2nd International Conference on Marine Science (ICMS 2017)

Better Insight for the Healthy Ocean

IOP Conference Series: Earth and Environmental Science
Volume 176

Bogor, Indonesia
6 - 7 September 2017

Editor:

Dietriech Geoffrey Bengen

ISBN: 978-1-5108-6819-9
ISSN: 1755-1307

Printed from e-media with permission by:

Curran Associates, Inc.
57 Morehouse Lane
Red Hook, NY 12571

Some format issues inherent in the e-media version may also appear in this print version.

Printed by Curran Associates, Inc. (2018)

Institute of Physics
Dirac House, Temple Back
Bristol BS1 6BE UK

Phone: 44 1 17 929 7481
Fax: 44 1 17 920 0979

techtracking@iop.org

Additional copies of this publication are available from:

Curran Associates, Inc.
57 Morehouse Lane
Red Hook, NY 12571 USA
Phone: 845-758-0400
Fax: 845-758-2633
Email: curran@proceedings.com
Web: www.proceedings.com

Table of contents

Volume 176

2nd International Conference on Marine Science: Better Insight for the Healthy Ocean

6–7 September 2017, Bogor, Indonesia

Accepted papers received: 3 July 2018
Published online: 30 July 2018

Preface

Papers

IOP Conf. Series: Earth and Environmental Science **176** (2018) 011001 doi:10.1088/1755-1315/176/1/011001

Preface

Ocean plays an important role in various aspects such as physical, chemical, and biological processes, marine-atmospheric dynamics, distribution of marine ecosystem diversity and marine resources. Marine ecosystem and resources have tremendous potential that will be able to utilize sustainably in the field of marine industry development such as food, pharmaceutical, cosmetics, chemicals and bio-energy. However, marine ecosystems and resources are also under serious anthropogenic pressure, exploitation and marine pollution. Another serious threat is the phenomena of global climate change. Researches and efforts have been made to examine and understand the phenomena that occur. Collaboration and sharing knowledge among researchers on the topic Marine Science and Technology are key to achieve these goals and interdisciplinary collaboration within marine sciences is expected to also trigger research and development activities, especially in the tropical regions of Indonesia.

Department of Marine Science and Technology, Faculty of Fisheries and Marine Science, Bogor Agricultural University (IPB) in collaboration with France Embassy Jakarta held the 2nd International Conference on Marine Sciences (ICMS) 2017 on September 6-7th, 2017 at the Aston Bogor Hotel and Resort, Indonesia. The theme of this conference was "Marine Science and Technology: better insight for the healthy ocean". The conference was a scientific gathering event that will provide a media for information sharing, ideas, and knowledge from various stakeholders related to marine sciences subjects, such as researcher, students, NGOs, and government. International conference is one of media for disseminating the research results from researchers and parties related to this topics and issues.

The conference was divided into three parts: plenary session, parallel session, and poster exhibition. The plenary session consisted of presentation by international and national keynote and invited speakers. The parallel session contained the papers presentations by participants divided into 5 different topics: hydrobiology, oceanography, marine observation system, marine technology, and special topics. Exhibition activities displayed posters from selected participants who took apart on poster session. The keynote and invited speakers presented in this conference were listed below.

Speaker	Presentation Tittle	Institution
Dr. Tubagus B. Haeru (Keynote speaker)	Contribution of Sciences and Technology to Maintain Sustainable Management of The Ocean Health	Coordinating Ministry for Maritime Affairs of Indonesia (Kemenkomar)
Prof. Ocky Karna Radjasa (Keynote speaker)	Marine Biotechnology for Sustainable Use of Marine Resources	University of Diponegoro Semarang; Ministry of Research, Technology and Higher Education of Indonesia
Dr. Michel de Jean (Keynote speaker)	Development of Operational Oceanography in Indonesian Seas: Lesson Learned from INDESO Project	Collecte Localisation Satellites (CLS) Argos France
Dr. Aditya R. Kartadikaria (Invited speaker)	Concurrent Simulation of Mesoscale Phenomena and Explicit Tides Toward a Real-time Operational Ocean Model for the Indonesian Seas	Bandung Institute of Technology

Speaker	Presentation Tittle	Institution
Dr. Dwiyoga Nugroho (Invited speaker)	Tidal Energy Dissipation in Indonesia Seas	LEGOS Laboratory, Paul Sabatier University – Toulouse III, France
Dr. Joana Dias (Invited speaker)	Genetic Diversity of a Hitchhiker and Prized Food Source in the Anthropocene: the Asian Green Mussel *Perna viridis*	Department of Fisheries, Western Australia, Perth Australia
Dr. Davin H.E. Setiamarga (Invited speaker)	An insight to octopodid phylogeny from transcriptomics-based mitochondrial phylogenomics	National Institute of Technology Wakayama College, Japan
Prof. Henry M. Manik (Invited speaker)	Performance of Multibeam Echosounder for Target Detection and Seafloor Mapping	Bogor Agricultural University
Dr. Fadli Syamsudin (Invited speaker)	Toward Establishment of Regional Ocean Climate Observing System on the Indonesia Maritime Continent	Agency for the Assessment and Application of Technology (BPPT)
Dr. I Nyoman Radiarta (Invited speaker)	Potential Fishing Model Using VIIRS and Satellite Based Oceanography Data in Western Indonesia Seas	Agency for Marine and Fisheries Research and Human Resources
Dr. R. Heru Triharjanto (Invited speaker)	Satellite Technologies for Maritime Application : LAPAN's Perspective	National Institute of Aeronautics and Space of Indonesia

The participants of conference came from various research institutions and universities such as Bogor Agricultural University (IPB), Gajah Mada University (UGM), Padjadjaran University, Bandung Institute of Technology (ITB), University of Indonesia (UI), Diponegoro University, Udayana University, Sultan Ageng Tirtayasa University, Halu Oleo University, Jenderal Soedirman University, Hasanuddin University, Syiah Kuala University, Batam State Polytechnics, Indonesian Naval Postgraduate School, and National Institute of Technology - Wakayama College (Japan). Furthermore, the participants also came from The Research Center for Fisheries - Ministry of Marine Affairs and Fisheries Republic of Indonesia (KKP), Marine and Coastal Data Laboratory - Ministry of Marine Affairs and Fisheries Republic of Indonesia (KKP), Agency for the Assessment and Application of Technology (BPPT), Coordinating Ministry for Maritime Affairs of Indonesia (Kemenkomar), National Institute of Aeronautics and Space of Indonesia (LAPAN), and Indonesian Institute of Science (LIPI). Beside from universities and governmental institutions, there were participants from NGO such as WWF Indonesia, World Conservation Society, and Bali Reef Check.

The committee have received 120 research abstracts from both oral and poster presentations, then selected 47 full papers to be published in this issue. These papers have been reviewed by ICMS 2017 scientific committee. The publication is expected to be beneficial for all parties in order to provide recent understanding on information and knowledge in marine science and technology. We are grateful to IOP Publishing, especially Ms. Anete Ashton (Conference Commissioning Editor) and Ms. Stephanie Gill (Conference Publishing Coordinator) for allowing us to publish full paper from the conference on IOP Conference Series: Earth and Environmental Science. We also would like to thank to the authors who have participated in the conference and submitted their papers. We highly appreciate to the sponsors who financially supported this event especially to The Ministry of Research, Technology and Higher Education (Ristekdikti), Coordinating Ministry for Maritime Affairs of Indonesia (Kemenkomar), Collecte Localisation Satellites (CLS) Argos, Koperasi Ben Makmur Muara Tawar, PT. PJB UP Muara Tawar, Direktorat Jenderal Perhubungan Laut, Kantor Penyelenggara Pelabuhan Kelas III Pamanukan, and others. This symposium was also supported partly by IPB World Class University Program scheme for fiscal year 2017.

IOP Conf. Series: Earth and Environmental Science **176** (2018) 011001 doi:10.1088/1755-1315/176/1/011001

Editor-in-Chief
Proceeding The 2nd International Conference on Marine Sciences (ICMS) 2017
Dietriech Geoffrey Bengen

Conference Chair
The 2nd International Conference on Marine Sciences (ICMS) 2017
Meutia S. Ismet

Editorial Committee
Agus S. Atmadipoera (Bogor Agricultural University, Indonesia)
Dietriech G. Bengen (Bogor Agricultural University, Indonesia)
I Wayan Nurjaya (Bogor Agricultural University, Indonesia)

Scientific Committee
Agus S. Atmadipoera (Bogor Agricultural University, Indonesia)
Ariane Koch-Larrouy (LEGOS Toulouse, France)
Charles P.H.S (Bogor Agricultural University, Indonesia)
Davin H. E. Setiamarga (National Institute of Technology, Wakayama College, Japan)
Dietriech G. Bengen (Bogor Agricultural University, Indonesia)
Hawis H. Madduppa (Bogor Agricultural University, Indonesia)
Henry M. Manik (Bogor Agricultural University, Indonesia)
Indra Jaya (Bogor Agricultural University, Indonesia)
I Wayan Nurjaya (Bogor Agricultural University, Indonesia)
Joana Dias (Department of Fisheries, Government of Western Australia)
Jonson L. Gaol (Bogor Agricultural University, Indonesia)
Neviaty P. Zamani (Bogor Agricultural University, Indonesia)
Peter J. Schupp (Oldenburg University, Germany)
Philippe Gaspar (CLS Toulouse, France)
Ren-Chieh Lien (University of Washington, USA)
Roza Yusfiandayani (Bogor Agricultural University, Indonesia)
Safri Burhanuddin (Coordinating Ministry for Maritime Affairs of Indonesia)
Sam Wouthuyzeen (Indonesian Institute of Science, LIPI)
Takahiro Osawa (Yamaguchi University, Japan)
Tri Prartono (Bogor Agricultural University, Indonesia)
Tohuru Mukai (Hokkaido University, Japan)
Vincentius P. Siregar (Bogor Agricultural University, Indonesia)

Assistant Editors
Dea Fauzia Lestari, Aditya Hikmat Nugraha, Ani Haryati, Sri Ratih Deswanti (Department of Marine Science and Technology, Faculty of Fisheries and Marine Science, Bogor Agricultural University, Indonesia)

IOP Conf. Series: Earth and Environmental Science **176** (2018) 011001 doi:10.1088/1755-1315/176/1/011001

Conference Photographs

ICMS 2017 Participants

Keynote Speaker
Prof. Ocky Karna Radjasa

Keynote Speaker
Dr. Tubagus B. Haeru

Keynote Speaker
Dr. Michel de Jean

Invited Speakers

Invited Speaker
Dr. Joana Dias

IOP Conf. Series: Earth and Environmental Science **176** (2018) 011001 doi:10.1088/1755-1315/176/1/011001

Invited Speaker
Dr. Fadli Syamsudin

Invited Speaker
Prof. Henry M. Manik

Invited Speaker
Dr. R. Heru Triharjanto

Discussion Session

Oral Presentation

Presentation Session

Poster Exhibition

Sponsor Banners

Best Presenters

ICMS 2017 Committee

Peer review statement

All papers published in this volume of *IOP Conference Series: Earth and Environmental Science* have been peer reviewed through processes administered by the proceedings Editors. Reviews were conducted by expert referees to the professional and scientific standards expected of a proceedings journal published by IOP Publishing.

IOP Conf. Series: Earth and Environmental Science **176** (2018) 012001 doi:10.1088/1755-1315/176/1/012001

Overview of physical oceanographic condition at Biawak Island: past achievement and future challenge

N P Purba[1,2*], I Faizal[2], I F Pangestu[2], P G Mulyani[2], M F Fadhillah[3]

[1] Marine Research Laboratory, Universitas Padjadjaran, Jatinangor, Indonesia
[2] Department of Marine Science, Universitas Padjadjaran, Jatinangor, Indonesia
[3] PARIMANTA, Padjadjaran University
*e-mail: noir.purba@unpad.ac.id

Abstract. Biawak Islands is one of Marine Protected Areas (MPA) in Java Seas and categorized as remote area. The purpose of this research is to assess data's from 2011-2017 and to understand about these island from oceanographic point of view. Due to its location in the center of Java Sea (JS), variations of oceanographic condition at Biawak Island is supposed to be similarity with JS. Seasonal variations in this shallow-flat topography of JS are mostly influenced by the monsoon due to the sea level difference between Pacific and Indian Ocean. Water masses from the South China Sea (SCS) and Pacific Ocean (PO) take turns to enter JS at two distinct monsoon seasons. The results showed that temperature ranges 26.28 - 34.3 °C, salinity ranges 27.5 - 34.5 psu, transparency ranges 0.7 - 20 m, dissolved oxygen ranges 4.1 - 6.5 mg/L, and pH 6.30-8.5. Tides with mixed tide prevailing diurnal. The wind pattern with monsunal variation and the surface current is about 0.2-0.8 m/sec. The dominant substrat in the Biawak is sand and coral. Further, there is a need long-term programs concentrate on delivering quantification data related to environmental issues.

1. Introduction

Biawak Island (before namely as Bonpies or Rakit) located at Java Seas. Biawak Island consists of two islands (Biawak and Gosong) and 1 atoll (Candikian). The biggest island named Biawak (other name is Rakit Island) is a small island located in the north of West Java, precisely in the north of Indramayu with its center coordinate at 5.930752 °S and 108.379548 °E. These islands is one of 4 remote islands in Java Sea, which have been intensively observed by Marine Science Department of Universitas Padjadjaran in the last six years by several research and projects. However, little is known about data and information that summarized all those six years observations, since the observations itself were performed by separate individuals or groups as well as at different time. Thus, the results that have been obtained are limited by short observations and the discussion of one component of oceanography, either physical or biological is also limited. Therefore, the information about the annual variability and the relationship between the physical and biological components can't be obtained.

Due to its location in the center of Java Sea (JS), variations of oceanographic condition at Biawak Island is supposed to be similarity with JS [1]. Seasonal variations in this shallow-flat topography of JS are mostly influenced by the monsoon due to the sea level difference between Pacific and Indian Ocean [2, 3]. Water masses from the South China Sea (SCS) and Pacific Ocean (PO) take turns to enter JS at two distinct monsoon seasons. The Northwest (NW) monsoon transports the SCS water through Karimata Strait from November to March and increasing the Sea Surface Temperature (SST) of JS, whereas the Southeast (SE) monsoon transports PO water through Makassar Strait and Banda Sea from May to September and decreasing the Sea Surface Salinity (SSS) of JS [3, 4]. The circulation that regulated between SCS-JS-PO suggested containing strong signal of El-Nino and Southern Oscillation

(ENSO) and Indian Ocean Dipole (IOD) that somehow altered the weather and climate regionally or globally [4, 5, 6]. Therefore, all those alterations to the JS may as well influencing the local condition of Biawak Island and vice versa.

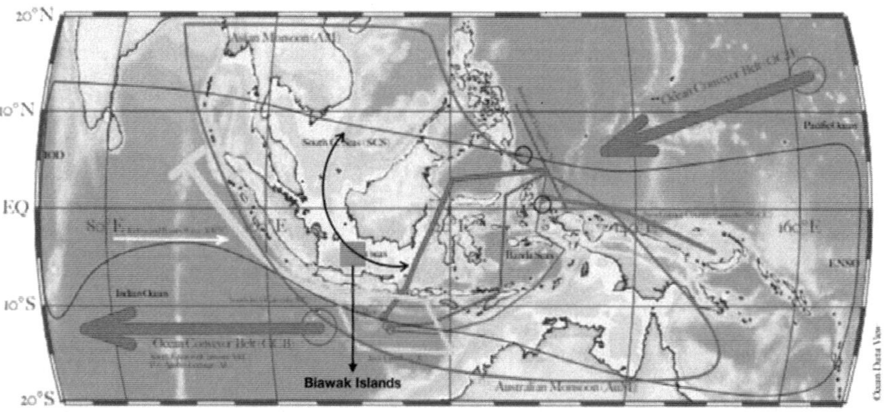

Figure 1. The location of Biawak Island between local and regional phenomenom [7, 8, 9].

The islands are remote areas and limited access to go, therefore oceanographic, geomorphological, and ecosystem data are still rare [10]. From the literature, not more than 3 studies were conducted before 2011 in this island. Since 2011, KOMITMEN (Study Group of Instruments and Marine Surveys) Unpad and FPIK-Unpad conduct research in this region intensively as well as by some researcher [11, 12]. Ecosystem that can be found in Biawak Island are mangroves, seagrasses and coral reefs. The density of mangrove ecosystem in this island ranging from low to high density. The percentage of seagrass beds coverages are very low (5-10%) with sand as bottom substrate and *Enhalus acoroides* as the main species. The coral reef was categorized as poor (6.25%) to good (63.75%) condition, but in the average, coral reefs are in the medium (34.69%) condition [10].

The objective of this paper is to provide an introduction of the local physical oceanographic data and information of Biawak Island based on previous observations. First, we briefly introduce and describe the physical condition in Biawak Islands. Second, analyzing the conditions with the characteristic of Java Seas and regional phenomenom. Finally, we discuss about future activities in Biawak Island that would lead to assess a comprehensive feature of oceanographic condition.

2. Materials and methods

2.1. Study sites

This study area chosen for this work, Biawak Islands, consists of three islands, the Candikian, Gosong and Biawak Islands itself (figure 1 and 2). These island is Marine Protected Areas because of its high diversity. The distance from mainland of Java Islands is about 36 km. All of data in the Biawak Island are primary data that collected from 2011 to 2017 from several research projects.

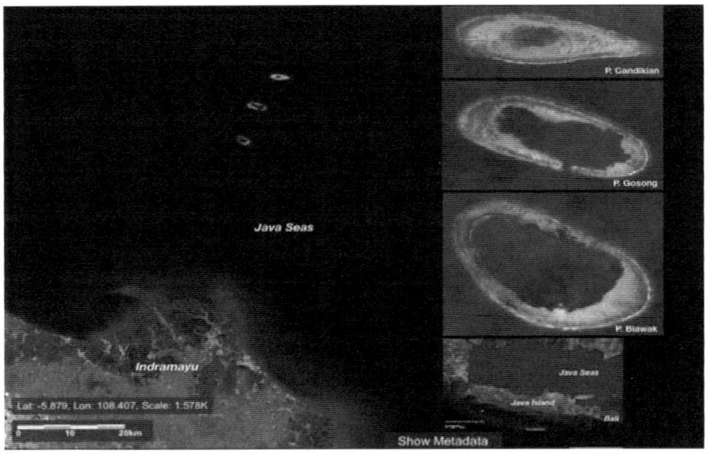

Figure 2. Position of Biawak Islands, Biawak as a main island (5.9310 S, 108.3795 W), Gosong (5.882 S, 108.4026 W), and Candikian (5.8076 S, 108.4394 W).

In this research, we use several portable instrument to measure insitu data. Hand optical refractometer to measure salinity, the temperature measure by two instrument, one in GARMIN and in pH meter combine, with the standard deviation is less than 1 degree Celsius. For sediment, the Ekman grab [13, 14,15] operated from the water surface by a line and messenger. Salinity measured by optical hand-refractometer, the tides with model combine and manual insitu measurement only in Biawak Island. The transparency measured by white secchi disc with 40 cm diameter.

Table 1. List of research/project in Biawak from 2011 to 2016.

No.	Date	Parameter measured
1	March and May 2011 [16, 17, 18]	Temperature, salinity, dissolved oxygen
2	June and November 2012 [19]	Temperature, salinity, pH, dissolved oxygen, water transparency, bathymetry, tides, current wind, substrate
3	June 2013	Temperature, salinity, dissolved oxygen, pH
4	October 2014	Temperature, salinity, pH, dissolved oxygen, water transparency, bathymetry, tides, current wind
5	April 2015	Temperature, salinity, dissolved oxygen, pH
6	May 2015	Temperature, salinity, pH, dissolved oxygen, water transparency, bathymetry, tides, current wind, substrate
7	May 2016	Temperature, salinity, pH, dissolved oxygen, water transparency, bathymetry, tides, current wind, substrate, and tides
8	April 2017 [20]	Temperature and salinity

Here, we measured the temperature with 4668 data, salinity from 187 stations, wind, current, tidal, and brightness from 124 stations, dissolved oxygen and pH from 209 stations. The tides are measured by using tidal pole that installed at the Biawak Island, while other parameters above are measured from the ship and also from the transect that conducted in the shallow waters. Depth profiles data are obtained by primary measurements from 4668 stations and an analog maps from DISHIDROS (2013). In addition, wind data is analyzed from the European Center for Medium-Range Weather Forecasts (ECMWF) (www.ecmwf.int) per 6 hours with spatial resolution 0.125° x 0.125°. Obtained data are then calibrated statistically, processed, visualized and interpreted using ODV (Ocean Data View) software [21], while tidal data are analyzed using Formzahl numbers to determine the type of tidal. We analyze the data with temporal and spatial distribution with focus on surface water column

3. Results and discussions

3.1. Tides

Tidal conditions in the area shows a mixed tide prevailing diurnal type with 1.7446 Formzahl number. This is similar to the type of tides in some areas such as Kalimantan and Java Sea in general [22, 23]. It means that there will be 2 peaks during high waters and 2 valleys during low waters in one day with differ high water level.

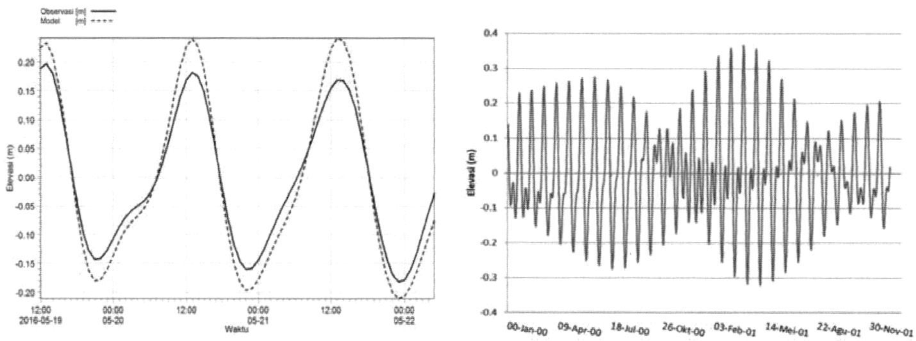

Figure 3. Three days tides pattern (left); combination between three days tide pattern and one month model (right).

Figure 3 shows that the primary data and the model have a similar pattern, which means the tidal corection is right, but there is a little different at maximum and minimum in height. Measurement data affected by many factors, such as condition of weather during measurement result a big deal in accuracy. These condition made a measurement more complex than model.

3.2. Bathymetry and substrate

Bathymetry of the Biawak Islands according to DISHIDROS Data (2013) which from analog map in 1970, ranged from 5 to 20 m. Bathymetry in Biawak Islands have similarity with Java Sea that have isodepth 20 m in open water [24]. Gosong Island has 25 m maximum depth located in the middle of island and 5 m depth in attol's reef flat during observation in 2006. This Island become a target for sand mining location in 1990's. The area changes from 2765 km^2 (1990) to 2257 km^2 (2016). Thus, the island has decreased about 0.508 km^2 in 26 years.

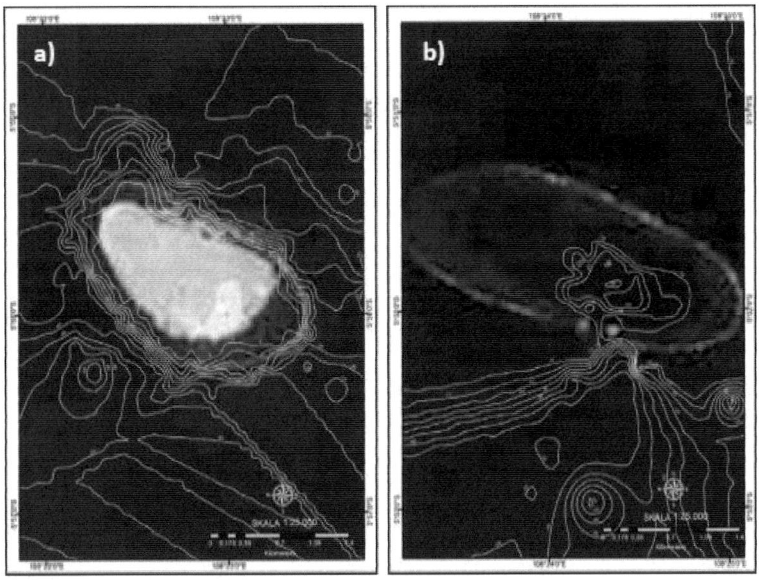

Figure 4. Bathymetry of Biawak Islands; (a) Biawak Island; (b) Gosong Island.

From survey data, depth of the Biawak Islands ranges from 0.9 to 75 m (figure 4a). The largest shallow areas of Biawak Island is located in the southeast of the island (0.9-1.2 m), which composed of corals, sand, and mud as dominant substrates, while the deepest waters are in the southern of the island. Around 500 m to the outside dominated by coral and sand substrate. In some parts (West) is dominated by mud. Depth at Gosong Island is about 4 - 28 m on the inside while on the outside it is around 50 m (figure 4b). The inside of the island of Gosong is a closed water. Gosong Island mostly dominated by sand and coral with percentage 70:30 on the inside or outside atoll. Depth in Candikian Island is about 1-2 m with the coral substrate. Outside island, the depth can reach 47 m. This is also similar to the data released by DISHIDROS in 2013. This depth contour also shows the shallow profile of Java Seas. Candikian Island dominated by sand and corals with a percentage of 30:70.

Java Seas is generally dominated by mud and sand. The most part of bottom substratum of the Java Sea is constituted by silt and formed by highly dense mud layer, with large muddy bed in the North-East and central area which are mixed with coral and shell debris [22].

3.3. Wind and currents

Wind patterns in the Biawak Island, following periodically monsoon pattern of the Java Sea that blows horizontally from east to west during the eastern monsoon and blows in the opposite direction during the west monsoon [22]. Western monsoon occurs in December to February, the Transitional Monsoon 1 and 2 occurs in March to May and in September to November, while the eastern monsoon occurs in June to August as shown in the line 1 and 2 of Figure 5. Wind data measurement in 4 m height above mainland (U^4).

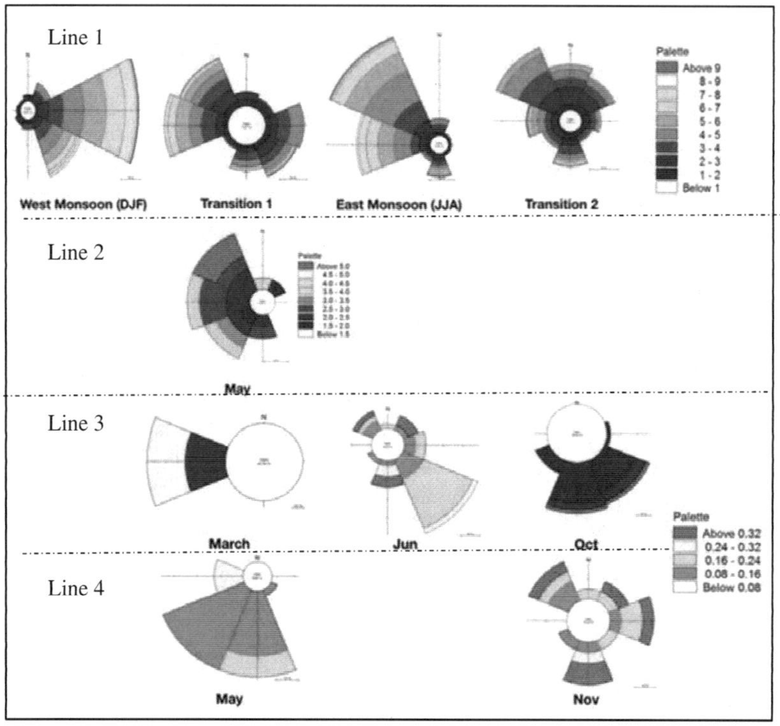

Figure 5. Wind (U^4: ~4 m from sea level) rose patterns in Biawak Island (line 1: wind from ecmwf; line: 2 from field survey), and current rose pattern (line 3 and 4) from field survey).

Monthly wind speed ranges from 1-7 m/sec throughout the year, while from the speed of the instantaneous measurement in May is 1-5 m/sec. Wind direction corresponds to the seasonal direction that controls the atmosphere in the Java Sea. For surface currents, the current velocity is below 1 m/sec and the direction is not always the same as the wind direction. Current Speed pattern in Biawak Island more complex than the other 2 islands, Gosong and Candikia Island. Besides, wind forces of monsoon will give effect to current speed pattern. Characteristically this region represents Java Sea waters with current speed of 0.15-1 m/sec, which mostly influenced by tidal [25] and monsoon [26].

3.4. Sea surface temperature and salinity

Sea Surface Temperature in Biawak Islands range 26.3 - 34.4 °C with an average of 30.7 °C (figure 6). The highest sea temperature in Biawak Islands occurs during NW Monsoon (34.4 °C) while the lowest sea surface temperature value 26.2 °C founded in May. The NW monsoon transports the South Cina Seas (SCS) water through Karimata Strait from November to March and increasing the Sea Surface Temperature (SST) of Java Sea (JS). Before NW Monsoon enter (Transition II), temperature in Biawak Island considerate in 29-30 °C. These Variation determined by bathymetry and water transparency of waters. In general, temperature in Biawak Island still consider as normal condition [27].

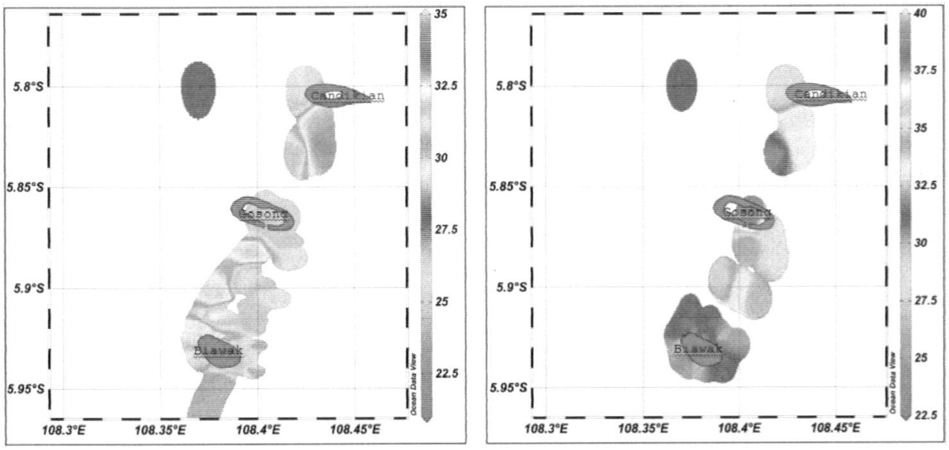

Figure 6. Distribution of temperature (left) and salinity (right) in Biawak Island.

The maximum temperature is in Biawak Island is 32 °C. North area of Biawak Island has the highest temperature than south area. Variability of temperature in Biawak Island is diverse by location. The minimum temperature is in west of Candikian Island, because this island faced to the open sea area. The waters nearest island has a highest temperature than in the open water area. This is because the type waters with a semi-closed water. SST at Gosong Island ranges from 29.6 - 30.2 °C.

Salinity distribution in the sea is influenced by various factors such as the pattern of water circulation, evaporation, rainfall and river flow [28, 29]. Salinity in Biawak Islands ranges 27.50-34.50 psu with an average of 30.88 psu (figure 6). Salinity in Biawak Island has range between 27.50-31.08 psu and in Gosong Island 32.00 – 34.00 psu. In Java Sea (JS), salinity in range 29.00 – 34.00 psu still categorized normal [30]. The fluctuations of these two parameters (SST and Salinity) are influenced by the water mass that coming from the South China Sea (TSS) [1] and also from Makassar Strait via ITF [24].

3.5. Dissolved Oxygen (DO) and pH

The content of dissolved oxygen (DO) in a waters is influenced by the diffusion speed of oxygen from the air as well as the rate of photosynthesis and respiration of aquatic organisms. DO needs in organisms varies greatly depending on type, phase and activity [31]. Phytoplankton as primer organism in sea-life, can live well at oxygen concentrations greater than 3.00 mg/L. DO in Biawak Island between 4.10 - 6.50 mg/L (figure 6) which support phytoplankton boundary life. DO in Gosong Island tends to be homogeneous and do not have much variability.

A certain range of pH values may indicate a change in water quality [32]. The value of pH in Biawak Islands has range 6.30 – 8.52 with averages 7.54 (figure 7). According to Quality Standart from Ministry of Environment and Forestry, pH values in Biawak Islands categorized as suitable for marine life.

IOP Conf. Series: Earth and Environmental Science **176** (2018) 012001 doi:10.1088/1755-1315/176/1/012001

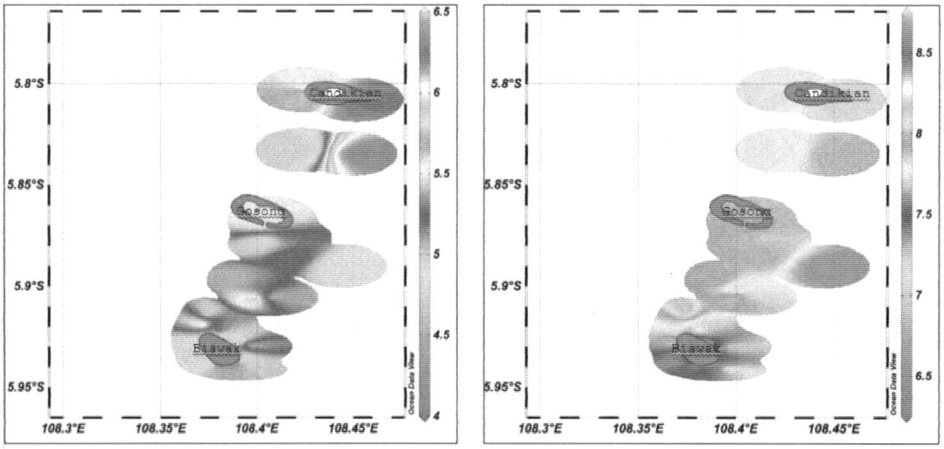

Figure 7. Distribution of dissolve oxygen (left) and pH values (right) in Biawak Island.

3.6. Water transparency

The water transparency of Biawak Island has wide ranges variability from 0.7 – 20 m with an average of 8.06 m. The transparency of ocean water affected by chemical and physical factors [33]. Water transparency also affected by bathymetry and the type of sediment in sea bed. The intensity of sunlight is often a very rapid limiting factor because it is affected by depth and turbidity [34]. Water transparency becomes an important factor in controlling the primary productivity of waters as it relates to the rate of penetration of sunlight that will determine the rate of photosynthesis and primary productivity of the waters [20].

3.7. Substrate

Substrate in the area of Biawak Island and its surroundings have morphological similarities. The basic material that built is sand and the coral fragments. Observations in November 2014 stated that, found four types of substrate in the shallow waters of Biawak are sand, rocky sand, muddy sand and mud. According to [35] the substrate of waters of Biawak Island consists of sand, Dead Coral (DC), live corals, sandy mud, and rubble. The DC area is located on the island of Biawak with of ± 216,000 m^2, while the coral reef area is ± 111,000 m^2. The sand substrate has an area of 525.000 m^2, then the fracture ranges from 348.000 m^2, sandy mud around 91.000 m^2 and mud with an area of 90.000 m^2 range. In the northern part of the island has a rubble substrate, while in the southeast part of the island is dominated by dead coral (DC). [36] states that in the intercidal zone Biawak Island consists of three basic substrate types, namely substrate of fine sand in the southeast to the east, rocky rock substrate (coral rubble) in the north and muddy sand substrate in the northwest to south. In the southern part of the island has sand and rocky sand substrate. The southern part of the shallow waters of Biawak Island is a habitat of coral massive coral reefs. Such coral reefs can only live in waters with sandy sediments. In the western part it has mud substrate and muddy sand. In the area is found many mangrove *Avicenia* sp. This type of mangrove generally grows shallow water with sediments of muddy sand. The substrate on Gosong Island consists of sand, live corals, dead coral, and grassy seagrass. The southern part of the island is dominated by sand grass substrate (sand with seagrass), which indicates that the island also has potential as a macrozoobenthos habitat. According [36] the most optimal substrate for sand cucumber habitat is sand or sand with seagrass-covered rocks. For the surroundings of the island there are many living coral substrates, but in the eastern part of the island is dominated by DC.

4. Future challenge

The study of hydro-oceanography in the islands of Biawak is need to develop. The Physical conditions are important parameters which can determined productivity of marine life. Continuous data and routine measurements can determine how the characteristics of the water masses in these islands primarily to see how regional phenomena such as the Indian Ocean Dipole (IOD) [8] and El Nino Southern Oscilation (ENSO) affect it [9]. Biawak Island that located in JS is affected by local, regional, and global phenomenom. Biawak Islands also as path of national and international ship.

Furthermore, that shallow sea waters located in Biawak are very rich in important micro-organisms that build food chains. On Gosong Island, morphological changes because sand dredging conduce sea bed profiles changes. Ecosystems degradation in this area which can reduce marine biota who assosiate with ecosystems. Research on the Gosong Island and Candikian Island also needs to be multiplied in the future to get complete and comprehensive data about the condition of Biawak Island waters. The Biawak Islands of the future will be determined by the physical oceanographic condition. Integrated research is required condition especially from biological and chemical aspects.

5. Conclusions

This paper deals with physical oceanographic condition on the Biawak Island. During the period of research, shows that Biawak has unique water characteristics based on physical conditions of the area. This is because the areas located in the Java Seas that have many variabilities from local to global scale impact (IOD, ENSO, Monsoon, ITF path, and climate change). The monsoonal system seems has a significant effect on physical parameters (SST, salinities, oxygen dissolved) and the substrate similar to Java Island than Borneo island. Based on the physical aspect, there are lack of data especially in temporal data collection. The next research in temporal scale and in inner of Gosong and Candikian is the priorities because of its geomorphology and local condition.

Acknowledgements

Thanks to Jaya Kelvin, Andreas Bastian, and Rahmi Nailah for first discussion in 2015. The authors would like to thanks for anonymous reviewer and committee of ICMS Bogor. Some of this projects received funding from Padjadjaran University and RISTEKDIKTI Grants. We also say thanks to Marine Coast Guard and Marine Transportation of Indramayu for administration use. Mr. Tasyid and crews as local assistants who has been helps our survey since 2011.

References

[1] Purba N P, Kelvin J, Annisaa M and Teliandi D 2014 Preliminary research of using ocean currents and wind energy to support lighthouse in small island, Indonesia *J. Energy Procedia* **47** 204–210

[2] Wyrtki K 1987 Indonesian through flow and the associated pressure gradient. *Journal of Geophsical Research* **92** 12941–12946

[3] Sadhatomo B 2006 Review of environmental features of the Java Sea *J. Ind. Fish Res.* **12**(2) 129-157

[4] Qu T, Du Y, Meyers G, Ishida A and Wang D 2005 Connecting the tropical Pacific with Indian Ocean through South China Sea *Geophysical Research Letters* **32**(L24609) 3–6

[5] Juneng L and Tangang F T 2005 Evolution of ENSO-related rainfall anomalies in Southeast Asia region and its relationship with atmosphere – ocean variations in Indo-Pacific sector *Climate Dynamics* **25** 337–350

[6] Wang D, Liu Q, Huang R X, Du Y and Qu T 2006 Interannual variability of the South China Sea throughflow inferred from wind data and an ocean data assimilation product *Geophysical Research Letters* **33**(L14605) 40–43

[7] Susanto R D, Moore T S and Marra J 2006 Ocean color variability in the Indonesian Seas during the SeaWiFS era *J. of The Earth Sciences* **7**(5) 1-16

[8] Hidayat R 2016 Interannual variability of rainfall over Indonesia: impacts of ENSO and IOD and their predictability *J. IOP Conf. Ser.: Earth Environ. Sci.* **31** 1-10

[9] Prasetyo Y and Nabilah F 2017 Pattern analysis of El Nino and La Nina phenomenon based on sea surface temperature (SST) and rainfall intensity using oceanic nino Index (ONI) in West Java Area *J. IOP Conf. Ser.: Earth Environ. Sci.* **98** 1-10

[10] Harahap S A, Sunarto, Riyantini I and Ihsan Y N 2013 Overview of Biawak waters in Indramayu regions *Report* **62** pp

[11] Fitriadi C A, Dhahiyat Y, Purba N P and Harahap S A 2017 Coral larvae spreading based on oceanographic condition in Biawak Islands, West Java, Indonesia *Journal of Biodiversitas* **18**(2) 681-688
[12] Pangestu I F, Purba N P and Syamsuddin M L 2016 Microplastic condition in Indramayu waters West Java *Proceeding National Seminar of Fisheries and Marine Science* **1** 475-478
[13] Ekman S 1911 New apparatus for the qualitative and quantitative research of soil fauna of lakes *Int. J. Hydrogr.* **3** 553–561
[14] Birge E A 1921 A second report on limnological apparatus *J. Acad. Sci. Arts Lett.* **20** 533–552
[15] Blomqvist S 1990 Sampling performance of ekman grabs insitu observations and design improvements *J. Hidrobiologia* **206**(3) 245-254
[16] Kristiadhi F 2011 Distribution and condition of coral reefs in the Waters of Biawak Island, Indramayu Regency *Report* pp 62
[17] Pramesthi D A 2011 Mapping of potential land suitability for diving and snorkeling tourism at utilization zone in Biawak Island, Indramayu Regency *Report* pp 65
[18] Sudrajat Z 2011 Distribution of makrozoobenthos in Island Biawak *Report* pp 70
[19] Hendrayana C 2013 The Phytoplankton community structure in relation to the distribution of nutrients in the Water of Biawak Island, Indramayu *Report* pp 65
[20] Widiastuti K S 2017 Nutrient status and primary productivity on waters with different coral reefs condition at Biawak Island, Indramayu District *Report* pp 80
[21] Schlitzer R 2016 *Ocean Data View* https://odv.awi.de
[22] Nagara G A, Sasongko N A and Olakunle O J 2007 *Introduction to Java Sea* (University of Stavanger) pp 14
[23] Ray R D, Egbert G D and Erofeeva S Y 2005 A brief overview of tides in the Indonesian Seas *J. Oceanography* **18**(4) 74-79
[24] Lubis E, Pane A B, Kurniawan Y, Chaussade J, Lamberts C and Potier P 2005 The atlas book of capture fisheres and fishing port in Java island Georaphy capture fisheries Indonesia Approach
[25] Purba N P and Harahap S A 2013 Small island in marine perspective, Biawak-Gosong-Candikian Edition Unpad Press pp 42
[26] Gaol J L and Sadhotomo B 2007 Charachteristic and variability of parameters oceanographic in Java Sea and the relationship with distribution of fish cath result *J. Lit. Fisheries Ind.* **13** pp 10
[27] Siregar S N, Yuliadi L P, Purba N P, Pranowo W S and Syamsuddin M L 2017 The water mass exchange in Java Sea due to periodicity of monsoon and ITF in 2015 *J. Depik* **6**(1) 44-59
[28] DKP 2006 Arrangement the academic scenario of marine conservation area management in Biawak Island, district Indramayu *Report* pp 120
[29] Larrouy A K, Madec G, Iudicone D, Atmadipoera A and Molcard R 2008 Physical processes contributing to the water mass transformation of the Indonesian Throughflow *J. Ocean Dynamics* **58** 275-288
[30] Nontji A 2005 *Marine Archipelago* (Jakarta: Djambatan) p 358
[31] Odum E P 1993 *The Basic of Ecology Third editon* (Yogyakarta: Gadjah Mada University Press) p 697
[32] Souhoka J and Patty S I 2013 Observation of hidrology condition and the relations with coral reef condition in Talise island, North Sulawesi *J. Platax* **1**(3) 138-147
[33] Denman K L and Gargett A E 1995 Biological-physical interactions in the upper ocean: the role of vertical and small scale transport processes *J. Ann. Rev. Fluid Mech.* **27** 225-256
[34] Yentsch C S, Yentsch C M, Cullen J J, Lapointe B, Phinney D A and Yentsch S W 2002 Sunlight and water transparency: cornerstones in coral research *J. Exp. Maret. Biol. Ecol.* **268** 171-183
[35] Yudha R A 2016 Distribution of Kima (Tridacnidae) based on substrates basic waters in Biawak Island, Indramayu *Report* pp 75
[36] Arohman 2015 Abundance and diversity of sea cucumber relation with the condition of the Island of Biawak Island, Regency of Indramayu, West Java *Report* pp 75

Influence of El Niño Southern Oscillation (ENSO) phenomena on Eddies Variability in the Western Pacific Ocean

Y A Simanungkalit[1], W S Pranowo[2,3], N P Purba[4], I Riyantini[1], Y Nurrahman[1]

[1]Department of Marine Science, Padjadjaran University, Dekanat Building, FPIK-UNPAD Jatinangor Km. 21 UBR 40600, West Java, Indonesia
[2]Marine and Coastal Data Laboratory, Marine Research Center, Indonesian Ministry of Marine Affairs & Fisheries, Pasir Putih II, East Ancol, Jakarta UBR 14430, Indonesia
[3]Department Hydrography Engineering, Indonesian Naval Postgraduate School (STTAL), Komplek Bina Samudera Jl. Pantai Kuta V, Ancol Timur, Jakarta Utara 14430, Indonesia
[4]Marine Research Laboratory (MEAL), Dekanat Building, Padjadjaran University Jatinangor Km. 21 UBR 40600, West Java, Indonesia
*e-mail: noir.purba@unpad.ac.id

Abstract. Sea Surface Height (SSH), seawater temperature, and ocean currents are an important oceanographic variables that were used to identify the ocean eddies variability. This research aims to investigate the influence of ENSO on eddies variability in Western Pacific Ocean (WPO), by using the Automated Eddy Detection (AED). The results showed that cyclonic eddies are mostly formed in WPO from a total of 2024 eddies. Eddies total increases within the amount of the sea's depth, at 200 m has the highest of eddy total. The temperature in eddy center mostly started changing at a depth of 150 m, which is at the cyclonic eddy has a lower value and at anticyclonic has a higher value. That has the large effect of changes SSH. During El Nino, sea level anomaly decreases down to -0.3 m, while sea level anomaly minimum decrease down to -0.1 m during La Niña. At the WPO, Mindanao Eddies formed along the year, whereas Halmahera Eddies were weakened in West Season. A periodic ENSO affects eddies variability, in general, the diameter, total and temperature center of eddies are smaller during El Niño, and greater during La Niña than normal condition.

1. Introduction

Western Pacific Ocean (WPO) in the last few decades has been a center for the attention of researchers meteorology and oceanography, because the waters had a very important role to incident El Niño and Southern Oscillation (ENSO) and is part the importance of "Great Conveyor Belt". The eddies and low-latitude westrern boundary currents (WBC) addressed here are important aspects of these.processes [1].

In Equator WPO between Mindanao and New Guinea there are two eddies [2], Mindanao Eddy (ME) and Halmahera Eddy (HE). The ME is a clockwise (or cyclone) eddy, while the HE is clockwise (or anticyclonic) eddy. In Northen hemisphere there are warm-core eddy and cold-core eddy. Warm-core eddy rotates clockwise or known as a anticyclonic eddy, while the cold-core eddy rotates counterclockwise or known as cyclonic eddy [put references]. The cyclonic eddy can drive upwelling in their interior and increasing marine primary production [3]. Ocean eddy can move at speeds of about 0.5 knots and may occasionally persist for many month [4]. Spatial distribution of eddy varies with the size. All eddies are categorized into two groups according to their size: a sub-mesoscale eddy is about 10-100 km, and a mesoscale eddy can reach more then 50-200 km [5].

El Nino Southern Oscilation (ENSO) is ocean-atmosphere phenomenon in equatorial Pacific Ocean that influences climate variability over Indonesia region, as well as regional and global influences. ENSO phenomenon consists of two events as El Nino and La Nina. The ENSO events can be identified from an index of Southern Oscillation Index (SOI) [6]. ENSO phenomenon affects some of Indonesia's eastern waters, where the region is also affected by the Indian Ocean Dipole (IOD). These two phenomena can further influence ocean dynamics such as eddies currents in these waters [7].

Research on eddy varability in the WPO has been conducted by several researchers [8-13]. However, this study focused to investigate the varability of eddy current during normal year and ENSO year, using the Automated Eddy Detection (AED) developed by Nencioli et al. (2010) to analyze and identify vertical structures and eddies characteristics.

2. Methods

2.1. Study area

Study area is located in the WPO with coordinate 3° S – 10° N and 125° E– 140° E (figure 1). Data processing was carried out in Marine and Coastal Data Laboratory, Marine Research Center, Indonesian Ministry of Marine Affairs, and in Marine Research Laboratory (MEAL), Faculty of Fisheries and Marine Science, Padjadjaran University.

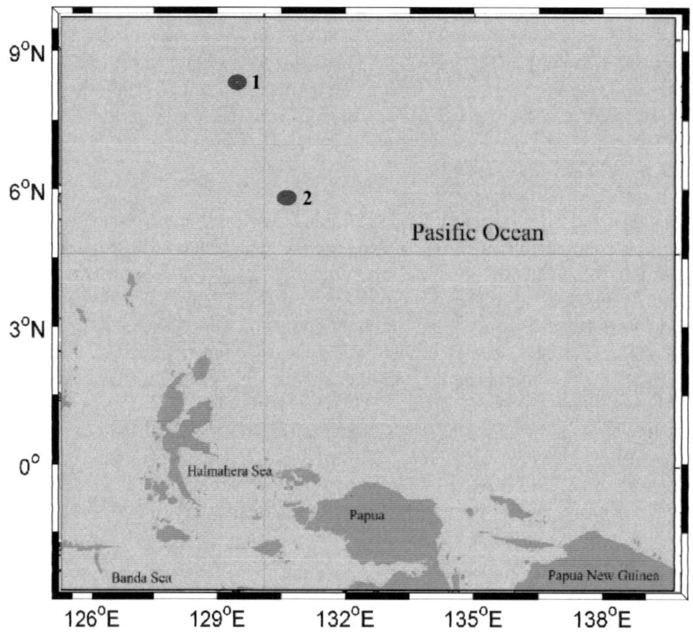

Figure 1. Study Area in Western Pacific Ocean and the point of ME and HE [8,9,10].

2.2. Material and method

The parameter that used to identify the eddy characteristic research are shown in table 1. The parameter distributions separated at 0 m, 75 m, 150 m, and 200 m. In detection the existence of eddy was using the Automated Eddy Detection algorithm which developed by Nencioli et al. (2010) [14].The method is based on some of the features that characterize the velocity field associated with mesoscale eddies, such as minimum velocities in the proximity of the eddy center, and tangential velocities that increase approximately linearly with distance from the center before reaching a maximum value and then decaying [15]. This method was used by Tussadiah 2015 [16] showed the eddy vertical characteristic in Southern Java Indian Ocean. Geostrophic Current data is used for AED method with software Matlab 2013b, then to strengthen the analysis eddy visualization was overlaid with SSH and temperature to obtained eddy characteristic.

Table 1. Research materials.

No	Dataset (2011, 2014-2016)	Data Resolution	Source
1.	Geostrophic Current	8.8 km	INDESO, HYCOM
2.	Sea Surface Height	8.8 km	INDESO, HYCOM
3.	Temperature	8.8 km	INDESO, HYCOM
4.	SOI	-	BOM

To know the condition of El Nino and La Nina, we used SOI from BOM . If the value of index greater than +7, then the condition is indicated La Nina year, but if index values smaller than -7 then indicated El Nino year. The period of time which indicated of ENSO phenomenon was used to compared the influence ENSO on Eddies Variability[17].

3. Results and discussion

3.1. Eddies variability

From the automated eddy detection analysis at four vertical levels (0 m, 75 m, 150 m, 200 m), it is shown that the number of eddies that have been detected in 2011 and between 2014-2016 were 2,024 eddies. Those are 1,119 cyclonic eddies and 892 anticyclonic eddies. Mostly the eddies formation increases at 200 m depth, which may be associated with differences of generating eddy vertically like wind stress, Kelvin and Rosby waves, topography, and baroclinic instability[18]. It is found eddies that can survive from depth 0 m to 200 m, but there is also a new eddy formed within the depth of 150-200 m. At the surface to bottom eddy has a variety diameter, dominated with the submesoscale eddy (30-500 km). The appearance of submesoscale, mesoscale and big eddy maybe come from drifter aggregation, maintained by converging Ekman currents and Coriolis force [18] and [19].

Eddies formation in October 2015, represented El Nino year (figure 1: right), shows 39 total eddies found at four depths, which are 23 cyclonic eddies and 16 anticyclonic eddies with variation of diameter ranges between 20-220 km. Based on the results of the eddies pattern in April 2011 which represents the La Niña event, it is known that the total of eddy cyclonic is 24 eddy, while the total eddy anticyclone type is 20 eddy, with varying diameters ranging from 30 - 450 km. In this month, it can be seen that Halmahera Eddy is formed at a depth of 75 m to a depth of 200 m.

In this month, it can be seen that Mindanao Eddy is only formed at a depth of 75 m and 200 m with a central point at 130-132 °E and 3.5 °N, while Halmahera Eddy is formed on four depth layers at 0 m, 75 m, 150 M and 200 m with the center point at 128.5 °E and 7.5 °N. The pattern of eddies formation will increases with the depth and when El Nino with total 39 eddies, but diameter of eddies become smaller. At normal eddy diameter measuring 30 - 300 km, while the occurrence of el nino phenomenon eddy diameter tends to decrease to 20-200 km. It may be caused at El Nino intensity, Mindanao currents as the caused of the formed of Eddy Mindanao weakened, otherwise North Equatorial Counter Current getting stronger marked with the movement of the warmpool of western Pacific to the eastern Pacific Ocean. Caused weakening of the wind moving toward the west in waters tropical Pacific.

During El Nino, showed that SSH decreased than normal with value ranges of 0.3-0.6 m. When cyclonic eddy formed at the coordinates of 130 °E and 7.5 °N has a lower value at eddy center of 0.3m. While on antisiclonic eddy has a higher value at the eddy center point with value ranges of 0.45 - 0.6 m. It shows that in the event of El Nino conditions in the Pacific Ocean sea surface height western part is lower than it would be in the eastern part of the Indian Ocean. It shows that in the event of El Nino conditions in the Pacific Ocean sea surface height western part is lower than it would be in the eastern part of the Indian Ocean [20].

IOP Conf. Series: Earth and Environmental Science **176** (2018) 012002 doi:10.1088/1755-1315/176/1/012002

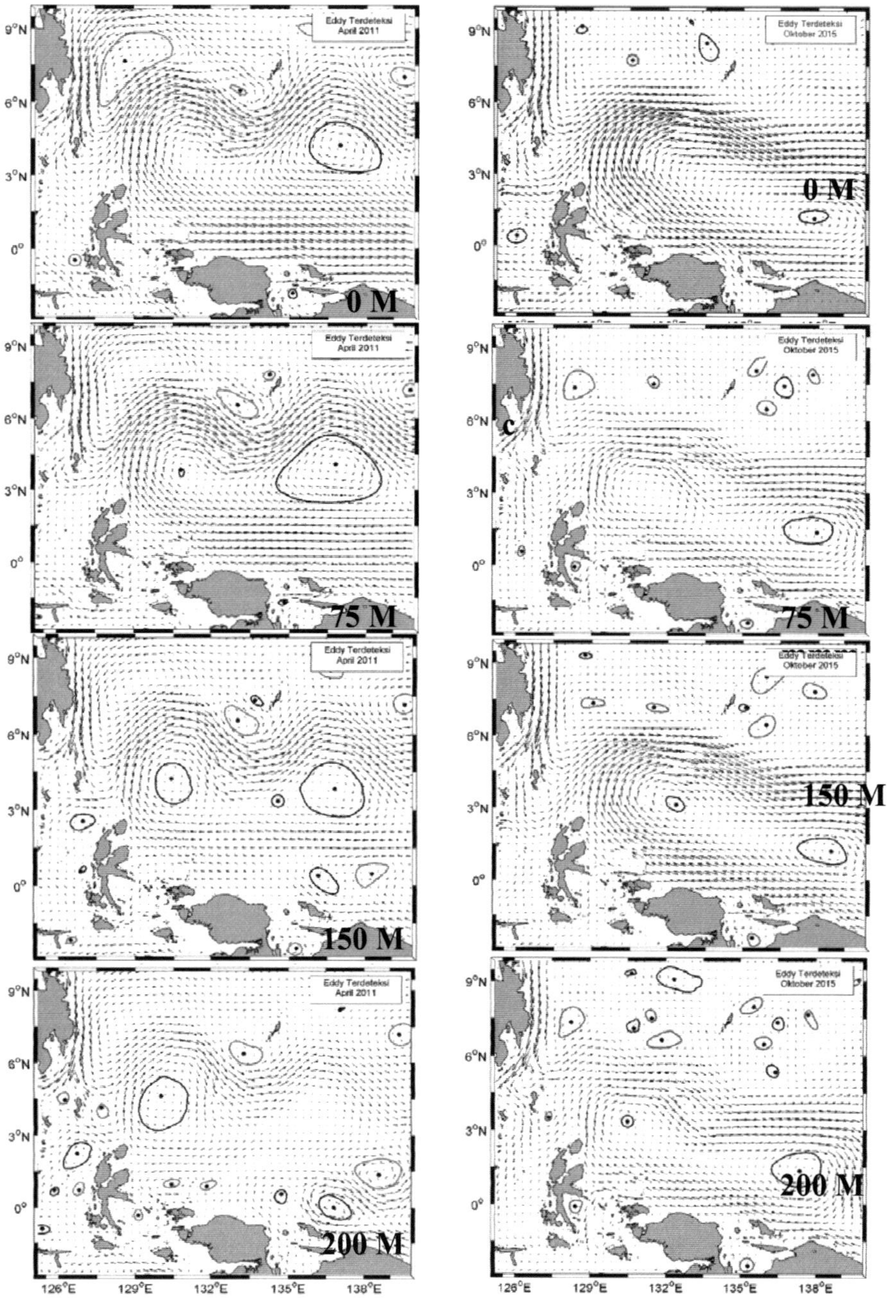

Figure 2. Eddy visualisation in April 2011 (left) and October 2015 (right). Red (black) circles depict cyclonic (anticyclonic) eddies.

At depth 0 m center anticyclonic eddy has a temperature value of 29.36 °C greater than the temperature value at center cyclonic eddy of 29,30 °C. This showed that at a depth of 0 m starts to show upwelling and downwelling. At depths of 75 m and 150 m in eddy cyclonic type still has lower temperature values than antisyclones eddy. The temperature value decreased drastically from 75 m to of 150 m, with value of 25.5 °C at a depth of 75 m and 16 °C at a depth of 150 m. It can be said that the depth of 150 m including the depth of the thermocline layer, because at this depth a drastic temperature drop reaches 7 °C. At depths of 75 m and 150 m it is seen that values at the center point of the cyclonic and antisyclonic eddy type have significant differences ranging from 2-3 °C, this becaused in eddy cyclonic there is a divergence on the surface caused the layer of thermocline in the eddy cyclones is higher, whereas at anticyclonic eddy occurs a surface congvergence that results in a deeper cyclonic layer of thermocline in eddy deeper.

The results of processing the formation of eddy in April 2011 represented La Nina condition shows that the total eddy at four depths of cyclonic eddy type are 24 eddies, while the total of antisyclonic eddy type are 20 eddy with total eddy formed of 44 Eddy. With variation diameter ranges from 30-450 km. In this month (figure 1: left) it can be seen that Mindanao Eddy is only formed at a depth of 75 m and 200 m with a central point at 130 °E and 5 °N, while Halmahera Eddy is formed on one depth layers at 0 m the center point at 128.5 °E and 7.5 °N. The pattern increases with the depth of the total eddy will tend to increase still occur at the time of La Nina phenomenon and, but during La Nina changed the size of diameter greater than the normal diameter. It maybe caused at El Nino intensity, Mindanao currents as the caused of the formed of Eddy Mindanao strengthened, otherwise North Equatorial Counter Current getting weakened marked with the movement of the warmpool of eastern Pacific to the western Pacific. Caused strengthened of the wind moving toward the west in waters tropical Pacific.

During La Nina, showed that minimum SSH decreased than normal with value of 0.4 m. When cyclonic eddy formed has a lower with value ranges of 0.4 – 0.6 m. While on antisiclonic eddy has a higher value at the eddy center point with value ranges of 0.4 - 0.85 m. During La nina phenomenon would be an increase in water mass volume, caused mass of water from the east is getting stronger to move western pacific marked with returned of the warmpool of western Pacific.

At depth 0 m center anticyclonic eddy has a temperature value of 29.97 °C greater than the temperature value at center cyclonic eddy of 29,12 °C. This showed that at a depth of 0 m starts to showed upwelling and downwelling. At depths of 75 m and 150 m in eddy cyclonic type still has lower temperature values than eddy antisyclones. The temperature value decreased drastically from 75 m to of 150 m, with value of 27.5°C at a depth of 75 m and 21.5 °C at a depth of 150 m. It can be said that the depth of 150 m including the depth of the thermocline layer, because at this depth a drastic temperature drop reaches 7°C. At depths of 75 m and 150 m it is seen that values at the center point of the cyclonic and antisyclonic eddy type have significant differences ranging from 3 ° C. During La nina conditions has an effect of increasing value temperatures eddy center started at 75 m to 200 m [21].

The correlation value between SOI with total cyclonik and antisiklonic eddy, it is known that the highest correlation value in both types is at depth 0 m (0.31) and 200 m (-0.36). This is assumed caused ENSO phenomenon has a influence to total eddy that formed up to depth 200 m. The correlation value between SOI and the temperature at center cyclonic and antisyclonic eddy is known that the highest correlation value of both types is at 200 m depth in eddy antisyclonic (0.98) and at 75 m depth in eddy cyclonic (0. 96). This is presumably because at the time of the ENSO phenomenon has a larger influence on the temperature at center eddy.

4. Conclusion

It can be concluded that ENSO conditions can affect on the eddies characteristics. Number of eddies increases with increase of depth, which has the large effect of SSH changes. The diameter, number of eddies and temperature center of eddies are decrease during El Niño, and increase during La Niña, compared to the normal year. The coefficient correlation between SOI and eddy diameter is high (-0.86), while coefficient correlation between SOI with total eddy is low (-0.36). The highest coefficient correlation is found between SOI and temperature center of eddy (0.98).

Acknowledgement

The author would like to thank Marine and Coastal Data Laboratory, Marine Research Center, Indonesian Ministry of Marine Affairs & Fisheries for providing the data and the facilities during the research process; Faculty of Fisheries and Marine Sciences, University of Padjadjaran for the facilities that have been provided; INDESO KP Project Research Agency which has been providing the dataset; and Komitmen Research Group.

References

[1] Arruda Z W and Nof D 2003 The Mindanao and Halmahera eddies–two eddies induced by nonlinearities *Journal of Physical Oceanography* **33** 2815-2830
[2] Wyrtki K 1961 *Physical Oceanography of the Southeast Asian Waters Naga Report California* pp 195
[3] McGillicuddy D J 2016 Mechanisms of Physical-Biological-Biogeochemical Interaction at the Oceanic Mesoscale *Annual Review of Marine Science* **8**(1) 125-159
[4] Hansen D V and Paul C A 1984 Genesis and effects of long waves in the equatorial Pacific *Journal of Geophysics Research* **89**(C6) 10431–40
[5] Klein P and Lapeyre G 2009 The Oceanic Vertical Pum Induced by Mesoscale and Submesoscale Turbulance *Annual Review of Marine Science* **1** 351-375
[6] Ropelweski C F and Halpert M S 1987 Global and Regional Scale Precipitation Pattern Associated with the El Niño /Southern Oscillation *Journal of* Monthly Weather Review **115** 1606-26
[7] Nuzula F, Syamsudin M L, Yuliadi L P S, Purba N P and Martono 2017 Eddies spatial variability at Makassar Strait – Flores Sea *IOP Conf. Series: Earth and Environmental Science* **54** 1-9
[8] Nencioli F, Dong C, Dickey T, Washburn L and McWilliams J C 2010 A vector geometry-based eddy detection algorithm and its application to a highresolution numerical product and high-frequency radar surface velocities in The Southern California Bight *Journal of Atmospheric and Oceanic Technology* **27**(3) 564-579
[9] Qu T D, Mitsudera H and Yamagata T 1998 On the western boundary currents in the Philippine Sea Journal of *Geophysi*cal *Res*earch *Oceans* **103**(C4) 537-548
[10] Wang F, Hu D and Bai H 1998 Western boundary undercurrents east of the Philippines *In*: He M X, Chen G eds. Proceedings of PORSEC'98-Qingdao, 28-31 July *Ocean Remote Sens. Inst.* (China: Ocean University of Qingdao) p 551-556
[11] Firing E, Kashino Y and Hacker P 2005 Energetic subthermocline currents observed east of Mindanao *Deep Sea Res. Part II Top. Study Oceanography* **52**(3-4) 605-613
[12] Qiu B and Chen S M 2012 Multidecadal sea level and gyre circulation variability in the northwestern tropical Pacific Ocean *Journal of Physical Oceanography* **42**(1) 193-206
[13] Kashino Y, Atmadipoera A and Kuroda Y 2013 Observed features of the Halmahera and Mindanau Eddies *Journal of Geophysical Research* **118** 6543-6560
[14] Chiang T L and Qu T D 2013 Subthermocline eddies in the western equatorial Pacific as shown by an eddy-resolving OGCM *Journal of Physical Oceanography* **43**(7) 241-253
[15] Dong C, Lin X, Liu Y, Nencioli F, Chao Y, Guan Y, Chen D, Dickey T and McWilliams J C Threedimensional Oceanic Eddy Analysis in the Southern California Bight from a Numerical Product *Journal of GeophysicalResearch* **117**(1) 1-17
[16] Tussadiah A, Syamsuddin M L, Pranowo W S, Riyantini I and Purba N P 2016 Eddy vertical structure in Southern Java Indian Ocean: identification using automated eddies detection *International Journal of Science and Research* (IJSR) **5**(3) 967-971
[17] Troup A J 1965 The southern oscillation *Journal of the Royal Meteorological Society* **91**(390) 490-506
[18] Kumar S P, Babu M T and Rao D P 1992 Energy and generating mechanism of a subsurface cold core eddy in the Bay of Bengal *Indian Journal of Marine Science* **21**(2) 140-142
[19] Maximenko N, Hafner J and Niiler P 2012 Pathways of marine debris derived from trajectories of lagrangian drifters *Marine Pollution Bulletin* **65** 51-62
[20] Zheng S, DuY, Li J and Cheng X 2015 Eddy Characteristics in the South Indian Ocean as inferred from Surface Drifters *Journal Ocean Science* **11**(3) 361-371
[21] Rejeki H A, Munasik and Kunarso 2017 The effect of ENSO to the variability of sea surface height in western Pacific Ocean and eastern Indian Ocean and its connectivity to the Indonesia Throughflow (ITF) *IOP Conf. Series: Earth and Environmental Science* **55** 1-7
[22] Martono 2008 *Studi Eddy Mindanao dan Eddy Halmahera Report thesis Bandung* pp50

IOP Conf. Series: Earth and Environmental Science **176** (2018) 012003 doi:10.1088/1755-1315/176/1/012003

The wave characteristics in Natuna Sea and its adjacent for naval operation base purposes

P D Anggara[1*], T M Alam[1], D Adrianto[2], W S Pranowo[1,3*]

[1] Department of Hydrography Engineering, Indonesian Naval Postgraduate School (STTAL), Jl. Pantai Kuta V No. 01, Ancol Timur, Jakarta 14430, Indonesia
[2] Hydrography and Oceanography Center, Indonesian Navy, Jl. Pantai Kuta V No.1 Ancol Timur, Jakarta Utara Indonesia 14430, Indonesia
[3] Marine Research Center, Indonesian Ministry of Marine Affairs and Fisheries, Jl. Pasir Putih II, Ancol Timur, Jakarta 14430, Indonesia
*e-mail: purrydjati@gmail.com; widodo.pranowo@gmail.com

Abstract. Natuna Sea is one economy strategic outer area, which connected into archipelagic international sea lanes (ALKI I). The research area is in that area using two specifics observation Naval base stations: (South) Dabo Singkep and (North) Ranai. Wave characteristics are needed to be conducting a secure the naval operation cruises. The statistical analysis is carried out using 2016 ECMWF 6 hourly datasets with 14 km grid spatial resolution. The variable of significant wave height (SWH), period and wave direction has been analyzed by monthly-yearly averages. Maximum SWH during rough wave period is range of 5.1 m (January) and 2.3 m (February). The minimum SWH is range of 0.2 m (April) and 0.1 m (July). The predominantly wave direction at Dabo Singkep is coming from southeast (135°; mostly in July), and at Ranai is coming from the northeast (45°; mostly in January). During the rough wave period, it is highly recommended using the KRI warship with length > 82 m and high hull > 2.3 m for surveillance cruise in the southern part of Natuna Sea. While in the northern part, would be needed KRI with a length > 146 m, and the high hull > 4.8 m.

1. Introduction

Land, water, and air are dimensions of space used to implement military operations. Every space dimension has parameters that can determine the success of a military operation. Parameters that affect military operations in the sea is an attribute of seawater, sea surface behavior, marine topography and representative naval ramifications. The uppermost layer of seawater is eternally dynamic in response to ocean climate interaction [1].

One important parameter of the hydrodynamic is wave. Wave is a mechanism of up and down sea surface in perpendicular which happened in a moment until it reaches equilibrium [2]. Sea surface height difference can be generated by some sources such as the wind, earth plate activity, due to the movement of the ship, tidal pumping and ocean currents [3]. In this study we are focussing on the wind waves.

Waves energy will be considered as a nuisance when used as a parameter in the field of navigation, port, coastal and offshore engineering [4]. Therefore, it is very important to study hydrodynamics for management and its control [5]. So that energy resulting from ocean waves do not give a loss in human life.

Nowadays, the maritime economy is one of the priority target sectors by Indonesian Government for the next 25 years [6]. Indonesia should take the sea as a media liaison between the islands. Indonesia have 16.056 named and coordinated islands [7]. The connectivity among those islands is expected to be optimalized, so that all development sectors can be well distributed. One important role is attainment with protecting stability and security in Indonesian Seas.

Various ways can be conducted to protect stability and security of Indonesian Seas. One of those, by providing information of ocean wave characteristics. It can be used as a reference of how to choose the class's warship (KRI) for surveillance cruise. Because protect cruise safety and maritime security is the responsibility of The Indonesian Government [8].

Indonesian archipelagic setting among the two oceans will have various and unique wave characteristics. The appropriate management system can reduce risk level of KRI accidents during the military operation, if the determination of KRI class is adjusted to the wave characteristics. In this study, we are using 5 (five) class of KRI.

2. Materials and methods

This study focuses on the waves generated by the wind, where there will be three different physical processes and eventually known as wind sea and swell [9]. The location of wind sources, fetch, wind speed, long wind blow and wind direction are parameters that can affect the height, direction and wave period.

The study area is located in Natuna Waters (table 1). Dataset used is significant wave height (SWH), mean wave direction (MWD) and mean wave period (MWP) throughout 2016, downloaded from ECMWF (European Centre for Medium-range Weather Forecasts) website with spatial resolution 14 km x 14 km (0.125° x 0.125°) and resolution temporal 6 hours [10]. Mean wave direction is defined using the meteorological convention that zero means "coming from the north" and 90 "coming from the east" [11]. Data is calculated monthly and seasonal average to see the characteristics [8].

Table 1. Classification of wave directions.

Wave direction	Class degree (°)
North	337.5 – 22.5
Northeast	22.5 – 67.5
East	67.5 – 112.5
Southeast	112.5 – 157.5
South	157.5 – 202.5
Southwest	202.5 – 247.5
West	247.5 – 292.5
Northwest	292.5 – 337.5

Ocean Data View (ODV) Software is employed to visualize ocean wave's propagations (2D SWH model), as a tool for analyze wave characteristics and its form in the research area. WRPLOT software is applied to visualize wave directions, in the form of a rose diagram, with 5 (six) colors as classification of wave velocity. Classifications of wave direction and velocity in WRPLOT are grouped in the table 1 and 2, respectively.

Table 2. Classification of wave velocity.

Color	Velocity (ms^{-1})
Orange	≥ 16
Yellow	12 – 16
Light green	8 – 12
Dark green	4 – 8
Blue	0 – 4

The research location is determined from the data station (●). Station 1 (◆) is located in the working area of the Dabo Singkep Navy Base around meeting ALKI 1, with branch 1A connecting the Singapore Strait to geographic coordinates 00° 45' 00" N 106° 7' 30" E. Station 2 (▲) is located in the working area of Ranai Navy Base around the boundary line of the Indonesian continental shelf in geographic coordinates 6° 30' 00" N 109° 15' 00" E [12]. From geographical view, station 2 has more open sea space than station 1 flanked by Kalimantan Island (east) and Bintan Island (west), see figure 1.

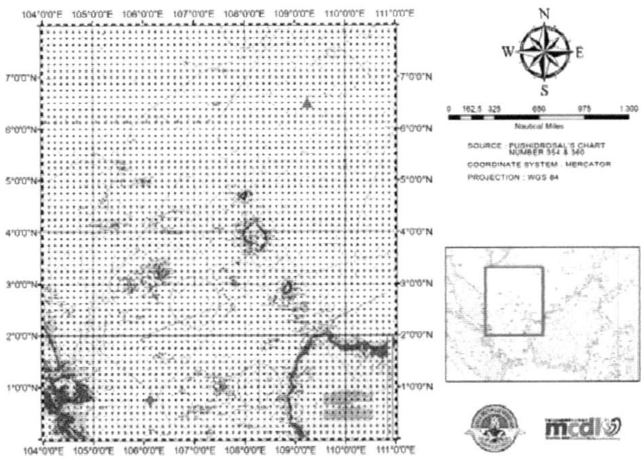

Figure 1. Location map of research station.

Ocean waves are the physical form of sea water where still water level (SWL) experience up and down movement resemble sinusoidal line so that it has wave height, wavelength and sea surface fluctuation with wave direction and wave velocity that affects the motion of water particles. Ocean waves have some parameters such as wave height (H), wavelength (L), wave velocity (C), wave period (T), wave amplitude (a) and water surface fluctuation (η). Significant wave height (Hs) is obtained from average of 1/3rd (one-third) highest individual wave height that occurs during a specific time period [13], see Figure 2.

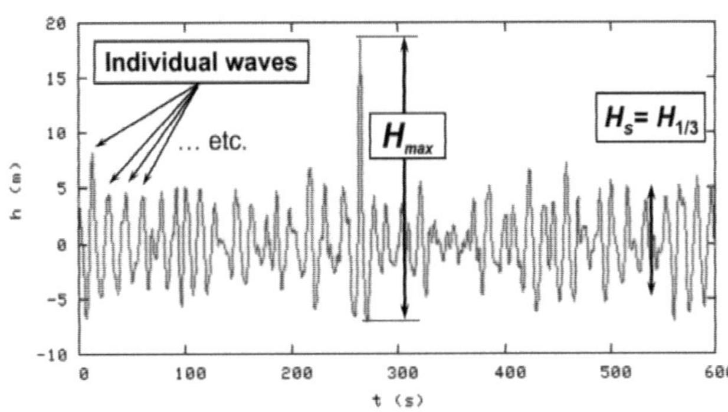

Figure 2. Time series of surface elevation of the Draupner platform in the North Sea [13].

Analyzing wave height can not be equated for all areas. The depth is a parameter must be considered. Wave characteristics will be affected by the friction of water particles with the seabed, when the depth is relatively shallow. The motion of water particles does not erode the seabed when the depth is relatively deep. Based on The Indonesian navigation chart number 354, station 1 has average depth of 200 meters [14]. The chart number 360 shows station 2 has average depth of 50 meters [15]. The length and wave velocity are determined using the equation of Airy wave theory [16], see equation (1) and (2). Wave period data can be used to compute the wavelength value (equation 2), which then used to compute the wave velocity value (equation 1).

$$C = LT^{-1} = gT(2\pi)^{-1} = 1.56 \, x \, T \quad (1)$$

$$L = gT^2 (2\pi)^{-1} = 1.56 \, x \, T^2 \quad (2)$$

Explanation: L: wavelength (m), g: Acceleration of gravity (9.8 ms^{-2}), T: wave period (s), C: wave velocity (ms^{-1}), π: 3.14.

The effect of ocean wave motion to ship stability will cause two motion types i.e translational motion and rotation motion [17]. According to them, motion forward/backward (surge), right/left side (sway) and up/down (lifted) is translation motion; while rotational motion is rotating motion to axis line as rotation shaft (figure 3).

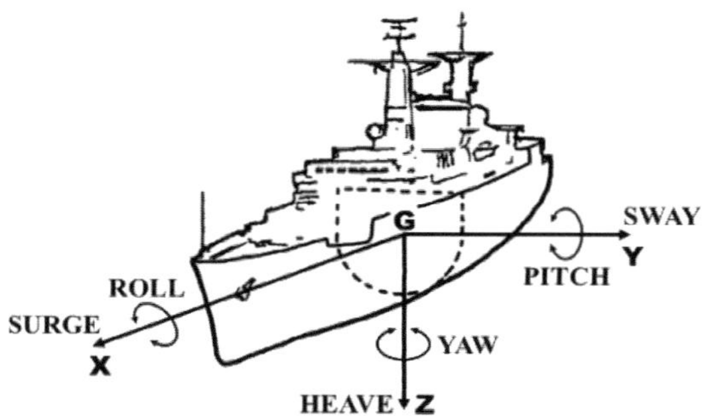

Figure 3. Effects of ocean wave motion on ship stability [17].

The safest motion KRI to ocean waves when navigating is surge translation motion and pitch rotation motion [18, 19]. Based on those motions, the line of water length of the ship (LWL) become a parameter to determine wavelength limit and hull height of prow (HH) is used as the parameter to determine wave height limit. The work flow for analyzing a wave characteristics is presented in figure 4.

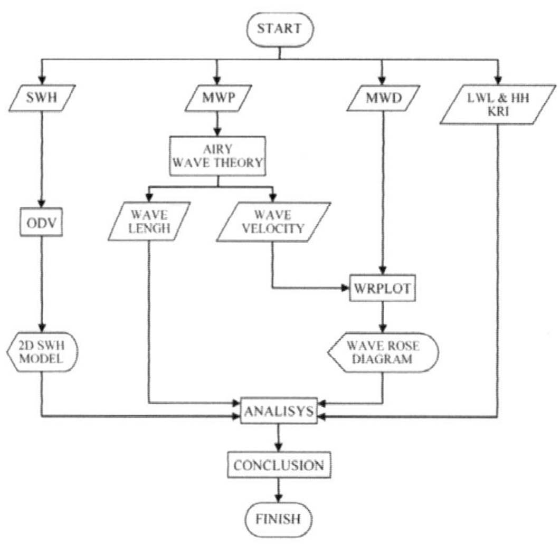

Figure 4. Flowchart analysis of wave characteristics for KRI recommendations.

3. Results and discussion

3.1. Significant wave height characteristics

In February 7, 2016 from 18.00 UTC to 24.00 UTC, maximum significant wave height in station 1 reached 2.3 meters. Minimum significant wave height reached 0.1 meters, occurred in July 01, 2016 from 06.00 UTC to 12.00 UTC. High wave dominant occurred on February and low wave dominant occurred on April (figure 5).

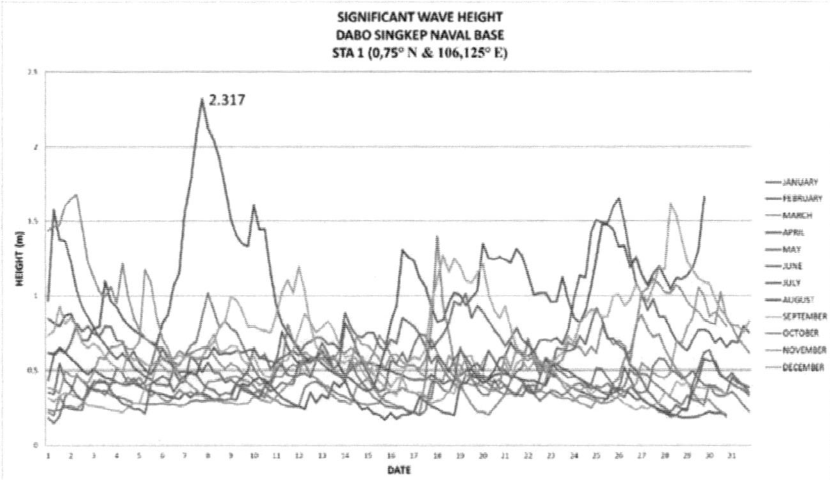

Figure 5. Significant wave height mareogram of 6 hours average in station 1.

In January 25, 2016, from 12.00 UTC to 18.00 UTC, maximum significant wave height in station 2 reached 5.1 meters. Minimum significant wave height reached 0.2 meters, occurred in April 14, 2016, from 18.00 UTC to 24.00 UTC. High wave dominant occurred in February and low wave dominant occurred in April (figure 6).

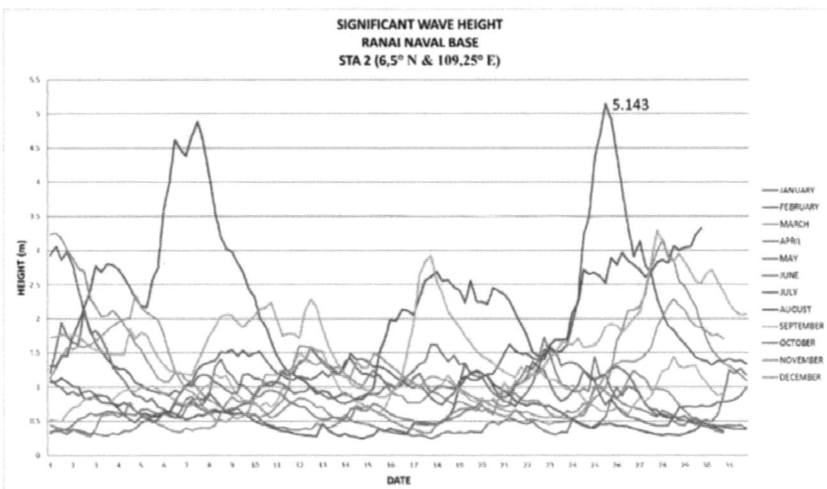

Figure 6. Significant wave height mareogram of 6 hours average in station 2.

3.2. Significant wave height propagation

The global ocean is strongly dominated by swell waves spreading away from the influence of the wind zone [20]. Based on the form of significant wave height contour lines can be known the location of influence of the wind zone and the direction of its propagation. The change in direction of the propagation of significant wave height influenced by the state of geographic and hydrographic.

Analyzes of wave propagation patterns are grouped into four seasons [21]. Natuna People distinguish four seasons based on wind direction i.e. north season on November to February, East season on March to June, the south season in July and August, west season on September and October [22]. In northern season, wave height reaches the maximum height and bad weather occured in Natuna Sea [23]. The wind can influence the length and significant wave height that occurs in Indonesian Seas [19].

The 2D model of significant wave height in Natuna Waters has 4 (four) propagation's colors to help analyzing wave heights (figure 7). Wave direction dominant changes from northeast to southwest during North season and East season. During North season, in February 7, 2016, from 18.00 UTC to 24.00 UTC in station 1 has maximum significant wave height reached 2.3 meters (green) and wave propagation direction came from Natuna Besar Islands (25°) with wave height ranged 2.3 to 3.2 meters (green). The area between Anambas Islands and Natuna Besar Islands has significant wave height ranged 3.3 to 3.4 meters (green). Wave propagation direction in station 2 came from the northeast (27°) toward Natuna Besar Islands with wave height ranged 3.3 to 4.6 meters (green-yellow).

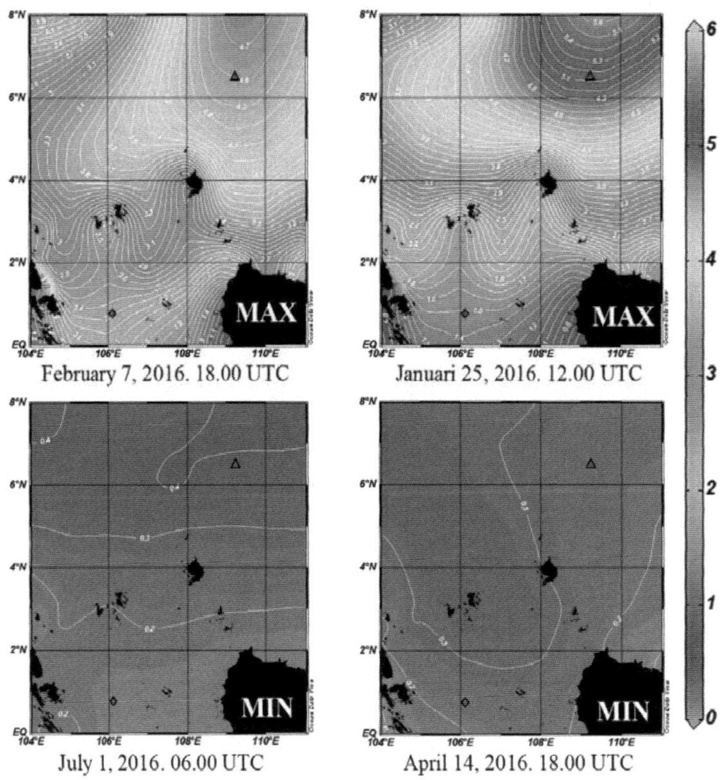

Figure 7. Significant wave height propagation in station 1 (1st column) and station 2 (2nd column).

During North season, in January 25, 2016, from 12.00 UTC to 18.00 UTC, in station 2 has maximum significant wave height reached 5.1 meters. The wave propagation direction came from the northeast (30°), toward Natuna Besar Islands, with significant wave height ranged of 2.9 to 5.1 meters (red-yellow-green). The area between Natuna Besar Islands and Anambas Islands has significant wave height ranged of 2.5 and 2.7 meters (green). Wave propagation direction, in station 1, is came from Natuna Besar Islands (34°), with significant wave height ranged of 1.5 to 2.1 meters (green-blue).

Minimum significant wave height in Natuna Sea in every season has almost evenly. It might because of wind speed is getting slower, and cannot generate significant waves. During South season, in 2016, station 1 has minimum significant wave height. Station 2 has minimum significant wave height during East season. Because in Natuna Waters experienced dry season where the wind blows from southern hemisphere toward northern hemisphere [24].

In July 1, 2016, from 06.00 UTC to 12.00 UTC, station 1 has minimum significant wave height reached 0.1 meters (light purple), with wave propagation direction is came from southeast (134°). The area between Natuna Besar Islands and Anambas Islands has significant wave height ranged 0.2 meters (purple). Station 2 has significant wave height reached 0.3 meters (dark purple), with wave propagation direction is came from the southwest (231°).

In April 14, 2016, from 18.00 UTC to 24.00 UTC, station 2 has minimum significant wave height reached 0.2 meters (purple), with wave propagation direction is came from southeast (161°). The area between Natuna Besar Islands and Anambas Islands has significant wave height reached 0.3 meters (dark purple). Station 1 has significant wave height reached 0.2 meters (purple), with wave propagation direction is came from the south.

In general, Indonesia territory based on geographical location has two monsoons i.e. east monsoon and west monsoon [25, 26]. East monsoon (April - September) brings a dry season causing by the wind blows from Australian Continent to Asian Continent [27]. West monsoon (October - March) brings a rainy season causing the wind blows from Asian Continent to Australian Continent [28]. In general, west monsoon a lot of sea transportation vehicles undergo accident due to bad weather condition [29]. Bad sea weather is induced by extreme wind including its ocean waves [30]. That is why very important to analyzing wave's characteristic, due to forecasting, to get the information about the safest periods for surveillance cruise activities. The results can also used as reference of decisions taken by policy (stake) holders to manage their marine zonation [31].

In Natuna sea, based on wave roses diagram in station 1 and station 2, during the north and east seasons are dominated by west monsoon. Wave propagation direction is came from the South China Sea. The highest mean wave velocity, ranging from 12 to 16 ms^{-1}, occurs ofently during the north season in station 2 (figure 8).

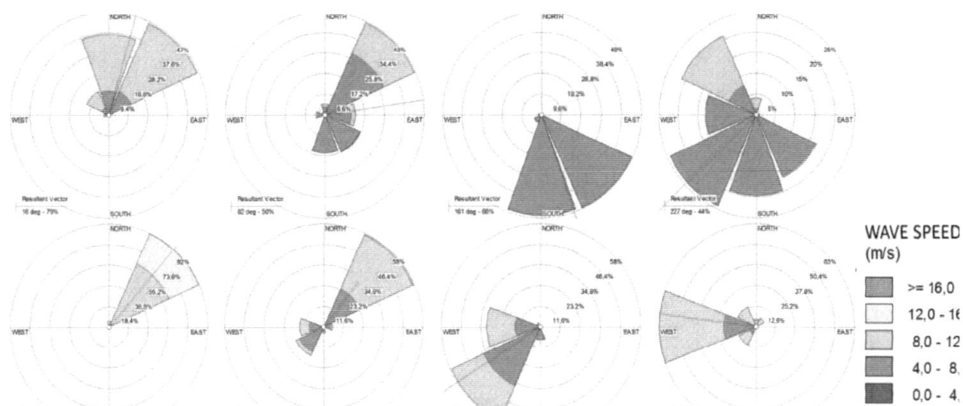

Figure 8. Wave rose chart of seasonal average in station 1 (top row) and station 2 (bottom row).

Monthly wave variations can be showed by computing a monthly average data of SWH [18]. Station 1 has range maximum SWH in February ranged of 0.4 to 2.3 meters. In April and May, minimum SWH has range of 0.2 - 0.7 meters. In January, February and December, wavelength is over 80 meters. Maximum wavelength reached 86.3 meters is founded in January. In general, during January-April and November - December, wave is propagate from northeren toward south hemisphere. Maximum wave velocity (8.6 ms^{-1}) is found in February (table 3).

Table 3. Monthly average of wave characteristics in station 1.

	Height (m)	Length (m)	Direction (^{0})	Velocity (ms^{-1})
January	0.4 – 1.7	24.8 – 86.3	41	8.3
February	0.4 – 2.3	21.2 – 82.1	31	8.6
March	0.4 – 1.7	18.9 – 61.4	39	7.5
April	0.2 – 0.7	11.5 – 47.8	62	6.5
May	0.2 – 0.7	14.2 – 50.3	100	6.5
June	0.2 – 1.4	11.8 – 44.7	164	5.8
July	0.3 – 2.3	12.8 – 35.4	159	5.7
August	0.2 – 1	13.3 – 36.5	162	6.2
September	0.2 – 0.9	15.6 – 50.3	207	6.5
October	0.2 – 0.8	13.3 – 47.7	240	6.3
November	0.3 – 1.2	19.4 – 73.3	353	7.7
December	0.3 – 1.6	22.3 – 80.1	349	8

Thousands of islands setting, in Indonesian seas, then creating various fetches, which is influences wind movement characteristic [32]. In general, monthly wave average of station 1 is lower than station 2. It might because of the station 2 is located more open sea than station 1. Then, Station 2 has significant wave heights over 3 meters, during January - March, and also in December. Beware of wave height in January because maximum significant wave height reached 5.1 meters. In April has minimum range significant wave height ranged 0.2 to 1.2 meters. But if we have a look into the wavelength characteristic, it is only December has a wavelength over 170 meters, with maximum length of 171.7 meters. The wave moves from the South China Sea to Karimata Straits in January - May, and also November – December. Maximum wave velocity (12.3 ms^{-1}) is found in December (table 4).

Table 4. Monthly average of wave characteristics in station 2.

	Height (m)	Length (m)	Direction (^{0})	Velocity (ms^{-1})
January	0.7 – 5.1	32.3 – 168.4	45	10.3
February	0.8 – 4.9	30.8 – 146.8	45	11.3
March	0.5 – 3.3	24.9 – 105.6	45	9.4
April	0.2 – 1.2	16 – 64.5	54	7.2
May	0.3 – 1.7	21 – 83.3	25	7.8
June	0.3 – 1.2	13.5 – 42.4	234	6.7
July	0.2 – 1.4	16.5 – 44.3	232	6.6
August	0.4 – 2.1	23.3 – 66.1	242	7.9
September	0.5 – 1.6	24.7 – 70.8	268	7.9
October	0.4 – 1.6	21 – 110	289	8.4
November	0.5 – 2.3	33.6 – 147.1	41	11.2
December	0.8 – 3.3	53.1 – 171.7	33	12.3

3.3. Analysis of wave height and length versus KRI warship dimension

Dabo Singkep Naval Base and Ranai Naval Base is work unit of the Naval Main Base (Lantamal) IV under Indonesian Navy Western Fleet (Koarmabar), which have type and class of KRI warship in table 5.

Table 5. The array of KRI under Koarmabar.

Type	Class	LWL (m)	HH (m)
Frosch I	Teluk Sibolga	82.5	10
Parchim	Wiratno	75.2	5.3
FPB 57 NAV IV	Barakuda	54.4	2.8
Kondor II	Pulau Romang	53	3.2
KCR-40	Kujang	40	2.7

The wave energy will be considered as the intrusion, when used as parameter of navigation [33]. The calm waves will intensively increase cruise activities. However, if strong winds and high waves then it is expected to have kind of an information of the proper KRI warship dimension to keep the safety during cruises. These information is also needed for public sea water transportations [18].

Analysis of wave height and length of the (war-) ships are important role for Naval policy holders concerned at sea, i.e. for cruise management strategy, and for marine infrastructure development [31]. Wave height data is used to determine hull height (HH) of KRI prow from sea water line (LWL). If prow hull of KRI higher than wave height, then ship equipment on the open deck is safe from overtopping waves. Wavelength data is also used to determine line water length of KRI. For minimize pitch rotation motion of KRI, the line water length KRI must be longer than the wavelength. Pitch rotation motion of KRI will influence the accuracy of cannon fire during cruise. The worst scenario (damaging and drown of KRI warship) will be happened when wave height and length higher than LWL and HH, in respectively.

3.4. Recommendation

For those who sail in ALKI I, a warning (SWH > 3 m; ~80% susceptibility) should be given during December - January to every ship before sailing [34]. Based on 2016 datasets, high waves often occur in February, then the recommended KRI, who will carry out surveillance cruise at station 1, is to have prow hull height more than 2.3 meters, and ship water line length more than 82 meters. For KRI, who will carry out surveillance cruise at station 2, is recommended to have prow hull height more than 4.8 meters, and ship water line length more than 146 meters. An example, KRI Kujang is not recommended to carry out surveillance cruise at station 2 because every month is formed wavelength exceeds line water length of class's KRI Kujang (see Table 8 and compared with Table 3 & 4).

4. Conclusion

In station 1, all class's KRI should be alert to ocean waves that occurred in January because wavelength exceeds ship line water length and wave height reached 1.7 meters so that it causes pitch rotation motion of KRI. In station 2, all class's KRI should be alert to ocean waves that occurred in the north season. And for class's KRI Kujang is not recommended to carry out surveillance cruise at station 2 because every month is formed wavelength exceeds line water length of class's KRI Kujang. Without losses of material and personnel than a military operation is declared successful. The success of military operation at sea must be supported by information on ocean wave characteristics. It might because of high waves and bad weather will threats a military operation at sea.

Acknowledgement

The authors would like to thank to Lieutenant Commander Johar Setiadi, M.Si. and Lieutenant Commander Eri Lesmana, S.T. (STTAL), who fully supported to publishing this article. All work has been done in Marine and Coastal Data Laboratory, Marine Research Center (Indonesian Ministry of Marine Affairs and Fisheries). This publication is funded by STTAL.

IOP Conf. Series: Earth and Environmental Science **176** (2018) 012003 doi:10.1088/1755-1315/176/1/012003

References

[1] Collins J M 1998 *Military Geography for Professionals and The Public* (Washington, DC: National Defense Univercity Press) part 1 pp 47-59
[2] Purba N P and W S Pranowo 2015 *Dinamika Oseanografi, Deskripsi Karakteristik Massa Air dan Sirkulasi Air Laut* (Bandung: Unpad Press) chapter 4 pp 148-163. [in Bahasa]
[3] Pranowo W S 2016 *Geology and Geophysics of Ocean:* Semester III S1/XXXVI-2016. Lecture Module. Department of Hidrography. Indonesian Naval Postgraduate School (STTAL) Jakarta
[4] Falcão A F de O 2010 Wave energy utilization: a review of the technologies renewable and sustainable energy *Reviews* **14** 899-918
[5] Mehdiabadi F E, Mehdizadeh M M and Rahbani M 2015 Simulating wind driven waves in the Strait of Hormuz using MIKE21 *J. Ilmu Kelautan* **20**(1) 1-8
[6] Saragih J F S 2016 *Prioritas Dalam Industri Maritim* Maritimnews (http://maritimnews.com/prioritas-dalam-industri-maritim/)
[7] Kemendagri 2017 *Sebanyak 16.056 Pulau Sudah Dibakukan dan Dilaporkan ke PBB* (http://www.kemendagri.go.id/news/2017/10/09/sebanyak-16056-pulau-sudah-dibakukan-dan-dilaporkan-ke-pbb)
[8] Pranowo W S 2016 *Adjustment Computation:* Semester II S1/XXXVI-2016. *Lecture Module.* Department of Hidrography. Indonesian Naval Postgraduate School (STTAL), Jakarta
[9] Stewart R H 2008 *Physical Oceanography* (Texas: A&M University Press)
[10] ECMWF 2016 *ERA Interim Daily* (http://apps.ecmwf.int/datasets/data/interim-full-daily/levtype=sfc/) [17-08-2017]
[11] ECMWF 2015 *What is The Direction Convention for Wave Fields?* (https://www.ecmwf.int/en/faq/what-direction-convention-wave-fields)
[12] Kasal Decree No. Kep/326/II/2016 *About The Degree Arrangement and Classification and Working Areas Boundaries of Indonesian Naval Bases*
[13] Bidlot J R 2016 *Ocean Waves* ECMWF p 50
[14] Pushidrosal 2012 *Indonesian Navy Chart No. 354: Natuna Sea, Anambas Islands and Natuna to Tanjung Datu*
[15] Pushidrosal 2014 *Indonesian Navy Chart No. 360: Malaka Strait, Natuna Sea to Java Sea*
[16] Triatmodjo B 1999 *Teknik Pantai* (Yogyakarta: Beta Offset) pp 14-20. ISBN: 979-8541-05-7 [in Bahasa]
[17] Gates P J and Lynn N M 1990 *Ships, Submarines and The Sea* (London: Brassey's)
[18] Wicaksana S, Sofian I, Pranowo W S, Kuswardani A R T D, Saroso and Sukoco N B 2015 Karakteristik gelombang signifikan di Selat Karimata dan Laut Jawa berdasarkan rerata angin 9 tahunan (2005-2013) *J. OmniAquatika* **11**(2) 33-40
[19] Taryono, Sofian I, Kuswardani A R T D and Alam T M 2016 Analisis panjang dan tinggi gelombang untuk operasi KRI TNI AL di Perairan Indonesia *J. Chart Datum* **2**(1) 72-87\
[20] Semedo A, Sušelj K, Rutgersson A and Sterl A 2011 A global view on the wind sea and swell climate and variability from ERA-40 *J. Climate* **24** 1461-1479
[21] Mustikasari E, Dewi L C, Heriati A and Pranowo W S 2015 Three-dimensional (3D) barotropic current circulation modelling of seasonal upwelling in Indonesian Seas *J. Segara* **11**(1) 25-35
[22] Regional Government of Natuna 2013 *Selayang Pandang Kabupaten Natuna* (http://www.natunakab.go.id/sekilas-natuna.html)
[23] Sagala S L, Bramawanto R, Kuswardani A R T D and Pranowo W S 2014 Distribution of heavy metals in Natuna Coastal Waters *J. Ilmu dan Teknologi Kelautan Tropis* **6**(2) 297-310
[24] Pranowo W S, Puspita C D, Adi R A, Kuswardani A R T D and Dewi L C 2013 *Atlas of Natuna Marine and Coastal Resources nad its Adjacent* First Ed 126 pages ISBN: 978-602-9086-37-9
[25] Siregar S N, Sari L P, Purba N P, Pranowo W S and Syamsuddin M L 2015 *The water mass exchange in Java Sea to periodicity of monsoon and ITF in 2015 J. Depik* **6**(1) 44-59
[26] Hong Z, Sannasiraj S A and Chan E S 2009 Wind wave effects on hydrodynamic modeling of ocean circulation in the South China Sea *The Open Civil Engineering Journal* **3** 48-61

[27] Pranowo W S, Phillips H and Wijffels S 2005 Upwelling event 2003 along South Java Sea and the Sea of Lesser Sunda Islands *J. Segara* **1**(3) 116-123
[28] Pranowo W S 2012 Upwelling and downwelling dynamics in Arafura and Timor Seas *J. Widya Riset* **15**(2) 415-424
[29] Pranowo W S, Adi T R, Makarim S and Hasanah N N 2012 *Marine & Climate Research Contributions to the National Program on Climate Change Adaptation & Mitigation* Procced The International Workshop on Climate Information Services in Supporting Mitigation & Adaptation to Climate Change in Transportation & Tourism 15-16 May 2012 pp 76-79 ISBN:978-602-19508-3-8
[30] Behrens J, Beisiegel N and Pranowo W S 2010 *Inundation modelling with adaptive triangular mesh refinement – state of the art numerical methods for accurate simulation* Proc. Abs., Storm Surges Congress 2010: Risk and management of current and future storm surges. University of Hamburg, 13th – 17th September, 2010
[31] Risandi J, Sagala S L and Pranowo W S 2015 The application of wave characteristic numerical model for site selection of marine aquaculture development in Situbondo, East Java *J. Kelautan Nasional* **10**(1) 21-31
[32] Aisjah A S, Arifin S and Danistha W L 2016 Sverdruv munk bretschneider modification (SMB) for significant wave height prediction in Java Sea British *J. of Applied Science and Technology* **16**(2) 1-8
[33] Pratomo Y, W S Pranowo, H Setiadi, G Harsono, Kamija, S M Simanjuntak and T M Alam 2016 Identifikasi penjalaran gelombang panjang Samudera Hindia ke Selat Lombok berdasarkan komponen harmonik arus *J. Omni Aktuatika* **12**(1) 22-29
[34] Kurniawan R and Khotimah M K 2015 Ocean wave characteristics in Indonesian Waters for sea transportation safety and planning *J. IPTEK* **26**(1) 19-27

IOP Conf. Series: Earth and Environmental Science **176** (2018) 012004 doi:10.1088/1755-1315/176/1/012004

Characteristic of eddies kinetic energy associated with yellowfin tuna in southern Java Indian Ocean

A Tussadiah[1*], W S Pranowo[1, 2], M L Syamsuddin[3], I Riyantini[3], B Nugraha[4], D Novianto[5]

[1] Marine and Coastal Data Laboratory, Marine Research Center, Agency of Research and Human Resources, Ministry of Marine Affairs & Fisheries Republic of Indonesia
[2] Department of Hydrography Engineering, Indonesian Naval Postgraduate School (STTAL)
[3] Padjadjaran University, Department of Marine Science, Fisheries and Marine Science Faculty, Padjadjaran University
[4] Fisheries Research Center, Ministry of Marine Affairs & Fisheries Republic of Indonesia
[5] Tuna Fisheries Research Institute, Ministry of Marine Affairs & Fisheries Republic of Indonesia
*e-mail: armyanda@gmail.com

Abstract. Eddies kinetic energy (EKE) is one of the most indicators in observed the phenomena of upwelling and down welling generated by circular current or eddies. This research aims are to determine the characteristics and intensity of eddies and EKE and its relation to the distribution of yellowfin tuna (*Thunnus albacares*) in Southern Java Indian Ocean. Based on Automated Eddies Detection (AED) method, it is succesfully discovered many eddies well formed in Southern Java Indian Ocean through 2014. Furthermore, EKE has a vary values to the depth, which tends to be higher at the surface layer than at thermocline layer. EKE in this region display a distinct seasonal cycle with maximum occurred in Southeast Monsoon and minimum in Northwest Monsoon. Increasing of EKE in Southeast Monsoon tend to be affected by currents system of Indonesian Throughflow (ITF) and South Equatorial Current (SEC). The highest of yellowfin tuna catches found in June 2014 and occurred in the current system with high EKE intensity. Correlation between eddies and yellowfin tuna found at thermocline layer (109 m) and occurred in cyclonic eddies ($R^2 = 0.59$).

1. Introduction

Indonesian archipelagos link the Pacific and Indian Oceans and play a crucial role in global ocean thermohaline circulation [1]. Furthermore, Tropical Indian Ocean interaction with the atmosphere plays an important role in shaping climate on either regional or global scales [2]. There are unique ocean currents that flow seasonally through Indian Ocean, among others are the Indonesian Throughflow (ITF) which links Pacific and Indian Ocean [3], the South Java Current (SJC) which flows along the south Java coast [4], and the South Equatorial Current (SEC) which flows westward from western Indian Ocean [5]. Despite this importance, many aspects of upper ocean dynamics are still unknown, especially of different types of motion and scales. Upper ocean variability is characterized by eddies, mostly geostrophic and generated by instabilities of ocean currents [6]. Eddies are important because it have so much kinetic energy, as well it is responsible for the irreversible mixing of waters with different properties. Eddies have a deep roots and can carrying energy and momentum to the seafloor [7].

Eddies are generally more energetic than the surrounding currents and are an important component of dynamical oceanography at all scales [8]. The fact that both cyclonic and anti-cyclonic eddies may embody either upwelling or downwelling in their centres [9]. *Note*: In Southern Hemisphere, "cyclonic eddies" rotate clockwise and "anti-cyclonic eddies" rotate counter clockwise (figure 2; red and black circles represent cyclonic and anticyclonic), vice versa in Northern Hemisphere. When cyclonic eddies occurred there may be nutrients in abundance, hence of the mixing water mass from bottom layer to the upper layer. This nutrient might be supports living organism through food chain start from plankton, small fish, up to pelagic fish, for instance tuna [4]. Tuna (*Thunnus.sp*) has a very high economic value in the world market. Hence, it is the most important fishery resources targeted by fisherman from various countries including Indonesia, also it is an important food source in the world [10].

In detection of eddies formation, a few method have been proposed, based either on the physical or geometrical characteristics of the flow field [11]. For instance by [12] using the higher resolution Sea Surface Height (SSH) fields afforded by the merged T/P and ERS-1 and ERS-2 satellite datasets, then by [13] using Sea Level Anomaly (SLA) maps from the multi satellite AVISO - product. Moreover, mesoscale eddies activities have been studied in the Eastern Indian Ocean associated with the South Equatorial Current (SEC). In Indian Ocean a high Eddies Kinetic Energy (EKE) intensity (around 150 to 250 cm^2/s^2) can be found around 25°S extending westward from the Australian coast (figure 1) [14]. In Indian Ocean, there are several researches about eddies has been done. For instance by [14] about the formation and mechanism of eddy kinetic energy; by [15] about eddies variability and the oceanography characteristics at eddy centre; and by [16] which clearly clarify eddies mechanism in the South-eastern Tropical Indian Ocean.

Most of eddies studies in Indian Ocean are focused on the sea surface and little is known about the impact on the fisheries, specifically to tuna catches [17]. In this paper, we apply an oceanography data based on NEMO Model to study eddies structures and its association with tuna catches. In detection eddies formation, here we use the automatic eddies detection algorithm for the high-resolution numerical product developed by [11]. First of all, eddies data is set up by identifying cyclonic and anti-cyclonic eddies from the numerical product and then statistical analysis is applied. This paper is composed of five sections: Section 2 describes the physical oceanography model from NEMO Model through INDESO – KKP Project. Section 3 clearly identified eddies characteristics and intensity of EKE. Section 4 emphasizes a series of statistical analysis is applied to compute the coefficient correlation of eddies with tuna catches data from LPPT Benoa, Bali. Section 5 concludes the paper with summary.

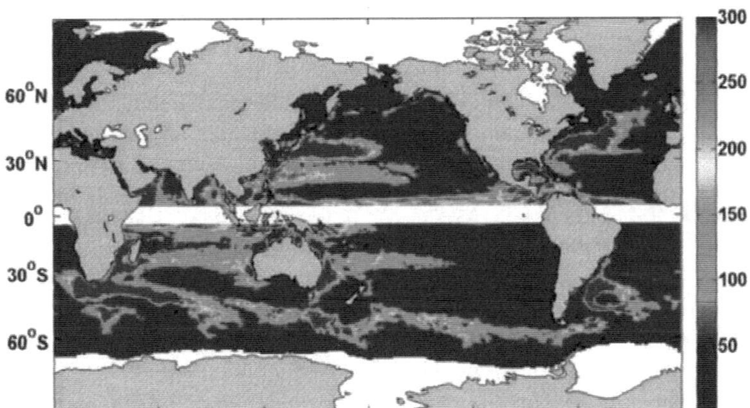

Figure 1. Global EKE (cm^2/s^2) calculated from the 15 years SLA [11].

2. Data

The data used in this study, including u and v components of sea current at two vertical level, which represents surface layer (mixed layer depth) at 5 m depth and thermocline layer at 109 m depth. This was generated by Nucleus for European Modeling of the Ocean (NEMO) Model through Physical Ocean Model of Infrastructure Development for Space Oceanography (INDESO) Project KKP [18]. INDESO-Project KKP is a collaboration project between Indonesia and France to improve the institutional capacity and human resources of oceanographic operational systems. Oceanographic data service has been running since 2014 through www.indeso.web.id [19].

NEMO model is a numerical simulation of the ocean. It have three components, which are: "blue-ocean" NEMO-OPA simulates the dynamics; the "white-ocean" NEMO-LIM simulates the sea-ice; and the "green-ocean" NEMO-TOP simulates the biogeochemistry [20]. The ocean physical model used for the INDESO projects is a regional configuration of the NEMO-OPA. These model is forced at the surface using 3-hourly ECMWF atmospheric analysis and forecast fields. [21].

The operational INDESO Physical Ocean Model system has a 1/12° or 9.25 km x 9.25 km horizontal resolution and 50 vertical layers with increased resolution near the surface, also provides 10 days of 3D ocean forecast with updated weekly. This product includes daily mean fields of atmospheric fluxes, physical parameters (temperature, salinity, currents, sea level), hourly surface variables (SST, SSH, currents), as well as hourly values of all fields at selected mooring sites and validation metrics [18]. The NEMO 2.3 version was used in INDESO Project and has been developed at Mercator-Ocean. This regional ocean physical model provides physical fields for the past two weeks (forced by a global 2-weeks analysis) and ocean forecast (forced by a global 10-day forecast) for the next 10 days [21].

3. Seasonal Cycle of Eddies and EKE Variation

3.1. Eddy Variation and Characteristic

In detection of eddies occurrences was using the automatic eddy detection algorithm developed by [11]. This method can be applied to any velocity field and also has been validated by [11]. Basically the method was based on eddies characteristics, such as for the minimum velocities were indicates in the proximity of the eddy center, and the velocity values that increase gradually linear from its center up to reach the maximum near the edges. Furthermore, four constraints were defined on this method to identifying eddies occurrence based on the characteristics of eddy velocity fields, which: (i) along east-west section, meridional velocity (V) has to reverse in sign across the eddy center and its magnitude has to increase away from it; (ii) Along north-south section, zonal velocity (U) has to reverse in sign across the eddy center, and its magnitude has to increase away from it; (iii) the velocity magnitude has a local minimum at the eddy center; and (iv) around the eddy center, the directions of the velocity vectors have to change with a constant sense of rotation [22].

Based on automatic eddies detection it is found in Southern Java Indian Ocean mostly generated by cyclonic eddies rather than anticyclonic. In total of 474 cyclonic and 442 anticyclonic, found 192 cyclonic and 181 anticyclonic at the surface layer; as well at the thermocline found 282 cyclonic and 261 anticyclonic eddies. Generally cyclonic eddies mostly occurred during northwest season (at surface layer) and southeast season (at thermocline layer); meanwhile anticyclonic mostly occurred during first transition season (at surface layer) and second transition (at thermocline layer) [22]. From figure 2a there were two coloured circles, which are indicates red circle as a cyclonic eddy and black circle as an anticyclonic eddy. During northwest season (figure 2a), eddies dominantly generate in the Southern Indian Ocean along 16 – 20°S, and there were several eddies occurred along the western Sumatera coast and reach to Southern Java. These eddies might be generated due to high shear velocity of SJC. As we can see in figure 2a, SJC flows eastward dominantly along the region with a strong velocity (red arrow) around 0.4 – 0.6 m/s. On the southeast season (figure 2b) the existence eddies near western Sumatera coast and southern Java decrease due to the weakness of SJC during this season. However, as seen in the figure 2b there were cyclonic eddies that existence in the southern coast Java during this season. This cyclonic eddies suggested was developed by the stronger ITF current system that flowing south-westward in this region [23]. Furthermore, from figure 2b showed that SEC current system flowing westward with a strong intensity (0.5 to 0.9 m/s). So thus, during southeast season eddies existences were moving southward to the offshore hence of the high intensity of SEC and ITF.

Eddies formation in the Southern Java Indian Ocean is not only generated by local wind but also by the remote forcing that generates SEC, ITF, and SJC [16]. The seasonally current flow in Southern Java Indian Ocean play an important role in eddies formation [24]. Thus, in the open ocean, mesoscale eddies that generate by the wind stress is not likely to be effective rather than by current flow pattern [9]. The mesoscale cyclonic eddies mostly appeared along the SJC current system near the southwest Sumatera coast to the southern Java, whereas the anticyclonic eddies mostly appear in the south of SEC current system. Occurrences of anticyclonic eddies also presumed generated by propagation of Rossby waves. Total eddies formation at two vertical layers predominantly rich at thermocline layer rather than surface layer. It is might be due to the high intensity of shear velocity at this depth, [23] stated that the eastward currents (SJC) form a large vertical shear in the depth range from 100 to 300 m in contrast with the westward current which is homogenized in the upper 300 m [24]. Furthermore, there were other factors in generating eddies vertically for instance local wind, topography, and propagation of Kelvin and Rossby waves.

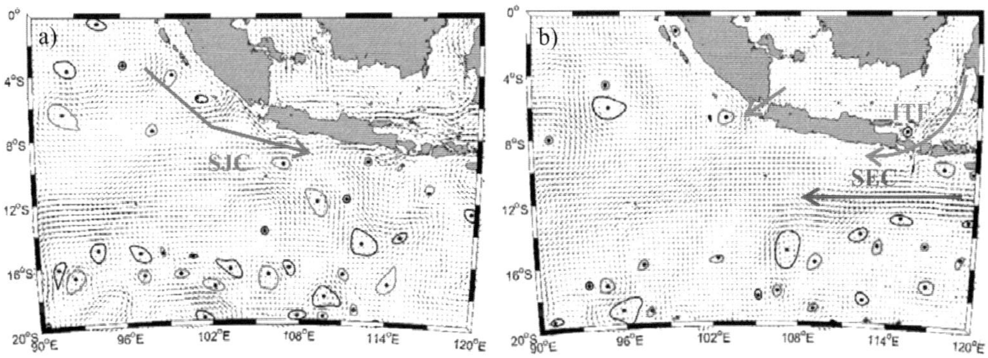

Figure 2. Eddies spatial distributions at surface layer in: a) Northwest Season and b) Southeast Season.

Histogram of eddies sizes at surface and thermocline layers are shown at figure 3. Both at surface and thermocline layers there are more cyclonic than anticyclonic eddies. The peak number at surface layer is located about 50 – 80 km, but for cyclonic eddies the frequencies may reach to 90 km. Whereas, at thermocline layer eddies sizes is greater than at surface layer, here the peak number is about 60 – 100 km. It has a similar results with what [25] found using surface satellite-tracked drifter data, which more sub-mesoscale eddies are densely over the entire Southern Indian Ocean (100 km > r > 60 km). Greater diameter at thermocline layer, presumed that eddies in this layer has a stronger intensity than at surface layer, which might be due to the high intensity of shear velocity at this depth as argued above.

The spatial distribution of eddies sizes is plotted at figure 4. Generally, eddies with larger diameter mostly found in the southern of Indian Ocean along 10° – 15°S. The spatial distribution of mesoscale cyclonic and anticyclonic eddies were different. As for the mesoscale anticyclonic eddies dominantly appears at the SEC pathways; whereas the mesoscale cyclonic eddies rich in the SJC pathways. It is consistent with [16] that mostly cyclonic eddies appears at the southern part of east Java coast; and for anticyclonic eddies at the south of SEC that flow westward, which are also as a part of Rossby waves. In the western Sumatera coast and southern Java, the shape of large eddies is restrained by difference in the velocity of SJC and the weaker currents around it as well as the morphology [16]. Furthermore, the submesoscale eddies generally occur below 15°S. It is consistent with [25] that medium eddies have a similar spatial distribution to that of large eddies but occur in a wider area. As well mention that the submesoscale eddies are densely distributed over the entire Indian Ocean, and for the appearance of submesoscale anticyclonic eddies might be due to the drifter aggregation maintained by converging Ekman currents.

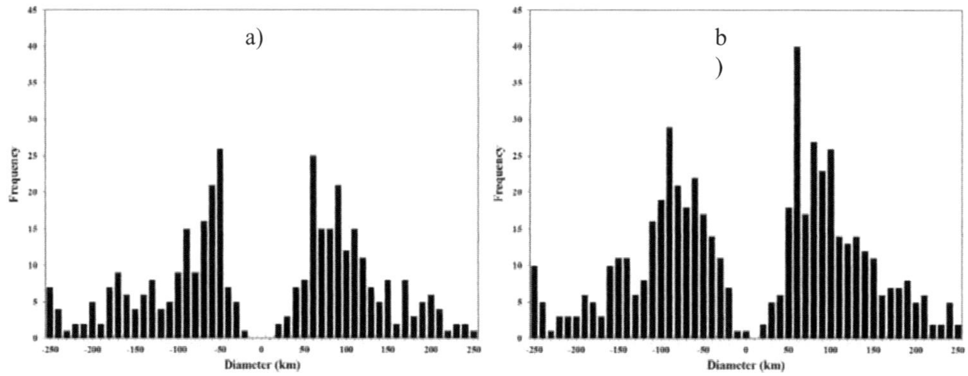

Figure 3. Histogram of eddies sizes at: a) surface layer; and b) thermocline layer; Positive (negative) eddy sizes represents cyclonic (anticyclonic) eddies.

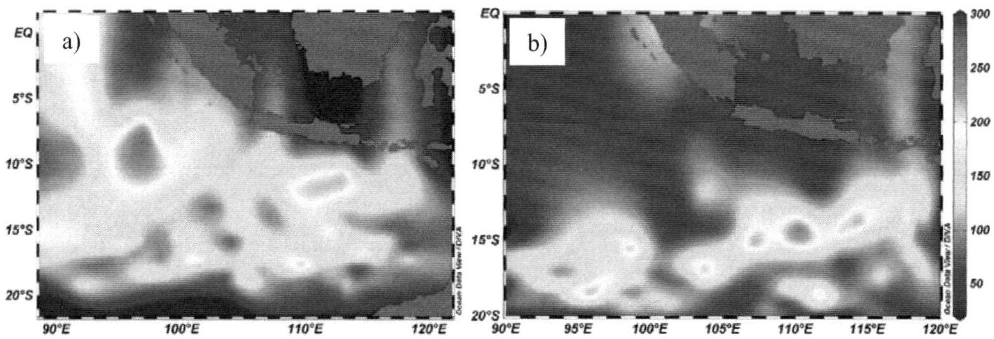

Figure 4. Eddies sizes distribution (km) at: a) cyclonic eddies; and b) anticyclonic eddies.

3.2. Seasonal Cycle of EKE Intensity
The EKE value is computed based on velocity components using the classical relation as follows:

$$\text{EKE} = \frac{1}{2}\,(U^2 + V^2) \qquad [8]$$

Where U and V are the zonal and meridional velocity components.

The EKE maps in the Southern Java Indian Ocean at four different seasons can be seen in figure 5. As shown in figure 5, EKE displays a distinct seasonal cycle with a maximum during southeast season (JJA) and 2nd transition season, and for a minimum during northwest season (DJF). The highest EKE (around 0.5 to 0.9 m^2/s^2) dominantly occurred along the ITF pathway near the Lombok Strait and in the south of SEC pathway along 12°S. It is might be indicating moderate and intensive eddies activity over these regions. To further demonstrate the EKE seasonal cycle, we compute time series of EKE averaged and maximum value. Seasonal variations can be clearly identified, with a mean EKE of 0.03 m^2/s^2, a maximum of 1.52 m^2/s^2 in September; and a minimum around 0.02 m^2/s^2 in December. [14] suggested that the seasonal variations of EKE in the southeast Indian Ocean are predominantly regulated by baroclinic instability associated with the mean flows. Moreover, [14] states that there is phase about 2-4 months for the vertical shear precedes the EKE variation. This indicates that due to the peak ITF signal in April through July and the maximum occurs in early June [26], then the EKE values reach its maximum in September. As well the high EKE signal during September also generated by instability currents system of SEC. [27] state that SEC flows throughout the year and have a strong intensity in the southeast season along 10° – 20°S [24]. This results also similar with [14] where the seasonal modulation of EKE in Southeast Indian Ocean is mediated by baroclinic instability associated with the underlying

IOP Conf. Series: Earth and Environmental Science **176** (2018) 012004 doi:10.1088/1755-1315/176/1/012004

westward-flowing SEC system and the surface eastward-flowing South Indian Ocean Countercurrent (SICC).

The mean EKE at cyclonic (anticyclonic) eddies centers could be calculated in the Southern Java Indian Ocean at surface layer giving average EKE 0.010 (0.012) m^2/s^2, and at thermocline layer giving average EKE 0.555 (0.313) m^2/s^2. From figure 6 can be seen the seasonal cycle of EKE at eddies centers. Generally, for both layers the cyclonic eddies have a larger EKE than anticyclonic. In the surface layer, EKE at cyclonic eddies reach its maximum on October, whereas the anticyclonic have a peak on November. However, in the thermocline layer larger EKE at cyclonic eddies found on June to October, while the anticyclonic on March and May. The high modulation of EKE at cyclonic eddies that occurred during southeast and 2nd transition season are mediated by geostrophic currents instability of SEC and ITF system. As argued above, that EKE modulation in this region is predominantly restrained by the instability mean flows of ITF and SEC. Furthermore, the results indicate that eddies in the thermocline layer has a higher intensity than in surface layer. As well the cyclonic eddies have a greater intensity in transporting water mass than anticyclonic. It is consistent with [28] that the energy of the downwelling generated by anticyclonic eddies tend to be weak.

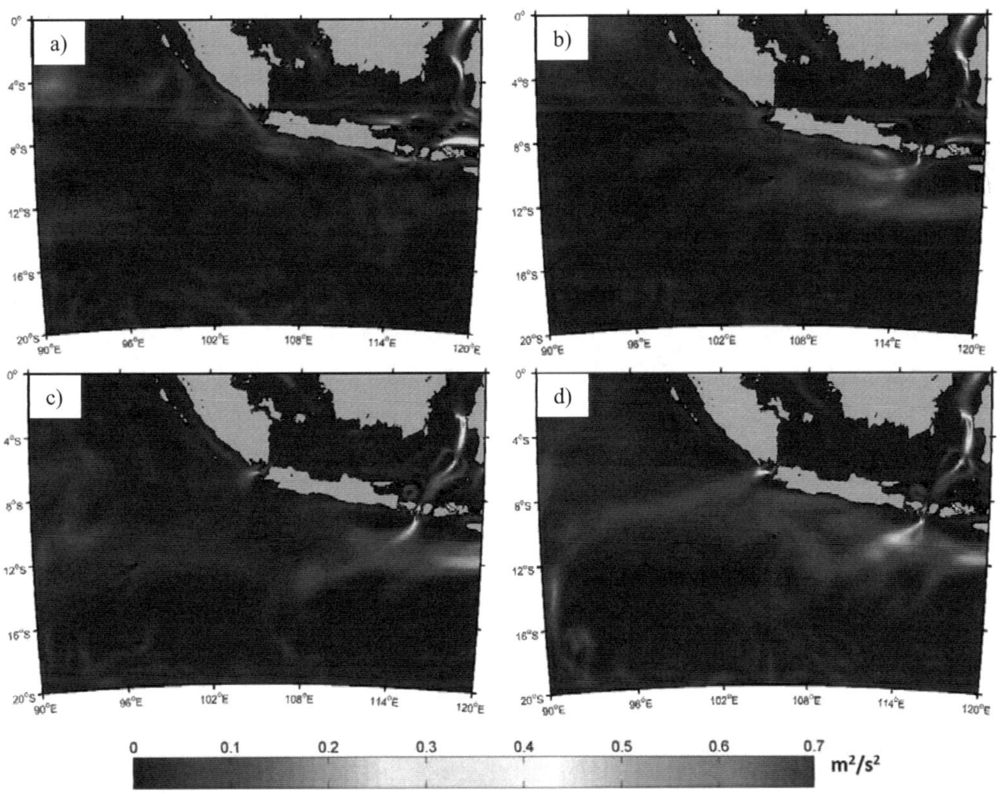

Figure 5. Seasonal distribution of EKE in: a) Northwest Season (Dec – Feb [DJF]); b) 1st Transition Season (Mar – May [MAM]); c) Southeast Season (Jun – Aug [JJA]); and d) 2nd Transition Season (Sep – Nov [SON]).

IOP Conf. Series: Earth and Environmental Science **176** (2018) 012004 doi:10.1088/1755-1315/176/1/012004

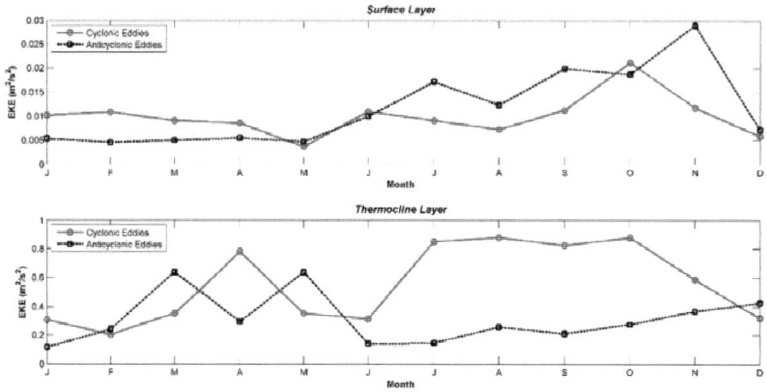

Figure 6. Mean seasonal of EKE at eddies centers.

4. Eddies and Yellowfin Tuna Catches

In analyse eddies and its association with the yellowfin tuna, we overlay eddies plot with the tuna catches, then used a statistical method which is Pearson's Correlation [22]. This correlation is to measure the strength of the association between two variables. Here we only compute the correlation for the thermocline layer, this refers to the yellowfin tuna catches data that predominantly appears in this layer [17]. The yellowfin tuna catches ware obtained from observation data by Loka Penelitian Perikanan Tuna (LPPT) Benoa in 2014 using a tuna longline. These tuna catches data represents as a frequency number of tuna caught in each month. The yellowfin tuna catches data can be seen on table 1.

Table 1. Yellowfin Tuna Catches on 2014.

Month	Yellowfin Tuna Catch
January	1
February	0
March	6
April	6
May	7
June	48
July	15
August	1
September	0
October	3
November	2
December	0

Based on the yellowfin tuna catches data, found the highest tuna catches was on June 2014; whereas the lowest on January 2014 [22]. Meanwhile, on February, September, and December no tuna catches were found. This is might be due because of the ships were not operated during these months. The overlay between eddies and tuna catches can be seen on figure 7 with the total of tuna caught represent as a yellow circle. On June 2014 (figure 7) yellowfin tuna catches found along in the edges either cyclonic or anticyclonic eddies. The yellowfin tuna distribution in this region probably due to the downwelling and upwelling generated by eddies. [9] mention that in the types of eddies where upwelling and divergence; downwelling and convergence often occur near the edges or in eddies outer edges. Thus, eddies motion in the ocean yield rich nutrients and further can support fisheries. Otherwise, yellowfin tuna catches not always found near in the edge of eddies. This indicates that there were other oceanographic parameters which are more influence on the abundance and distribution of yellowfin tuna, such as sea surface temperature, fronts, and depth of thermocline [29].

As the results of Pearson's Correlation found the highest correlation is in the cyclonic eddies (r = 0.59) rather than anticyclonic eddies (r = 0.14). A highest correlation between cyclonic eddies and yellowfin tuna might be due to the fishery abundance, hence the mixing water mass from bottom layer to upper layer [22]. Therefore, nutrients from bottom layer will be rising up and might be generated the primary productivity [17]. Furthermore eddies presumably localize tuna forage and then create a good feeding opportunity [30]. Moreover, the primary production increased within cyclonic rings in the Southern Hemisphere. The Coriolis effects force deflects water movement to the left away from the eddy center, and thus the depth of thermocline would be decreases and nutrients-rich waters closer to the surface layer and increasing primary production [31].

Based on the Pearson's correlation between EKE at eddies center and yellowfin tuna also found the highest correlation on the cyclonic eddies (r = 0.37). Strong correlation on EKE at cyclonic eddies is due to the larger EKE at this type rather than at anticyclonic eddies. This result also shows that EKE distribution has an influence in distribution of yellowfin tuna, besides the oceanographic parameters as mentioned above. Predominantly yellowfin tuna distributed in the area with the larger EKE. Moreover [32] state that high yellowfin tuna catches were most likely in locations where the surface ocean characteristics were: increased current shear (EKE), relatively shallow MLD [22], and increased of phytoplankton concentrations. [30] also found the association between EKE and other tuna species, where large catches of albacore were mostly concentrated in area with relatively high EKE and geostrophic velocities. This suggests that tunas are associated with eddies fields.

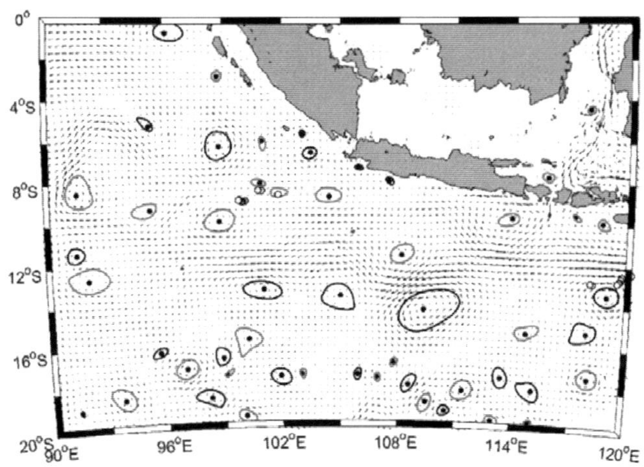

Figure 7. Eddies spatial distribution and yellowfin tuna catches.

5. Conclusion

Eddies appearances have been investigated by using the automatic eddies detection algorithm for the high resolution numerical product at two vertical layers in Southern Java Indian Ocean. In total of 916 eddies were detected, among 192 (181) are cyclonic (anticyclonic) eddies at the surface layer; as well at the thermocline 282 (261) are cyclonic (anticyclonic) eddies.

Predominantly for cyclonic eddies appear in the northwest season, whereas anticyclonic eddies appear in the 1st and 2nd transition season. These cyclonic and anticyclonic eddies appearances tend to followed the pattern of shear velocity current along in the Indian Ocean. As for the cyclonic eddies dominantly appear when a high intensity of SJC occurred. Meanwhile for the anticyclonic eddies suggest were associated with the appearance of SEC current system, as well with other forcing like Rossby wave. In addition, other local or nonlocal processes including local wind stress, kelvin waves, and topography may also modulate the activity of eddies in the Southern Java Indian Ocean.

Mesoscale cyclonic (anticyclonic) eddies populate along the SJC (SEC) system current, along 10°S (15°S). The spatial distribution of mesoscale eddies corresponds to the large EKE region. The EKE in this region shows a dis.tinct seasonal cycle with a maximum during southeast season (with maximum 1.52 m^2/s^2), and for a minimum during northwest season (with maximum 0.39 m^2/s^2). It is found that this seasonal modulation of EKE is mediated by the vertical velocity shear associated with the SEC and ITF currents system. Furthermore, by using the mean EKE as a background the mean EKE at eddies centers could be calculated. For cyclonic (anticyclonic) eddies at surface layer giving average EKE 0.010 (0.012) m^2/s^2, and at thermocline layer giving average EKE 0.555 (0.313) m^2/s^2. These results indicate that eddies in the thermocline layer has a higher intensity, as well the cyclonic eddies have a greater intensity in transporting water mass rather than anticyclonic eddies.

By overlying and using Pearson's Correlation the association between eddies, EKE, and yellowfin tuna have been investigated. It is found that the yellowfin tuna catches not constantly found along eddies formation. Otherwise, the cyclonic eddies have a higher correlation (r = 0.59) with the yellowfin tuna rather than anticyclonic eddies. It is might be due to the nutrients in abundance in the cyclonic eddies, hence the mixing of water mass from bottom layer to the upper layer. As well for the EKE at cyclonic eddies center also has the highest correlation with yellowfin tuna (r = 0.37).

Acknowledgments
This work mostly conducted in the Marine and Coastal Data Laboratory (MCDL), Marine Research Center, Ministry of Marine Affairs & Fisheries Republic of Indonesia; and the Department of Marine Science, Padjadjaran University. The u and v sea surface current is provided by Physical Ocean Model at INDESO – Project KKP (http://www.indeso.web.id). The tuna catches data 2014 is provided by observers from Tuna Fisheries Institute, Benoa. Special thanks to the Fisheries Research Center and the Tuna Fisheries Research Institute, Ministry of Marine Affairs & Fisheries Republic of Indonesia; to Sharing Data Team from INDESO KKP; and also to KOMITMEN Research Group.

References
[1] Kartadikaria AR, Miyazawa Y, Nadaoka K and Watanabe A 2012 Existence of eddies at crossroad of the Indonesian Seas *Ocean Dyn.* **62** 31–44
[2] Schott FA, Xie SP and Jr. JPM 2009 Indian ocean circulation and climate variability *American Geo Union.* **47** 1–46
[3] Gordon AL, Sprintall J, Van Aken HM, Susanto D, Wijffels S, Molcard R, Ffield A, Pranowo Wm and Wirasantosa S 2010 The Indonesian throughflow during 2004-2006 as observed by the INSTANT program *Dyn Atmos Ocean.* **50** 115–28
[4] Pranowo WS, Phillips H and Wijffels S 2005 Upwelling event 2003 along south Java Sea and lesser Sunda Islands *J Segara.* **1** 119–26
[5] Bray NA, Wijffels SE, Chong JC, Fieux M, Hautala S, Meyers G and Morawits ML 1997 Characteristics of the Indo Pacific throughflow in the eastern Indian Ocean. **24** 2569–72
[6] Griffa A, Lumpkin R and Veneziani M 2008 Cyclonic and anticyclonic motion in the upper ocean *Geophys Res Lett.* **35**
[7] Tussadiah A, Syamsuddin ML, Pranowo WS, Purba NP and Riyantini I 2016 Eddy vertical structure in southern Java Indian Ocean : Identification using automated eddies detection. *Int J Sci Res.* **5** 967–71
[8] Chaigneau A, Gizolme A and Grados C 2008 Mesoscale eddies off Peru in altimeter records : Identification algorithms and eddy spatio-temporal patterns *Progress in Oceanography.* **79** 106–19
[9] Bakun A 2006 Fronts and eddies as key structures in the habitat of marine fish larvae : opportunity, adaptive response *Scientia Marina.* **70** 105–22
[10] Sukresno B, Hartoko A and Sulistyo B 2015 Empirical cumulative distribution function (ECDF) analysis of thunnus.sp using ARGO float sub-surface multilayer temperature data in Indian Ocean south of java *Procedia Environ Sci.* **23** 358–67 Available from: http://dx.doi.org/ 10.1016/j.proenv.2015.01.052
[11] Nencioli F, Dong C, Dickey T, Washburn L and Mcwilliams JC 2010 A vector geometry-based eddy detection algorithm and its application to a high-resolution numerical model product and high-frequency radar surface velocities in the Southern California Bight *J Atmos Ocean*

Technol. **27** 564–79

[12] Chelton DB, Schlax MG, Samelson RM and de Szoeke RA 2007 Global observations of large oceanic eddies *Geophys Res Lett.* **34** 1–5

[13] Chaigneau A, Texier M Le, Eldin G and Grados C 2011 Vertical structure of mesoscale eddies in the eastern South Pacific Ocean : A composite analysis from altimetry and Argo profiling floats *J Geophysical Research.* **116** 1–16

[14] Jia F, Wu L and Qiu B 2011 Seasonal modulation of eddy kinetic energy and its formation mechanism in the southeast Indian Ocean *J of Physical Oceanography.* **41** 657–65

[15] Pranowo WS, Tussadiah A, Syamsuddin ML, Purba NP and Riyantini I 2016 Karakteristik dan variabilitas eddy di Samudera Hindia selatan jawa *J Segara.* **12** 159–65

[16] Hanifah F, Ningsih NS and Sofian I 2016 Dynamics of eddies in the southeastern tropical Indian Ocean *J Phys Conf Ser.* **739** 12042

[17] Pranowo WS, Kuswardani ARTD, Nugraha B, Novianto D, Muawanah U, Prihatno H and Yu W 2016 Ocean-climate interaction of south eastern Indian Ocean for tuna fisheries & its socio-economy impacts *Int J Sci Res.* **5** 1956–61

[18] Purba NP and Pranowo WS 2015 *Dinamika oseanografi, deskripsi karakteristik massa air dan sirkulasi air laut* UNPAD Press Bandung ISBN : 978-602-0810-20-1 276 p

[19] Tussadiah A, Subandriyo J, Novita S and Pranowo WS 2017 Verification of PISCES dissolved oxygen model using in situ measurement in Biak, Rote, and Tanimbar Seas, Indonesia *Int J Remote Sens Earth Sci.* **14** 37–46

[20] Levy C and Benshila R 2012 NEMO for Dummies *Mercat Ocean Q Newslatter.* **2** 4–7

[21] INDESO - KKP 2015 Product user manual – physical model outputs 1–17

[22] Teliandi D, Djunaedi OS, Purba NP and Setiyo W 2013 Relationship between variability mixed layer depth $\Delta T=0.5\ ^{o}C$ criterion and distribution of tuna in the eastern Indian Ocean **2** 162–71

[23] Syamsudin F and Kaneko A 2013 Ocean variability along the southern coast of Java and Lesser Sunda Islands *J Oceanogr.* **69** 557–70

[24] Mustikasari E, Dewi LC, Heriati A and Pranowo WS 2015 Pemodelan pola arus barotropik musiman 3 dimensi (3D) untuk mensimulasikan fenomena upwelling di Perairan Indonesia *J Segara.* **11** 25–35

[25] Zheng S, Du Y, Li J and Cheng X 2015 Eddy characteristics in the South Indian Ocean as inferred from surface drifters *Ocean Sci.* **11** 361–71 Available from: http://www.ocean-sci.net/11/361/2015/

[26] Potemra JT 1999 Seasonal variations of upper ocean transport from the Pacific to the Indian Ocean via Indonesian Straits *J Phys Oceanogr.* **29** 2930–44

[27] Utamy RM, Purba NP, Pranowo WS and Suherman H 2015 The pattern of south equatorial current and primary productivity in south Java Seas *Int Proc Chem Biol Environ Eng.* **51** 139–42

[28] Oki D and Putri MR 2014 Identifikasi awal eddies di perairan Laut Jawa **1** 12-20

[29] Rajapaksha JK, Nishida T and Samarakoon L 2010 Environmental preferences of yellowfin tuna (*Thunnus albacores*) in the northeast Indian Ocean : an application of remote sensing data to longline catches 1-6

[30] Zainuddin M, Kiyofuji H, Saitoh K and Saitoh SI 2006 Using multi-sensor satellite remote sensing and catch data to detect ocean hot spots for albacore (*Thunnus alalunga*) in the northwestern North Pacific *Deep Res Part II Top Stud Oceanogr.* **53** 419–31

[31] Abigail W, Zainuri M, Tisiana A, Kuswardani D and Setiyo W 2015 Distribution of nutrient, light intensity, chlorophyll-a and water quality in Badung Strait, Bali during southeast monsoon *Depik.* **4** 87–94

[32] Dell J 2012 Fisheries oceanography of yellowfin tuna (*Thunnus albacares*) in the Tasman Sea *Tasmania Univ.*

Homotrema rubrum taphonomy as an indicator of sedimen transport of the Bangka Island Waters, North Minahasa

M Alam[1*], R Zuraida[2], L P Sari Yuliadi[3], W Lili[4], I Nurruhwati[5]

[1] Marine Science Major, Faculty of Fisheries and Marine Science, University of Padjadjaran, Indonesia
[2] Marine Geological Institute, Jl. Junjunan 236, Bandung City 40174, Indonesia
[3] Marine Science Department, Faculty of Fisheries and Marine Science, University of Padjadjaran, Indonesia
[4]Aquaculture Department, Faculty of Fisheries and Marine Science, University of Padjadjaran, Indonesia
[5] MSP Department, Faculty of Fisheries and Marine Science, University of Padjadjaran, Indonesia
*e-mail: mujizat13001@mail.unpad.ac.id

Abstract. The determination of current direction from sedimentary deposits were based on silisiclastic deposits that behave differently than biogenic fragment-rich deposit. The objectives of this research are to determine biogenic sediment transport pathway of Bangka Island water, North Minahasa, by the taphonomy of benthic foraminifer *Homotrema rubrum* and to verify the reconstructed pathway with sea surface currents data. This study used twelve surface sediment samples that were acquired by grab sampler in 2016 and 2014-2016 annual average sea surface current data from INDESO that was generated by OPA/NEMO model. The result of this study shows that samples with higher degree of taphonomy are located closer to coral reef which forms the source of *H. rubrum* while samples with lower degree of taphonomy are located further from the source. Sediment transport pathway reconstructed from *H. rubrum* taphonomy shows inflow of water from the north and west part of the islands and West Likupang Strait to the east and flowing out of East Likupang Strait. This direction corresponds to annual sea surface current derived from INDESO. This study also shows that *H. rubrum* taphonomy can be used as sediment transport indicator in areas that are dominated by biogenic sediments.

1. Introduction

Bangka Island, a part of islands cluster in North Minahasa District, North Sulawesi Province, are susceptible to abrasion due to marine forcing such as ocean currents and waves (Regional Regulation of North Minahasa Regency No. 1 of 2013). Current movement will affect beach profile leading to coastal abrasion or accretion [1,2]. This longshore current transports sediment along the coast, and hence the reliable information of current patterns and sediment transport is required to develop coastal management. This information can be used to reduce the impact of coastal abrasion and accretion.

Determination of sediment transport direction can be conducted by observing sediment deposited to the updrift of a jetty or any other coastal construction [11]. Identification of past sediment transport direction from sedimentary deposits have been carried out by studying grain-size distribution. Finer grains are found deposited further from the source and vice versa. However, that method can only be applied to detrital sediment as it was developed based on experiment conducted on siliceous sediments, and cannot be applied on sediment dominated by biogenic fragments, i.e. biogenic sediments, that are found covering the seafloor in areas where coral reefs are present. Considering that Indonesia is in the center of Coral Triangle that resulted in biogenic sediments, it is imperative to find another method to identify current direction from sedimentary deposits.

One of the most recent method in establishing sediment transport direction is by using benthic foraminifera from seabed sediments [3]. The taphonomy of *Homotrema rubrum* was successfully applied to understand sediment transport in Sioba/Rio do Fogo, Brazil [4]. Taphonomy is able to reveal information on sediment source and also sedimentation process [5]. *Homotrema rubrum* (Lamarck) is benthic foraminifera that lives attached to the bottom part of corals [6].

The application of this method in Bangka Island is determined based on the position of Bangka Island within Coral Triangle Scientific Boundary that houses over 500 species of coral *zooxanthellae* [7]. To date, there is no information on benthic foraminifer from the study area. However, considering the association of *H. rubrum* with corals, it is expected tat the surface sediment might contain abundant *H. rubrum* that could be used for the study. This study aims to determine sediment transport pathway of Bangka Island water, North Minahasa, by observing *H. rubrum* taphonomy and to verify the reconstructed pathway with sea surface currents data.

2. Methods

A total of 3715 *H. rubrum* specimens from 12 sediment samples in the waters of Bangka Island, North Minahasa and surrounding areas (table 1). The samples were selected to represent each straits in the island cluster of which the currents flow in and out. The sediment samples were acquired in 2016 by Marine Geological Institute (MGI). The selected samples were split until it was estimated to contain approximately 300 individus of *H. rubrum.* Picking activity of *H. rubrum* was performed in Micropaleontology and Petrology Laboratory of MGI in Bandung from February to April 2017.

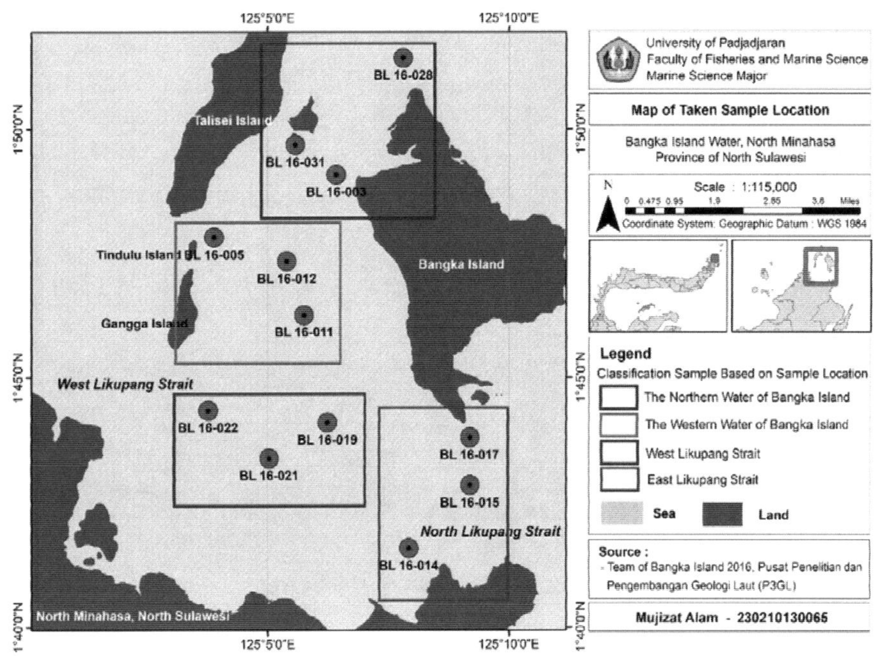

Figure 1. Location map of samples collection.

The observation of *H. rubrum* taphonomy was conducted by using Nikon SMZ 800 binocular microscope. Parameters observed in this study included color, shape and chamber structure [8] that determined the degree of preservation of *H. rubrum*. Plotting those data into a map developed sediment transport pathway. The developed pathway was then compared to surface current pattern of Bangka Island waters [9]

This study used five levels of *H. rubrum* degree of of preservation instead of seven groups [8]. The modification was conducted due to different characteristics of *H. rubrum* that were observed in Bangka Island waters: white, angular with intact chamber structure (table 2 and figure 2). Those characteristics are assigned as *well to moderately preserved.*

Table 1. Sample locations.

Sample	Longitude (°)	Latitude (°)
BL 16-031	125.093	1.828
BL 16-003	125.107	1.818
BL 16-005	125.065	1.797
BL 16-022	125.063	1.739
BL 16-017	125.153	1.730
BL 16-015	125.153	1.714
BL 16-028	125.130	1.857
BL 16-021	125.084	1.723
BL 16-019	125.104	1.735
BL 16-011	125.096	1.771
BL 16-014	125.132	1.693
BL 16-012	125.090	1.789

Table 2. Degree of preservation *Homotrema rubrum* (modified from [8]).

Degree of Taphonomy Preservation	Test Colour	Fragment Angularity	Chamber Structure
Exceptionally Preserved (EP)	Red	Angular	Intact
Well Preserved (WP)	Pink	Angular	Hollowed
Well to Moderately Preserved (WMP)	White	Angular	Hollowed
Moderately Preserved (MP)	Pink	Rounded	Hollowed
High Altered (HA)	Light Pink to White	Highly Rounded	Hollowed

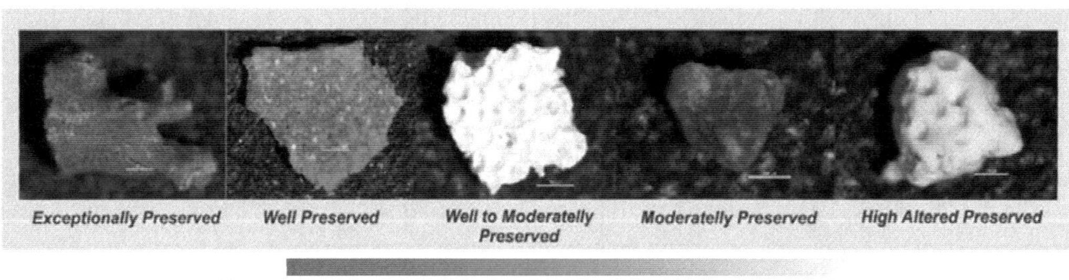

Figure 2. *H. rubrum* taphonomy indicated by color, angularity and chamber structure (modified from [8]).

3. Results

3.1. Taphonomy Homotrema rubrum

3.1.1 The Northern Bangka Island. Taphonomy observations on BL 16-003, BL 16-028 and BL 16-031 samples shows that higher preservation was found in BL 16-028 (figure 2). *H. rubrum* BL 16-028 is characterized by red colour (3%), pink (73%), and light pink to white (24%). The degree of angularity in this location ranges from angular and sub-angular (89%) and sub-angular and sub-rounded and rounded (11%). The sample exhibit intact chamber structure (3%) and hollowed chamber (97%).

H. rubrum in BL 16-003 shows lower preservation than BL 16-028 but higher compared to BL 16-031 (figure 2). *H. rubrum* in BL 16-003 is characterized by red colour (2%), pink colour (43%), and light pink to white (56%). The majority of observed specimens (97%) are angular and sub-angular while 3% are sub-rounded to rounded. Only 1% of observed specimens have intact chamber while 99% show hollowed chamber.

The degree of preservation of *H. rubrum* observed in sample BL 16-031 is lower compared to that in BL 16-003 and BL-16-028 (figure 2). *H. rubrum* in BL 16-031 is characterized by 0% red, 23% pink, and 77% light pink to white. The shape *H rubrum* in this sample consists of 89% angular to sub-angular and 10% rounded to sub-rounded with 0% shows intact chamber structure.

H. rubrum in study area can be classified into 5 degree of preservation (table 2, figure 3). The degree of preservation in BL 16-028 sample is 3% exceptionally preserved (EP), 63% well preserved (WP), 24% well to moderately preserved (WMP), 11% moderately preserved (MP), and 0% high altered (HA). BL 16-003 shows the following degree of preservation: 2% EP, 39% WP, 56% WMP, 3% MP, and 0% HA. BL 16-031 shows the following preservation levels: 0% EP, 12.7% WP, 76.6% WMP, 10.4% MP, and 0.3% HA.

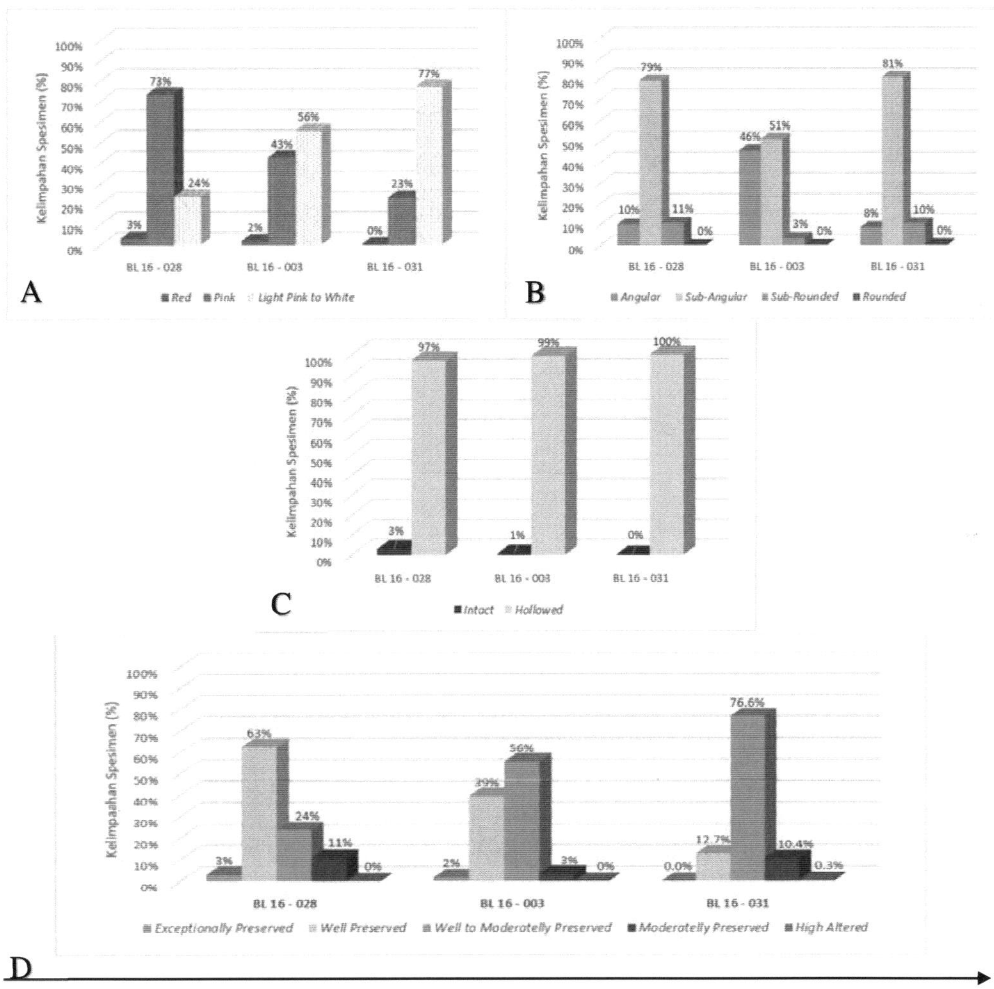

Decreasing Preservation

Figure 3. *H rubrum* taphonomy of the study area; (A) test color (%), (B) angularity (%), (C) chamber structure (%), (D) preservation degree (%).

The best level of preservation in sample is characterized by red colour, angular shape and intact chamber structure and has been used as proximity indicator to coral reef as the source of *H. rubrum* [4]. Degree of *H. rubrum* preservation shows decreasing trend from BL 16-028 that is located in the entrance of the strait, to BL 16-003 that is located near Bangka Island and BL 16-031 that is located off Talisei Island. Higher degree of preservation indicates nearest site to coral reef. Thus, BL 16-028 is closest to the reef, followed by BL 16-003 and BL 16-031. Therefore, it can be inferred that sediment transport in this area is from BL 16-028 to BL 16-003 to BL 16-031.

3.1.2 The Western Bangka Island. The highest preservation sample from western Bangka Island is BL 16-005 (figure 4). The sample is characterized by 0% red, 88% pink, and 12% light to pink white in colour; 77% angular to sub-angular and 22% rounded to sub-rounded in shape; and 0% intact chamber.

The degree of preservation of sample BL 16-012 is lower than BL 16-005 but higher than BL 16-011 (figure 4). BL 16-012 sample is characterized by 0% red, 84% pink, and 16% light to pink white in colour; 59% angular to sub-angular and 41% sub-rounded to rounded in shape; and shows 0% intact chamber structure. *H. rubrum* observed in BL 16-011 has lowest degree of preservation (figure 4) and characterized by 0% red, 14% pink, and 86% light pink to white; 89% angular to sub-angular and 11% of rounded to sub-rounded in shape; and 0% intact chamber.

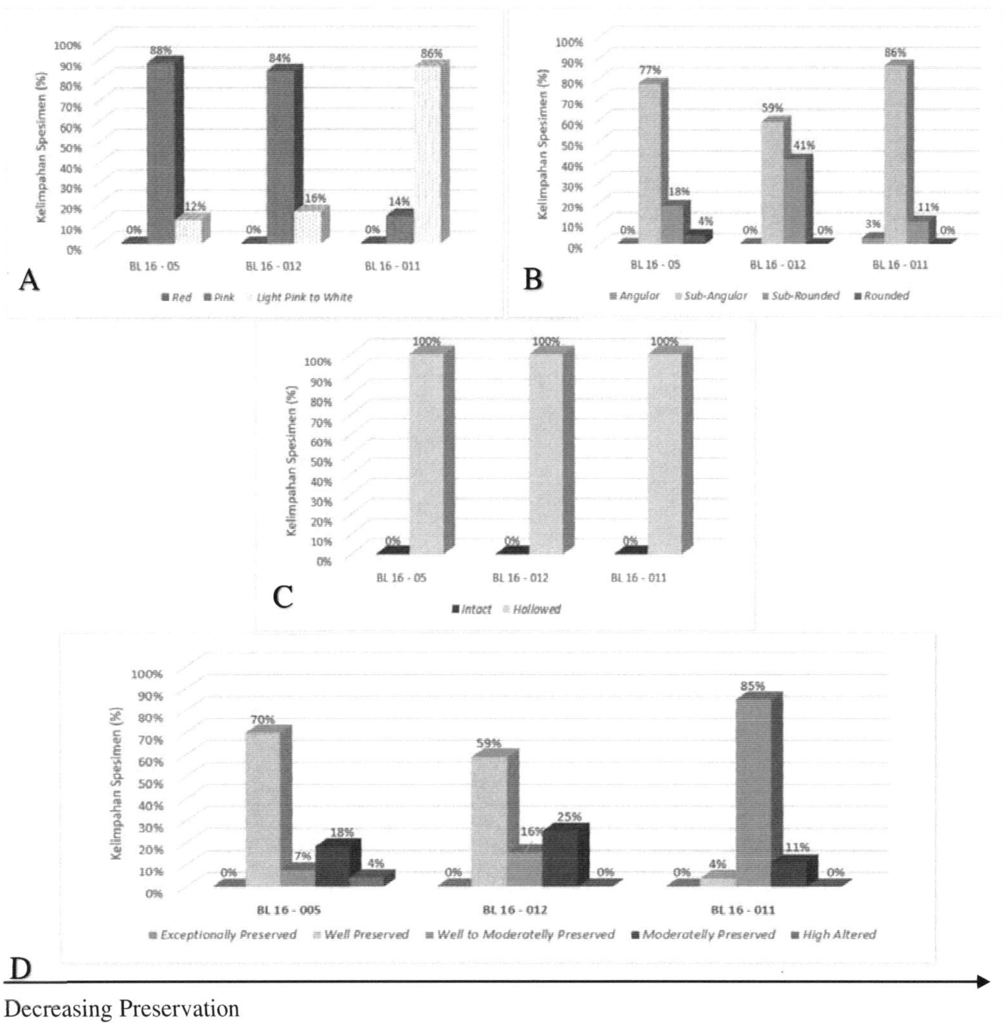

Figure 4. *H. rubrum* taphonomy in Western Bangka Island; (A) test color (%), (B) angularity (%), (C) chamber structure (%), (D) preservation degree (%).

The three samples show considerable amount of angular to sub-angular shape (>50%), but also exhibit 100% hollowed chamber structure. The abundance of pink *H. rubrum* in these samples indicates that *H rubrum* were deposited near or in the proximity of the source. The degree of *H. rubrum* preservation in these samples indicate that BL 16-005 and BL 16-012 are located near coral reefs, while BL 16-011 lies furthermost of the coral reefs. The shape and chamber structure that were observed in BL 16-005 and BL 16-022 reveal the possibility of shell abrasion during sediment transport. The abrasion might involve scouring of *H. rubrum* shells by other sediment particles during transport

Combination of observed parameters yields 5 levels of *H. rubrum* preservation (figure 4). Sample in BL 16-005 exhibits 0% EP, 70% WP, 7% WMP, 18% MP, and 4% HA. Sample BL 16-012 consists of 0% EP, 59% WP, 16% WMP, 25% MP, and 0% HA. The lowest level of *H. rubrum*'s preservation is observed in BL 16-011 with 0% EP, 4% WP, 85% WMP, 11% MP, and 0% HA.

Decreasing preservation is shown from BL 16-05 to BL 16-012 to BL 16-011 indicating that BL 16-05 is closest to the reef followed by BL 16-012 and BL 16-011. Thus, the direction of sediment transport in western Bangka Island is from BL 16-005 to BL 16-012 to BL 16-011.

This area shows different preservation degree compared to Northern Bangka that shows higher percentage of WP and WMP compared to this area that shows a mixture of WP and MP. The distinction might be related to bathymetry of Bangka waters that is deeper in the central part of the islands (figure 5). This central part forms a depositional basin for sediments that are transported into the central part of the islands from the west or the north. Because the cental part of the islands are further from coral reef, *H, rubrum* that is accumulated in site BL 16-011 is mostly less preserved (MP) compared to BL 16-05 and BL 16-012.

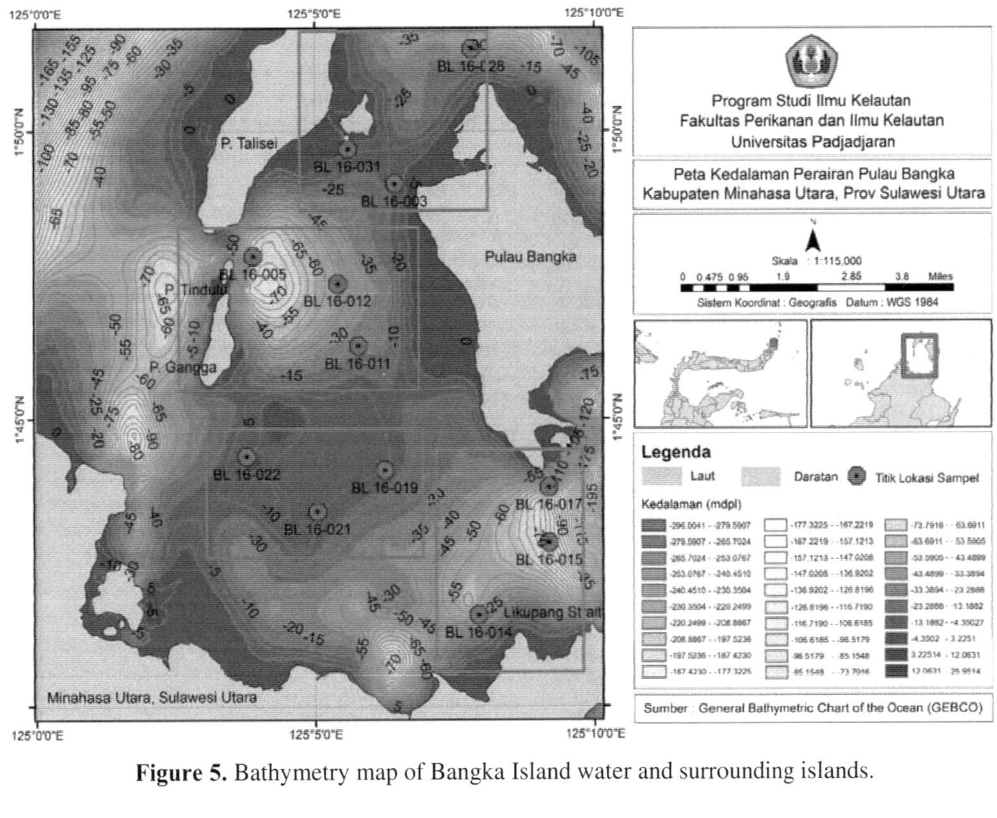

Figure 5. Bathymetry map of Bangka Island water and surrounding islands.

3.1.3 West Likupang Strait. The taphonomy of *H. rubrum* from samples acquired from West Likupang Strait shows highest preservation in BL 16-022 (figure 6). *H. rubrum* in BL 16-022 sediment is characterized by 0% red, 25% pink and 55% light pink to white in colour; 56% angular to sub-angular and 44% rounded to sub-rounded in shape; and 19% have intact chamber structure while 81% have hollowed chamber.

IOP Conf. Series: Earth and Environmental Science **176** (2018) 012005 doi:10.1088/1755-1315/176/1/012005

H. rubrum observed in BL 16-021 have lower preservation than those in BL 16-022 but higher than BL 16-019 (figure 6). *H. rubrum* in BL 16-021 is characterized by 1% red, 34% pink and 65% light pink to white in colour; 22% angular to sub-angular and 78% rounded to sub-rounded in shape; and 1% with intact chamber and 99% with hollowed chamber. The lowest preservation degree, BL 16-019, shows 0% red, 2% pink, and 98% light pink to white *H. rubrum*; 7% angular to sub-angular and 93% rounded to sub-rounded in shape; with 0% intact chamber.

The degree of *H. rubrum* preservation in West Likupang Strait are as follows (figure 6). Preservation level of sample BL 16-022 consists of 20% EP, 33% WP, 12% WMP, 24% MP, and 12% HA. Sample BL 16-021 comprises of 1% EP, 21% WP, 55% WMP, 13% WP, and 10% HA. BL 16-019 shows the following level of preservation: 0.3% EP, 1.3% WP, 6% WMP, 28.9% MP, and 63.5% HA.

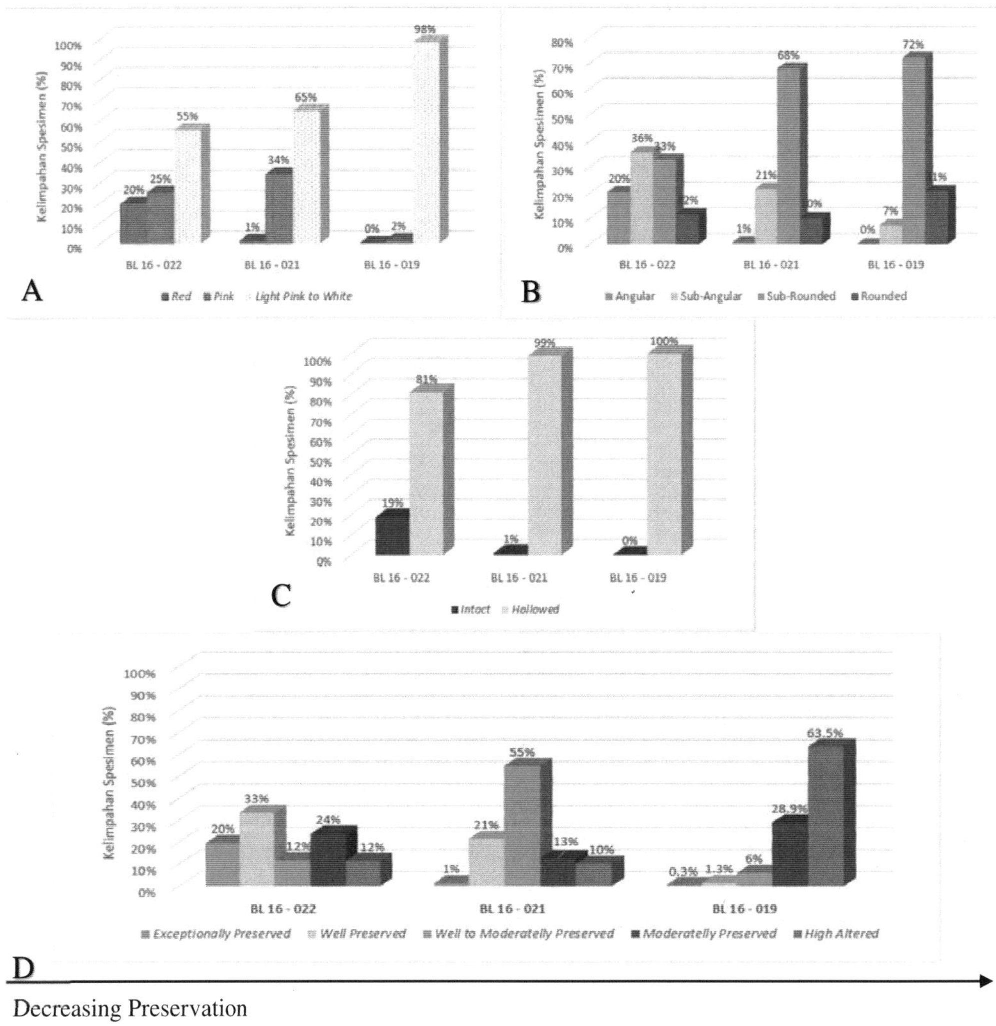

Figure 6. *H. rubrum* taphonomyin West Likupang Strait; (A) test color (%), (B) angularity (%), (C) chamber structure (%), (D) preservation degree (%).

Red colour and angular *H. rubrum* can be used as indicator of source (coral reef) proximity [4]. Those characteristics were observed in BL 16-022 suggesting that this sample has not been transported far from or deposited near the coral reefs. Declining preservation trend from BL 16-022 to BL 16-021 to BL 16-019 points to the direction of sediment transport from BL 16-022 to BL 16-021 to BL 16-019.

High percentage of HA in BL 16-019 might be related to its position in the central part of the islands. As was discussed in pervious section, the central part of the islands is deeper than the rest forming depositional basin for sediments flowing through. Moreover, site BL 16-019 is further from coral reef that will lead to less preserved H. rubrum compared to BL 16-022 and BL 16-021. Thus, sediments that are accumulated in this site will be less preserved than the others.

3.1.4 East Likupang Strait. H. rubrum observed in BL 16-014 have higher preservation than those in BL 16-015 and BL 16-017 (figure 7). The characteristics of *H. rubrum* in BL 16-014 are 5% red, 27% pink, and 67% light pink to white in colour; 35% are angular and sub-angular and 65% rounded and sub-rounded in shape; with 3% have intact chamber and 97% have hollowed chamber.

H. rubrum in BL 16-015 has lower preservation than those in BL 16-014, but higher than those in BL 16-017 (figure 7). The taphonomy of *H. rubrum* in BL 16-015 is as follows: 2% red, 52% pink and 46% of light pink in colour; 89% are angular and sub-angular and 11% are rounded and sub-rounded in shape; with 0% show intact chamber structure. Sample BL 16-017 has lowest preservation and is characterized by 0% red, 65% pink, and 35% light pink to white in colour; 89% are angular and sub-angular and 11% are rounded and sub-rounded in shape; with 0% shows intact chamber structure.

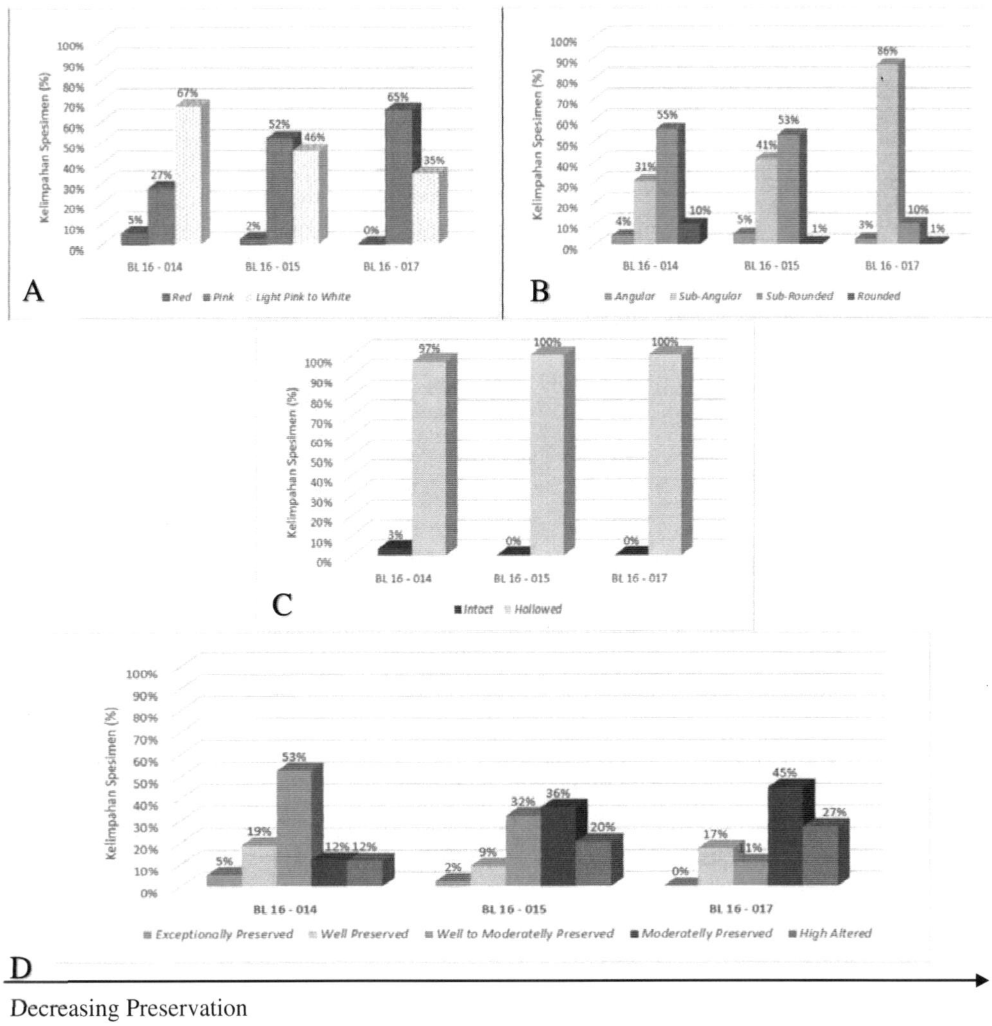

Figure 7. *H. rubrum* taphonomy in East Likupang Strait. (A) Test Color (%), (B) Angularity (%), (C) Chamber Structure (%), (D) Preservation Degree (%).

The observed parameters of the taphonomy then used to categorize *H. rubrum* in samples from East Likupang Strait into 5 levels of preservation (figure 7). Sample of BL 16-014 shows the following preservation level: 5% EP, 19% WP, 53% WMP, 12% MP and 12% HA. The preservation level of BL 16-015 is as follows: 2% EP, 9% WP, 32% WMP, 36% MP and 20% HA. BL 16-017 comprises of 0% EP, 17% WP, 11% WMP, 45% MP and 27% HA.

In general, preservation level decrease from BL 16-014 to BL 16-015 to BL 16-017. It indicates that sediments in Eastern Likupang Strait are transported from BL 16-014 to BL 16-015 to BL 16-017. This area shows distinc trend compared to the rest due to relatively higher percentage of MP and HA. The presence of MP and HA might be the result of the accummulation of H. rubrum that is transported from Northern and Western Bangka Island as well as West Likupang Strait to East Likupang Strait that forms the outlet of the water flowing through Bangka Islands.

3.2 Reconstruction of sediment transport based on H. rubrum taphonomy

As demonstrated in Brazil, *H. rubrum* with red color and sharp points were mostly deposited near the source while white and rounded specimens were found further from the source [4]. A study from Wallis Island, Polynesia, show that *H. rubrum* taphonomy can be used to identify sediment transport trends in tropical island [8]. Sediment reddish and angular specimens of *H. rubrum* indicates that the sample location is close to source [4]. The further from source, the dominant taphonomy of *H, rubrum* woud decrease from EP to WP to WMP to MP and HA indicates location furthest from the source [9]. The modification of H, rubrum taphonomy (color, angularity and chamber structure) is the result of sediment transport and/or energetic hydrological conditions [4]. Surface current in study area is estimated to be between 0 t 0.35 m/s that can be classified as low to medium current velocity [12], therefore the taphonomy the taphonomy is mostly caused by sediment transport.

Plot of *H. rubrum* preservation degrees demonstrate that the higher degree of preservation is observed in samples located in the inflow areas (figure 9) where most coral reefs in this area are located (figure 8) while H, rubrum with lower preservation degree is observed in samples from the outflow area (figure 9). The lowest preservation degree is observed in the central part of study area (BL 16-019) that is interpreted as the preferential site for sediment accummulation.

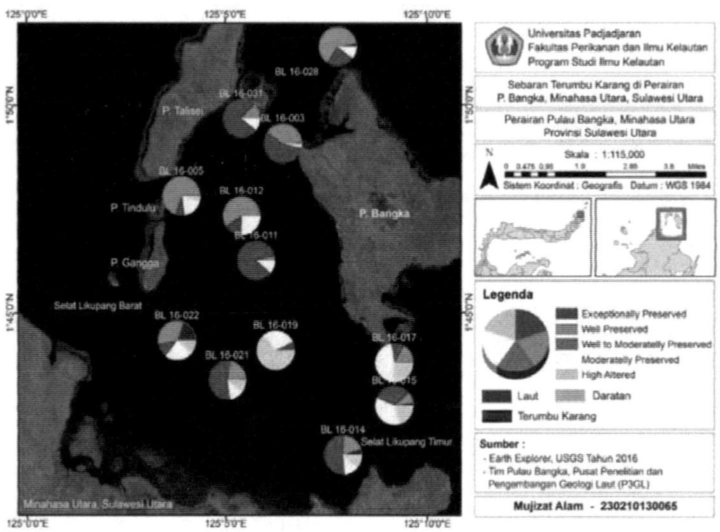

Figure 8. Distribution of Coral Reefs in Bangka Island, North Minahasa.

Identification of current direction can be reconstructed by following the taphonomy of H, rubrum in the study area. Tracing of the degree of preservation from EP to HA points to the direction of sediment transport that indicates current direction. Figure 9 shows that sediments are transported from the north and west into the central part of the island cluster and then flow out from the East Likupang Strait. The predominant sediment transport direction is Bangka Island water is eastward.

Because sediments reflect net transport flux as was demonstrate in Fly River, Papua New Guinea, we can infer that net transport direction in study area is eastward [10]. To validate the reconstructed sediment transport pathway, we use annual annual average surface current data. The annual average sea surface current is obtained from Infrastructure Development and Space Oceanography (INDESO) within the period of 2014-2016 and has been discussed in detail [9]

In general, the reconstructed sediment transport pattern shows similar direction with 2014 - 2016 annual average sea surface current from INDESO (figure 10) detailed sediment transport within Bangka Island waters is not visible in the INDESO map (figure 10). The discrepancy is caused by the size of study area (18x11 km^2) that is smaller to INDESO resolution (1/12°)[9]. Thus, *H. rubrum* taphonomy can be used to reconstruct detail net sediment transport in areas that are dominated by coral reefs [9].

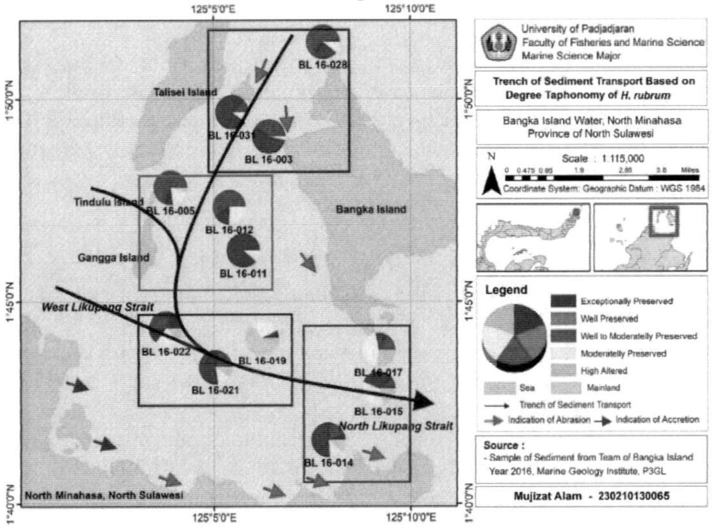

Figure 9. Direction of sedimen transport in Bangka Island waters constructed from *Homotrema rubrum* taphonomy.

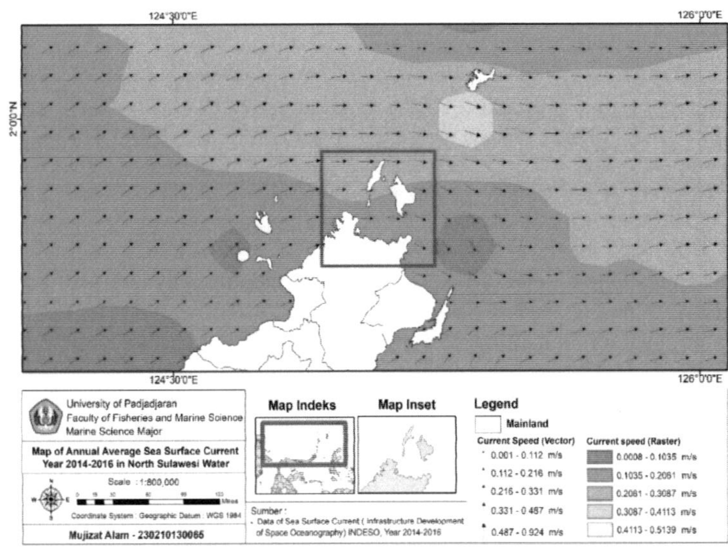

Figure 10. Map of annual sea surface current 2014-2016 from INDESO In North Bangka Island water [9].

IOP Conf. Series: Earth and Environmental Science **176** (2018) 012005 doi:10.1088/1755-1315/176/1/012005

4. Conclusions

Reconstruction of sediment transport pathway based on preservation degree of *H. rubrum* in Bangka Island waters reveal predominant eastward transport. Water from the open sea flows into study area through the straits between islands and flows out through East Likupang Strait. The eastward movement corresponds with annual average sea surface current for 2014-2016 period from INDESO. Our result shows that *Homotrema rubrum* taphonomy can be used to reconstruct sediment transport pathway in coral reef region.

Acknowledgments

I would like to thank the head of the marine geology research team of Bangka Island and its surroundings, Ir Catur Purwanti, M.T., for the use of samples and data that used in this study. I also want to thank the head of Pusat Penelitian dan Pengembangan Geologi Kelautan (PPPGL) that allowed me to do the research there. The greatest appreciation I convey to committee of International Conference of Marine Science (ICMS) that accepted me to can presentation this research.

References

[1] Triatmodjo B 1999 *Teknik Pantai* (Yogyakarta: BETA OFFSET)

[2] Cristina J 2014 Studi perubahan garis pantai selatan Teluk Ambon luar dengan metode komar dan bikjer *J. Teknologi*. **2**(1) 14-19

[3] Dawson J Q H and Smithers S 2012 Benthic foraminifera: their importance to future reef island resilience. *Yellowless D, Hughes T P [Eds] the 12th International Coral Reef Symposium* (p 7). Townsville, Cairns, Australia: Debenay, J.-P., Payri,.

[4] Machado A and Moraes S 2002 A note on the occurence of the encrusting foraminifera *Homotrema rubrum* in reef sediments from two distinctive hydrodynamic settings. *Anais da Academia Brasileira de Ciências*. **74**(4) 727-735

[5] Lawrence D R 1979 Taphonomy, in Fairbridge, R., W., Jablonski, D., *eds., The Encyclopedia of Paleontology* (Stroudsburg, Pennsylvania, Dowden, Hutchinson and Ross) p 793–799

[6] Emiliani C 1951 On the species *Homotrema rubrum* (Lamarck), Crushman Found, Foram. *Cushman Found Foram Res*. **2**(1) 143–14

[7] Veron J M L, Devantier, Turak E and Green 2009 Delineating the Coral Triangle. *J. of Coral Reef Studies*. **11**(1) 91-100

[8] Pilarczyk J E, Goff J, Mountjoy J, Lamarche G, Pelletier B, and Horton B P 2014 Sediment transpor trends from a tropical Pasific lagoon as indicated by *Homotrema rubrum* taphonomy in Wallis Island, Polynesia *Elsevier: Marine Micropaleontology*. **109** 21-29

[9] Alam M, Zuraida R, Yuliadi L, and Lili W 2017 Validasi arah arus data INDESO menggunakan pola sebaran preservasi taphonomy *Homotrema rubrum* di perairan Pulau Bangka, Minahasa Utara. *Kumpulan Abstrak SNSA 2017*.

[10] Harris PT, Hughes M G, Baker E K and Dalrymple R 2004 Sediment transport in distributary channels and its export to the pro-deltaic environment in a tidally dominated delta: Fly River, Papua New Guinea. *Continent. Shelf Res*. **24** 2431–2454

[11] Berman G 2011 *Longshore Sediment Transport* (Cape Cod Massachusetts: Woods Hole Sea Grant)

[12] Hutabarat, S. & S.M. Evans. 1985. Pengantar Oseanografi. Jakarta: UI Press.

IOP Conf. Series: Earth and Environmental Science **176** (2018) 012006 doi:10.1088/1755-1315/176/1/012006

Foraminifera spatial analysis as bioindicator in Java Sea to Flores Sea

B Valentina[1*], R Zuraida[2], B Marhaeni[1]

[1] Fisheries and Marine Science Faculty Jenderal Soedirman University
[2] Marine Geological Institute
*e-mail: debernval@gmail.com

Abstract. Foraminifera is potential bioindicator to understand the modern and ancient marine environment and have been widely used as bioindicator. Java Sea to Flores Sea are productive area influenced by physical and chemical reactions from nature and human activity, so that their characteristics are important to recognize. The aim of this research is to determine oceanography condition of Java Sea to Flores Sea covers the depth, temperature, thermocline, and upwelling according to foraminifera analysis. The method used were inventory and species analysis from 12 stations along study area. The analysis was conducted based on species and abundance. Foraminifera species which are warm water bioindicators are *Globigerinoides rubra* and *Globigerinoides sacculifera*, upwelling bioindicators are *Globigerina bulloides* and *Globorotalia menardii*, and shallow thermocline bioindicators are *Pulleniatina obliqueloculata* and *Neogloboquadrina dutertrei*. The result show that Java Sea has more benthonic species and abundance because of its shallow depth while Flores Sea has more planktonic species and abundance because of its deep depth and part of open ocean. In Flores Sea the abundance of warm water species bioindicators fall significantly and high abundance of upwelling and shallow thermocline species bioindicators indicates that there are upwelling phenomenon with shallow thermocline, thin mix layer, and high productivity.

1. Introduction

Java Sea and Flores Sea are located in between Java, Kalimantan, and Sulawesi and play important roles in human activities such as transportation. The seas are also part of the pathways that allow the passage of water mass from Pacific Ocean to Indian Ocean that forms the Indonesia Throughflow (ITF). The bathymetry of Java Sea varies with an average of 120 m, while the bathymetry of Flores Sea reaches 5000 m [1, 2]. Population explosion as well as human activity in land and ocean produced substantial waste that might influence the waters in the area.

The complex bathymetry of the Indonesian Seas, such as narrow channel and sills, is a result of geological processes working in the archipelago involving transform faults and subduction zone in plate boundaries. Those architectures mix inflowing the Pacific water mass with freshwater to produce Indonesian water mass [3] that in turn affect the type and distribution of organisms living in the area and also influence productivity of the area [4]. Indonesian Throughflow (ITF) transport became determinant factor for marine organism distribution [5]. According to [6] Indonesian Archipelago has high foraminifera diversity so that foraminifera can be used as potential bioindicator to recognize oceanography characteristics.

The characteristics of Java Sea and Flores Sea are influenced by both natural and anthropogenic-induced physical and chemical reactions that are important to recognize. One of the method to understand them is by studying the foraminifera assemblage. Foraminifera is a useful tool and easy to handle bioindicator. They reflect general and accurate environmental condition, thus changes in their assemblage would reflect environmental changes. After their deaths, foraminifera sink to the bottom, and accummulate for million of years to then form 75% of seafloor sediment [6]. Some of foraminifera characteristics are small size, simple body structure, hard test, widely distributed in marine environment and high ability to react to environmental changes [7, 8].

Water mass movement yields physical and chemical reactions such as as upwelling and changes in thermocline depth. One of the movement is surface current that changes semiannually in the Indonesian Seas. During different monsoon seasons, water mass from the Indian Ocean and Pacific Ocean flow to the Indonesian Seas and cause variation in temperature, salinity, and others [9]. During Southeast Monsoon, cold surface water mass from Banda and Flores Seas flow into the eastern part of Java Sea to the North Natuna Sea through Karimata Strait [10].

When the ITF reaches the southern part of Makassar Strait, it flows to two directions: straight to the Indian Ocean via Lombok Strait and flows to the east (Flores and Banda Sea) before going to the Indian Ocean through Timor Sea and Ombai Strait. Those pathways are very complex because of topographic variation of the three Straits (Makassar, Malaka, and Lombok) that yield the mixing and reversal of water mass. Local process working in the Indonesian Seas that is connected to local monsoonal wind such as upwelling and downwelling along with high tide, ocean heat flux, precipitation volume and river runoff change the Pacific water mass temperature and salinity stratification to become specific Indonesian Seas profile [9]. The aim of this study is to determine the depth, temperature, thermocline, and upwelling condition of Java Sea and Flores Sea based on foraminifera analysis.

2. Materials and methods

Map of sample locations and their depths show in figure 1. The method of this research including sampling, sample preparation, species identification, and data analysis. The samples that are existing samples that stored at the Core Laboratory of the Marine Geological Institute (MGI). Some of the samples were obtained during MAJAFLOTE cruise in 2015 onboarg Geomarin III. The cruise is a collaboration of Marine Geological Institute and Bogor Agriculture University. The samples are acquired by multicorer in 4 stations in Java Sea and gravity corer in 1 stations in Makassar Strait and 7 stations in Flores Sea. This study used surface sample (0-2 cm) for foraminifera analysis. Sample preparation was conducted at Marine Geological Institute Core Laboratory, Cirebon. The samples were washed and dried in the oven before sieved in multilevel sieves (-0.5; 0; 0.5; 1; 1.5; 2; 2.5; and 3ø in diameter). The dry weight of each size fraction then weighed separately. Foraminifer identification was carried out between February 1 to March 11, 2016 at the Petrology and Micropaleontology Laboratory of Marine Geological Institute in Bandung. Size fraction from 0,5 to 2ø was split until the smallest fraction is estimated to consist of ±300 foraminifera individus.

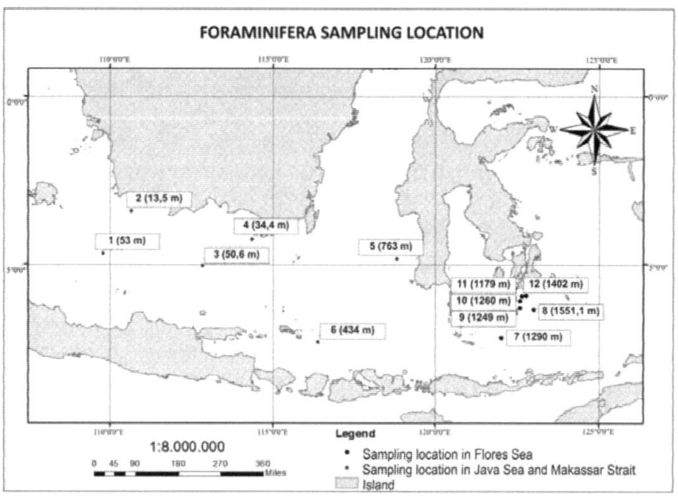

Figure 1. Map of sample locations and their depths (data from Marine Geological Institute).

Foraminifera which had been picked and separated from other sediment and tests fractions was identified below electronic microscope SMZ 800r following key identification of Saito *et al* (1981), identification book Postuma [11], Loeblich and Tappan [12], and Maryunani [13]. A collection of foraminifera specimens from 12 stations were put together in chartman slide showing the dorsal and ventral view of each species. The documentation of the specimens was conducted using binocular microscope connected to Nikon camera and NIST Element software. Data analysis involved calculation of P/B ratio as well as abundance of specific foraminifera that serve as bioindicator for temperature, upwelling, and thermocline. Identification of bioindicator species follows previous studies such as [14]. The calculation of P/B ratio uses the following equation:

$$\text{P/B ratio} = \frac{\sum planktonic\ species}{planktonic\ and\ benthonic\ species}$$

3. Results and discussion

Identification of foraminifera yields 59 species from 12 samples that compose of 31 benthonic species (1,111 individuals) and 28 planktonic species (2,240 individuals). Benthonic species show higher diversity than planktonic species, although planktonic species show higher abundance than benthonic species (figures 2 and 3).

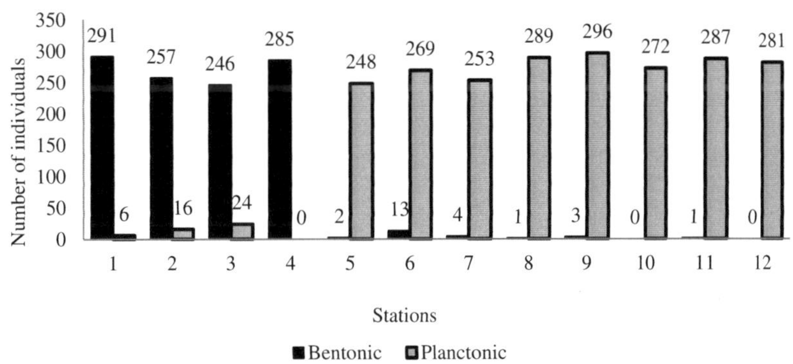

Figure 2. Planktonic and benthonic foraminifera number of individuals for each stations.

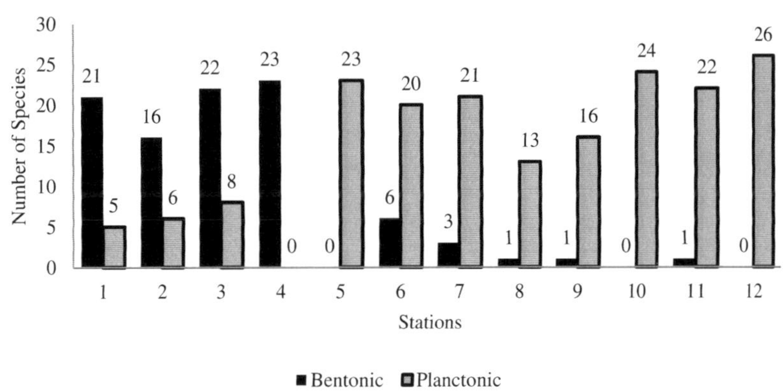

Figure 3. Planktonic and benthonic foraminifera number of species for each stations.

Our results showed that samples from the shallow Java Sea has higher benthonic species diversity and number of individuals while from deeper Makassar Strait and Flores Sea have higher planktonic species diversity and number of individuals. The result shows that planktonic foraminifera are more abundant in deeper water while benthonic foraminifera are more abundant in shallow water. This pattern is reflected in P/B ratio.

3.1. P/B ratio

The P/B ratio of the Java Sea show low P/B ratio value compare to Makassar Strait and Flores Sea. All four stations from Java Sea which is categorized as inland and shallow seas (±13-50 m) have low P/B ratio. All stations from Makassar Strait and Flores Sea which is categorized as open and deep seas (±400-1500 m) have higher P/B ratio up to 1 (figure 4).

The results show correlation between P/B ratio and water depth. P/B ratio has been traditionally used to estimate water depth on continental margins. Water depth often shows a relationship to the distribution patterns of living planktonic foraminifera, but depth-correlated factors such as temperature, salinity, dissolved oxygen, photic conditions, and nutrient levels probably exert direct control over foraminiferal depth habitats [15]. P/B ratio will increase along with the distance from shore and type of the water body toward the open ocean [18]. It can be concluded that Java Sea with shallow depth is an inland sea while Makassar Strait and Flores Sea with deeper water depth is part of open seas that is close to oceanic condition.

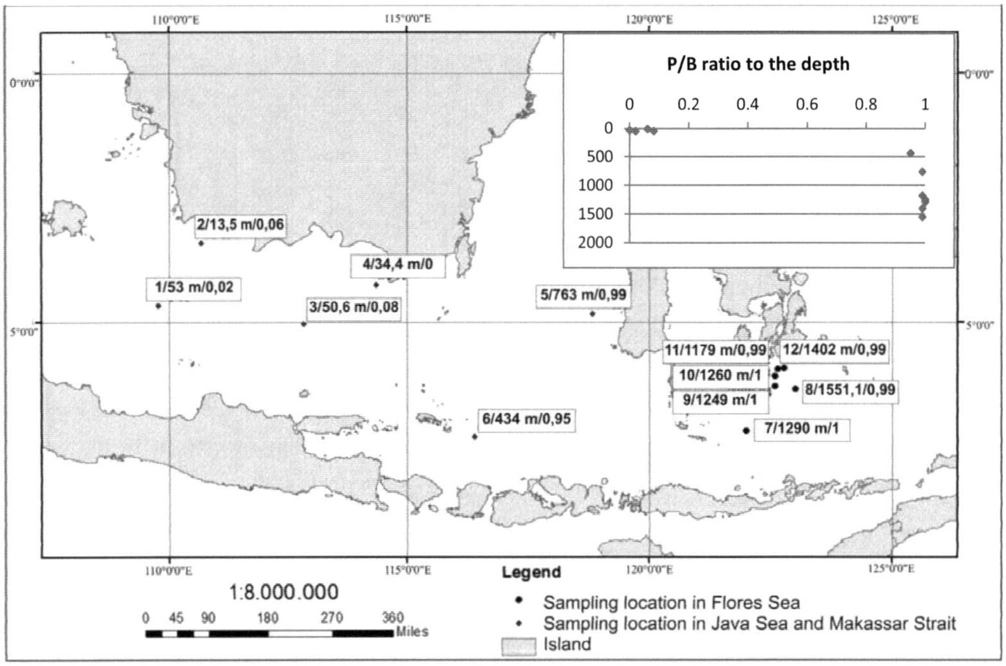

Figure 4. Map of P/B ratio and chart that show relation of P/B ratio to water depth.

3.2. Environment characteristics based on foraminifera assemblage

Environmental characteristics such as temperature, thermocline, and upwelling are observed from foraminifera assemblage. The analysis involved species identification, counting, and literature review of bioindicator species that can be applied to understand the oceanography of Java Sea and Flores Sea. Previous studies show that *Globigerinoides rubra* and *Globigerinoides sacculifera* can be used as warm water bioindicators considering their living distribution in tropical ocean; *Globigerina bulloides* and *Globorotalia menardii* as upwelling bioindicators; *Pulleniatina obliqueloculata* and *Neogloboquadrina dutertrei* as shallow thermocline indicators. Table 1 shows the references for bioindicator species.

Table 1. Foraminifera that serve as bioindicator species for warm water, upwelling, and shallow thermocline.

No	Indication	Species	References
1	Warm water	*Globigerinoides sacculifera* and *Globigerinoides rubra*	[6, 15, 17, 18]
2	Upwelling	*Globigerina bulloides* and *Globorotalia menardii*	[6, 17, 19, 20]
3	Shallow thermocline	*Pulleniatina obliqueloculata* and *Neogloboquadrina dutertrei*	[6, 17, 18, 20, 21]

The abundance of *Gs. sacculifera* and *Gs. rubra* in 4 samples from Java Sea are quite low while no *G. bulloides*, *Gr. menardii*, *P. obliqueloculata*, and *N. dutertrei* were found. All of bioindicators use planktonic foraminifera that are rarely found in Java Sea. The abundance of *Gs. sacculifera* and *Gs. rubra* in Makassar Strait was high and decrease toward the Flores Sea. The lowest abundants of the six species in Flores Sea was observed in Station 10.

High abundance of *G. bulloides* and *Gr. menardii* abundance was observed in Makassar Strait and some stations in Flores Sea. The highest abundance of those species were found at Station 10 and Station 11 in Flores Sea. *N. dutertrei* and *P. obliqueloculata* are abundant in Flores Sea with highest abundance were observed at Station 11 and Station 12 (figure 5).

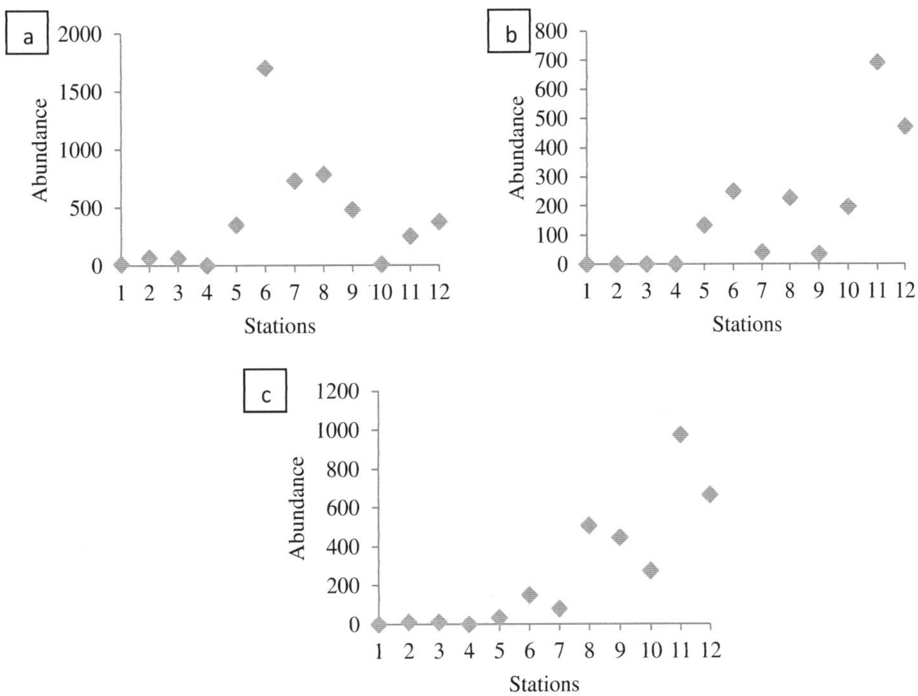

Figure 5. Species of foraminifera that are used as bioindicators for a) *Gs. rubra* and *Gs. Sacculifera* abundance as warm water bioindicators b) *G. bulloides* and *Gr. menardii* as upwelling bioindicators, and c) *N. dutertrei* and *P. obliqueloculata* as shallow thermocline bioindicators.

Higher abundance of *N. dutertrei* and *P. obliqueloculata* was observed in Flores Sea. High abundance of those species in Stations 11 and 12 indicate that shallow thermocline layer in those stations. This interpretation is supported by high abundance of *G. bulloides* and *Gr. menardii* in Station 11 that show upwelling activity. Low abundance of *Gs. sacculifera* and *Gs. rubra* as warm water bioindicators was observed at Stations 10, 11 and 12. This interpretation is also supported by highest abundance of *N. dutertrei* and *P. obliqueloculata* at Stations 11 and 12.

[20] studied that *Gr. menardii* lived within the upper thermocline (UTH) during the late Miocene, but Fairbanks and Wiebe (1980) *in* [20] found that today, *Gr. menardii* is known to occur in greatest abundances in association with the chlorophyll maximum zone, similar to *N. Dutertrei* [20]. Therefore, their position in the upper thermocline is likely controlled by the availability of food rather than temperature. The abundance of *Gr. menardii* and *N. dutertrei* at Stations 11 and 12 indicate upwelling phenomenon that was accompanied by shallow thermocline, thin mixed layer, and high primary productivity. The mechanism behind high surface productivity in the Eastern Indonesia is monsoon-related upwelling. From foraminiferal tests, we know that monsoon system disturbance will lead to decrease on marine organism productivity [6].

4. Conclusion

A study on foraminifera assemblage from Java Sea, Makassar Strait and Flores Sea show the relationship between P/B ratio and water depth. In general, benthonic species show higher diversity but lower abundance compared to planktonic species. Java Sea is characterized by higher abundace and diversity of benthonic species due to its shallow water depth and inland position. Flores Sea is characterized by higher abundance and diversity of planktonic species because of its deeper water depth and relatively open sea. Analysis of foraminifera as bioindicators of environmental condition show that upwelling phenomenon was observed in Flores Sea. The upwelling phenomenon was accompanied by shallow thermocline, thin mixed layer, and high primary productivity.

Acknowledgeements

The field and laboratory work was supported by Marine Geological Institute, Bandung, Indonesia which providing core GLWPL and 2208 available for this study and all of laboratory stuff. We also thank Agriculture Bogor University as host organizers for 2nd International Conference of Marine Science providing the means for this research to be published.

References

[1] Romimohtarto K and Juwana S 1999 *Biologi Laut: Ilmu Pengetahuan tentang Biota Laut* (Jakarta: Pusat Penelitian dan Pengembangan Oseanologi-LIPI) p 540
[2] Mulyana W and Salahuddin M 2010 *Morfologi Dasar Laut Indonesia* http://www.mgi.esdm.go.id/ [22 January 2016]
[3] Gordon A L 2005 Oceanography of the Indonesian Seas and their throughflow *Oceanography.* **18**(4) 14-27
[4] Nagara G A, Sasongko N A and Olakunle O J 2007 *Introduction to Java Sea* (University of Stavanger) p 18
[5] Safitri M, Cahyarini S Y and Putri M R 2012. Variasi Arus Arlindo dan Parameter Oseanografi di Laut Timor sebagai Indikasi kejadian ENSO *J. Ilmu dan Teknologi Kelautan Tropis.* **4**(2) 369- 377
[6] Tomascik T, Mah A J, Nontji A and Moosa M K 1997 *The ecology of the Indonesian Seas Part One. The ecology of Indonesia* Series VII p 746.
[7] Loeblich A R and Tappan H 1988 *Foraminiferal Genera and Their Classification* (New York: Van Nostrand Reinhold) p 2031
[8] Nurdin J and Afrizal S 2013 Kepadatan dan Keanekaragaman Foraminifera Di Perairan Laut Teluk Bayur Padang Sumatera Barat. *Prosiding Semirata FMIPA Universitas Lampung* p 7
[9] Thorpe, 2009
[10] Tubalawony S, Kusmanto E and Muhadjirin 2012 Suhu dan salinitas permukaan merupakan indikator upwelling sebagai respon terhadap angin muson tenggara di perairan bagian utara Laut Sawu. J. *Ilmu Kelautan* **17**(4) 226-239
[11] Postuma J A 1971 *Manual Planktonic Foraminifera* (New York: Elsevier Publishing Company) p 422.
[12] Loeblich A R and Tappan H 1988 *Foraminiferal Genera and Their Classification* (New York: Van Nostrand Reinhold) p 2031
[13] Maryunani K A 1999 *Panduan Praktikum Foraminifera* (Bandung: Laboratorium Mikropaleontologi Jurusan Teknik Geologi ITB) p 12
[14] Ding X, Bassinot F, Guichard F, Li Q Y, Fang N Q, Labeyrie L, Xin R C, Adisaputra M K and

Hardjawidjaksana K 2006 Distribution and ecology of planktonic foraminifera from the Seas around the Indonesia Archipelago *J. of Marine Micropaleontology*. **58** 114-134

[15] Gupta B K S 2002 *Modern Foraminifera* (Dordrecht: Kluwer Academic Publishers) p 384

[16] Demchuk T D and Anthony C G 2004 *Geologic Problem Solving with Microfossils: A Volume in Honor of Gary D Jones* SEPM Special Publication (93) p 347.

[17] Boltovskoy E and Wright R 1976 *Recent Foraminifera* (Netherland: The Haque) p 515

[18] Spooner M I, Barrows T, Deckker P D and Paterne M 2005 Paleoceanography of the Banda Sea, and Late Pleistocene initiation of the Northwest Monsoon *Elsevier J. Global and Planetary Change*. **49** 28-46

[19] Clift P D, Kroon D, Gaedicke C and Craig J 2002 *The Tectonic and Climatic Evolution of the Arabian Sea Region*. Geollogical Society Special Publication no 195 (London: Geological society of London special publication)

[20] Nathan S A and Leckie R M 2009 Early history of the Western Pacific Warm Pool during the middle to late Miocene (~13.2 5.8 Ma): Role of sea-level change and implications for equatorial circulation *Elsevier J. Palaeogeography, Palaeoclimatology, Palaeoecology*. **274** 140-159

[21] Xiang R, Sun Y, Li T, Oppo D W, Chen M and Zheng F 2014 *Paleoenvironmental change in the middle Okinawa Trough since the last deglaciation: Evidence from the sedimentation rate and planktonic foraminiferal record*. Chinese Academy of Sciences p 18

IOP Conf. Series: Earth and Environmental Science 176 (2018) 012007 doi:10.1088/1755-1315/176/1/012007

Dynamic changes of organic content over time in the sea sludge collected from the Gobo/Hidaka coastal area in Wakayama, southwest Japan

H Hirano[1*], D Semura[1], K Sakamoto[1], T Toyama[2], DHE Setiamarga[3*]

[1]Department of Civil Engineering and Urban Planning, National Institute of Technology, Wakayama College, Gobo City, Wakayama, Japan 644-0023
[2]Department of Materials and Applied Chemistry, College of Science and Engineering, Nihon University, Tokyo, Japan 101-8308
[3]Department of Applied Chemistry and Biochemistry, National Institute of Technology, Wakayama College, Gobo City, Wakayama, Japan 644-0023
*e-mail: hirano@wakayama-nct.ac.jp; davin@wakayama-nct.ac.jp

Abstract. The Gobo/Hidaka coastal area is known for its pristine nature and biodiversity, but recent urban developments might have affected its environmental condition. In previous study, we found a relatively low amount of organic matter from the Hidaka Port samples. In this study, we determine the possible cause of such an anomalous low organic matter content. Sludge samples were collected from the bottom of the sea of Hidaka Port monthly for one year (July 2016 to June 2017). The result showed the lowest content of organic matter in July 2016 (0.89%) and almost repeated in November 2016 (1.47%), February 2017 (1.08%), and May 2017 (1.42%). Meanwhile, the highest record was in June 2017 (3.93%). The fluctuating organic contents are in congruent with the water temperature data. The possible interactions among the seawater temperature, monthly rainfall amount of the corresponding months, possible abundance of marine microbes, and farming in the surrounding area might have caused the fluctuations. Our result would certainly be useful to infer the interrelationships of anthropogenic activities and other biotic and abiotic factors to the sea sludge, and eventually, the oceanic environment. Future studies needed to analyze inorganic content and data collections over multiple years.

1. Introduction

Sedimentary sea sludge is an important component of the seafloor environment. For example, it plays an important role as one of the matrices for living organism habitats. Therefore, information on the chemical compositions of sea sludge (e.g. the organic and inorganic contents) and their dynamics might bring an understanding about the interaction between environments and organisms living in the matrix. Furthermore, such data might bring possible insights about the effect of anthropogenic influences on seafloor environment.

Sea sludge is formed through the deposition of organic and inorganic matter at the seafloor, causing viscous mud-like sediment to pile at the bottom of the sea. By definition, sea sludge is different from a mud because of the lack of organic matter and sulfate in the latter. However, it is very difficult to differentiate the two seafloor sediments simply through its outer appearance. In Japan, sea sludge is always associated with bad smell, causing it to be called “hédoro” (flatulence (= hé); and mud (= doro)) or "hénatsuchi" (a wet darkly mud). Because of this compositional difference the organisms inhabiting mud and sea sludge are presumably different.

Some previous research has studied the chemical characteristics of the sea sludge [5, 6]. However, such studies have concentrated only on elucidating the compositional aspect of the sea sludge. Moreover, the sea sludge used as the objects of those studies were collected from different sampling localities, and located at different times. Since sea sludge composition is influenced greatly by small difference in the environs of the location, an objective, standardized comparison of the studies difficult [1]. In our previous study, we reported that samples collected from different coastal areas in Japan (Funabashi-port in Chiba vs. Hidaka-port in Wakayama) showed different compositional values in organic and inorganic contents, and thus in their characteristics (table 1) [4, 6]. The generalization of the results obtained from such studies, however, have allowed for a standard-but-vague definition of a sea sludge based on its chemical composition. The definition of sea sludge can thus be given as follow: Mud-like viscous sedimentary substance composed of (1) organic matter partially derived from the activities and carcasses of living organisms including microorganisms, (2) inorganic matter derived from the deposition of mud, silt, and sand, and (3) high sulfur content (which gives the sea sludge its distinct "rotten egg-like smell" and dark color).

The lack of a more specific definition has caused a detailed classification of seafloor sludge non-existent. Such lack of a classification system based on the differences and dynamic changes in compositions of the different types of sludge, and their relations to differences and changes of the environments, have hindered a detailed understanding of the position and importance of sludge in seafloor environment. Meanwhile, the inter-relationships and interactions among natural ecosystem changes, climate and weather dynamics, biological activities, and other external environmental changes, with the dynamic chemical compositions changes of the sea sludge, are rarely studies. In fact, to our knowledge, there is no report about sea sludge compositional changes collected in a time series from one particular sampling point.

In this paper, we report the changes of organic content of the coastal sea sludge in natural conditions collected from a single location at the Hidaka Port in Gobo/Hidaka coastal area in the Kichuu region of the Wakayama Prefecture, southwest Japan (figure 1). The Gobo/Hidaka coastal area is known for its pristine nature and biodiversity, but recent urban developments might have affected its environmental condition. In order to obtain an insight on the dynamic changes of the organic content of the sea sludge collected from one particular location, we collected sea sludge samples in a time-series for 13 months. We also surveyed and collected relevant local weather and water temperatures in order to infer possible correlations among the three data sets. We concentrated on the changes in organic content because it is relatively easy to detect and analyze with relative accuracy. Moreover, since changes in organic content are usually influenced by external environmental factors such as living organism activities including microbes and humans, the data might enable us to discuss about the possible dynamics of such external factors.

Table 1. Difference in organic content of the sea sludge in different collecting sites.

Collecting sites	Measured days	Organic contents [%]
Chiba (Funabashi Port)	15/9/2011	23.6
Wakamaya (Hidaka Port)	15/6/2015	7.3

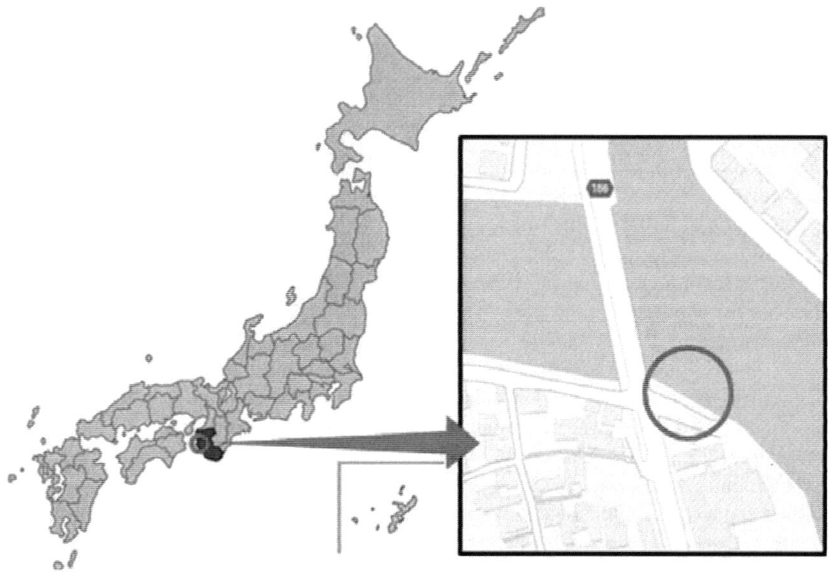

Figure 1. Sea sludge collection sites for this study (33°52'55.3"N, 135°09'10.7"E).

2. Materials and Methods

2.1. Sea sludge sampling and weekly water temperature data

Wet sea sludge from Hidaka Port in Gobo City in Wakayama Prefecture (figure 1) was collected manually. The sea sludge was obtained from its natural environment. To see the dynamic changes of the organic content of the sludge from one geographic locus, we collected sea sludge samples once a month for 13 months from July 2016 to July 2017. Samples were collected from one sampling location at three different loci, and the sludge from the three loci were mixed together for further analyses, in order to avoid compositional bias. Sampling dates for each month are shown in table 2, along with the information of the mean temperature of water for the week of the sampling date. Seawater temperature data of the week of the sampling dates, of Wakayama region, were also referenced from the homepage of the 5th Regional Coast Guard Headquarters (http://www1.kaiho.mlit.go.jp/ KAN5/).

Then, the wet sea sludge samples were oven-dried, in 100 °C for 10-15 min to evaporate the moisture content. Afterward, one gram of the dried sample was extracted for future analyses.

Table 2. Sample sludge collection dates and corresponding weekly water temperature.

Collected days	Water temperature [°C]	Collected days	Water temperature [°C]
7 July 2016	28	26 February 2017	16
8 August 2016	30	29 March 2017	17
27 September 2016	27	25 April 2017	19
31 October 2016	25	24 May 2017	21
29 November 2016	22	28 June 2017	24
22 December 2016	20	25 July 2017	27
25 January 2017	18		

2.2. Analysis of organic content

We conducted Thermo-Gravimetric (TG) analysis [6] on the oven-dried samples to obtain the information of organic content, in three conditions: (1) the original dried sludge sample, (2) dried-up sample further burnt at 100 °C in order to obtain information about both inorganic and organic contents, and (3) dried sample burnt twice, first at 100 °C and later at 600 °C, to get rid of the organic matter completely. The organic content of samples was then deduced by calculating the lost mass caused by burning at 100 °C and 600 °C, considering sample (1) (the solid/dried soil sample) as the standard.

2.3. Weather condition data

We collected weather data of three consecutive days before, three consecutive days after, and on the day of sludge collection, provided by the Japan Weather Association (http://tenki.jp/).

3. Results

3.1. Analysis of organic content

Heat weight loss analyses (TG analysis) on the samples with burning temperature between 100 °C to 600 °C indicate a fluctuation of the content of organic matters in the sludge samples over time in the course of one year. Our observation suggested that the fluctuations are random but not drastic, with the "usual" organic content in a given month as around 1.00% to 2.00% (Table 3). The lowest content in July 2016 (0.89%) and was almost repeated in November 2016 (1.47%), February 2017 (1.08%), and May 2017 (1.42%). However, data of some of the months show drastic fluctuations of organic content. The value in June 2017 was 3.93%, and in September 2016 was 2.68%, which were twice than the "usual" value of 1 - 2%.

Table 3. Monthly organic contents.

Collected days		Organic contents [%]
2016	July	0.89
	August	1.92
	September	2.68
	October	1.78
	November	1.47
	December	1.47
2017	January	1.93
	February	1.08
	March	1.65
	April	1.89
	Mei	1.42
	June	3.93
	July	1.53

3.2. Correlation among the weather, water temperature data, and the organic content of sea sludge

Environmental data such as water temperature and weather conditions were collected from relevant public institutions (e.g. the Japan Weather Association), in order to see their influence on the dynamic fluctuations of organic content of the sea sludge. Although sample collections were mainly done on days with fair weather condition, the weather of previous days might affect the composition of the sludge. Meanwhile, changes in water temperature might also influence the composition of the sludge.

Interestingly, we found that the temperature and the preceding weather of September 2016 (9/27) and June 2017 (6/28), as well as the mean water temperature of the relevant week, were drastically increased. Another fluctuation in mean water temperature of the week and/or preceding weather was also observed surrounding the December 2016 (12/22) sampling, with seawater temperature at 20 °C, and the weather was cloudy with temporary rain, but sunny or cloudy then sunny for the preceding three days (table 4).

Meanwhile, a sunny weather condition was observed on the sampling dates where the organic contents of samples were categorized as "usual". This result indicates a possible effect of external environment to the compositional fluctuations of the sea sludge. The results of organic content changes over time were plotted on figure 2, together with seawater temperature changes. Weather conditions are shown on table 4.

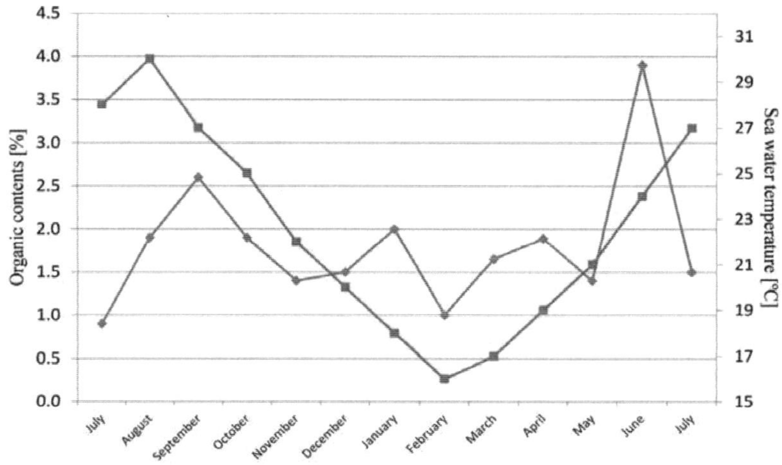

Figure 2. Change of organic contents with water temperature.
◆ : Organic content ■ Water temperature

4. Discussion

The composition of sea sludge, including the organic matter composition, varies from one locality to another. Such composition is affected by many external factors, including, but not limited to, living organisms inhabiting the area, water temperature, salinity, local climate, terrestrial debris flowing in from the river [1]. This has caused difficulties in defining the sea sludge based on a standardized chemical composition [6]. Meanwhile, research on local variations of sea sludge in an area has been meager. To our knowledge, there has been no study on the chemical compositional changes over time of sea sludge collected from a single locality. The lack of such information hampers an objective discussion about standardized compositional changes of sea sludge.

In this study, we followed the organic content of sea sludge collected from one sampling locality, and discussed our result together with the seawater temperature fluctuations and changes in weather conditions.

4.1. The possible role of microbial activities

A data plot of weekly seawater temperature and sea sludge organic matter composition over time (figure 2) shows that the two changes are congruent, indicating possible correlations. For example, during the months of July 2016 to November 2016, organic content and temperature showed similar fluctuation patterns, with the organic content lagged behind for one month. Water temperature peaked in August then decreased, while the organic content's peak was in September, and then decreased.

Such changes in organic content, when discussed with changes in water temperature, can be explained by considering the presence of microbes. Water temperature increased above 25 °C during the summer months (June to August), and decreases afterwards. Such warm temperature is conducive for microbial activities. We thus suggest that microbial activities might be strongly related to the increase of organic matter in the seafloor sludge. The one-month lag of the organic content peak could also be explained with the microbial activities. The organic matters produced by the microbes accumulate in a process that peaked a month after the actual peak of microbe activities. When the season gradually enters autumn, water temperature drops, and microbe activities slowed down, causing the production of organic matters and its inclusion in the sea sludge, to also gradually drop down.

Table 4. Weather conditions on and up to three days before the collection day.

Year	Date	Conditions	Year	Date	Conditions
2016	7/17	Cloudy then sunny	2017	1/22	Cloudy the sunny
	7/18	Cloudy then sunny		1/23	Sunny then rain
	7/19	Sunny		1/24	Snow later rain
	7/20*	Sunny		1/25*	Sunny
	8/16	Cloudy then sunny		2/23	Rain then cloudy
	8/17	Cloudy then sunny		2/24	Cloudy and sometimes sunny
	8/18	Cloudy then sunny		2/25	Sunny
	8/19*	Cloudy then sunny		2/26*	Cloudy
	9/24	Cloudy then rain		3/26	Sunny the rain
	9/25	Cloudy then rain		3/27	Cloudy then sunny
	9/26	Rain then cloudy		3/28	Cloudy then sunny
	9/27*	Mostly sunny and cloudy		3/29*	Cloudy and sometimes sunny
	10/28	Cloudy then rain		4/22	Cloudy then sunny
	10/29	Cloudy then sunny		4/23	Sunny
	10/30	Sunny		4/23	Sunny
	10/31*	Cloudy		4/23*	Moatly sunny and cloudy
	11/26	Sunny then rain		5/21	Cloudy and sometimes sunny
	11/27	Rain then cloudy		5/22	Sunny
	11/28	Cloudy then sunny		5/23	Mostly sunny and cloudy
	11/29*	Sunny		5/24*	Rain
	12/19	Cloudy then sunny		6/25	Rain then cloudy
	12/20	Cloudy then sunny		6/26	Cloudy then rain
	12/21	Cloudy with temporary rain		6/27	Cloudy then rain
	12/22	Cloudy with temporary rain		6/28*	Rain then cloudy
				7/22	Cloudy
				7/23	Cloudy
				7/24	Cloudy and sometimes sunny
				7/25*	Cloudy

*) Collection day

Figure 3. Collecting sites: (A) for samples from February 2017, (B) for samples from July 2017. Notice algal growth and disappearance on the collection sites.

4.2. The possible influence of algal life cycle

While a peak and gradual decrease in organic content were explicable with an increase in microbial activities caused by high water temperature, a deep drop in organic content, such as what was observed in February 2017, could not be explained similarly. Water temperature drops down gradually, and starting from November 2016 and all the way to June 2017, the temperatures stayed below 25 °C despite small fluctuations. Since such low seawater temperatures could not support microbial activities, other environmental factors must be considered to explain the steep drop in the organic content of sea sludge. Interestingly, we also observed the emergence of algal-bed (figure 3) at the sample-collecting locality in January 2017.

Taking everything into account, we therefore suggest that drastic fluctuations in organic content for the non-summer months could be explained in two steps. First, conducive water-temperature during summer triggered microbial activities, causing organic content to be released into the sea sludge. When water temperature dropped down after summer, the organic content of the sea sludge was still high, and it was then used as nutrients by the alga to grow into an algal-bed (figure 3(A)). Because microbial activity significantly decreased during the non-summer months, the organic content was not replenished, causing a gradual decrease in organic matter presence during the months of November 2016 to April 2017.

We observed *in situ* that the algal-bed disappeared by March 2017 (figure 3(B)), while the organic content gradually increased during the months of February and April 2017, although it did not peak. We hypothesized that the life cycle of the alga could explain this observation. The alga might reach the end of the life cycle, which could partially be caused by the decrease of organic matter nutrients contained in the sea bottom sludge. Despite the non-optimal temperature in February to April, a slow decomposition process by microbes might still happen. Dead alga then decomposed, causing a gradual increase of organic matter during the months of March and April 2017.

4.3. The correlation between weather conditions and changes in organic content

Other non-biological factor might also affect the changes of organic content of the sea floor sludge collected from Hidaka Port in Wakayama. We observed that between April and July 2017, the organic content decreased to 0.47% in May 2017, and increased to 2.51% in June 2017. This sudden and drastic change could also be explained by non-biological causes, such as local weather on the week of sampling date. Our previous study has shown that heavy rains could cause nutrients, salts, and organic matter contained in coastal sea sludge to decrease, probably due to washing–away caused by the sudden increase of freshwater [2]. However, we also did not rule out the possibility of a sudden increase of organic content caused by heavy rains happened inland. Such rain could cause organic materials from terrestrial agents brought down by the river to the sea, and deposited on the coastal sea floor sludge.

Data on the three-days leading to and on the sample-collecting day showed that there was a heavy rain for four days, around the sampling date of May 2017 (table 4). This might have caused the decrease of organic content. Meanwhile, several days before the sampling day of June 2016, heavy rain also happened inland. This might have caused terrestrial organic matters to be brought in to the sea (figure 4). Then, several days of calm weather might have let the terrestrial organic components to settle at the bottom, causing a sudden increase in organic content of the sea sludge.

Figure 4. Effects of heavy rain in collected point (June 2017).

5. Conclusion and Future Directions

Our result reported here indicates possible explanations for the changes of organic content in natural sea sludge:

1. Change of organic content matches the change of seawater temperature with one-month lag, especially during summer. High water temperature during summer might have caused increased microbial activities.
2. The decrease of organic content during the months of autumn and winter could be caused by the algal lifecycle, which might include feeding on organic materials in the sludge.
3. Weather changes might affect the organic content in the sea sludge, since heavy rains might leach the organic matters inside the sludge. The heavy rains occurred inland might brought organic matter into the coastal waters, and its deposition might cause an increase in the sea floor sludge organic content.

The elucidation of organic contents also might allow for future studies aiming at the detection of changes in other chemical components such as inorganic matter. Future studies on (1) detailed analyses of chemical compositional changes for the course of multiple years, (2) similar studies on several other localities will be needed to not only reconfirm our result presented here, but also to gain a more detailed picture of the interaction between sea sludge chemical composition dynamics and the external environment. Moreover, studies on microbial activities are also crucial, in order to decisively elucidate the actual mechanisms of the changes. For example, in this study only three possible causes of the dynamics of the sea sludge organic content were discussed. However, there are other possible reasons to explore, such as farming activities. Accordingly, future studies in a more comprehensive manner spanning multiple years, involving samples of different sites and multiple loci are warranted.

Acknowledgement

HH would like to thank Gobo City municipal government for their cooperation during sampling collection. HH and DHES were partially supported by the National Institute of Technology, Wakayama College Competitive Internal Research Grant for Education and Research 2016.

References

[1] Yoshida T A 1974 About generation and the removal of the sludge - In bottom sampler examination of the Gulf of Minamata *J. Hedoro* 11-27 (In Japanese)

IOP Conf. Series: Earth and Environmental Science **176** (2018) 012007 doi:10.1088/1755-1315/176/1/012007

[2] Hirano H, Sasahara R, Morita Y, Okamoto K, and Hotta K 2011 Purification of Sludge by Micro-Bubble and Microorganisms Activator-Case of Using Enzyme, Collection of papers: Annual Conference in CST 723-724 (In Japanese)
[3] Nakao A 2012 Cesium adsorption and fixation on soils trend in the science **17**(10) 10_40-10_45
[4] Hirano H, Toyama T, Nishimiya N, and Okamoto K 2015 Removal of cesium from sea sludge through decomposition of organic matter with aqueous hydrogen peroxide *International Journal of GEOMATE* **8**(1) 1203-1206 (In Japanese)
[5] Hirano H, Uragaki Y, and Morita S 2016 Preparation of prototype artificial sludge based on compositional information of sea sludge Part 2: Improvement idea about content 50th Annual Conference of Japan Society on Water Environment p.190
[6] Hirano H, Toyama T, Nishimiya N, Setiamarga DHE, Morita S, Uragaki Y, and Okamoto K 2017 Artificial sludge based on compositional information of a natural sea sludge *International Journal of GEOMATE* **12**(31) 95-99

IOP Conf. Series: Earth and Environmental Science **176** (2018) 012008 doi:10.1088/1755-1315/176/1/012008

Impact of reclamation on monsoonal circulation changes in Jakarta Bay

D Adhyatma[1*], A S Atmadipoera[2]

[1] Graduate Program in Marine Sciences, Bogor Agricultural University (IPB)
[2] Department of Marine Science and Technology, Bogor Agricultural University
*e-mail: dewa.adhyatma@gmail.com

Abstract. The Jakarta Bay plays an important role in supporting maritime activities such as marine transportation, tourism, industry, and coastal communities. To meet the needs of the people in the future, it is planned to build islands (giant sea wall) through huge coastal reclamation. In oceanography-point-of-view, new islands formation is believed to modify natural coastal circulation in the region. Here, we investigated the impact of new islands formation on Monsoonal circulation and passive tracer distribution by performing a validated numerical model experiment with ROMS-AGRIF. The results show that natural Monsoonal circulation is characterized by eastward (westward) flow during the Northwest (Southeast) monsoon period, but mean circulation flows eastward. Water mass in the inner bay is flushed partially by water from western/eastern peninsula. With the presence of reclaimed islands, the Monsoonal surface flows are partitioned into channels in-between the islands. An intensification of the flows is found around a narrow channel. Furthermore, passive tracer of water, released in Ciliwung River estuary, indicated Monsoonal reversal spreading along the channels closed to the coastal mainland. Further experiment by changing channels width to be much smaller, the flows revealed are much weaker, was resulting in an accumulation of passive tracer around the estuary.

1. Introduction

The Jakarta Bay is a strategic area for fisheries, transportation, and tourism sector. This area is the central of coastal communities' activity and produces economical profit to the surrounding regions. However, the number of human activities also generate a high environmental pollution. Based on previous research, Jakarta Bay has elevated contaminants of 537 tons per day [1]. Organic pollutants could cause an eutrophication that gives impact to the area ecologically. A study by [2] about the variability of Monsoonal currents in Jakarta Bay, showed that sea surface currents in June and September 2003 flowed broader from the southwest to northwest and from the near-bottom to the coastal area, except for towards the sea. Additionally, sea surface currents pattern in May 2004 generally came from northeast to northwest.

Based on [3], a giant sea wall (GSW) will be built in Jakarta Bay and the construction of the master plan will take one and a half year with the physical development takes about 10-25 years. Physical development is targeted to be complete in 2025. The presence of the GSW construction will presumably influence various aspects of ecology, including oceanographic aspect in Jakarta Bay that causes a controversial issue.

Numerical modeling method was used to assess possible changes of the structure of the coastal area that could be affected due to reclamation activities by indicating current circulation pattern changes. Numerical modeling and computational advance in oceanography have improved the capacity in oceanographic conditions prediction on the offshore and the onshore area. One of modeling tools that was already developed is Regional Ocean Modelling System (ROMS) with its nesting version ROMS-

AGRIF. ROMS uses primitive equations (Navier-Stokes equation) for hydrodynamics model and can be applied in regional scale [4]. Certain numerical modeling using ROMS in Indonesian territory were also implemented, i.e. simulation of Monsoonal currents in Indonesian waters [5] and study of upwelling mechanism using numerical modeling in Southern Sulawesi [6].

As sea current acts as the carrier of sediments, nutrients, and pollutants, a study of current patterns in Jakarta Bay is needed. The pattern of flow was examined to explain the distribution of physical and chemical parameters as well as the distribution of marine organisms. Given the above context, we aim to investigate the annual cycle of sea surface currents pattern and its change as the impact of reclamation, as well as the distribution of passive tracer released from Ciliwung River.

2. Methods

2.1. The data

Data used for model were from research centers and World Database [7]. Atmospheric climatology (wind stress) was obtained from QuickSCAT, SST from AVHRR (Advanced Very High Resolution Radiometer)-Pathfinder Observations 1985-1997, bathymetry used ETOPO data with 0.5° resolution, the atmospheric flux used data from COADS05, the lateral boundary conditions (Drakkar Simulation of INDO-ORCA05), coastline (Coastline GHSS Folder), and seawater properties (World Ocean Atlas 2009).

Domain of model ranges from 106°18'E to 107°12'E and 5°18'S to 6°6'S. Domain for visualization was zoomed only between 106°40'E–107°5'E and 5°52'S–6°8'S. Jakarta Bay (visualization domain) consists of shallow water with average of depth approximately 32 m.

2.2. Model configuration

Ocean circulation model was built using ROMS-AGRIF configuration. ROMS uses primitive equations and free surface (short time-step for barotropic dynamics and long time-step for baroclinic dynamics) to calculate atmospheric or oceanic state at the initial time and after.

ROMS requires horizontal grid data input (position of grid point, size of grid dimension), topography of sea floor, addition of surface forcing (wind stress, heat flux, surface freshwater flux), initial (temperature, salinity, currents, sea level height) and lateral (temperature, salinity, currents, sea level height) boundary conditions.

Equation of motion in ROMS applies the second Law of Newton where the change of momentum against time equals to the total forcing [8]. [9] describe the equation of motion of water in 3D form and Cartesian coordinates:

$$\frac{du}{dt} + \vec{U}.\nabla u - fv = -\frac{1}{\rho}\frac{dP}{dx} + k\frac{d^2u}{dz^2} + A_h\Delta u \qquad (1.a)$$

$$\frac{dv}{dt} + \vec{U}.\nabla v + fu = -\frac{1}{\rho}\frac{dP}{dy} + k\frac{d^2v}{dz^2} + A_h\Delta v \qquad (1.b)$$

$$\frac{dw}{dt} + \vec{U}.\nabla w + g = -\frac{1}{\rho}\frac{dP}{dz} + k\frac{d^2w}{dz^2} + A_h\Delta w \qquad (1.c)$$

Where u, v, w are the velocity of fluid at x, y, z axes; t is time; f is Coriolis parameter (=2Ωsinϕ); $\vec{\Omega}$ is the angular velocity of the Earth rotation; Φ is latitude coordinate; $\frac{1}{\rho}$ is specific volume; p is pressure; k is vertical Eddy viscosity coefficient; A_h is horizontal Eddy viscosity coefficient; $\Delta = \frac{d^2}{dx^2} + \frac{d^2}{dy^2}$; $\nabla = \hat{\imath}\frac{d}{dx} + \hat{\jmath}\frac{d}{dy} + \hat{k}\frac{d}{dz}$ (del operator); and g is Earth gravity.

a

b

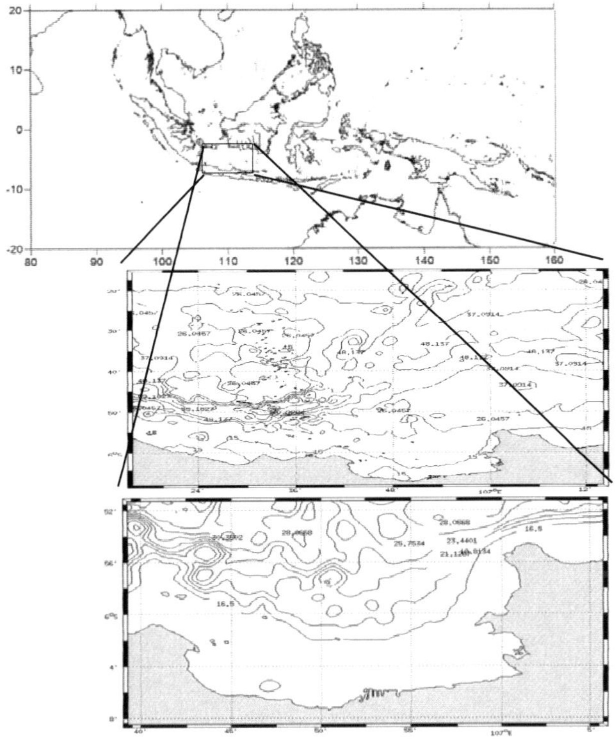

Figure 1. (a) Domain of model and bathymetry and (b) Domain of visualization and bathymetry of Jakarta Bay.

ROMS uses Navier Stokes primitive equations and ignores the acceleration among vertical speeds (w = 0) [10], thus, the equation of momentum conservation is simply referred as momentum conservation on the direction of x-axis and y-axis:

$$\frac{du}{dt}+\vec{U}.\nabla u - fv = -\frac{1}{\rho_0}\frac{dP}{dx}+\nabla_h\,(K_{Mh}.\nabla_h u)+\frac{d}{dz}(K_{Mv}\frac{du}{dz}) \quad (2.a)$$

$$\frac{dv}{dt}+\vec{U}.\nabla v + fu = -\frac{1}{\rho_0}\frac{dP}{dy}+\nabla_h\,(K_{Mh}.\,\nabla_h v)+\frac{d}{dz}(K_{Mv}\frac{dv}{dz}) \quad (2.b)$$

where the parameter k is transformed into K_{Mv} and A_h is replaced by K_{Mh} as the vertical and horizontal mixing water mass coefficient.

ρ is the density function of the temperature (T), salinity (S), and pressure gradient (P). However, density component is ignored in advection of equation (2) (the fluid is incompressible) so that on the equation of vertical hydrostatic equilibrium, the value of density can be described in equation (3) with the value of pressure on the depth equals to the negative of density and gravity:

$$0 = -\frac{dP}{dz} - \rho g \quad (3)$$

Change of horizontal density in continuity equation was neglected or fluid is incompressible ($\rho = \rho_0$) (Boussineq approximation) so that density value is constant for horizontal pressure gradient. Continuity equation for incompressible fluid can be written as:

$$0 = \frac{du}{dx} + \frac{dv}{dy} + \frac{dw}{dz} \quad (4)$$

ROMS can analyze fluid movement due to the advection and mixing of water masses. The equation used is tracer conservation, with temperature and salinity are included. Conservation equations of temperature (5) and salinity (6) can be written mathematically as:

$$\frac{\mathrm{d}T}{\mathrm{d}t} + \vec{u}\,.\nabla T = \nabla_h(K_{Th}.\nabla_h T) + \frac{\mathrm{d}}{\mathrm{d}z}(K_{Tv}\frac{\mathrm{d}T}{\mathrm{d}z}) \tag{5}$$

$$\frac{\mathrm{d}S}{\mathrm{d}t} + \vec{u}\,.\nabla S = \nabla_h(K_{Sh}.\nabla_h S) + \frac{\mathrm{d}}{\mathrm{d}z}(K_{Sv}\frac{\mathrm{d}S}{\mathrm{d}z}) \tag{6}$$

where $\frac{dT}{dt}$ and $\frac{dS}{dt}$ show the transform rate of temperature and salinity over time, $\vec{u}\,.\nabla T$ and $\vec{u}\,.\nabla S$ are advection, K_{Th} is horizontal mixing coefficient and K_{Tv} is vertical mixing coefficient. In this model, passive tracer calculates the debit of river in arbitrary unit.

Forcing of fluid on the surface and bottom are described using surface and bottom boundary conditions [11]. Surface waters are driven by kinematic (high sea level), surface wind stress and temperature-salinity flux. Mathematically, the driving force of the fluid in the surface boundary conditions can be written as:

$$\frac{\mathrm{d}\eta}{\mathrm{d}t} = w \tag{7.a}$$

$$K_{Mv}\frac{\mathrm{d}u}{\mathrm{d}z} = \frac{\tau_x}{\rho_0} \tag{7.b}$$

$$K_{Mv}\frac{\mathrm{d}y}{\mathrm{d}z} = \frac{\tau_y}{\rho_0} \tag{7.c}$$

$$K_{Tv}\frac{\mathrm{d}T}{\mathrm{d}z} = \frac{Q}{\rho_0 C_p} \tag{7.d}$$

$$K_{Sv}\frac{\mathrm{d}S}{\mathrm{d}z} = \frac{S(E-P)}{\rho_0} \tag{7.e}$$

Equation 7.a shows driven fluid by transforming sea level elevation ($\frac{\mathrm{d}\eta}{\mathrm{d}t}$). Wind as forcing is figured by equation 7.b in x-axis and 7.c in y-axis. The equations describe mixed fluid (K_{Mv}) driven by a stretch of wind (τ_x and τ_y) through the coastal area. Equation 7.d describes the change in temperature due to heat flux (Q) on a mixed layer and heat transfer coefficient is C_p. Equation 7.e is the change of salinity due to evaporation-precipitation S(E-P) on a mixed layer.

Bottom force on the boundary condition waters consists of basic kinematic, friction (friction) and bottom temperature and salinity flux. Mathematically it can be written as:

$$\vec{u}\,.\nabla(-H) = w \tag{8.a}$$

$$K_{Mv}\frac{\mathrm{d}u}{\mathrm{d}z} = \frac{-C_d|\vec{u}|u}{\rho 0} \tag{8.b}$$

$$K_{Mv}\frac{\mathrm{d}v}{\mathrm{d}z} = \frac{-C_d|\vec{u}|v}{\rho 0} \tag{8.c}$$

$$K_{Tv}\frac{\mathrm{d}T}{\mathrm{d}z} = 0 \tag{8.d}$$

$$K_{Sv}\frac{\mathrm{d}S}{\mathrm{d}z} = 0 \tag{8.e}$$

Equation 8.a describes the fluid motion due to advection process in bottom layers. Equation 8.b and 8.c explain the friction on x-axis and y-axis where the linear drag coefficient (C_d) has a value of $3\text{x}10^{-4}$ m/s. Thus, the temperature and salinity flux in the bottom is zero.

2.3. Description of experiment

Model was made in three scenarios. Scenario 1 is the normal condition of Jakarta Bay which is exist before recalamation, scenario 2 is the condition of Jakarta Bay after reclamation with GSW blueprint modified, and scenario 3 is the condition of Jakarta Bay including reclaimed artificial islands in accordance with the original GSW blueprint plan. The addition of reclaimed land of artificial islands in scenarios 2 and 3 was performed by masking technique on the grid of model for scenario 1.

Tidal component was calculated as a forcing for the simulation model on the open boundary condition. This was done since Jakarta Bay circulation is influenced by tides. However, the results of the simulation were saved as daily average to focus on the annual cycle of circulation. Moreover, this simulation model also set river discharge point to be made constantly. This parameter is required to study Monsoonal pattern distribution based on passive tracer of Ciliwung River.

3. Results

3.1. Model validation

A numerical model of Jakarta Bay circulation on scenario 1 was validated using sea surface height (SSH) data from altimetry satellite (AVISO) 2005-2010. Altimetry satellite data was averaged weekly in monthly climatology. Sea level height data result in climatology (zeta) from ROMS model (scenario 1) in the 5th year was also used as a comparison. Both of data was calculated for anomaly values (MSL references) hence the sea surface height data could be compared. The following presented the result of SSH data comparison between model and satellite altimetry (figure 2).

Correlation value between model data with the altimetry on the SSH is 0.7, which means that the model represents allegedly against the data of observation. RMSE was neglected as the values of the two signals are highly different, therefore the scale used were adjusted. Nevertheless, the patterns formed in the time series were similar from one another. The value of altimetry data shows that the sea level is below zero from January to March, and above zero from April to August. Sea level increases in September and decreases significantly in December. Similar result is also shown on the model results. Yet, model data was averaged daily which creates the sea level fluctuation was occurred more on the model data. Generally, the increasing of sea level in monthly period follows the pattern of SSH data altimetry.

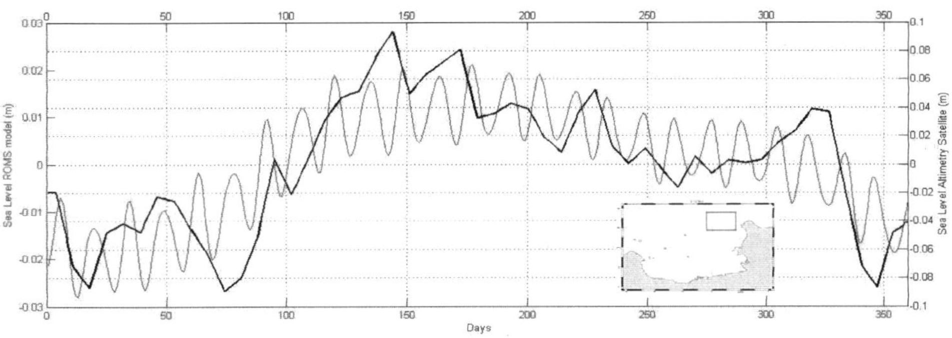

Figure 2. Annual cycle of sea level from satellite altimetry (black line) and model output (red line) in the Northeast side of Jakarta Bay. Sea level correlation between data and model output is 0.7.

3.2. Surface circulation pattern in Jakarta Bay

Simulation model in Jakarta Bay on scenario 1 (normal condition) was run for 5 years and averaged daily; hence tide signal was already eliminated. Thus, the current pattern in Jakarta Bay is only affected by wind with each grid on the model contains current vector. However, due to high resolution and very small distance of each grid on the visualization, simplification is necessary. Display grid of flow vector was given in every six grids with 0.1 m/s scale vector. Flow magnitude was not simplified or values are displayed on the entire grid.

Highest current circulation takes place in February 16 (West Monsoon) where the water flows eastward. Subsequently, sea current in May 16 (Transitional I Monsoon) moves slowly towards westward likewise in August 16 (East Monsoon). In November 16 (Transitional II Monsoon), transitional water flows from east to west and moves slowly until the West Monsoon sets in when the current speed increases eastwardly. Sea current pattern of the scenario 1 is presented in figure 3.

The development of the cape in Jakarta Bay causes the water flow from the western side is divided into two directions, into the bay area until near coastal and free flowing to the east or west. Then the two streams will converge in the eastern or western side of the mainland. Based on a research by [2] concerning the variability of currents in Jakarta Bay from field observation, the circulation pattern of model is less in accordance with the results of the study, most likely due to the current plot vector from field observation still consisted of tidal currents signal. However, different field measurement results show different circulation patterns, and it is indicated on the results of the model.

Simulation model of Jakarta Bay on scenario 2 (reclaimed GSW modified) has similar time step with scenario 1 (daily average). Current circulation pattern in Jakarta Bay on scenario 2 is also dominated by zonal components, but the presence of reclamation (mainland) configuration stimulates current partitions into the channels. Channels are the slits where increasing flow occurs in narrow gap. figure 4 is presented to describe the pattern of currents in Jakarta Bay on scenario 2.

Simulation model of Jakarta Bay on scenario 3 (reclaimed GSW blueprint) shows the current circulation pattern in Jakarta Bay is complicated and isolated, thus a small amount of water flows and enters the bay and turned over the GSW land reclamation. The presence of GSW reclamation is presented at Tg. Priok harbor expansion which causes the current enters inside (August 16th and November 16th) and outside (February 16th and May 16th) the bay. Current intensification seems weaker than scenario 2, which is expected as the result of a little channel that parallel towards the coastline. Current zonal component also appears weaker compared to scenario 1 and 2.

Vertical structure of sea current in Jakarta Bay was made for describing current pattern based on its depth whether Jakarta Bay current has robust stratification or not. Typically, Jakarta Bay is dominated by zonal components in the surface, then the flow is shown by the components of the zonal flow (u) in transect A-B that extends vertically. In general, the flow at the surface has a powerful magnitude. This is shown in figure 6, 7 and 8 where the magnitude of the flow is way more than zero (positive or negative). Afterwards, the current pattern is getting into the bottom and facing an attenuation. Vertical structure of current on scenario 1 follows the surface current pattern on which the direction reversal occurred twice (in West and East Monsoon) based on Monsoonal pattern. Reversal of flow is indicated by a positive or negative value on the u component. Based on four pictures in different Monsoons (figure 6), sea current magnitude in the bottom has a value close to zero. However, in May 16, August 16, and November 16, the magnitude represents a strong current at the bottom. This is due to the vertical transect that intersects with islands.

Scenario 2 shows the difference of the vertical structure of current from scenario 1 where the intensification of current magnitude occurred. The Monsoonal pattern of vertical flow on scenario 2 also adheres to its surface. The intensification of currents in the bottom prevails at the channel of reclamation land. Vertical transect on zonal current component reveals a similar pattern on scenario 2. However, sea current near Jakarta coastline shows a negative value throughout the year. Blank area at figure 7 and 8 arises due to the vertical transect through reclamation land.

IOP Conf. Series: Earth and Environmental Science **176** (2018) 012008 doi:10.1088/1755-1315/176/1/012008

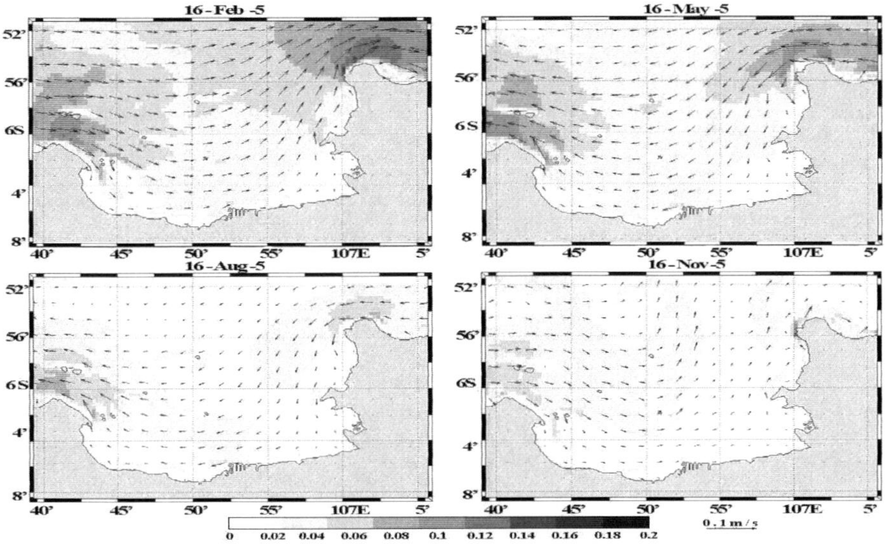

Figure 3. Current pattern in Jakarta Bay on scenario 1 at 10 m depth overlapped with flow magnitude (color) (ms^{-1}), presented in four representative Monsoons.

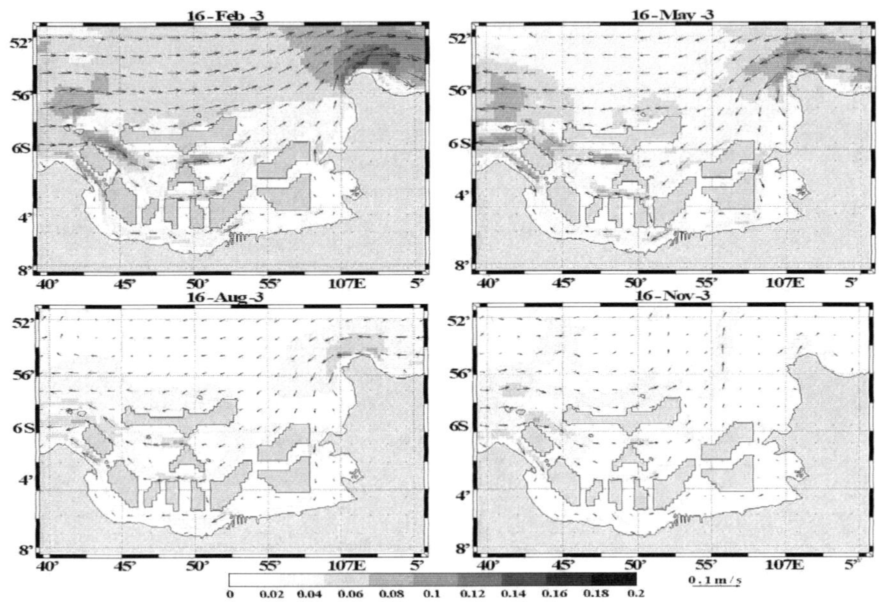

Figure 4. Current pattern in Jakarta Bay on scenario 2 at 10 m depth overlapped with flow magnitude (color) (ms^{-1}), presented in four representative Monsoons.

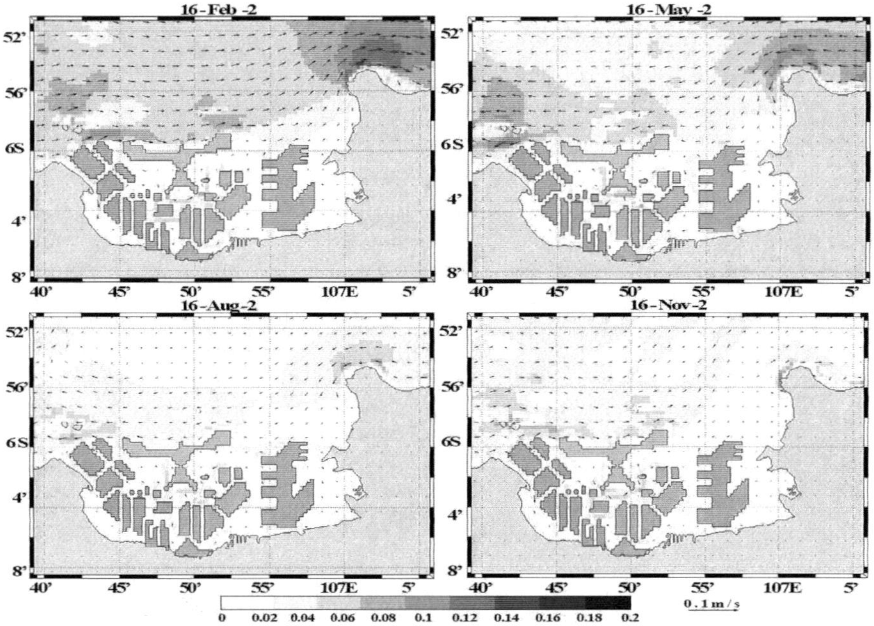

Figure 5. Current pattern in Jakarta Bay on scenario 3 at 10 m depth overlapped with flow magnitude (color) (ms^{-1}), presented in four representative Monsoons.

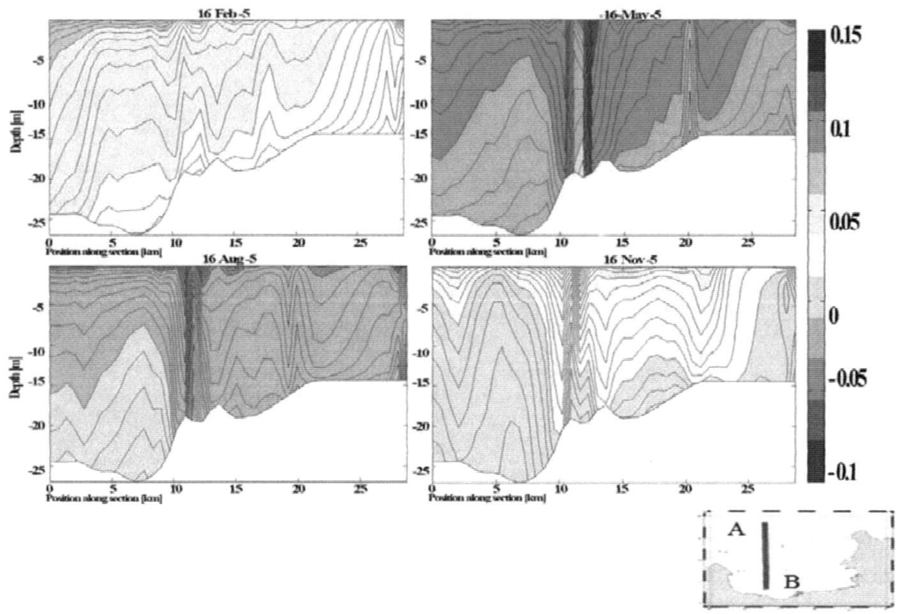

Figure 6. Structure and pattern of vertical current in Jakarta Bay on scenario 1, indicated by zonal component (u) (ms^{-1}) at transect A-B in four representative Monsoons.

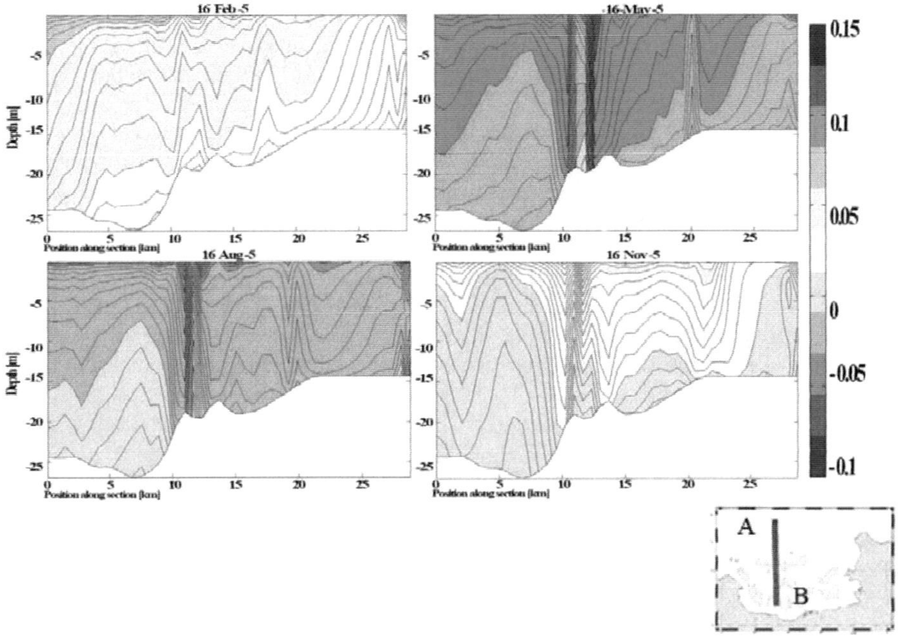

Figure 7. Structure and pattern of vertical current in Jakarta Bay on scenario 2, indicated by zonal component (u) (ms^{-1}) at transect A-B in four representative Monsoons.

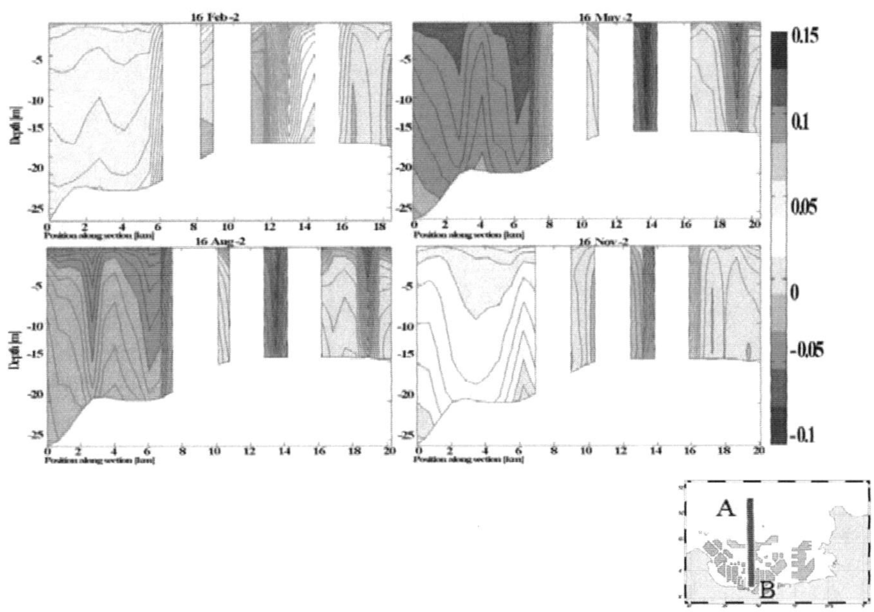

Figure 8. Structure and pattern of vertical current in Jakarta Bay on scenario 3, indicated by zonal component (u) (ms^{-1}) at transect A-B in four representative Monsoons.

3.3. Freshwater dispersal from Ciliwung River discharge

Components of the river were included in the model to see the distribution of water output from the river mouth into the bay. The river source was determined in the middle of Jakarta mainland configuration (the mouth of Ciliwung River) and river discharge value was made as constant along the year.

Scenario 1 shows that the waters movement of the river in West Monsoon head eastward along the flow. River discharge that comes out relatively high and dispersed uniformly along the coast. In Transitional I Monsoon, river discharge were carried away to the west side. However, the dispersion is higher and tends to gather in the coastal region (west of Jakarta). River discharge still appears smaller which is marked by thin blue color on the east coast of Jakarta. Then in East Monsoon, the river discharge output seems reduced to centralize the distribution closed to the river source in Transitional II Monsoon.

The process of river flow intrusion into the Jakarta Bay water system on scenario 2 has similar pattern with scenario 1. However, relatively low river discharge occurs and tends to disperse more widely than scenario 1. This happens due to the intensification of currents on the channels paralleled to the coastline so that the river discharge is carried out faster and causes an increase on the dispersion. The presence of GSW reclamation modification causes a quicker and wider movement of river input, which fills the channel of the reclamation land. Based on the model on figure 10, the movement of the input river occurs due to the intensification of current, while dispersion after the reclamation is caused by the formation of channels in the reclamation area.

Scenario 1 shows that the passive tracer from the river input in coastal area encounters a change of direction in Monsoonal period. The presence of modified GSW reclamation on scenario 2 (figure 10) causes a current discharge is carried away by the river input faster than scenario 1, hence it is assumed that the pollutants are disposed faster from Jakarta Bay. Model result of GSW reclamation on scenario 3 (figure 11) illustrates a high level of dispersion in the channel but isolated. This causes a closed current circulation so that the output streams are difficult to escape from the bay system. The distribution of output river discharge on scenario 3 has two river sources. The position of source point located on the west side (Cisadane River) and on the central (Ciliwung River) of Jakarta Bay. Transitional I and Transitional II Monsoons have a contribution to accumulate the river discharge in the western side of GSW reclamation Jakarta Bay.

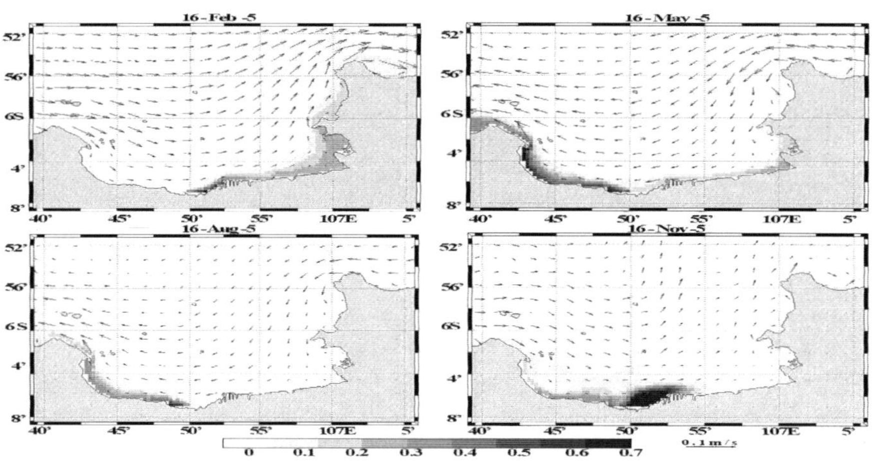

Figure 9. Passive tracer movement pattern of the river (the river source point is set on the mouth of Ciliwung River) assuming constant all the year with the river discharge in the Jakarta Bay on scenario 1 (normal condition) overlapped with the vector of current in four representative Monsoons.

IOP Conf. Series: Earth and Environmental Science **176** (2018) 012008 doi:10.1088/1755-1315/176/1/012008

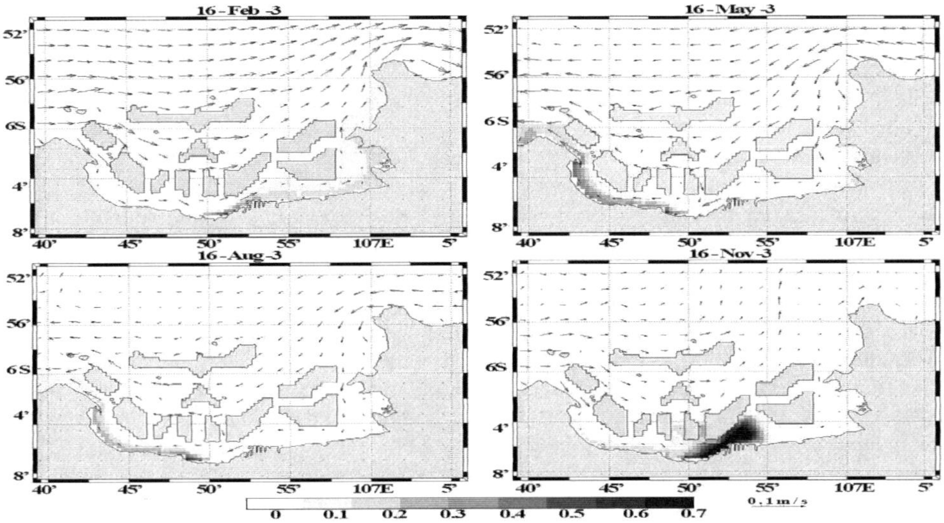

Figure 10. Passive tracer movement pattern of the river (the river source point is set on the mouth of Ciliwung River) assuming constant all the year with the river discharge in the Jakarta Bay on scenario 2 (modified reclamation condition) overlapped with the vector of current in four representative Monsoons.

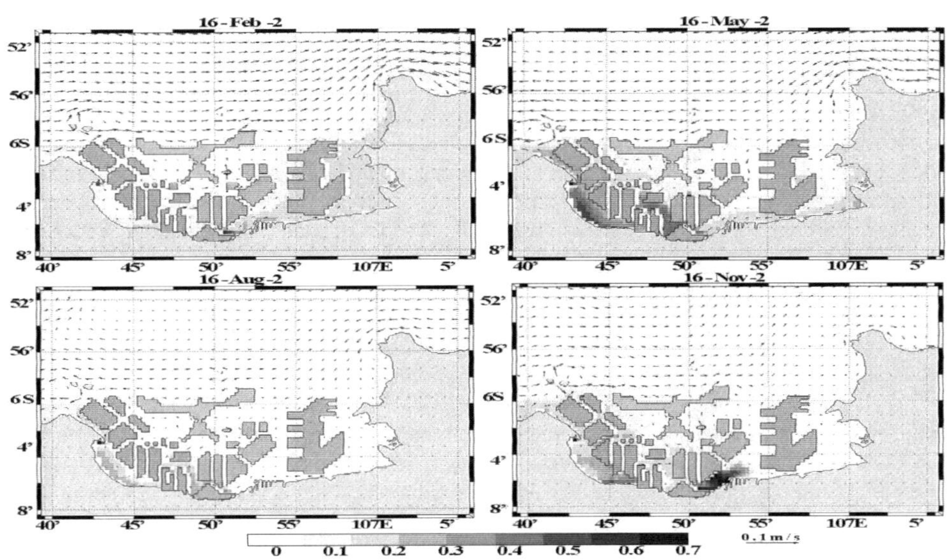

Figure 11. Passive tracer movement pattern of the river (the river source point is set on the mouth of Ciliwung River) assuming constant all the year with the river discharge in the Jakarta Bay on scenario 3 (reclamation with GSW blueprint) overlapped with the vector of current in four representative Monsoons.

4. Conclusion

Current circulation pattern in Jakarta Bay varies Monsoonally. Zonal component dominates the current movement of the scheme. Current circulation pattern simulation in scenario 1 (normal condition) shows that the water flows freely through the side of western or eastern of the cape. However, scenario 2 (GSW reclamation modified) shows that the water flow enters through the side of the cape and is divided and flowed through the channels. On each channel, there are slits where current intensification occurs and develops an eddy.

The highest magnitude of currents occurs in the West Monsoon where the water flows eastward for scenario 1 and scenario 2. The pattern of currents in scenario 3 (in accordance with GSW reclamation) appears complex, isolated, and weaker than scenario 1 and 2 (almost no intensification of currents).

The presence of the reclamation (scenario 2) also leads to the formation of eddy due to current intensification. Moreover, eddy formation is caused by the current intensification and reclamation configuration forms a shape of 'Garuda bird'. On the results of passive tracer in the river input, the discharge is influenced by current circulation pattern and in scenario 2, it moves relatively faster than scenario 1 and scenario 3.

Acknowledgments
The authors are grateful to the Bogor Agricultural University colleagues, Prof. Mulia Purba and Dr. I Wayan Nurjaya for the support, discussion, and comments on this work. We also wish to thank to the reviewers for their advice on how to improve this manuscript.

References

[1] Suriwati L M, Osawa T and Mahendra M S 2009 Study of total suspended matter transportation and circulation in Jakarta Bay using numerical simulation and satellite data *Ecotrophic* **5** 51-56
[2] Hadikusumah 2007 Variabilitas musiman arus di Teluk Jakarta *J. ITKT*. P 305-9
[3] Badan Pengelola Lingkungan Hidup Daerah DKI Jakarta 2011 *Laporan status lingkungan hidup daerah Provinsi Daerah Ibukota Jakarta* (Jakarta: Pemerintah Provinsi DKI Jakarta)
[4] Marta-Almeida M, Ruiz-Villarreal M, Otero P, Cobas M, Peliz A, Nolasco R, Cirano M and Pereira J 2010 OOFε: A Python engine for automating regional and coastal ocean forecasts *EMS*. **26** 680-2
[5] Prihatiningsih I 2014 *Simulasi Arus Musiman di Perairan Indonesia* (Bogor: Institut Pertanian Bogor)
[6] Atmadipoera AS and Widyastuti P 2015 A numerical modelling study on upwelling mechanism in Southern Makassar Strait *J. ITKT*. **6** 355-371
[7] Penven P Debreu L Marchesiello P McWilliams JC 2006 Evaluation and application of the ROMS 1-way embedding procedure to the Central California upwelling system *Oce. Mod.* **12** 157-158
[8] Pond S and Pickard G L 1983 *Introductory to Dynamical Oceanography* (New York: Pergamon Press)
[9] Ramming H G and Kowalik Z 1980 *Numerical Modelling of Marine Hydrodynamics: Application to Dynamic Physical Processes* (New York: Elsevier Scientific Publishing Company)
[10] Marchesiello P and Estrade P 2010 Upwelling limitation by geostrophic onshore flow *J. Mar. Res.* 68: 37-62
[11] Stewart R H 2008 *Introduction to Physical Oceanography* (Texas A & M University (USA): Department of Oceanography)
[12] Lathuiliére C, Echevin V, Lévy M and Madec G 2010 On the role of the mesoscale circulation on an idealized coastal upwelling ecosystem *J. Geophys. Res.* **115** C09018

IOP Conf. Series: Earth and Environmental Science **176** (2018) 012009 doi:10.1088/1755-1315/176/1/012009

Building up the database of the Level-2 Java Sea Ecoregion based on physical oceanographic parameters

J E Simanjorang[1*], W S Pranowo[2,3], L P Sari[1], N P Purba[4], M L Syamsuddin[1]

[1]Department of Marine Science, Padjadjaran University, Dekanat Building, FPIK-UNPAD Jatinangor Km. 21 UBR 40600, West Java, Indonesia
[2]Marine and Coastal Data Laboratory, Marine Research Center, Indonesian Ministry of Marine Affairs & Fisheries, Pasir Putih II, East Ancol, Jakarta UBR 14430, Indonesia
[3]Department Hydrography Engineering, Indonesian Naval Postgraduate School (STTAL), Komplek Bina Samudera Jl. Pantai Kuta V, Ancol Timur, Jakarta Utara 14430, Indonesia
[4]Marine Research Laboratory (MEAL), Dekanat Building, Padjadjaran University Jatinangor Km. 21 UBR 40600, West Java, Indonesia
*e-mail: noir.purba@unpad.ac.id

Abstract. The Java Sea (JS) is one of the sea with unique characteristic with especially in biodiversity and oceanographic condition. The purpose of this research is to identify the EL into level 2 in Java Sea based on oceanographic condition. The Java Sea, which covers an area of 467,000 Km^2 with an average depth of 50m, is located on the Southeastern part of the Sunda Shelf. This research uses the temperature and salinity data provided by the INDESO, the sea surface height (SSH) data downloaded from the HYCOM website, and the level 2 Java Sea ecoregion shape file data from the Indonesian seas delineation team. The physical oceanographic conditions of the level 2 Java Sea ecoregions are influenced by the moonsun and their respective geographical positions. The results showed that the distribution of temperature, sea current, salinity, and Sea Surface Height differed in each Java Sea Ecoregion level 2 because it is influenced by the geographical and astronomical location of each Java Sea Ecoregion level and influenced by the monsoon that exists over Indonesia. In the east of Java Sea influence from Makassar Strait and in the west affected by Karimata Strait. Within the region, EL 6.1. (Sunda Strait), the physical oceanography condition is slightly different from other Java EL because it is strongly influenced by the Indian Ocean.

1. Introduction

According to Geospatial Information Agency (BIG), the vast waters of Indonesia covers 6.376.744 km^2. With such vast waters, the diversity of flora, fauna, environmental conditions, and other parameters will surely be diverse. With such vast waters and the biological resources contained in it, a thorough and detailed inventory is needed. One of the database that exist now is the Indonesia Sea Ecoregion (EL) which is arranged by ministry of Environment (KLH) in 2013.

Sea ecoregion is a vast water territory that is filled with species, communities of nature, and environmental conditions that are really united in one geographic scope. Ecoregion is a geographic ecosystem, which means that the patterns of the various ecosystems and the processes between those ecosystems are bound in a geographical unit [16]. The delineation of ecoregion zones is a very important step in the process to build area of conservation [4], and Indonesia has applied this scheme.

At the present time, the Indonesian seas have been divided into 18 marine ecoregions and 11 fisheries management areas. One of the important EL Indonesia areas is the Java Sea located in the ecoregion zone 2. The Java Sea Ecoregion zone is characterized by shallow sea, whose oceanic conditions are heavily influenced by mixed-type tides predominantly diurnal, as well as Asian-

Australian Monsoon system. The Java Sea Ecoregion is also situated with relatively low bottom slope (gradient of 1-3°), although there are some areas that have a steep slope class (3-10°) [17]. The Java Sea is also located on continental exposure with a depth of 60 m, so that with the Java Sea conditions and its location within the territory of Indonesia, the Java Sea is relatively safe against earthquakes and tsunami [3]. The determination of the ecoregion zone is still at the stage of level 1 (global), which is still with the national map scale of 1: 1.000.000. On that scale, the information obtained is less detailed. Due to the marine environment, the zone classification on the global scale (level 1) is still within a limited spatial resolution, so that the accuracy of the desired information is still imperfect [12]. Zone downscalling is required to increase the EL level to level 2 with the scale of 1: 250,000 to obtain bathymetry information, current, sediment, temperature, freshwater affect, nutrient, isolation level, upwelling, and morphostructure in Java Sea accurately and specifically. The objective of this study was to identify the EL into level 2 in Java Sea based on oceanographic condition.

2. Methods

2.1. Time and area of study

This study was conducted from November 2016 to February 2017. The study area is in the EL-6 zone of Java Sea, covering an area of 437,978 km^2 (figure 1). Data processing and analysis will be performed at the Computer Center, Faculty of Fisheries and Marine Sciences, Padjadjaran University (UNPAD) and at the Sea and Coastal Data Laboratory, Marine Research Center, Human Resource Research Agency, Ministry of Marine Affairs and Fisheries, North Jakarta.

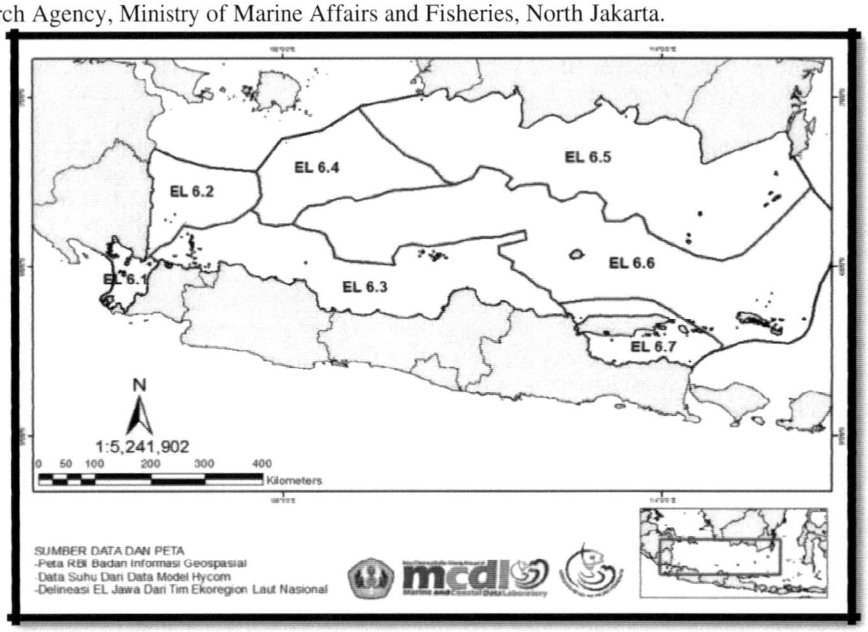

Figure 1. Indonesian sea ecoregion (EL) in Java Sea Level-2.

Figure 1 shows a zone of Sea Ecoregion 6 (EL-6) located within the Java Sea. Administratively, EL-6 is not only composed of the Java Sea, but also some parts that are directly intersection with Karimata Strait and Makassar Strait, which is the same characteristic as the Java Sea zone. It will be incorporated into the group of EL 6 zones because the creation of an Indonesian EL zonation is based on similarity between parameters that have been set up. Location around the waters of the Jakarta Bay is not detailed on EL Level 2 with a scale of 1: 250.000, because the provincial government of DKI Jakarta has set ecoregion zone at level 3 in the territorial waters of DKI Jakarta Province.

2.2. Data collection
The dataset used in this research are seawater temperature, salinity, current, sea surface height (SSH), and shape files from delineation of Java Sea Level Ecoregion 2. The data were obtained from the Infrastructure Development of Space Oceanography (INDESO) (http://www.indeso.web.id/) and HYCOM (https://hycom.org) sites, downloaded directly from the official website. The spatial resolution used for Monsoon and Arlindo data contained in INDESO is 1/12° or about 9.25 km. The shape data file of the 2nd level Java Sea Ecoregion was obtained from the Indonesian Geospatial Information Agency (BIG). The data were taken with an average of 0-50 m.

2.3. Data analysis
The method used in this study is a method of spatial analysis, temporal in a descriptive way. The analysis method is spatial and temporal, which is to process the secondary spatial data obtained from the official site of the institution in the field of oceanography, where the data come from the satellite measurement which are then processed using ArcGis software and the results are spatial maps and overlaid spatial maps between parameters. Spatial analysis method is used to process sea current, salinity, and seawater temperature data in the Java Sea. The MATLAB software is used to average the data that have been downloaded, and the ArcGis software is used to process the data that have been averaged. The results of the processed data are maps of complete distribution with color gradation and a map legend, which then be overlaid with the Java Sea Ecoregion level 2 delineation data shape file.

3. Results and discussion

3.1. Monthly mean of temperature distribution in Java Sea Ecoregion
The result of this study shows that the temperature in the areas of EL 6.1, EL 6.2, EL 6.3, and EL 6.7 located in the southern part of Java Sea is warmer than that the areas of EL 6.4, EL 6.5, and EL 6.6 regions in the north region (figure 2).

In the area of EL 6.1 the temperature range is 27-31.6 °C, the lowest temperature of 27 °C, occurs at all seasons in Indonesia, and the highest temperature of 31.6 °C occurs during the Monsoon Transition I (March-April-May) and East Monsoon (June-July-August). The dispersion of water temperature in EL 6.1 is not evenly distributed as shown in figure 2, as there seems to be 2 temperature groups. In areas closer to Java mainland, its temperature is warmer, which ranges between 30.5 - 31.6 °C, and in areas slightly farther from the Java mainland, the temperature is around 28.5-29 °C. This is because the areas closer to Java's mainland is still directly affected by Java's mainland activities, it is also affected by the warmer state of EL 6.2. While the position rather farther from the mainland is influenced by the Indian Ocean that has a cooler water temperature. This causes the water temperature dispersion in EL 6.1 to be uneven. The temperature distribution in EL 6.1 differs slightly from the other EL-6 sub, since EL 6.1 is mostly influenced by the Indian Ocean. The temperature distribution in EL 6.1 in November is different compared to other Java EL areas. There was a decrease in temperature when other regions experienced an increase in temperature in November. This is because of the positive IOD phenomenon that occurred in 2012 in the Indian Ocean that also affected the temperature along the Java coast; and the peak of this phenomenon is in November [10]. This phenomenon causes the pressure in Indian waters to be higher than the waters in Africa, so that the warm water mass in the mixed layer moves towards the waters in Africa [14]. The displacement of the mix-layer layer from the slough in the Indian Ocean resulted in a low temperature thermocline layer rising to fill the void, resulting that in the positive IOD, the temperature of waters in the Indian Ocean decreasing. This is what causes the temperature at EL 6.1 to decrease in November.

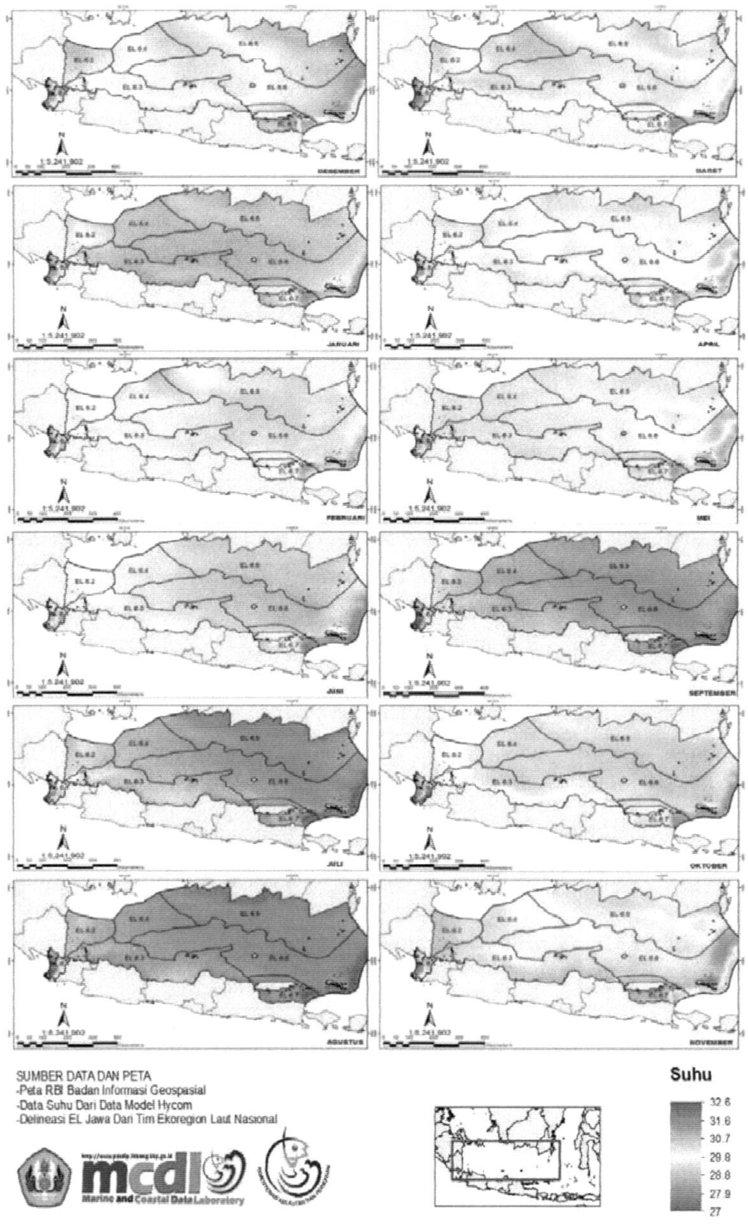

Figure 2. Temperature distribution in the Java Sea Ecoregion in 2012.

In the areas of EL 6.2, EL 6.3, EL 6.4, and EL 6.5, the temperature distribution is also uniformed, because the area is not directly in contact with the oceans. Area EL 6.2 has a temperature range of 28.5-30.5 °C, and EL 6.3 also has the same temperature distribution range with EL 6.2 with a value of 28.5-30.5 °C. This happens because EL 6.2 and EL 6.3 affect each other. In EL 6.6, the temperature spread has a value of 27-31.6 °C, but the temperature distribution in EL section 6.6 is uneven. The eastern part of EL 6.6 has a lower temperature spread than other EL 6.6 regions. It can be seen from the figure that during East Season, the eastern region of EL 6.6 has the lowest temperature. The time of the East Season showed the same thing, that is getting a cooler Water Mass from the East [5].

The lower temperature distribution conditions in the eastern part of EL 6.6 are also influenced by Arlindo [15]. At the peak of Arlindo (East Monsoon, May-October), the western part of the Java Sea Temperature ranges from 28-30.5 °C, while in the east it varies between 27-29 °C, especially in the eastern part of EL 6.6. At the time of Arlindo, Java Sea always gets influence from South China Sea Water Mass in West Monsoon and also get influence of Water Mass from East (in North of Makassar Strait).

3.2. Monthly mean of salinity distribution in Java Sea Ecoregion

The value of salinity during the Western Monsoon is lower than at the time of the East Monsoon, ranging from 29.5 to 33 psu. This is in agreement with [8], which stated that the Java Sea experiences the rainy season due to the water vapor that the wind takes as it passes the Pacific Ocean and South China Sea, resulting in a decrease in salinity (low salinity) due to the entering of fresh water and precipitation in large quantities. The distribution of salinity in the Java Sea Ecoregion at the time of West Monsoon and the Monsoon transition I is shown in figure 3.

According to [6], at the time of the West Monsoon, the water mass with low salinity of the Java Sea is pushed back to the southern Makassar Strait. This water mass exchange causes the pressure of water mass to go to the northern Makassar Strait. At the time of the East Monsoon, the winds on the Java Sea carry a surface water mass that is higher than the Banda Sea to the Makassar Strait. [17] argued that the Southern Equatorial Current (SEC) water supply comes from the southern Indian Ocean, an upwelling region between Indonesia and Australia, the Timor Sea and Indonesian Waters, and Equatorial Counter Current (ECC) that turn southwards in the loose waters of the western Sumatra coast. According to [6], at the time of the West Monsoon, the water mass with low salinity of the Java Sea was pushed to the South of the Makassar Strait. This water mass exchange causes the pressure of water mass to go to the northern Makassar Strait. At the time of the East Monsoon, the wind on the Java Sea carries the surface water mass that is higher than the Banda Sea to the Makassar Strait.

Region of EL 6.1 shows that its range of exchanges is almost constant in every turn of seasons in Indonesia, but the areas in EL 6.1 that is close to the mainland of Java and Sumatra have lower salinity. The salinity of the waters close to the mainland is about 30.4-31 psu. This happens because the waters close to the mainland have been mixed with fresh water coming from the river that empties into the shore. Influenced by several factors, which includes precipitation, evaporation, freshwater input (runoff), mixing process, and changes in the current due to the changing seasons. The areas of EL 6.1 which are directly linked to the Indian Ocean has high salinity, even occupying the highest range in each season, 34 psu. The salinity distribution conditions in EL 6.1 at the time of March are different from the conditions of distribution of salinity of other EL 6 areas. By March, the salinity distribution conditions in EL 6.1 area increased, while in other regions decreased. This is because the location of EL 6.1 located in the Sunda Strait is strongly influenced by the characteristics of the Indian Ocean. Research by [1] says that at the time of the West Monsoon, the Sunda Strait received mass water input from the Indian Ocean that has high temperatures and salinity. The condition is also influenced by the negative IOD phenomenon, which occurred in 2012 in the waters of the Indian Ocean. The negative IOD phenomenon and its peak occurred in March [10], which brings warmer water masses and higher salinity to the waters of the Indian Ocean. The South Java Current (SJC) entering EL 6.1 carrying a mass of water from the Indian Ocean affects the distribution of salinity in EL 6.1, so that the salinity in March at EL 6.1 increases.

IOP Conf. Series: Earth and Environmental Science **176** (2018) 012009 doi:10.1088/1755-1315/176/1/012009

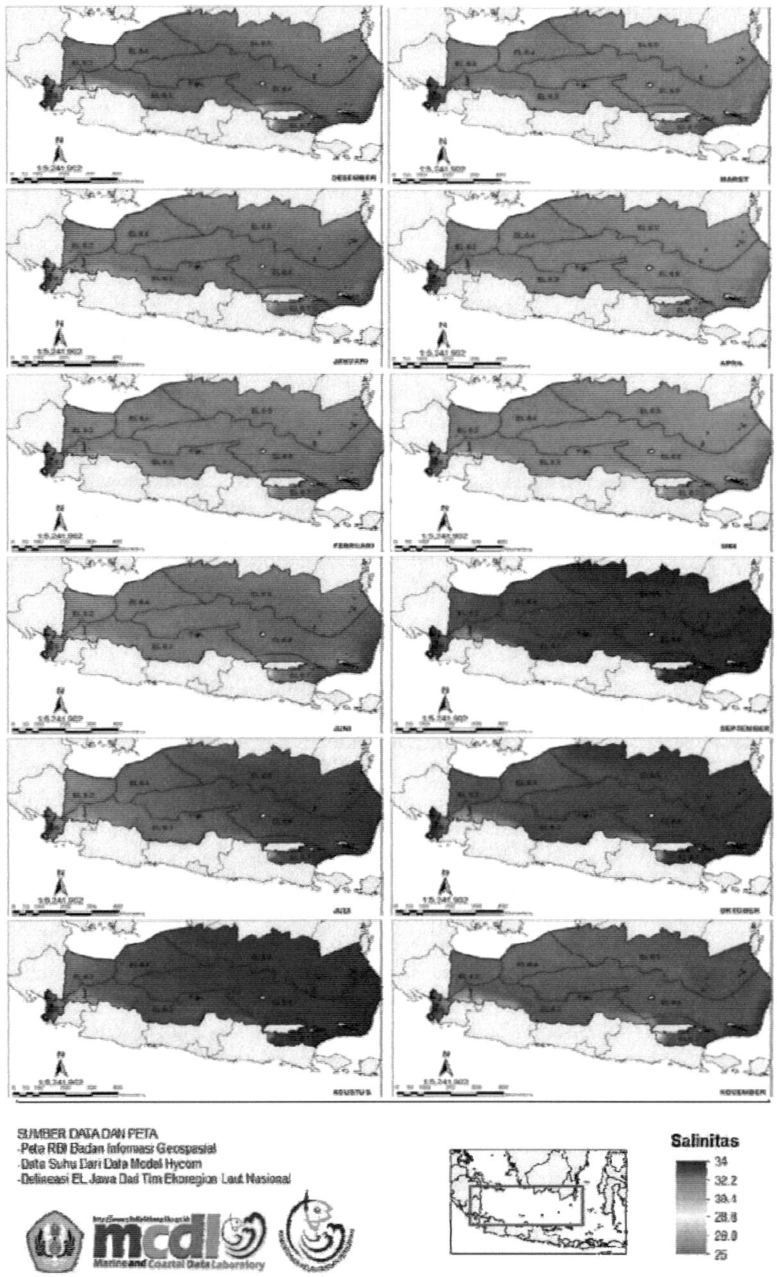

Figure 3. Distribution of Salinity in the Java Sea Ecoregion in 2012.

In the area of EL 6.2, the distribution of salinity is more evenly distributed, as is in the EL region 6.4. In area EL 6.2 has the lowest salinity with 31 psu, and in area EL 6.4, the lowest salinity level is 32 psu. The situation occurred during the West Monsoon and the Transitional Monsoon. At the time of the East Monsoon, both areas experienced the highest salinity level, which is 34 psu. This result is consistent with the research by [15] which found that the lowest salinity was during the West Monsoon, and the highest salinity was during the East Monsoon.

The areas of EL 6.3, EL 6.5, and EL 6.7 have unequal distribution of salinity distribution. The area around the mainland has a lower salinity than the areas further away from the mainland. The lowest value of 28.5 psu is during the West Monsoon. This is especially true in the West Monsoon, in which the mainland of Borneo and Java has its rainy season, so that rivers on the land carry large amounts of freshwater, as freshwater and seawater are mixed in the coastal areas, resulting in salinity reduction. In the East Monsoon and the Monsoon Transition II, the distribution of salinity was almost evenly distributed, either at a point close to the land, or further away from the land. Its salinity distribution is 33.5-34 psu. This occurs because during the East Monsoon, the dry season occurs in the mainland, so that the mass of water carried by rivers in the land is a few, so it does not affect the distribution of salinity in the area.

The area of EL 6.6 also has a uniform distribution of salinity in each season. The lowest salinity distribution in EL 6.6 occurred during Transition Season I on a scale of 31.5-33 psu. The highest salinity distribution occurred during East Monsoon and Monsoon Transition II on a scale of 32.2-34 psu. The eastern conditions of the EL region 6.6 shown in figures 3 have the highest salinity values for each season. The salinity of the area is 34 psu. This is due to the current flowing through Indonesia passing through the Lombok Strait which brings the masses of water with high salinity. The same can be seen in the easternmost areas of EL 6.7. The currents passing through the areas of EL 6.6 and EL 6.7 carry the mass of water from the South China Sea, so the distribution of salinity is influenced by the salinity conditions of the waters of the South China Sea. During the West Monsoon, currents move from west to east, because in the West Monsoon, the wind blows from west to east. The depth of the Java Sea is not deep, so that the current that moves in the Java Sea is, on average, affected by the Monsoon wind. The water mass of the West Monsoon is a mass of water from the South China Sea coming from the Pacific Ocean through the Luzon Strait [7]. Then it goes through the Karimata Strait, whose depth is below 45 meters, so that West Monsoon greatly affect the flow from the Karimata Strait to the Java Sea. The water vapor, carried by the Monsoon wind as it passes through the South China Sea, causes a decrease in the value of Salinity (low Salinity) due to large quantities of freshwater input and precipitation [8]. The current moves from the Java Sea to the east in the Western Monsoon, because at the time of the West Monsoon, the Ekman current suppresses the surface current with a low Salinity from the Java Sea to the south of the Makassar Strait [6].

The Java Sea has always been influenced by the mass of South China Sea waters in the West Monsoon and is influenced by the Water Mass from the East (South of Makassar Strait) in the East Monsoon. As seen in figure 3, at the time of West Monsoon, EL 6.1, 6.2, 6.3, and 6.4, which are located on the western part of EL Java, were affected by lower Water Mass. This happens because the wind blows from the West to the East. The time of the East Monsoon shows the same thing, that is getting the influence of water masses from the East [5].

3.3. Monthly mean of ocean current distribution in Java Sea Ecoregion

The mass movement of the seawater arises from the effect of the resultant forces acting and the factors that influence it [2]. The movement of the current direction and its velocity on the Java Sea Ecoregion at the time of West Monsoon and Transition I Monsoon is shown in figure 4.

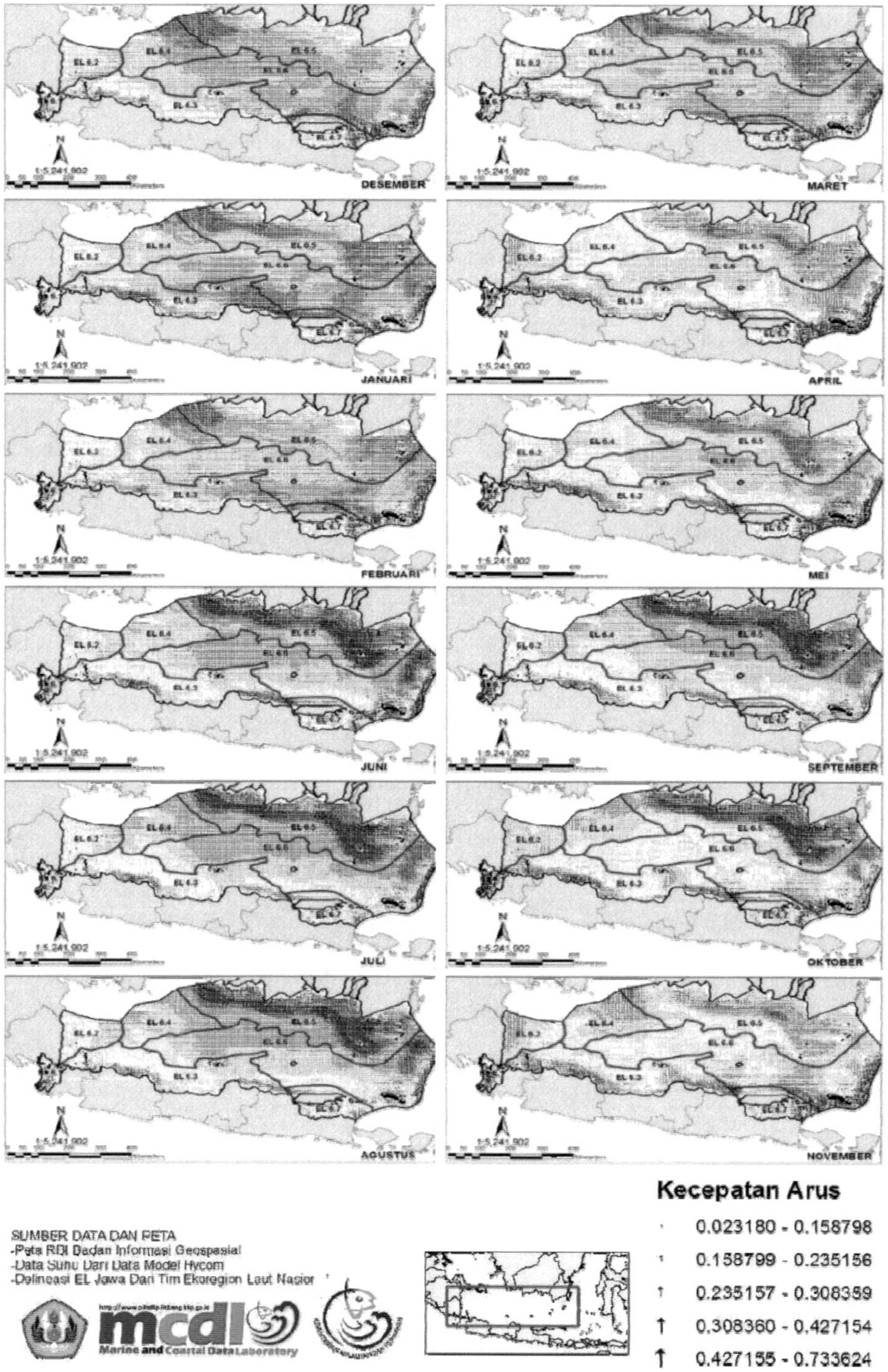

Figure 4. Distribution of Salinity in the Java Sea Ecoregion in 2012.

EL 6.1 is an area directly linked to the Indian Ocean. The direction of current movement and velocity in EL 6.1 is influenced by the movement of currents from the Indian Ocean, as well as the Monsoon. At the time of the West Monsoon, the current velocity in EL 6.1 was 0.158-0.235 m/s, and at the time of East Monsoon, 0.158-0.308 m/s. The pattern of current movement in EL 6.1 is irregular. This occurs because the current in EL 6.1 is affected by the topographic condition of EL 6.1 topographic shape of the ocean floor and the surrounding islands. Some of the major ocean systems in the world are limited by land masses from three sides and islands by equatorial current counters on all four sides. These limits produce a nearly closed system of flow and tends to make the flow lead into a circular shape or makes the direction of the current to become irregular [2].

The current velocity in EL 6.3 is 0.023-0.15 m/s. This is in accordance with the opinion of which states that the area around the north coast of Java is not directly influenced by Monsoon. In coastal areas, currents move slightly faster. The further away toward the sea, the flow rate will be smaller. The Monsoon affects only the direction and speed of the wind on the Javanese soil, so that local wind affects the current velocity in EL 6.3. The direction of the land wind, influenced by Monsoon, also influences the direction of the current movement in EL 6.3. It is seen in the picture that during the West Season, the direction of the current is headed south, while at the time of East Monsoon, the direction of the current movement is to the north.

EL 6.6 is of part of Java EL area that is in direct contact to EL 6.3 on the north. In EL 6.6. the motion pattern and current velocity are not significantly different from those in EL 6.3. The current velocity in EL 6.6 is 0.158-0.308 m/s, and it is the same in each monsoon. This happens because in this area, the wind season that blows has many buffer factors. The Javanese mainland will be a buffer to the wind that blows on East Monsoon, so it cannot have direct wind influence on East Monsoon, while the mainland of West Borneo and Bangka Belitung Island become buffers in the wind that blows in the West Monsoon. Current velocity in EL 6.6 is influenced by the local wind. The current and current velocity patterns in EL 6.4 are similar to those in EL 6.6.

In EL 6.5, the current velocity patterns in vary each season. During the West Monsoon, the current velocity in EL 6.5 is lower than during East Monsoon. At the time of the West Monsoon the current velocity in EL 6.5 is 0.158-0.235 m/s. This happens because of the inhibition of the wind that blows on the sea surface. But it is different with the western part of EL 6.5 in the same season, in which the current velocity in the area is higher, which is about 0.235-0.308 m/s. This is because the situation in that area is directly influenced by the moving currents of the South China Sea, and the area also lacks buffers, so the wind that blows as a result of the Asian Monsoons directly affects the velocity of the ocean currents. During the East Monsoon, the current velocity at EL 6.5 is 0.308-0.733 m/s. This speed is the highest current velocity found in EL areas in Java. This occurs because the wind that blows during East Season directly affects the current surface conditions in EL 6.5. The wind gives rise to viscosity at sea level. The viscosity force at sea level is caused by the movement of the wind on the surface of the sea, causing the exchange of adjacent water mass periodically, this is due to the pressure difference in the fluid [15].

3.4. Spatial and Temporal Analysis of Sea Surface Height (SSH) in the Java Sea Ecoregion

Sea level can change on variety of timescales due to number of factors. Usually, the seasonal changes are influenced by tides and current movements. Monthly distribution of sea surface height (SSH) in Java EL in 2012 is shown in figure 5.

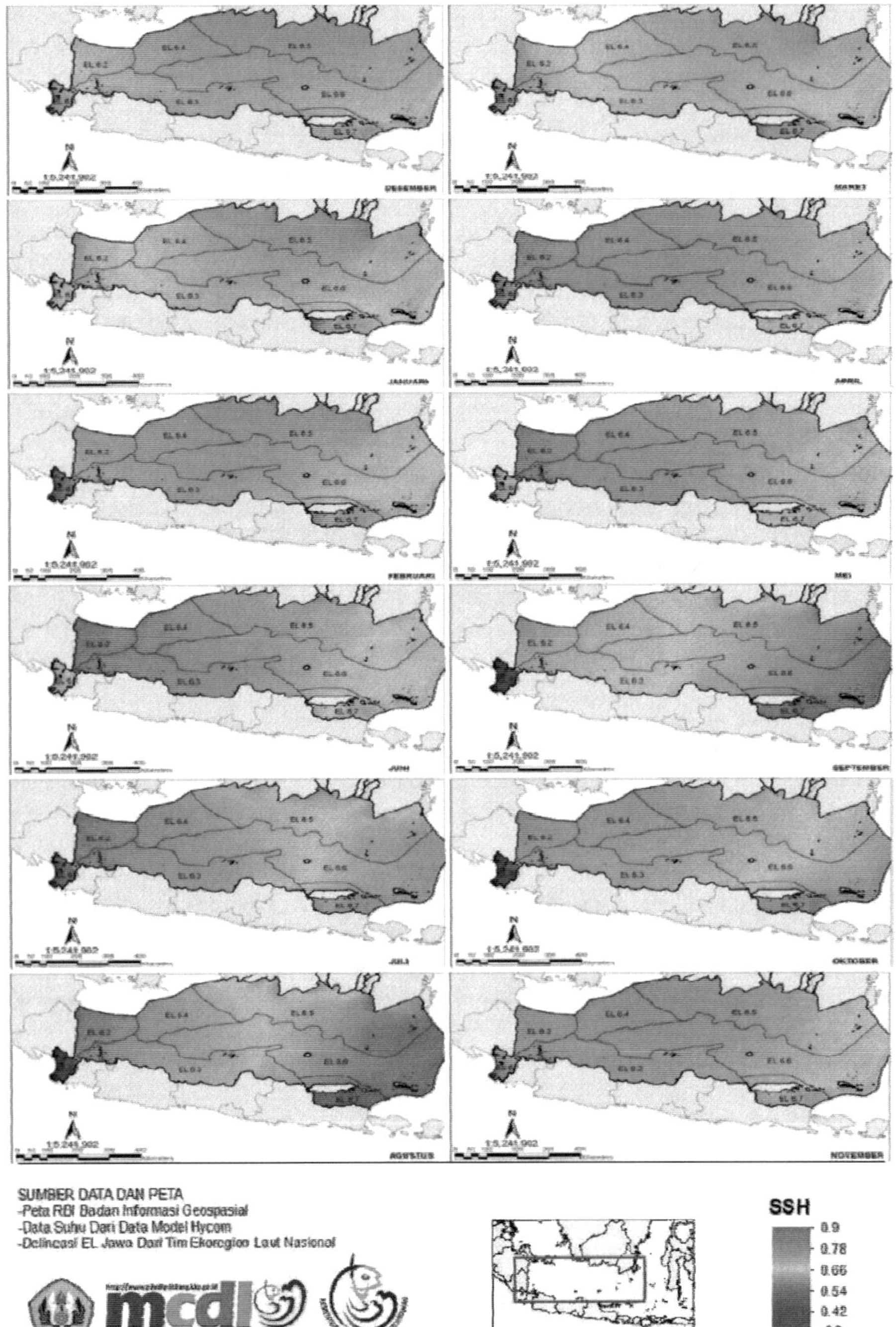

Figure 5. The distribution of Sea Surface Height in the Java Sea Ecoregion in 2012

Sea Surface Height is affected by surface wind and movement of water masses [11]. In Java EL, the movement of water mass is influenced by temperature, pressure, and current. And it all directly related to the four seasons contained in Indonesia. The Water Mass Exchange in Java Sea can be caused by

Monsoon periodicity. The Water Mass Exchange is characterized by differences in Temperature and Salinity values based on Monsoon periodicity. The difference in sea level that is affected by the position of the sun also affects the height of sea level during East Monsoon and Monsoon Transition II to be different from during West Monsoon and Monsoon Transition I.

EL 6.1 is part of the Java EL which is located geographically flanked by the island of Java and the island of Sumatra. EL 6.1, as shown in the image, has the lowest average SSH each season compared to other parts of EL Java. At the time of the West Season, the SSH of EL 6.1 is 0.42-0.7 m, at the time of the Monsoon Transition I it is 0.54-0.66 m, during East Monsoon, 0.3-0.6 m, and during the Monsoon Transition II, 0.3-0.54 m. From these results, it can be seen that the highest SSH values in the EL 6.1 area are during the West Monsoon, and the lowest SSH values occurred during East Monsoon and Monsoon Transition II. This is especially true during the West Monsoon where a high air pressure center is developed over the Asian Continent and the center of low air pressure occurred over the Australian Continent so that the wind blows from the northwest to the southeast in the Java Sea in the West Monsoon [18]. In the West Monsoon, most of the Sea Water Mass in the Java EL gets the Water Mass from the South China Sea and the Karimata Strait. Because the location of EL 6.1 is flanked by the Java Island and the Sumatra Island, the water mass carried from the South China Sea and the Karimata Strait did not exert a great impact on SSH in EL 6.1. Figure 5 also show that the SSH conditions in the Northern part of EL 6.1 are higher than that of the Southwest EL 6.1. This happens because of the topographical differences in the area. Amri and friends says that the depth of the Sunda Strait in the North is about 40 m, then towards the southwest of the strait, the depth grows to about 75-100 m; at the time of his research, Amri and friends also found the upwelling phenomenon on the month of August and September [1]. Upwelling phenomenon occurs because of the mass water void in the water column, so the water level becomes lower than usually.

In EL 6.2, the highest SSH average occurred during the East Monsoon. Figure 5 shows that at the time of West Monsoon, it has a value of SSH 0.6-0.7 m. During Monsoon Transition II, the SSH values were 0.6-0.78 m, at the East Monsoon, 0.78-0.8 m, and during the Monsoon Transition II, the SSH value was 0.7-0.78 m. The highest SSH value in EL 6.2 area occurred during the East Monsoon, that is 0.8 m, and the lowest SSH value occurred at West Monsoon and Monsoon Transition I with 0.6 m. This is because the wind that blows during the East Monsoon has a much stronger effect than the wind that blows during the West Monsoon. During the West Monsoon, the mass of water carried by the current flowing from the Karimata Strait and the South China Sea is blocked by the islands of Bangka and Belitung and the island of Sumatra. While during East Season, moving water masses of EL 6.3 and 6.6 directly affect SSH in EL 6.2.

In figure 5, there is a difference in sea level (SSH) in EL 6.3. During the West Monsoon, SSH in the western part of EL 6.3 is lower than the eastern part. While in the East Monsoon, the East part is lower than the West in EL 6.3. In the West Monsoon, the SSH value in EL 6.3 in the West part is 0.6-0.7 m, while in the East of EL 6.3 is 0.7-0.8 m. At the time of the Transition Monsoon I, SSH value in the West is 0.6-0.8 m and the eastern part 0.7-0.8 m. In the Eastern Monsoon, the western part of EL 6.3 has a SSH value of 0.6-0.85 m and in the east 0.66-0.8 m. During the Monsoon Transition, the SSH value in the western part of EL 6.3 is 0.66-0.78 m, whereas the eastern part of EL 6.3 has a SSH value of 0.54-0.75 m. At the time of the West Monsoon, the sun on the Southern earth causes the wind to move southward. The wind that moves towards the South affect the movement of seawater, so that the sea water also moves to the south [7].

In EL 6.4, its SSH values also vary. At the time of West Monsoon, the SSH value is 0.66-0.8 m. At the time of Convertible Monsoon Transition I, 0.66-0.78 m. At the East Monsoon, 0.66-0.85 m. And at the time of the transition Monsoon II, 0.66-0.78 m. From that result, it is seen that the highest SSH value occurs at the time of the East Monsoon with the SSH value that reach 0.85 m, and the lowest SSH value occurs in all four seasons with the value of 0.66 m. At the time of the West Monsoon, the current moving from the South China Sea led the mass of water into the Java EL passing through EL 6.3 and 6.4, so that at the time of West Monsoon, the SSH of EL 6.4 region, which is directly tied to the Karimata Strait, has a higher SSH value compared to the areas covered by Bangka and Belitung Islands. In contrast, at the time of the East Monsoon, the highest SSH value reached is found in the areas covered by the island of Bangka and Belitung.

Figure 5 show that in area 6.5, the highest SSH value in each season occurs in the waters closer to the land. This is consistent with those of Yang and Liu who said that in closed waters, the farther from the coast, the lower the water level [11]. At the time of the West Monsoon, the value of SSH reached 0.88 m, while the value in areas far from the mainland of Borneo is 0.66 m. At the time of the West Monsoon, the current that moved in the area of EL 6.5 is not as strong as during the East Monsoon. This resulted in the movement of water masses to be not too fast in EL 6.5 at the time of the West Monsoon. While at the East Monsoon, the SSH value is 0.5-0.7 m. The lowest SSH value occurs in the areas far from land. At the time of the East Monsoon, the current that moved faster than at the time of West Monsoon, resulted in the mass of water that also moves faster, so that the value of SSH in EL 6.5 at the time of the East Monsoon is lower than the SSH at the time of West Monsoon. The highest average SSH value occurred at the time of the Monsoon Transition I. The SSH value at the time of Transition Monsoon I was 0.85 m.

In the EL region of 6.7, during the West Monsoon, the SSH value in EL 6.7 is 0.6-0.66 cm, and at the time of the Monsoon Transition I is 0.6-0.7 m. While at the East Monsoon is 0.4-0.6 m, and during the Transitional Monsoon is 54-0.6 m. In the area of EL 6.7, the average SSH values were lower than in the other EL 6 regions. This occurs because the location of EL 6.7 is geographically located in the bay, so that the mass of water that moves due to the seasonal wind blow does not directly affect it, whereas [13] says that the Monsoon affects the circulation of seawater movement. Because the location of EL 6.7 which is embraced by the current movement that occurs in EL Java is not too affected. figure 5 also shows that the SSH in the area is low, and because of the conditions that are embraced, the current is obstructed by the land to enter into the area so that the water mass exchange is also not very significant and the SSH value in EL 6.7 is relatively low.

4. Conclusion

Based on the result of the research, it can be concluded that the distribution of temperature, sea current, salinity, and Sea Surface Height differed in each Java Sea Ecoregion level 2 because it is influenced by the geographical and astronomical location of each Java Sea Ecoregion level and influenced by the monsoon that exists over Indonesia. Because the position of EL 6.1 is in the Sunda Strait, the physical oceanography condition of EL 6.1 is slightly different from other Java EL because it is strongly influenced by the Indian Ocean.

References

[1] Amri K, Priatna A and Suprapto 2014 Oceanographical characteristics and phytoplankton abundance in Sunda Strait Waters in East Monsoon *Bawal* 11-20

[2] Baumgart, Jennerjahn and Pranowo 2006 Nutrient distribution in the Indian Ocean off Java and Sumatra, Indonesia, related to coastal upwelling *Spice/Loicz/Atsef/Seacorm Southeast Asia Coastal Governance and Management Forum Bali*

[3] Budiono, Suprapto, Kristanto, Raharjo and Noviandi 2003 *Peta Wilayah Rawan Bencana Tsunami Indonesia* (Bandung: Marine Geological Institute)

[4] Forbs E 1856 Map of the distribution of marine life *The Physical Atlas of Natural Phenomena* 99-102

[5] Gaol J and Sadhotomo B 2007 Karakteristik dan variabilitas parameter-parameter oseanografi Laut Jawa hubungannya distribusi hasil tangkapan ikan *J. of Indonesian Fisheries Research* 201-211

[6] Gordon A 2005 Oceanography Indonesian Seas and their throughflow *Oceanography* **4** 15-18

[7] Gordon A, Huber B, Metzger E, Susanto R and Hulburt A 2012 South China Sea throughflow impact on the Indonesian throughflow *Geophysical Research Letters* **39**

[8] Ilahude A G and Nontji A 1999 *Oseanografi dan Perubahan Iklim Global (El Nino dan La Nina)*

[9] Ilahude A G 1996 Thermocline stratification within the Indonesian Seas *Journal of Geophysical Research* **101**(C5) 12401-12409

[10] Lan M 2012 *Analysis of 2012 Indian Ocean Dipole Behavior* (Tokyo: University of Tokyo)

[11] Liu Z and Yang H 2000 Regional dynamics of seasonal variability in the South China Sea *Journal of Physical Oseanography* **31** 273-23
[12] Mark, Spalding, Davidson N, Finlayson M, Robertson J 2007 A bioregionalization of coastal and shelf areas *Marine Ecoregions of the World* 573-582
[13] Nontji A 1993 *Laut Nusantara* (Jakarta: Grafindo)
[14] Rao A S, Behera, Masumoto Y and Yamagata 2002 Interannual variability in the subsurface Tropical Indian Ocean *Deep Sea Res. I.* **49** 1549-1572
[15] Siregar S 2016 The water mass exchange in Java Sea due to periodicity of monsoon and ITF 2015 *Depik* 1-16
[16] Spalding, Fox, Allen, Davidson, Ferdana, Finlayson and Robertson 2007 Marine ecoregions of the world: a bioregionalization of coastal and selfareas *Bioscience* **57**(7) 573-583
[17] Sulistiyo B and Triyono 2009 *Atlas Kelautan dan Atmosfer. Balai Riset Kelautan dan Perikanan* (Jakarta: Kementerian Kelautan dan Perikanan)
[18] Wyrtki K 1961 Physical oceanography of the Southeast Asian Waters (California: Naga Report, California Digital Library) pp 195

IOP Conf. Series: Earth and Environmental Science **176** (2018) 012010 doi:10.1088/1755-1315/176/1/012010

Seasonally physical characteristics of ROFI zone (Region of Freshwater Influence) in Pelabuhan Ratu Bay

Z Zibar[1*], I W Nurjaya[2], N M N Natih[2]

[1] Department of Marine Science, Graduate School of Bogor Agricultural University, Agatis Road, Bogor 16680, Indonesia
[2] Departemen of Marine Science and Technology, Bogor Agricultural University, Agatis Road, Indonesia
*e-mail: zanzibar301@gmail.com

Abstract. Coastal waters have many oceanographic phenomena are still being studied. One such phenomenon is how the influence of freshwater to the water mass in coastal waters. This study aims to calculate freshwater fraction, volume transport of freshwater and water mass stability. Physical water characteristics were measured by CTD (Conductivity, Temperature and Depth) deployed in 15 locations for temperature, salinity and density. The results of freshwater fraction analysis show that the west season had a greater on water fraction than the east season with value ranging from 0.40 % - 0.70 % with the amount 18.07 %, while the east season had a value of 0.02 % - 0.28 % with the amount is 7.30 %. The transport of freshwater volume in west season is 1.429.244 m^3s^{-1} while in the east season it is 1.066 m^3s^{-1}. Water mass stability of water column in the research location has static stability with its average N^2 in west and east monsoon are 0.008 cycl/s and 0.7 x 10^3 cycl/s, respectively. Those analyses above infer that freshwater input hugely influences physical water mass of ROFI zone in research location.

1. Introduction

Pelabuhan Pelabuhan Ratu Bay is located geographycally at 6°-57'-7°07'S and 106°22'-106°-23'E and its coastal line is about 105 km length. It is also being the mouth for some big rivers such as Cimandiri and Cimaja [1]. The freshwater inputs influences on coastal water trough estuaries are verily important information to coastal areas. Predicted transport, distribution, total pollutants, suspended materials are very important matters for various coastal information because the dynamic mixing as a result of some physical process [2,3]. The importance of physical processes in coastal areas, caused by freshwater inputs from rivers that have been studied by several researchers, Some research such as the alongshore transport of freshwater in a surface-trapped river plume [4], A simple theory for the fate of buoyant coastal discharges [5], Relating river plume structure to vertical mixing [6], SST observations of upwelling induced by tidal straining in the Rhine ROFI [7], observation of coastal front and circulation in the northern Java, Indonesia [8].

Region of freshwater influence (ROFI) is a zone between sea exposure and estuary where as local freshwater bouyancy from coastal resource is equal, or exceeds seasonally bouyancy inputs as heat happening in all sea exposures [9]. The effect of overflowing suspended material moving in the water column (plume) in coastal waters is a way to understanding the behavior of physical (plume) and front because plume and front are strongly influenced by river flow, tidal and meteorological conditions [10]. The main characteristic of the ROFI system is the significant buoyancy of the freshwater source, which has important implications for the structure and dynamics of the water column. If spatial scales are constrained by coastal features and weak stirring, freshwater inputs tend to encourage parallel beach densities where the effects of the Earth's rotation (Coriolis force) act to limit spread and form a baroclinic (density driven) that can affect the hydrodynamics of hundreds kilometres of coastal waters [9].

Other factors effect ROFI zone are convergence current, bathymetry, solar heating, and other oceanographic vactors [11]. The happening system in ROFI zone is very important and it is needed to be accounted practically. Freshwater input from estuaries that are mixed with suspended materials will give any impact to sea exposure environments [12]. Run-off input input maintain highly nutritive concentration in ROFI zones which induces blooming phytoplankton and in some cases it can endanger or faze [13]. Combination between tides and river buoyancy produces sea organism groups, and higly dispersive water current affects to sea creatures dynamic [14,15]. [16] reveals that biota live coastal areas are higly affected by chemic-physical condition. This is figured by coastal community activity of Pelabuhan Ratu Bay in each month collect *impun*, or called by local as "*ngala impun*". *Impun* are small fish/juveniles that are captured in some river estuaries such Citepus, Sukawayana, Cibuntu and Cimandiri. The caught impun by local abandontly happens on July, and their abudance found in two estuaries, Cimandiri and Citepus, with 12.6 kg/people/day and 4.2 kg/people/day, respectively.

There are many researches conducted in Cimandiri's estuary, such as [17] studied up-welling in Southern Java shows that sea water temperature of Palabuhan Ratu bay on September to October (end east monsoon) and November to Desember (early west monsoon) are 26 °C are 28 °C, respectively. Salinity the bay was affected by seasonality which the main factor is freshwater input from rivers. Freshwater input from rivers especially in west monsoon result lowering the water salinity of the bay, however the central bay have relative small different surface salinity between two season, west and east monsoon. Salinity measurement conducted on two different season, August to October period and May to July are 32.96 ‰ and 32.33 ‰, respectively. Other researchers such [18] chemical characteristics and water fertility of Pelabuhan Ratu Bay on west and East monsoon, [1] two-dimensional simulation water current pattern of the bay, [19] oceanographic condition as the basis for fish resource preservation in the bay.

So far, research about seasonally physical characteristics of ROFI zone in Cimandiri estuary Pelabuhan Ratu Bay is still not conducted yet. This research were aimed to calculate salinity anomaly, freshwater fraction, volume transport of freshwater and water mass stability in two different season. This research was expected resulting beneficial information about physical process of the water in the bay which can be used for further management, both government bodies and research.

2. Methods

This research was conducted in Cimandiri estuary of Pelabuhan Ratu Bay at two different field observational surveys namely January and August, 2016. Observation locations consisted of three section which are directed divergently or a whole station directed away against to the estuary mouth. First section consisted of five stations (st. 1 to st. 5) which is spread diagonally between the estuary mouth and the second section. Second section was consisted of 5 stations (st. 6 to st. 10) that was perpendicular to the estuary mouth. The last section was stretches diagonally between second section and the estuary mouth. All data sampling sites were assumed to be representatives for research locations at all. Research map is displayed in figure 1.

Equipment were used in the data sampling consisted of GPS MAP for determining observation location, boat track directions and other supported information, CTD (Conductivity, Temperature and Depth) has a role for collecting salinity, temperature and density data, Surface current data in January and August 2016 is derived from the measurement data directly in the field as well as the observation model data derived from INDESO (Infrastructure Development for Space Oceanography) and survey boat from local fisherman was operated to carry out research phases.

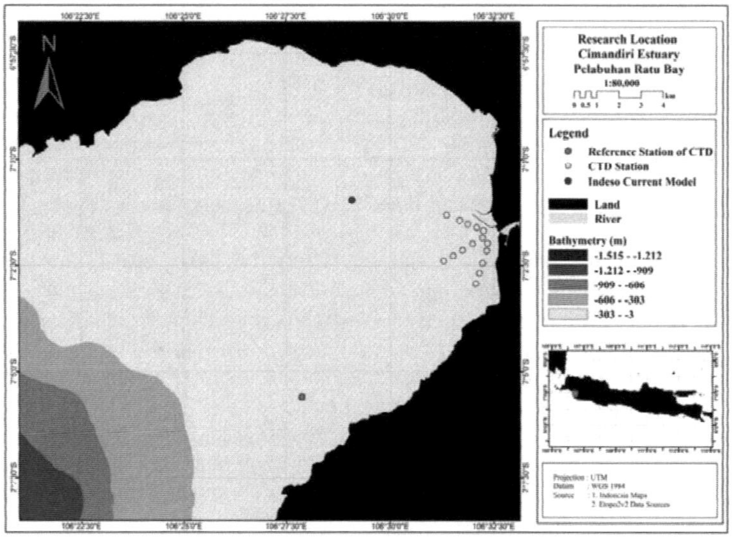

Figure 1. Research location.

Calculating anomaly salinity is based [20,21] as the following formula:

$$S'(x,z) = S(x,z) - S_{ref}(z) \quad (1)$$

Where S'(x,z) is salinity anomaly value (PSU), S (x,z) is salinity value that is measured at research location and $S_{ref}(z)$ is reference salinity is averaged at depth z (PSU).

Freshwater fraction is in volume unit which is difined according to [20,21] formula as follow.

$$F(x,z) = \frac{S_{ref}(z)-S(x,z)}{S_{ref}(z)} \quad (2)$$

where F(x,z) is freshwater fraction value, $S_{ref}(z)$ is the salinity reference at depth z (PSU) and $S(x,z)$ is the salinity value measured at each station x and the depth z (PSU). The reference salinity used in this study is 33.76 PSU which is the salinity value of a research station located far from the mouth of the river mouth, in other words have not gained influence from the fresh water supply flowing from the mouth of the Cimandiri River. The average value of freshwater fractions along the section can be defined:

$$Fr = \frac{1}{h}\int_{h}^{o} \frac{S_{ref} - S}{S_{ref}}\, dz \quad (3)$$

Fr is the average value of freshwater fractions along the transect, h is the depth of the waters of each depth of the station (m).

The volume transport of freshwater is estimated [20,21].

$$V_{fw} = \int_A F_r u dA \quad (4)$$

Where V_{fw} is volume transport of freshwater (m^3s^{-1}), F_r is the average of freshwater fraction, A is the area of low salinity water band (m^2) and u is the average alongshore velocity across section A (ms^{-1}).

IOP Conf. Series: Earth and Environmental Science **176** (2018) 012010 doi:10.1088/1755-1315/176/1/012010

The stability of the water mass can be determined by the stability equation (E) [22]:

$$E = -\frac{1}{\rho}\frac{\partial p}{\partial z} \quad (5)$$

where ρ is the water density (kg m^{-3}) and z is the depth (m). Fluid is stable if E > 0, neutral if E = 0 and unstable if E < 0. If the difference in density value to the depth is greater, then the water layer will be more stable. Movements rise and fall of the fluid to reach a stable position known with bouyancy frequency *or Brunt Vaisala frequency*. Counting Brunt Vaisala Frequency is generated by salinity data, temperature per dept data which is extracted from CTD of research station. N^2 value is calculated to determine stratified value [23].

$$N^2 = -\frac{g}{\rho_0}\frac{\partial \rho}{\partial z} \quad (6)$$

where N^2 is Brunt Vaisala frequency, g is earth's gravitational acceleration (9.8 ms^{-2}), ρ_0 is background density is the average density of the measurement results, ∂ρ changes (gradient) density to depth ∂z (1 m).

3. Results and discussion

3.1. Salinity anomaly

Dispersed salinity anomaly values on west season at all station section are in ranged of -7.44 to -1.406 (section 1), -1.397 to -23.701 (section 2), and -19.983 to -1.546 (section 3). The lowest value is the nearest stations to the estuary mouth and the salinity values go up paralelly to the water depth increases and getting further away of the estuary mouth. In this research, changes in section on depth are section 1 (-5.47) with 1.5 meter depth from the surface, and section 2 (-21.8) with 2 meter depth from the surface and it still decreases until sea bottom surface. The distribution of salinity anomaly is influenced by salinity change. Salinity at the mouth of the Cimandiri River is influenced by tides and seasons. Towards the mouth of the estuary, the lower salinity anomaly. Flavor during the dry season when the river flow decreases, sea air entering further towards the mouth of the estuary cause the salinity. Compound in the rainy season, fresh water flows from the ocean into larger amounts so that the anomalous salinity in the estuary decreases. Field observation carried out [24] in Mesoamerican Barrier Reef System states that salinity has low value in the surface layer with more prominent in 5-10 meter depth which is an effect of freshwater input from mainland. The difference in salinity of seawater with river water that meets in the rofi zone causes both to be mixed. Because the sea salt content is greater, the seawater tends to move in the bottom of freshwater waters in the surface layer.

Furthermore, dispersed anomaly salinity on east season in all research station section are -1.124 to -0.886 (section 1), -1.114 to -0.881 (section 2), -9.571 to -0.908 (section 3). Those values exhibits less freshwater input to the seawater column, so that anomaly salinity almost disperse evenly in the water column. The salinity values is so less compare with the salinity anomaly on January measurement. [25] Seawater salinity under goes change causing by evaporation and presipitation, run off from rivers, coldness or ice melting. Areas close to the river estuary will have lower salinity about 20‰ and it is varied depends on depth.

IOP Conf. Series: Earth and Environmental Science **176** (2018) 012010 doi:10.1088/1755-1315/176/1/012010

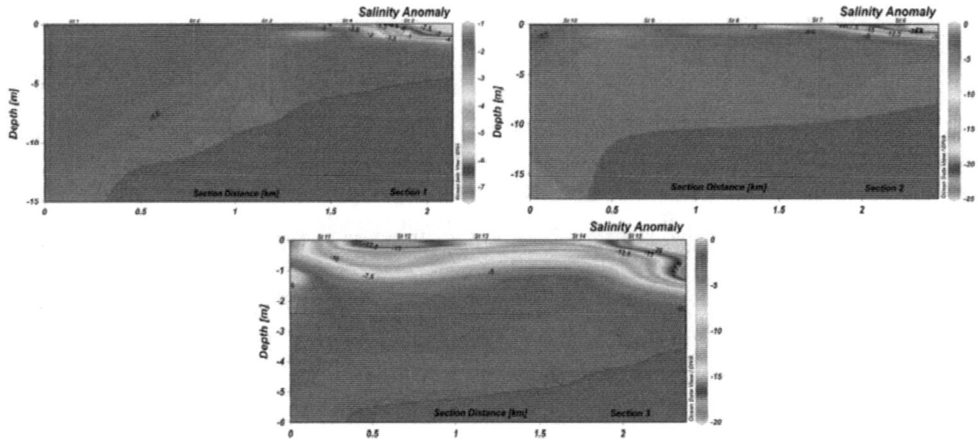

Figure 2. Vertical cross section salinity anomaly section 1, 2 and 3 on West Season.

The anomaly salinity on east season distribution in section 3 has an extremely high salinity anomaly value in the water columns compared to section 1 and 2 so it provides information that the freshwater supply coming from more rivers is carried along the coast towards southeastern south and south of the bay.

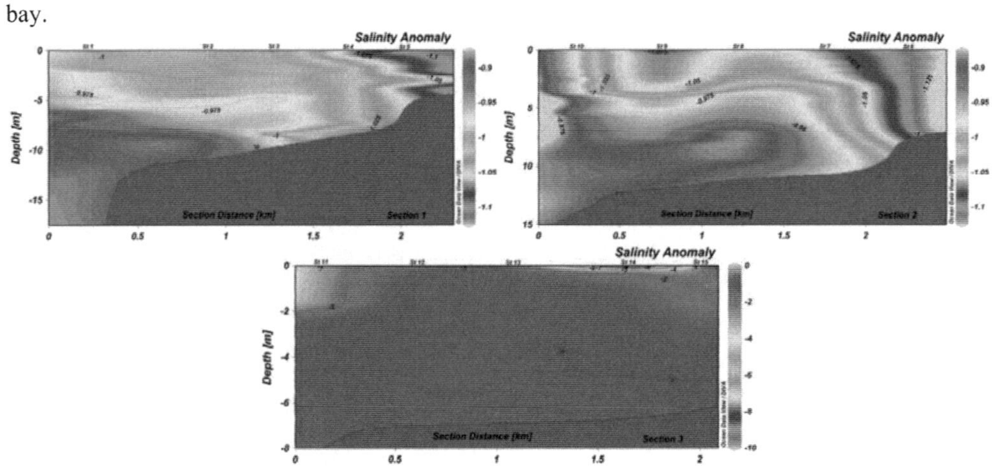

Figure 3. Vertical cross section salinity anomaly section 1, 2 and 3 on East Season.

3.2. Freshwater fraction

Sampling on west season 2016, the averaged freshwater fractions enter sea water column in three station section are 0.05 % leading away as far as 1.7 km from the estuary mouth (section 1), 0.07 % taking for 2 km away from the estuary mouth (section 2), and 0.12 % bring out for 2.5 km from the estuary (section 3). The distribution of freshwater fractions is more carried along the coast, toward the southeast and south of the bay. This is influenced by fresh water discharge, so the value of freshwater fraction varies for each line available at each research station. [26] freshwater discharge has an important effect on the estuary waters, thus providing salinity value. [27] freshwater fraction values are influenced by water mass mixing process due to the oscillation of tidal estuaries and solubility of salinity in estuary waters. Pelabuhan Ratu Bay waters have tidal phase 6 hours period just before its up and down with the type of tide mixed tide prevailing semidiurnal with the value of the Formzahl number is equal to 0.44. The tides that occur in Pelabuhan Ratu Bay are more influenced by the tidal propagation that occurs in the open sea (Indian Ocean) so that with the tidal period that occurs for 6 hours and stirring by the wave as it approaches the waters of the estuary with bathymetry conditions at the location of data collection that varies from depth of 3.5 - 16.3 m can influence the distribution of high fresh water fraction value in water column. the occurrence of up and down the sea surface regularly is one of the important factors affecting the current around the coast as well as various oceanographic processes around the coastal waters, as well as a major influence on coastal geomorphology.

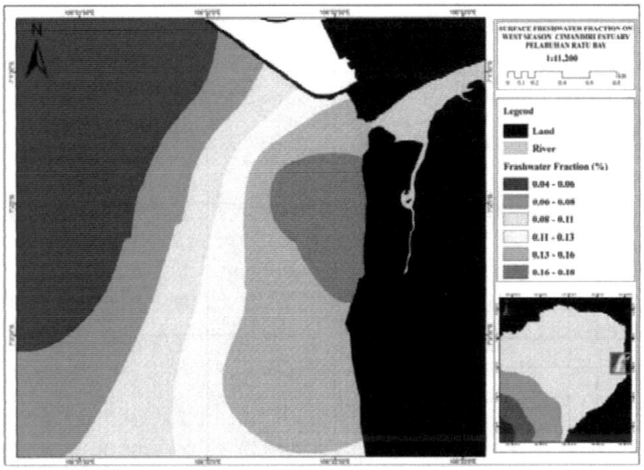

Figure 4. Surface freshwater fraction on West Season.

Sampling on east season 2016, the averaged fresh water fraction enter sea water column in three station section are 0.028 % leading away as far as 0.5 km form the estuary mouth (section 1), 0.029 % taking for 0.5 km away from the estuary mouth (section 2), and 0.038 % bring out for 2 km from the estuary (section 3). [28] that the amount of freshwater mass derived from the precipitation as well as freshwater input from the river may affect the value of the freshwater fraction in the waters.

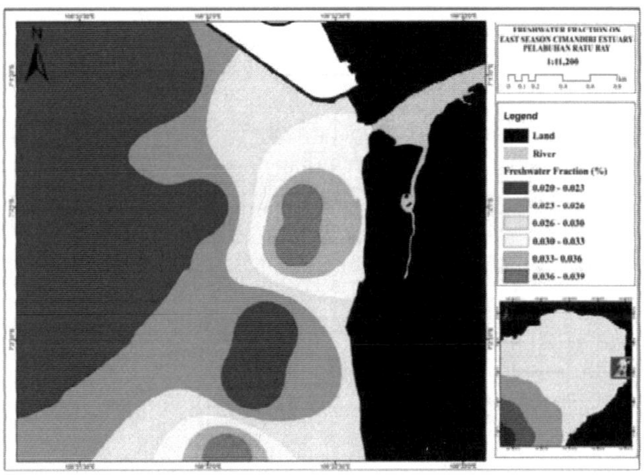

Figure 5. Surface freshwater fraction on East Season.

3.3. Volume Transport of Freshwater

The volume transport of freshwater at the mouth of the Cimandiri River in the west season is 1,429,244 m^3s^{-1} while volume transport of freshwater at the mouth of the Cimandiri River in the east season is 1,066 m^3s^{-1}. This is caused by freshwater discharge in the west season is greater than east season. The result of this research in line with Sanusi [18] conducting research in the whole of Pelabuhan Ratu Bay on two seasons (east and west) where the salinity of the water (river discharge 95.19 m^3s^{-1}) was recorded 5 ‰ and 26 ‰ (river discharge 35.84 m^3s^{-1}), respectively.

Tabel 1. Current velocity, freshwater fraction, area of low sanity water and volume transport of fershwater.

Month	Freshwater fraction (%)	Current velocity (m/s)	Area of low salinity water (m^2)	Volume transport of freshwater (m^3s^{-1})
January	18.07	0.21	375.142	1.429.244
August	7.30	0.09	1.494	1.066

Moreover, other factors affecting the differences of water mass transport are river's mouth channel from, bathymetry slopes, freshwater currents entering the bay, and some oceanography factors. Others studies related to water mass transport influencing by inlet and outlet of currents in certain bay [3]. [29] express about structure of mouth river channel and estuary slope it is important to understand the transition of estaury circulation.

3.4. Water mass stability

Calculating of Brunt-Vaisala (N^2) values results that all research station section (Figure 10) on west season 2016 condition are in ranged of -0. 697 x 10^{-3} – 0.128 Cycl/S (section 1), -0.00131 – 0.473 Cycl/S (section 2), and -0.043 – 0.336 Cycl/S (section 3), respectively.

The all values indicate that freshwater inputs to the seawater column of Pelabuhan Ratu Bay will give a result to be static stability of sea surface layer. It is proven by the water mass to be unstable condition with N^2 is being negative when those water mass spread away the estuary to enter the center of the Bay. The higher value of a water layer, the morely static stability of the layer (something conditions in which the mass of water with low density is above the water masses with high density) and reversly when the value is going to be negative, the water column condition is going to be unstable or static un-stability condition [30]. However, in water column of station 12, 13 and 14 (section 3) undergo un stability.

Moreover, counting of Brunt-Vaisala values on August 2016 condition at all research section (Figure 17) are in ranged of -0.00996 – 0.010812 Cycl/S (section 1), -2.05 x 10^{-3} – 1.94 x 10^{-3} Cycl/S (section 2), -0.00336 – 0.4324 Cycl/S (section 3). The gained average value shows that the research location on August (east monsoon/dried season) is still on stable condition. However, the water mass stability is more less than January which is characterized by almost any negative value in each station. The less water column stability indicates that there is any influence of northwest monsoon wind blowing in southern equator which mixes water mass potentially disturbing water column stability. [31,32] state that, in Indonesia territory, on December to March the northeast monsoon wind blows in the northern and the northwest monsoon wind blows in the southern of equator, while June to August the southwest monsoon wind blows.

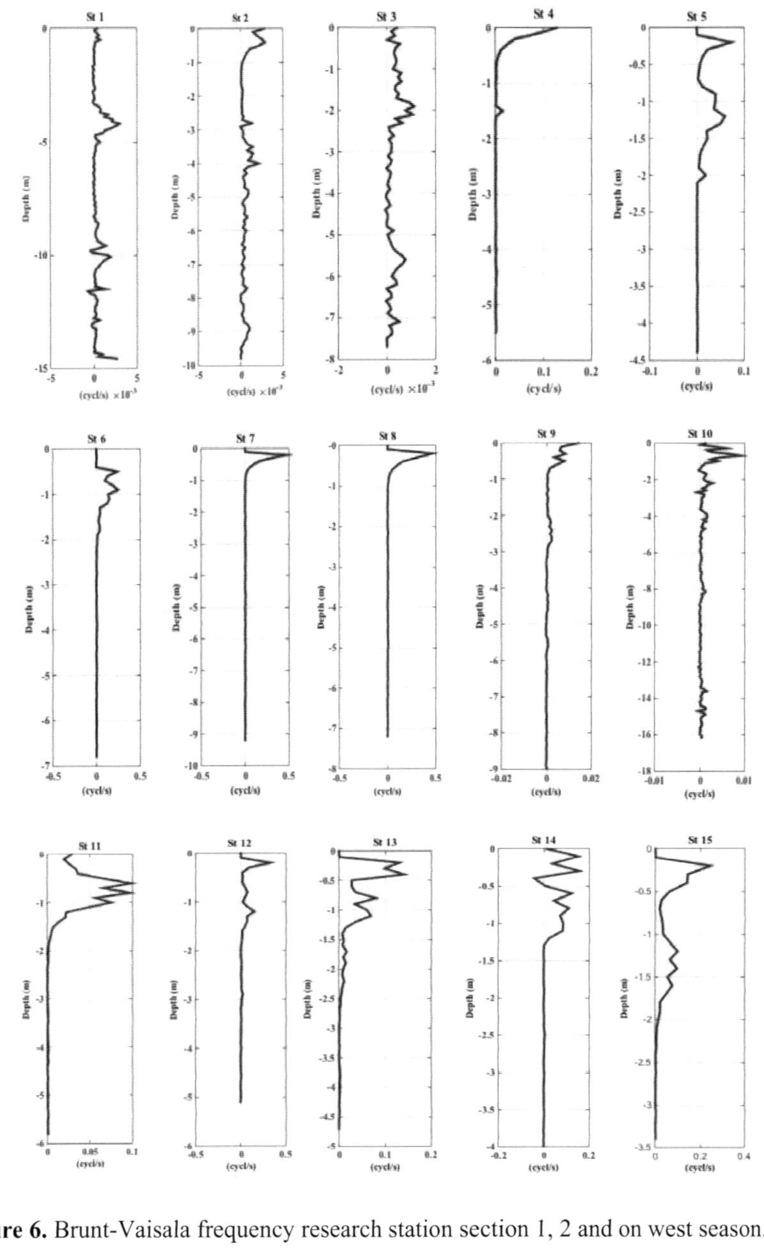

Figure 6. Brunt-Vaisala frequency research station section 1, 2 and on west season.

IOP Conf. Series: Earth and Environmental Science **176** (2018) 012010 doi:10.1088/1755-1315/176/1/012010

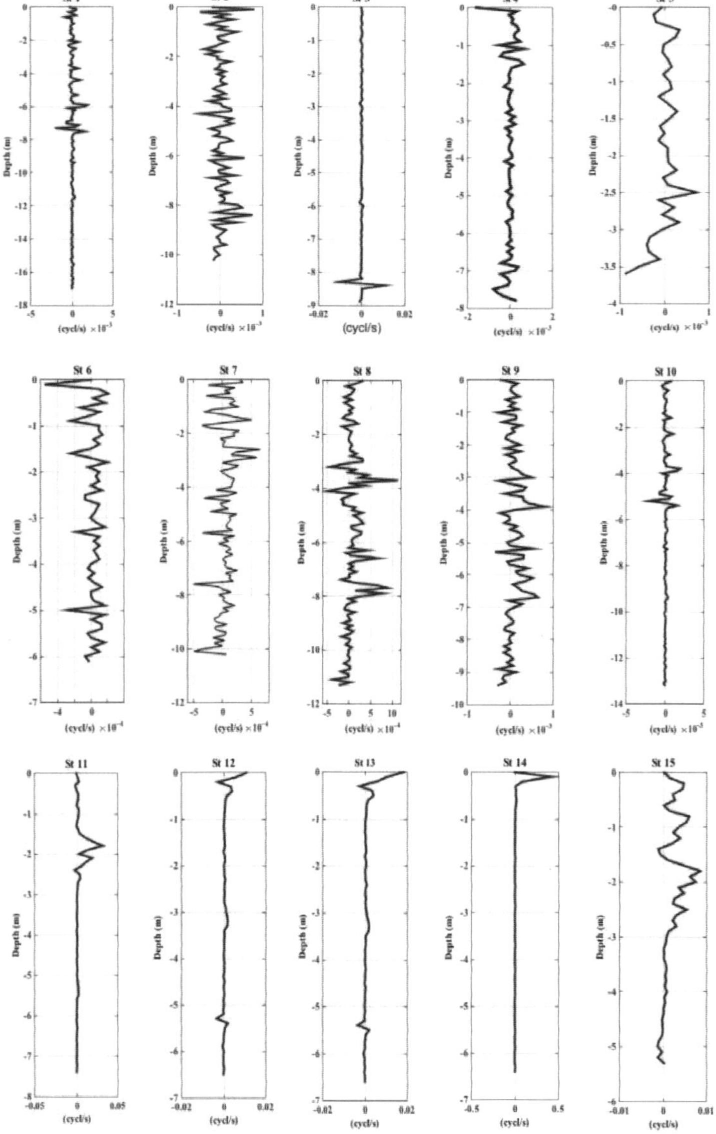

Figure 7. Brunt-Vaisala frequency research station section 1, 2 and on east season.

4. Conclusion

The main characteristic of ROFI (Region of freshwater influence) at the Cimandiri estuary is significant of buoyancy input from freshwater which has important effect for structure and water column dynamic. ROFI on the Cimandiri estuary are more influenced by freshwater supply that entering the sea, and this interaction could change the water mass characteristic. On the west and east season, fraction of freshwater with volume transport carried away from the river along the coast towards southeast and south direction of the bay. Brunt Vaisala frequency on west season at all line of research station was ranged between -0.043 – 0.473 Cycl/S with the highest stable water column formed at freshwater flow from the estuary to the water column while the instability of the column of the waters increasingly occurred at station 10 which getting away from the mouth of the estuary was marked with a negative value and characterized by value 0 as a sign of area of the intersection (front). The calculation of Brunt Vaisala frequency in east season was ranged from -0.00996 – 0.4324 Cycl/S so that the water column entered the condition of unstable waters marked by almost negative value in each station at the research location. The occurrence of water mass instability in water column indicates the influence of the southwest monsoon blowing in the south of the equator, so it cause mixing of water masses in the water column. This is hypothesis confirmation that freshwater supply from Cimandiri River and its location facing the open sea causes the structure and dynamics of the water column.

Acknowledgments

Thanks to supervisor Biological Oceanography Laboratory Marine Science and Technology Mr. M. Tri Hartanto for field assistance and valuable discussion along with the research. This research would not be implemented without supporting by the Laboratory of Oceanography for research tolls and Data Processing.

References

[1] Nugroho R B A and Surbakti H 2009 Two dimensional simulation flow patterns in Pelabuhan Ratu Bay during September 2004 *Nation Marine Science Journal* **4**(1) 48-55
[2] Zimmerman J T F 1986 The tidal whirlpool: a review of horizontal dispersion by tidal and residual currents *Netherlands Journal of Sea Research* **20** 133-154
[3] Van Dam G C, Ozmidov R V, Korotenko K A and Suijlen J M 1999 Spectral structure of horizontal water movement in shallow seas with special reference to the North Sea, as related to the dispersion of dissolved matter *Journal of Marine Systems* **21** 207-228
[4] Fong D A and Geyer W R 2002 The alongshore transport of freshwater in a surface-trapped river plume *Journal of Physical Oceanography* **32** 957-72
[5] Yankovsky A E and Chapman D C 1997 A simple theory for the fate of buoyant coastal discharges *Journal of Physical Oceanography* **27** 1386-401
[6] Hetland R D 2005 Relating river plume structure to vertical mixing *Journal of Physical Oceanography* 35 1667-688
[7] Boer de J G, Pietrzak D J and Winterwerp C J 2009 SST observations of upwelling induced by tidal straining in the Rhine ROFI. *Continental Shelf Research* **29** 263-277
[8] Atmadipoera S A, Kusmanto E, Purwandana A and Nurjaya I W 2015 Observation of coastal front and circulation in the Northeastern Java Sea, Indonesia *JITKT* **7**(1) 91-108
[9] Simpson J H 1997 Physical processes in the ROFI regime *Journal of Marine Systems* **12** 3-15
[10] Ge J, Ding P and Chen C 2014 Low-salinity plume detachment under non-uniform summer wind off the Changjiang Estuary *Estuarine Coastal and Shelf Science* **30** 1-10
[11] Mann K H and Lazier J R N 1996 *Dynamics of Marine Ecosystems Biological–Physical Interactions in the Oceans 2nd Ed* (Cambridge: Blackwell Science)
[12] Nedwell D B, Dong L F, Sage A and Underwood G J C 2002 Variations of the nutrients loads from the Mainland UK estuaries: correlation with catchment areas, urbanization and coastal eutrophication *Estuarine Coastal and Shelf Science* **54** 951-970
[13] Desprez M, Rybarczyk H, Wilson J G, Ducrotoy J P, Sueur F, Olivesi R and Elkaim B 1992 Biological impact of eutrophication in the bay of Somme and the induction and impact of anoxia *Netherlands Journal of Sea Research* **30** 149-159
[14] Sentchev A and Korotenko K 2004 Stratification and tidal current effects on larvaltransport in the Eastern English Channel: observations and 3D modeling *Environmental Fluid Mechanics* **4** 305-331
[15] Sentchev A and Korotenko K 2005 Dispersion processes and transport patterns in the ROFI of Eastern English Channel derived from a Particle Tracking Method *Continent Shelf Res* **25**(4) 2293-308
[16] Nastiti A S, *et al.* 2010 Resource assessment of fish larvae in the Gulf Palabuhanratu, Sukabumi, West Java. Technical Report. Research colaboration BRKP- KKP with the Research and Technology [Unpublished]

[17] Pariwono J I, Eidman M, Rahardjo S, Purba M, Prartono T, Widodo R, Juariah U and Hutapea J H 1988 *Study Up Welling in The Waters of The South Island of Java* (Bogor: Faculty of Fisheries and Marine Science, IPB)

[18] Sanusi H S 2004 Chemical characteristic and fertility of Pelabuhan Ratu Bay Waters at East and West Monsoon *Journal of Aquatic Sciences and Fisheries Indonesia* **2** 93-100

[19] Nastiti A S 2011 Oceanographic conditions as the Basis for Preservation of Fish Resources in the Gulf of Pelabuhan Ratu, Sukabumi West Java *Proceedings of the National Forum of Fish Resources Pacing III*

[20] Gilbert P S, Lee T N and Podesta G P 1996 Transport of anomalous of low salinity waters from the missisippi River flood of 1993 to the Straits of Florida *Con. Shelf Res.* **16**(8) 1065-085

[21] Nurjaya I W 2002 *Behavior of Low Salinity Water Near The Mouth of Tokyo Bay* (Tokyo: Marine Science and Technology, Tokyo University of Fisheries)

[22] Stewart R H 2002 *Introduction to Physical Oceanography* (Texas: Texas A and M University)

[23] Griffiths R W 1986 Gravity currents in rotating system *An. Ref. of Fluid Mech.* **18** 59-89

[24] Carillo L, Johns E M, Smith R H, Lamkin J T and Largier J L 2016 Pathways and hydrography in the Mesoamerican Barrier Reef System Part 2: Water masses and thermohaline structure *Continental Shelf Research* **120** 41-58

[25] Bhatt J J 1978 *Oceanography Exploring The Planet Ocean* (New York: D.Van Nostrand Company)

[26] Song I J and Woo S B 2015 Study on variability of residual current and salinity structure according to river discharge at the Yeoungsan River Estuary, South Korea *Procedia Engineering* **116** 1002-008

[27] Regnier P and O'Kane J P 2004 On the mixing processes in estuaries: the fractional freshwater method revisited *Estuaries* **27**(4) 571-582

[28] Gustafsson B and Stigebrandt A 1996 Dynamics of the freshwater-infuenced surface layers in the Skagerrak *Journal of Sea Research* **35** 39-53

[29] Lee J and Valle-Levinson A 2016 Influence of bathymetry on hydrography and circulation at the region between an estuary mouth and the adjacent continental shelf *Continental Shelf Research* **41** 77-91

[30] Pond S and Pickard G L 1983 *Introductory Dynamical Oceanography 2nd Edition* (Oxford: Pergamon Press)

[31] Wyrkti K 1961 *Physical Oceanography of South East Asian Water*. Naga Report. Vol 2. Scripps Institution of Oceanography. The University of California. La Jolla. California. Naga Rep. **2** 1-195

[32] Tomczcak M and Godfrey J S 1994 *Regional Oseanography: An Introduction* (Oxford:PP)

IOP Conf. Series: Earth and Environmental Science **176** (2018) 012011 doi:10.1088/1755-1315/176/1/012011

Indonesian Throughflow (ITF) variability in Halmahera Sea and its coherency with New Guinea Coastal Current

M C Wattimena[1*], A S Atmadipoera[1], M Purba[1], I W Nurjaya[1], F Syamsudin[2]

[1] Department of Marine Science and Technology, FPIK IPB Bogor
[2] The Agency for the Assessment and Application of Technology (BPPT) Jakarta
*e-mail: marlinchrisye_wattimena@yahoo.com

Abstract. The Indonesian Throughflow passing through the Halmahera Sea (H-ITF) is supposed to be supplied by the western boundary current system, particularly from the New Guinea Coastal Current (NGCC). However, characteristics of H-ITF and NGCC is not well understood yet. This paper aims to investigate vertical structure, transport volume estimate, and transport variability of H-ITF and its coherency with NGCC in the upper 200 m depth. The dataset are obtained from daily INDESO model output (2008-2014). H-ITF across latitude of 0.3°S is indicated by a strong intensified southward flow in upper 150 m with 67.2 km width, and mean transport volume 2.35 (±1.94) Sv (southward). Furthermore, NGCC across 134°E is characterized with 291.2 km width, strong velocity core (>0.4 m/s) in the upper 200 m depth, and transport volume 13.47 (±6.6) Sv (westward). The model indicated that H-ITF is supplied by 17.45% of NGCC. Variability of H-ITF is dominated by intraseasonal period. High coherence (0.85) between H-ITF and NGCC transport is revealed on intraseasonal (45-day) period, with NGCC signal is 4.86 days leading to H-ITF. This sinyal may be associated with a passage of westward equatorial Pacific waves into Halmahera Sea.

1. Introduction

The main inflow of ITF is known as the western path from the Mindanao Current (MC) that flows over the top of the thermocline carrying North Pacific Subtropical Water (NPSW) and the North Pacific Intermediate Water (NPIW). This branch fills the Indonesian Sea through the Sulawesi Sea and then flows through the Makassar Strait. This ITF branch flows out to the Indian Ocean through the Lombok Strait and eventually reaches the Flores Sea and the Banda Sea to finally exit through the Ombai Strait or the Timor Passage [1,2].

In addition to the western path, the ITF inflow also passes the eastern path with two routes available. The first route is via Maluku Sea, bringing the South Pacific Intermediate Water (SPIW) from the Maluku Sea to the Lifamatola Strait to the Banda Sea and then through the Ombai Strait or the Timor Passage to the Indian Ocean. The second route via Halmahera Sea (H-ITF) brings the South Pacific Subtropical Water (SPSW) through Halmahera and the Seram Sea and eventually joins the first eastern route of the Banda Sea. Shallower waters from the South Pacific enter through the passages that line the Halmahera Sea [3,4].

Water mass in Halmahera Sea (HS) is different from other Indonesian waters because it is influenced by the presence of equatorial currents dynamics and warm pools that have a major effect on ocean-atmosphere interactions. Warm pools are caused by a year-round blowing Trade winds that drags the masses of tropical warm water to the western tropical Pacific Ocean and then accumulates in the Tropical Western Pacific region covering the waters of Halmahera and Northern Papua. These territorial waters also have complex current dynamics that may be caused by complex configuration of bathimetry. There are several currents that play a role in transporting water masses in this area such as

Mindanao Current (MC), New Guinea Coastal Current (NGCC), New Guinea Under Current (NGCUC), North Equatorial Counter Current (NECC), Halmahera Eddies (HE), and Mindanao Eddies (ME) [5, 6, 7, 8, 9].

The circulation and transport within the Indonesian seas has a large seasonal variation due to the influence of the reversing annual winds fileds associated with the Asian-Australian monsoon system [10,11,12]. Several studies of the ITF transport which pass through the HS from the measurement and modeling have been reviewed previously. The results of measurements made by [6] noted of 1.5 Sv (1 Sv=10^6 m^3 s^{-1}) transport volume leads to the south from the average measurement of one year mooring. Measurement results by [13] in Lifamatola showed that this path brings about 2.5 Sv mass of water coming from the Southern Pacific Ocean from the deeper layers (South Pacific Subtropical Lower Thermocline Water / SPSLTW) through the Maluku Sea to the Banda Sea.

However, it is not yet possible to estimate the amount of water mass carried through this eastern path due to other water mass inputs on the eastern path, through the HS [14,15]. Apart from the measurement results, the estimated transport passing through the HS from the modeling analysis is also done by Morey *et al.* [16] using the Navy Layered Ocean Model (NLOM), found the 3 Sv transport within the thermocline and the intermediate layer; [17] using the Parallel Ocean Climate Model (POCM), 2.2 Sv flow to the south in the HS; [18] using NEMO Ocean General Circulation Model, 0.8 Sv.

Although there is already a study of transport in the HS, there is no specific study of the dynamics and variability of Halmahera ITF (H-ITF) and its relation to NGCC/NGCUC. The purpose of this study is to investigate the structure and characteristics of H-ITF, and the variability of the H-ITF transport and its coherence with NGCC.

2. Data and data analysis

2.1. Study area

This study was conducted from March to June 2017 in the Physical Oceanography Laboratory, Department of Marine Science and Technology (ITK) - FPIK IPB Bogor. The study area includes one transect in HS (0.3°S, 128°E- 132°E) and one transect in the North of New Guinea (134°E, 2.2°N - 0.6°S) (figure 1).

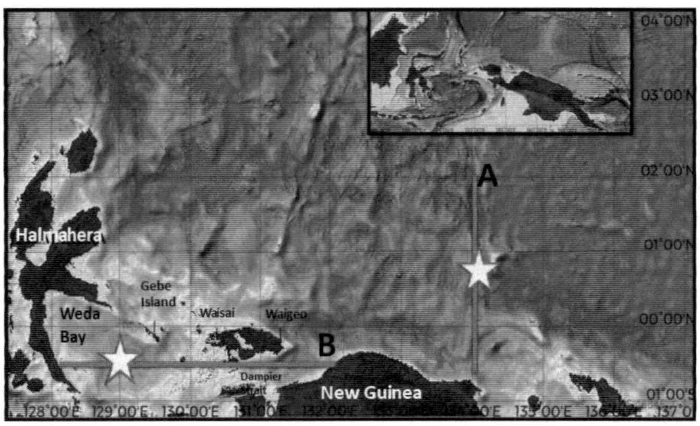

Figure 1. Study area in the Halmahera Sea and its adjacent waters in western equatorial Pacific Ocean. Lines that perpendicular (A) and parallel (B) denote sections for NGCC/NGCUC and H-ITF transport calculations. The location of the vertical current profiles (denoted by the star symbols).

2.2. The data

2.2.1. INDESO model outputs. Model outputs from a series of simulation of NEMO ocean general circulation model of *Infrastructure Development of Space Oceanography* (INDESO) configuration, performed by the CLS France & BPOL Bali, were used in this study. The outputs of data-series span from January 1st 2008 to December 31th 2014. INDESO is a scientific cooperation program at the Ministry of Marine Affairs and Fisheries with a consortium of joint European Marine Research Institute (MERCATOR-Ocean and Collecte Localisation Satelittes - CLS) France. Daily average of 3-dimensions ocean current fields were used for this study.

2.3. Data analysis

Characteristics of circulation and current variability in the study area were assessed by applying time-series analysis, i.e. calculating the mean value of zonal component current (134°E, 0.6 °S - 2.2°N) and meridional (128 °E – 132°E, 0.3°S) from 2008-2014, Continuous Wavelet Transform (CWT) transformation analysis, cross-PSD (Power Spectrum Density) analysis, and EOF (Empirical Orthogonal Function) analysis, according to [19] and [20]. The mean flow aims to analyze general circulation patterns in the study area in spatial form. The climatological method is used to analyze the annual cycle of variable current and temperature. The EOF method is used to analyze the spatial and temporal variability of H-ITF and NGCC/NGCUC.

The estimated transport volume (0-200 m) of H-ITF passing the transect line of east Halmahera and estimated transport volume of NGCC were calculated by integrating the velocity of the zonal component current (*u*) for the estimated transport volume of NGCC (*Qu*) and meridional components (*v*) to estimate the transport volume of H-ITF (*Qv*) against transect length and depth, following [19].

$$Qu_{NGCC} = \int NGCC \int_z^0 udydz \quad (1)$$

$$Qv_{Halma} = \int Halma \int_z^0 vdxdz \quad (2)$$

where Qu_{NGCC} is the total volume of NGCC that intersects line A (transect line in North Papua); Qv_{Halma} (Sv unit) is the total transport volume of H-ITF cutting line B (line in eastern Halmahera); z (200 m) is a lower limit of integration velocity to surface depth (0 m), and *u* is zonal current component, while *v* is meridional current component (m/sec).

2.4. Data validation

To assess a performance and reliability of the model results, a snapshoot of current vector data from shipboard acoustic Doppler current profiler (SADCP) data from the INDOMIX cruise 2010 and the INDESO model output is overlaid on July 06 and July 10, 2010 at 50 m depth level (figure 2).

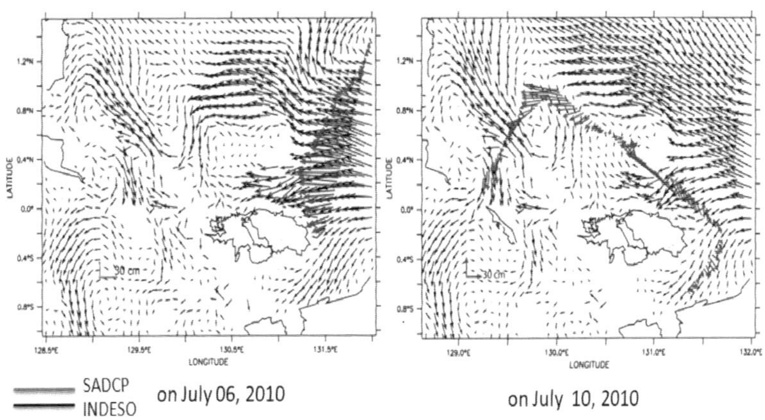

Figure 2. Comparison between observed current vectors from INDOMIX data (red) and modeled current vectors of INDESO model (black) at 50 m depth level on July 6, 2010 (left) and July 10, 2010 (right) in Halmahera Sea and adjacent waters.

SADCP measurements at 50 m depth level in along the northern coast of New Guinea show NGCC flows northwestward and turns eastward to Halmahera Island. Current vectors of INDESO model (black) simmilar with the observed current vectors from INDOMIX data (red), although model velocities being generally slightly weaker than observations. At some locations, magnitudes of current vector amplitude of INDESO model look a little weakened compared with observations such as the northern coast of New Guinea and the eastern of Halmahera. This is allegedly due to the difference in the resolution of the data where the INDESO model is the average spatial resolution of 9 km x 9 km, but the current observation INDOMIX data is the average 1 km. It can be said that INDESO model output can reproduce pretty good current vector of a the spatial pattern, so that the quality of the outcome a simulation model indeso can be used to further analysis in this study.

Evaluation of an operational ocean model configuration at 1/12 degrees spatial resolution for the NEMO (NEMO2.3/INDO12) physical ocean model covering the whole Indonesian Exclusive Economic Zone (IEEZ) in the framework of the INDESO has also been done through comparisons with various data sets including outputs of the parent model, climatologies, in situ temperature and salinity measurements, and satellite data [4].

3. Results and discussion

3.1. Characteristics of H-ITF

The averaged vertical structure of the meridional current component (v) of INDESO model from 2008-2014 at 0.3°S in the Halmahera Sea is presented in figure 3. ITF Halmahera is indicated by a southward strong current. The existence of H-ITF is characterized by a southward flow that is found from the surface to a depth of 200 m, with a speed magnitude exceeding 0.3 m/s, and it weakened as it approached the deeper depth (0.1 m/s). The magnitude velocity of the ITF within its core layer is 0.3 m/s, confined between the surface layer and a depth above 100 m. The width axis of ITF Halmahera was 72 km in the surface and narrowed further reaching a depth of 200 m. The southward flow in the Halmahera Sea was found from the surface to a depth of 700 m. The main driving force of the ITF flow in the upper 200 m layer was the differences in the strong sea level pressure between the Pacific Ocean and Indian Ocean so that the ocean current flows to the south throughout the year [21].

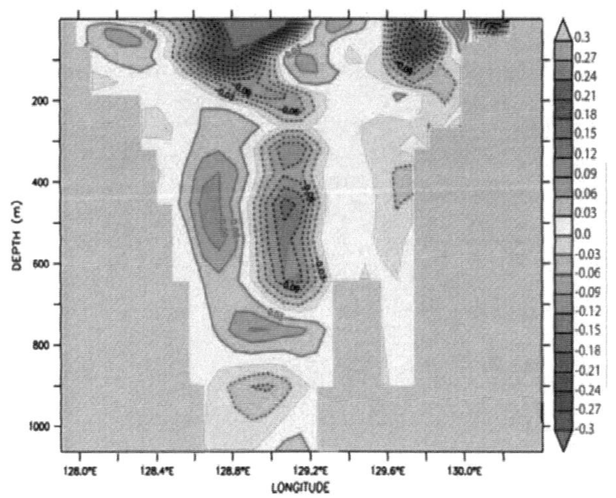

Figure 3. The cross-section of the averaged meridional current component (v) of H-ITF at 0.3°S; 129°E. The colors in meridional currents show the speed scale (m/s). The location of the current transect (figure 1).

3.2. The Annual Cycle of Meridional Component Current of H-ITF

The annual cycle of the meridional current component of ITF Halmahera (figure 4) shows that the current velocity became much stronger during the Southeast Monsoon-SEM (Jul-Aug-Sept) than that of Northwest Monsoon-NWM (Nov-Dec-Jan) period. In September, the intensification of ITF strengthens in the surface layer to almost 200 m (> 0.4 m/s). At a depth of 300 m - 650 m, the movement of current towards the south with the speed magnitude of 0.04 - 0.12 m/s in Halmahera Sea is found. The fluctuations of ITF during the NWM shows clearly a weaker intensity of ITF (on the surface to a depth of 200 m) when compared to that of the SEM.

The main current of ITF flowing during the SEM shows a stronger current velocity when compared to that of NWM. This implies that the transport volume of ITF was greater in the SEM compared to that in NWM, which is in good agreement with previous studies [22, 17, 12]. In the Transition Monsoon Break-TMB 2 (Oct-Nov), the current velocity of ITF Halmahera weakens. This is due to the reversal of the direction of the SEM winds fields into the northwest winds, so that Ekman transport was heading to the north [23].

The description of physical characteristics of water masses (temperature, salinity and current) from the observational data sources in the Halmahera Sea is still challenging. Previous research conducted in the Halmahera Sea [6], after measuring with the mooring current meter at a depth of 400 m, 700 m and 900 m revealed that currents at each depth have different velocities and directions. However, current changes in the Halmahera Sea were dominantly controlled by NGCC and NGCUC [6].

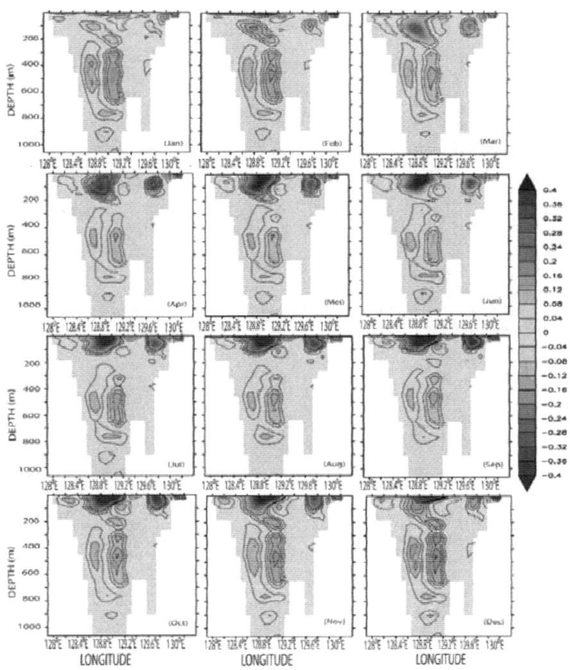

Figure 4. The annual cycle of meridional current component of H-ITF (0.3°E). Negative (positive) values indicate southward (northward) flows.

3.3. Spatial Structures and Temporal Variations of H-ITF

The spatial structure of H-ITF can be interpreted from the EOF analysis results for meridional current component, expressed by the four largest percentage of explained variances (EOF Mode 1 - Mode 4 (figure 5).

The largest signal amplitude of EOF Mode 1 was found at the entrance of ITF in the northeastern Halmahera Sea with a value of 0.58 and a 75% explained variance. The pattern in EOF mode 1 shows that the main flow of ITF Halmahera dominantly flowed towards the south indicated by a positive value, while the negative value indicates the recirculation of the flow to the north. EOF Mode 2 with a 7% explained variance revealed a smaller variability with a positive coefficient on the surface, but below surface, the dominance was negative. EOF 3 with a 5% explained variance had a positive coefficient value on the surface, but with the increasing depth, the variance value was more dominantly negative. The pattern in EOF Mode 4 with 3% explained variance showed a negative coefficients. Spatially, the surface EOF value was higher at the middle and west branches of the eastern HS. Surface flow tended to pass through the central and western branches while in the western branches, fluctuations were quite significant due to seasonal changes.

The temporal variation of the time-series data corresponding to EOF Mode 1 - Mode 4, and the results of the wavelet transformation and power spectral density (PSD) analysis are presented in figure 6. EOF Mode 1 shows high intraseasonal variability (25 and 78 days) with a CWT coefficient > 0.6. Based on the result of density energy analysis, some dominant peaks occurred during the period of intraseasonal (i.e. 25 - 78 days), in addition to the peaks period of 14 day and 214 days. High intraseasonal variability was found approximately in January 2008, June 2008, January 2009, August 2009, April 2010 and July 2012.

Based on the temporal spectral analysis of EOF Mode 1 on the surface (figure 5), it is shown that the intraseasonal period of 25 and 78 days had greater energy density values of 63 and 60 $(m/s)^2$/cpd. The 78-day period is thought likely to be influenced by the arrival of Rossby waves which has a period of 70-100 days [24, 25]. The signals propagated southwards through the HS into the Seram Sea and then southwestward around the Banda Sea to pass through Ombai Strait in the annual period (214 days) and biweekly period (14 days) were also found and had a reasonably strong energy of 53 $(m/s)^2$/cpd, but they were weaker than the intraseasonal period.

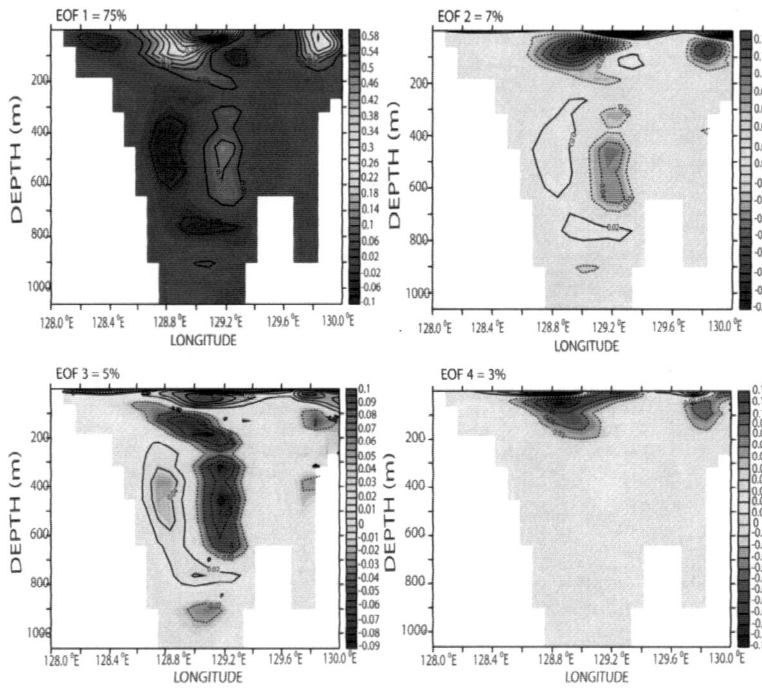

Figure 5. Four largest EOF modes, showing vertical structures of H-ITF, acrossing 0.3°S in Halmahera Sea.

EOF Mode 2 in the surface depth has a spatial distribution with a positive predominant value. Temporally, through CWT analysis, it is known that intraseasonal and biweekly periods (14 days) are still visible but with lesser CWT coefficients than the semi-annual (180 days) period. The temporal variability of EOF Mode 3 of the CWT and PSD coefficients shows the fluctuations of meridional current with the peaks of annual period (373 days). In EOF Mode 4, the opposite site was found. In spectral analysis, the intraseasonal peak with the periods of 18, 41 and 84 days becomes more visible.

Visually, the fluctuations of ITF Halmahera are dominated by the fundamental periods of intraseasonal time-scales. The expedition result of Internal Tides and Mixing in Indonesia Throughflow (INDOMIX) in a moored bottom ADCP for about 2.5 years in the Halmahera Sea revealed fluctuations in current flows on the intraseasonal, seasonal and interannual periods [25]. On the intraseasonal scale of near-bottom current in the Halmahera Sea, power spectral density energy of current peaks at 14, 21, 23 and 27 days periods [26]. Period of 14 days scale is expected to be influenced by the tide at dominant components lunar monthly (MM) and diurnal fortnightly (MF). Moreover, 27 days periods is expected to be driven by remotely equatorial zonal winds through the propagation of Rossby waves in the tropical Pacific Ocean.

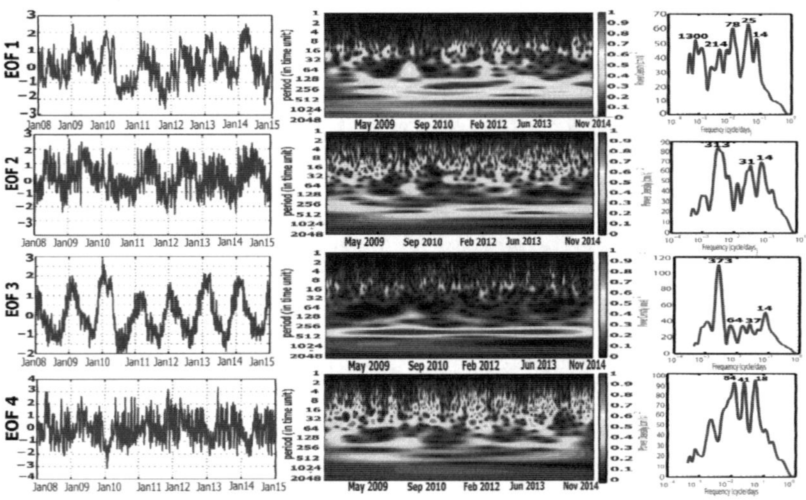

Figure 6. The principal component of EOF mode 1-mode 4 (left panel), showing temporal variation of meridional component of H-ITF from the results of EOF analysis. Middle and right panels are related to CWT and PSD analysis of ti1me series.

3.4. Characteristics of the NGCC and NGCUC

The averaged cross-section of the zonal component (*u*) of the INDESO model output 2008-2014 accrosing 134°E in Northern Papua shows the presence of three layers of core of NGCC and NGCUC, indicated by a strong westward flow with magnitude exceeds 0.4 m/s was near 0.4°N - 1.0°N and could be identified until a depth of 200 m. The NGCC and NGCUC width dimension that intersected at 134°E is 291.2 km. Vertical extent of NGCC was seen down to a depth of 200 m. In fact, NGCUC was found beneath the NGCC at about 0.4°N - 2.0°N to a depth of 1142 m with a current magnitude of about 0.1 m/s (figure 7a).

The NGCC and NGCUC flow along the New Guinea (Papua) coast, and they are part of the South Equatorial Current. NGCC flows northwestward [11], and NGCUC flows northwestward and then turns eastward to Halmahera Island, joining the Mindanao Current and flows eastward as the North Equatorial Counter Current. The cyclonic ME and anticyclonic HE were found at the confluence region of Mindanao and NGCUC [8].

NGCC is a surface current caused by seasonal influences [27]. In the boreal summer characterized by the southeasterly monsoon, westward currents of over 60 cm/s were dominant in the surface layer. In the boreal winter, an eastward surface current developed to 100 cm/s extending down to 100 m depth in response to the northwesterly monsoonal winds. During the Southeast Monsoon, NGCUC flows strongly northwestward [8].

3.5. Annual Cycle of Zonal Component Current of NGCC and NGCUC

The annual cycle of NGCC/NGCUC during the NWM period was characterized by strong mainstream velocity in December, but it began to weaken in January to February (figure 8). In the transitional period I (March, April, May), NGCC was still visible with a strong amplitude of speed. However, NGCUC magnitude of velocity became weaker, and shallower. The main current velocity of the NGCC and NGCUC was weakening towards the West, and part of the current entered the Halmahera Sea. This is because the direction of the wind changed from the northwest monsoon to the southeast. Thus, surface circulation patterns in the northern waters of Papua have a strong seasonal variability [23].

During the SEM period (JJA), the magnitude of NGCC was on the surface and NGCUC beneath NGCC began to increase. In the second Transitional Monsoon Break periods (October - November), the NGCC mainstream velocity was intensified, reaching a current velocity of 0.45 m/s in almost all depth levels and still visible until the end of Monsoon Break.

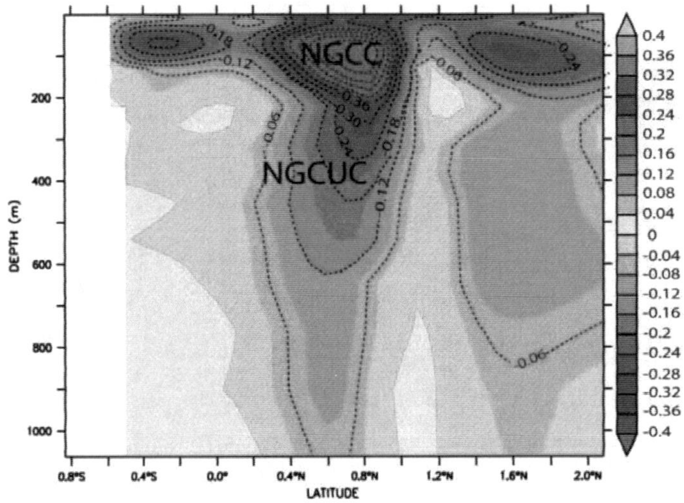

Figure 7. The average cross section of the zonal component current (u) of NGCC and NGCUC (a). The color in the zonal current shows the speed scale (m/s). The location of the current transect (figure 1).

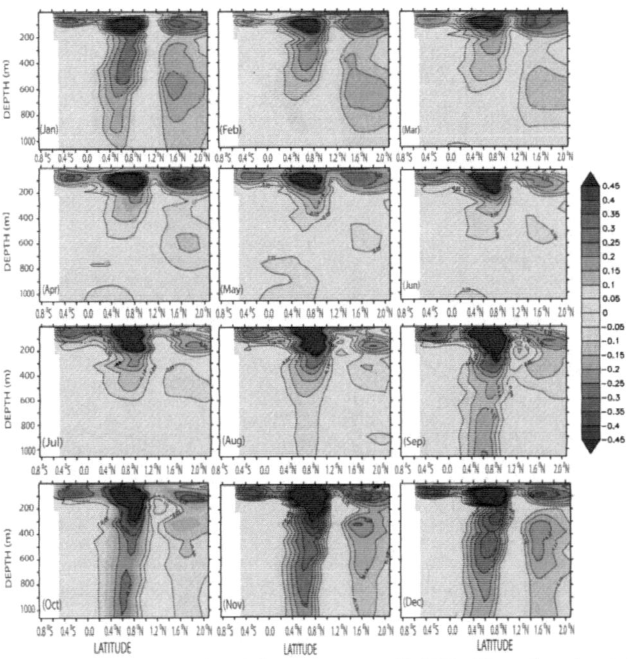

Figure 8. Annual cycle of NGCC and NGCUC (across 134°E). Negative (positive) values indicates westward (eastward) flow direction.

3.6. Spatial Structures and Temporal Fluctuations of NGCC/NGCUC

The spatial structure of NGCC/NGCUC can be interpreted from the results of EOF analysis for the zonal current expressed by the largest percentage of explained variances for EOF Mode 1, Mode 2, Mode 3 and Mode 4 (figure 9).

Amplitude of the largest signal of EOF Mode 1 with a value reached 0.62 and percentage of variance 53%. This is in line with the higher standard deviation of the zonal current (u) in the region. EOF Mode 2 with a 15% variance percentage had a smaller variability on the surface to a depth of 30 m. EOF Mode 3 with a 7% percent variance percentage had a negative dominant variance value. The pattern in EOF Mode 4 with a 4% variance percentage had a negative dominant variance value on the surface, but with the increasing depth, the variance value is more dominantly positive.

The temporal variation of time series data corresponding to four largest EOF Modes and the results of the wavelet transformation and power spectral density (PSD) analysis are presented in figure 10. EOF Mode 1 shows the high intraseasonal variability (84 days) with a CWT coefficient value of > 0.8 (figure 10). Based on the result of density energy analysis, some dominant peaks occurred during the intraseasonal periods of 24, 39 and 84 days (figure 10). High intraseasonal variability was found in June 2008, August 2009, November 2010 and July 2012.

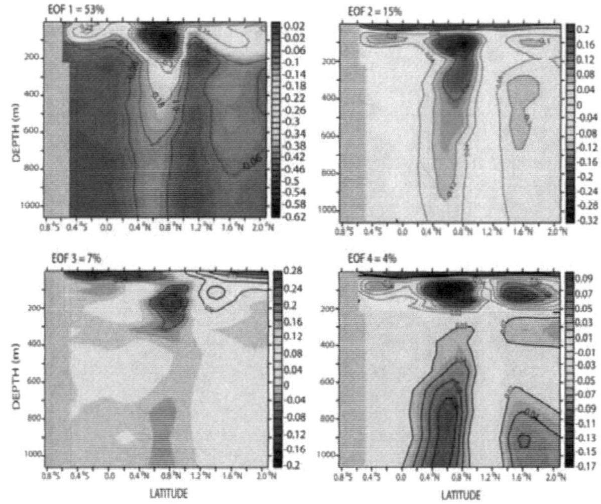

Figure 9. Four largest EOF modes, showing vertical structures of NGCC and NGCUC, across 134°E in northern New Guinea (Papua).

Temporally, through the CWT analysis (figure 10), it is known that intraseasonal periods were still visible but with a considerable greater coefficient of CWT (> 0.8). The distributions of the CWT coefficient and the temporal graph of EOF Mode 2, Mode 3 and Mode 4 shows the fluctuations of meridional currents have an annual dominant cycle (figure 10). Meridional current become strong during the Southeast monsoon, and weaken during the Northwest Monsoon. The process occurring during the annual period was described earlier in figure 9. Using the 1,024 segment length on the spectral analysis, it is clear that the annual period with the peak period was 387 days in EOF Mode 2, 361 days in EOF Mode 3, and 373 days in EOF Mode 4.

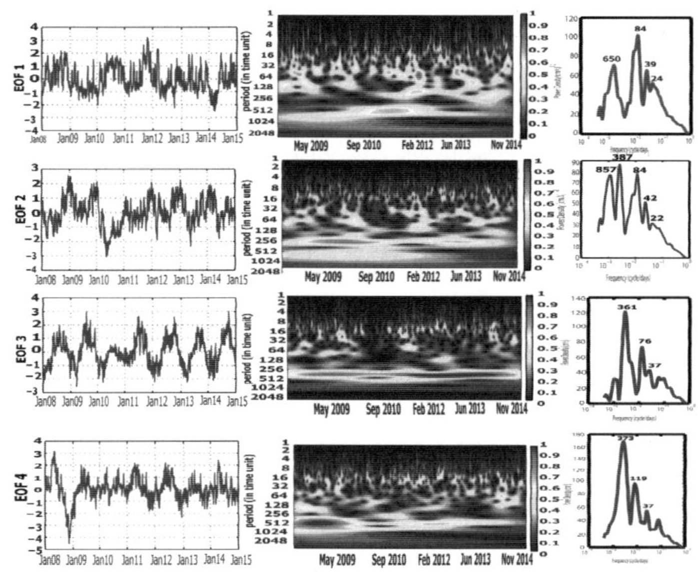

Figure 10. The principal component of four largest EOF modes (left panel), showing temporal variation of zonal component of NGCC and NGCUC at 134°E from the results of EOF analysis. Midle and right panels display results of CWT and PSD analysis of time series.

3.7. Variability of Transport Volume at the H-ITF and the NGCC and NGCUC

The average time series data of ITF Halmahera transport volume from January 2008 to December 2014 show the ranges of transport fluctuations between -7.8 Sv (to the south) and 4.0 Sv (to the north), with an average value of approximately -2.2 Sv and standard deviation of approximately 1.8 Sv (figure 11a). The annual cycle of transport of ITF Halmahera shows the transport strengthened during the Southeast Monsoon toward the south and obtained its maximum in October (3.4 Sv) in the final period of the Season, whereas it became relaxed at 0.4 Sv in January in the Monsoon period (figure 11b). Visually, fluctuations in the transport of ITF Halmahera are dominated by fundamental periods of intraseasonal time scales.

In the northern region of Papua, the average transport time series of NGCC and NGCUC shows fluctuations in the high range between -77 Sv (toward the west) and 10 Sv (towards the east), with an average of approximately -27.5 Sv and a standard deviation of approximately 16 Sv (figure 11c). The annual cycle of NGCC and NGCUC transport shows the ITF transport became strong during the Southeast Monsoon and reached its maximum in November (50 Sv), and subsequently, it relaxed to 3 Sv in May during the first transitional season (figure 11 d). Visually, NGCC and NGCUC transport fluctuations show an annual fundamental period. This means the annual variation of transport becomes dominant in this region.

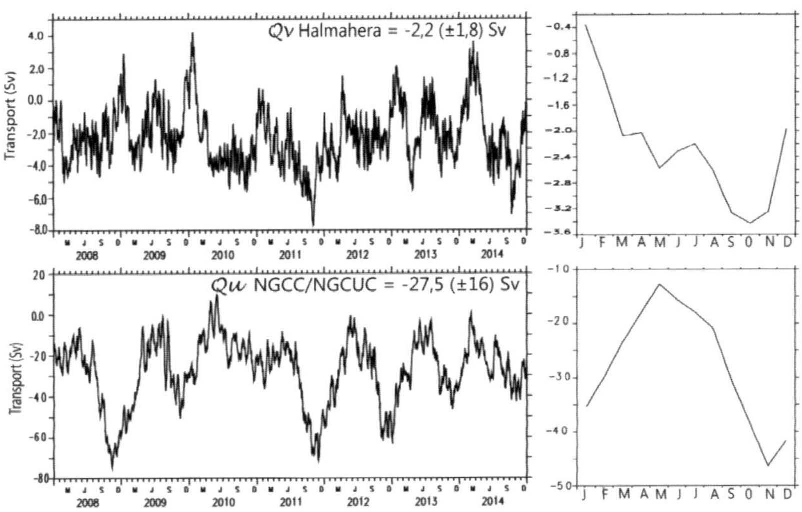

Figure 11. Time-series of transport volume in the upper 1000m in the H-ITF regions and its annual cycle (upper panels), and at NGCC and NGCUC region (lower panels).

The relationship between the western edge of the Pacific and ITF is very complex, especially when the ENSO event occurs. The results of observations and modeling studies show that on average, the proportion of ITF water masses from the North and South Pacific is dependent on the ENSO phase [28]. Climatologically, approximately 80% of ITF water masses come from the North Pacific and 20% from the South Pacific through NGCUC. As La Nina develops, ITF transport increases because of the strong eastern trade wind, which induces a gradient of pressure between the western Pacific and the eastern Indian Ocean. Climate models show larger amounts of ITF water masses originating from the South Pacific during La Nina than during the El Nino occurrence.

3.8. Coherence of Transport Fluctuation between H-ITF and NGCC

The relationship between ITF Halmahera transport and NGCC transport (above 200m) can be seen from the time-transport data of ITF Halmahera transport and NGCC transport from January 2008 to December 2014 (figure 12).

The mean transport volume of ITF Halmahera is around 2.35 (± 1.94) Sv (southward), while the mean transport volume of NGCC is about 13.47 (± 6.6) Sv (westward). The annual cycles of the average transport of ITF Halmahera and NGCC transport show similar fluctuation pattern with the maximum transport was found from October to November, and the minimum transport was found in January. NGCC is found along 134°E 0.4°N - 2.0°N (figure 2 and 7a) with a strong flow intensity westward across northern Halmahera with an average transport 13.47 Sv. The mean transport volume of ITF Halmahera is around 2.35 Sv. In regards to the similarity of fluctuation pattern of H-ITF transport and NGCC transport and based on the ratio calculation, the percentage of NGCC entering through the Halmahera Sea above 200 m depth was 17.45% from the total transport volume of NGCC. This shows that NGCC transport contributes by about 17.45% to the ITF transport entering through the Halmahera Sea.

The result of cross-PSD analysis between the ITF Halmahera transport time series and the NGCC transport time series revealed significant coherence over several periods, i.e. 28 days, 38 days, 45 days, 79 days, 128 days and 341 days periods (figure 13). The highest coherence value (0.85) occurred in the 45-day period, with a positive phase difference (4.86 days). This indicates that the NGCC signal fluctuation at the 45-day periods appeared first in transect of 134°E (in Northern Papua) and 4.86 days later, the signal propagation appeared in transect of 0.3°S of ITF Halmahera. The values and phase differences between the NGCC signals at transect of 134°E of Northern Papua with the signal of ITF Halmahera at transect of 0.3°S are presented in table 1.

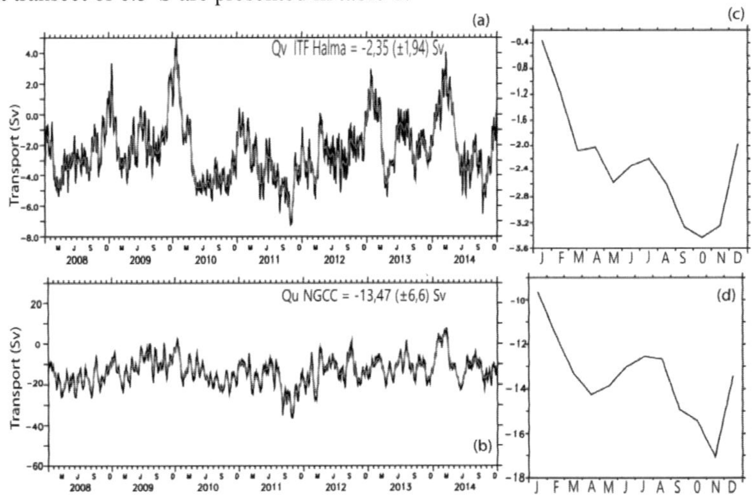

Figure 12. Time-series of transport volume in the upper 200 m in the H-ITF regions and its annual cycle (upper panels), and at NGCC and NGCUC region (lower panels).

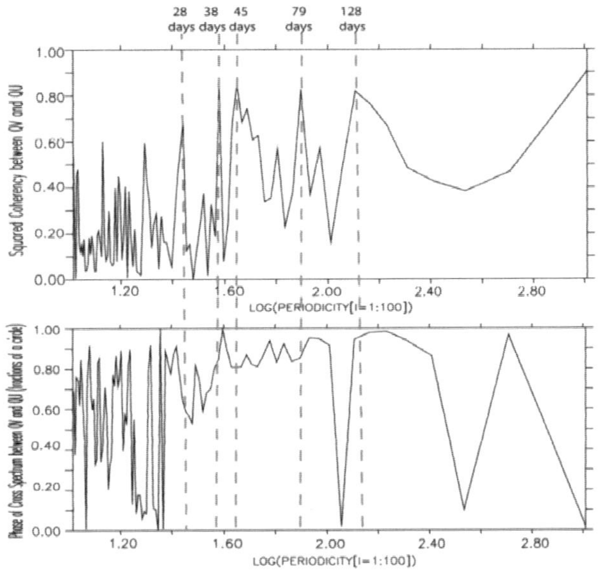

Figure 13. Coherence (a) and phase (b) between the H-ITF transport and the NGCC transport series.

Coherence analysis results revealed that the phase difference between signal fluctuations in NGCC and ITF Halmahera ranged from 2.56 days to 15.4 days. This indicates that NGCC at the transect of 134°E (in Northern Papua) flowing into the ITF Halmahera region at transect of 0.3°S needs between 2.56 days to 15.4 days, indicating the presence of propagation of the Rossby wave signal entering the HS.

Table 1. Period, coherence and phase between H-ITF transport and NGCC transport.

No	Period (day)	Coherency	Phase (day)
1	28	0.6835	2.56
2	38	0.8474	4,25
3	45	0.8519	4.86
4	79	0.8256	8.8
5	128	0.8192	15.4
6	341	0.3826	5.46

The oceanic response to wind forcing is often accomplished through wave processes that propagate along the equatorial and coastal wave guides within the Indonesian Archipelago, and impact the water properties, thermocline and sea level on all timescales. The equatorial winds that produced free equatorial Rossby waves whose signals indicated reached the study region. Equatorial Pacific Rossby waves excited coastally trapped waves off the western tip of New Guinea that propagated poleward along the Arafura/ Australian shelf break. As well, Pacific energy radiated westward into the southeast Indian Ocean via the Banda Sea [29, 30].

4. Conclusion

The circulation pattern in the eastern part of the Halmahera Sea is characterized by the strong axis of ITF Halmahera on the middle lane of the Halmahera Sea. The average of ITF Halmahera transport at transect of 0.3° SL is -2.2 (± 1.8) Sv (negative: southward). On the northern side of Papua, NGCC/NGCUC transport is approximately -27.5 (± 16) Sv (positive: westward).

The upright structure of ITF Halmahera at 0.3° LS transect is characterized by the wide size of around 67.2 km, the depth of the current from the surface to the depth of 150 m with a maximum current layer of >0.3 m/s from the surface to 100 m. The annual cycle of ITF Halmahera shows the width and speed of ITF Halmahera are fully developed in the Southeast Season with transport of about 3.4 Sv, which is in contrast to the minimum ITF Halmahera (about 0.4 Sv) in the Northwest season. The size of the NGCC NGCUC cross section of the transect at134°BT is 291.2 km in width, 0-1142 m in depth with the maximum speed of > 0.4 m/s) in the depth of 50-200 meters.

Variability of ITF Halmahera is dominated by fluctuations with the subsequent ISV of annual and biweekly periods where as variability of NGCC/NGCUC is dominated by annual period. Significant coherence between NGCC and ITF Halmahera occurs between 28 days to 341 days. The highest coherence (0.85) occurs in the 45-day period with a phase difference of about 4.86 days. This means that the propagation of the Rossby wave signal from NGCC into the Halmahera Sea region take approximately 4.86 days.

Acknowledgments

INDESO project funded by the Indonesian Ministry of Marine Affairs and Fisheries. SADCP current data provided by INDOMIX cruise using RV Marion Dufresne. We appreciated comments from reviewers to improve the quality of this paper.

References

[1] Arief D and Murray S P 1996 Low-frequency fluctuations in the Indonesian throughflow through Lombok Strait *J Geophys Res* **101** 2455–12464
[2] Gordon A L and Fine R 1996 Pathways of water between the Pacific and Indian oceans in the Indonesian Seas *Nature* **379**(6561) 146-149
[3] Sprintall J and Revelard A 2014 The Indonesian throughflow response to Indo-Pacific climate variability *J. Geophys. Res. Oceans* **119** 1161–1175
[4] Tranchant B, Reffray G, Greiner E, Nugroho D, Koch-Larrouy A and Gaspar P 2016 Evaluation of an operational ocean model for the Indonesian seas – Part 1 *Geosci. Model Dev.* **9** 1037–1064
[5] Fine R A, Lukas R, Bingham F, Warnar W and Gammon R 1994 The western equatorial Pacific: a water mass crossroads *J. Geophys. Res.* **99** 25063–80
[6] Cresswell G R and Luick J L 2001 Current measuranment in the Halmahera Sea *J. Geophys. Res.* **106**(C7) 13953-58
[7] Atmadipoera A, Kuroda Y, Pariwono J L and Purwandani A 2004 Water Mass Variation in the Upper Layer of the Halmahera Eddies Region Observed from a TRITON buoy IEEE
[8] Kashino Y, Ueki I, Kuroda Y and Purwandani A 2007 Ocean variability north of New Guinea derived from TRITON buoy data *J. Oceanogr.* **63** 545–559
[9] Kashino Y, Atmadipoera A S, Kuroda Y and Lukijanto 2013 Observed feature of the Halmahera and Mindanao Eddies *J. Geo. Ress. Ocean.* **118** 6543-60
[10] Clarke A J and Liu X 1993 Observations and dynamics of semiannual and annual sea levels near the eastern equatorial Indian Ocean boundary *J. Phys. Oceanogr.* **23** 386–399
[11] Masumoto Y and Yamagata T 1996 Seasonal variations of the Indonesian Throughflow in a general circulation model *J. Geophys. Res* **101** 12287–93
[12] Sprintall J, Wijffels S E, Molcard R and Indra J 2009 Direct estimates of the Indonesian throughflow entering the Indian Ocean: 2004–2006 *J. Geophys. Res.* **114** C07001
[13] van Aken H M, Brodjonegoro I J and Jaya I 2009 The deepwater motion through the Lifamatola passage and its contribution to the Indonesian throughflow *Deep-Sea Res.* **56**(1) 1203–16
[14] Ilahude A G and Gordon A L 1996 Thermocline stratification within the Indonesian seas *J. Geophys. Res.* **101**(12) 401-409
[15] Gordon A L 2005 Oceanography of the Indonesian seas and their throughflow *Oceanography* **18** 14–27
[16] Morey S L, Shriver J F and O'Brien J J 1999 The effects of Halmahera on the Indonesian Throughflow *J. Geophys. Res.* **104** 23281–96
[17] Potemra J T 2005 Indonesian throughflow transport variability estimated from satellite altimetry

Oceanography **18**(4) 98–107
[18] Koch-Larrouy A G, Madec B, Blanke R and Molcard 2008 Water mass transformation along the Indonesian throughflow in an OGCM *Ocean Dyn.* **58** 289-309
[19] Emery W and Richard E Thomson 2014 *Data Analysis Methods in Physical Oceanography 3rd Edition* Elsevier Science **eBook ISBN:** 9780123877833
[20] Bendat J S and Piersol A G 1971 *Random Data Analysis and Measurement Procedure* (New York: John Wiley and Sons Inc.)
[21] Gordon A L, Ffield A and Ilahude A G 1994 Thermocline of the Flores and Banda seas *J. Geophys. Res.* **99** 18235–42
[22] England M and Huang F 2005 On the interannual variability of the Indonesian throughflow and its linkage with ENSO *J. Clim.* **18** 1435–44
[23] Wyrtki K 1961 *Physical Oceanography of the Southeast Asian Waters* NAGA Rep. 2 Scripps Inst. Oceanography
[24] Qiu B, Mao M and Kashino Y 1999 Intraseasonal variability in the Indo-Pacific throughflow and the regions surrounding the Indonesian *Seas J. Phys. Oceanogr.* **29** 1599–1618
[25] Atmadipoera A, Koch-Larrouy A, Cuypers Y, Molcard R, Jaya I and Harsono G 2013 New current measurement in the Halmahera Passage, Indonesia *Poster Sessions, Asian Oceania Geosciences Society Symposium Australia, 24-27 June 2013*
[26] Wattimena M C, Atmadipoera A S, Purba M **and** Koch-Larrouy A 2014 Intraseasonal **v**ariability **of** **n**ear-**b**ottom **c**urrent **in** **t**he Halmahera Sea *JITKT* 6(2) 267-282
[27] Kuroda Y 2000 Variability of the currents off the north coast of New Guinea. *J Oceanogr.* 56 103–106
[28] Gordon A L and Fine R 1996 Pathways of water between the Pacific and Indian oceans in the Indonesian seas *Nature* **379**(6561) 146-149
[29] Sprintall J and Revelard A 2014 The Indonesian throughflow response to Indo-Pacific climate variability *J. Geophys. Res. Oceans* **119** 1161–75
[30] Wijffels S and Meyers G 2004 An intersection of oceanic wave guides: variability in the Indonesian throughflow region *J. Phys. Oceanogr.* **34**(5) 1232–53

Coherence of transport variability along outer Banda Arcs

V C Masoleh[1*], A S Atmadipoera[2*]

[1] Graduate Program in Marine Sciences, Bogor Agricultural University (IPB)
[2] Physical Oceanography Laboratory, Dept. of Marine Science and Technology-IPB
*e-mail: atmadipoera_itk@ipb.ac.id, vanessamasoleh@yahoo.com

Abstract. Previous study hypothesized that the Pacific equatorial waves could enter into the interior Indonesian Seas and influenced the Indonesian Throughflow (ITF) along its eastern pathway. Here we investigate the variability of volume transport in the upper 500 m depth and its coherency along the outer Banda Arcs (OBA) channel from Ceram Sea (CS) to Aru Basin (AB) to western Timor Passage (TP) using daily INDESO model output (2008-2014). As expected, mean volume transport along OBA is in clock-wise direction: +1.04 (±1.54) Sv eastward in CS, -0.61 (±0.79) Sv southward in AB, and -8.42 (±2.05) Sv westward in TP, respectively. The vertical structure of currents is prominently intensified in the upper 300 m layer. Scales of transport variability in CS and AB are dominated by annual period with secondary in intraseasonal (ISV) period. In contrast, TP dominant periodicities are on the semiannual and ISV scales. Transport variability on ISV 30-43 days period among CS - AB - TP shows a significant coherence (>0.8) and phase difference (2-5 days) with signal in CS leading to AB and TP. This suggests a presence of Pacific equatorial waves which propagate along OBA channel.

1. Introduction

The Indonesian Throughflow (ITF) is a current system influenced by the thermocline of the North Pacific Ocean, which flows into Indonesian waters through two main routes, namely, the west path through the Sulawesi Sea and the eastern path through the Halmahera Sea and the Maluku Sea, where the components of ITF comprise water masses from the North Pacific and South Pacific [1, 2]. The ITF current system plays an important role in the formation of ENSO, which affects the global climate in channeling the Pacific Ocean to the Indian Ocean, where the stronger ITF transport occurs in the La Nina period (West Pacific is warmer) and weakens in the El Nino period (West Pacific is cooler) [3, 4]. The ITF varies seasonally, with the highest transport rates being found during the Southeast Monsoon (SEM, June to August) while the lowest transport during the Northwest Monsoon (NWM, December to February) [5, 6, 7].

The North Pacific water mass flow into Indonesian waters from Mindanao Current through the Sulawesi Sea to the Makassar Strait and then 20% of the water mass passes through the Lombok Strait and the rest follows the Flores Sea and Banda Sea [8]. Previous studies show that upon entering Indonesian waters, characteristics of ITF's water masses are changed into a considerable transformation, in areas closed to Sangihe undergoing a transformation of 40%, in Makassar Strait and the Flores Sea 30%, in the Halmahera Sea, Maluku Sea and Ceram Sea at 26 %, and in the Strait of Ombai and Timor by 26% while in the Banda Sea, Java Sea, and in the Indian Ocean, the masses are transformed in the opposite direction with the total transformation of each region are -15%, -2%, and -3%, respectively [9, 10]. According to [11], the propagation of the Rossby Pacific wave signal trapped offshore from the western tip of New Guinea spreads, which popagate poleward along west coast of Australia, in response to the Pacific zonal winds. Also visible is the propagation of the Rossby wave to the west across the

Banda Sea and into the subtropical south Indian Ocean within of 1500 km, in response to the Equatorial Pacific wind energy. This paper aims to investigate the variability of volume transport in the upper 500 m depth and its coherence along the outer Banda Arcs (OBA) channel from Ceram Sea (CS) to Aru Basin (AB) to western Timor Passage (TP).

2. Data and data analysis

2.1. Study area

This study was conducted in Banda Sea and its adjacent seas, which focused on outer Banda Arcs (Ceram Sea, Aru Basin, and Timor Passage) (figure 1). Data processing and analysis was carried out in Physical Oceanography Laboratory, Marine Science and Technology Department, Bogor Agricultural University.

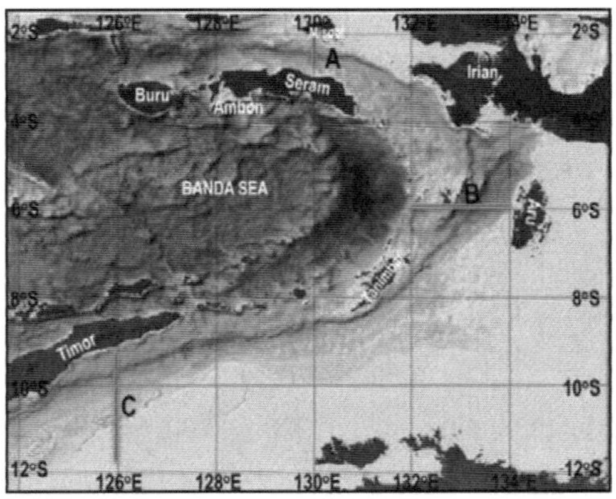

Figure 1. Study area in Banda Sea and its adjacent seas by emphasizing in outer Banda Arcs.

Arcs from eastern Ceram Sea to Aru Basin to Timor Passage. Lines A, B and C are cross section for volume transport calculation in Ceram Sea (130°E), Aru Basin (6°S) and Timor Passage (126°E), respectively.

2.2. The data

2.2.1. INDESO model outputs. The data source of this research was obtained from the INDESO (*Infrastructure Development for Space Oceanography*) project of the Ministry of Fisheries and Marine Affairs. The INDESO model used the NEMO ocean general circulation model with specific configuration within the domain model in Indonesian Seas and adjacent waters (90°E-145°E; 25°S-20°N), including detailed bathymetry, explicit tidal forcing and mixing parameterization. Model simulation was performed at the Marine Observation and Research Institute (BPOL) Perancak Bali and at CLS (*Collecte Localisation Satellites*) Toulouse France. Model configuration was described in [12]. The model output data consist of daily zonal and meridional current components, seawater temperature, and salinity time-series data spanning from January 2008 to December 2014 (7 years). The data could be accessed through the website *http://www.indeso.web.id/.* The INDESO data is averaged daily simulation data results with horizontal resolution 1/12° (9.25 km) and 50 depth levels.

2.2.2. Model validation. The validation of INDESO model output with satellite observation and field current measurement data in the study area (Banda Sea) was performed by [13, 14]. They reported that correlation between model sea surface temperature (SST) and satellite SST data was reasonably high (r=0.95). Fluctuation of model SST and the satellite data was in phase [13]. Comparison of model current vector and observed shipboard Acoustic Doppler Current Profiler (ADCP) data during the INDOMIC cruise in July 2010 showed a good agreement between the model and observation, where upper-surface eddies in Banda Sea and strong persistent Indonesian Throughflow current vectors in Ceram and Ombai

Straits were well reproduced by the model [14]. In addition, general model validation of INDESO model output was also described in [12].

2.3. Data analysis

Analyses of time-series data of zonal and meridional components at each section in Ceram, Aru and Timor Passage were performed using Cross-Power Spectral Density (Cross-PSD) method [15], and Continuous Wavelet Transform (CWT) method [16] to examine periodicity from current component and volume transport time-series.

Volume transport estimate at each section (A, B and C in Fig.1) was calculated using the equation below [17]:

$$Qu_A = \int_A^B \int_z^0 u \, dy \, dz \quad (1)$$

$$Qv_B = \int_B^C \int_z^0 v \, dx \, dz \quad (2)$$

$$Qu_C = \int_C^A \int_z^0 u \, dy \, dz \quad (3)$$

where Qu_A, Qv_B, and Qu_C are volume transport estimates (in Sverdrup Sv; 1Sv=10^6 m^3/s) at Ceram section (A), Aru section (B), and Timor section (C) by integral operation taking account of a length of section (m), depth integration (0-500 m), and zonal and meridional current components (u and v).

The wavelet analysis aims to find out time-frequency or time-period distribution within time-series data, i.e. how the power changes over [16]. In this research, CWT analysis was applied to volume transport times-series data in Ceram, Aru and Timor sections. Following [17], the CWT equation is expressed as follows:

$$\mathrm{C(scale, position)} = \int_{-\infty}^{\infty} \mathrm{f(t)}\psi\mathrm{(scale, position, t)dt} \quad (4)$$

which C is wavelet coefficients; $\mathrm{f(t)}$ is time series data; ψ is mother wavelet. The CWT Analysis is using wavelet as *band pass filter* which grouped in time and scale, so $\eta = s.t$and normalized in unit of energy.

Power Spectral Density (*PSD*) Analysis was used to find out the significant energy peaks for respective periods within times-series data with 95% confidence interval. In this research, length of the time-series data (N) is 2557 and segment length used for PSD calculation is 1024. Following [15], the PSD of each volume transport time-series was estimated using the equation, as follows:

$$X_i(f_k) = \Delta t \sum_{n=0}^{N-1} x_{in} \exp\left[\frac{-j2\pi kn}{N}\right] \quad (5)$$

Which $X_i(f_k)$ is Fourier components; x_{in} is data values, $n = 0, 1, \dots, N-1, i = 1,2, \dots, n_d$; $k = 0,1, \dots, N/2$.

The *cross-PSD* (*cross-Power Spectral Density*) analysis method was used to find out shared periodicity from both time-series data. Their degree of correlation is indicated by its normalized coherence value, which varies from minimum (0) to maximum (1). In this research, *cross-PSD* analysis was used to obtaine correlation between time-series data of volume transport in Ceram Sea (Q_{uc}) vs Aru (Q_{va}); in Aru (Q_{va}) vs in Timor (Q_{ut}); and in Ceram (Q_{uc}) vs in Timor (Q_{ut}).

The result of *cross-PSD* analysis is *co-spectrum* energy that shows magnitude of energy fluctuations at specific frequency within both of times-series data; *coherency* and *phase* show correlation and phase difference (lag or lead) at specific frequency within both of times-data series. Positive phase means one time-series leads (or proceed) to another series in contrary to negative phase to be lag to another series. According to [15], co-spectrum estimate is calculated by using equation:

$$G_{XY}(f_k) = \frac{2\Delta t}{T} \left| X(f_{(k)}) * Y(f_{(k)}) \right| \quad (6)$$

IOP Conf. Series: Earth and Environmental Science **176** (2018) 012012 doi:10.1088/1755-1315/176/1/012012

where $G_{XY}(f_k)$ is density spectrum of cross energy in frequency to-*k* (f_k); f_k is the discretized frequencies, where f_k = k/Nh; k = 0, 1, 2,..., N-1; $X(f_{(k)})$ = *Fourier* component from (xt); $Y(f_{(k)})$ is *Fourier* component from (yt); Δt is time cross sampling data (1 day); *T* is data period.
Coherence value is calculated by using this equation [15]:

$$\gamma^2{}_{XY}(f_k) = \frac{|S_{xy}(f_k)|^2}{S_x(f_k)S_y(f_k)} \tag{7}$$

where $\gamma^2{}_{XY}(f_k)$ coherence is value in frequency to-*k* (f_k); $S_{XY}(f_k)$ is density spectrum of cross energy in frequency to-*k* (f_k); $S_x(f_k)$ is density spectrum energy *X* (f_k); $S_y(f_k)$ is density spectrum energy *Y* (f_k) in frequency to-*k* (f_k).
Phase different value is calculated by using this equation [15]:

$$\theta_{XY}(f_k) = tan^{-1}\left|\frac{Q_{xy}(f_k)}{C_{xy}(f_k)}\right| \tag{8}$$

where $Q_{xy}(f_k)$ imaginer is value from $G_{XY}(f_k)$; $C_{xy}(f_k)$ is real value from $G_{XY}(f_k)$.

3. Results and discussion

3.1. Mean structure of ITF along outer Banda Arcs

Figure 2 shows the circulation pattern of ITF average transport along the outer Banda Arcs within the three study areas: Ceram Sea, Aru Basin, and Timor Passage. In the Ceram Sea, the average flow of ITF is dominant to the East with a current velocity of about 0.20 m/s at depth above 150 m. The average flow rate of ITF in Aru Basin is about 0.12 m/s at depth above 110 m, with the dominant flow direction to the South.

According to [5], the flow of ITF at depth above 200 m is caused by the strong gradient pressure difference between the Pacific and Indian Oceans. The circulation pattern in Timor Passage has a unique pattern in which the average speed of ITF is about 0.40 m/s flowing Westward from the surface (0 m) to the depth of 380 m, and the reversal of the current flow direction to a depth below 400 m with the average current velocity by 0.04 m/s (figure 2c). The reversal of the currents to the East is the response of Kelvin wave occurring due to the zonal wind anomalies in the equatorial Indian Ocean [18, 19].

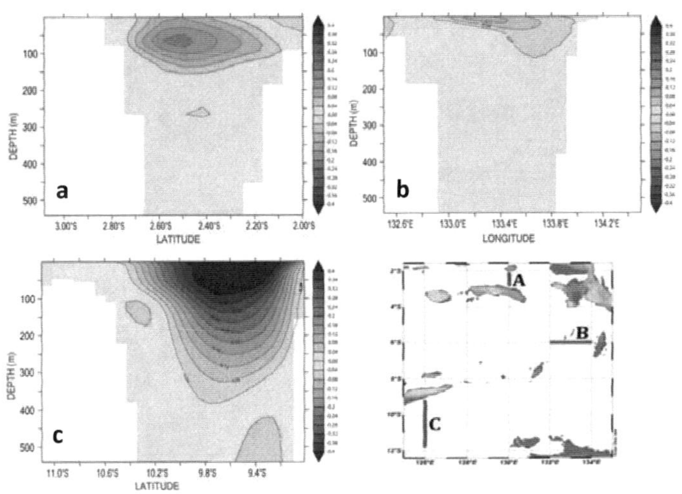

Figure 2. Mean structure of ITF along outer Banda Arcs; (a) mean structure in Ceram Sea at 130°E, (b) mean structure in Aru Basin at 6°S, (c) mean structure in Timor Passage at 126°E.

3.1.1. Annual cycle. The annual cycle of marine circulation in the study area varied on a seasonal basis. In the Northwest Season (December-February), the direction of current flow in the Ceram Sea is predominantly Westward with a current velocity of about 0.28 m/s at depth above 200 m (figure 3) while in the transition season I (March-May), there is a reversal of the current direction at a depth below 100 m to the East, so that in the Southeast (June-August) the direction of the dominant current flow to the East with the current velocity is stronger than that of in the Northwest Season which is about 0.40 m/s. In addition, water mass propagation is indicated in South Pacific flowing into the Banda Arcs through Ceram Sea by bringing in high-energy water mass [1].

The direction of current flow in the Aru Basin in the Northwest Season is dominant to the North at a depth above 200 m, with a current velocity by 0.08 m/s. This is expected due to the mass water input from the Indian Ocean while in the Southeast, the current flow is dominant to the South with a current velocity by 0.16 m/s (figure 4). In the Southeast Season, the mass of Northern Subtropical Lower Water is indicated in the Pacific through the Ceram Sea and Irian Jaya [20].

In the Timor Passage, the flow of currents tends to have the same pattern, with a current velocity of about 0.40 m/s continuously moving westward (figure 5). However, in the Southeast Season, there is a reversal of current flow with a current velocity of 0.80 m/s at a depth below 400 m in accordance with previous research [21]. The direction of maximum ITF flow occurs during the Southeast Season as a response of local winds occurring in the study area.

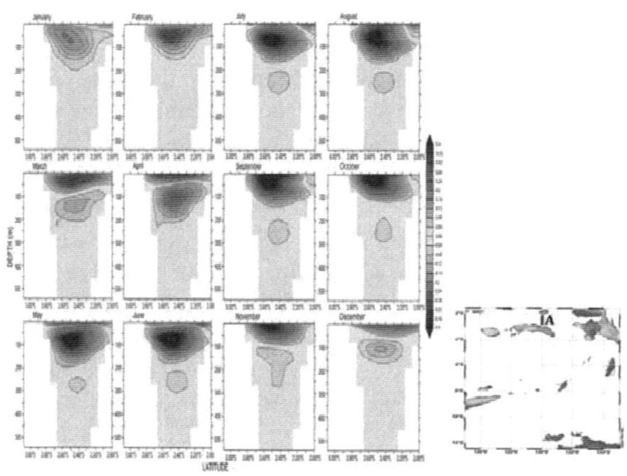

Figure 3. Annual cycle of zonal component (Ceram Sea at 130°E).

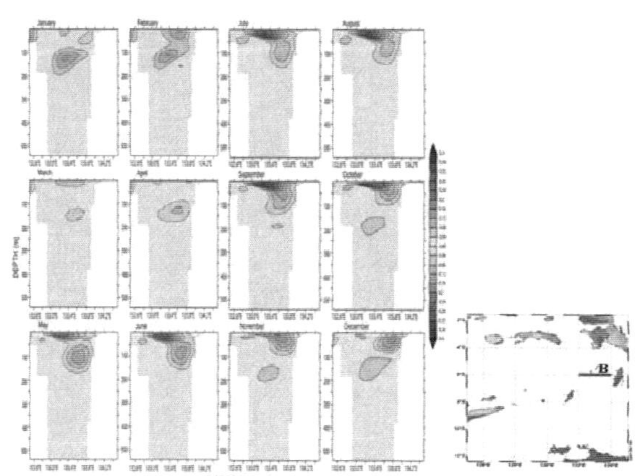

Figure 4. Annual cycle of meridional component (Aru Basin at 6°S).

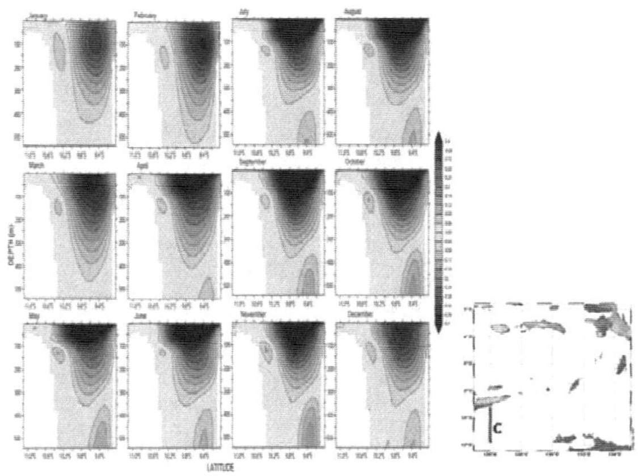

Figure 5. Annual cycle of zonal component (Timor Passage at 126°E).

3.2. Transport variability of ITF along outer Banda Arcs

The ITF volume transport of Ceram Sea (CS) from January 2008 to December 2014 is presented in figure 6a. The transport fluctuation ranges between -3 Sv to + 2 Sv, with a mean of 1.04 Sv and a standard deviation of 1.54 Sv. In general, the volume transport of CS indicates its direction to be dominant to the West. Alindo transport dominant to the West occurs during the Southeast Season period, while in the North West Seasons the transport of ITF changes to the East (figure 6b).

In the Aru Basin (AB) area, the volume transport of ITF shows the range of fluctuations between -2 Sv to +1.5 Sv, with a mean of about -0.61 Sv and a standard deviation of about 0.79 Sv (figure 7a). In general, the maximum transport of ITF is more dominant to the South, with fluctuations in transport on an annual time-scale (figure 7b). The volume transport of ITF Timor Passage (TP) shows the range of -13 Sv to -3 Sv, with transport rate of around -8.42 and standard deviation of 2.05 Sv (figure 8a). The transport movement of ITF ST is dominated by an intra-seasonal time-scale, transport to the South.

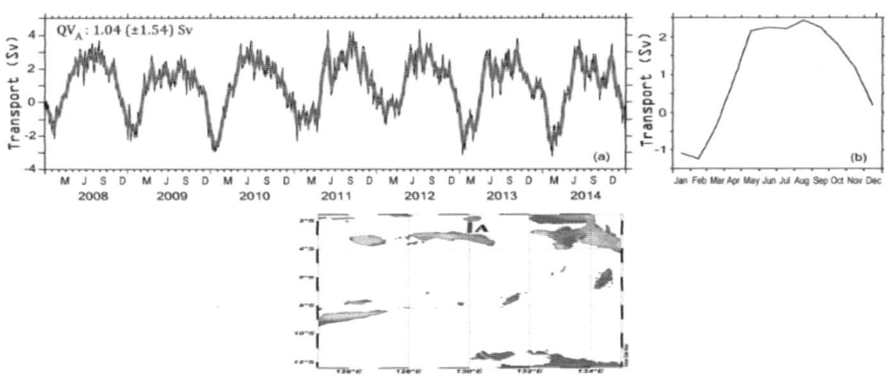

Figure 6. Time series of volume transports in the Ceram Sea with *band-pass filter* 31 days (a), and volume transport climatology (b).

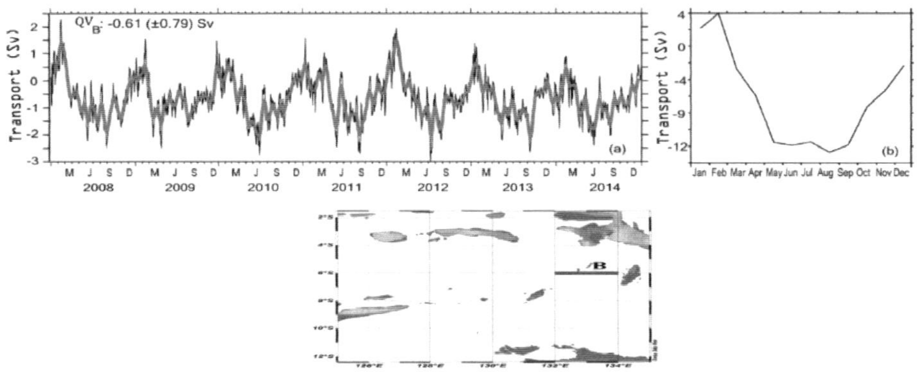

Figure 7. Time series of volume transports in the Aru Basin with *band-pass filter* 31 days (a), and volume transport climatology (b).

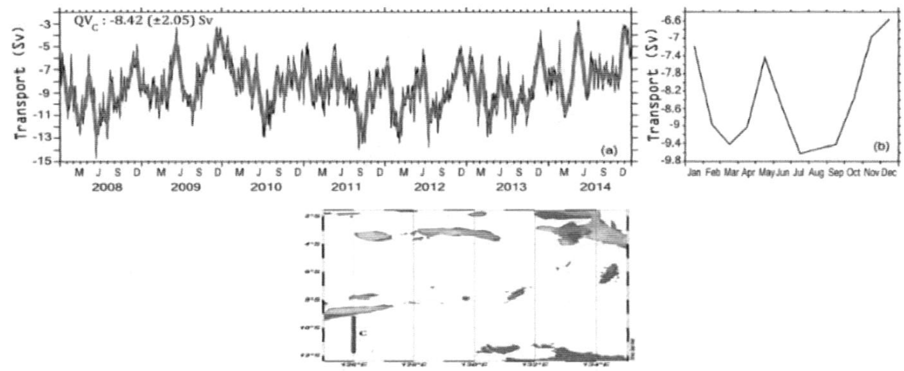

Figure 8. Time series of volume transports in the Timor Passage with *band-pass filter* 31 days (a), and volume transport climatology (b).

The results of time series analysis using continuous wavelet transform (CWT) method for transport of ITF CS, AB, and TP from 2008 to 2014 show 3 periodicities of variability with intra-seasonal time-scale (20-90 days), semi-annual time-scale (180 days) and annual time-scale variability (360 days) (figure 9). The results of the CWT analysis in the CS region show that transport variability is dominated by annual time-scale fluctuations occurring from January 2008 to December 2014, with relatively high CWT coefficients (figure 9a). The CS transport variability also occurs on the semi-annual, which occurred in 2009, 2010, 2013 and 2014 with relatively small coefficient CWT.

The transport variability of ITF AB shows fluctuations on a more dominant annual time-scale in the region, starting from 2008 to 2014, with high CWT coefficient values occurring in 2010 to 2013 (figure 9b). AB transport fluctuations on a semi-annual scale appear in the first 3 years of 2008-2010 with relatively small CWT coefficients. AB transport variability on inter-seasonal time-scales occurs in the transitional period 2 (September-November).

The transport variability of ITF TP is dominated by fluctuations at semi-annual time-scales, with fairly strong CWT coefficients (figure 10c) while fluctuations in the intra-seasonal time-scales occurred during the Southeast of the period of 2008 to 2014.

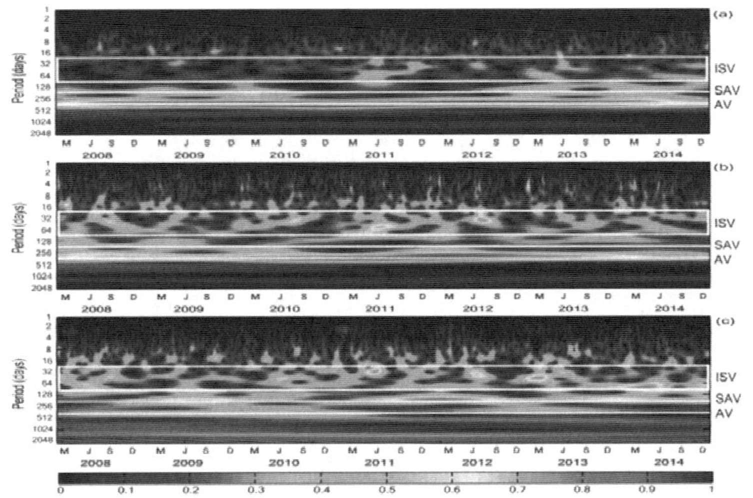

Figure 9. Results of Continous Wavelet Transform analysis (CWT). (a) Time series of Arlindo volume transport in the Ceram Sea, (b) time-series of Arlindo volume transport in Aru Basin, (c) time-series of Arlindo volume transports in the Timor Passage.

3.3. Power spectral density time-series of transport ITF along outer Banda Arcs

Spectrums of the energy densities for the zonal and meridional components in the three study areas are presented in figure 10. Visually, the signal propagations among the Ceram Sea, the Aru Basin and the Timor Passage can be seen from the peak of the spectrum of significant energy densities found in the periods of 20, 24, 26, 43, 171, 341 days, where the current time-series data are dominated by annual, semi-annual, and intra-seasonal fluctuations.

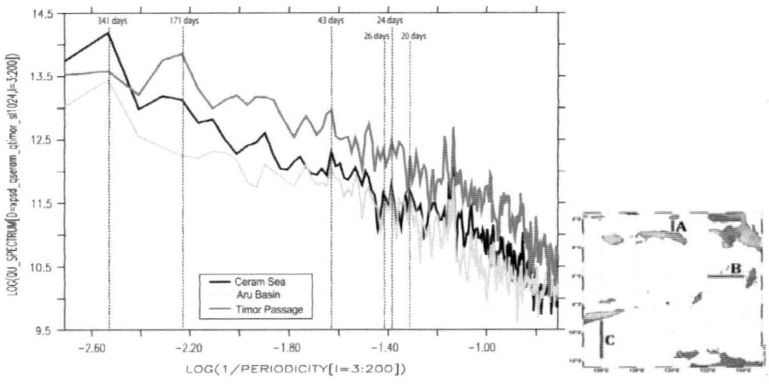

Figure 10. Power spectral density time-series of ITF along outer Banda Arcs.

3.4. Coherence and phase time-series transport variability ITF along outer Banda Arcs

3.4.1. Coherence and Phase Time-Series Transport Variability between Ceram Sea and Aru Basin. The analysis results of Cross-PSD time series of ITF transport between the Ceram Sea and the Aru Basin show that the significant coherence values occurred on annual, semi-annual, and intra-seasonal time-scales, in 341 days, 171 days, 43 days, 26 days , 24 days, and 20 days (figure 11). The highest coherence value (0.9689) occurs in a 341 day period with a 26.5 days different phase difference, which means the signal fluctuation of the Ceram Sea appears earlier than the Aru Basin signal. The second largest coherence value is 0.9127 with a phase difference of 2 days. Furthermore, for the coherence value from the periods of 171 days, 43 days, 24 days and 20 days respectively is 0.8168 with a phase difference of 13.8 days, 0.8373 (2.5 days).

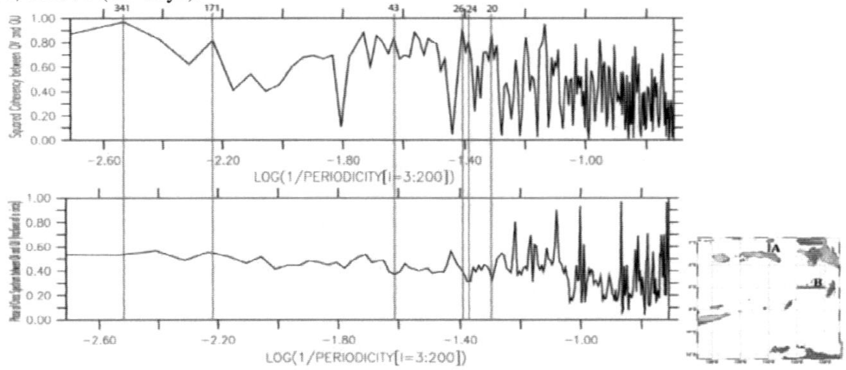

Figure 11. Coherence and phase time-series volume transport between the Ceram Sea and the Aru Basin.

Table 1. Coherence and time series phase of transport variability between the Ceram Sea and the Aru Basin .

Period (days)	Coherence	Phase (days)
341	0.9689	26.5
171	0.8168	13.8
43	0.8373	2.5
26	0.9127	2
24	0.8089	1.3
20	0.8624	1.3

3.4.2. Coherence and phase time-series transport variability between Aru Basin and Timor Passage. Figure 12 shows the results of Cross-PSD analysis of the ITF transport time series between Aru Basin and Timor Passage having significant coherence values occurring on an intra-seasonal, semi-annual and annual time-scales. The highest coherence value of 0.8682 occurs in a period of 43 days with a 5-days phase difference. This suggests that the Aru Basin propagation signal emerged earlier than that of the Timor Passage. In the periods of 341 days, 171 days, 26 days, 24 days and 20 days, coherence values of 0.8285 (2.6 days), 0.6214 (8.4 days), 0.7025 (3 days), 0.7962 (2.7 days) and 0.7931 (2.3 days) were obtained. This means that the Aru Basin signal flowing into the Timor Passage takes 2.3 days to 8.5 days (table 2).

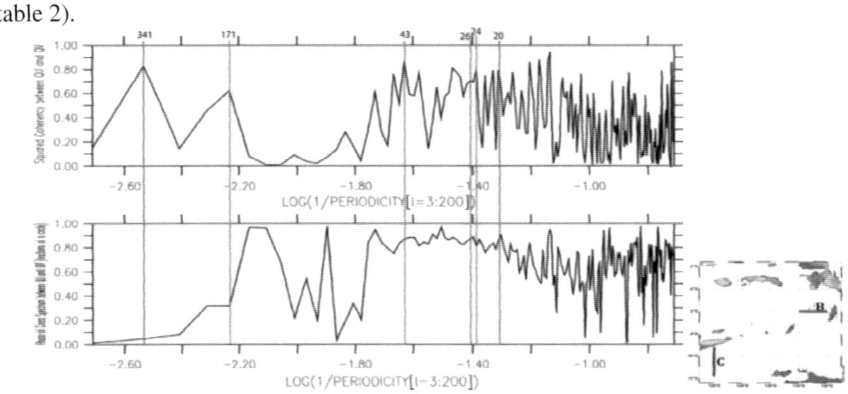

Figure 12. Coherence and phase time-series volume transport between the Aru Basin and Timor Passage.

Table 2. Coherence and time series phase of transport variability between the Aru Basin and Timor Passage.

Period (days)	Coherence	Phase (days)
341	0.8285	2.6
171	0.6214	8.4
43	0.8682	5
26	0.7025	3
24	0.7962	2.7
20	0.7931	2.3

3.4.3. Coherence and phase time-series transport variability between Ceram Sea and Timor Passage. The coherence analysis of zonal component shows the highest coherence value (0.8917) in the 20 days period (table 3), in which the propagation signal in Timor Passage preceded the Ceram Sea with a difference of 1 day phase. The coherence values on the intra-seasonal scale in the periods of 43 days, 26 days, and 24 days respectively are 0.6798 (1.8 days), 0.8537 (1.5 days), and 0.5969 (0.7 days). This indicates that the Timor Passage signal is earlier than that of the Ceram Sea.

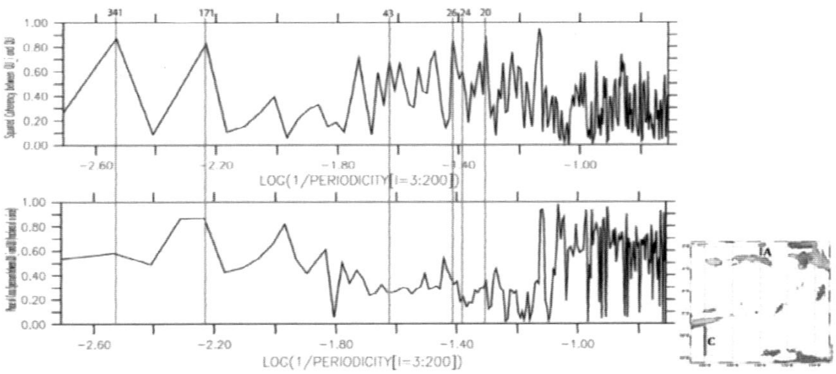

Figure 13. Coherence and time-series phase of volume transport between the Ceram Sea and the Timor Passage.

Table 3. Coherence and time series of transport variability between the Ceram Sea and the Timor Passage.

Period (days)	Coherence	Phase (days)
341	0.8696	28.5
171	0.8253	19.4
43	0.6798	1.8
26	0.8537	1.5
24	0.5969	0.7
20	0.8917	1

4. Conclusion

The annual cycle of ITF along the Banda Sea bow is characterized by a stronger flow rate in the Southeast Season compared to that of the North West Season. The volume of transport of ITF of the Ceram Sea ranges from -3 Sv to + 2 Sv, with an average of 1.04 Sv (± 1.54 Sv) to the East. In the Aru Basin, the average volume transport is about -0.61 Sv and the standard deviation is about 0.79 (± 0.79) Sv Southward and -8.42 (± 2.05) Sv Westward in the Timor Passage.

Transport variability in the Ceram Sea and the Aru Basin is dominated by fluctuations in semi-annual and annual time-scales, whereas in the Timor Passage, it is dominated by intra-seasonal and semi-annual time-scales. A significant coherence exists between the Ceram Sea and the Aru Basin occurring over a 1.3 day period up to 26.6 days, with the a high coherence value of 0.9127 between the Ceram Sea and the Aru Basin with a phase difference of about 2 days.

The highest coherence value between the Aru Basin and Timor Passage is 0.8682 occurring in the 43 day period with 5 days phase difference. In the period of 20 days, the highest coherence value between Ceram Sea and Timor Passage with phase difference about 1 day was obtained. This indicates that there is an indication of the propagation of the equatorial wave signal from the Ceram Sea to the Timor Passage along the outer Banda Arcs.

Acknowledgments
The Authors would like to thank INDESO Project, especially to Dr. Philippe Gaspar, Dr. Benoit Tranchant, Dr. Ariane Koch-Larrouy and Berny Subki MSc. The Authors also to thank BPOL Perancak Bali and CLS Toulouse France. Suggestions and comments from Reviewers for the refinement of this article are greatly appreciated.

Reference

[1] Gordon A L and Fine R A 1996 Pathways of water between the Pacific and Indian oceans in the Indonesian seas *Nature* **379** 146–149 doi:10.1038/379146a0

[2] Atmadipoera A, Molcard R, Madec G, Wijffels S, Sprintall J, Koch-Larrouy A, Jaya I and Supangat A 2009 Characteristics and variability of the Indonesian Throughflow water at the outflow straits *Deep-Sea Res I* **56** 1942-1954

[3] Godfrey J S 1996 The effect of the Indonesian Throughflow on ocean circulation and heat exchange with the atmosphere: a review *J Geophys Res.* **101**(C5) 12.217-37 doi: 10.1029/95JC03860

[4] England M H, Huang F 2005 Notes and Correspondence on the interannual variability of the Indonesian Throughflow and its linkage with ENSO *J. Climate* **18** 1435-1444

[5] Wyrtki K 1987 Indonesian Throughflow and the associated pressure gradient *J. Geophys Res.* **92**(C12):12.914-12.946 doi:10.1029/JC092iC12p12941

[6] Gordon A L, Susanto R D, Ffield A 1999 Throughflow within Makassar Strait. *J. Geophys. Res. Lett.* **26**(21) doi:10.1029/1999GL002340

[7] Shinoda T, Han W, Mezger E J and Hurlburt H E 2012 Seasonal variation of Indonesian Throughflow in Makassar Strait *J. Phys. Oceanogr.* **42**(7) 1099-23 doi:10.1175/JPO-D-11-0120.1

[8] Murray S P and Arief D 1988 Throughflow into the Indian Ocean through Lombok Strait, January 1985–January 1986 *Nature* **333** 444–447

[9] Koch-Larrouy A, Madec G, Iudicone D, Atmadipoera A and Molcard R 2008 Physical processes contributing to the water mass transformation of the Indonesian Throughflow *Ocean Dyn.* doi:10.1007/s10236-008-0154-5

[10] Hautala S L, Reid J L and Bray N 1996 The distribution and mixing of Pacific water masses in the Indonesian seas *J. Geophys. Res.* **101**(C5) 12.375-89 doi:10.1029/96JC00037

[11] Wijffels S and Meyers G 2003 An intersection of oceanic waveguides: variability in the Indonesian Throughflow Region. *J Phys Res.* **34** 1232-1253

[12] Tranchant B, Reffray G, Greiner E, Nugroho D, Koch-Larrouy A and Gaspar P 2015 Evaluation of an operational ocean modal configuration at 1/12° spatial resolution for the Indonesian seas - Part 1 *Oce. Phys. Geosci. Model Dev.* **9** 1037-1064

[13] Horhoruw S M, Atmadipoera A S, Nanlohy P and Nurjaya I W 2016 Anomaly of surface circulation and Ekman transport in Banda Sea during 'normal' and ENSO episode (2008-2011) *IOP Conference Series: Earth and Environmental Science* **54** 012041 doi:10.1088/1755-1315/54/1/1/012041

[14] Atmadipoera A S 2017 The Indonesian Throughflow in Banda Sea (IOP cenference proceeding)

[15] Bendat J S and Piersol A G 2010 *Random Data Analysis and Measurement Procedure* (New York: John Wiley and Sons Inc.)

[16] Torrence C and Compo G P 1998 *A Practical Guide to Wavelet Analysis. User Guide* (Colorado University)

[17] Emery W J and Thomson R E 2014 Data Analysis Methods in Physical Oceanography Waltham (US): Elsevier

[18] Qiu B, Mao M and Kashino Y 1999 Intraseasonal variability in the Indo-Pacific Throughflow and regions surrounding the Indonesian Sea *Am Meteorol Soc.* **29** 1599-1618

[19] Sprintall J, Gordon A L, Koch-Larrouy A, Lee T, Potemra J T, Pujiana K and Wijffels S E 2014 The Indonesian seas and their role in the coupled ocean-climate system *Nature Geoscience* **7** 487-492 doi:10.1038/NGEO2188.

[20] Zijlstra J J, Baars M A, Tijssen S B, Wetsteyn F J and Witte J I J 1990 Monsoonal Effects on the Hydrography of the Upper Waters (<300 m) of the Eastern Banda Sea and Northern Arafura Sea, with Special Reference to Vertical Transport Processes *Netherlands Journal of Sea Research* **25** (4) 431-447

[21] Sprintall J, Wijffels S E, Molcard R and Jaya I 2009 Direct estimation of the Indonesian Throughflow entering the Indian Ocean: 2004-2006 *J. Geophys Res.* **114** 1-9

IOP Conf. Series: Earth and Environmental Science **176** (2018) 012013 doi:10.1088/1755-1315/176/1/012013

The beach slopes and grain size distribution at Anoi Itam and Pasir Putih Beaches, Sabang City

S Purnawan[1]*, N A Mailala[1], S Karina[1], Muhammad[1], I Setiawan[1], Y Ilhamsyah[1,2]

[1]Department of Marine Sciences, Faculty of Marine and Fishery Sciences, Syiah Kuala University, Banda Aceh 23111, Indonesia
[2] Applied Climatology, Graduate School of Bogor Agriculture University, Bogor, Indonesia
*e-mail: syahrulpurnawan@unsyiah.ac.id

Abstract. Since ocean and land greatly influence on coastal processes, and continually affected by these dynamics, the coastal zone become the most dynamical region of the ocean. Coastal geomorphological changes principally in a more specific of beach processes and made of are found from the characteristics of the sediments over certain coastal waters. This research examines the relationship between beach slope and mean grain size. The objective of the research is to explain the mechanisms of sediment distribution associated with different beach slopes. The sampling locations were situated at Anoi Itam and Pasir Putih Beaches of Sabang City on October 2016 and April 2017. Coring method was used in this research. There were six collection sites with two repetitions of treatment in each location. The distribution of sediment based on grain size was measured using wet sieving procedure extracted by stratified sieving. The results showed that medium sand are dominant at Anoi Itam and Pasir Putih Beaches while beach slopes vary from nearly flat to gently sloping.

1. Introduction

Hydrological processes play a major role in sediment transport. Coastal dynamics such as: waves, tidal currents, stream flow affect sediment contents greatly [1]. Grain size is one of the fundamental properties in sedimentology that provides information about depositional environment and transport processes [2]. The relationship between sediment and coastal morphodynamics have been examined by some studies [3,4,5]. The characteristics of sediment are associated with variations of beach profiles. Thus, variations in grain size are described by the changes of beach profiles [6,7].

Seasonal variability influences the composition of sediment over certain waters [8], where transport processes and sediment deposition from one to another location cause the water bottom to change [9,10]. One of the constraints in this study, however, is the availability of long-term field data. Sabang city is administratively located at Weh Island, far northern Sumatera Island, Indonesia. Hydrodynamics condition of Weh Island waters is influenced by such adjacent waters as the Andaman Sea, the Malacca Strait, and the Indian Ocean [11]. Two sampling locations are chosen for the study, i.e., Anoi Itam and Pasir Putih Beaches, situated in the eastern and southern part of the Weh Island, respectively.

The beach sediment characterizes environmental influences and abundant mineral composition from surrounding areas [12,13]. Based on their given names, Anoi Itam is associated with dark sand with magnetite content while Pasir Putih is named after bright white sand with carbonate-silicate contents. Sediment characteristics in these two locations are different marked by coarser sediments found at Pasir Putih.

IOP Conf. Series: Earth and Environmental Science **176** (2018) 012013 doi:10.1088/1755-1315/176/1/012013

Field measurements of beach profiles, in this case the beach slope and sediment properties are important to conduct to achieve a better description of beach dynamics. This study focused on the relationship between seasonal beach slopes and grain size at Anoi Itam and Pasir Putih. The objective of the study is to obtain advanced understanding on seasonal variability of sediment dynamics over Weh Island.

2. Methods

Data were collected at Pasir Putih and Anoi Itam on October 2016 and April 2017. Each location consists of 6 stations, separated by a distance of 200 meters covering sediment samples and beach profiles (Figure 1). Sediment samples were collected by coring method to a depth of 5 centimeters beneath the ground surface. The extraction was done using wet sieving method with the following The Krumbein *phi* (φ) scale: -1, 0, 1, 2, 3, 4, and 5φ. The graphic mean was calculated based on the formula of Folk and Ward. Folk triangle was used to describe the types of sediment.

The characteristics of the beach is represented by beach slopes with the following expression:

$$\arctan \alpha\ (°) = \frac{y}{x} \tag{1}$$

$$\text{Slopes}\ \% = \frac{\alpha}{90°} \times 100\% \tag{2}$$

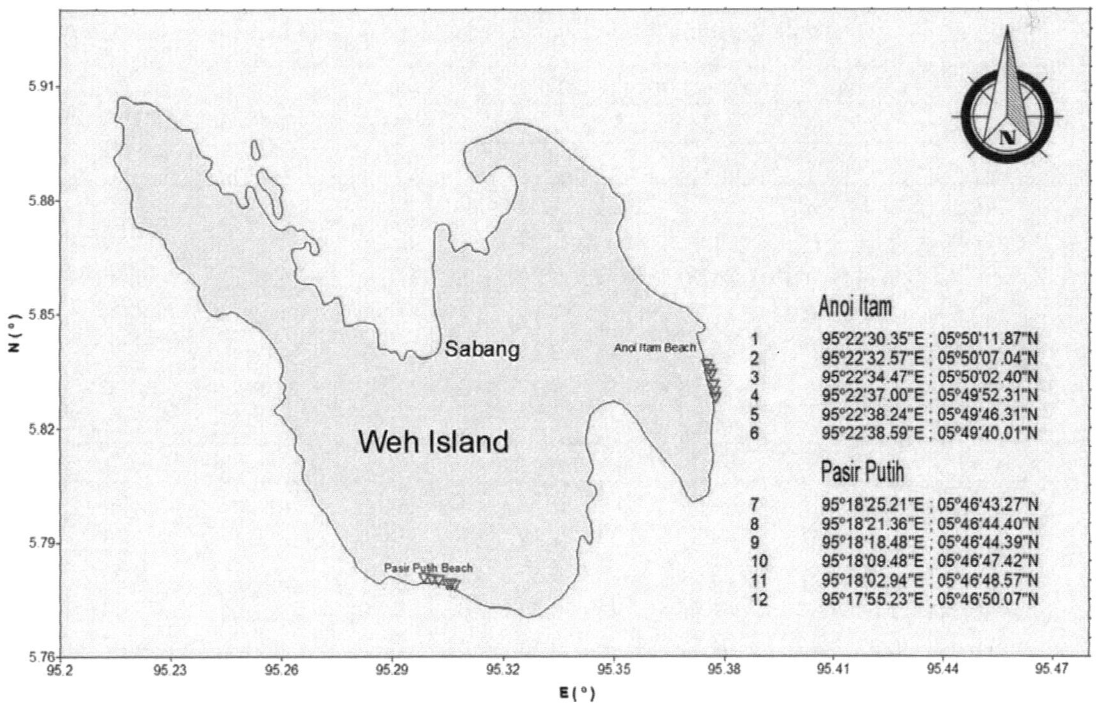

Figure 1. Sampling location in Pasir Putih Beach and Anoi Itam Beach, Weh Island, Aceh Province.

3. Result and discussion

3.1 Result

Seasonal variability during October (west season) and April (east season) lead sediment types to change at 10 of 12 stations, both Anoi Itam and Pasir Putih beaches. It is shown that sediment textures are coarser from October to April at Anoi Itam (Table 1). Anoi Itam's grain size range from 0.59φ to 1.82φ and has averaged of 1.37φ and 1.23φ during west and east season, respectively. Several stations at Pasir Putih showed coarser types of the sediment, but the average of grain size is less during east than west season (Table 2). Pasir Putih has mean grain size (Mz) ranged from -0.09φ to 2.09φ and has averaged 1.19φ and 1.35φ for west and east season, respectively. This is consistent with the presence of increasing find sand fraction during east season at Pasir Putih. An overview of the the grain-size shift that has occurred on the Pasir Putih Beach and Anoi Itam Beach can be seen in Figure 2.

Anoi Itam and Pasir Putih have typical gently slopes. It is found that the beach slopes at Anoi Itam and Pasir Putih change in response to season. In general, the slope is steeper during east than west season (Figure 3). The slope averages change from 2.63 to 3.40 during west and east season at Anoi Itam, respectively and 1.01 to 2.48 during west and east season at Pasir Putih, respectively.

The relationship betweeen slope variations and grain size are depicted from seasonal linear relationship in each sampling locations. The pattern showed the changes of the grain size (Mz) which are coarser as the slope increase (Figure 4 and Figure 5)

Table 1. Sediment textures and beach slope at Anoi Itam.

St	Sediment percentage of class (phi)							Mz	Sediment type	Beach Slope (%)
	-1	0	1	2	3	4	5			
	October									
1	9.30	13.51	42.11	29.23	3.34	2.53	0.00	0.60	gS	2.50
2	0.00	0.08	12.45	51.03	35.17	1.08	0.19	1.79	S	1.50
3	0.32	0.74	40.02	44.40	13.48	0.89	0.15	1.18	(g)S	4.60
4	0.39	1.60	0.22	82.17	15.46	0.15	0.00	1.58	(g)S	3.90
5	0.88	0.97	27.72	66.70	3.48	0.25	0.00	1.21	(g)S	2.20
6	3.71	0.41	5.73	52.18	37.48	0.47	0.02	1.82	(g)S	1.10
	average							1.37		2.63
	April									
1	0.07	0.42	18.78	44.09	31.79	4.39	0.45	1.72	(g)S	1.60
2	0.18	2.71	41.89	49.42	4.37	1.19	0.24	1.07	(g)S	3.00
3	5.89	12.63	56.45	22.61	1.80	0.52	0.10	0.59	gS	5.00
4	0.32	1.54	34.13	61.06	2.45	0.44	0.07	1.14	(g)S	3.60
5	0.00	0.27	30.31	64.50	4.75	0.13	0.04	1.22	S	4.80
6	5.78	0.19	8.01	62.89	22.91	0.19	0.03	1.64	gS	2.40
	average							1.23		3.40

References: Sediment types were determined based on "Folk triangle". Gravelly sand (gS); Sand (S); Slightly gravelly sand ((g)S).

Table 2. Sediment textures and beach slope at Pasir Putih.

St	Sediment percentage of class (phi)							Mz	Sediment type	Beach Slope (%)
	-1	0	1	2	3	4	5			
	October									
7	3.82	3.37	26.00	40.22	24.98	1.26	0.35	1.39	(g)S	0.80
8	0.47	1.02	13.39	55.20	26.77	2.86	0.28	1.73	(g)S	0.70
9	11.57	12.41	30.59	23.67	18.45	3.32	0.00	0.84	gS	1.30
10	22.86	3.66	19.16	45.15	8.07	0.71	0.39	0.55	gS	0.95
11	9.66	2.61	13.74	63.29	9.31	1.02	0.36	1.19	gS	1.00
12	1.55	3.02	20.48	54.13	19.24	1.21	0.37	1.42	(g)S	1.30
	average							**1.19**		**1.01**
	April									
7	5.46	4.63	25.81	31.33	31.23	1.54	0.00	1.41	gS	3.30
8	0.14	0.23	2.93	40.36	48.56	7.73	0.05	2.09	(g)S	1.20
9	1.69	0.06	1.33	40.16	53.36	3.16	0.24	2.07	(g)S	2.50
10	39.47	8.51	31.71	17.03	2.94	0.26	0.07	-0.09	sG	3.10
11	36.12	2.03	7.98	34.19	18.12	1.50	0.06	0.59	sG	2.40
12	11.29	0.59	1.52	24.98	58.60	2.98	0.05	2.03	gS	2.40
	average							**1.35**		**2.48**

References: Sediment types were determined based on "Folk triangle". Slightly gravelly sand ((g)S); Gravelly sand (gS); Gravelly sand (gS)

IOP Conf. Series: Earth and Environmental Science **176** (2018) 012013 doi:10.1088/1755-1315/176/1/012013

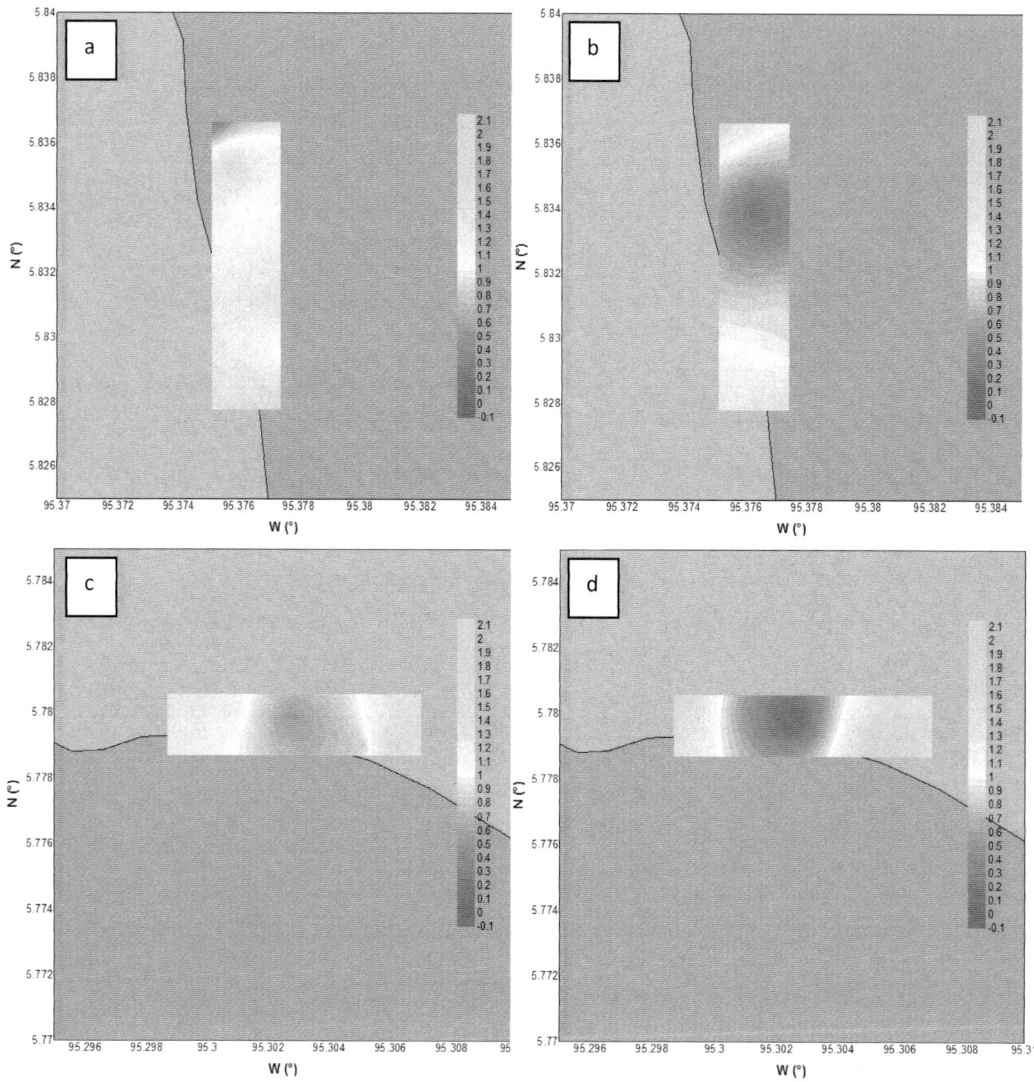

Figure 2. Changes in sediment grain size: a) Anoi Itam Beach in October 2016; b) Anoi Itam Beach in April 2017; c) Pasir Putih Beach in October 2016; d) Pasir Putih Beach in April 2017.

IOP Conf. Series: Earth and Environmental Science **176** (2018) 012013 doi:10.1088/1755-1315/176/1/012013

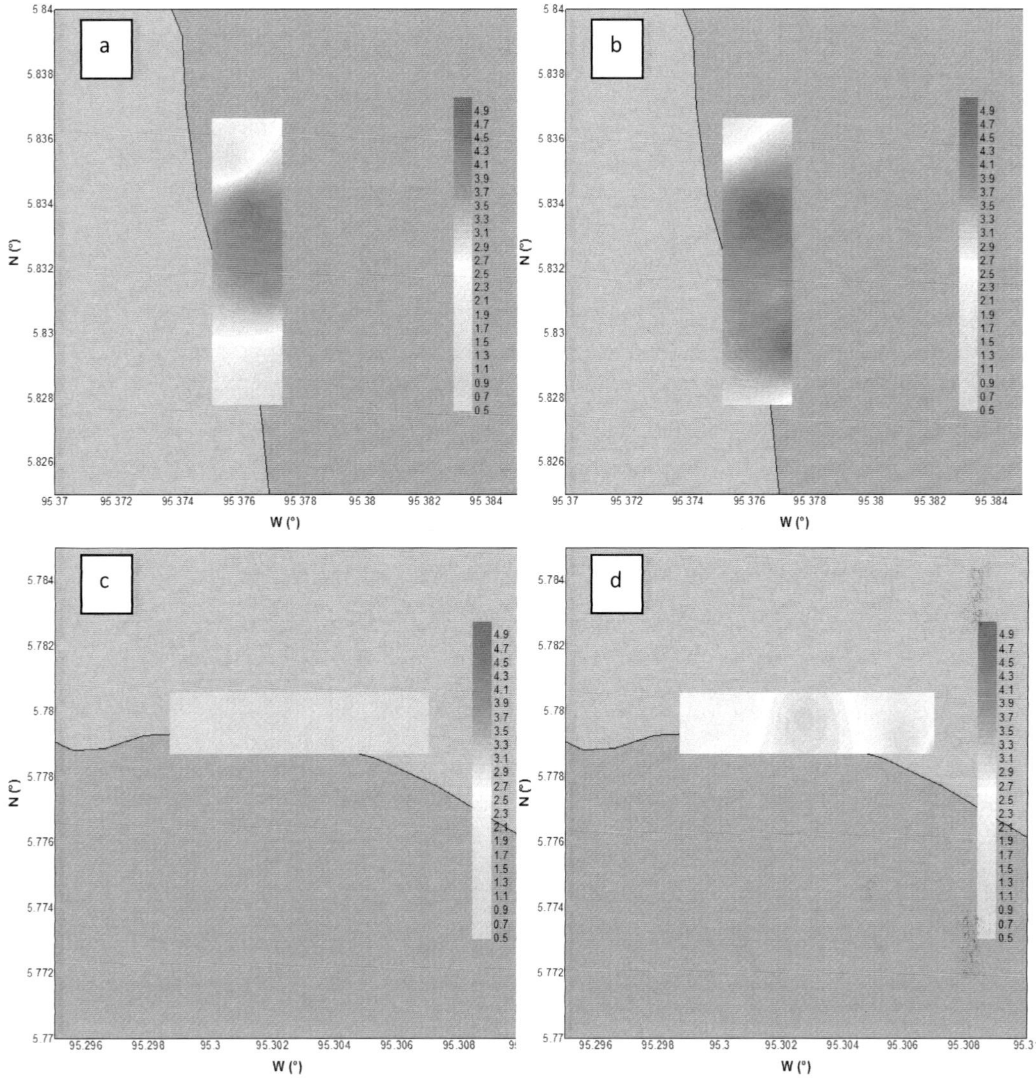

Figure 3. Changes in Beach slope: a) Anoi Itam Beach in October 2016; b) Anoi Itam Beach in April 2017; c) Pasir Putih Beach in October 2016; d) Pasir Putih Beach in April 2017.

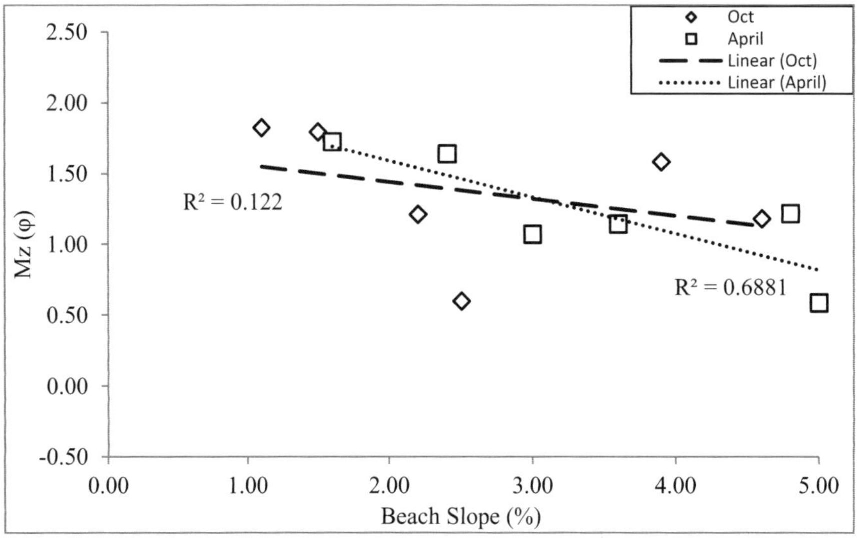

Figure 4. The relationship between beach slope and mean grain size (Mz) at Anoi Itam.

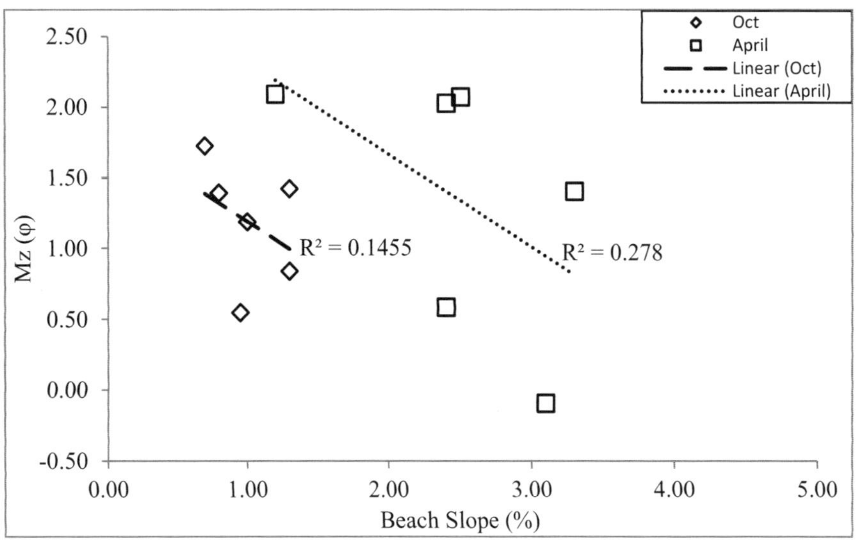

Figure 5. The relationship between beach slope and mean grain size (Mz) at Pasir Putih.

3.2 Discussion

The variations of grain size obtained at foreshore are associated with energy received from wave and tidal activity [1], where coarser grain size characterized by high intensity of energy exposure [14,15]. Seasonal variability affects to the changes of sediment size distribution. On October, sediment was highly distributed at medium sand fraction (phi=2), while in April the distribution of sediments was more evenly distributed on the other fractions. This condition leads the average value to change where coarser grain size is found on April instead of October.

Generally, changes of the beach slope to the steep profiles are associated with coastal erosion due to intensive energy in that season. It is shown that a steeper profile occurred on April as confirmed by the changes of sediment roughness, in particular at Anoi Itam where coarser than averages were found. Although the average grain sizes were finer on April at Pasir Putih, but the additional gravel fraction was found during east season which is consistent with earlier statement where the erosion occurred on April. Seasonal variability plays a major role on the changes of beach characteristics and sediment at two locations. Wind circulation at Weh Island is strongly influenced by monsoonal pattern, i.e., southwest and northeast monsoon during August and February, respectively. Based on NCEP long-term data, averages wind velocity of 4 m/s flow over Weh Island [11]. Meanwhile, the ocean current move to the west from the Andaman Sea and the Malacca Strait, during northeast or southwest monsoon as confirmed by advanced model study by Rizal *et al.* [11,16].

Based on wind and current profiles around Weh Island, it is concluded that strong current toward west occurred on April, driven by strong wind circulation from east as well as current from the Andaman Sea and the Malacca Strait. These conditions have an impact on sediment variations at Anoi Itam and Pasir Putih on April where beach slope increase and coarser grain size are found in many stations. High energy during east season is suspected to transport finer grain size leading the slope to become steeper [17]. On the other hand, steeper changes of the beach slope influence the beach hydraulic conductivity, swash mechanism, which also correlated to grain size [18].

The relationship between the slope and grain size show nearly similar pattern in every condition during east and west season at Anoi Itam and Pasir Putih. It is shown that the steeper the beach, the coarser the grain size. The R-value in each location is likely to be lower if October (west season) and April (east season) data are combined. Thus, seasonal changes give different pattern and these might be expected to be separated [6].

4. Conclusion

Seasonal variability affects on the changes of sediment roughness and beach slopes, specifically steeper profile is found during east season. Despite the correlation between beach slope and grain size in this study is not significant, yet a slight tendency that finer sediment established at a gentle beach slope.

References

[1] Curtiss G M, Osborne P D and Horner-Devine A R 2008 Seasonal patterns of coarse sediment transport on a mixed sand and gravel beach due to vessel wakes, wind waves, and tidal currents *Marine Geology* **259** 73–85

[2] Wilson A M, Huettel M and Klein S 2008 Grain size and depositional environment as predictors of permeability in coastal marine sands *Estuarine, Coastal and Shelf Scienc* **80** 193–199

[3] Marino-Tapia I J, Russel P E, O'Hare T J, Davidson M A and Huntley D A 2007 Cross-shore sediment transport on natural beaches and its relation to sandbar migration patterns: 1. Field observations and derivation of a transport parameterization *Journal of Geophysical Research* **112** C03001 doi:10.1029/2005JC002893

[4] Austin M J and Buscombe D 2008 Morphological change and sediment dynamics of the beach step on a macrotidal gravel beach *Marine Geology* **249** 167-183

[5] Reis A H and Gama C 2010 Sand size versus beachface slope- An explanation based on the Constructal Law *Geomorphology* **114** 276–283

[6] Sajeev R, Sankaranarayanan V N, Chandramohan P and Nampoodiripad K S N 1996 Seasonal changes of the sediment size distribution and stability along the beaches of Kerala, south west coast of India *Indian Journal of Marine Sciences* **25** 216-220

[7] Stauble D K 2005 A review of the role of grain size in beach nourishment projects National Conference on Beach Preservation Technology February 2 - 4, 2005. Destin, Florida

[8] Staub J R, Among H L and Gastaldo R A 2000 Seasonal sediment transport and deposition in the Rajang River delta, Sarawak, East Malaysia *Sedimentary Geology* **133** 249–264
[9] Charkin A N, Dudarev O V, Semiletov I P, Kruhmalev A V, Vonk J E, Sanchez-Garcia L, Karlsson E and Gustafsson O 2011 Seasonal and interannual variability of sedimentation and organic matter distribution in the Buor-Khaya Gulf: the primary recipient of input from Lena River and coastal erosion in the southeast Laptev Sea *Biogeosciences* **8** 2581-2594
[10] Mao Z, Pan D, Tang C L, Tao B, Chen J, Bai Y, Chen P, He X, Hao Z, Huang H and Zhu Q 2016 Dynamic sediment model based on satellite-measured concentration of the surface suspended matter in the East China Sea *Journal of Geophysical Research Oceans* **121** 2755–68 doi:10.1002/2015JC011466
[11] Rizal S, Damm P, Wahid M A, Sundermann J, Ilhamsyah Y, Iskandar T and Muhammad 2012 General circulation in the Malacca Strait and Andaman Sea: a numerical model study *American Journal of Environmental Science* **8**(5) 479-488
[12] Saniah, Purnawan S and Karina S 2014 Karakteristik dan kandungan mineral pasir pantai Lhok Mee, Beureunut dan Leungah, Kabupaten Aceh Besar *Depik* **3**(3) 263-270
[13] Purnawan S, Alamsyah T P F, Setiawan I, Rizwan 2016 Sediment distribution analysis in Balohan Bay, Sabang *Jurnal Ilmu dan Teknologi Kelautan Tropis* **8**(2) 531-538
[14] Purnawan S, Setiawan I, Haridhi H A, and Irham M 2018 Granulometric analysis at Lampulo Fishing Port (LFP) substrate, Banda Aceh, Indonesia *IOP Conf. Ser.: Earth Environ. Sci.* **106** 012070
[15] Purnawan S, Setiawan I and Muchlisin Z A 2015 Sediment grain-size distribution in the Lake Laut Tawar, Aceh Province, Indonesia *AACL Bioflux* **8**(3) 404-410
[16] Rizal S, Setiawan I, Iskandar T, Ilhamsyah Y, Wahid M A and Musman M 2010 Currents simulation in the Malacca Straits by Using three-dimensional numerical model *Sains Malaysiana* **39**(4) 519–524
[17] Buscombe D and Masselink G 2006 Concepts in gravel beach dynamics *Earth Science Reviews* **79**:33-52
[18] Kulkarni D C, Levoy F, Monfort O and Miles J R 2004 Morphological variations of a mixed sediment beachface (Teignmouth, UK) *Continental Shelf Research* **24** 1203–18

IOP Conf. Series: Earth and Environmental Science **176** (2018) 012014 doi:10.1088/1755-1315/176/1/012014

How do El Niño Southern Oscillation events impact on small pelagic fish catches in the west Java Sea

M Syamsuddin*, Sunarto, L Yuliadi

Department of Marine Sciences, Faculty of Fishery and Marine Science, University of Padjadjaran, Jl. Raya Bandung–Sumedang KM. 21 Jatinangor 45363 Indonesia
*e-mail: vegha16@gmail.com

Abstract. We analyse how oceanographic factors included sea surface temperature (SST), sea surface height anomaly (SSHA), and chlorophyll-*a* influence on small pelagic fish catches during El Niño Southern Oscillation (ENSO) event. The small pelagic fish catches and oceanographic factors derived from satellite imagery were analysed for the 5 years datasets (2010-2014). In this study, we demonstrate two species small pelagic fish consist of Eastern Little Tuna (*Euthynus affinis*) and Spanish Mackerel (*Scomberomorus commerson)* as representative of dominant catch in the west Java Sea. The results showed that Eastern Little Tuna and Spanish Mackerel catches had increment during El Niño than during La Niña, with an average catches of 839.6 t (Eastern Little Tuna) and 273,7 t (Spanish Mackerel) during El Niño event. The average catches during La Niña event were significantly decreased of 602.6 t (Eastern Little Tuna) and 210.3 t (Spanish Mackerel). Further analysis performs the habitat optimum related to higher small pelagic fish catches were correspond with oceanographic conditions of SST ranged from 28-29°C, SSHA from 0-8 cm and chlorophyll-*a* ranged from 0.3 - 0.5 mg.m^{-3}. This inferred that different oceanographic conditions during ENSO events might cause different of small pelagic fish catch rates in the study area.

1. Introduction

The ocean climate variability of ENSO causes warm event (El Niño) and cool event (La Niña) in the tropical Pacific, which have been known to affect the tuna fisheries in some part of the world [1]. The trade winds during El Niño weaken along the equator as atmospheric pressure rises in the western Pacific and falls in the eastern Pacific Ocean. In contrast, the trade winds during La Niña were stronger than normal and the ocean temperature were colder than normal in the tropical Pacific ocean. Impacts of La Niña on oceanographic conditions are opposite to those of El Niño events [2].

Previous studies showed that the oceanographic variability are affected directly by the variability of ENSO as well as the monsoon system [3]. The oceanographic conditions in Java Sea mostly affected by monsoon that driven ocean circulation in the surface layer. During mature phase of El Niño the relatively cold water mass of South China Sea through the Karimata Strait was maximum on inter-annual time scale [4]. The Java Sea's surface water seasonally travels according to monsoonal winds. Those make the regions very dynamic and conducive to the migration of small pelagic fish. Climate variability of ENSO has demonstrable impacts on the tuna catches and distribution [5]. The effects of climate variability on oceanographic conditions and pelagic fish catches in the eastern Indian Ocean have been reported widely [6,7,8], but less studied in the Java Sea.

The Eastern Little Tuna represents a dominant catch in the Java Sea, followed by the Spanish Mackerel. The Eastern Little Tuna is found in ocean temperatures ranging from 18 °C to 29 °C and tend to form multispecies schools by size. The Eastern Little Tuna is largely confined to continental shelves and islands of the Indian Ocean [9]. The Eastern Little Tuna is a highly migratory species and form schools with other species such Scombrid. There are seasonal spawning peaks varying according to regions: i.e. In Indonesia, from August to October; from the middle of the North West monsoon period to the beginning of the South East monsoon (January to July) off East Africa and in Philippine waters from March to May [10].

The behaviour and survival of Eastern Little Tuna and Spanish Mackerel are related to a range of oceanographic factors, necessitating the need to combine several oceanographic factors to investigate the effects of oceanographic factors on the Eastern Little Tuna catch in the west Java sea. As proxies of potential fishing grounds, sea surface height (SSH) measurement which could indicate oceanic features, as well as measurements of ocean temperature can be used to identify frontal area as a productive area [11]. A high productivity area can also be seen from thermal satellite [12]. Chlorophyll-*a* concentrations have been used as indicator for water mass boundaries that can influence small pelagic fish distribution in the study area.

On the other hand, there seems to be only a few studies of the oceanographic variability impacts on Eastern Little Tuna and Spanish Mackerel distributions in the Java Sea. The satellite data included SSHA, SST and chlorophyll-*a* are used to understand ENSO impacts on two species of small pelagic fish in west Java Sea. Therefore, understanding the effect of ENSO on Eastern Little Tuna and Spanish Mackerel will enhance our ability to minimize the risk of ocean climate impacts and to have sustainable management of small pelagic fish in the Java Sea.

2. Research area

The research area was in the west Java Sea in the Indonesian Sea region focusing on geographical coordinates of 3 °S to 7 °S and 108 °E to 110 °E (figure 1).

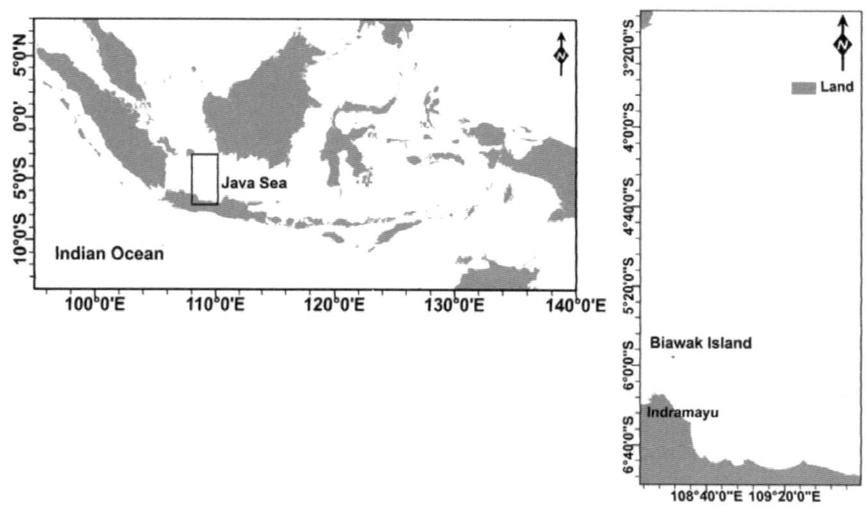

Figure 1. The study area in the west Java Sea, Indonesian Seas.

3. Data and methods

3.1. Data

In this study, we utilized fish catch and oceanographic parameter derived from remote sensing data. The fish catch data provided by the Fishing Port Indramayu, Indonesia. Data consist of the fishing position and total monthly catches. The remotely sensed data consist of SST, SSHA and chlorophyll-*a* concentrations. The SST and chlorophyll-*a* were derived from satellite imagery of Aqua Modis and downloaded from http://oceancolor.gsfc.nasa.gov. The SST and chlorophyll-*a* data had spatial resolutions of 4 km and temporal resolutions of monthly data. The SSHA data derived from the TOPEX and Poseidon ERS-1/2 altimeters which are downloaded at http://aviso.altimetry.fr. All oceanographic parameter resampled into 9 km spatial resolution and were analyzed for the 5 years datasets from January 2010-December 2014 during ENSO events. We used the climatic index of Niño 3.4 as indicator of ENSO based on SST anomaly in the Niño 3.4 region.

3.2 Methods

In this study, the continous wavelet transform (CWT) analysis was applied on the monthly of small pelagic catches, SST, SSHA, and chlorophyll-*a* data. The wavelet transform analysis (WT) was performed to obtain the temporal distributions the ocean variability during 2010-2014. Wavelet analysis performs a time frequency analysis of the dominant signal and decompose the signal in different time scales and identify how these signals vary in time. The wavelet transform has been successfully applied in climate studies [13].

High energy variances are represented in red and low values are in blue as shown in the colorbars that represent different wavelet amplitude. The black line (solid contour) shows the 95% confidence level and a thin black line shows the cone of influence (COI) [14].

4. Result and discussion

4.1. Climate feature of oceanographic conditions

We have chosen the month that represents the La Niña (October 2010) and El Niño (November 2014) events based on the Niño 3.4 index values during ENSO events in 2010 to 2014. The result on figure 2 shows snapshots of monthly mean of SST represented the El Niño and La Niña.

During La Niña in 2010, the SSHA value tends to be higher in the range of 1-25 cm. In contrast, during El Niño, the SSHA showing a lower value range from (-15) to (-10) cm (figure 2a). In general, the SST along the coast and offshore region shows a contrast conditions during El Niño and La Niña events. SST during La Niña in October 2010 showed warmer SST (30 °C to 31 °C) concentrating along the coast and offshore of the Java Sea due to changing monsoon from southeast to northwest, the cold SST changing to warm SST. SST during El Niño (November 2014) showed in contrasting condition with La Niña events where cold SST ranged from 27 °C to 29 °C concentrated in the offshore and 29 °C to 30 °C in the coastal area (figure 2b). Our result is similar with the research finding of Qu *et al.* [3] who reported there was maximum inflow of cold relatively fresh South China Sea water during the mature phase of El Niño event. It seems that the magnitude of El Niño affects the cold water distribution in the region.

Figure 2c shows snapshot of mean monthly of chlorophyll-*a* concentration during November 2010 (La Niña) and November 2014 (El Niño). The chlorophyll-*a* concentrations were much higher in the coastal area (1 to 1.5) mg.m^{-3} than in the offshore (0.1-0.9) mg.m^{-3}.

IOP Conf. Series: Earth and Environmental Science **176** (2018) 012014 doi:10.1088/1755-1315/176/1/012014

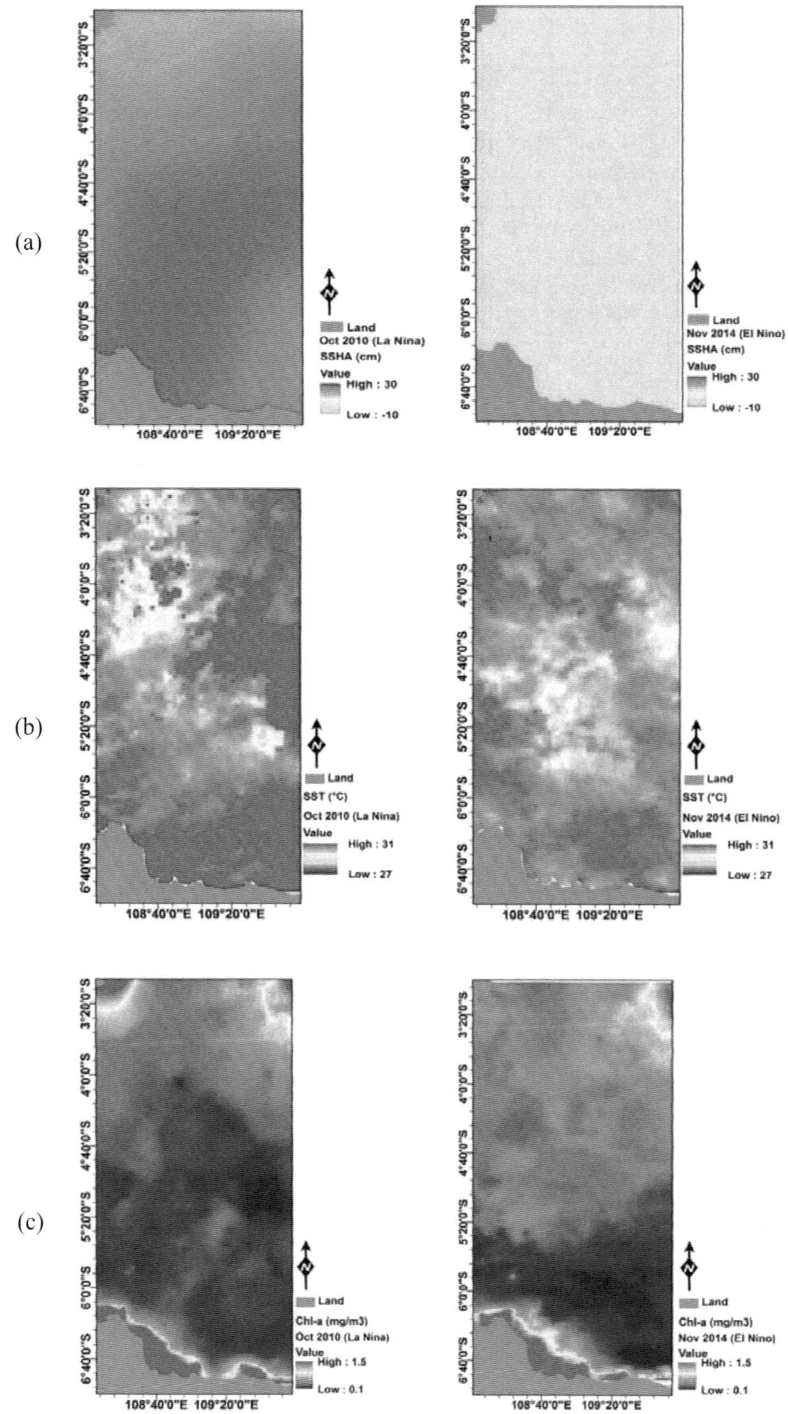

Figure 2 Mean Monthly of oceanographic factors of (a) SSHA, (b) SST and (c) Chlorophyll-*a* in October 2010 (La Niña) and November 2014 (El Niño).

4.2. Oceanographic factors associated with small pelagic fish cathes

This study has shown the effect of monsoon inducing oceanographic condition in the west Java Sea. Seasonal change features were dominant for all the selected oceanographic parameters of SSHA, SST and chlorophyll-*a*, and also small pelagic fish catches, respectively. The time series plots on figure 3 showed that the Eastern Little Tuna catch rates have the peak season in March and October (900 to 1000) ton that corresponded with the value of SSHA from 4 to 8 cm, SST from 29 °C to 30 °C following the decreasing chlorophyll-*a* concentrations in September to November (0.4 to 0.5 mg.m^{-3}). Spanish Mackerel catch peaks in December with relatively high SSHA conditions 12 cm, SST ranging from 29-30°C, and chlorophyll-*a* concentration from 0.4 - 0.5 mg.m$^{-3.}$

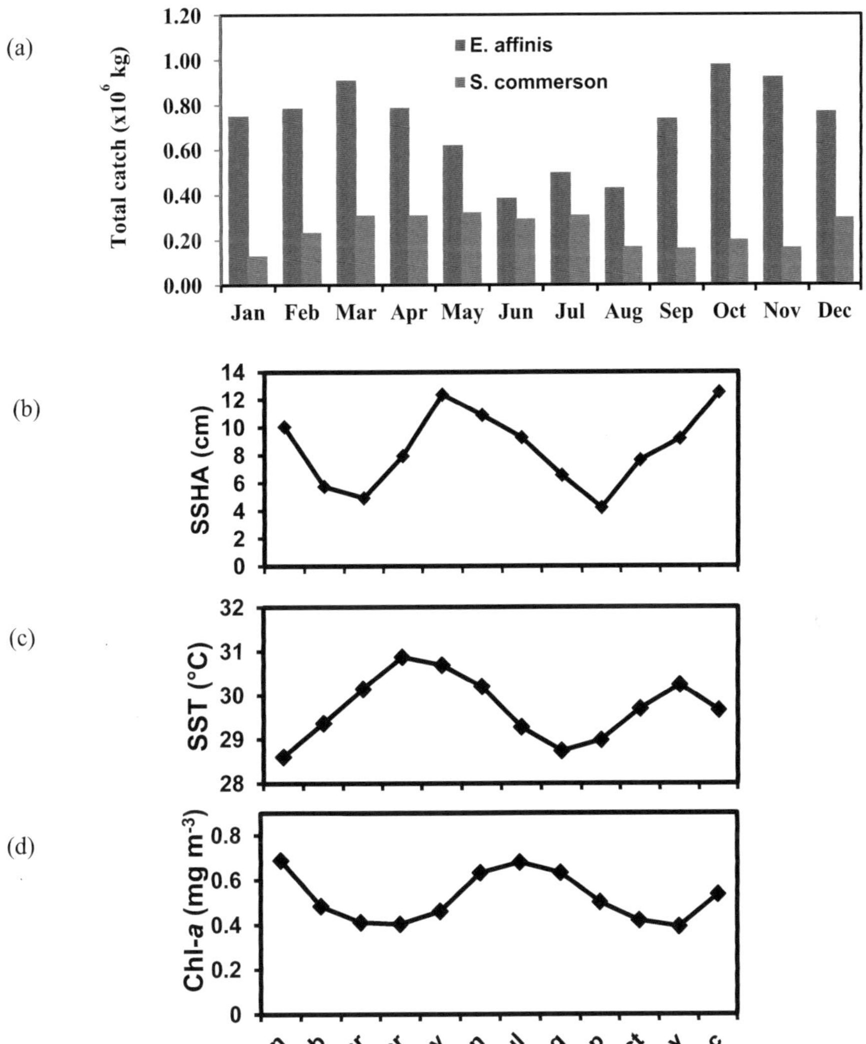

Figure 3 Mean monthly of (a) catch rate of Eastern Little Tuna and Spanish Mackerel, (b) SSHA, (c) SST, (d) Chlorophyll-*a* concentration during 2010-2014.

4.3. Variability of small pelagic fish and oceanographic conditions during ENSO events

The annual catch rates of small pelagic fish catches (Eastern Little Tuna and Spanish Mackerel) and Niño 3.4 index during 2010 to 2014 are shown in figure 4a. Catch rates varied temporally relatively significant over year-round. The oceanographic condition are change during ENSO events and associate with the Eastern Little Tuna and Spanish Mackerel catches, with higher catch rates during El Niño event. Catches are higher during El Niño (January to April 2010 and October to December 2014) compared to during La Niña events (July 2010 to April 2011 and September 2011 to March 2012).

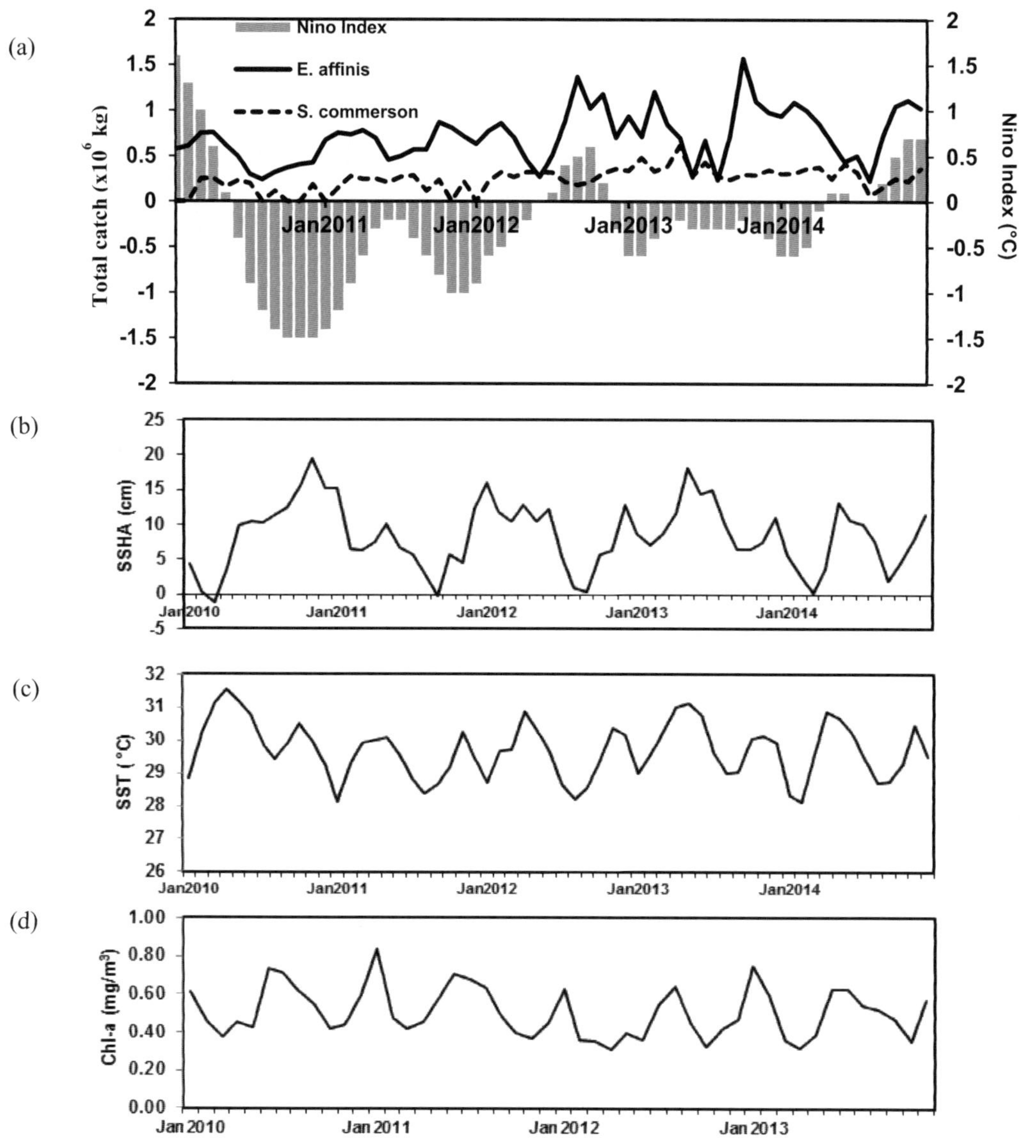

Figure 4 (a) Eastern Little Tuna (solid line) and Spanish Mackerel (dash line) catch rates and Niño 3.4 index during 2010 to 2014 (grey bars); (b) SSHA; (c) SST; (d) Chlorophyll-*a*.

IOP Conf. Series: Earth and Environmental Science **176** (2018) 012014 doi:10.1088/1755-1315/176/1/012014

Figures 4b-d showed the temporal variability of the oceanographic parameters of SSHA, SST and chlorophyll-*a*, respectively. During La Nina (July 2010-April 2011 and September 2011-March 2012), positive values of SSHA ranging from 2-20 cm following with increasing SST value ranging from 29.5-31°C, and decreasing Chl-*a* concentration from 0.4-0.7 mg · m^{-3}. The oceanographic variables in January 2010-April 2010 and October 2014-December 2014 (El Nino) showed in contrast condition with La Nina events where negative and lower SSHA -2 to 9 cm corresponding with colder SST of 29-30.5, and Chl-*a* concentration ranging from 0.4-0.6 mg · m^{-3}.

The conducive oceanographic conditions during El Niño resulted in increasing of fish catches, with an average catches of Eastern Little Tuna of 839.6 t and Spanish Mackerel of 273,7 t. The catch rates of both species were decreased with an average catches of 602.6 t (Eastern Little Tuna) and 210.3 t (Spanish Mackerel) during La Niña event. We also found that the peak season of total catch of small pelagic fish varied from year to year during ENSO events. The maximum peak of Eastern Little Tuna catch rates were found in April 2010 (757.5 t), October 2011 (869.4 t), September 2012 (1 368.5 t), October 2013 (1 570.9 t) and November 2014 (1 113.9 t).

The peak season of Eastern Little Tuna occurred in September-December and the maximum catches occurred in April 2010 and October to December 2014 during El Niño events. It seems that the oceanographic conditions was less favorable for small pelagic catches during La Niña than during El Niño. Our research concluded that ENSO events might cause different oceanographic conditions favourable to Eastern Little Tuna and Spanish Mackerel catches in the Java Sea.

4.4. Wavelet spectrum analysis

The dominant periods related to the amplitude and phase time series were computed by the wavelet analysis. The wavelet analysis confirmed dominant signals related to amplitude (variance) and phases of small pelagic fish catches, SSHA, SST and chlorophyll-*a* during 2010-2014 (figure 5).

The wavelet analysis revealed the seasonal (6 months) signal of Eastern Little Tuna, Spanish Mackerel, SST (period of 2010-2014), SSHA (middle of 2012-2014) and chlorophyll-*a* (April-October 2010, and May-July 2011). Strong interannual signal (16 months) was observed in Eastern Little Tuna catches, SSHA and Chl-a. The seasonal signal of Spanish Mackerel indicated that this species does not have direct correlation with ENSO because inter-annual signal could only be seen on Eastern Little Tuna catches.

(a)
(b)

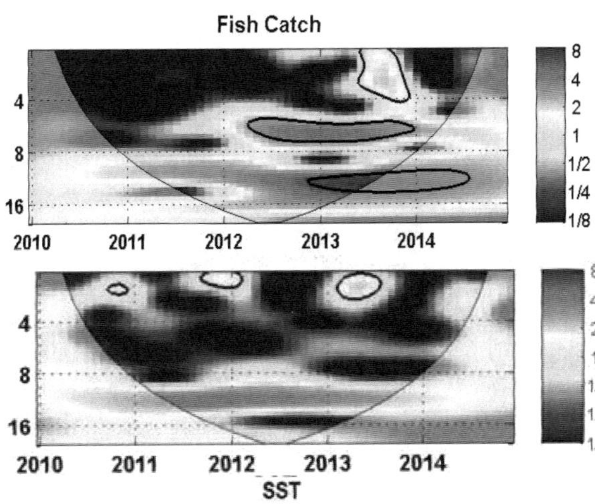

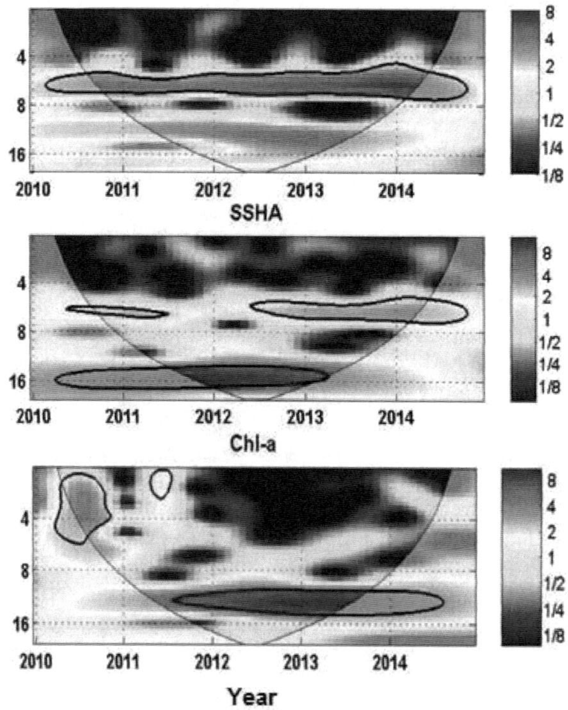

Figure 5. Wavelet power spectrum for small pelagic fish catches (a) Eastern Little Tuna; (b) Spanish Mackerel (c) SST; (d) SSHA; and (e) Chlorophyll-*a* based on raw data (monthly mean) during 2010-2014. The y-axis is represented a period frequency (monthly) and x-axis showed the time (year). Colorbars represented different wavelet amplitude, the black line (solid contour) showed the 95% confidence level and a thin black line showed the cone of influence.

5. Conclusion

The results of present study further highlight both seasonal and interannual (ENSO) patterns affecting the the small pelagic fish catches and oceanographic conditions in the study area. Higher catch rates of Eastern Little Tuna and Spanish Mackerel catches occurred during El Niño event. The wavelet analysis confirmed that Eastern Little Tuna, SSHA and Chl-*a* affected by the ENSO in inter-annual signal.

We recommend to have further investigations by using long-term time series datasets to predict fishing ground locations and an emerging need to improve our climate understanding and forecast skill to conserve small pelagic catches in the region.

Acknowledgements

The authors thank the anonymous reviewers. We also thank Padjadjaran University, Marine Affairs and Fisheries Indramayu, the GSFC NASA, PODAAC, NOAA Climate Prediction Center, ICMS 2017 Committee and also CSPACE Project Team.

References

[1] Lehodey P, SeNiña I, Sibert J, Bopp J L, Calmettes B, Hampton J and Murtugudde R 2010 Preliminary forecast of Pacific Bigeye tuna population trends under the A2 IPCC Scenario *Progress of Oceanography* **86** 302–315

[2] McPhaden M J 1999 Genesis and evolution of the 1997–1998 El Niño *Science* **283** 950–954

[3] Goelzer H, Levermann A and Rahmstorf S 2009 Two way coupling of an ENSO model to the global climate model CLIMBER-3α *Ocean Modelling* **29** 94-101

[4] Qu T, Du Y, Strachan J, Meyers G and Slingo J 2005 Sea surface temperature and its variability in the Indonesian Region *Oceanography* 18 50–123\
[5] Howell dan Kobayashi 2006 El Nino effects in the Palmyra Atoll region: oceanographic changes and bigeye tuna (*Thunnus obesus*) catch rate variability *Fisheries Oceanography* **15** 477-489
[6] Syamsuddin M, Saitoh S, Hirawake T, Syamsudin F and Zainuddin M 2016 Interannual variation of bigeye tuna (*Thunnus obesus*) hotspots in the eastern Indian Ocean off Java *International Journal of Remote Sensing*
[7] Sartimbul A, Rohadi E, Yona D, Yuli E H, Sambah A B and Arleston J 2016 Change in species composition and its implication on climate variation in Bali Strait: case study in 2006 and 2010 In the 3rd International Conference on Fisheries and Aquaculture, Negombo, Sri Lanka 1-7
[8] Lumban-Gaol J, Leben R R, Vignudelli S, Mahapatra K, Okada Y, Nababan B, Mei-Ling M, Amri K, Arhatiin R E and Syahdan M 2015 Variability of satellite-derived sea surface height anomaly, and its relationship with Bigeye tuna (*Thunnus obesus*) catch in the Eastern Indian Ocean *European Journal of Remote Sensing* **48**(1) 465-477
[9] Pepperell J 2010 *Fishes of The Open Ocean* (Chicago: The University of Chicago Press) p 75-80
[10] FAO 2017 Species Fact Sheet Eastern Little Tuna Food and Agriculture Organization of the United Nations *FAO Fisheries and Aquaculture Department*
[11] Zainuddin M, Saitoh S and Saitoh K 2004 Detection of potential fishing ground for albacore tuna using synoptic measurements of ocean color and thermal remote sensing in the northwestern North Pacific *Geophysical Research Letter*
[12] Saitoh S, Chassot E, Dwivedi R, Fonteneau A, Kiyofuji H, Kumari B, Kuno M, Matsumura S, Platt T, Raman M, Sathyendranath S, Solanki H and Takahashi F 2009 Remote sensing applications to fish harvesting. *In* Remote sensing in fisheries and aquaculture. *Reports of the International Ocean-Colour Coordinating Group (IOCCG)*, No 8 (M H Forget, V Stuart, and T Platt, eds.) IOCCG, Dartmouth Canada p 57–76
[13] Menard F, Marsac F, Bellier E and Cazelles B 2007 Climatic oscillations and tuna catch rates in the Indian Ocean: a wavelet approach to time series analysis *Fisheries Oceanography* **16**(1) 95-104
[14] Torrence C and Compo G P 1998 A practical guide to wavelet analysis *Bulletin of the American Meteorological Society* **79** 61-78

IOP Conf. Series: Earth and Environmental Science **176** (2018) 012015 doi:10.1088/1755-1315/176/1/012015

Chemical characteristics of the surface sediment in the Java Sea

Owanda Alam Pugung[1], Tri Prartono[1], Rina Zuraida[2*]

[1] Department of Marine Science and Technology, Faculty of Fisheries and Marine Science, Bogor Agricultural University, Bogor, West Java, Indonesia
[2] Marine Geological Institute, Ministry of Energy and Mineral Resources, Bandung, West Java, Indonesia
*e-mail: rina@mgi.esm.go.id

Abstract. Sediments have been used as an indicator of environmental pollution because of its role as a final sink for most of materials. This study aims to describe the chemical characteristics of 13 metal contents in surface sediments and to predict the source of metal using enrichment factor (EF) approach. During the Java-Makassar-Flores (JMF) Triangle experiment in August 2015, about 11 surface sediment samples were collected using Multicore Octopus. Here we analyzed only three sediment samples at sampling station close to the coast of Java (MC-09), at mid-Java Sea (MC-11) and close to the coast of Kalimantan (MC-13). The result show that surface sediments consist of olive loose sand to silty clay containing organic materials, foraminifera shells, fragments of litic and quartz with dominant color of olive. The result of total metal analysis yields *Al* as the most dominant element in all stations. The enrichment factor (EF) revealed that the surface sediments of Java Sea enriched in *As*, *Zn* and *Co* at MC-09, wtih *Zn* and *Ni* MC-11 and with *Cu*, *As*, *Zn*, *Ni* and *Cr* at MC-13. All surface sediments have indicated medium contamination of *Zn*.

1. Introduction

Marine sediment is an accumulation of particulate, organic or inorganic materials that settles in the bottom of water body [1]. Marine sediment serves as a sink for material deposited from the seawater and produces varying characteristics that reflect their sources and sedimentation processes. Deposition of marine sediment involves weathering, erosion and transportation from the land by wind, water and ice into the sea before they accumulate on the bottom of the sea. As a result, marine sediments show distinct physical, chemical and biological characteristics that differ from basin to basin. Physical characteristics of marine sediments include grain size that ranges from fine to coarse grain, while biological characteristics comprise of marine organisms and terrestrial organic matter that are accumulated in the sediments. Chemical characteristics of marine sediments include pH, potential of reduction-oxidation (redox) (Eh), organic matter (OM) content and metal content [2].

The Java Sea that composes 7% of Indonesian Seas is a shelf sea that is affected by monsoon winds and its dimension can be described as rectangles with an average depth of 50 m [3] and [4]. Java Sea is located between Kalimantan, Java and Sumatra and it is connected to the Natuna Sea in the northwest through Karimata Strait, to the Indian Ocean in the south through Sunda and Bali Straits, and is bordered by Flores Sea in the east [5]. Human activities in surrounding area of Java Sea would affect water quality of the sea which then would be recorded in marine sediments.

Marine sediments play important role in understanding sources of pollutants due to their function as sink for heavy metals. Previous studies show that marine sediments provide information on the source of heavy metals suspected as natural and non-natural pollutants [6, 7, 8, 9, 10], bioaccumulation [11] and benthic habitat quality monitoring [12].

To date, information on chemical characteristics of Java Sea surface sediments is limited to studies that focus on coastal sediments. One such studies is conducted by [13] in estuaries and coasts in the northern part of East Java province and the information on the chemical characteristics of the surface sediment in terms of metal content in these locations are still limited. Considering Java Sea is also an important fishing ground for surrounding population, including sediment feeder mollusks and crustaceans, it is imperative to understand the characteristics of heavy metals in Java Sea sediments. This study aims to identify chemical characteristics of metals within the Java Sea sediments using of enrichment factor (EF) approach.

2. Materials and methods

2.1. Materials

This research is a part of multi-disciplinary study titled "JMF Triangle Experiment" in Java Sea, Makassar Strait and Flores Sea using R.V. Geomarin III. Thanks to fruitfull collaboration between Marine Geological Institute (PPPGL) of the Ministry of Energy and Mineral Resources (ESDM), Department of Marine Science and Technology (ITK), Bogor Agricultural University (IPB) and Marine Fisheries Research Institute (BRPL) of the Ministry of Marine Affairs and Fisheries (KKP). The cruise involved field sampling campaign in Java Sea using multicore Octopus as well as gravity corer including the three sampling sites used in this study (figure 1): MC-09 (-6.565°S and 114.404°E, 32 m water depth) that represents mainland Java; MC-11 (-5.034°S and 112.849°E, 58.6 m water depth) that represents the middle part of Java Sea; and MC-13 (-4.241°S and 114.358°E, 34.4 m water depth) that represents Kalimantan. Those samples were acquired on 7th and 8th August 2015. The samples were obtained by cutting the top 10 cm of the sediments and covered with aluminum foil before stored in 4° C cold storage. Metal analysis was conducted from February to April 2016 at the Laboratory of Marine Chemistry, Natural Resources and Environment, Center of Isotope and Radiation Application, National Agency for Nuclear Power (BATAN), Jakarta.

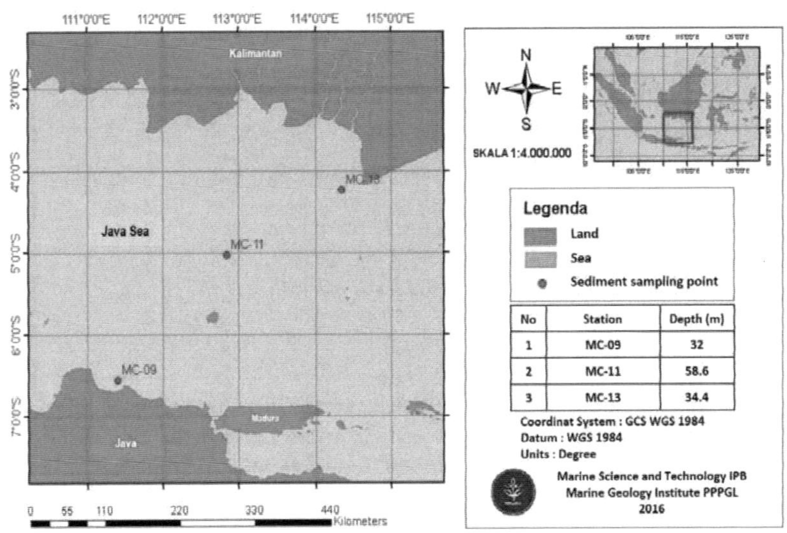

Figure 1. Map of sediment sampling at three locations of the research in the Java Sea.

2.2. Methods

Sample preparation started by subsampling the middle part of the sample to avoid aluminum contamination. The subsamples then were dried in the oven at 100°C for 24 hours followed by grounding the samples. The powder then weighed before being destructed following [14]. Sample destruction involved dissolving samples with strong acid and heated them in a closed system using microwave digestion system to accelerate the dissolution process. The destruction steps used approximately 0.2 grams of samples that were put into Teflon vial and dissolving reagents composed of 9 ml HNO3 65% and 3 ml HCl 37%. Samples were then put into Milestone Start D Microwave labstation at 150° C and 700 watts for 30 minutes. After cooling the samples into 40°C, samples were transferred into 50 ml vial to be centrifuged. An aliquot of 2 ml was than obtained from the resulting filtrate and diluted by *aquabidest* by a factor of 5. After homogenization, the samples were measured using the Inductively Couple Plasma-Optical Emission Spectrometry (ICP-OES) instrument following [15] procedures. The measurement result (C) is then converted into metal concentration in sediment sample (Cs) by taking into account the total volume of solution in ml (V), dilution factor (D, which is 5) and weight of dry sediment sample in grams (W) with the following equation:

$$\mathbf{Cs} = \frac{(\mathbf{C\ x\ V\ x\ D})}{\mathbf{W}} \quad (1)$$

Measurement results were calculated using spreadsheet program to obtain elemental ratio as well as enrichment factor (EF). Elemental ratios was calculated to understand the influence terrigenous material and clay minerals to metal content in the sediments. The ratio used in this study is ratio of each element to Al, Fe and Ti. The determination of these three elements aimed to illustrate the influence of some indicators, such as Al can be used as a clay minerals indicator [16] as well as to determine the environmental conditions of sediment deposition [17]; Fe can be used as an indicator of material origin of terrigenous and minerals that formed *in situ* [18] as well as describes the condition of redox in marine sediments [19]; and Ti as an indicator terrigenous material influx [18, 20].

The enrichment factor (EF) is an indicator to assess the influence of anthropogenic sources of contaminants on the surface of soil or sediment. Enrichment Factor can be calculated by normalizing the metal concentrations in sediment layers to reference metal concentration [21]. The reference is usually element, preferable metal, that is stable in the soil or sediment and essentially linked to sediment grain (fine particles) and its concentration is unaffected by anthropogenic influences [21]. The metals that are often used in previous studies include Al, Fe and Mn. In this study, we used Al because it is conservative, thus its concentration is stable or not affected by chemical and biological processes [16] [22]. The calculation of EF value for this study follows [21]:

$$\mathbf{EF} = \frac{\mathbf{X_{i(s)}/Al_{(s)}}}{\mathbf{X_{i(b)}/Al_{(b)}}} \quad (2)$$

Where:

X(s) : concentration of X_i metal in this study (μg/g),
Al(s) : concentration of Al in this study (μg/g),
X(b) : concentration of X_i metal in the background value (μg/g),
Al(b) : concentration of Al in the background value (μg/g)
i : Al, Fe, Mn, Ti, Ba, Cu, Pb, Ni, As, Zn, Cr, Co and Cd

Background value is natural elemental concentration in soil or sediment that depends on mineral characteristics and composition of soil and sediment source. The value that is usually applied as background value is based on elemental concentration in the Earth's crust or soil. This study used elemental concentration in marine mud as background value following Wedepohl (1960) in [16] as Java Sea sediments consists mostly of mud. Classification of elemental enrichment in sediments follow [21] as presented in table 1.

Table 1. Classification of EF values, after [21].

EF value	Classification
EF <2	Minimal enrichment
2 <EF <5	Medium enrichment
5 <EF <20	Moderate enrichment
20 <EF <40	High enrichment
EF> 40	Very high enrichment

3. Result and discussion

3.1. Characteristic of surface sediments

Java Sea surface sediments are dominated by soft, olive silty clay to loose sand compose of shell fragments, organic matter, lithic fragments and quartz. Sediment of MC-13 contains coarser grained material compares to the other two sites with sediment of MC-09 that was acquired off Rembang, Central Java, exhibits the finest-grained sediment among the three samples. Detail descriptions of each samples are given in table 2.

The northeastward trend of coarsening grain size is considered to be related to the distance of each sample locations to their source. Particle size of sediment is influenced among other by ocean current while settling of particular fragments with larger size would be faster than smaller ones. Generally, larger particles would be deposited near sediment source while finer particles would be deposited far from the source [23]. Thus, the source of MC-13 sediments is closer than the sources of MC-09 and MC-11 sediments.

Shell fragment contents in MC-09 and MC-13 that are closer to Java and Kalimantan are relatively higher than MC-11 that indicate higher productivity off Java and Kalimantan compare to the middle part of Java Sea. One of productivity indicator is organic matter content in sediments that might be originated from the land as well as the sea [24] [25]. Quartz and lithic fragments in marine surface sediment signify terrigenous input in sample location. Highest quartz content in MC-13 points to higher terrigenous input to the site compare to MC-09 and MC-11.

Table 2. Characteristics of Java Sea Surface Sediments.

Locations	Depth	Characteristics of Java Sea Surface Sediment	Microscopic photos
MC-09	32 m	Silt to silty clay, olive 5Y, soft. Sediment consistency is higher in lower part which is composed of fine sands containing angular shell fragments.	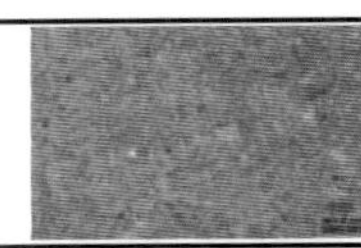
MC-11	58.6 m	Olive sandy silt, soft, composed of silt to fine sand grained shell fragments and organic matter. Sediment consistency is higher in the lower part, while grain size show coarsening upward and shell fragment content is decreasing upward.	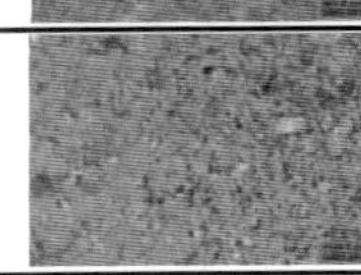
MC-13	34.4 m	Loose medium to fine sand, olive, soft with high water content, composed of shell fragments including foraminifera >50%, quartz 25%, and lithic fragments 25%. Grain size becomes finer downward to form silty sand to clayey sand, olive, soft and relatively drier than the upper part. The sand fraction is composed of coarse to very fine sand, that consist of shell fragments including foraminifera 50-60%, lithic fragments 25-30% and quartz 15-20%. Shell fragments decrease upward while quartz fragmets increase upward.	

Source: [26]
*Magnification: 40x - 1000x

3.2. Metal content in sediment

The result of ICP-OES measurement show that in general the area close to Java (MC-09) and Kalimantan (MC-13) have relatively higher metal contents compare to MC-11 that was acquired from the middle Java Sea (figure 2). The result yields Al as the element with the highest concentration in the measured Java Sea sediment samples with concentration ranges of 15184.73-28799.50 μg/g. High concentration of Al is considered to be related to the fact that Al is one of major rock forming elements which natural abundance in nature is 8.1% and ranked third after oxygen (47%) and silicon (28%) [27].

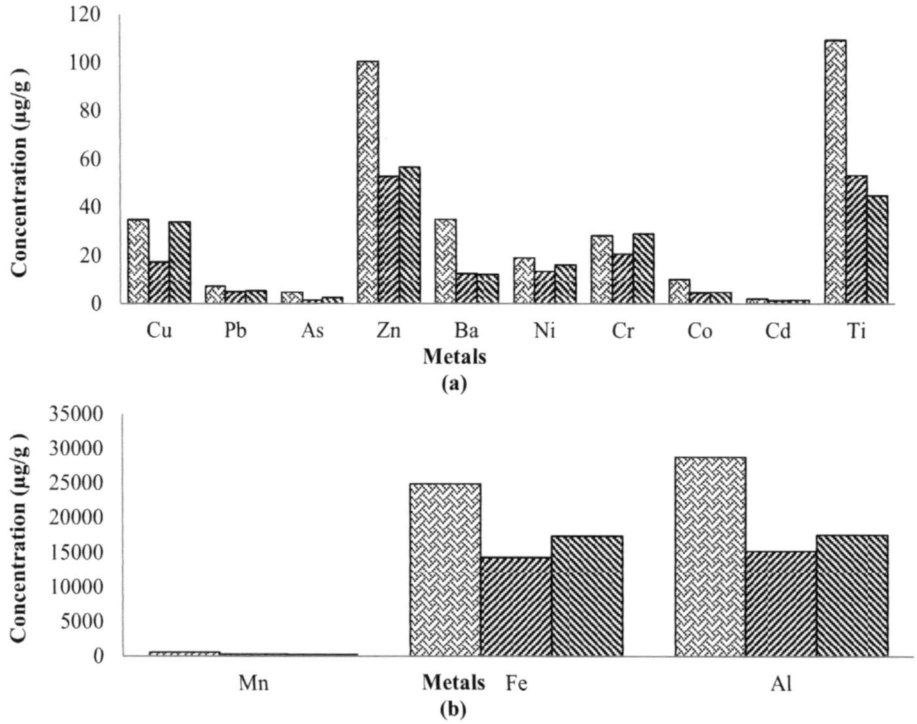

Figure 2. Concentrations of (a) Cu, Pb, As, Zn, Ba, Ni, Cr, Co, Cd, Ti and (b) Mn, Fe, Al on surface sediments in areas close to Java (: MC-09), in the middle of the Java Sea (: MC-11) and in areas close to Kalimantan (: MC-13).

A closer look on figure 2 show that metal contents in the site off Java (MC-09) are higher than in the site off Kalimantan (MC-13). The high level of metal concentration in the surface sediments in this study was thought to be due to several factors, including metallic characteristics and surface sediments observed, the distance of the metal source to the study site and oceanographic conditions. The physical characteristics of sediment in the form of fine grain particle (MC-09: silt to silty clay) have greater metal binding ability than coarser grain particle (MC-11: sandy silt and MC-13: sand) due to larger surface area of fine grain particle in metal binding [28, 29, 30]. It can be said that the high metal content in the MC-09 was suspected to be associated with fine grain particle (clay).

Surface circulation in Java Sea plays a major role in metal distribution. Java Sea is influenced by monsoon cycle that is generated by different air pressure cells over Asia and Australia [31] that initiates surface current flows in the Java Sea [32, 33]. During west season, monsoonal currents flow eastward from Karimata Strait to the Java Sea while weaker east monsoonal currents flow from the Flores Sea to the Java Sea [34]. Those currents influence the distribution of metal-bearing particles from their source to study locations and also control particle size that are deposited in the seafloor. Friction due to near-bottom current between suyrface sediment and water mass would release metals from surface sediment into the water column and resulted in lower metal content in surface sediment and higher metal content in water column.

The average current vectors obtained from INDESO model in 2015 show little seasonal variations (figure 3). During west season (February), the current flows eastward with stronger current flowing off Java and weaker currents flows in the middle of Java Sea and off Kalimantan (figure 3a). Transitional I season in May (figure 3b) is indicated by westward flowing current with stronger current in the middle part of Java and weaker currents off Java and Kalimantan (figure 3b). East season (August) is also characterised by westward flowing current although relatively weaker current is observed in the middle Java Sea compared to Transitional I season (figure 3c). Transitional II season shows similar eastward flowing currents as west season (figure 2d) with weaker currents flowing off Java.

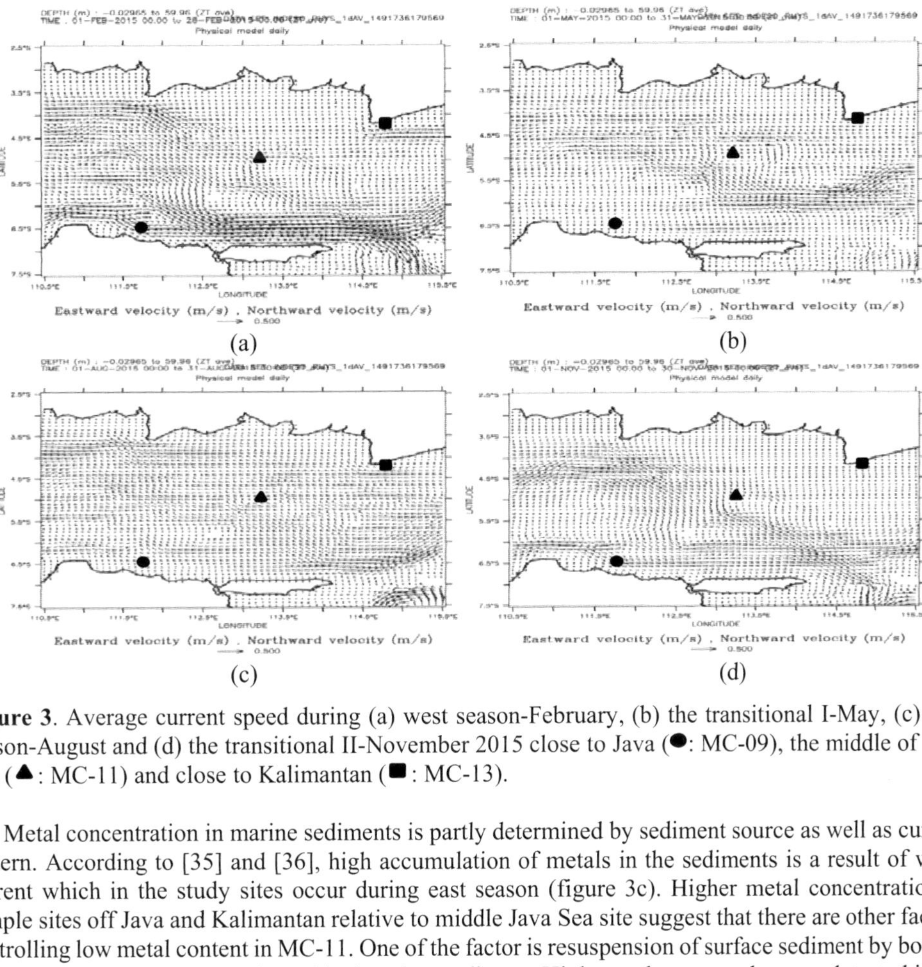

Figure 3. Average current speed during (a) west season-February, (b) the transitional I-May, (c) east season-August and (d) the transitional II-November 2015 close to Java (●: MC-09), the middle of Java Sea (▲: MC-11) and close to Kalimantan (■: MC-13).

Metal concentration in marine sediments is partly determined by sediment source as well as current pattern. According to [35] and [36], high accumulation of metals in the sediments is a result of weak current which in the study sites occur during east season (figure 3c). Higher metal concentration in sample sites off Java and Kalimantan relative to middle Java Sea site suggest that there are other factors controlling low metal content in MC-11. One of the factor is resuspension of surface sediment by bottom which reduces metal contents in residual surface sediment. High metal contents that was detected in off Java and Kalimantan sites are suspected to be related to anthropogenic activities that include both land-based and marine-based activities such as shipping, mining, agriculture, household and industrial wastes [37, 38, 10].

Even though human activities in Java and Kalimantan might release heavy metal into Java Sea, the scope of this study is too broad to understand the source of metal input. Nevertheless, it is worth mentioning that anthropogenic activities in study area include fishing, shipping, ports, oil fields and power plant on northern coast of Java and mining activity in Kalimantan. In order to understand the impact of anthropogenic activities to metal content in marine sediments, we have to determine natural metal content in marine sediments that are mostly influenced by terrigenous material derived from weathering of rock formations in Java, Kalimantan, Sumatra and Sulawesi. Metal contents measured from study area are compared to background value of nearshore mud presented by Wedepohl (1960) in [16] as well as previous studies from around the world to identify the level of contamination in study area (table 3).

Table 3. Comparison of heavy metals concentrations in surface sediments of the Java Sea with the background value and other studies (µg/g).

Metal	Java Sea	Background Value[1]	Other studies
Cu	33.73 - 34.65	56	9.6 - 35.6 [a]
Pb	4.97 - 7.18	22	22.6 - 43.7 [a]
As	2.48 - 4.70	5	3.4 - 13.6 [a]
Zn	52.71 - 100.50	92	106 [b]
Ba	12.15 - 34.90	-	42-596 [c]
Ni	13.43 - 19.06	35	3.5 - 35.8 [a]
Cr	20.64 - 29.02	60	11.6 - 76.2 [a]
Co	4.48 - 10.15	13	11.2 [b]
Cd	1.24 - 1.98	-	0.02 - 0.24 [a]
Ti	44.89 - 109.41	5000	1700 - 5600 [c]
Mn	246.53 - 538.61	850	168 [b]
Fe	14276.98 - 24893.56	65000	29000 [b]
Al	15184.73 - 28799.50	84000	26000-108000 [c]

[1] nearshore mud, Ba and Cd undetected Wedepohl (1960) in [16]
[a]Luanhe River Estuary [39]
[b]Vigo Ria [40]
[c]Messolonghi Lagoon Complex [41]

Table 3 shows that the contents of all metals, except Zn, measured from Java Sea sediments are below background value. Zn concentration in Java Sea sediments (52.71-100.5 µg/g) is slightly higher than the background value (92 µg/g) indicating that Java Sea sediments received excessive Zn from surrounding landmass. The source of Zn is believed to be from anthropogenic activities in islands surrounding Java Sea.

Comparison of metal contents in Java Sea sediments to previous studies reveals that Cu, Cd and Mn measured in this study exceed the concentrations of corresponding metals in Luanhe River estuary [39] and Vigo Ria [40]. That condition indicates that Java Sea sediments receive higher level of Cu, Cd and Mn compare to Luanhe River and Vigo Ria. The higher concentratons of Cu, Cd and Mn are considered to be related to the particle size of Java Sea sediments (table 2). Fine-grained sediments containing organic matter that predominate Java Sea sediments would affect absorption of Cu, Cd and Mn due to larger surface area of fine-grained particle to bind metal or organic matter [29, 30]. Another factor that might influence higher concentration of Cu, Cd and Mn in Java Sea sediments is anthropogenic activity from surrounding area.

3.3. Ratio of metal elements in sediments

Elemental ratios of each metal show slight variation between sample sites (table 4). Ratios of elements to Ti yield higher values than ratios to Al and Fe that suggest higher terrigenous influx in sample sites, particularly MC-13. Titanium in this study is used as terrigenous indicator following [18] and [20]. The site off Kalimantan (MC-13) has the highest ratios except Ba/Ti which highest value is observed from MC-09. In general, elemental Ti ratios among sample sites are observed to be higher in MC-13 followed by MC-11 and MC-09. This result indicate that terrigenous influx from Kalimantan is higher than from Java and the influx from Kalimantan might reach MC-11. Maximum value of Ba/Ti that is observed off Java suggests high productivity in MC-09 that might be related to terrigenous nutrient supply considering Ba/Ti can be used as an indicator of productivity [19].

Table 4. Comparison of elemental ratios to Al, Fe and Ti measured from Java Sea sediments to the background value of nearshore mud presented by Wedepohl (1960) in [16].

		Al		
Elements	MC-09	MC-11	MC-13	Background value[1]
Cu	0.0012	0.0011	0.0019	0.00066
Pb	0.00024	0.00032	0.00029	0.00026
As	163×10^{-6}	98×10^{-6}	141×10^{-6}	59.5×10^{-6}
Zn	0.00348	0.00347	0.00322	0.00109
Ba	0.0012	0.0008	0.0069	-
Ni	0.00066	0.00088	0.00091	0.00042
Cr	0.00098	0.00135	0.00165	0.00071
Co	0.00035	0.00029	0.00026	0.00015
Cd	69×10^{-6}	82×10^{-6}	85×10^{-6}	-
Ti	0.0037	0.0035	0.0025	0.0595
Mn	0.018702	0.018781	0.014049	0.010119
Fe	0.8643	0.9402	0.9895	0.77

		Fe		
Elements	MC-09	MC-11	MC-13	Background value[1]
Cu	0.0013	0.0012	0.0019	0.0008
Pb	0.00028	0.00034	0.0003	0.00033
As	189×10^{-6}	104×10^{-6}	143×10^{-6}	76.9×10^{-6}
Zn	0.004	0.0036	0.0032	0.0014
Ba	0.0014	0.0008	0.0007	-
Ni	0.00076	0.00094	0.00092	0.0005
Cr	0.0011	0.0014	0.0016	0.0009
Co	0.000408	0.000313	0.000271	0.000402
Cd	80×10^{-6}	87×10^{-6}	86×10^{-6}	-
Ti	0.0043	0.0037	0.0025	0.076
Mn	0.0216	0.0199	0.0141	0013
Al	1.1569	1.0635	1.0105	1.2923

Ti				
Elements	MC-09	MC-11	MC-13	Background value[1]
Cu	0.316	0.322	0.751	0.011
Pb	0.065	0.093	0.116	0.004
As	0.042	0.028	0.055	0.001
Zn	0.918	0.99	1,259	0.018
Ba	0.319	0233	0.27	-
Ni	0.174	0.252	0.359	0.007
Cr	0..257	0.387	0646	0.00092
Co	0.092	0:08	0.1	0.0026
Cd	0.018	0.02	0.03	-
Ti	1	1	1	1
Mn	4,923	5,359	5,491	0.17
Fe	227.53	268.32	386.8	13
Al	263.23	285.38	390.89	16.8

The ratios of elements to Al and Fe are distributed evenly in the three sites (table 3). This study uses Al as an indicator of clay minerals following [16] and Fe as an indicator of terrigenous material, *in situ* mineral [18] as well as reduction-oxidation condition of the sediments [19]. The ratios show that elemental to Al ratios are slightly higher tha elemental to Fe ratios in sites off Java and Kalimantan. These results indicate higher terrigenous influence associated with clay minerals in MC-09 and MC-13 in comparison to MC-11. Higher elemental to Fe ratios in MC-11 might suggest different redox condition in the middle part of Java Sea as oppose to the sites of Java and Kalimantan.

3.4. Enrichment factor

Enrichment factor (EF) is calculated to understand the source of metal contamination. Calculation was performed for all elements except Ba and Cd which background values are below detection limit. The calculated EF of Java Sea sediments yields the following EF range: off Java site (MC-09) 0.06 – 3.18; central part of Java Sea (MC-11) 0.05 – 3.16 and off Kalimantan (MC-13) 0.04 - 2.94 (figure 4).

Generally, Java Sea sediments can be classified into minimal enrichment (EF<2) and medium enrichment (2<EF<5). Minimal enrichment is observed for Cu, Pb, Ni, Cr, Ti, Mn and Fe in MC-09; Cu, Pb, As, Cr, Co, Ti, Mn and Fe in MC-11; and Pb, Co, Ti, Mn and Fe in MC-13. Medium enrichment is observed for As, Zn, Co in MC-09; Zn and Ni in MC-11; and Cu, As, Zn, Ni and Cr in MC-13.

EF values (EF<5) from the three sites suggest that most of metal contents in Java Sea sediments are derived from natural sources. This interpretation follows [42] that classified metals with EF<2 are predominantly derived from natural sources, while EF > 5 are considered to be contaminated by anthropogenic activities. Because most of metals measured from Java Sea sediments yield EF values between 2 and 5, there is a possibility that metal enrichments in study area are mixture of natural and anthropogenic sources.

The highest EF values (2.94-3.18) are observed in Zn in all sites ranging from, while the lowest EF values are observed in Ti (0.04-0.06) in all sites. The results point to a mixture of natural and anthropogenic sources of Zn, particularly in MC-09 site which is located off northern coast of Java that is home to various industries. This interpretation is supported by ratios of Zn to Ti (table 4) and comparison between Zn from off Java to background value (table 3). Ratios Zn/Ti of all samples suggest that Zn is derived from terrigenous material. The fact that Zn slightly exceed background value which is interpreted as the result of anthropogenic input to sample site.

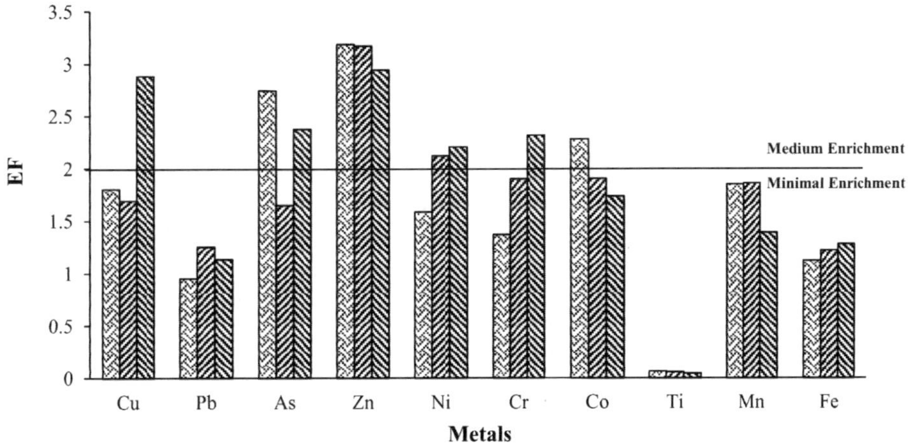

Figure 4. EF values of Cu, Pb, As, Zn, Ni, Cr, Co, Ti, Mn and Fe in Java Sea sediments from off Java (: MC-09), the central part of Java Sea (: MC-11) and off Kalimantan (: MC-13)

4. Conclusion

The study of chemical characteristics of Java Sea sediments was conducted in three sites: off Java Sea (MC-09), central part of Java Sea (MC-11) and off Kalimantan (MC-13). Java Sea sediments are characterized by olive silty clay to sand containing lithic fragment, quartz, organic matter and shell fragments. Thirteen metals (Al, Fe, Mn, Ti, Ba, Cu, Pb, Ni, As, Zn, Cr, Co and Cd) were measured using ICP-OES and the result is calculated to obtain ratios of elements to Al, Ti and Fe and enrichment factor (EF). The EF values show minimal enrichment on Cu, Pb, Ni, Cr, Ti, Mn and Fe in sample from off Java (MC-09); Cu, Pb, As, Cr, Co, Ti, Mn and Fe in sample from central part of Java Sea (MC-11); and Pb, Co, Ti, Mn and Fe in sample from off Kalimantan (MC-13). Medium enrichment is observed in As, Zn and Co in MC-09; Zn and Ni in MC-11; and Cu, As, Zn, Ni and Cr in MC-13. The highest EF values are observed in Zn from all sites that are interpreted as derived from natural sources with additional input from anthropogenic activities.

References

[1] Ilahude A G I 2013 *Oceanographic dictionary English-Indonesian* Jakarta (ID): PT. Trias Cemerlang
[2] Rahman A A 2008 *Vertical distribution of Pb, Cu, Zn, Cd and Ni concentration in sediment in Parin Island, Northern Thousand Island* [undergraduate thesis] Institut Pertanian Bogor
[3] Damayanti W Z 2016 *Mixed vertical in Matasiri waters, Java Sea* [undergraduate thesis] Institut Pertanian Bogor
[4] Koropitan A F and Ikeda M 2016 Influences of physical processes and anthropogenic influx on biogeochemical cycle in the Java Sea: numerical model experiment *in* The 2nd International Symposium on LAPAN-IPB Satellite for Food Security and Environmental Monitoring 2015, LISAT-FSEM 2015 *Proc Environ Sci.*, 33: 532 – 552
[5] Lubis E, Pane A B, Kurniawan Y, Chaussade J, Lamberts C and Potier P 2005 *Atlas of Fisheries and Fishery Ports in Java Island An Approach of Indonesian Capture Fisheries Geography Fishery Port and Maritime Transportation Study Program* Lembaga Penelitian Institut Pertanian Bogor
[6] Rabee A M, Al-Fatlawy Y F, Najim A and Nameer M 2011 Using pollution load index (PLI) and geoaccumulation index (I-geo) for the assessment of heavy metals pollution in Tigris River sediment in Baghdad region *J Al-Nahrain Univ.*, 14(4): 108-114

[7] Ahmad F 2013 *Distribution and Prediction of Heavy Metal Pollution Level (Pb, Cd, Cu, Zn, And Ni) in Sediments in Bangka Island Waters using Pollution Load Index and Geoakumulation Index JITKT* **5**(1): 170-181
[8] Edward 2014 *The content of heavy metals in sediments in the waters of Teluk Wawo Batu, Kendari, Southeast Sulawesi Depik* **3**(2): 157-165
[9] Najamuddin, Prartono T, Sanusi H S and Nurjaya I W 2016 Seasonal distribution and geochemical fractionation of heavy metals from surface sediment in a tropical estuary of Jeneberang River, Indonesia *Mar Poll Bull.***111**(1-2): 456-462
[10] Ramadhan M H 2016 *Geochemical partition of Pb and Cu in sediment of Sunda Strait of Banten* [undergraduate thesis] Institut Pertanian Bogor
[11] Davies O A, Allison M E and Uyi H S 2006 Bioaccumulation of heavy metals in water, sediment and periwinkle (*Tympanotonus fuscatus var radula*) from the Elechi Creek, Niger Delta *African J Biotech.* **5**(10): 965-973
[12] Naser H A 2013 Assessment and management of heavy metals pollution in the marine environment of the Arabian Gulf: a review *Mar Poll Bull.* 72:6-13
[13] Everaats J M 1989 Heavty metals (Cu, Zn, Cd, Pb) in sediment of the Java Sea, estuarine and areas of East Java and some deep-sea areas *Netherlands J Sea Res.* **23**(4): 403-413
[14] Milestone Application Note 2009 HPR-EN-33: Sea Sediment, SK-10 Application Book (Rev. 5) Italy (IT): Milestone SRL
[15] [USEPA] United State Environtmental Protection Agency. 2001. Method 200.7:Trace elements in water, solids, and biosolids by inductively coupled plasma-atomic emission spectrometry [internet]. [diunduh 2016 April 28]; EPA-821-R-01-010. Tersedia pada: https://www.epa.gov/sites/production/files/2015-06/documents/epa-200.7.pdf
[16] Chester R and Jickells T 2012 *Marine Geochemistry 3rd ed.* London (UK): Willey-Blackwell
[17] Schnetger B, Brumsack H J, Schale H and Hinrichs J 2000 Geochemical characteristics of deep-sea sediments from the Arabian Sea: a high-resolution study *Deep-Sea Res.* **2**(47): 2735-2768
[18] Calvert S E and Pederson T F 2007 Chaper Fourteen Elemental proxies for palaeoclimatic and palaeoceanographic variability in marine sediments: interpretation and apllication *Develop Mar Geol.* **1**: 567-643 doi: 10.1016/S1572-5480(07)01019-6
[19] Croudace I W, Rindby A and Rothwell R G 2006 ITRAX: description and evaluation of a new multi-function X-ray core scanner in: Rothwell RG (ed.) *New Techniques in Sediment Core Analysis* Geology Society London (GB): Special Publications **267**: 51-63
[20] Tjallingii R, Stattegger K, Wetzel A, and Van Phach P 2010 Infilling and flooding of the Mekong River incised valley during deglacial sea-level rise *Quarter Sci Rev.,* **29**: 1432-1444
[21] Barieri M 2016 The importance of enrichmment factor (EF) and geoaccumulation index (I_{geo}) to evaluate the soil contamination *J Geol Geophys.* **5**:237
[22] Herut B and Sandler A 2006 Normalization methods for pollutans in marine sediments: review and recommendations for the Mediteranian *Israel Ocean Limnol Res.* p 1-23
[23] Wibisono M S 2005 *Introduction to Marine Science* Jakarta PT Gramedia Widiasarana Indonesia
[24] Yun P S, Ariffin J, Siang H Y and Tahir N M 2015 Influence of monsoon on the distribution of organic carbon in inner continental shelf core sediments, South China Sea, Malaysia. *Sains Malay.* **44**(7): 941-945
[25] Permanawati Y 2016 *Reconstruction of the marine environment of Java-Makassar-Flores Sea Flores Sea* [thesis] Institut Pertanian Bogor
[26] Marine Geology Institute 2015 *Cruise Report of Oceanic and Sediment JMF Triangle Seas Experiment* Center of Marine Geology Research and Development
[27] Hefferan K and O'Brien J 2010 *Earth Materials* London (UK): Willey-Blackwell
[28] Sanusi H S, Prartono T and Supriyono E 2006 *Marine Chemistry: The Physical Process of Chemistry and its Interaction with the Environment* Department of Marine Science and Technology Faculty of Fisheries and Marine Science Institut Pertanian Bogor

[29] Afriansyah A 2009 *The concentration of cadmium (Cd) and copper (Cu) in water, seston, shells and fractionation in sediments in the waters of Delta Berau, East Kalimantan* [undergradute thesis] Institut Pertanian Bogor

[30] Maskulah L 2013 *The relationship between concentrations of heavy metals Pb, Cd, Cu, Zn with organic matter and particle size in sediments in West Flood Canal Estuary, Semarang. Bul Osean Mar.* **2**: 55-62

[31] Mulyana E 2002 *The relationship between ENSO and the variation of rainfall in Indonesia JSTMC.* **3**(1): 1-4

[32] Yulianto I 2001 *Variability of surface currents in the Java Sea generated by wind (Wind Driven Currents)* [undergradute thesis] Institut Pertanian Bogor

[33] Manoppo A K S 2003 *Study of a-chlorophyll concentration in the waters of the Java Sea, Flores Sea, Makassar Strait and Indian Ocean from Fengyun and SeaWifs satellite imagery: a comparison* [undergradute thesis] Institut Pertanian Bogor

[34] Widyastuti R, Handoko E Y and Suntoyo 2010 *The modeling of surface ocean current patterns in Indonesian waters uses Jason-1 altimetry satellite data. Teknik Kelautan* ITS*:* Surabaya

[35] Sagala S L, Bramawanto R, Kuswardani A R T D, Pranowo W S 2014 *Distribution of heavy metals in Natuna waters JITKT,* **6**(2): 297-310

[36] Ma'rifah A, Siswanto A D and Romadhon A 2016 *Characteristics and effect of current on heavy metal accumulation of lead (Pb) on sediment in Kalianget waters of Sumenep regency In Proceedings of National Marine Seminar*, Universitas Trunojoyo Madura [2016 Juni 27] 82-88

[37] Rochyatun E, Kaisupy M T and Rozak A 2006 *Distribution of heavy metals in water and sediments in the waters of the Cisadane River estuary Makaira Sains* **10**(1): 35-40

[38] Permanawati Y, Zuraida R and Ibrahim A 2013 *The content of heavy metals (Cu, Pb, Zn, Cd and Cr) in water and sediments in the waters of Jakarta Bay J Geol Kel.* **11**(1): 9-16

[39] Liu J, Yin P, Chen B, Song H and Li M 2016 Distribution and contamination assessment of heavy metals in surface sediments of the Luanhe River Estuary, northwest of the Bohai Sea *Mar. Poll Bull*

[40] Segarra M J B, Prego R, Wilson M J, Bacon J and Santos-Echeandia J 2008 Metal speciation in surface sediments of the Vigo Ria (NW Iberian Peninsula) *Sci. Mar.* **72**(1): 119-126

[41] Karageorgis AP, Sioulas A, Krasakopoulou E, Anagnostou CL, Hatiris GA, Kyriakidou H, Vasilopoulos K. 2012. Geochemistry of surface sediments and heavy metal contamination assessment: Messolonghi lagoon omplex, Greece. *Environ Earth Sci.,* 65: 1619-1629

[42] Werorilangi S, Tahir A, Noor A and Samawi M F 2013 The pollution status and potential bioavailability of metals in the coastal waters of Makasar *Semnaskan Proceedings*: 1-18

IOP Conf. Series: Earth and Environmental Science **176** (2018) 012016 doi:10.1088/1755-1315/176/1/012016

Study of current circulation in the Northern Waters of Aceh

I Setiawan[1,2*], S Rizal[1,2], Y Haditiar[1,2], Y Ilhamsyah[1,3], S Purnawan[1], M Irham[1,4] S M Yuni[5]

[1]Department of Marine Sciences, Faculty of Marine and Fisheries, Syiah Kuala University, Banda Aceh, 23111, Indonesia
[2]Graduate School of Mathematics and Applied Sciences, Syiah Kuala University, Banda Aceh, 23111, Indonesia
[3]Applied Climatology, Graduate School of Bogor Agriculture University, Bogor, Indonesia
[4]Center for Marine and Fishery Studies, Syiah Kuala University, Banda Aceh, 23111, Indonesia
[5]Department of Mathematics, Faculty of Mathematics and Natural Sciences, Syiah Kuala University, Banda Aceh, 23111, Indonesia
*e-mail: ichsansetiawan@unsyiah.ac.id

Abstract. The northern waters of Aceh are directly adjacent to the Malacca Strait, the Andaman Sea, and the Indian Ocean. These waters are an important route in the activities of shipping, fishing and others. The study of current circulation was studied by using the 2D hydrodynamic models. The purpose of this study is to determine the current patterns during the Northeast (February 2016) and southwest (August 2016) monsoons. To generate the models, we use the wind force obtained from European Center for Medium-Range Weather Forecast (ECMWF) 2016. The results show that the current circulation generally follows the bathymetry pattern and monsoonal direction.

1. Introduction

The Northern Waters of Aceh is geographically located at 5.13° – 5.97° North and 94.97° – 95.77° East (figure 1). These waters border Sumatra Island, Indian Ocean, the Gulf of Bengal, the Andaman Sea, and the Strait of Malacca. Earlier studies on current circulation have been done in the Strait of Malacca, the Andaman Sea and the Indian Ocean that divides the regions into two seasonal circulation, i.e., northeast and southwest monsoons [1-3]. The activity in northern Aceh and surrounding waters is quite active, which include shipping, mining, mariculture, capture fisheries, tourism and climate outdoor activities [4-8].

Based on these activities, the study of current dynamic is essential to study further on wave, sediment transport and coastal morphology changes, and pollutants distribution [9-13]. Thus the investigation of current dynamic of northern of Aceh waters aims to examine the current circulation process during the northeast monsoon on February 2016 and southwest monsoon on August 2016 by employing numerical model simulation.

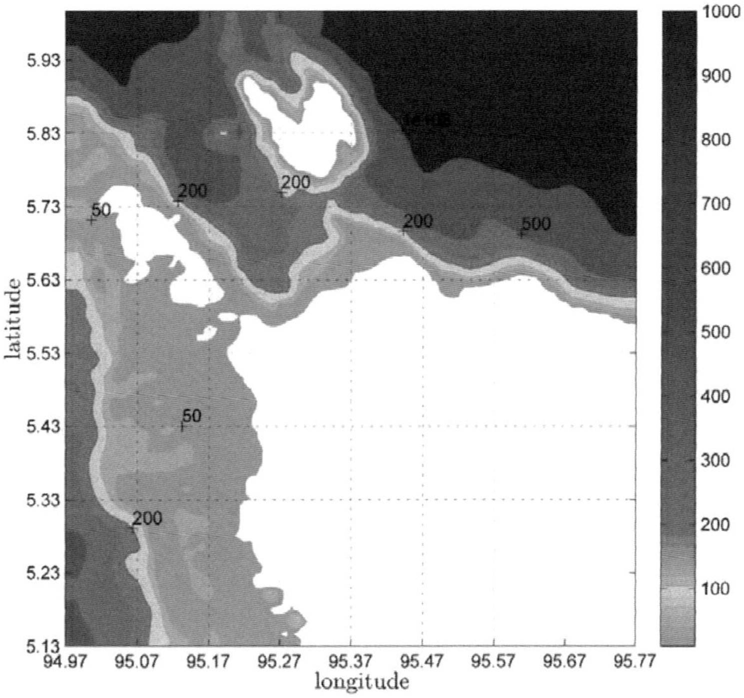

Figure 1. The bathymetry of the Aceh Waters (in meters). It is obtained from the Shuttle Radar Topography Mission (SRTM).

2. The methods of the research

2.1. The data

The bathymetric data used for the model was obtained from Shuttle Radar Topographic Mission (SRTM). The data is a digital elevation model (DEM) supplied globally free of charge by the National Aeronautics and Space Administration (NASA) and the United States Geological Survey (USGS). The resolution of the data considerably high that reaches 1 arc second (~ 30 meters) to region of America and 3 arc-second (~ 90 meters) to the rest. However, the data still has a gap to high resolution, especially in a region with detail profile. Therefore, some of the SRTM data is based on 30 arcs second resolution [14]. To prevent a good result, we are expected to use SRTM with 30 arcs second resolution by modifying the data and correcting the error. The data is projected uniformly to 30 seconds Cartesian resolution.

Surface wind speed play a dominat role to drive the sea current. Generally, the wind speed produces tangential friction toward the sea. The wind speed data used to model the current dynamic is the wind speed data analysis from 10 meters high above the sea surface based on interpolated measurement of the high resolution of European Center for Medium-Range Weather Forecast (ECMWF) data (figure 2 and figure 3) [15].

Absolute Dynamic Topography (ADT) is the data used to determine geostrophic current. These data obtained from the reduction of geoid altitude from Mean Sea Surface (MSS) height on an ellipsoid reference. These data also originate from the calculation of Mean Dynamic Topography (MDT) and Sea level Anomaly (SLA). The accuracy of the data can reach several centimeters, with a resolution up to 15 minutes. Currently, the ADT is a measurement of sea surface data from some satellites including Envisat, TOPEX / Poseidon, Jason-1 and OSTM / Jason-2 referring the geoid [16, 17]. These data used as the input of open boundaries for the two dimensional (2D) numerical model.

The current models are validated with [18]. The result will also be validated using surface current average data of Simple Ocean Data Assimilation (SODA) v2.2.4 from 2006 to 2010 [19].

2.2. Hydrodynamical model

The model derived from numerical finite different method from the Navier-Stokes 2D dynamics of seawater [20]. This equation involves pressure gradient force from sea level elevation, wind friction at surface sea, sea bottom friction, advection and diffusion components in two dimensions and Coriolis force.

The study area covers 5.13° N – 5.97° N and 94.97° E – 95.77° E. In the Cartesian coordinate, it is discretized at dx = dy = 30 seconds. For the stability of the model and to avoid the overflow calculation, the time-step is determined with Courant-Friedrich-Lewy (CFL) method. On the western part of the domain area, the depth of the water range from 50-100 meters. While on northern part, the water depth is about 50-1400 meters. The maximum depth is located on the eastern part of the sea, which is about 1400 meters. The time-step (dt) used is 5 seconds. For other coefficients such as bottom friction (r), wind friction (Cd) and viscosity (A_h) are 0.038, 1.285×10^{-3} and 500 m^2/s, respectively. To generate current, wind force from monthly analysis of ECMWF and sea level elevation of ADT is applied at the begining of the simulation.

Equations (1-6) this is equation dynamics of seawater two dimensions [20] and had been applied by [21] in the Gulf of Thailand and in the Malacca Strait by [2]. Where u(t, x, y), v(t, x, y) speeds current sea in direction west-east and north-south, respectively. Adv_h (u, v) is component horizontal advection, $Diff_h$ (u, v) is component horizontal diffusion. While $-fv$ and $+fu$ is style Coriolis. Component style Coriolis completed semi-implicit manner. So that on early calculations, Coriolis calculated from estimation speed (u, v). Style gradient pressure on field two dimensions is $-g\frac{\partial \eta}{\partial x}$ and $-g\frac{\partial \eta}{\partial y}$. While the sum of friction and friction under the sea breeze is $\frac{\tau_x^{wind}-\tau_x^{bottom}}{(\rho_0 h)}$ and $\frac{\tau_y^{wind}-\tau_y^{bottom}}{(\rho_0 h)}$. Seawater depth is h and depth maximum is h_{max}, while g is the force gravity.

$$\frac{\partial u}{\partial t} + Adv_h(u) - fv = -g\frac{\partial \eta}{\partial x} + \frac{\tau_x^{wind}-\tau_x^{bot}}{\rho_o} + Diff_h(u) \quad (1)$$

$$\frac{\partial v}{\partial t} + Adv_h(v) + fu = -g\frac{\partial \eta}{\partial y} + \frac{\tau_y^{wind}-\tau_y^{bot}}{\rho_o} + Diff_h(v) \quad (2)$$

$$\frac{\partial \eta}{\partial t} + \frac{\partial (uh)}{\partial x} + \frac{\partial (uh)}{\partial y} = 0 \quad (3)$$

$$Adv(\psi) = u\frac{\partial \psi}{dx} + v\frac{\partial \psi}{dy} \quad (4)$$

$$Diff(\psi) = \frac{\partial}{\partial x}\left(A_h\frac{\partial \psi}{dx}\right) + \frac{\partial}{\partial x}\left(A_h\frac{\partial \psi}{dx}\right) \quad (5)$$

Courant-Friedrich-Lewy (CFL) criterion for stability:

$$\Delta t \leq \frac{\min(\Delta x, \Delta y)}{\sqrt{2gh_{max}}} \quad (6)$$

3. Results and discussions

3.1. The wind circulation

February 2016 and August 2016 are months of northeast monsoon and southwest monsoon, respectively. On this situation the pressure and the influence of monsoon winds is relatively large. In general the wind circulation on northern Aceh water is depended on the monsoon system. In northeast monsoon (figure 2), the wind circulation on northern waters of Aceh flow from east to west with the velocity magnitude of 3-4 m/s. On the western part of Aceh waters, the wind tends to deflect toward southwest, however, the speed of wind is relatively constant and no change in magnitude. On this occasion, the wind visibly blows along the eastern part Sabang and Breueh Islands and form lee wind along the the east coast region, covering Sabang, Island of Breueh and Aceh Besar. On northern coast of Aceh Besar, wind tends to flow down the coast toward east. On the southwest monsoon (figure 3), the dominant wind circulation flows from southwest toward northeast with the speed of 4-6 m/s.

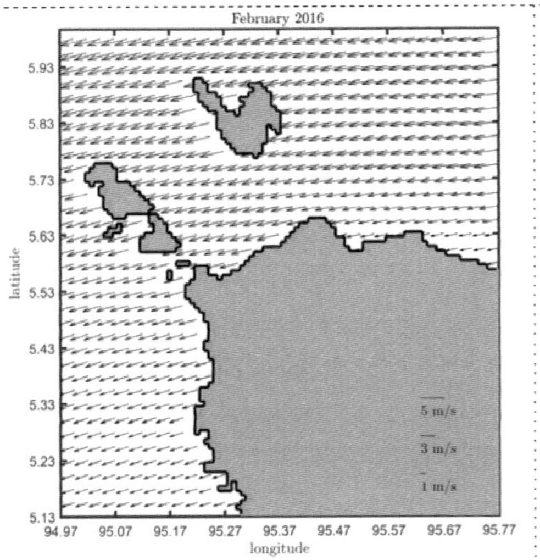

Figure 2. Wind circulation during Northeast monsoon in the northern of Aceh waters (m/s).

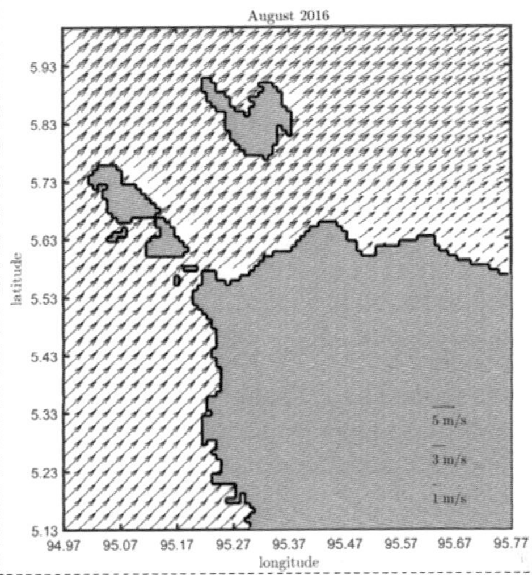

Figure 3. Wind circulation during Southwest monsoon in the northern of Aceh waters (m/s).

3.2. The current circulation

The 2D flow simulation results are displayed in figure 4 and 5, respectively. The circulation pattern on the month of February 2016 (figure 4) follows the wind direction and Coriolis force. The current circulation, it is seen to flow toward north. However, current circulation on Sabang water is relatively weak, while offshore of western of Aceh Besar, relative strong flow occurs on this area as well as on the north-eastern of Aceh Besar. On the western of Aceh Besar, current flows toward north follow the topography contour. The current circulation is relatively fast offshore and linearly decrease toward shoreline. The current circulation of west offshore is about 0.25 m/s, while near the coast is approximately 0.1 m/s.

In the month of August 2016 the current flows faster on north-eastern part compared to the western part (figure 5). On north-eastern part the current flows to the northwest with speed of 0.3 m/s. The currents than split and accelerate due to the effect of Sabang island topography. Meanwhile, western offshore part of Aceh Besar, the currents flow toward South follow the wind force and Coriolis force direction.

Based on the result of numerical model of current produced in this research, it has a good agreement with SODA (figure 6 and 7), the work of [16,22,23]. The influence of monsoon toward the current circulation can be clearly described as the season change, especially on the western part of Aceh waters. Generally the northeast monsoon current can be seen during the month of February 2016 and the southwest monsoon current is visible during the month of August 2016. On the western part, the northeast monsoon current has stronger flow than the southwest monsoon current. However, the north-eastern bordering with Malacca Strait and the Andaman Sea, the southwest monsoon current is visible greater (figure 5).

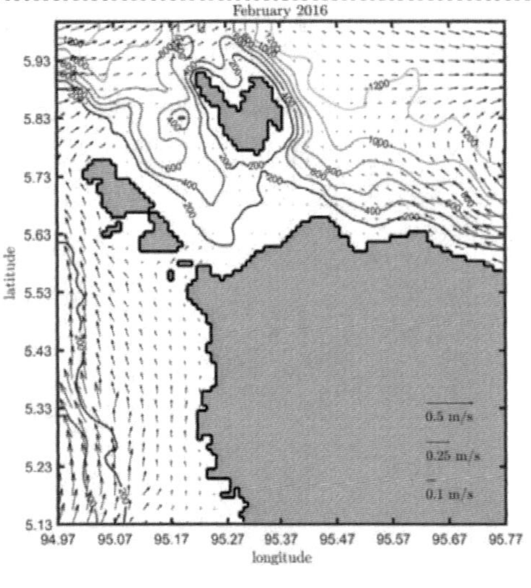

Figure 4. Current circulation during Northeast monsoon in the northern of Aceh waters. Current is depicted by vectors (m/s) and levels of bathymetry are shown by colour contours (m).

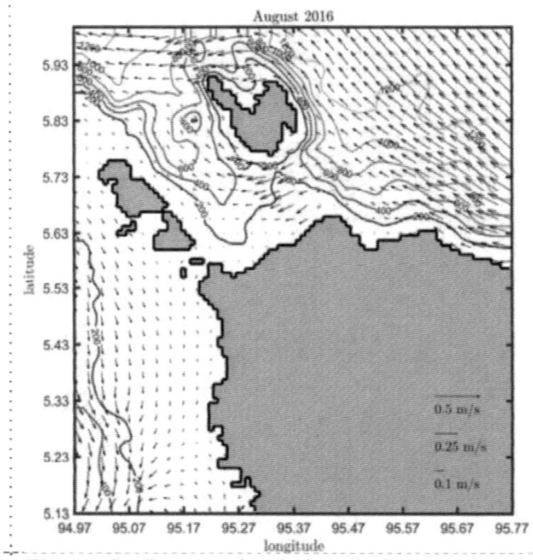

Figure 5. Current circulation during Southwest monsoon in the northern of Aceh waters. Current is depicted by vectors (m/s) and levels of bathymetry are shown by colour contours (m).

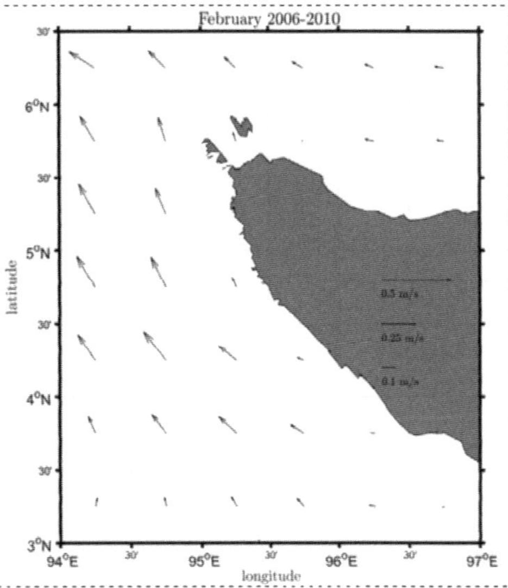

Figure 6. Average surface current circulation during Northeast monsoon (2006-2010) based on SODA (m/s).

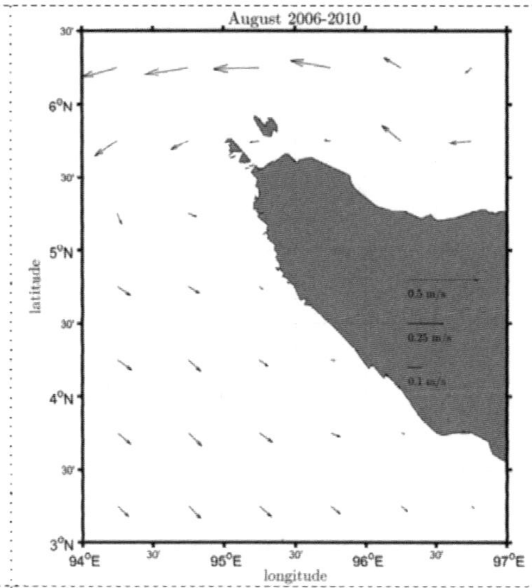

Figure 7. Average surface current circulation during Southwest monsoon (2006-2010) based on SODA (m/s).

4. Conclusions

In general, the results show that the dynamic of current circulation follows the bathymetry pattern. In the western part of the model, the direction of current heading toward north during February 2016 and change toward south during August 2016. The model has good agreement with the other results for verification.

Acknowledgments
Authors express gratitude to the Ministry of Research, Technology and Higher Education of Indonesia for financial assistance under 'Fundamental Research Grants', with contract number: 105/SP2H/LT/DPRM/IV/2017. We also thank to Syiah Kuala University through facilities support at Ocean Modelling Laboratory during the research.

References
[1] Rizal S, Damm P, Wahid M A, Sundermann J, Ilhamsyah Y, Iskandar T and Muhammad 2012 General circulation in the Malacca Strait and Andaman Sea: A numerical model study *American Journal of Environmental Science* **5** 479-488
[2] Haditiar Y, Rizal S and Abdullah F 2016 Current simulation in the Malacca Strait and part of South China Sea due to wind *12th International Conference on Mathematics, Statistics, and Their Applications (ICMSA)* pp 47-50 doi: 10.1109/ICMSA.2016.7954306
[3] Joseph S and Ravichandran M 2013 Validation of 0.25^{O} x 0.25^{O} Indian Ocean HYCOM. INCOIS Report. 43 pp. http://incois.gov.in/documents/hycom/hycom_0.25x0.25_tech_rep.pdf
[4] Rizal S, Haridhi H A, Wilson C R, Hasan A and Setiawan I 2013 Community Collection of Ocean Current Data: An Example from Northern Aceh Province, Indonesia *SPC Traditional Marine Resource Management and Knowledge, Information Bulletin* **31** 3-11
[5] Haridhi H A, Nanda M, Wilson C R and Rizal S 2016 Preliminary study of the sea surface temperature (SST) at fishing ground locations based on the net deployment of traditional purse-seine boats in the northern waters of Aceh — A community-based data collection approach *Regional Studies in Marine Science* **8** 2352-4855
[6] Haridhi H A, Nanda M, Haditiar Y and Rizal S 2018 Application of Rapid Appraisals of Fisheries Management System (RAFMS) to identify the seasonal variation of fishing ground locations and its corresponding fish species availability at Aceh waters, Indonesia *Ocean & Coastal Management* **154** 46-54
[7] Ilhamsyah Y, Fadli N, Setiawan I, Haridhi H A 2014 Notes: coral reef bleaching in Weh Island, Indonesia, a natural climate variability or global climate change impact? *AACL Bioflux* **7**(6) 508-515
[8] Rizwan T, Dewiyanti I, Haridhi H A, Setiawan I, Ilhamsyah Y and Alirudin J 2014 Analysis of fish catches by traditional purse seine boat in Aceh waters based on setting and hauling duration *AACL Bioflux* **7**(2) 63-67
[9] Setiawan I, Yuni S M, Mariana and Ilhamsyah Y 2017 Modelling of waste concentration distribution at Iskandar Muda Fertilizer Company *Proceding on Nasional Seminar of Post Graduate of Syiah Kuala University*
[10] Irham M and Setiawan I 2017 The study of flow resulting from wave on Lhonga beach, Aceh Besar *OmniAkuatika* **13**(1) 5-12
[11] Chen, W -B, Liu W -C, Hsu M -H and Hwang C -C 2015 Modeling investigation of suspended sediment transport in a tidal estuary using a three-dimensional model *Applied Mathematical Modelling* **39**(9) 2570–2586 https://doi.org/10.1016/j.apm.2014.11.006
[12] Irham M, Setiawan I 2018 The numerical model of the sediment distribution pattern at Lampulo National fisheries port *IOP Conf. Ser.: Earth Environ. Sci.* 106 012067 doi: 10.1088/1755-1315/106/1/012067
[13] Purnawan S, Setiawan I, Haridhi H A and Irham M 2018 Granulometric analysis at Lampulo Fishing Port (LFP) substrate, Banda Aceh, Indonesia *IOP Conf. Ser.: Earth Environ. Sci.* 106 012070 doi: 10.1088/1755-1315/106/1/012070
[14] Jarvis A, Reuter H I, Nelson A and Guevara E 2008 Hole-filled SRTM for the globe Version 4 available from the CGIAR-CSI SRTM 90m Database (http://srtm.csi.cgiar.org)
[15] Balsamo G, Albergel C, Beljaars A, Boussetta S, Brun E, Cloke H, Dee D, Dutra E, Muñoz-Sabater J, Pappenberger F, de Rosnay P, Stockdale T and Vitart F 2015 ERA-Interim/Land: a global land surface reanalysis data set *Hydrology and Earth System Sciences* **19**(1) pp 389-407 doi: 10.5194/hess-19-389-2015

[16] Rio M -H, Mulet S and Picot N 2014 Beyond GOCE for the ocean circulation estimate: Synergetic use of altimetry, gravimetry, and in situ data provides new insight into geostrophic and Ekman currents *Geophys. Res. Lett.* **41** doi:10.1002/2014GL061773
[17] Ablain M, Cazenave A, Larnicol G, Balmaseda M, Cipollini P, Faugère Y, Fernandes M J, Henry O, Johannessen J A, Knudsen P, Andersen O, Legeais J, Meyssignac B, Picot N, Roca M, Rudenko S, Scharffenberg M G, Stammer D, Timms G and Benveniste J 2015 Improved sea level record over the satellite altimetry era (1993–2010) from the Climate Change Initiative project *Ocean Science* **11**(1) pp. 67-82 doi:10.5194/os-11-67-2015
[18] Wyrtki K 1961 *Physical Oceanography of the Southeast Asian waters* Scripps Institution of Oceanography UC San Diego: Scripps Institution of Oceanography Retrieved from: http://escholarship.org/uc/item/49n9x3t4
[19] Giese, B S and Ray S 2011 El Niño variability in simple ocean data assimilation (SODA), 1871–2008 *J. Geophys. Res.***116** C02024 doi:10.1029/2010JC006695
[20] Kämpf J 2009 *Ocean Modelling For Beginner Using Open-Source Software* Springer Verlag, Heildelberg
[21] Adliansyah, Haditiar Y, Muhammad and Rizal S 2016 Current simulation in the Gulf of Thailand due to wind and sea level elevation *12th International Conference on Mathematics, Statistics, and Their Applications (ICMSA)* pp 55-58 doi: 10.1109/ICMSA.2016.7954308
[22] Diansky, N A, Zalesny V B, Moshonkin S N and Rusakov A S 2006 High Resolution Modeling of the Monsoon Circulation in the Indian Ocean *Oceanology* **46**(5) 608–628 https://doi.org/10.1134/S000143700605002X
[23] Schott F A, Xie S P and McCreary J P 2009 Indian Ocean circulation and climate variability *Reviews of Geophysics* **47**(1) 1-46

IOP Conf. Series: Earth and Environmental Science **176** (2018) 012017 doi:10.1088/1755-1315/176/1/012017

Deep water masses exchange induced by internal tidal waves in Ombai Strait

A S Atmadipoera[1*], Y Suteja[2]

[1]Dept. of Marine Sciences and Technology, Faculty of Fisheries and Marine Sciences, Bogor Agricultural University, Bogor Indonesia.
[2]Dept. of Marine Science, Faculty of Marine and Fisheries, Udayana University, Bali Indonesia
*e-mail: agus.itk@gmail.com

Abstract. Yoyo CTD/LADCP measurement for a 24-h tidal period and along trackshipboard ADCP, has been carried out in the northern entrance of Ombai Sill during the INDOMIX Cruise in July 2010 to investigate current structure and associated near-bottom water mass exchange induced by internal tidal waves. The results show that a strong meridional current fluctuation is observed on quasi-semidiurnal period with much stronger southward flow, related to the existence of southward Ombai Indonesian Throughflow (Ombai ITF). The vertical extent of the current deepens to the sill depth (1600m). Under 500 m depth to the sill, the southward ebb flow brings cooler and less salty water Flores water to southern Ombai Strait in Savu Sea. In contrast, the northward flood flow supplies Flores/Banda Sea with much warmer and saltier water. Thus, on tidal period there exists deep water masses exchange induced by internal tidal waves. Strong baroclinic instabilities appear when induced northward tidal flow against Ombai ITF, resulting highest vertical diffusivity.

1. Introduction

The Indonesia Throughflow (ITF) is a main branch of the global thermohaline circulation, connecting between the Pacific and the Indian Oceans [1,2]. The ITF flow in interior Indonesian seas is mainly via western pathway, which conveys about 11.6 Sv (1 Sv = 10^6 m^3/s) of water masses that consist of thermocline and sub-thermocline water masses from North Pacific Subtropical Water (NPSW) and North Pacific Intermediate Water (NPIW). The smaller contribution through eastern route which consists of South Pacific Subtropical Lower Thermocline Water (SPSLTW) and only drawn 2.5 Sv water masses [1,3,4,5,6,7]. Most of the western and eastern routes passes to Indian Ocean through the Timor Passage (7.6 Sv) and the rest exits through Ombai Strait (4.9 Sv) and Lombok Strait (2.6 Sv) [8,9].

Besides getting input from the Pacific Ocean, the ITF also gets water input from Indian Ocean. Previous research observed that there is an upper layer of North Indian Intermediate Water in northern Sawu Sea and Ombai Strait, as well as Indian Ocean Deep Water (IDW) in Timor Passage [7]. Ombai Straits is located closed to Timor Passage and its water masses are considered to be influenced from the Indian Ocean through the straits in the western part of Sawu Sea. INSTANT data in Ombai Strait indicated that there are flows into interior Indonesian seas through Ombai Strait drawn by the South Java Current (SJC) and South Java Under Current (SJUC), and also arrival of equatorial Kelvin waves from the equatorial Indian Ocean penetrate through the Savu Sea [10,9,11]. In addition to the coastally trapped Kelvin waves, water mass transport also strongly affected by internal waves generated by rough topography and sills [12].

Ombai Strait is one of the main ITF outflow straits to the Indian Ocean which constrains to deep sill of about 3250 m depth [13,9]. From numerical simulation showed that interaction between sill and

IOP Conf. Series: Earth and Environmental Science **176** (2018) 012017 doi:10.1088/1755-1315/176/1/012017

strong barotropic tidal current can generate internal waves [14]. Ombai Strait is one of the spot areas in Indonesia seas, which displays a high internal tidal energy, transferring from barotropic tides to baroclinic tides [15,16] and has strongest internal tidal current velocity (up to 0.5 ms^{-1}), although it is only generated by M2 tidal component [17]. Internal tide is one of the main forcing for generating turbulent mixing in the ocean [18], including in the deep ocean [19].

Indirect method of turbulent mixing measurements using CTD showed that Ombai Strait revealed a higher turbulent mixing than other regions in Indonesia such as Banda Sea and showed highest vertical diffusivity of turbulent mixing in the deep layer but lowest in thermocline layer [20]. Vertical diffusivity estimates in Ombai Strait vary between 1.2–9.7 x 10^{-4} m^2s^{-1} with its maximum of 2.3 x 10^{-1} m^2s^{-1} [21,22]. The high value of tubulent mixing was thought by strong internal tidal energy and its interaction with deep sill in Ombai Strait [20].

Intense internal wave activity as seen from high energy transfer from barotropic tides to baroclinic tides, as well as high vertical diffusivity of turbulent mixing may influence strongly stratification and exchange of different water masses in Ombai Strait. The objectives of this study attempt to investigate vertical current structure during a tidal cycle (24-hour) and its associated deep water mass exchange and internal tidal waves in Ombai Strait, acquired from the conductivity-temperature-depth (CTD) and lowered acoustics Doppler current profiler (LADCP), and hull-mounted shipboard (SADCP) measurement during the INDOMIX cruise in July 2010.

2. Methods

The Hydrographic data measurement in Ombai Strait has been carried out during the *Indonesian Mixing* (INDOMIX) 2010 cruise. During two weeks cruise, the INDOMIX voyage was started in Sorong, then to western equatorial Pacific Ocean, to Halmahera Sea, to Seram Sea, to Banda Sea, to Ombai Strait, to northern Sawu Sea, to Lombok Strait and finished at Tanjung Perak port in Surabaya. The CTD/LADCP "yoyo" measurement in Ombai Strait has been conducted in northern entrance of Ombai Strait (between Alor and Ataura, figure 1), collecting 9 CTD/LADCP casts for 24 hours (one tidal cycle) measurement on July 15-16, 2010. Measurement was conducted in the sill slope in the northern entrance of Ombai Strait with its width of about 17.5 km. Instrument depths during repeated CTD/LADCP measurement varied from 409 m to 1549 m depth. The CTD data processed by standard processing procedure using *the SBE data processing software* and CTD profile data were derived from downcasting data only. CTD data from INSTANT program on June 2005 in South Ombai Sill were used as a comparison to CTD profiles from INDOMIX data. The ocean current profiles measured by 2 Lowered Acoutics Doppler Current Profiler (LADCP) RDI 300kHz that installed in CTD frame, one is looking-upward and the other looking downward, where absolute current is obtained from bottom track of LADCP looker downward. The bin size for each water column is 8 m. The LADCP data processing follows the guidelines of [23], [24] and [25]. Ocean currents also measured using hull-mounted shipboard ADCP, known as the shipboard ADCP (SADCP). The SADCP data processing and analysis follows the guidelines from [26].

Data processing, analysis and visualization were done by using MATLAB and spreadsheet software. The Thorpe displacement (*d*) is calculated from seawater density profile that has been re-ordered under stable stratification, i.e., conditions in which higher density water mass is below low density water mass. For example, if we have *n* samples at depth z_a with density ρ_n and to obtain static stability conditions samples at depth z_a are moved to depth z_b then the value of *d* is calculated [20] by using equation of :

$$d = z_a - z_b$$

A positive d value indicates when a high density water mass is above a low density water mass, so to make static stability low density water mass must be moved up. In contrary, when the value of d is negative indicates high density mass must be moved down replacing the low density water mass to achieve static stability conditions.

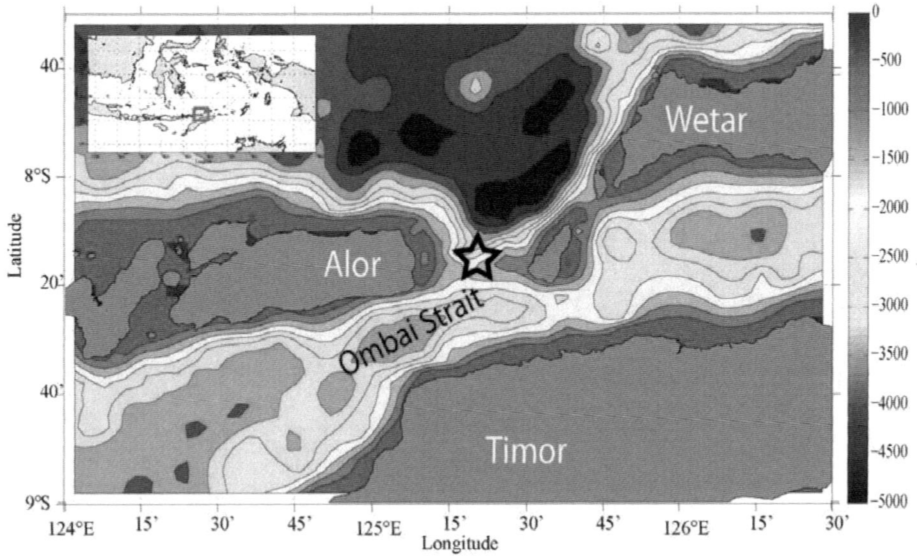

Figure 1. Map of study area in Ombai Strait. The CTD/LADCP measurement during the INDOMIX cruise has been conducted in the sill slope of the nothern entrance between Alor and Wetar. Star denotes for 24-hour repeated "yoyo" CTD/LADCP measurement on 15-16 July 2010. Bathymetry interval is 500.

3. Results and discussion

Averaged current vectors at different depths in Ombai Strait during first SADCP measurement is presented in Figure 2. From stick plot in East Ombai it is found that dominant current flow is southward and southwestward. However, there are variations in magnitude and direction at different averaged depths. Current direction average between 50-150 m depth is dominant southward with a slight zonal deflection. At 150-300 m depth, current direction spreads eastward and westward, but the southward flow are still dominant (figure 2b). Current velocity at 300 - 500 m depth flows dominantly southward, but the eastward and westward currents are getting stronger. This result also has good agreement with previous measurements using ADCP during November 1995 - November 1996 that show net flows current to Savu Sea in southern Ombai [9], but in December 1995 was found that above 160 m depth there is a current move towards the east [8].

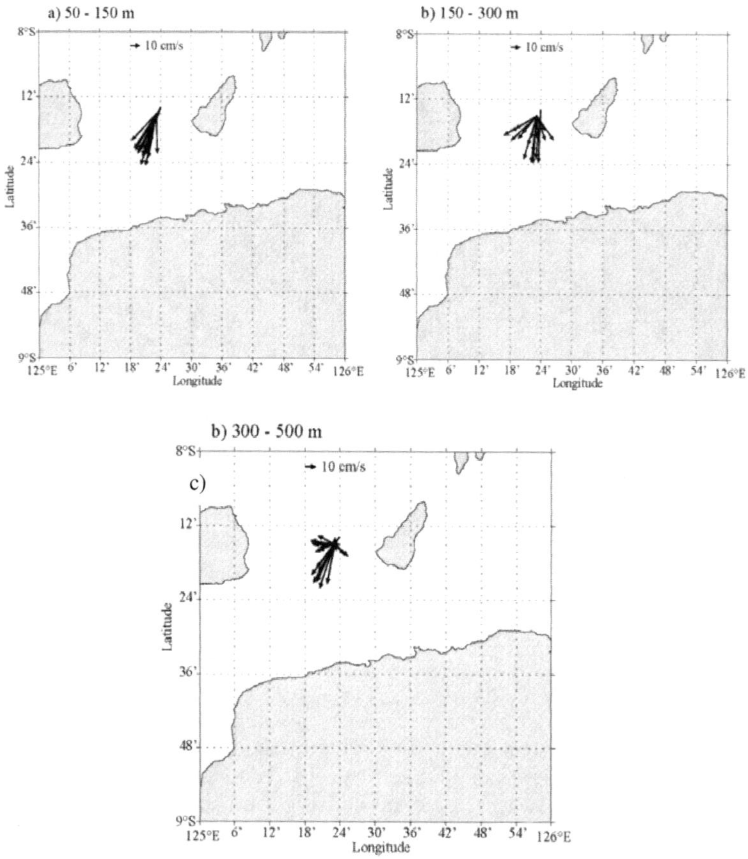

Figure 2. Current vector plots from first SADCP measurement in Ombai Strait; (a) averaged current vector between 50-150 m depth, (b) between 150-300 m depth, (c) between 300-500 m depth.

Averaged current vector at different depths in Ombai West (figure 3) show that the flow pattern of current at three different depths in Ombai west is similar to those observed in Ombai east, where the dominant flows are southward. However, the magnitude of current velocities are much stronger and the current directions are much variable in Ombai west. At 50-150 m depth the direction of current is southward (figure 3a), but some current vectors flow westward. In Ombai east, at the same depth, there is no current flowing westward (figure 2a). In lower layer (150-300 m depth), the northward velocity current is stronger than that in 50-150 m depth, but the current leading south remains dominant. The currents direction and velocity average between 150-300 m depth in western Ombai stronger and more northward then eastern Ombai. Northward current flow decreases between 300-500 m depth and the southward current increases. The current velocity in western Ombai is larger (zonal or meridional current) than the eastern Ombai at the same depth. The increased in vertical flow is supposed due to the maximum ITF transport in thermocline layer compared to surface layer [13,27]. From current measurements using SADCP at two locations near the eastern and western sides of Ombai Strait (figures 2 and 3), it is clear that the dominant currents flowing southward and southwestward represent the ITF in Ombai Strait, where the current are stronger and intensive in the thermocline layer. Although current measurements are made only snapshots, but time measurement (in July 2010) has been part of the Southeast Monsoon period when seasonal variation of ITF maximum. From this current measurement, one interesting novel is revealed a western intensification (strengthening) flow in the northern entrance of Ombai Strait that is apparently the main pathway of ITF originating from the Flores Sea.

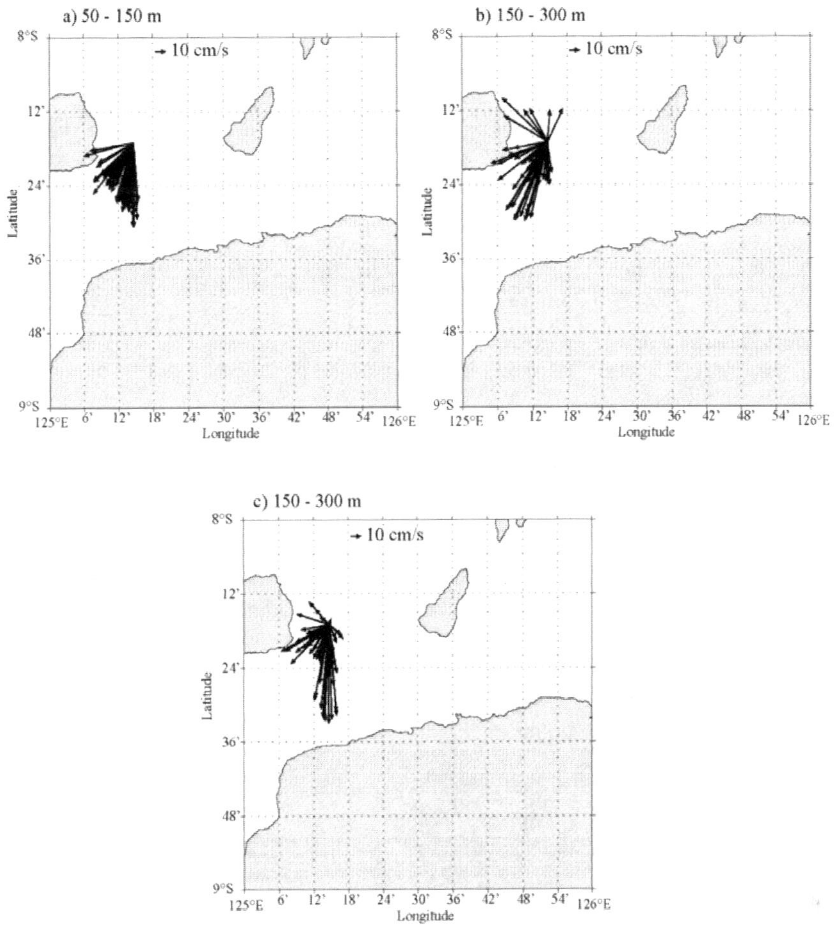

Figure 3. Current vector plots from second SADCP measurement in Ombai Strait; (a) averaged current vector between 50-150 m depth, (b) 150-300 m depth, (c) 300-500 m depth.

The depth-time plot of the zonal and meridional current components from 24-hour SADCP measurements shows clearly temporal (tidal scale) and vertical (different depth) variations of the flow (Figure 4). The meridional current component (north-south component) is much stronger than zonal (east-west component) component, associated to configuration of the strait in north-south direction. The flow direction of meridional current component indicates alternating flow northward during certain period (yellow-red color) and southward during the opposite period (green-blue color). During 24-hour current measurement, it is found twice current reversal, which is strongly related to arrival of barotropic semi-diurnal tidal waves in Ombai Strait. Semi-diurnal tides propagate from the Indian Ocean entering through the Savu Sea and into the Ombai strait [28].

Distribution of meridional current component shows that magnitude of southward velocity is much stronger than the northward flow (Figure 4, bottom left panel), indicating an additional velocity quasi-constant derived from the ITF component that flows southward. It is also seen that, the southward flows are more consistent and much stronger between 0-500 m depths. The averaging of 24-hour current to eliminate tidal signal, indicating that the magnitude flow of meridional velocity is much stronger than the cross-strait velocity (Figure 4, right panel). The maximum velocity (about 75 cm/s) of meridional current is found near 25 m depth, then the velocity decreases to 35 cm/s near 100 m depth. Magnitude of meridional current velocity between 100 m and 650 m depth is relatively persistent and quasi-constant (approximately 25 cm/s). From meridional velocity profile, it is found that Ombai ITF is relatively stable from sub-surface to about 650 m depth, which is good agreement with [8] and [9], where the vertical extent of Ombai ITF reach 800 m depth. A similar pattern of stronger along-strait flows is also found on Timor passage [29,30,31].

The zonal current component (cross-strait velocity) indicates that the currents moving westward are more dominant than those moving eastward (figure 4, upper panel). In contrast to the meridional component, the average of zonal component flow profile is much weaker and fluctuates with the maximum speed of 25 cm/s near 75 m and 250 m depth. The average flow to the west has a maximum speed of 25 cm/s near 75 m depth. The meridional current component is much more dominant than the zonal current flow with the highest speed near 25 m depth. The average meridional current also shows that the dominant current is moving southward.

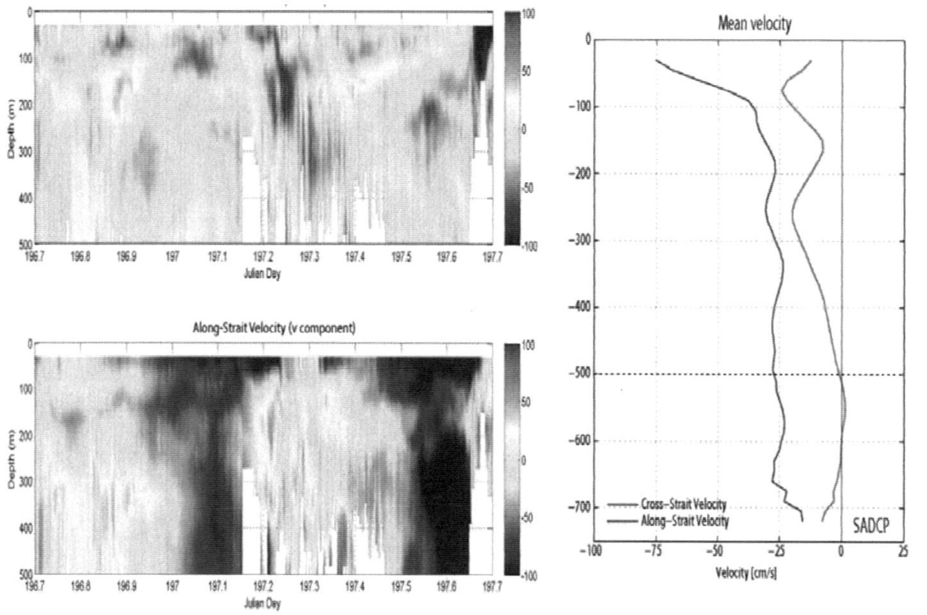

Figure 4. Zonal and meridional current components (left panel) and mean zonal (cross-strait) and meridional (along-strait) current profiles (right panel) from SADCP measurement on July 15-16, 2010 in northern entrance of Ombai Strait.

The zonal and meridional current components from LADCP measurement at varying depths according to the CTD casting (500 m to 1600 m) show a consistent pattern with SADCP measurement (figure 5). It is revealed two reversals northward and southward flows within approximately 24 hours. The meridional current component is much stronger than the zonal component (Figure 5, left). However, the depth-time plot of zonal and meridional current components with LADCP is much more detailed with the presence of strong meridional current cores northward (red) and southward (blue).

The zonal component is weaker than the meridional component (figure 5). In zonal currents it is seen that the variation of east-west direction has no significant, compared to north-south variation of meridional current. It is also show that there is a very strong flow in meridional component that changes two times in quasi-semidiurnal periods, where the current flowing northward at a particular time (18:00-21:00 and 06:00-11: 00) and southward at other times (00:00-04:00 and 11:00-15:00). The northward current dominant occurs at depth between 400 m - 1200 m, and between 400- 1400 m depth.

The mean zonal and meridional profile, averaged for 24 hours to eliminate the tidal current signal, show that the meridonal current is much stronger than the zonal current. The maximum speed of zonal current is 20 cm/s near 250 m depth, while the meridional current reveals a magnitude velocity more than 60 cm/s near the surface layer. The southward flows are related to the existence of southward Ombai ITF as one of the main outflow straits to the Indian Ocean [8,7,9]. The mean meridional profile also shows that the current is weakening with increasing depth (Figure 5, right), even currents moving to the north near bottom (more than 1600 m).

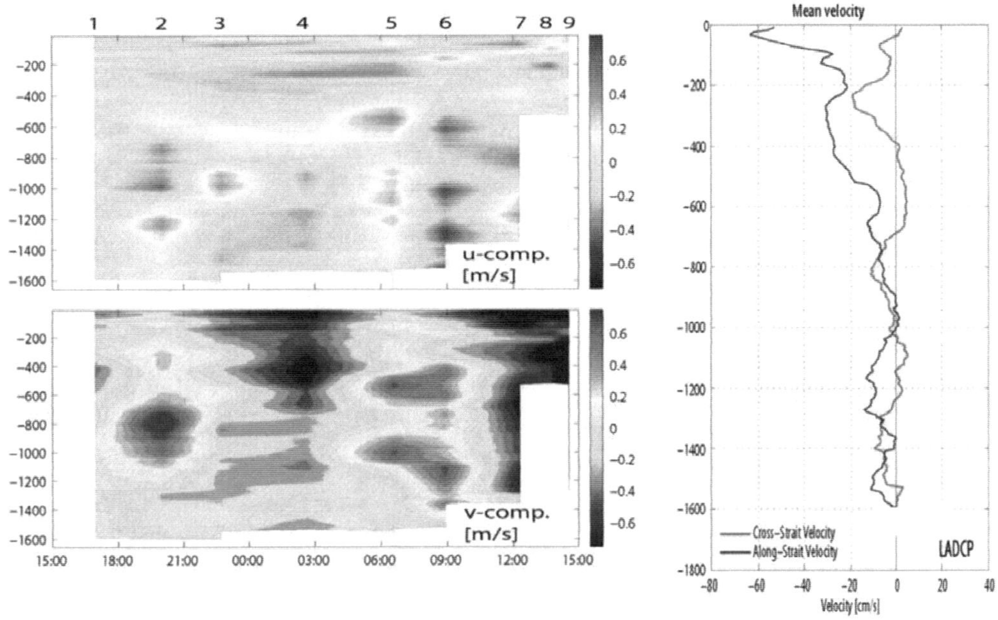

Figure 5. Zonal and meridional current components (left panel) and mean flow profiles (right panel) from lowered ADCP measurement in Ombai Strait

Stratification of water masses, expressed by its depth-time plot of potential temperature, salinity and potential density anomaly (figure 6, upper panel) revealed a quasi-semidiurnal oscillation of water masses within 600-1500 m depth water column. It is seen that the internal waves in the Ombai Strait prevails semidiurnal, because it has two peaks and two valleys during measurement (24 hours). This result also has good agreement with [32] that Indonesian waters are generally dominated by internal tides M2 (semidiurnal tidal component). More detail also we can see that the first internal wave height is lower than the second internal wave. The first internal wave has about 550 m height with the first internal wave crest at about 700 m depth and the wave trough is around 1250 m depth. The second internal wave has around 1000 m height with a wave crest at 500 m depth and a wave trough at about 1500 m depth.

Corresponding different water masses appear during one-tidal-cycle (24-hour) in Ombai Strait, one relatively warmer, less salty, and less dense water mass occur two times, for example around 197.8 and 198.4 Julian-day, and the other relatively colder, saltier and denser water mass appear alternately around 198.1 and 198.6 Julian-day (figure 6, a-b-c). The changes of water masses stratification, as indicated by abrupt changes of their isotherm, isohaline and isopycnal (Figure 6, a-b-c), indicate that the internal tidal waves control predominantly on structures of water masses and stratification. The internal wave activity in a water column affect greatly on the vertical profile (temperature, salinity, and density) of waters masses [33].

The presence of relatively warmer, less salty and less dense water mass around 197.8 and 198.4 Julian-day (figure 6, a-b-c) is associated with northward flow. Modeling tidal current showed that semi-diurnal tidal component propagates from the Indian Ocean, into Sawu Sea, then pass into Ombai Strait entering Flores Sea and Banda Sea [28,32]. Hence, the northward flow around 197.8 and 198.4 Julian-day corresponds to flood tide (high tide), which brings distinct water mass from Sawu Sea into

Flores/Banda Sea. On the other hand, the southward flow around 198.1 and 198.6 Julian-day (figure 6c) is related to ebb tide (low tide), plus quasi-constant of southward Ombai ITF that draws water mass from Flores/Banda Sea into Sawu Sea.

The internal wave pattern in deep layer (near bottom) affects the exchange of water masses in its layer. The southward ebb flow brings cooler and much salty water mass, which is derived from deep Banda/Flores Water [1,34,13]. In contrast, the northward flood in Ombai Strait brings less salty and much warmer water from Savu Sea, which may influence Flores or Banda water mass. Further research is needed to determine the origin of this water mass. Previous study showed that North Indian Intermediate Water (NIIW) at depth between 600-800 m depth is dominant in Sawu Sea, and at deeper layer upper Indian Deep Water (IDW) also occupies in the region [35,8,13,7].

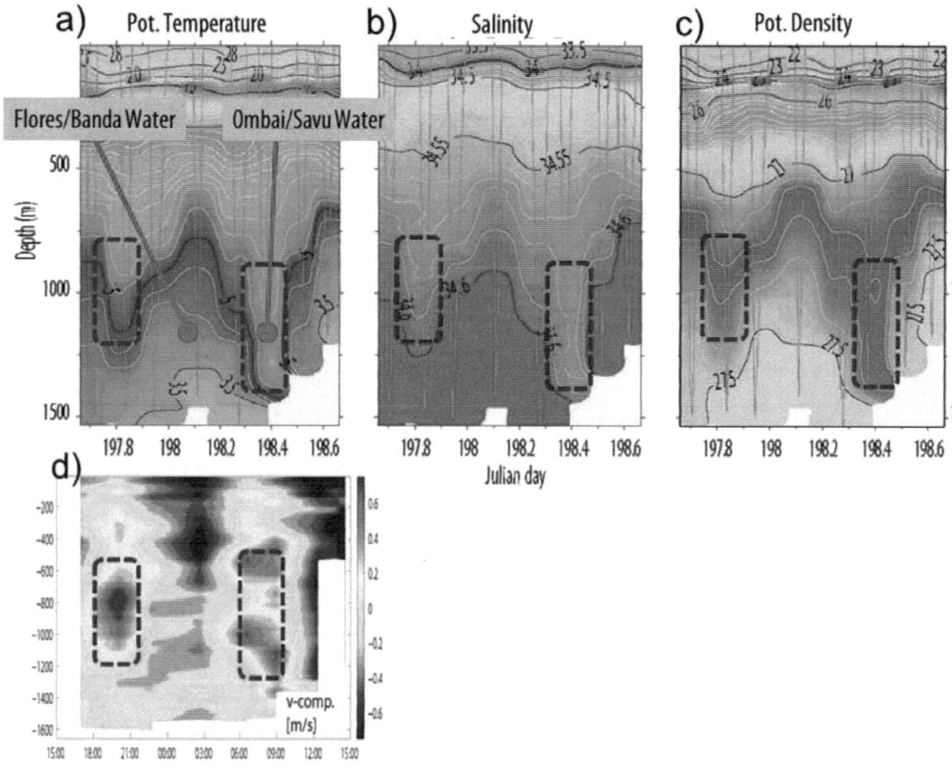

Figure 6. Fluctuation of seawater properties (potential temperature, salinity and potential density anomaly (a-c), and meridional current component (d) on quasi-semidiurnal time-scale in Ombai Strait.

Profiles of potential temperature, salinity and potential density anomaly (sigma-t) from INSTANT CTD measurements in the north Ombai side (Flores/Banda Sea) and south of Ombai (Savu Sea) between 1000-1500 m depth are separate (Figure 7). The Flores/Banda water masses in north side of Ombai is characterized by much cooler, saltier, and denser water, while Savu water mass in south side of Ombai is characterized by warmer, less salty and less dense water. Intense and strong internal tidal activities in deeper layer in Ombai Strait are strongly suspected to cause a mixing of water masses at semidiurnal tidal-scale.

Water masses in deeper layer before passing through Ombai Strait has a characterized by much cooler, saltier and denser water, but after passing through the Ombai Strait the characteristics of the water mass changes to be warmer, less salty and less dense water. This indicated that the vertical mixing is strong enough in Ombai straits to change the characteristics of the water mass. The intens vertical mixing in Ombai Strait is strongly suspected due to the deep sill at the bottom of the waters interacting with the internal waves [16,20]. Baroclinic tides in Ombai Strait are generated by Ombai sill, generating strong internal tides waves [32], as confirmed by field measurement [21,20] and by the model [32,36].

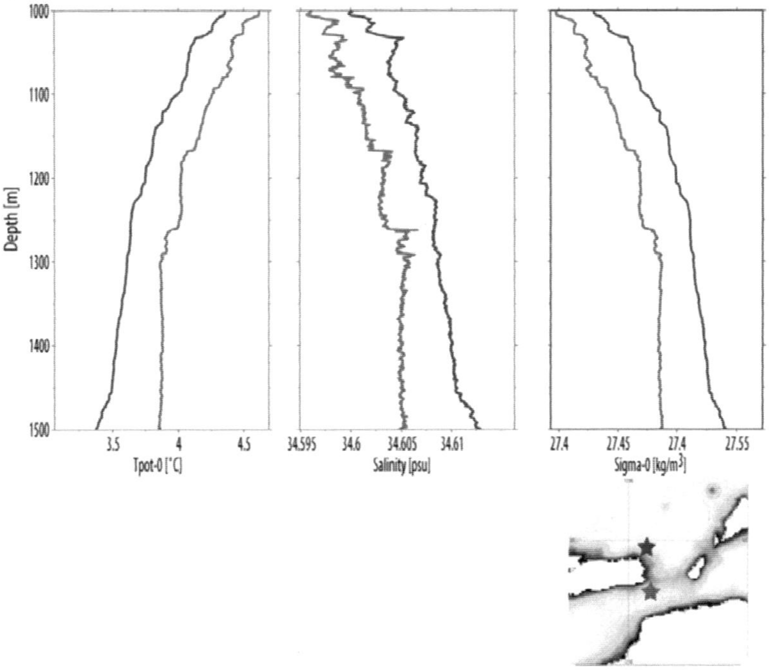

Figure 7. Comparison of potential temperature and salinity profiles in Ombai/SavuSea side and Flores/Banda Sea side at deeper layer between 1000 m and 1500 m depth. Colder, saltier and denser water mass is found in Flores/Banda Sea, compared to warmer, less salty and less dense water in Savu Sea. Data source: the CTD archive data from INSTANT measurement in June 2005 [7].

Depth-time plot of potential temperature, overlayed with Thorpe displacement (d) profiles during a tidal cycle (Figure **8**a) reveals that Thorpe displacement appears to be maximum at deeper layer between 600 m and 1500 m depth, in contrast, its minimum is found in the upper thermocline layer (0-500 m depth). The high static stability at the thermocline layer causes the low Thorpe displacment [37]. During the northward flood tide (yellow-red color in figure 8b), the Thorpe displacement increases in deeper layer, and conversely during the southward ebb tide. A strong Thorpe displacement occurs twice in 24-hour measurement, when reversal meridional current flows northward (yellow-red color, figure 8b). This is assumed because the northward propagation of the tidal current against the dominant southward flow of the Ombai ITF, so it can cause baroclinic instability of water column during this tidal period. High Thope displacement corresponding to the northward tidal current (for example around 20:00 and 09:00 local time), charaterized by much warmer water mass. On the other hand, when the southward meridional current (for example around 03:00 and 13:00), Thorpe displacement are relatively low and is dominated by much cooler water mass. So it can be assumed that there is an exchange and mixing of water masses in the deep layer in Ombai between Flores/Banda water mass and Savu sea water mass.

Deep homogeneous layer in Ombai Strait revealed much larger Thorpe displacement that varies between 20 m and 107 m, compared to those observed in the mixed surface and thermocline layers [20]. The high value of Thorpe displacement at deeper layer may be due to the interaction between currents and rough topography [38]. The Ombai Strait is a region with a high current velocity (up to 0.5 m/s) [9]. The high Thorpe displacement in the Ombai Strait indicates that there is a strong mixing of the water mass on deeper homogenous layer, so that, it is very reasonable that internal tidal waves influence strongly the characteristics of water mass when passing through Ombai Strait.

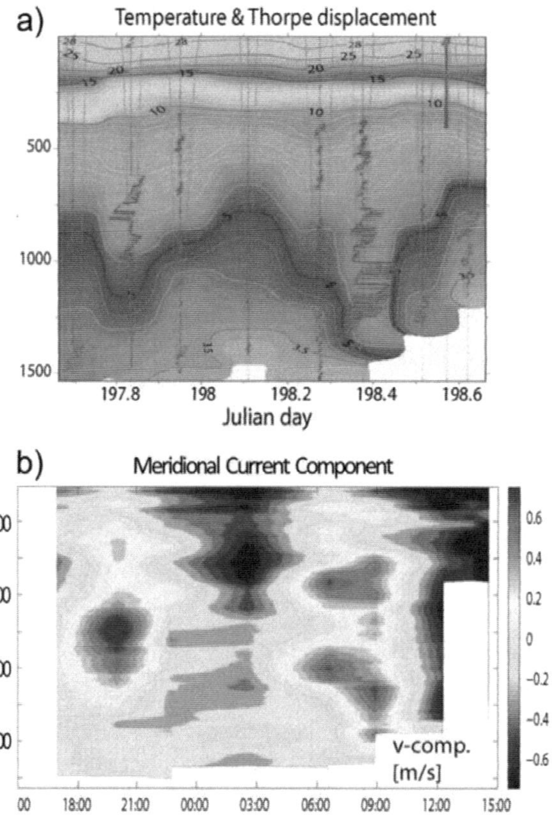

Figure 8. (a) Depth-time plot of potential temperature, overlayed with 9 Thorpe displacement profiles during a tidal cycle (24-hour); (b) Structure of meridional current component obtained from LADCP on July 15-16 2010.

4. Conclusion

Vertical structure of zonal and meridional current components during a tidal cycle (24-hour) measurement in Ombai Strait has been described in its associated with possible deep water masses exchange induced by internal tidal waves. Since the north-south orientation of Ombai Strait, it is revealed that meridional (north-south) current component is predominant. It is found two reversals of meridional flow during a tidal cycle, expressed by northward flood current (high tide) and southward ebb current (low tide), associated with a propagation of semidiurnal tidal waves originated from the Indian Ocean. During the low tide, a quasi-persistent southward Ombai ITF modulates southward ebb current, resulting in asymetric pattern of flood/ebb currents. Ombai ITF is characterized by a strong persistent southward flow that extends vertically to about 800 m depth. The northward flow (flood current) brings water mass from Savu Sea into Flores/Banda Sea, characterized by warmer, less salty and less dense water. On the other hand, the southward flow is associated with the arrival of cooler, much salty and much dense water from Flores/Banda Sea. A drastic increase of the Thorpe displacement is found during the northard flood current, when the flood current against the southward persistent Ombai ITF. This suggests that an intense of deep water mass mixing and exchange between Flores/Banda water and Savu Sea water may take place on semidiurnal tidal scale in Ombai Strait.

Further hydrographic and microstructure measurement is needed to quantify turbulent mixing in both sides of Ombai Strait.

5. Acknowledgment

The authors say many thanks to ministry of Ristekdikti who has funded this research. The authors also express great gratitude to Ariane Koch-Larrouy (INDOMIX PI), all INDOMIX cruise team and Marion Dufresne crew for their cooperation.

References

[1] Wyrtki K 1961 *Physical Oceanography of the Southeast Asian Waters* Naga Report Volume 2 (California: University of California) p 225
[2] Gordon A L 1986 Interocean exchange of thermocline water *J. Geophys. Res.* **91** 5037-5046
[3] Ffield A and Gordon A L 1992 Vertical mixing in the Indonesian thermocline *J. Phys. Oceanogr.* **22** 184-195
[4] Gordon A L 2005 Oceanography of the Indonesian Seas and their Throughflow *Oceanography* **18** 14-27
[5] Gordon A L, Susanto R D, Ffield R D, Huber B A, Pranowo W and Wirasantosa S 2008 Makassar Strait Throughflow, 2004 to 2006 *Geophys. Res. Lett.* **35** 1-5
[6] van Aken H M, Brodjonegoro I S and Jaya I 2009 The deep-water motion through the Lifamatola Passage and its contribution to the Indonesian Throughflow *Deep-Sea Res.* **56** 1203-1216
[7] Atmadipoera A, Molcard R, Madec G, Wijffels S, Sprintall J, Koch-Larrouy A, Jaya I and Supangat A 2009 Characteristics and variability of the Indonesian throughflow water at the outflow straits *Deep-See Res. I.* **56** 1942-1954
[8] Molcard R, Fieux M and Syamsudin F 2001 The throughfow within Ombai Strait *Deep-Sea Res. Pt. I* **48** 1237-1253
[9] Sprintall J, Wijffels S E, Molcard R and Jaya I 2009 Direct estimation of the Indonesian Throughflow entering the Indian Ocean: 2004-2009 *J. Geophys. Res.* **114** 1-19
[10] Hautala S L, Sprintall J, Potemra J, Ilahude A G, Chong J C, Pandoe W and Bray N 2001 Velocity structure and transport of the Indonesian Throughflow in the major straits restricting flow into the Indian Ocean *J. Geophys. Res.* **106(19)** 527-546.
[11] Sprintall J, Wijffels S E, Molcard R and Jaya I 2010 Direct evidence of the South Java current system in Ombai Strait *Dyn. Atmos. Oceans* **50**(2) 140–156
[12] Shea E R and Broenkow W W 1981 The rule of internal tides in the nutrient enrichment of Monterey Bay, California *Estuar. Coast. Shelf. S.* **15** 57-66
[13] Gordon A L, Susanto R D and Vranes K 2003 Cool Indonesian Throughflow as a consequence of restricted surface layer flow *Nature* **425** 824–828
[14] Nakamura T, Awaji T, Hatayama T, Akitomo K, Takizawa T, Kono T and Fukasawa M 2000: The generation of large-amplitude unsteady lee waves by subinertial K1 tidal flow: A possiblevertical mixing mechanism in the Kuril Straits *J. Phys. Oceanogr.* **30** 1601–1621
[15] Carrere L and Lyard F 2003 Modeling the barotropic response of the global ocean to atmospheric wind and pressure forcing - comparisons with observations *Geophys. Res. Lett.* **30**(6) 1275
[16] Koch-Larrouy A, Madec G, Bouruet-Aubertot P and Gerkema T 2007 On the transformation of Pacific Water into Indonesian Throughflow water by internal tidal mixing *Geophys. Res. Lett.* **34** 1-6
[17] Robertson R and Ffield A 2005 M2 baroclinic tides in the Indonesian Seas *J. Oceanogr.* **18** 62-73
[18] Garrett C 2003 Internal tides and ocean mixing *Science* **301**(5641) 1858-1859
[19] Laurent L S and Garrett C 2002 The role of internal tides in mixing the deep ocean *J. Phys. Oceanogr.* **32** 2882-2899
[20] Suteja Y, Atmadipoera A S and Purba M 2015 Turbulent mixing in Ombai Strait (in Bahasa) *Jurnal Ilmu dan Teknologi Kelautan Tropis* **7**(1) 71-82

IOP Conf. Series: Earth and Environmental Science **176** (2018) 012017 doi:10.1088/1755-1315/176/1/012017

[21] Koch-Larrouy A, Atmadipoera A, van Beek P, Madec G, Aucan J, Lyard F, Grelet J and Souhaut M 2015 Estimates of tidal mixing in the Indonesian Archipelago from multidisciplinary INDOMIX in-situ data. *Deep-Sea Res. Pt. I.* **106** 136-153

[22] Bouruet-Aubertot P, Cuypers Y, Ferron B, Dausse D, Ménage O, Atmadipoera A and Jaya I 2018 Contrasted turbulence intensities in the Indonesian Throughflow: a challenge for parameterizing energy dissipation rate Ocean Dynamics (in press).

[23] Thurnherr A M 2008 How to process LADCP data with the LDEO software (last updated for version IX.5)

[24] Visbeck M 2002 Deep velocity profiling using lowered acoustic Doppler Current Profilers: bottom track and inverse solutions *J. of Atmos. and Oceanic Tech.* **19** 794-807

[25] Thurnherr A M 2014 How to process LADCP data with the LDEO software (Versions IX.7 – IX.10)

[26] Firing E and Hummon J M 2010 Shipboard ADCP measurements. The GO-SHIP repeat hydrography manual: a collection of expert reports and guidelines. IOCCP report 14, *ICPO publication series* **134**

[27] Song Q and Gordon A L 2004 Significance of the vertical profile of the Indonesian Throughflow transport to the Indian Ocean *Geophys. Res. Lett.* **31** 1-4

[28] Hatayama T, Awaji T and Akitomo A 1996 Tidal currents in the Indonesian seas and their effect on transport and mixing *J. Geophys. Res.* **101**(12) 353-373.

[29] Cresswell G, Frische A, Peterson J and Quadfasel D 1993 Circulation in the Timor Sea *J. Geophys. Res.* **98**(C8) 14379-14389

[30] Molcard R, Fieux M, Swallow J C, Ilahude A G, and Banjarnahor J 1994 Low frequency variability of the currents in Indonesian Channels *Deep-Sea Res.* **41** 1643-1661

[31] Molcard R, Fieux M and Ilahude A G 1996 The Indo-Pacific throughflow in the Timor Passage *J. Geophys. Res.* **101**(C5) 12411-12420

[32] Robertson R and Ffield A 2005 M2 Baroclinic tides in the Indoensian Seas. *J. Oceanogr.* **18** 62-73

[33] Shanmugam G 2014 Review of research in internal-wave and internal-tide deposits of China:: Discussion. *J. Palaeogeogr.* **3**(4) 332-350

[34] Van Bennekom A J 1988 Deep-water transit times in the eastern Indonesian basins, calculated from disolved silica in deep and interstitial water *Neth. J. Sea Res.* **22** 341-354

[35] Fieux M, Molcard R and Ilahude A G 1996 Geostrophic transport of the Pacific-Indian Oceans throughflow *J. Geophys. Res.* **101**(C5) 12421-12432

[36] Nugroho D, Koch-Larrouy A, Gaspar P, Lyard F, Reffray G and Tranchant B 2017 Modelling explicit tides in the Indonesias seas: an important process for surface water properties *Mar. Poll. Bull.* doi: 10.1016/j.marpolbul.2017.06.033.

[37] Thorpe S A 2007 *An introduction to ocean turbulence* (Cambridge: Cambridge University Press) p 235

[38] Polzin K L, Toole J M, Ledwell J R and Schmitt R W 1997 Spatial variability of turbulent mixing in the Abyssal Ocean *Science* **276** 93-96

Upwelling characteristics during El Nino 2015 in Maluku Sea

A S Atmadipoera[1*], Z Khairunnisa[1], D W Kusuma[2]

[1]Department of Marine Science and Technology, Bogor Agricultural University [2]Marine Observation and Research Center (BPOL-KKP), Bali
*e-mail: atmadipoera_itk@ipb.ac.id

Abstract. Investigation of upwelling characteristics in Banggai-Maluku Sea (BMS) during super El Nino 2015 event and its interannual variation related to ENSO used validated INDESO model output and satellite imagery datasets between 2008 and 2015. The result shows the upwelling episode occurs during the Southeast Monsoon period from June to October, and its maximum in September. It is mainly forced by fully developed southerly monsoon winds, dragging surface water northward similar to the wind direction since the effect of Coriolis vanishes near the equator. Warm surface water in the center upwelling is replaced by upwelled colder water from about 60 m, which evolves from southern to northern region then curving to northeastern due to the boundary of Sulawesi mainland. Upwelling characteristics in 2015 event is indicated by drastic changes in ocean-atmosphere variables, such as increased northward meridional winds/ infrared heat flux/ transport volume, decreased sensible heat flux/ mixed layer depth, upwelled 26°C isotherm from 60 m depth to surface, and blooming surface chlorophyll-a. Between 2008 and 2015, upwelling intensity in 2015 was the most powerful (index: -5.35), in contrast to upwelling intensity during 2010 strong La Nina event (+0.27). Hence, ENSO influences significantly on fluctuation of upwelling intensity in BMS.

1. Introduction

Upwelling is an oceanographic phenomenon of which deeper and colder water mass rises towards the surface, replacing warmer surface water mass [1]. There are three types of upwelling: coastal, equatorial, and polar upwelling. Coastal upwelling happens when water depletion occurs in upper layer and space is vacated by offshore drift, which is affected by wind and Ekman drift [2]. The trade winds generate equatorial upwelling, which leads to water mass uplifting to both northern and southern hemispheres, while polar upwelling happens near ice edge. There are several factors affected upwelling, such as wind, topographic condition, length and curvature of the coast [20].

Rising water mass in upwelling affects water mass characteristics. Upwelled water mass has cooler temperature with higher salinity, thus affecting surface temperature and salinity distribution in the region. It is also rich of nutrients, related to higher primary productivity [19]. Therefore, upwelling drives important biological consequences, and from there related to marine fisheries [12]. It also alters weather as [19] stated that weather onshore of upwelling region tend to be foggy with low stratus clouds, along with stable stratified atmosphere, less convection, and less rain.

Several methods are be able to understand ocean phenomenon such as upwelling. In-situ hydrographic measurement can be carried out to provide information on oceanographic conditions, such as temperature-salinity, nutrients, and chlorophyll-a distribution, conducted by [20] and [21] in Makassar upwelling region, based on MAJAFLOX Cruise 2015. Recently, remote sensing and numerical model has been widely used to complement field measurement, due to its inexpensive cost and larger spatial data retrieval, which depicts actual condition. Using remote sensing data, [3] and [4], studied upwelling in southern Makassar Strait and Gulf of Tomini, respectively. Both studies show SST cooling and chlorophyll-a blooming occurred during Southeast Monsoon, associated with the southerly winds fields. Numerical ocean model is also used to study upwelling, such as research on southern Makassar Strait by [5] using numerical model to analyze main upwelling generating force, and [6] using carbon-based productivity model to identify upwelling effect to primary productivity in Indonesian Sea. Recent hydrographic observation in southern Makassar reported upwelled water layer was found from subsurface (60 m) layer to surface as indicated by surfacing 25°C isotherm, 34.5 isohaline [21].

According to [17] and [27], Indonesian archipelago is the only region near the equator that connects two major oceans, the Pacific and Indian Oceans. This complicates the variability of ocean dynamics in the region, along with its complex topography, coastlines' geometry, and monsoonal winds. These factors may as well affect oceanography phenomena happening in the area. There are several known and indicated upwelling area in Indonesia, such as in southern Makassar Strait [5,7,18,21,22], in southern coast of Java and Sumatera [8,9], in Banda Sea [10,23], and in Timor Sea [11,12], as well as in Banggai waters in western Maluku Sea [4].

Maluku Sea located in the eastern Sulawesi Island and bordered with North Maluku Islands to the east. The deepest depth of Maluku Sea reaches 4,500 m, in contrast with shallow coastal water of Banggai Islands on its southwestern boundary, and opened passage to the Seram-Banda Sea in the southern boundary. There is still limited research on upwelling in Maluku Sea. There are SST cooling and chl-a blooming in Gulf of Tomini, affected by strong winds passing Maluku Sea, which is in good agreement with [4,13]. However, previous upwelling studies in this region have only provided a simplified physical mechanism analysis and only focused on surface features of upwelling. The objective of this study is to investigate physical processes and dynamics of upwelling, particularly changes in ocean-atmosphere variables before-during-after upwelling, mechanism of upwelling generating force during 2015 super El Nino event, and to analysis interannual variation of upwelling intensity related to ENSO between 2008-2015.

2. Methods

2.1. Study area

Study area is located off northeast Banggai waters in western Maluku Sea (figure 1). The southern boundary located in a shallow passage between Banggai and Taliabu Islands, connecting directly to northwestern Banda Sea. The eastern boundary is central deep (>2000 m) Maluku Sea, while western and northern boundaries are linked with outer Tomini Gulf and Gorontalo Sulawesi mainland. Peleng Strait is a narrow and shallow strait, located just in the northern Banggai Islands.

The sampling box (black rectangle) denotes for extraction of model/data validation. Black line represents *Hovmoller* analysis of ocean-atmosphere parameters. Yellow-line represents cross-section for seawater temperature, while red line indicates cross-section for transport-volume calculation. The daily time-series data analysis of model output and satellite imagery data comprised from 1 January to 31 December 2015, and interannual data analysis spanned from 2008 to 2015.

IOP Conf. Series: Earth and Environmental Science **176** (2018) 012018 doi:10.1088/1755-1315/176/1/012018

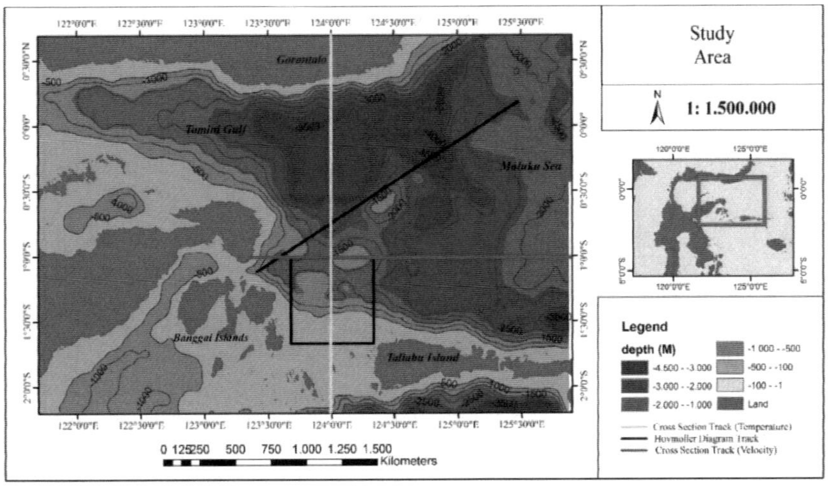

Figure 1. Study area off Banggai waters in western Maluku Sea.

2.2. *The data*

This study used several keys atmospheric and oceanic variables derived from satellite observation and INDESO model output. Those are sea surface temperature (SST), seawater temperature, wind speed, zonal and meridional wind stress components, latent heat flux, sensible heat flux, infrared heat flux, relative humidity, air temperature, instantaneous SST, salinity, and zonal and meridional current components. Model seawater temperature, salinity, and current are 3-dimensional data sets. These data acquired from the INDESO model. Configuration of this model described in detail [17]. Briefly, this model is based on NEMO (Nucleus for European Modeling of the Ocean) 9.0 [14] with 1/12° horizontal resolution. The horizontal grid is an extraction of global ORCA (global tripolar grid used in NEMO) grid at 1/12° developed at Mercator Ocean. The vertical grid spreads over 50 depth levels that a depth-dependent resolution (1 m at surface to 450 m near the bottom). Atmospheric forcing fields acquired from European Centre for Medium-Range Weather Forecasts (ECMWF).

Data accuracy of INDESO model output are validated using satellite imagery data. Satellite derived SST and chl-a data are used to map spatial distribution of upwelling indicated area. Satellite sea surface height anomaly (SSHA) and SST data were used to validate INDESO model output. The SSHA data with horizontal resolution of 1/4° or 27-28 km were obtained from the Copernicus Marine Environment Monitoring Service (http://marine.copernicus.eu/). The data based on assimilated multi-satellite data, such as Jason-3, Jason-2, and Cyrosat-2.

The SST and surface chl-a data are used to map spatial distribution data, were obtained from http://www.indeso.web.id/indeso_wp/index.php as remote sensing observation data of INDESO Project. SST data are derived from MODIS/AQUA and MODIS/TERRA satellite with horizontal resolution of 0.02°. Surface chl-a data are derived from Suomi-NPP satellite using VIIRS sensor with spatial resolution of 0.02°.

2.3. *Data analysis*

Data analysis of upwelling ocean-atmosphere variables follows procedure of time-series analysis methods [15]. Model outputs are validated using satellite data to estimate its accuracy. Both satellite observation data and INDESO model outputs reanalyzed and visualized. We computed and visualized monthly mean along the 2015 year to map the spatial evolution of upwelling from satellite imagery data. Model outputs reanalyzed to understand a mechanism of generating force of upwelling, to estimate transport volume, and fluctuation of atmospheric and oceanic variables. To analyze generating force of upwelling, winds fields and currents fields are visualized as monthly average. Transport volume estimation was calculated following [18]. The depth and area of upwelling are analyzed with vertical structure of meridional current component, leading to estimate upwelling transport volume. We analyzed and visualized fluctuation of atmospheric and oceanic variables were *Hovmoller* diagram throughout 2015. Time-series ocean-atmosphere data are extracted from sampling box (figure 1) from 2008-2015 to analyses interannual variation of the variables related to El Nino Southern Oscillation (ENSO). The comprehensive analysis on physical processes and dynamics of upwelling and its variability related to ENSO in western Maluku Sea used various methods.

2.4. Data validation

INDESO model outputs were validated using satellite data, by comparing sea surface height model with observed sea level anomaly and sea surface temperature with observed SST retrieved from Copernicus, Marine Environment Monitoring Service. The plotting data along 2015 within a sampling box area used for validation, as shown in figure 1. Correlation coefficient determine the degree to which two variables are significantly correlated [15].

Sea level anomaly from both data shows the maxima during peak of the Northwest Monsoon (NWM) period (January-February), which decreases gradually from the first Monsoon Break (MB1), from March to June. While the minima occur during the Southeast Monsoon (SEM) period (June-August) (figure 2). Amplitude of fluctuations reveals that sea level anomaly is high during the NWM period and is remarkably low during the SEM period, with much higher frequency fluctuation appeared in the model due to higher temporal resolution. Correlation coefficient between model and observed SSHA is highly good (0.8713). Comparison of model and observed SST data is significantly high (0.9149) (figure 3). Both data-series show similar fluctuation pattern throughout the year. According to [15], both values show that accuracy of model is pretty good agreement and highly correlated with the data. Thus, INDESO model depicts observed condition and the model data sets are suitable for further analysis.

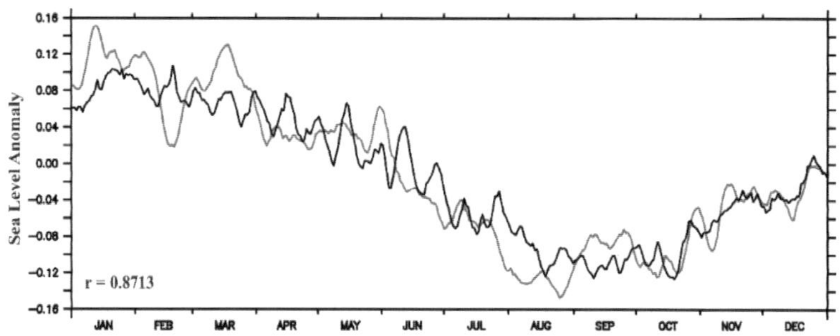

Figure 2. Comparison between INDESO model sea surface height anomaly (black line) and observed satellite data (red line) in 2015. Correlation coefficient between the data is 0.8713.

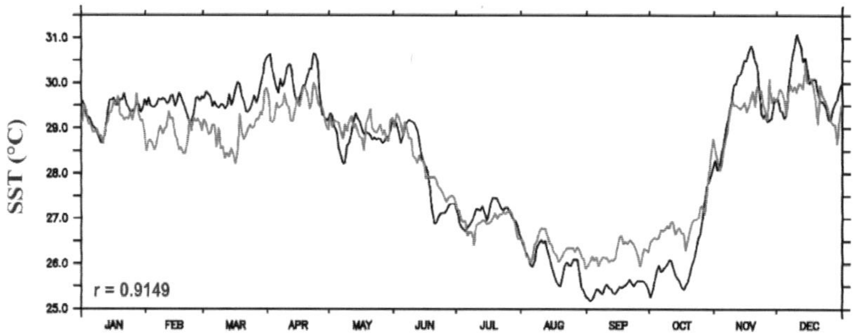

Figure 3. Comparison between sea surface temperature (black line) model and observed satellite data (red line) in 2015. Correlation coefficient is 0.9149.

3. Results and discussion

3.1. Surface features of upwelling

The first step to inspect upwelling event in the study area is from monthly mean SST and chl-a data from satellite imagery data. It reveals SST cooling and chl-a blooming related to upwelling that occurs during the SEM period from June to October 2015 (figure 4 and 5).

The SST during January-June varies from 27-31°C, with lower temperature is mostly seen in the southwestern region, close to central Sulawesi. In January, highest temperature can be seen around Banggai Island. Later in February and March, the SST distribution is similar, which shows much warmer water in the western part, while cooler water is seen in the northeastern region and closed to Taliabu Island (figure 4). As in May begins, marking the end of the MB1 period, SST starts to decrease, which vary from 27.4-30.4°C. Much cooler water is seen between Banggai and Taliabu waters with temperature of 27. 4 °C. This area of low SST evolves larger in June, the beginning of the SEM period, with SST ranges from 26.5-29.5°C. Low SST is distributed in most part of the region, except in Tomini Gulf. The lowest SST is found in between Banggai and Taliabu waters.

Contrast to first half of 2015, SST during the second half is lower, as shown cyan color dominated with hints of green and yellow, which ranges from 26-30°C. Highest temperature area can be seen around Tomini Gulf waters, while SST of Maluku Sea only reaches 27°C at most. High SST can be affected by region characteristics, as Tomini Gulf is semi enclosed waters (figure 4).

Low SST appears in the southern region, a pathway of water mass intrusion from northwestern Banda Sea. SST gradually increase towards the north, which changes to northeast direction near northern Gorontalo Sulawesi mainland (see figure 1). SST is generally minimum during the SEM period. Drastic changes appear during the SEM period which only reaches 26.2°C at most. Largest low SST can be seen in September, which then slightly weaken in October. This pattern changes drastically in November and December, which shows the lowest value of 28°C (figure 4).

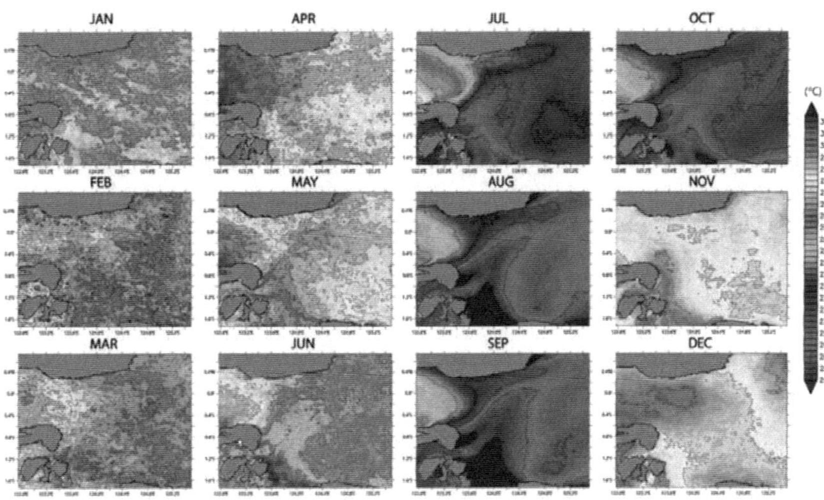

Figure 4. Evolution of monthly sea surface temperature in 2015, indicating surface features of upwelling off Banggai waters in Maluku Sea during the SEM period (June-October).

Chlorophyll-a (chl-a) in Maluku sea waters ranges from 0.02 – 12.1 mg/m^3 along 2015. High chl-a can be found in coastal area, which may be related to nutrient input from land which affects primary productivity. During the NWM period, surface chl-a ranges from 0.02 – 0.98 mg/ m^3, the highest chl-a exists in coastal area, while in Maluku Sea only reaches 0.68 mg/ m^3 with similar the coastal area (figure 5). It is in good agreement with [4], reporting that the chl-a in Maluku Sea ranges from 0.1 – 1.1 mg/ m^3 during the NWM period.

From January to April, there is no significant changes in spatial distribution of chl-a. But as the MB1 ends in May, high chl-a is found not only in coastal area but also off the Banggai waters. Tip of Tomini Gulf and Banggai waters and Taliabu waters reveal higher chl-a. It evolves into bigger area in June, which is shown by larger area (in green), which represents higher compared to its surroundings. The spatial distribution shows similar pattern with SST cooling (figure 5).

An increase of chl-a continues during the SEM period, reaching 0.98 mg/m^3 and above, with the highest chl-a is seen around Banggai Island and its eastern region. Area with high chl-a enlarges, evolve in northeast direction to northern Sulawesi, with the lower chl-a is seen in Peleng Strait, just north of Banggai Island.

Large seasonal variation of SST and chl-a in Maluku Sea is consistent with previous study e.g. [4]. Seasonal peaks of chlorophyll-a blooming and SST cooling in Maluku Sea happens in August [13]. High chl-a is gradually decreasing in October, and can no longer be seen in November and December, with chl-a only reaches 0.32 mg/m^3.

Past study (e.g. [4]) proposed that mechanism regarding chl-a blooms in Maluku Sea is associated with increased alongshore wind fields that drag cold water from beneath mixed layer, which support to phytoplankton growth in the region. Following section, we discussed more detail seasonal changes of surface wind fields and near-surface current in the region related to upwelling event.

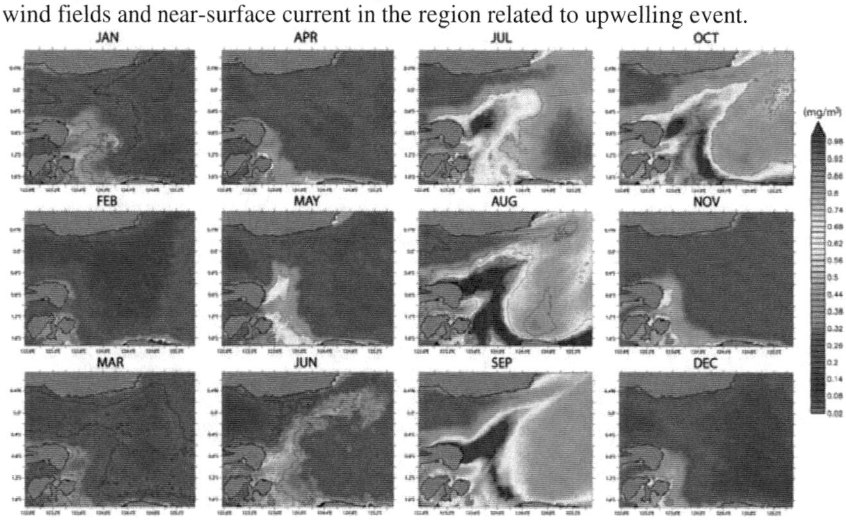

Figure 5. Evolution of monthly surface chl-a concentration in 2015 that indicates the upwelling features during the SEM period in Banggai waters in Maluku Sea.

3.2. Generating mechanism of upwelling in Banggai waters

Monthly mean of wind stress distribution is shown in figure 6. The arrows represent wind stress vectors, and its length represents wind stress magnitude. Spatial distribution of wind stress magnitude is also shown, ranging from 0 to 0.06 N/m^2.

During the NWM period, intensity of wind stress is rather high. During the first three months, northwesterly and northerly wind fields blow from Tomini Gulf and Maluku Sea to Banda Sea, which weaken as the MB1 period begins in March. In April, wind stress significantly weakens, as seen on the figure 6, the color shown in April is different, and wind stress only reaches 0.008 N/m^2. As May begins, southerly reversal wind stress occurs Banda Sea, heading northern Sulawesi. The wind fields intensify in June, and get much stronger when the SEM period begins, with wind stress reaching 0.048 N/m^2.

During the SEM period southerly monsoonal wind stress fields are fully excited. Wind stress with highest intensity tend to come from Banda Sea, which then head to Tomini Gulf and Maluku Sea. This high intensity wind stress reaches its peak in August. In September and October, wind stress gradually weakens, while in November it reaches almost calm over the region, as shown with significant changes of color, with wind stress only reaches 0.008 N/m^2. As the NWM period begins in December, the

IOP Conf. Series: Earth and Environmental Science **176** (2018) 012018 doi:10.1088/1755-1315/176/1/012018

northerly-northwesterly wind fields appear with only 0.032 N/m^2, heading to Banda Sea from Maluku Sea and Tomini Gulf.

Wind stress gradually increase of wind stress from June to August, as the southerly monsoon winds begin to intensify in July. This strong alongshore wind stress exceeding 0.06 N/m^2 is clearly seen from the passage between Banggai and Taliabu Islands, to the tip of Tomini Gulf, causing SST to drop by 2°C and increasing the surface chlorophyll-a of about 0.7 mg/m^3(figure 6). This pattern is related to the changes of SST distribution in the study area. SST shows significant decrease during the SEM period, which can be seen in the area between Banggai and Taliabu Island to its north, until the mouth of Tomini Gulf. Area with low SST widens to the northeastern waters.

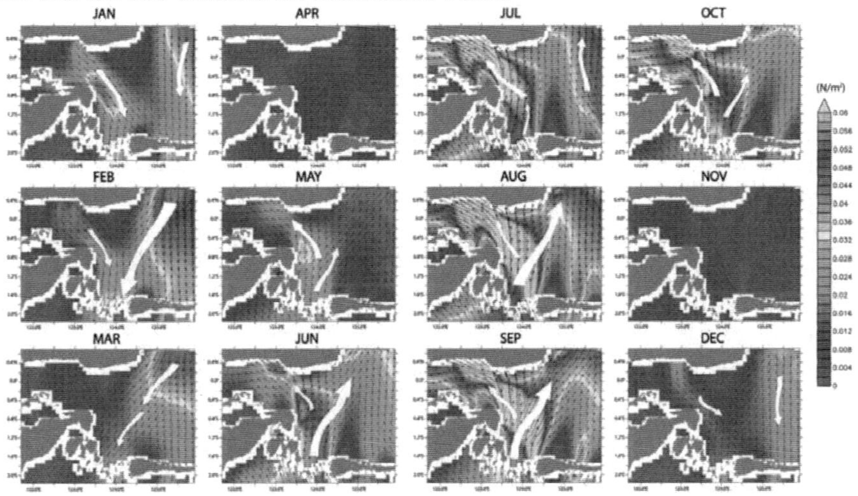

Figure 6. Evolution of monthly mean wind stress vectors in 2015.

Normal Ekman upwelling event tends to be affected by Coriolis effect, deflecting surface currents from wind direction, to the right of the wind direction in the northern hemisphere, and to the left in the southern hemisphere. Following Ekman theory, Ekman transport due to wind stress drag, net water movement is directed perpendicular to wind direction, to the left of the wind stress in southern hemisphere and to the right in the northern hemisphere [2]. The illustration of upwelling mechanism can be seen in figure 7.

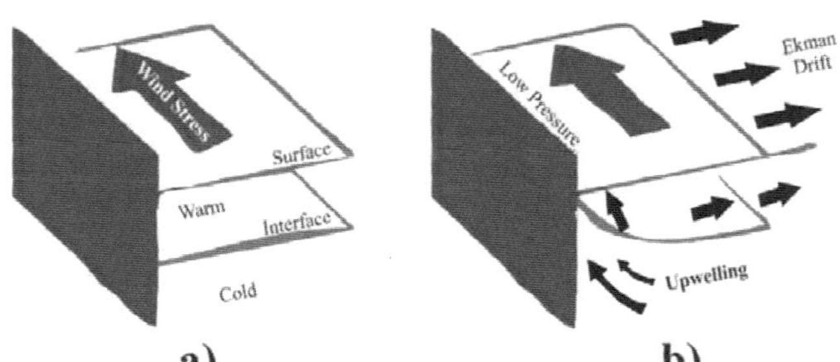

Figure 7. Schematic Ekman upwelling, (a) initial condition and (b) upwelling condition. In northern hemisphere, southerly winds stress result Ekman drift (transport) to the right of the wind direction. (after [2]).

Upwelling area in western Maluku Sea lies from the shallow passage between Banggai and Taliabu Islands to its northern waters. If we consider the Ekman theory with included Coriolis Effect in southern hemisphere, the Ekman transport due to southerly winds must be directed westward. However, Ekman drift in this study area is directed northward rather than westward. Since the study area is closed to the equator (2°S - 0.5°N), then the Coriolis Effect can be neglected. This implies that strong persistent southerly monsoonal winds fields during the SEM period drag directly surface water quasi-northward. Shallow passage between Banggai and Taliabu Islands is the initial path of colder water mass flow. Through the mechanism of wind-driven coastal upwelling by neglecting Coriolis Effect, surface water mass is dragged by the wind directly northward. The vacant surface water column is then replaced by colder water mass from deeper layer. Due to the boundary of Gorontalo Sulawesi mainland, the flow direction then changes to the northeast. During the MB2 in October and November, the wind stress weakens and upwelling event can no longer be seen, as seen from appearance of warmer SST and low surface chl-a in the southern area.

3.3. Near-surface circulation

As the wind stress alters the surface layer, surface current is established in upper-layer which is similar to the monsoonal wind field direction (figure 8). During the NWM period, surface circulation flows southward, though current vectors heading northeast can also be seen. It may be affected by secondary ITF Maluku pathway which recirculates in eastern part of Maluku Sea. Similar pattern can be seen during the MB1 period with lower magnitude. At the end of the MB1 period, sea surface temperature and surface circulation change and start flowing southward.

Entering the SEM period, the northward flow from northwestern Banda Sea can be seen in the passage between Banggai and Taliabu Islands. Strong northward currents from Banda Sea is associated with a development of colder STT. Spatial distribution of low SST also develops to the north direction. Surface current in the near-surface layer is directly affected by wind fields. The model shows no deflection on current direction relative to the wind, since Coriolis Effect vanishes in the equator area.

Due to this reason, current vectors during the SEM period flows quasi-northward. Shallow water between Banggai Islands and Taliabu Island as the pathway eases dragging of water mass. Strong southeast monsoonal wind generates stress on the surface. This causes dragging of water mass to certain water depth, and due to non-existent Coriolis' effect, water mass is dragged straight to the north, causing water masses on shallower area replaced by colder deeper water mass, commonly known as upwelling (figure 8).

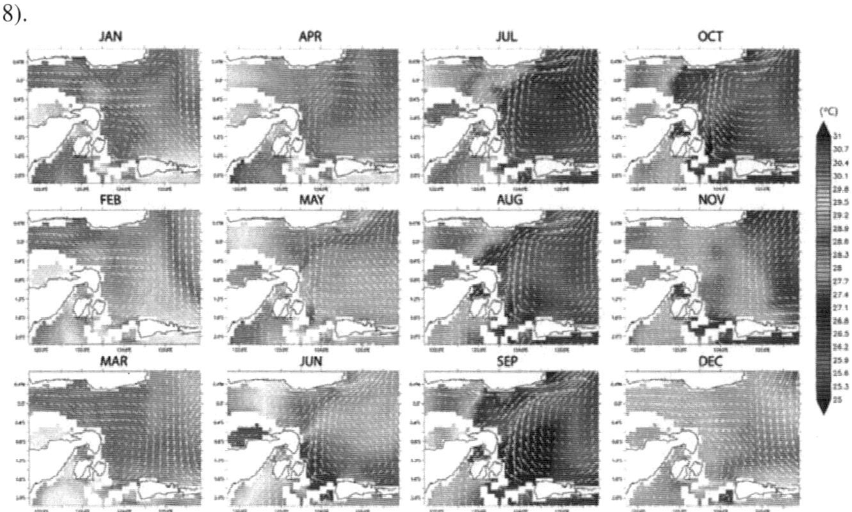

Figure 8. Monthly mean of model current vectors in 25 m depth and its seawater temperature in 2015

Vertical extent of seawater temperature and meridional current component during the NWM and SEM periods are shown in figure 9 and figure 10. It is seen clearly that near-surface warmer water with temperature above 28°C is found during the NWM period, in contrast to that the SEM period in which colder water above 24 °C is predominant. So that, temperature difference between two monsoon period is about 2 °C (figure 9). In general, seawater temperature ranges between 10 and 31°C during the NWM. During the SEM period, seawater temperature reaches 10°C in deeper layer, but the upper layer has lower temperature compared to the NWM period. Highest temperature only reaches 28°C, and can be seen along the track. In mixed layer, the temperature difference reach 2°C cooler during the SEM period, while in deeper layer, temperature difference is quite small. This suits the spatial distribution of SST, which lower during the SEM period, and evolve from southern boundary, in shallow passage between Banggai and Taliabu Islands towards the north. As seen in figure.9, isotherm of 26 °C is outcropped to surface layer during the SEM "upwelling" period. Coldest water with temperature 25-26 °C can be seen along 1.8°S to 1°S. This is the center of upwelling in Banggai waters. Colder surface water is developed due to upwelled water from deeper layer. During the NWM period isotherm of 26 °C lies at 80 m depth, but it uplifts to surface layer during the SEM "upwelling" period (figure 9).

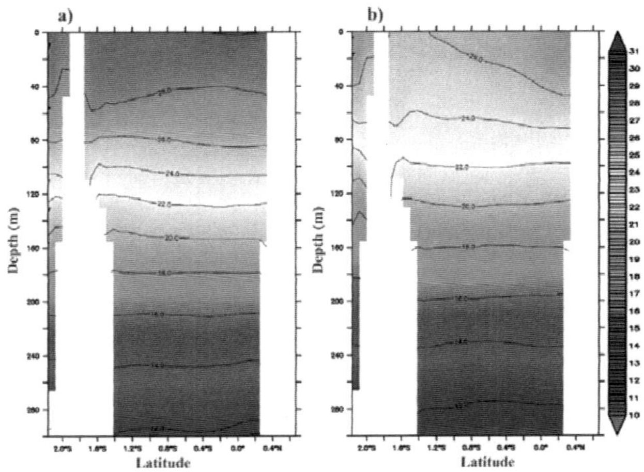

Figure 9. Vertical section of seawater temperature along 124°E (a) during the NMW period in February, and (b)during the SEM period in August 2015. Upwelling is indicated by outcropped isotherm of 26°C to the sea surface (b).

Vertical section of meridional current component during different monsoon period shows large seasonal variation of northward surface flow associated with upwelling event. During the NWM period, strong southward flow near surface layer is found (figure 10a), but closed to coastal area reveals weak northward flow. However, during the SEM period along with the occurrence of upwelling event, much stronger northward flow occurs (figure 10b). This northward current transports COLDER surface water mass from the center of upwelling in the shallow passage between Banggai and Taliabu Islands.

Vertical distribution of meridional current component varies with depth, which much stronger at the surface and much weaker at deeper depth. During the NWM period (represented by February), meridional velocity component reaches above 0.5 m/s that flows southward in surface layer and quasi-vanishes at deeper depth (figure 10a). During the SEM period (August), meridional velocity component reaches 0.78 m/s northward in surface layer and weakens at 80 m depth. At deeper depth (e.g. 160 m depth) meridional velocity component southward is seen, exceeding magnitude of 0.3 m/s.

Upwelling event in Banggai waters indicated chl-a blooming and SST cooling, along with much stronger southerly wind stress and northward surface current. The upwelling appears during the SEM period, commenced in June and terminated in October with its peak in August. During the peak of upwelling in August, isotherm of 26 is outcropped at the sea surface, and seawater temperature drops to about 2 °C, compared to that the NWM period. Vertical extent of upwelling depth is estimated to about 60 m depth. Upwelling indicated region is shown by the area marked by red square (figure 10b), which shows strong northward flow. This region reaches about 60 m deep, with area approximately 6.72 km^2. From this area, transport volume estimation can be known, with the value of 0.4 Sv. Intensity of transport volume of upwelling water mass may differ from area to area, depending on magnitude of the current, and transport calculation approach. One of the external factors may contribute to upwelling in Maluku

Sea is associated with interannual climate anomaly of El-Nino Southern Oscillation (ENSO), where Maluku Sea showed significant correlation with ENSO [5].

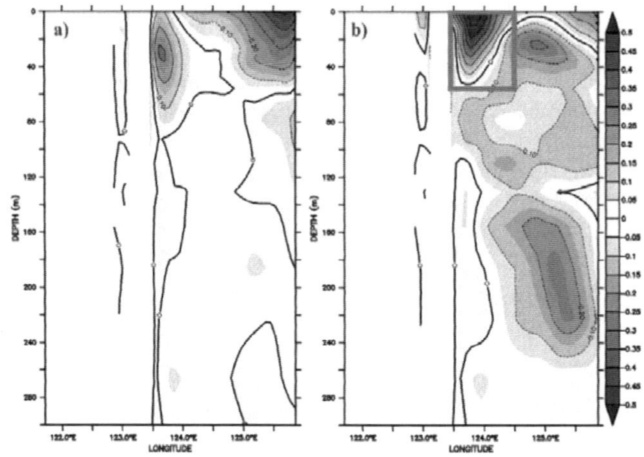

Figure 10. Vertical section of meridional current component in February (a) and August (b) 2015. Note: red rectangle denotes upwelling extent during the SEM period.

3.4. Evolution of ocean-atmosphere changes related to upwelling episode

Heat fluxes and instantaneous SST changed during the SEM "upwelling" period. While instantaneous SST cools, latent heat flux increases. Infrared heat flux weakens during the SEM period and no significant changes can be seen in relative humidity (not shown). Heat flux intensification may lead to heat loss in the mixed layer of atmosphere, and when heat loss exceeds, air temperature decreases. This may explain the relation between increased sensible heat flux intensification and decreased air temperature. Wind stress intensification may also affect heat flux intensification [16].

Strong seasonal variation is found from the ocean-atmosphere variables, contrasting between two monsoon periods. During the SEM period, modulus of wind speed is strongest, coincides with wind stress. Compared to zonal wind stress component, meridional wind stress component is significantly stronger. Air temperature cools during the SEM period. According to [16], wind stress intensification may affect heat flux intensification. Heat flux intensification may lead to heat loss in the mixed layer of atmosphere, and when heat loss is bigger than heat gain, temperature will be lower. This explains the changes of sensible heat flux that intensifies while air temperature is cooling.

Stronger wind stress drags surface water masses northward, which replaced by colder water mass from water column, generally known as upwelling. While simultaneously, wind stress is also affecting heat flux and heat loss in the study area, which explains low air temperature (not shown).

Time-series of ocean-atmosphere variables, extracted from a rectangle sampling box in figure 1, is shown in figure 11. Rapid inspection from the series revealed that negative anomaly of SST is commenced from June to October 2015. This seasonal change of SST is consistent with others variables (figure 11), suggesting that upwelling onset begins in June and is terminated in October. Upwelling affects on altering water mass as response to water mass uplifting, along with changes on atmospheric variables. Large seasonal fluctuation is found from the series. SST is minimum in June, then increased gradually in October. Contrary to SST, chl-a increased during the same period in the year. It increased significantly during June-October, as it reached maxima of 12.1 mg/m^3. Ekman drift due to strong persistent southerly monsoon winds field during the SEM period plays a role as the uplifted colder water mass from deeper water column (60 m depth). From there, higher nutrient colder water contributed to higher primary productivity, hence higher chl-a.

As discussed previously, wind stress generated upwelling in the area. Upwelling area evolved northward from southern boundary of shallow passage between Banggai and Taliabu Islands. Thus, fluctuation of meridional wind stress is examined (figure 11c). Started from June, meridional wind stress increased, reached its peak in September, and decreased again in October. Estimated transported volume of water mass also shows similar fluctuation (figure 11d). Northward transport (positive anomaly) appeared during the SEM period that transport surface water mass from southern boundary, while during the NWM period negative anomaly means southward transport volume. Much stronger southward transport is found in November-December 2015.

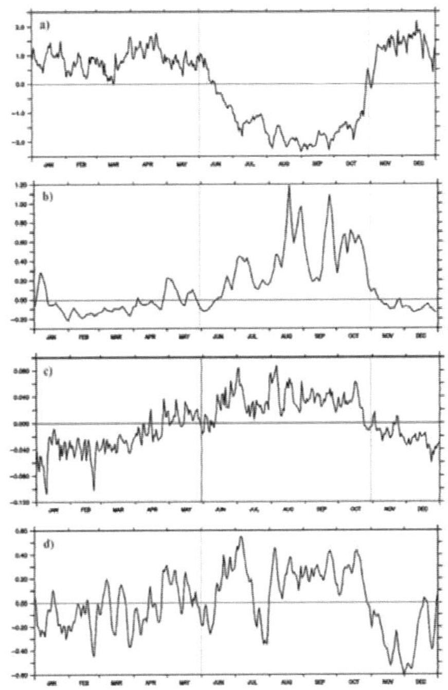

Figure 11. Anomaly of a) SST, b) chlorophyll-a, c) meridional wind stress, and d) transport volume estimate in 2015

Seasonal variation of atmospheric variables is also found (figure 12). Infrared heat flux and air temperature are minima during the SEM period, while sensible heat flux increases. This changes may occur due to heat loss, as heat flux intensifies when wind stress intensifies (figure 12). This is consistent with increased meridional wind stress during the SEM (figure 11).

As the atmospheric characteristics are changing, the changes on water mass also occur. Northward transport volume is higher during the SEM period, resulting in changes of sea surface temperature and chlorophyll-a (figure 11).

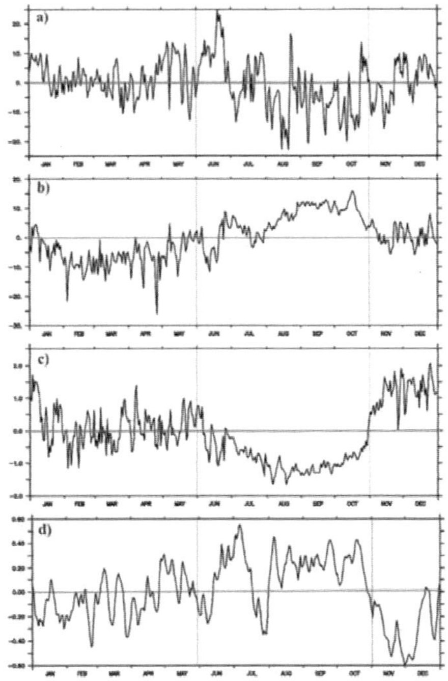

Figure 12. Anomaly plot of a) infrared heat flux, b) sensible heat flux, c) air temperature, and d) transport volume estimate in 2015

Seasonal fluctuation as mentioned above, can be seen as the effect of upwelling. Wind stress generates the upwelling in Maluku Sea, and induced the changes on oceanic and atmospheric variables. Initial condition before onset of upwelling, strong persistent southerly wind induced surface flow northward. This then generated the water mass transport. The initial upwelling pathway in shallow passage between Banggai and Taliabu Islands, possessed vacant surface layer due to transported water mass. This leads uplifting deeper colder water mass to replace the warmer surface transported water, which caused SST cooling in certain area, as shown with blue, includes the waters between Banggai and Taliabu Islands, to the tip of Tomini Gulf (figure 13).

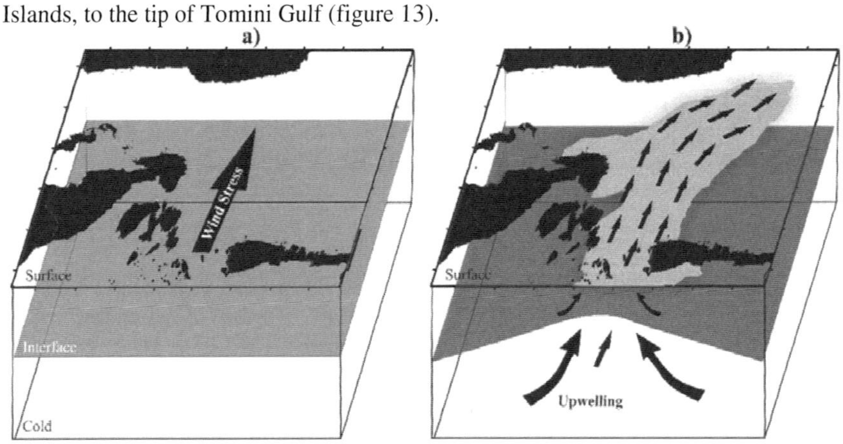

Figure 13. Schematic of upwelling process in Maluku Sea, a) initial condition and b) upwelling condition

The changes can also be seen quantitatively, as upwelling alters the condition of both oceanic and atmospheric parameters. As the upwelling onset in June begins, strong persistent southerly wind changes the wind stress magnitude. From there, northward wind-driven surface current strengthens and leads to water mass transport. It then alters SST, which cools down, as well as chl-a blooms. In addition to that, atmospheric variables also alter. These upwelling characteristics can be summarized in table 1 below.

Table 1. Characteristics of ocean-atmosphere variables during upwelling period in 2015.

Ocean-Atmosphere variables	Range (2015)	Range (Jun-Oct 2015) during upwelling period
SST (°C)	26 – 31	26 – 28
Chlorophyll-a (mg/m^3)	0.02 – 12.10	0.98 – 12.10
Wind Stress ((N/m^2)	0 – 0.06	0.032 – 0.060
Air Temperature (°C)	25.3 – 29.5	25.3 – 27.0
Sensible Heat Flux (W/m^2)	-42 – 10	-6 – 10

A proposed mechanism of upwelling in Maluku Sea is induced by southerly winds passing through Maluku Sea during the SEM period (figure 14). Wind stress shows that there is strong southerly wind blow during the SEM period. The wind stress generates wind-driven upwelling in western Maluku Sea. SST cooling and chl-a blooming are induced by the mechanism of wind-driven upwelling, the alongshore wind may drag the cold water from beneath the mixed layer to the surface in the Banggai tip waters, and from there, support phytoplankton growth in the region [4].

Wind intensification caused much stronger wind stress, affecting released heat flux. This causes an increase on seawater density, thus water mass sinks. In addition, the wind stress affects surface current in the mixed layer, dragging surface water mass northward. This is proposed mechanism of upwelling in Maluku Sea.

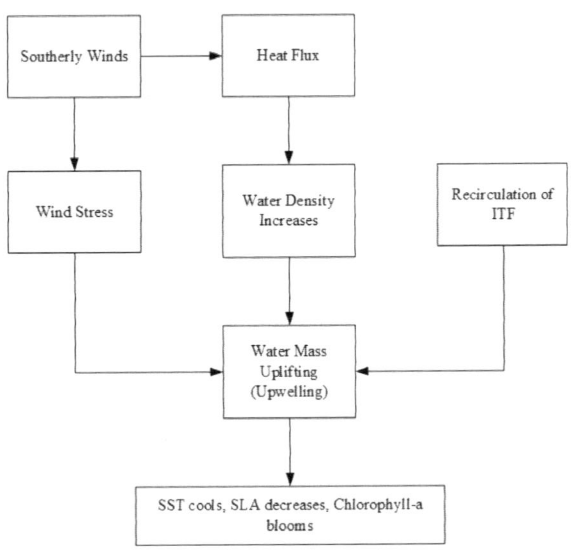

Figure 14. Proposed mechanism of upwelling of Banggai waters in Maluku Sea.

In the northern part of study area, recirculation of ITF Maluku plays a role on upwelling intensity during the SEM period. The existence of ITF's second pathway and Gorontalo Sulawesi mainland shape deflected upwelling area to the northeast. From the tip of Tomini Gulf, the upwelling area evolves toward northeast as Sulawesi mainland acts as solid boundary, as well as recirculation of ITF on the northeastern part of study area. Upwelling then causes SST cooling, decreasing sea level, as well as chl-a blooming.

3.5. Interannual variation of upwelling intensity related to ENSO

Analysis of monthly mean of satellite imagery and model output data sets in 2015 revealed clearly upwelling event off Banggai waters in Maluku Sea. Upwelling mechanism is mostly forced by the strong persistent southerly monsoon winds during the SEM period. Furthermore, initial upwelling region is located in southern boundary of a shallow passage between Banggai and Taliabu Islands, where it is directly linked with northwestern Banda Sea. It is found that upwelling event is commenced in June with its peak in August and is terminated in October. Ocean response to the upwelling processing is upwelled seawater temperature cooling with outcropped isotherm of 26 to the sea surface, chl-a blooming, and increased near-surface northward transport volume. These ocean responses propagate northward from initial upwelling area in the southern boundary. Meridional wind stress and current components show similar evolution.

In term of interannual variability, it is interesting to examine upwelling intensity during the 2015 upwelling event in Maluku Sea and to compare with past years of upwelling intensity, since the occurrence of 2015 upwelling event was coincident with the interannual climate anomaly of El Nino Southern Oscillation (ENSO) 2015/16. The 2015/2016 El Nino event is one of the strongest on record, comparable to the 1982-1983 and 1997-1998 events that triggered widespread ecosystem change in the northeast Pacific [23]. The 2015-2016 El Nino event is categorized as super El Nino [24]. Upwelling intensity along the southern Java Sumatera waters was found much stronger during the El Nino year [8]. However, study of interannual variation of upwelling in interior Indonesian seas, such as in Banggai waters Maluku Sea, needs to be advanced.

Time series of standardized southern oscillation index (SOI) obtained from [22], shows ENSO events that appeared between 2008-2015 (figure 15a). Positive anomaly (red) of the SOI represents La Nina events, while negative anomaly (blue) is for El Nino events. The ENSO events are defined if the index anomaly exceeds ±2. Over that period, there are weak La Nina 2008/2009, strong La Nina 2010/2011, and weak El Nino 2009/2010 and strong El Nino 2014/2015.

Time series of anomaly of ocean-atmosphere variables (extracted from sampling box in figure 1) between 2008 and 2015 is shown in figure 15(b-g). It is apparent that the variables indicate interannual variability related to ENSO (figure 15a), which show two major events, La Nina 2010/2011 and El-Nino 2014/2015. Peak of La Nina can be seen in 2010 while El Nino in 2015. Anomaly of ocean-atmosphere variables present on figure 15(b-g), with annual mean removed to seek better understanding on interannual variability.

The ocean-atmosphere variables show interannual variability, in which SST, SSH, and seawater temperature are well positive correlated, while meridional wind stress, salinity (0-25 m), and meridional current (0-25 m) are well negative correlated with SOI indices. Meridional wind stress (figure 15b) exhibits positive (negative) anomaly during El Nino (La Nina). The positive anomaly during El Nino 2015 is not as significant as the negative anomaly during 2010 La Nina, but it changes to great negative anomaly during the end of El Nino 2015. SST shows different anomaly during La Nina and El Nino. During El Nino (La Nina) event, negative (positive) anomaly of SST occurs, with SST anomaly significantly increased as El Nino 2015 ends. Similar event shown in figure 15d of SSH, where negative (positive) anomaly can be seen during El Nino (La Nina), with more significant anomaly changes seen during El Nino 2015.

Aside from the surface variables, oceanic parameters in the mixed layer alters as well. According to [13], ENSO modified the upwelling's magnitude in Maluku Sea. The western Pacific surface temperature plays role on controlling monsoon, with higher (lower) surface temperatures during La Nina (El Nino) years, leading to stronger monsoon [19]. Stronger southeast monsoon then leads to stronger southerly winds in the study area. This southerly winds impact on enhanced (reduced) the wind stress during El Nino (La Nina), shown by positive anomaly of meridional wind stress. Thus, alters the SST and SSH anomaly as well. The enhanced upwelling intensity during El Nino is indicated by the anomaly on mixed layer (figure 15d-g).

During the El Nino (La Nina) event, negative (positive) anomaly of seawater temperature (0-25 m) occurs. Greater anomaly can be seen during El Nino 2015. On contrary, salinity on the upper 25 m depth displays positive anomaly during El Nino. The increased southerly meridional wind stress during El Nino generates greater northward meridional current, shown by positive anomaly during El Nino 2015, which is much stronger in comparison with negative anomaly during La Nina. The greater meridional current drags surface water mass, letting a vacant area on the surface then filled with water mass from deeper colder water mass. The latter water mass has different characteristics, hence the increase of salinity and seawater temperature on the upper 25 m depth during El Nino.

Figure 16 summarizes interannual variation of ocean-atmosphere variables, related to upwelling intensity, during each upwelling period from 2008-2015. In general, figure 16a shows similar variability throughout 2008-2015, where positive anomaly of each variables can be seen during 2010 La Nina event, while the strongest negative anomaly is seen during 2015 El Nino event. Less magnitude of change is found in air temperature, while the other three variables do not differ as much. Changes on SST as well as seawater temperature in mixed layer (T25m and T50m) are bigger as it is directly affected by the uplifting water mass during upwelling.

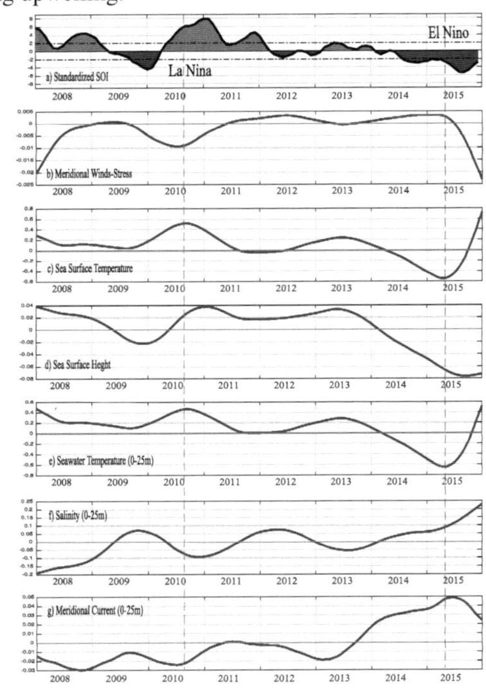

Figure 15. Time-series of ENSO indices (represented by SOI indices, a) and anomaly of ocean-atmosphere variables (b-g) in Banggai upwelling area in Maluku Sea. Note: daily ocean-atmosphere variables (b-g) are removed annual mean.

During the 2015 El Nino event positive anomaly of meridional wind stress is found, which is similar to that meridional current component in the upper 25 m depth (V25m) (figure 16b). During 2010 La Nina event, the zonal current component reaches positive anomaly while the others have the strong negative anomaly. For the meridional current component, it shows strong positive anomaly during 2015 El Nino event and significantly differs from the other three variables. This may occur due to enhanced upwelling intensity as impacted by El Nino event.

Table 2 summarizes anomaly of ocean-atmosphere variables during upwelling period (averaged between June-October) for each each year. Upwelling intensity (UI) is also shown in the last column, which is defined as the sum of anomaly of 9 ocean-atmosphere variables. As discussed previously, anomaly of ocean-atmosphere variables between 2008-2015 differs significantly during 2 specific periods: 2010 La Nina and 2015 El Nino events (figure 15 and 16). Meridional wind component acts as the local forcing of the upwelling, which was very weak during La Nina event, while the strongest value can be seen during El Nino event. This leads to much stronger meridional current component in the upper 25 m depth as well. Direction of meridional component flow southward during La Nina event, and in contrast, flows northward with much stronger value during El Nino 2015.

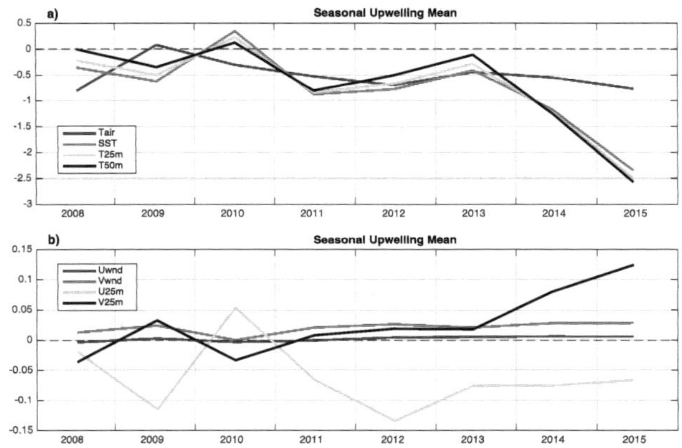

Figure 16. Interannual fluctuation of ocean-atmosphere variables in the upwelling region (off Banggai waters) between 2008 and 2015. The variables are averaged from June to October, representing upwelling intensity for each upwelling period; (a) air temperature (Tair), sea surface temperature (SST), seawater temperature 0-25m (T25m), seawater temperature 0-50m (T50m); (b) zonal wind component (Uwnd), meridional wind component (Vwnd), zonal current component 0-25 m (U25m), and meridional current component 0-25m (V25m).

Finally, it is evident that on interannual time-scale, fluctuation of upwelling intensity (UI) in Banggai waters exhibits positive anomaly (+0.27) during 2010 La Nina event, in contrast to 2015 super El Nino with the strongest negative anomaly of UI (-5.35) (table 2). Thus, it can be summarized that upwelling intensity during the super 2015 El Nino event is the most powerful upwelling intensity, compared to others ENSO events between 2008-2015. It is also shown that the weakest upwelling intensity is noted during the strong 2010 la Nina event. This suggests that upwelling intensity in Banggai waters are significantly impacted by external factors such as ENSO.

Table 2. Anomaly of ocean-atmosphere variables associated with *Upwelling Intensity* (UI). Each variable is averaged between June and October for each upwelling period.

Year	Tair	SST	SSH	uwind	vwind	T25m	S25m	U25m	V25m	Upwelling Intensity
2008	-0.791	-0.355	0.003	-0.003	0.013	-0.220	-0.019	-0.020	-0.035	-1.43
2009	0.082	-0.623	-0.040	0.003	0.024	-0.499	0.317	-0.115	0.033	-0.82
2010	-0.302	0.346	0.025	-0.003	0.001	0.231	-0.045	0.053	-0.033	+0.27
2011	-0.526	-0.877	-0.043	-0.009	0.021	-0.842	0.107	-0.065	0.008	-2.22
2012	-0.696	-0.779	-0.018	0.004	0.027	-0.665	0.238	-0.134	0.019	-2.00
2013	-0.443	-0.408	0.021	0.005	0.021	-0.284	0.051	-0.076	0.018	-1.10
2014	-0.550	-1.169	-0.072	0.006	0.028	-1.215	0.233	-0.075	0.080	-2.73
2015	-0.767	-2.335	-0.144	0.006	0.028	-2.486	0.291	-0.066	0.124	-5.35

4. Conclusion

From satellite imagery data, upwelling event is indicated by SST cooling in June in southern boundary where it extends northward, and terminates in October. The lowest SST (<26°C) is found in September. The cooling episode is followed by chl-a blooming one-month later from July to October. Concentration of surface chl-a reaches its maximum in August (>1 mg/m^3). Generating mechanism of upwelling in the study area is forced by strong and persistent southerly monsoon winds stress that exists from June to October during the Southeast monsoon (SEM) period. Following upwelling theory, in high-latitude southern hemisphere, southerly winds stress drag surface water northward and Ekman drift deflected to the left from the wind direction. However, since the study area is just over the equator, then the Coriolis effect may be neglected. So that the Ekman drift must be similar to the wind direction. This "special coastal-equator upwelling case" can be proven from surface circulation pattern, where the current vectors flow quasi-northward along the eastern Banggai Waters.

During upwelling episode, northward transport volume is drastically increased from surface to about 60 m depth. Shallow passage between Banggai and Taliabu Islands provides an initial upwelling region, where dragged warm surface water is replaced by upwelled colder water from sub-surface (80 m depth) layer. The upwelling area then evolves northward. Recirculation of Maluku ITF in southern boundary may contribute to the northward flow of upwelling. Vertical section of seawater temperature shows that during upwelling episode, isotherm of 26°C is outcropped to the sea surface. In contrary, during non-upwelling (the NWM period) this 26°C isotherm lies deeper at 80 m depth and temperature of 28-29°C is dominant at surface. Ocean-atmosphere variables changes in response to the upwelling event.

On interannual time-scale related to ENSO between 2008-2015, the most powerful upwelling intensity in Banggai waters was revealed during super El Nino 2015 with index of (-5.35), in contrast to weak upwelling intensity during strong La Nina 2010 (+0.27). Hence, Pacific origin of ENSO controls significantly on physical processes and dynamics of upwelling in the study area.

Acknowledgements

We would like to thank INDESO project Indonesia for providing and accessing to the model output data sets. We are indebted to CLS France, Mercator-Ocean, and Copernicus operational oceanography of Europe for providing satellite imagery data sets. We wish to thank Reviewer(s) for comments and critics to improve the manuscript.

References

[1] Duke D L and Sinz P D 2010 *Exact Solutions for Wind-Driven Coastal Upwelling and Downwelling Over Sloping Bathymetry* (California: California Polythecnic State University)
[2] Cushman-Roisin B and Beckers J 2011 *Introduction to geophysical fluid dynamics: physical and numerical aspects* Academic Press
[3] Habibi A, Setiawan R Y and Zuhdy A Y 2010 Wind-driven coastal upwelling along south of Sulawesi Island *Ilmu Kelautan* **15**(2) 115-118
[4] Setiawan R Y and Habibi A 2011 Satellite detection of summer chlorophyll-a bloom in the Gulf of Tomini *IEEE Journal of Selected Topics in Applied Earth Observations and Remote Sensing* **4**(4) 944-948
[5] Atmadipoera A S and Widyastuti P 2014 A numerical study of upwelling mechanism in Southern Makasssar Strait *JITKT* **6** 355-371
[6] Kemili P and Putri M P 2012 Pengaruh durasi dan intensitas upwelling berdasarkan anomali suhu permukaan laut terhadap variabilitas produktivitas primer di perairan Indonesia *JITKT* **4**(1) 66-79
[7] Nontji A 2007 *Laut Nusantara* (Jakarta: Djambatan)
[8] Susanto R D, Gordon A L and Zheng Q 2001 Upwelling along the coasts of Java and Sumatra and its relation to ENSO *Geophys. Res. Letters* **28**(8) 1599-1602
[9] Kuswardani R T D and Qiao F 2014 Influence of the Indonesian throughflow on the upwelling off the east coast of South Java *Chin. Sci. Bull.* **59**(33) 4516–4523

[10] Gordon A L and Susanto R D 2001 Banda Sea surface-layer divergence *Ocean Dynamics* **52** 2-10

[11] Siswanto and Suratno 2008 Seasonal pattern of wind induced upwelling over Java-Bali sea waters and surrounding area *International J. of Remote Sensing and Earth Science* **5** 46-56

[12] Kunarso, Hadi S, Ningsih N S and Baskoro M S 2011 Variabilitas suhu dan klorofil-a di daerah upwelling pada variasi kejadian ENSO dan IOD di perairan Selatan Jawa sampai Timor *Ilmu Kelautan* **16**(3) 171-180

[13] Wirasatriya A, Setiawan R Y and Subardjo P 2017 The Effect of ENSO on the Variability of Chlorophyll-a and Sea Surface Temperature in the Maluku Sea *IEEE Journal of Selected Topics in Applied Earth Observations and Remote Sensing* 1-6

[14] Madec G, Delecluse P, Imbard M and Levy C 1998 *OPA 8.1 Ocean General Circulation Reference Manual* (Paris: Institut Pierre Simon Laplace des sciences de l'Environnement Global)

[15] Emery W J and Thomson R E 2014 *Data analysis methods in physical oceanography* (Waltham: Elsevier)

[16] Sterl A and Hazeleger W 2003 Coupled variability and air-sea interaction in the South Atlantic Ocean *Clim. Dyn.* **21** 550-571

[17] Tranchant B, Reffray G, Greiner E, Nugroho D, Koch-Larrouy A and Gaspar P 2015 Evaluation of an operational ocean model configuration at 1/12° spatial resolution for the Indonesian seas-part 1:Ocean physics *Geoscient Mod. Develop.* 8:6611-6668.

[18] Atmadipoera A S and Hasanah P 2017 Characteristics and Variability of Flores ITF and its coherence with the South Java coastal current. *J. Ilmu dan Teknologi Kelautan Tropis* **9**(2) 537-556

[19] Steinke S, Prange M, Feist C, Groeneveld J and Mohtadi M 2014 Upwelling variability off southern Indonesia over the past two millennia *Geophys. Res. Lett.* **41** 7684–7693

[20] Steward R H 2008 *Introduction to Physical Oceanography* (Texas: Department of Oceanography Texas A & M University)

[21] Utama F G, Atmadipoera A S, Purba M, Sudjono E H, Zuraida R 2017 Analysis of Upwelling event in Southern Makassar Strait *IOP Conference Series: Earth and Environmental Science* **54** 012085 doi: 10.1088/1755-1315/54/1/012085

[22] Rosdiana A, Prartono T, Atmadipoera A S and Zuraida R 2017 Nutrient and chlorophyll-a distribution in Makassar upwelling region: from MAJAFLOX Cruise 2015 *IOP Conference Series: Earth and Environmental Science* **54** 012087. doi: 10.1088/1755-1315/54/1/012087

[23] Horhoruw S M, Atmadipoera A S, Nanlohy P and Nurjaya I W 2017 Anomaly of surface circulation and Ekman transport in Banda Sea during normal and ENSO episode (2008-2011) *IOP Conference Series: Earth and Environmental Science* **54** 012087. doi: 10.1088/1755-1315/54/1/012041

[24] Australian Bureau of Meteorology (BOM) Southern Oscillation Index (SOI) 2018 http://poama.bom.gov.au/climate/enso/enlist/index.shtml

[25] Jacox M G, Hazen E L, Zaba K D, Rudnick D L, Edwards C A, Moore A M and Bograd S J 2016 Impacts of the 2015-2016 El Nino on the California Current System: Early assessment and comparison to past events *Geophys. Res. Lett* **43**(13) 7072–7080

[26] L'Heureux M L, Takahashi K, Watkins A B, Barnston A G, Becker E J, Di Liberto T E, Gamble F, Gottschalck J, Halpert M S, Huang B, Mosquera-Vasquez K and Wittenberg A T 2017 Observing and Predicting the 2015/16 El Nino *Bulletin of the American Meteorological Society (BAMS)* 1363-1382 doi: 10.1175/BAMS-D-16-0009.1

[27] Atmadipoera A S, Molcard R, Madec G, Wijffels S, Sprintall J, Koch-Larrouy A, Jaya I and Supangat A 2009 Characteristics nad variability of the Indonesian Throughflow water at the outflow straits *Dee Sea Research I* **56** 1942-1954

IOP Conf. Series: Earth and Environmental Science **176** (2018) 012019 doi:10.1088/1755-1315/176/1/012019

Water mass dynamics in Balikpapan Bay, Eastern Kalimantan Indonesia

I Wayan Nurjaya[1*], H Surbakti[2], M T Hartanto[1], J L Gaol[1], A Sulardi[1]

[1] Department of Marine Science and Technology, IPB Bogor Indonesia
[2] Marine Science Program, University of Sriwijaya, Indonesia
*e-mail: i.wayan.nurjaya@apps.ipb.ac.id

Abstract. Balikpapan Bay (BB) located 90 km southwest of Delta Mahakam is one of estuarine systems in eastern Kalimantan directly flows into Makassar Strait. Dimension of BB is about 36 km length and between 1-14 km width, with mouth is wider than head of the bay. Water depth is much deeper (18 m) at the middle part of the bay. More than 10 secondary rivers flow into the bay to contribute large amount of freshwater mass flux. To understand physical properties and water mass dynamics, field measurement during high and low tides were conducted. The results shows that the warmer and saltier water mass are dominant at the entrance of the bay at flood period. Warmer and less salty water mass remain in the inner bay. However, during ebb, salinity is slightly decreased near the entrance and freshwater are observed in the inner bay. During both high and low tides condition, vertical stratification of water mass is very weak. The model demonstrates that this is related to large seawater flux brought by flood current from coastal waters, as well as small freshwater influx from river runoffs. Hence, the tidal dynamics significantly controls stratification and dynamics of water mass in the bay.

1. Introduction

Balikpapan bay is a semi-enclosed bay where the exchange of water masses within the bay with the mass of water outside the bay still occurs, mainly generated by tidal forcing. At high tide the flow of water that comes out of the bay is retained by the tidal masses even most of the water mass is pushed into the bay upstream [1]. In the upper reaches there are many freshwater inputs through small rivers so the salinity of the bay water at the head of the bay is lower than the salinity of the water masses at the mouth of the bay.

Well known that main function of bay is as transportation media, recreation and potential area for fishing ground, spawning ground and feeding ground for several kinds of fish [2], [3]. On the contrary, bay often undergoes sedimentation due to relatively sheltered [4], contaminated area from pollutant [5]. Freshwater from the river flow into the bay and mix with water mass from the Makassar Strait, formed salinity gradient (vertically or horizontally) from the mouth to the head of bay, on the contrary. Mixing type of water mass within bay depends on bay depth, opening of the bay's mouth, freshwater volume flow into the bay and tidal range [6], [7], [8], [9]. Several studies have been conducted in Balikpapan Bay, for example related to the heat distribution pattern of Pertamina UP V activities in Kampung Baru [10]; mangrove vegetation [11]. Study on erosion and sedimentation in Balikpapan Bay waters of East Borneo. The bathymetry of Balikpapan Bay can be seen in figure 1. In the vicinity of bay mouth the depth can reach 22 m, the further down into the bay the depth decreases to 9.6 m in front of the Wain River, when closer to the shore the depth diminishes Up to less than 2 m [12]. [13] said that the propagation of the tidal currents from the waters of the Strait of Makassar enters through the mouth of the bay and then moves into the head of the bay during flood and oppositely the water mass moves toward the mouth of the bay with the flow of freshwater mass coming from the tributaries that empty into the bay during ebb.

There are still many things that have not been studied in Balikpapan Bay related to some fundamental issues that must be answered in order to be able to manage the bay well. Why are some parts of the Bay more contaminated than others? What is the capacity of the Bay to assimilate watershed and in-Bay sources of contaminants and attain water quality objectives? [14]. Knowledge of the fundamental physical processes governing hydrological structures is necessary for a better understanding of complex physical-biological interactions [15]. On this paper water mass dynamics processes within Balikpapan Bay will be described. The objective of this study is to understand water mass dynamics within the Balikpapan Bay by analysing temperature, salinity and density structure as well as its currents pattern.

2. Data and data analysis

2.1. Study area

Study area is situated in the Balikpapan estuarine eastern Kalimantan Indonesia (figure 1).

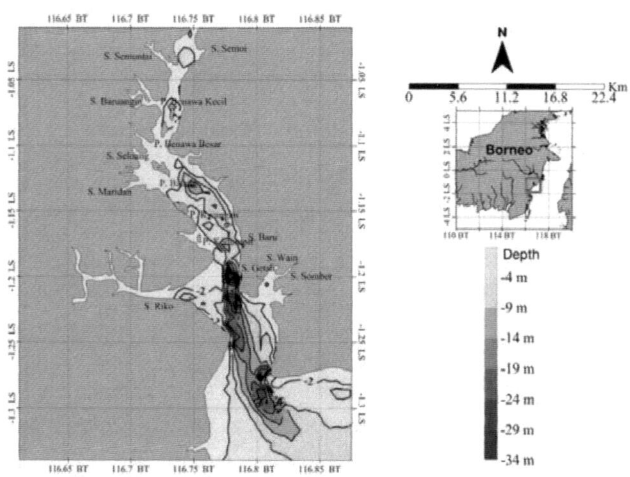

Figure 1. Bathymetry condition of Balikpapan Bay.

2.2. Material and methods

Field measurement of sea level fluctuation (tide) and water circulation by tidal gauge and current meter at mooring point (figure 2) was used to calibrate the model. Salinity and water temperature were measured by CTD sensor lowered at 12 stations located along the bay (figure 2). Three stations in front of the bay mouth are obtained from INDESO (Infrastructure Development of Space Oceanography) managed by the Agency for Marine and Fisheries Research & Development under the Ministry of Marine Affairs and Fisheries of Indonesian Government.

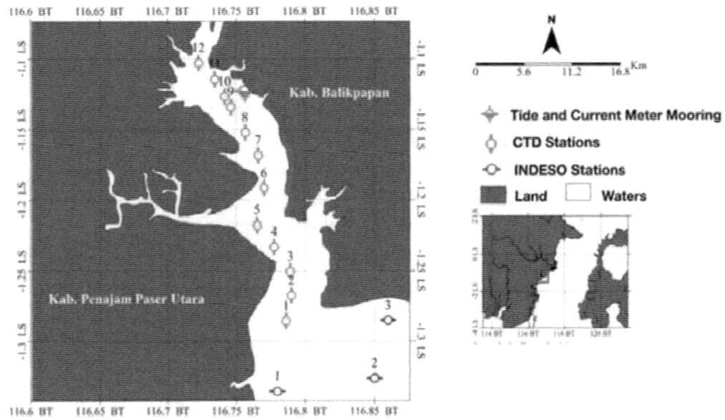

Figure 2. CTD stations, tide gauge and current meter mooring as well as INDESO stations.

The hydrodynamics model has been performed by using the MIKE Version 2005 software developed by Danish Hydraulics Institute (DHI) Water and Environment, Denmark. The equations used in this model are the continuity equation and the momentum equation with the average depth. This model uses a different method approach to (finite difference method) to solve the equations used. The following description of the equations used in the above model [16]. Details formulation described on [17].

Domain models are created in 4 stages, ranging from large, medium and small domains (figure 3). Models with broader domains are intended to look at external influences such as tidal propagation and wind forces.

2.3. *Model verification*

Verification of hydrodynamics model has been done by comparing the current model and field measurement. The result of verification between the model data and the measurement data is depicted on the scatter diagram on figure 4.

In figure 4 shows that the pattern of measurement current movement (blue triangle mark) is northwest-southeast direction. The pattern of current movement of the hydrodynamic model (orange quadrilateral mark) also shows the same direction, but the magnitude between the two results is still different. The measurement current data is measured at 5 m depth layer, while the model flow data is the result of the average of the depth, so that the measurement current velocity tend to be larger.

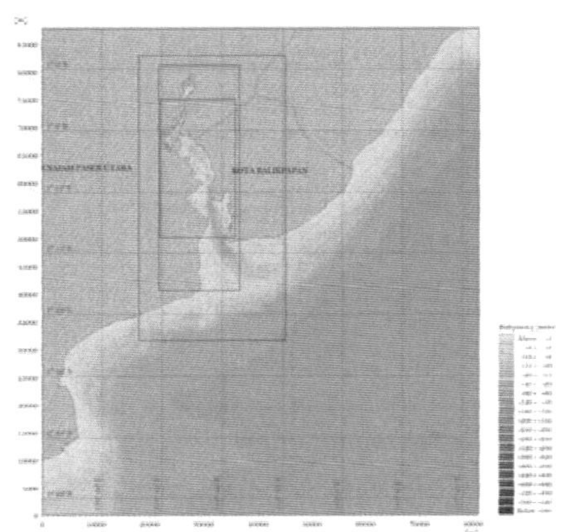

Figure 3. Hydrodynamics model domain.

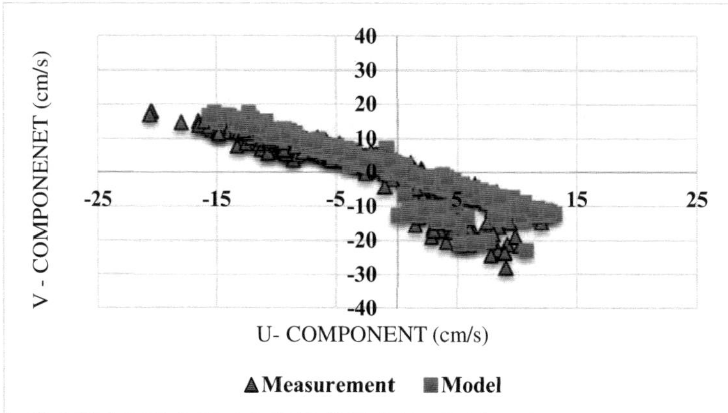

Figure 4. Model validation by field measurement.

3. Results and discussion

3.1. Properties of seawater within the bay

The temperature and salinity of seawater are two parameters of oceanography that are relatively easy to measure and important in studying the dynamics of water masses. The range of temperature and salinity of certain seawater is also indispensable for marine organisms to live. Significant temperature differences within a certain area can be used to identify the upwelling phenomenon or the increase of cooled water mass from the deeper to the upper layer. Generally characterized by low temperatures compared to around it. figure 5 is a plot of temperature and salinity data along Balikpapan Bay during tidal conditions, the color scale on the right shows the temperature and salinity values, the right-hand side image representing the transect line connecting stations as measured by their temperature and salinity values to CTD sensor (Conductivity Temperature Depth). On the left side the drawing draws toward the mouth of the bay, while the right side describes the condition toward the head of the bay.

The temperature of the seawater in the bay toward the bay head is relatively warmer than >30 °C compared to the water mass temperature around the bay mouth (<30 °C). Conversely, the salinity of the water mass around the bay mouth is higher (>29.75 psu) than the salinity of the water mass around the head of the bay (<27.75 psu).

The temperature and salinity at ebb conditions are presented in figure 6. The bay water temperature is higher in the bay waters (30.4-30.6 °C) than in the bay mouth (30.2 °C) and bay head (<30.2 °C). Compared to the tide (>29.75 psu), the salinity of water masses within the bay at low tide is smaller, i.e. <29 psu, and its value decreases towards the head of the bay (figure 5, bottom panel). There appears to be a Low Salinity Water (LSW) trapped upstream bay with a salinity value of 27.75 psu at flood time and < 26 psu at ebb time.

The distribution of water density within Balikpapan Bay is more controlled by the salinity of seawater than temperature one. At flood tide there is a mass of water with a higher water density flow into the middle part of the bay, this can be understood because at flood tide, higher density of water mass pushed towards the middle of the bay (figure 7, above panel). At ebb period there is a color degradation that indicates that the salinity decreases towards the head of the bay.

IOP Conf. Series: Earth and Environmental Science **176** (2018) 012019 doi:10.1088/1755-1315/176/1/012019

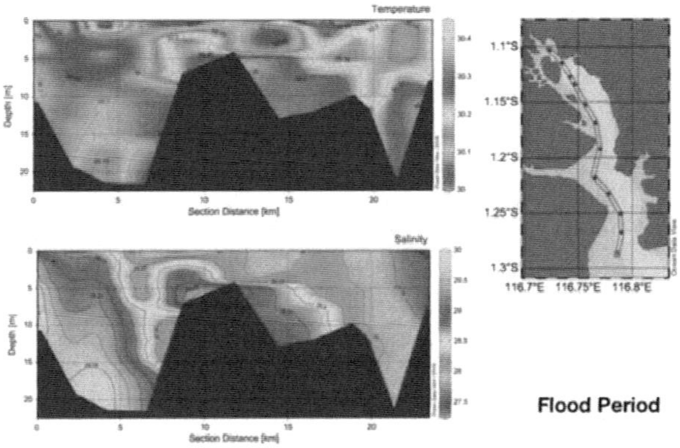

Figure 5. Longitudinal section of temperature and salinity along Balikpapan Bay at flood period.

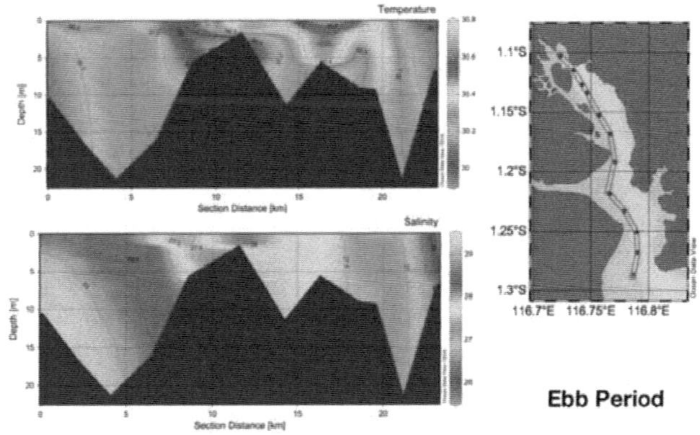

Figure 6. Cross section of temperature and salinity along Balikpapan Bay at ebb period.

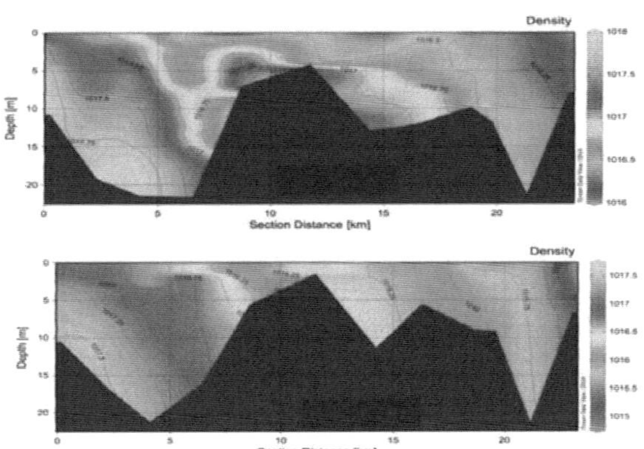

Figure 7. Cross section of Water Density along Balikpapan Bay at flood period (upper panel) and ebb period (lower panel) calculated from temperature and salinity data at the same.

3.2. Tidal Condition

One of the waves that has a long period is tidal or tidal wave. The phenomenon of the rise and fall of the sea surface periodically caused by the attraction between space objects, especially the moon and the sun against the surface of water on the surface of the earth. The existence of the tensile force causes the sea level towards the moon and / or the sun to rise higher, with the earth's rotation and the gravitational force of the disturbance to the sea surface can generate long-period waves of the ups and downs. The tidal waves that propagate into the bay come from the propagation of the tidal waves that spread from the Pacific Ocean and parts of the Java Sea through the Sulawesi Sea and Makassar Strait.

Table 1 is the main tidal constant values calculated from sea level observation data within 30 days in Balikpapan Bay. If the main constants are included in the formula Formzhall (F) = (K + O) / (M + S), then the value of F = 0.39 means that the type of tide that occurs in Balikpapan Bay is mixed tide prevailing semi-diurnal type (type mixed coupling tends to double daily). In 4 hours there are times the tide and two times the tide, but the amplitude is not the same. The range of tidal range in Balikpapan Bay can reach 2.7 m [13].

Table 1. Tidal Constant at Balikpapan Bay.

Tidal Component	M2	S2	N2	K2	K1	O1	Q1	MS4
Amplitude	3.1	600.1	40.3	5.8	10.9	22.3	17.2	7.4
Degree	244.3	127.6	176.8	141.3	176.8	258.2	239.3	258.2

3.3. Sea current pattern

The movement of water masses (currents) in a waters is generated by various forces of current generation such as wind, tidal, water density difference, and hydrostatic pressure of waters [6]; [18]; [19]. The magnitude of the influence of each force on the strength and direction of the current flow it generates depends on the type of water (beach. bay or sea off) and its geographic state.

Water circulation within the Balikpapan Bay is dominated by currents generated by the tides. Wind is the next important factor. especially in open water. The effect of the difference in density. hydrostatic pressure is generally small so it can be ignored. In the north-east monsoon from November to February the direction of the current is relatively small due to the dominant wind direction from the north where the land of Borneo. In the southeast monsoon the influence of wind and tidal propagation can strengthen inflows into the bay.

The tidal current dominates the circulation within the bay at low tide the current velocity range from 0.5 to 0.8 m/s with the dominant direction to the south (to the mouth of the bay), while at the tide 0.4-0.8 m/s northward (to the head of the bay).

To understand the more intact current pattern, the simulation of hydrodynamic model using bathymetry data input. Tidal and wind as the current generator. Simulations are performed on two differences monsoon. The northeast monsoon and southeast monsoon. Figure 8-15 are the snapshot of current results at different sea level position. i.e.: at the MSL (Mean Sea Level) point towards the highest tide. At the highest tide, when the sea level at the MSL point towards the lowest tide and when the sea level at low tide Lowest. The current pattern on these four positions of sea level is performed both in the north and south seasons.

Figure 8 is the current pattern towards the flood tide. The water mass is pushed into Balikpapan Bay. The average current velocity at flood tide can reach 10-15 cm/s. Due to the geographic conditions of the bay where the mouth is wider than the body of the bay and the head of the bay the current velocity when entering the bay looks stronger. On the contrary at the time of ebb tide the pattern of current appears to be reversed where the water mass flowed out from the bay into the Makassar Strait (figure 9).

The current pattern at the highest tide and lowest tide is presented in figures 10 and 11. From these two patterns the stream looks calmer when the sea level is at its highest tide point and there is a difference of direction especially in the southern mouth of the bay. When the flood tide the current moves counter clockwise. On the contrary, at the lowest ebb the direction of the current moves in a clockwise direction. Figure 12 through figure 15 are the results of current simulations in Balikpapan Bay and surrounding areas for the Southeast Monsoon. Similar with the result of current simulation at the MSL lead to the highest tide point in the northeast monsoon that the water mass circulation in the north-western Makassar Strait moves northward and a small part of the water mass is seen entering the bay of Balikpapan. The current velocity at the mouth of the bay ranges from 20-30 cm/s. the current is still strong and extending to the centre of the bay waters (figure 12).

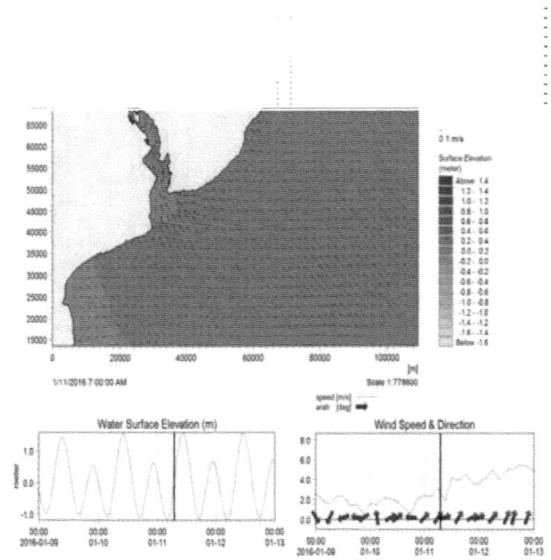

Figure 8. The current pattern during the northeast monsoon, when the sea level at MSL leads to the highest tide (the top panel is the current pattern presented in vector form. the lower left panel is the sea level (tide) position. and the lower right panel is the wind vector (velocity and direction).

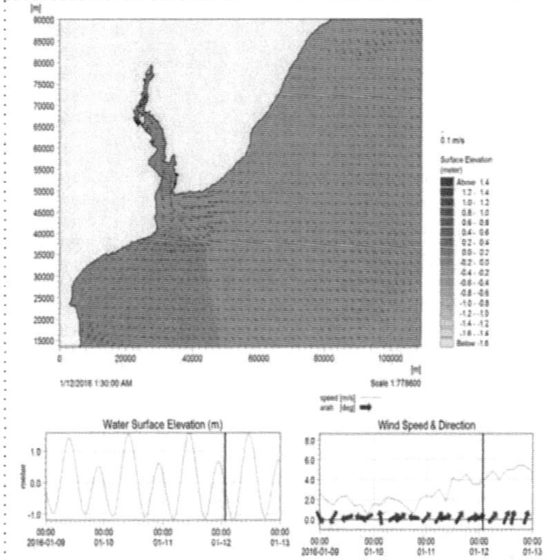

Figure 9. The current pattern during the northeast monsoon, when the sea level at MSL leads to the lowest tide (the top panel is the current pattern presented in vector form. the lower left panel is the sea level (tide) position. and the lower right panel is the wind vector (velocity and direction).

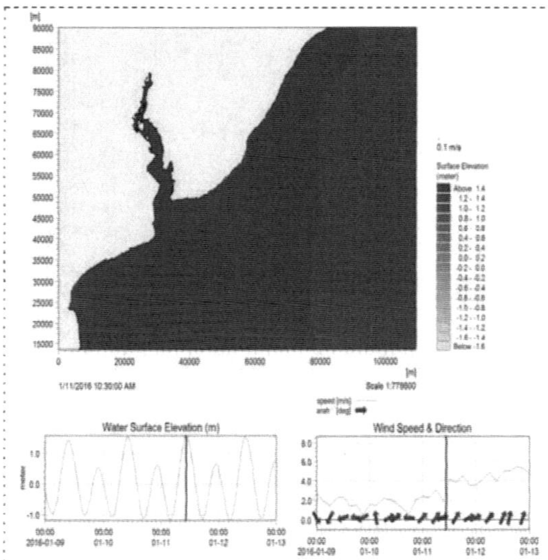

Figure 10. The current pattern during the northeast monsoon, when the sea level at the highest tide (the top panel is the current pattern presented in vector form. the lower left panel is the sea level (tide) position. and the lower right panel is the wind vector (velocity and direction).

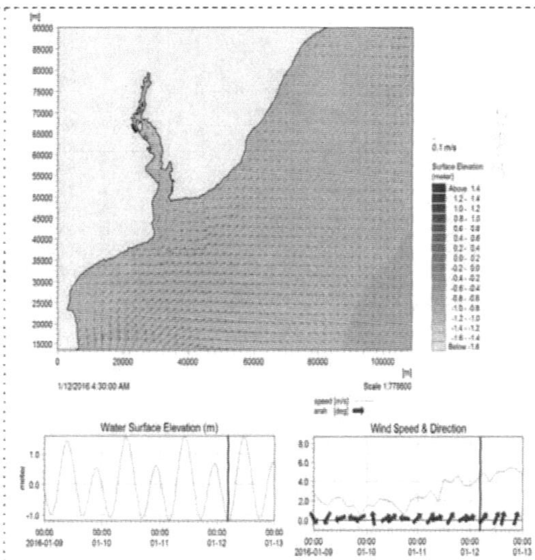

Figure 11. The current pattern during the northeast monsoon, when the sea level at the lowest tide (the top panel is the current pattern presented in vector form. the lower left panel is the sea level (tide) position. and the lower right panel is the wind vector (velocity and direction).

IOP Conf. Series: Earth and Environmental Science **176** (2018) 012019 doi:10.1088/1755-1315/176/1/012019

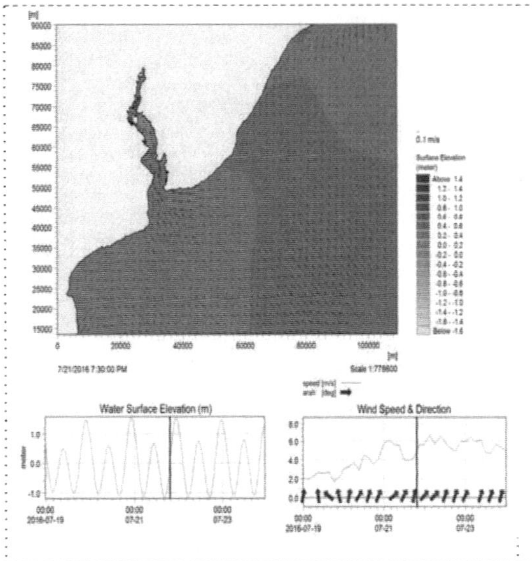

Figure 12. The current pattern during the Southeast monsoon. when the sea level at MSL leads to the highest tide (the top panel is the current pattern presented in vector form. the lower left panel is the sea level (tide) position. and the lower right panel is the wind vector (velocity and direction).

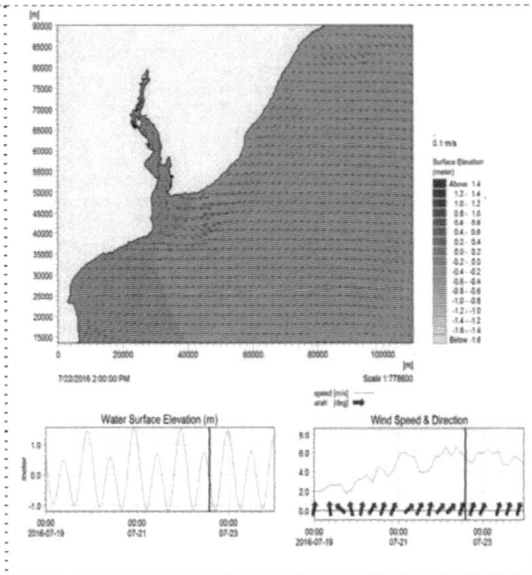

Figure 13. The current pattern during the Southeast monsoon. when the sea level at MSL leads to the lowest tide (the top panel is the current pattern presented in vector form. the lower left panel is the sea level (tide) position. and the lower right panel is the wind vector (velocity and direction).

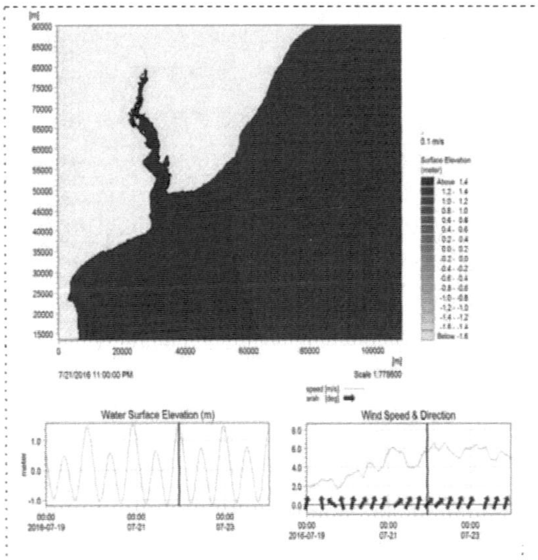

Figure 14. The current pattern during the Southeast monsoon. When the sea level at the highest tide (the top panel is the current pattern presented in vector form. the lower left panel is the sea level (tide) position. and the lower right panel is the wind vector (velocity and direction).

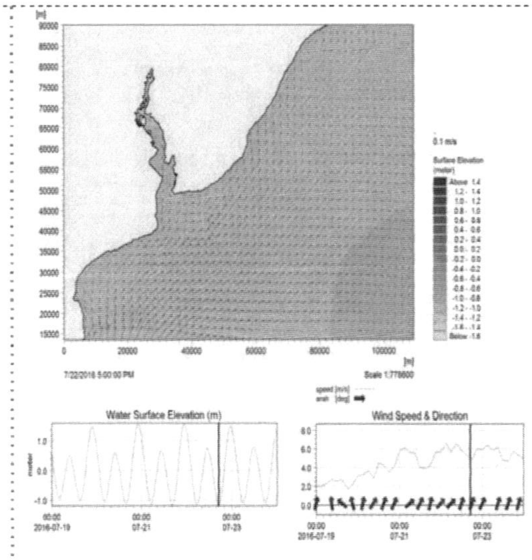

Figure 15. The current pattern during the Southeast monsoon. When the sea level at the lowest tide (the top panel is the current pattern presented in vector form. the lower left panel is the sea level (tide) position. and the lower right panel is the wind vector (velocity and direction).

Figure 13 is the current pattern when the sea level at the MSL flow to the lowest ebb in the southeast monsoon. The water mass within the bay is pushed out at the speed of 0.15-0.20 cm/s. when it outside of the bay the water mass is pushed eastward to the southeast as it forms a frontal zone which is the boundary of current flowed from the bay and the coastal current of the Makassar Strait. Plume of coastal current pattern is clearly visible from the difference of current direction and magnitude (current vector).

When the tidal conditions are highest during the south season. the water mass is seen moving northward in the north and east of the bay's mouth. while in the south-southwest of the mouth the water bay moves to rotate counter-clockwise. Current velocity appears to be stronger offshore compared to areas located near the coast (figure 14).

The current pattern is seen changing at the lowest tide (figure 15). almost at all model domain current flowing toward the northeast along the coast of Balikpapan. only slightly seen turning southwards. A stronger current is seen in the transition zone. there is between the coast waters of Balikpapan and the offshore. In the south-eastern part of the bay especially at the offshore part the current velocity decrease to be 10 cm/s.

3.4. Stratification of water mass

The gravitational stability of an adiabatically displaced air parcel in the atmosphere depends on the vertical stratification of pressure and density [20]; effects of continental slope and variable Brunt-Vaisla frequency [21]. There are three parameters usually used to calculated Brunt-Vaisala Frequency i.e.: vertical profile of temperature. salinity and density [22].

Figure 16 and 17 are longitudinal section according to water depth and parallel to bay length. respectively in periods of high and low tides. Generally. the water mass at low tide looks more stable compared to the high tide. This is indicated by the higher Brunt-Vaisala value at low tide compared to the high tide. The maximum value of Brunt-Vaisala Frequency was 22.5 cycle/hour. In the ocean where salinity is important. or in fresh water lakes near freezing. where density is not a linear function of temperature. Brunt-Vaisala frequency depend on both temperature and salinity [23]. [24]. [25].

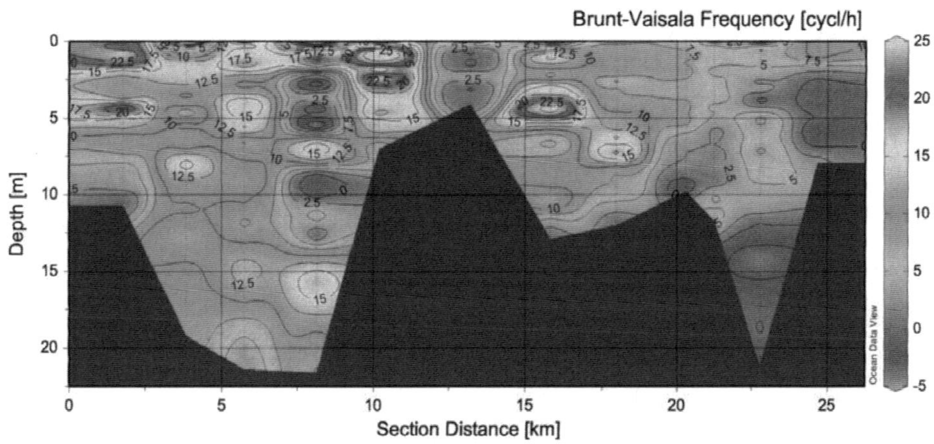

Figure 16. Longitudinal Section of Brunt-Vaisala Frequency on Flood Period along the Balikpapan Bay.

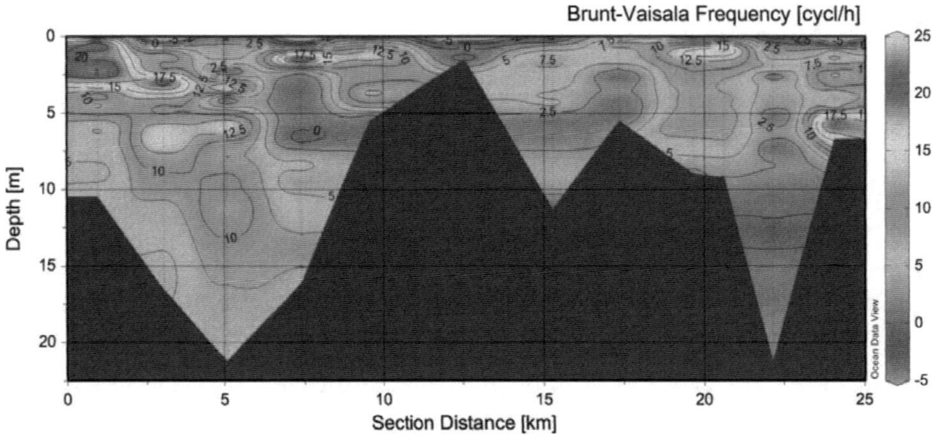

Figure 17. Longitudinal Section of Brunt-Vaisala Frequency on Ebb Period along the Balikpapan Bay.

4. Conclusion

Water mass stratification was found. It looks stronger in the center of the bay compared with the area near by the head and mouth of bay. Salinity more dominant to control water density within the bay compared to the water temperature. Tide is one of forces to withhold water mass which flow into and flow out the bay. Stability of water mass at along Balikpapan Bay at low tide looks more stable than when high tide.

Acknowledgements

I would like to say thank PT. Kretea Api Borneo Indonesia and PT. Hatfield Indonesia have been facilitated us during field survey.

References

[1] Kang X. Xia M. Pitula J S and Chigbu P 2017 Dynamics of water and salt exchange at Maryland Coastal Bays *Estuarine. Coastal and Shelf Science* **189** 1-16
[2] French P W 1997 *Coastal and Estuarine Management* (London: Routledge)
[3] Mitra A and Zaman S 2016 *Basics of Marine and Estuarine Ecology* (India: Springer)
[4] Najamuddin. Prartono T. Sanusi H S and Nurjaya I W 2016 Seasonal distribution and geochemical fractionation of heavy metals from surface sediment in a tropical estuary of Jeneberang River. Indonesia *Mar. Pollut. Bull. Elsevier* **111** 456-462
[5] Najamuddin. Prartono T. Sanusi H S and Nurjaya I W 2016 Distribution and behaviour of dissolved and particulate Pb. and Zn in Jeneberang Estuary. Makassar *J. Ilmu dan Teknologi Kelautan Tropis* **8**(1) 11-28
[6] Dyer K R 1973 *Esturaies: A Physical Introduction* (Great Britain: John Wiley and Sons Ltd.)
[7] Liu W C. Chen W B. Kuo J T and Wu C 2008 Numerical determination of residence time and age in pastially mixed estuary using a three-dimensional hydrodynamic model *Contiental Shelf Research* **28**(8) 1068-1088
[8] Casares-Salazar R and Marino-Tapia I 2016 Influence of remote Forcing and local winds on the barotropic hydrodynamics of an Elongated Coastal Lagoon *J. Coast. Res.* **32**(1) 116-130
[9] Nurjaya I W 2016 Salinity structure within the estuary of Bintuni Bay. at the southern part of bird head of West Papua. Indonesia *J. Segara* **12**(2) 73-80
[10] Rizkiyah D. Nugroho and Purwanto 2015 Studi pola sebaran buangan panas PT. Pertamina UP V Balikpapan. di Perairan Kampung Baru. Teluk Balikpapan *Bull. Oseanigrafi Marina* **6**(1) 1-13
[11] Warsidi and Endayani 2017 Komposisi vegetasi mangrove di Teluk Balikpapan. Provinsi Kalimantan Timur *J. Agrifor* **16**(1) 115-124

[12] Working Group of Erosion and Sedimentation 2002 *Kajian erosi dan Sedimentasi pada DAS Teluk Balikpapan. Kalimantan Timur* CRC/URI CRMP. Proyek Pesisir Kaltim Balikpapan p 51
[13] Sarwono. Mursidi. Abdunnur. Malik R and Audrie 1999 *Kondisi Hidrooseanografi Perairan Teluk Balikpapan. Proyek Pesisir technical Report TE-99/16-I* Coastal Resources Center. University of Rhode Island. Jakarta. Indonesia p 27
[14] Schoellhamer D H. Mumley T E and Leatherbarrow J E 2007 Suspended sediment and sediment-associated contaminants in San Fransisco Bay *Envirenmental Research* **105** 119-131
[15] Koutsikopoulos C and Cann B L 1999 Physical processes and hydrological structures realted to the Bay of Biscay *Scientia Marina* **60** 9-19
[16] Abbott M B. Petersen H M and Skovgard O 1978 On the numerical modelling of short waves in Shallow Water *Journal of Hydraulic Research* 16(3)
[17] Nurjaya I W and Surbakti H 2010 Thermal dispersion model of cooling water of gas and steam power plants at Cilegon CCPP discharge into Margasari coastal waters at the western coast of Banten Bay *J. Ilmu dan Teknologi Kelautan Tropis* **2**(1) 31-49
[18] Dyer K R 1977 *Lateral Circulation Effects in Estuaries. Esturaries. Geophysics and the Environment* (Washington D C: National Academy of Sciences) p 22-29
[19] Fischer H B. List E J. Koh R C Y. Imberger J and Brooks N H 1979 *Mixing in Inland and Coastal Waters* (New York: Academic Press) p 483
[20] Prandle D 2004 How tides and river flows determine estuarine bathymetries *Progress in Oceanography* **61** 1-6
[21] Prinsenberg S J and Rttray J M 1975 Effects of continental slope and variable Brunt-Vaisala frequency on the coastal generation of internal tide *Deep-Sea Res.* **22** 251-263
[22] Imberger J and Ivey GN 1991 On the nature of turbulence in a stratified fluid. Part II: application to lakes *Journal of Physical Oceanography* **21** 659–680
[23] Imberger J 1998 *Physical Processes in Lakes and Oceans Coastal Estuarine Studies Vol. 54* (Washington DC: American Geophysical Union)
[24] Thorpe SA 2007 *An Introduction to Ocean Turbulence* (Cambridge: Cambridge University Press)
[25] Wolanski E 2007 *Estuarine Ecohydrology* (Amsterdam: Elsevier)

IOP Conf. Series: Earth and Environmental Science **176** (2018) 012020 doi:10.1088/1755-1315/176/1/012020

Trophic interaction of fish larvae, phytoplankton, and zooplankton in Pari Island lagoon

R Puspasari[1*], A Damar[2], M M Kamal[2], N N Wiadyana[1]

[1] Centre for Fisheries Research, Indonesia
[2] Aquatic Living Resources Management, Bogor Agricultural University, Indonesia
*e-mail: reny.paksi@gmail.com

Abstract. Fish larvae are one of the critical periods in fish life cycle due to the availability of convenient prey. The aim of this research was to describe the dynamic and variation of fish larvae and its prey which could predict fish larvae survival and recruitment success of fish population in Pari Island lagoon. The research was conducted at Pari Island lagoon, Seribu Islands in Jakarta Province since June – November 2010. Fish larvae, zooplankton and phytoplankton samples were taken ten times in 6 months. Zooplankton and fish larvae were identified, and their biomasses were measured. A feeding experiment was done to analyze the food preferences of fish larvae. Results show that there were two peaks of fish larvae biomass indicated the post spawned season of fishes, occurred at the beginning of July and October. The high biomasses of fish larvae at the beginning of July was supported by the high growth rate of zooplankton and phytoplankton and resulted in the opportunity for recruitment succeed. The next spawned season occurred at the beginning of October was not supported by the high number of zooplankton biomass, resulted in the collapse of fish larvae biomass.

1. Introduction

Lagoons are a potential area for fish spawning and nursery ground regarding to its morphological structure which refers to semi-closed ecosystem. Semi-closed ecosystem could protect the organisms inside from any hydrodynamic exposure.

Fish stock abundance is apparently influenced by recruitment success of the fish population. Fish recruitment is determined by the high survival rate of fish larvae through the critical period in fish larvae cycle. Predation and starvation are thought to be the main causes of larval fish mortality.

Limitation of convenient prey during critical developmental stage would result in massive mortality. The availability of convenient prey determined to feed success of the first exogenous feeding larvae. The compatibility between fish larvae and its prey is determined by size, species composition [1] and production timing [2]. Planktonic organisms are the main prey for fish larvae in consequence of their small size. Nauplii copepod is the most frequent found in the gut content of fish larvae [3-4] adult copepod, polychaeta, mollusc larvae and phytoplankton could also found in gut content of fish larvae [4].

Production time of convenient prey plays an important role in the occurrence of match or mismatch condition between fish larvae and its prey. When convenient prey production occurs at the same time with the starting of exogenous feeding, it will support the food availability for fish larvae. The high abundance of food supply could increase fish larvae survival rate and minimize massive mortality. In the other time, while production rate of a planktonic organism is high, however, the species and their size are not convenient with fish larvae, it would not be eaten. The inconvenient planktonic organism could not support the high survival rate of fish larvae, such that the mismatch condition occurred. In this paper, the variability in fish larvae and its prey biomass is described as a prediction for recruitment success of fish population in Pari Island lagoon.

2. Methods

There were two kinds of research conducted that were experimental research and field sampling. Experimental research was carried out to analyze the trophic interaction among sized fractionated planktonic organism through grazing experiment and gut content analysis. The dynamic of several size planktonic organisms in Pari Island lagoon was observed by plankton sampling around Pari Island lagoon in Seribu Islands. Both experiment and field survey were conducted in 2010.

Table 1. The combination of planktonic size fraction in grazing experiment (*R = repeat).

	Treatment combination	*R1	*R2	*R3
1	F			
2	N			
3	F + N			
4	Mi + F			
5	Mi + N			
6	Mi + F + N			
7	Me + F			
8	Me + N			
9	Me + Mi			
10	Me + F + N			
11	ME + F + Mi			
12	Me + N + Mi			
13	Me + F + N + Mi			
14	L + F			
15	L + N			
16	L + Mi			
17	L + Me			
18	L + F + N			
19	L + F + Mi			
20	L + F + Me			
21	L + N + Mi			
22	L + N + Me			
23	L + Mi + Me			
24	L + N + F + Me			
25	L + N + F + Mi			
26	L + N + Me + Mi			
27	L + F + Me + Mi			
28	L + N + F + Mi + Me			

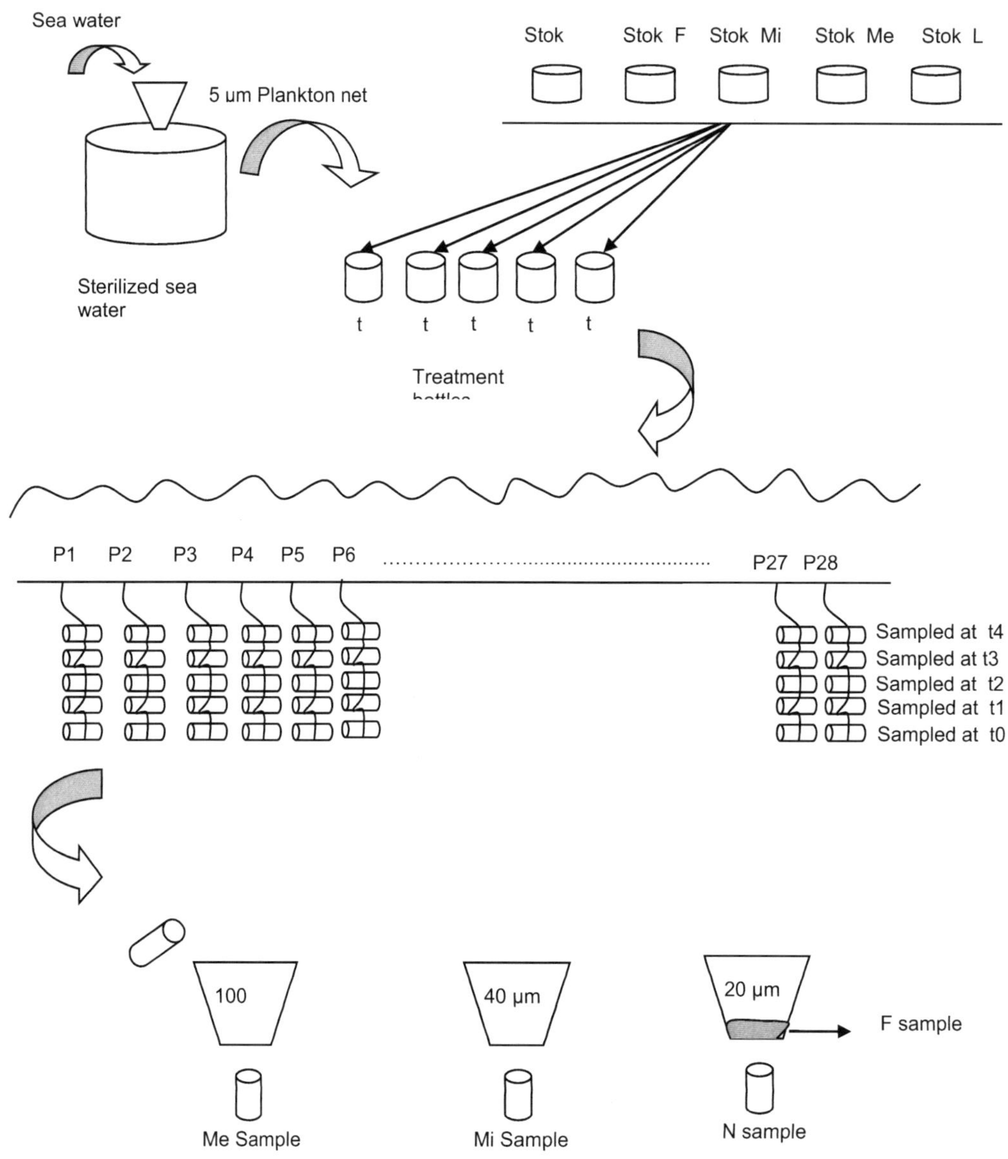

Figure 1. Grazing experiment design.

2.1. Grazing experiment

Grazing experiment was conducted to analyze the selection of prey by three fractionated planktonic organisms (predator), there were zooplankton that restrained by 40 µm mesh size net (Mi), zooplankton that restrained by 100 µm mesh size net (Me) and sea bass larvae that restrained by 500 µm mesh size net (L). Sea bass larvae used were the larvae came out from the hatchery. The larvae were in ages of 3 days, and had never been feed. Prey are the two size fractionated of phytoplankton; there was phytoplankton that wriggles out from 20 µm mesh size net (N) and phytoplankton that restrained by 20 µm mesh size net (F).

These three groups of predator and their prey then put in a 1500 ml transparent bottle with a certain concentration and combination of planktonic organisms. The bottles than were incubated in the lagoon for 26 hours. The first two hours was for acclimatization period, and the observation started at the third hours (t_0). There were 28 combinations of planktonic organisms (treatment) to be observed to measure the grazing rate of every zooplankton fraction to their prey. The combination of treatment is as described in table 1. The design method of grazing experiment described in figure 1. Every treatment combination has 5 sampling units, which would be taken one by one every 6 hours (t1 = 6 hr; t2 = 12 hr; t3 = 18 hr; and t4 = 24 hr).

The biomass of each fraction then measured by using wet weight methods (Downing and Rigler, 1984). Grazing rate of predator was measured by subtracting the growth rate of controlled F or N with the growth rate of F or N which has been combined with a predator, for example, the grazing rate of Mi are F – (F-Mi).

2.2. Food habit of fish larvae

Food habit of fish larvae was analyzed by surgery observation of gut content of Sea bass fish larvae that used in grazing experiment. The gut surgery was carried out using a micro dissecting set and a 100 x magnification of the binocular microscope. It was about 170 individuals of sea bass larvae observed.

2.3. Field survey

A field survey carried out in June – November 2010. The sample was taken in 5 stations around lagoon area and islands, i.e., St. 1 next to Tikus Island, St.2 next to Burung Island (backward), St. 3 next to Kongsi Island, St. 4 above the leeward and St. 5 next to Pari Island (figure 2).

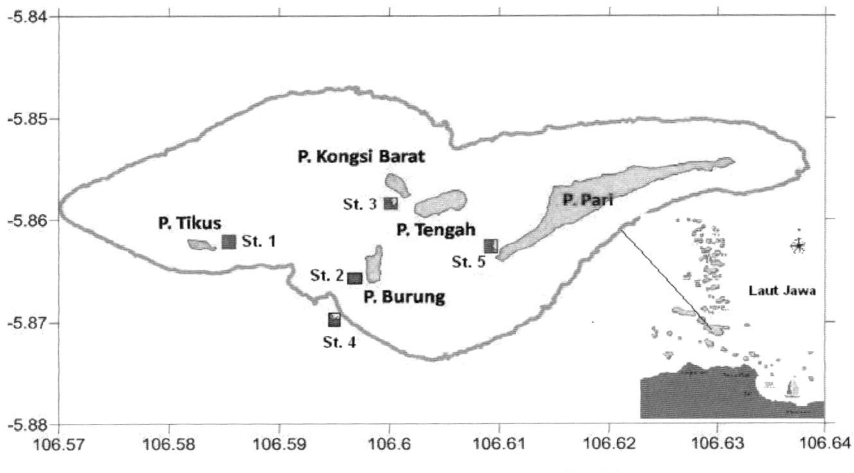

Figure 2. The study area in Pari Island lagoon.

2.3.1. Fish larvae, phytoplankton, and zooplankton. Fish larvae sample was collected by using the larval net with mesh size 500 μm, the net towed at 1 – 2-knot speed for 10 minutes. Samples were sorted to separate fish larvae from other materials. Fish larvae were then identified [5]. Wet weight biomass was done to quantify the biomass of fish larvae.

2.3.2. Phytoplankton. Phytoplankton sample was collected in the afternoon between 09.00-12.00 web, using Van Dorn water sampler at a depth of 5 m. Phytoplankton sample was then identified [6]. Phytoplankton biomass was measured by chlorophyll-a concentration measurement using a Spectrophotometer (Genesys U20).

2.3.3. Zooplankton. Zooplankton sample was horizontal hauled at night between 19.00 – 23.00 pm by using plankton net with mesh size 40 µm. The net was towed at 1 – 2-knot speed for 2 x 3 minutes. Zooplankton sample was then identified [6]. Dry weight biomass measurement was done [7]

3. Results and discussion

3.1. Grazing experiment

Grazing experiment resulted in the specific growth rate of every planktonic fraction and the grazing rate of different size zooplankton & fish larvae to other planktonic organisms. The measured specific growth rate and grazing rate are showed in the table 2.

Table 2. The specific growth rates and the estimation of grazing rates of the different sized planktonic organism.

Treatment	Specific growth rate x 10^{-2} (mg/m^3/hr)	Estimation of grazing rate x 10^{-2} (mg/m^3/hr)
Estimation grazing rate to N fraction		
Only N	8.5	
Mi – N	5.0	3.5
Me – N	8.0	0.5
L – N	7.5	1.0
Estimation grazing rate to F fraction		
Only F	10.2	
MI – F	9.0	1.2
Me – F	5.8	4.4
L – F	5.0	5.2
Estimation grazing rate to Mi fraction		
Mi – N	-3.5	
Me - Mi – N	-5.1	1.6
L - Mi – N	-3.9	0.4
Estimation grazing rate to Me fraction		
Me – F	8.7	
L - Me – F	-1.2	9.9

Table 2 showed that there were some negative growth rates found in experiment, especially from the treatment of Mi fractions and one of Me fraction treatment. Negative growth rate showed that there was a death of zooplankton Mi sized fraction feed by N sized fraction, Me fraction feed by Mi and N and also L feed by Mi and N, resulting in decreasing biomass at $_{an}$ compared to t_0. The death of organisms in the experiment unit could be caused by several factors such as the stress of organisms due to treatment. Negative growth rates indicating that the experiment failed for Mi unit treatment, and due to the three-repeating unit all were failed, so the experiment did not repeat anymore. According to this condition, we decided to sign the data from the failed experiment units as *"underestimate data."*

Mapping the grazing rate from the bigger fraction of sized fractionated planktonic organism into the smaller one, resulting in the grazing diagram as described in figure 3.

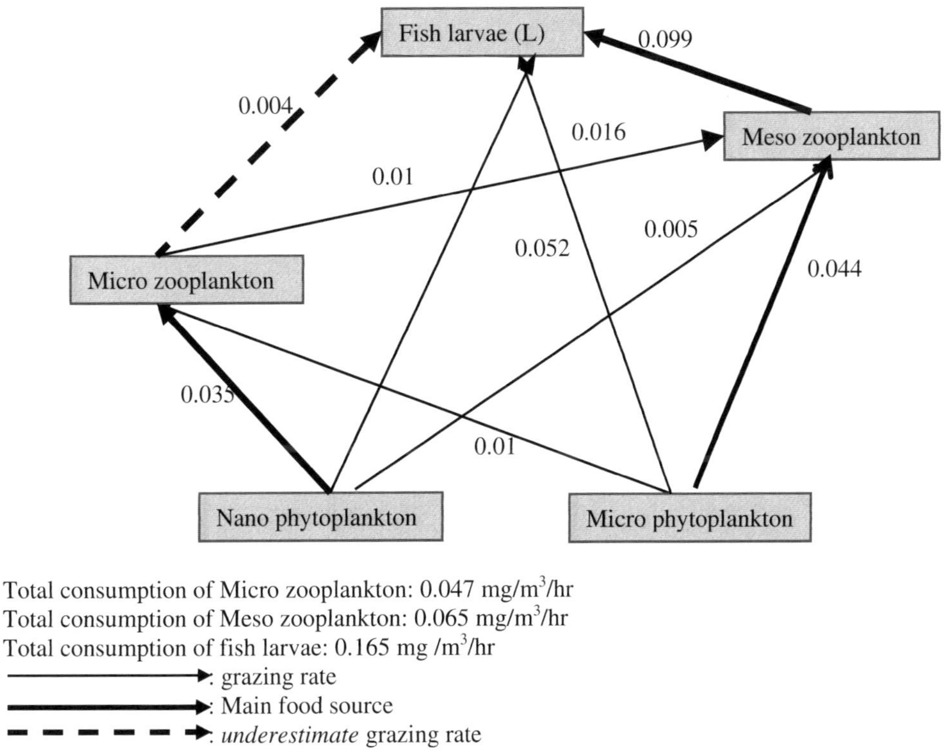

Figure 3. The diagram of grazing rates of the sized fractionated planktonic organism.

The grazing diagram showed that N sized fraction was consumed by Mi, Me and L, yet the main grazer is Mi, it was shown by the highest grazing rate value (0.03 mg/m^3/hr). The N sized fraction becomes the main food source for Mi, showed by the grazing rate of Mi was higher to N compare to F. The F sized fraction were consumed by Mi, Me and L, with the grazing rate of every zooplankton fraction size, were 0.01 mg/m^3/hr, 0.04 mg/m^3/hr, and 0.052 mg/m^3/hr respectively. Grazing rate of L into F sized fraction were the highest, yet F was not the main food source food L because L preferred Me more to consume than F and N sized fraction. The grazing rate of L to Me sized fraction was 0.09 mg/m^3/hr. According to the diagram, L was consumed all fraction sized of planktonic organisms, however, due to the failed treatment of Mi sample unit, we could not conclude that I was the main food source of L, L just consumed Me more than N and F.

Bigger sized zooplankton consumed smaller sized planktonic organisms. Bigger sized zooplankton consumed more food source than smaller one. Figure 3 showed there was a trophic interaction among sized fractionated planktonic organisms, which mentioned "predator-prey" interaction.

3.2. Food habit of fish larvae

Feeding experiment to sea bass (*Lates calcarifer*) larvae in the microcosm showed that the gut content of *sea bass* composed of 73.5% microzooplankton, 20.6% mesozooplankton and 5.9% are other fish larvae (figure 4). Brachionus and nauplius were microzooplankton due to its size less than 200 nm. Nauplius is a larvae phase of crustacean that could not be identified at the species level. The experiment showed that fish larvae consume microzooplankton more than other planktonic size fraction [8]. That copepod nauplius was dominated the gut content of fish larvae [3-4].

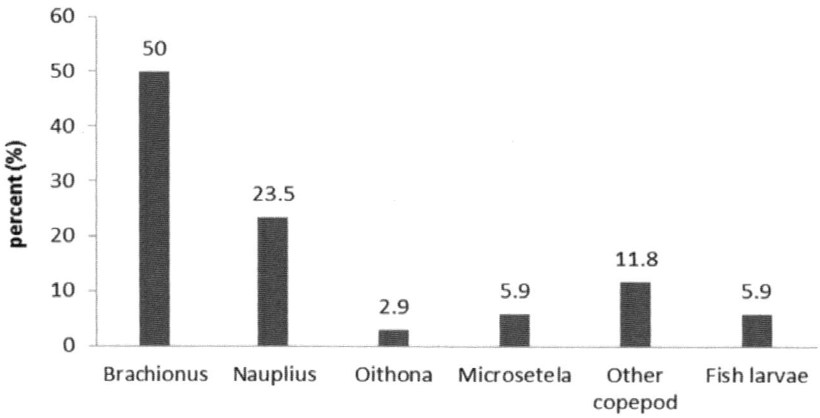

Figure 4. Gut content composition of sea bass (*Lates calcarifer*) larvae.

3.3. The temporal distribution of fish larvae biomass

Pari Island lagoon is the combination of a coral reef and seagrass ecosystem. Many fishes utilize the ecosystem as spawning and nursery ground showed by the high abundance of fish larvae found in this area. There were 79 families of fish larvae in Pari Island lagoon, dominated by Pomacentridae (20.9%), Aulostomidae (9.6%), Blenniidae (7.6%), Engraulidae (6.4%) and Pinguipedidae (5.7%) [9].

Seasonal biomass rate of fish larvae was 4.5 x 10^{-2} mg/m^3 ranged between 5.5 x 10^{-3} mg/m^3 at the end of July in St.5 to 3.6.10^1 mg/m^3 at the end of July in St.1. The dynamic of fish larvae biomass shows three peaks, occurred at the beginning of July, August, and October. The highest biomass was 8.7 x 10^{-2} mg/m^3 at the beginning of July. Fish larvae biomass peak found at the beginning of August was lower than July but more upper than October. There was an anomaly in the 21st of July, fish larvae biomass found in very high value (36 mg/m^3) exceeded the seasonal rate of biomass. Figure 2 shows the dynamic of average fish larvae biomass in Pari Island lagoon.

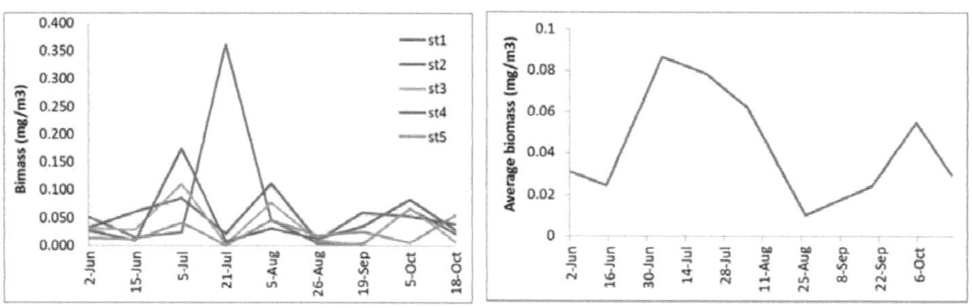

Figure 5. The dynamic of fish larvae biomass in every station (left), at the average number (right).

The number of fish larvae biomass was determined by the size of the larvae. The bigger size of fish larvae contributed more to total fish biomass. Fish larvae in Pari Island lagoon at that time was dominantly composed by small size larvae which were the beginning phase of larvae development [10]. The condition indicated that spawning seasons occur at the period before high larvae biomass found.

3.4. Variability in prey biomass

Fish larvae are consumed planktonic organisms which have a smaller size than mouth size of larvae [3-8]. Fish larvae consumed zooplankton more than phytoplankton. Phytoplankton biomass played an important role as primary producer and food source for zooplankton. Trophodynamic between phytoplankton zooplankton determine the survival rate of fish larvae.

There was a variation in phytoplankton biomass in Pari Island lagoon, shown by the dynamic of chlorophyll-a concentration. Chlorophyll-a concentration in Pari Island lagoon composed by 87,2% nanophytoplankton and 12,8 % microphytoplankton. The average of Chlorophyll-a concentration shows fluctuate value every two weeks. At the beginning of June, July, August, and middle of October the chlorophyll-a concentration was high, while two weeks later the chlorophyll-a concentration was dropped (figure 3). The two weeks cycle of phytoplankton corresponding to the dynamic of nutrient concentration and zooplankton biomass as a predator.

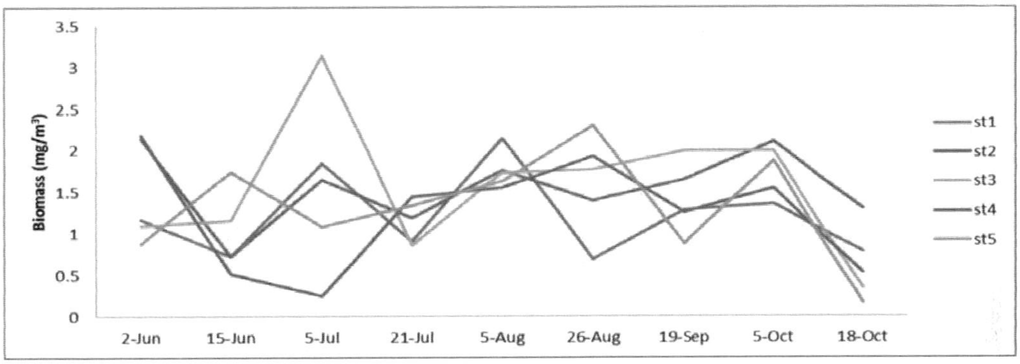

Figure 6. Dynamic of phytoplankton biomass.

Variation in zooplankton biomass showed a different pattern of phytoplankton. There were two biomass peaks during a sampling period that were at the beginning of August and in the middle of September, except for station 5, the first biomass peak occurred at the end of July (figure 6). Total zooplankton biomass range between 8.6 x 10^{-5} mg/m^3 in the middle of June to 1.3 x 10^{-2} mg/m^3 at the beginning of August, monthly mean for total zooplankton biomass was 4.6 x 10^{-3}. Approximately 63, 9 % of zooplankton biomass was microzooplankton, and the other was mesozooplankton.

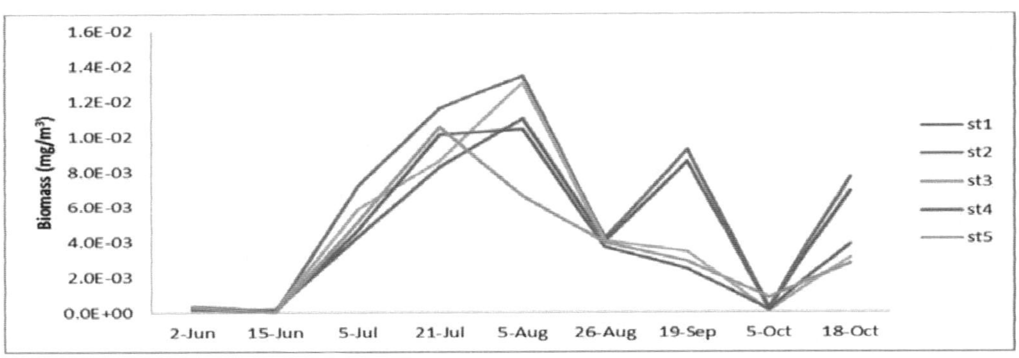

Figure 7. Dynamic of zooplankton biomass.

The high peaks of zooplankton biomass were corresponding to the species composition of zooplankton. Crustacea have dominated zooplankton composition during the research period, and reach the highest abundance at the beginning of August and middle of September while biomasses were high. The leading group of crustacean was nauplii copepod, which was composed of 54,5% of total zooplankton [9].

3.5. Match and mismatch hypothesis between fish larvae and phytoplankton and zooplankton

There were high variations in biomass distribution of planktonic organisms in Pari Island lagoon. The dynamic of phytoplankton, zooplankton and fish larvae showed match and mismatch timing between fish larvae and its prey. Figure 8 shows the dynamic of planktonic organisms biomass in Pari Island lagoon.

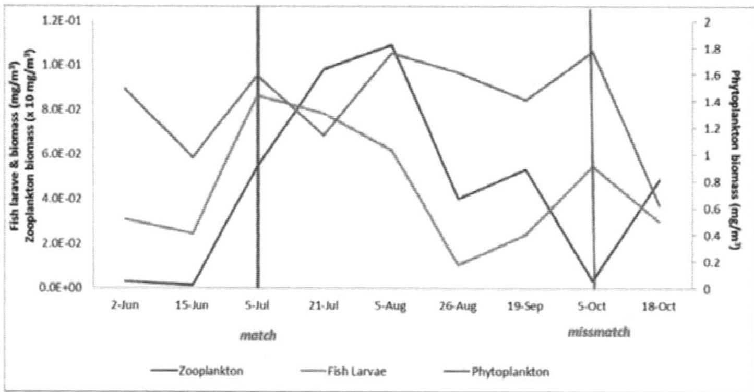

Figure 8. Dynamic of planktonic organism biomass in Pari Island lagoon.

The peak of fish larvae biomass occurred in July and October, indicating larval production season. At the beginning of July, when fish larvae had the highest biomass, phytoplankton was in the highest biomass (peak condition), at the same time zooplankton was in an exponential growth phase. When one population is in the exponential stage, they will have highest growth rate, such that can balance the predation rate by a predator. As a result, the population will not collapse or will be in a stable condition [11].

Figure 8 shows that in the first fish larvae production season (July), the high biomass of fish larvae was coinciding with high prey growth rate, particularly for zooplankton. One of the fish reproductive strategies is spawned at the time when the availability of food supply was high [12] therefore it could support the larvae survival [13]. This condition is convenient with the third Cushing hypothesis scenario [14], which larval production occurred while prey stock was in high abundance, such that fish larvae could survive and growth although it is not in a maximum rate. The match between fish larvae and its preys, could support the survival of fish larvae and be expected to the recruitment success.

The next larval production time occurred at the beginning of October. In the same time, zooplankton biomass was decreased suddenly, and phytoplankton biomass was highest during the research period. This condition was significantly different with the first larval production in July. At this time, steep increase in fish larvae biomass occurred while zooplankton population was at the negative exponential phase, means that their communities were in collapse. The increasing number of fish larvae biomass added the predatory pressure for zooplankton population; therefore, the zooplankton biomass was dropped.

Dropped in zooplankton biomass could not support the fish larvae survival due to the limited food source. The fish larvae had collapsed in the next two weeks observations. A small number of fish larvae and high phytoplankton biomass at the previous periods generate the increasing number of zooplankton biomass.

The interaction between fish larvae and its prey occurred in October did not include one of Cushing's hypothesis [14]. Production time of fish larvae coinciding with the collapse of zooplankton biomass, such that mismatch between fish larvae and their prey. Zooplankton could not support the survival of fish larvae due to limited abundance; eventually, the recruitment of fish population is failed. The dynamic of the planktonic organism as described in October shows that a match and mismatch between fish larvae and its prey may depend more on the dynamic of zooplankton. Fortier [15] found the same phenomena in South-eastern Hudson Bay where the momentum of fish larvae was more rely on the dynamic of zooplankton than that of phytoplankton. These results also support the hypothesis that fish larvae consume more zooplankton than phytoplankton [3-8].

4. Conclusion

The current work shows that the fish larvae production occurred in July was match with the availability of their prey and might result in recruitment success of the fish population.

Acknowledgments

We thank Prof. Djamar T.F Lumban Batu for his guidance during research conduct and M. Taufik and Koderi for their assistance in laboratory analysis.

References

[1] Jolley J C, Willis D W and Holland R S 2010 Match–Mismatch regulation for Bluegill and Yellow Perch Larvae and their prey in Sandhill Lakes *Journal of Fish and Wildlife Management* **1**(2) 73–85

[2] Siddon E C, Kristiansen T, Mueter F J, Holsman K K and Heintz R A 2013 Spatial match-mismatch between juvenile fish and prey provides a mechanism for recruitment variability across contrasting climate conditions in the Eastern Bering Sea *Plos One* **8**(12) e84526

[3] Rossi A, Sabates A, Latasa M and Reyes E 2006 Lipid biomarkers and trophic linkages between phytoplankton, zooplankton, and anchovy (*Engraulis encrasicolus*) larvae in the NW Mediterranean *Journal of Plankton Research* **28**(6) 551-562

[4] Voss R 2002 Recruitment Processes in The Larval Phase: The Influence of Varying Transport on Cod and Sprat Larval Survival (Kiel: Mathematisch-Naturwissenschaftlichen Fakultät der Christian-Albrechts-Universität zu Kiel)

[5] Leis J M and Carson-Ewart 2000 *The Larvae of Indo-Pacific Coastal Fishes: An Identification Guide to Marine Fish Larvae: Fauna Malesiana Vol 2* (Leiden: Brill)

[6] Yamaji I 1984 *Illustration of the Marine Plankton of Japan* (Japan: Hoikusha)

[7] Downing J A and Rigler F H 1984 *A Manual on Methods for the Assessment of Secondary Productivity in Fresh Water 2nd Edition* (Oxford: Blackwell Scientific Publication) pp 501

[8] Pepin P and Dower J E 2007 Variability in the trophic position of larval fish in a coastal pelagic ecosystem based on stable isotope analysis *Journal of Plankton Research* **29**(8) 727-737

[9] Puspasari R 2013 Size fractionation of zooplankton biomass and species composition in the lagoon of Pulau Pari, Seribu Islands *Widyariset* **16**(3) 1-10

[10] Puspasari R, Damar A, Kamal M M and Wiadnyana N N 2013 Fish larvae dynamic as a basis of Pulau Pari Lagoon area management *Jurnal Kebijakan Perikanan Indonesia* **5**(1) 1-7

[11] Emmel T C 1976 *Population Biology* (New York: Harper & Row Publisher)

[12] Sherman K, Smith W, Morse W, Berman M, Green J and Ejsymont L 1984 Spawning strategies of fishes in relation to circulation, phytoplankton production and pulses in zooplankton off the Northeastern United States *Marine Ecology Progress Series* **18** 1-19

[13] Santos A J P, Nogueira J and Martins H 2005 Survival of sardine larvae off the Atlantic Portuguese coast: a preliminary numerical study *ICES Journal of Marine Science* **62** 634-644

[14] Chusing D H 1975 *Marine Ecology and Fisheries* (Oxford: Cambridge University Press) pp 277

[15] Fortier L, Ponton D and Gilbert M 1995 The match/mismatch hypothesis and the feeding success of fish larvae in ice-covered Southeastern Hudson Bay *Marine Ecology Progress Series* **120** 11-27

The outbreak of *Chochlodinium* sp.: the red tide maker in the coastal of Lampung Bay

R Puspasari[1*], Y Sugianti[2], A Rustam[3], R A Adi[3], S S Sagala[3], W S Pranowo[4]

[1]Center for Fisheries Research, The agency of Marine and Fisheries Research and Human Resource Development, Ministry of Marine Affairs and Fisheries, Indonesia
[2]Research Institute for Fish Recovery and Resources Conservation, The Agency of Marine & Fisheries Research and Human Resource Development, Ministry of Marine Affairs & Fisheries, Indonesia
[3]Marine Research Center, The Agency of Marine and Fisheries Research and Human Resource Development, Ministry of Marine Affairs and Fisheries, Indonesia
[4]Marine & Coastal Data Laboratory, The Agency of Marine and Fisheries Research and Human Resource Development, Ministry of Marine Affairs and Fisheries, Indonesia
*e-mail: reny.paksi@gmail.com

Abstract. Red tide is a common phenomenon and almost present every year in Lampung Bay. In 2012, the red tide in Lampung Bay showed a different characteristic compared to the previous ones. The research was conducted to analyze the characteristics of organisms causing the red tide and the factors affecting its occurrence. Assessments were done by identifying the dominant phytoplankton and observing the environmental condition using rapid assessment method. Seawater sampling was conducted in 15 stations in December 2012 to monitor the phytoplankton abundance, nutrient, and physical condition. The results showed that the organism dominating the phytoplankton population is *Cochlodinium* sp., comprising of 85.89% of the total phytoplankton suggesting it was the red tide organism in the Bay. The average abundance of this species in the stations was 2.0 x 10^4cells/L. *Choclodinium* sp. was present agglomerated and distributed along the west coast of the bay. The highest concentration of the organism was found around the tip of the bay. The outbreaks of *Choclodinium* sp. were induced by the high concentration of nitrogen in the coastal water with N:P ratio of 35.5. Moreover, the physical condition of the coastal water strongly influenced the distribution and abundance of *Cochlodinium* sp. in the bay.

1. Introduction

Red tide phenomenon (harmful algal bloomings, HABs) commonly occurs in Lampung Bay. It is a phenomenon in which a few species of algae/phytoplankton accumulate rapidly in the water column, and the bloom takes on a red or brown color depending on the pigment-containing in the phytoplankton [1]. Algae blooms have occurred in Lampung Bay nearly every year since 2005. This bloom is caused by different species of microscopic algae/plankton. The most common species growing out of control, thus inducing red tides in the bay was *Noctiluca* sp. (2005-2010), while some other causative organisms were *Pyrodinium* sp. (2008), *Phaeocystis* sp. (2009) *Dynophysis* sp. (2010, 2011) *Trichodesmium* sp. (2002 – 2012), *Ceratium* sp. (2011), *Pseudonitzchia* sp. (2009 – 2012) [2].

In 2012 an outbreak of algae bloom showed different characteristic from that occurred in previous years. It turned the seawater red brownish and caused massive fish kills in a culture which mortality was characterized by redfish fin and gills. The outbreak lasted for a longer time than usual, starting in October 2012 until the beginning of March 2013 (Fisheries and Marine Affairs Service, Lampung

Province, personal communication). This outbreak resulted in adverse effects on the environmental conditions and may become a threat to mariculture activities in Lampung Bay.

In some other coastal regions, red tides with similar characteristics above were caused by the bloom of *Cochlodinium polykrikoides*. This spesies was the most common species of phytoplankton identified upon blooming in Korea coastal waters [3], Jepang [4] dan Chesapeake Bay [5]. This species was widely sp. read distributed from Atlantik, Europe and to Asia coastal waters. However, no *Cochlodinium* bloom was ever observed in Indonesia. Previous research had identified *C. fulvescens* phytoplankton in December 2003 in Lampung Hurun Bay which was an area for mariculture in floating net cages; however, this species was found not to induce red tides as that occurred in 2012 [6].

The present study was aimed to analyze the characteristics of organisms inducing red tides in Lampung Bay in December 2012. Additionally, identification of factors affecting the outbreaks of red tides in the bay was conducted as well.

2. Materials and method

2.1. Study location and sampling period

The present study and monitoring were conducted in Lampung Bay in December 2012. Automatic Weather Station (AWS), Tide Master, Acoustics Doppler Current Profiler (ADCP), and Temperature Logger is stationarily deployed to monitor the hydrodynamics characteristics [7-9]. Samples were collected from 15 selected stations, based on stratified sampling method (figure 1). The sampling stations were spread from the tip to the mouth of the bay and covering the red-tide area and the normal one. Sampling location with its geographical position was presented in table 1.

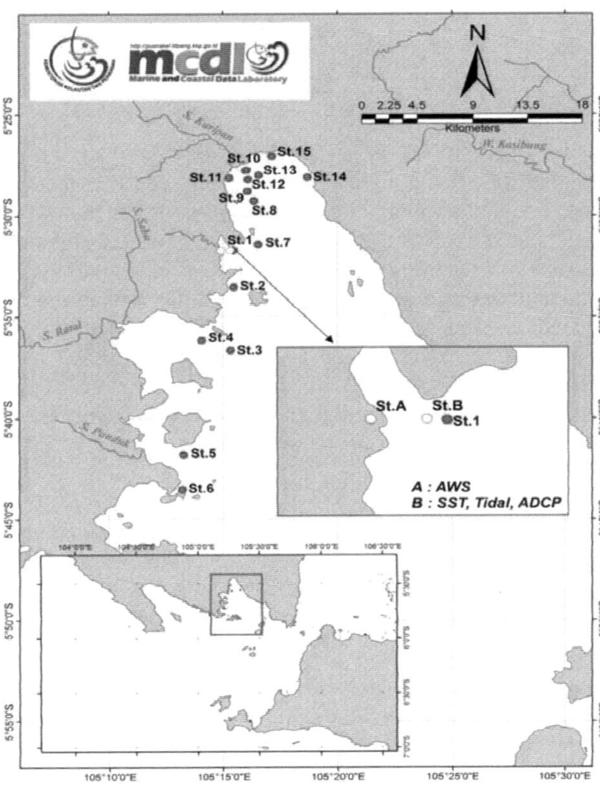

Figure 1. Sampling stations along the coastal area of Lampung Bay.

Table 1. Sampling location for red-tide monitoring in Lampung Bay.

No	Name of sampling location	Geographical position	
1	Kyoko (mouth of Hurun Bay)	05°31.704' LS	105°15. 456' BT
2	Ringgung	05°33. 529' LS	105°15, 474' BT
3	Adjacent to Maitam Islands	05°36. 661' LS	105°15.333' BT
4	Kelagian	05°36. 170' LS	105°14.057' BT
5	Puhawang Island	05°41. 800' LS	105°13. 252' BT
6	Putus Cape	05°43. 519' LS	105°13. 208' BT
7	Kyoko (Mutiara)	05°31. 405' LS	105°16.5359' BT
8	Adjacent Lempasing Fishing Port	05°29. 259' LS	105°16.353' BT
9	Floating Net Cages of Marine and Fisheries Service, Kota	05°28. 773' LS	105°16.072' BT
10	Kota Karang Estuary	05°27. 727' LS	105°16.026' BT
11	Keteguhan Embouchure	05°28.124' LS	105°15.269' BT
12	Pasaran Bay	05°28.192' LS	105°16.088' BT
13	Middle of the bay	05°27. 976' LS	105°16.561' BT
14	Panjang Port	05°28.085' LS	105°18.691' BT
15	Sukaraja	05°27. 027' LS	105°17.147' BT

2.2. Materials
Phytoplankton samples were collected using *Kemmerer Water Sampler* and then filtered through plankton net of bolting silk no. 25. Seawater filtration was conducted to concentrate the phytoplankton, particularly in the coastal area with no blooming. Filtered plankton samples were then fixed directly with Lugol solution and stored for later examination. Phytoplankton abundance was analyzed according to [10], and plankton was identified based on [11].

Physical parameters monitoring in the study were temperature, salinity, water current, and wind profile. Seawater temperature was measured using HOBO instrument, while the water current was measured using ADCP. Wind sp.eed and direction were obtained from AWS instrument. Meanwhile, the chemical parameters measured were pH, dissolved oxygen, chlorophyll-a, nitrate (NO_3-N), nitrite (NO_2-N), orthophosp.hate (PO_4-P), dan silica (Si). To determine the concentration of nutrients and chlorophyll, laboratory analysis using sp.ectrophotometer was conducted.

2.3 Data analysis
The samples were taken into Sedgwick Rafter counting Cell (SRC) and identification of plankton was carried out, and counting was done according to the procedure outlined by [10]. The abundance of phytoplankton was estimated by counting their presence per focus of the microscopic field. The abundance of phytoplankton was expressed using the following formula:

$$N = \frac{n}{100} x \frac{A}{B} x \frac{C}{D} x \frac{1}{E}$$

where,

N = Total number of plankton (cells/L)
n = Number of organisms per number of counting cells observed
A = Volume of SRC (mm^2).
B = The area of counting cells observed (mm^2)

C = Volume of concentrated plankton sample (mL)
D = Volume of 1-drop plankton sample underneath the cover glass (mL)
E = Volume of total water filtered (L)

3. Results and discussion

3.1. Composition and abundance of phytoplankton

The results indicated the phytoplankton population in Lampung Bay in December 2012 composed of four classes which were Bacillariophyceae, Dinophyceaea, and Cyanophyceae. Dinophyceae dominated the entire population in the coastal water, and it formed 92.88% of the total phytoplankton population. *Cochlodinium* sp. was the most dominant species of phytoplankton and it constituted 85.89% of the total phytoplankton abundance (figure 2), with the average abundance of 2.0 x 10^4 cells/L (table 2).

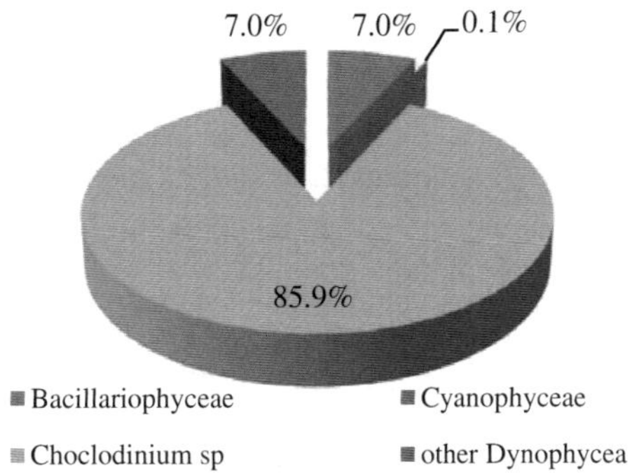

Figure 2. Phytoplankton abundance.

Table 2. The phytoplankton abundance in Lampung Bay.

Phytoplankton Classes	The range of abundance	The average abundance
Cyanophyceae	0.0 x 10^0 - 5.5 x 10^2	1.3 x 10^1
Bacillariophyceae	3.0 x 10^1 - 9.9 x 10^3	1.6 x 10^3
Dinophyceae	1.1 x 10^2 - 7.4 x 10^5	2.2 x 10^4
Cochlodinium sp.	0.0 x 10^0 - 7.4 x 10^5	2.0 x 10^4

Generally, in normal condition, no species *Cochlodinium* sp. was observed in Lampung Bay. The sudden occurrence of this dominant species in the bay suggested the blooming of this population. The average population abundance of the species was estimated at 2.4 x 10^4 cells/L. The value was indicated to be lower than the real population abundance because sample collection was conducted using a smaller mesh size of plankton net (60µm). However, the value is still in confirmation with the range value of abundance for *Cochlodinium* sp. blooms in Philippines, Malaysia, Korea, and Pakistan. *Cochlodinium* sp. bloom in Iligan Bay, South Philippine, was recorded as 3.1x 10^4 – 3.8 x 10^4 cells/L [12]. In Yeosu Korea, *Cochlodinium polikrikoides* bloom was found to occur 1 – 8.6 x 10^4 cells/L [3], whereas in Sabah Malaysia waters, average *Cochlodinium* sp. abundance could reach 6 x 10^6 cells/L [13]. Moreover, the occurrence of *Cochlodinium fulvescens* bloom in Pakistan was recorded at the abundance of 6x10^2 – 9 x 10^2 cells/L [14]. [15] reported that *C. polykrikoides* abundance at a density of 10^3 – 10^4 cells/L could lead to fish mortality, while the abundance of >5 x 10^4 cells/L could kill all fish (100%) within a week.

3.2. Morphology of Cochlodinium sp. *cell*

Visual observation during survey presented that during blooms, seawater turned into red-brownish. It was observed that upon water filtration through plankton net, the phytoplankton in the filtered water was sticking and coagulating each other leading to net clogging. Furthermore, when the filtered water was collected in a container, within less than 30 minutes, those phytoplanktons would coagulate and precipitate at the bottom of the container.

Based on morphological comparison results using microscopic imagery analysis, it was identified the bloom was caused by the high occurrence of the phytoplankton species closely related to *Cochlodinium* sp. [16] reported that *Cochlodinium* sp. had sub-sp.herical cells and made chains (consisting of 2-8 cells or more). The anterior part of the episode is conical, and the hypocone is subsp.herical forming epitheca. The cells have spiral cingulum, and the cell size ranged from 20 – 30 μm.

The observation also showed that the phytoplankton cells during bloom events in Lampung Bay had round shape, composing of two-to-four-cell chains and a single cell separated from the chains. The anterior of the cell is more rounded with a smaller diameter than the middle of the cell. The results of microscopic images with 400x enlargement showed that a circular line pattern in the middle of the cells was observed suggesting sp.iral cingulum commonly found in *Cochlodinium* sp. Morphological comparison suggested the cells had a closer shape to *C. fulvescens* cells than *C. polykrikoides* cells [17], however, with a smaller size. *Cochlodinium fulvescensh* with the average length of 45.8 μm (37.5–57.5 μm) and 35.3 μm wide (30.0–42.5 μm) have been previously found in Hurun Bay of Lampung in December 2003. It was also observed that the size of *Cochlodinium* sp. cells in the present study was 7-10 μm. This size is much smaller than that of *Cochlodinium polykrikoides* commonly found in bloom events in Korea coastal waters [3] and Persia Bay [18]. Figure 3 presented the morphology of the cells during bloom events in Lampung Bay.

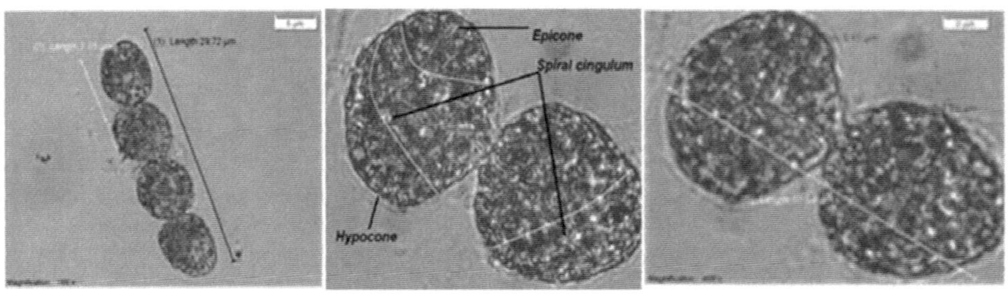

Figure 3. Morphology of phytoplankton forming red tides in Lampung Bay: (a) a four-cell chain, 38.65 μm long; (b) spiral cingulum encircling the cell (c) the cell size at 400x enlargement.

Morphological comparison among *Cochlodinium* sp. occurred in Lampung Bay, *C. fulvescens* [6, 17] and *C. polykrikoides* [6] is described below (table 3).

Table 3. Morphological comparison between *C. fluvescens, C. ploykrikoides* and *Cochlodinium* sp.

Morphological characters	*C. fluvescens*	*C. polykrikoides*	*Cochlodinium* sp.
Size	50 - 60 μm	20 - 30 μm	7 - 10 μm
Shape	Subspherical - ellipsoidal	Ellipsoidal	Subspherical
Sulcus	Branched	Shallow	Unclear
Cingulum	Encircle the cell twice	Encircle the cell twice	Encircle the cell twice

3.3 Distribution of Cochlodinium sp.

Cochlodinium sp. has a clumped distribution forming spots in particular spaces with different densities and coverage areas. This distribution type is typical for *Cochlodinium* bloom events [12,15,17]. It was observed that *Cochlodinium* sp. was distributed in the western coast of Lampung Bay, starting from Sukaraja Beach to Putus Bay located in southern part of Lampung Bay (figure 4). No bloom event of *Cochlodinium* sp. was found in the eastern coast of Lampung Bay (Panjang Harbour-Kalianda).

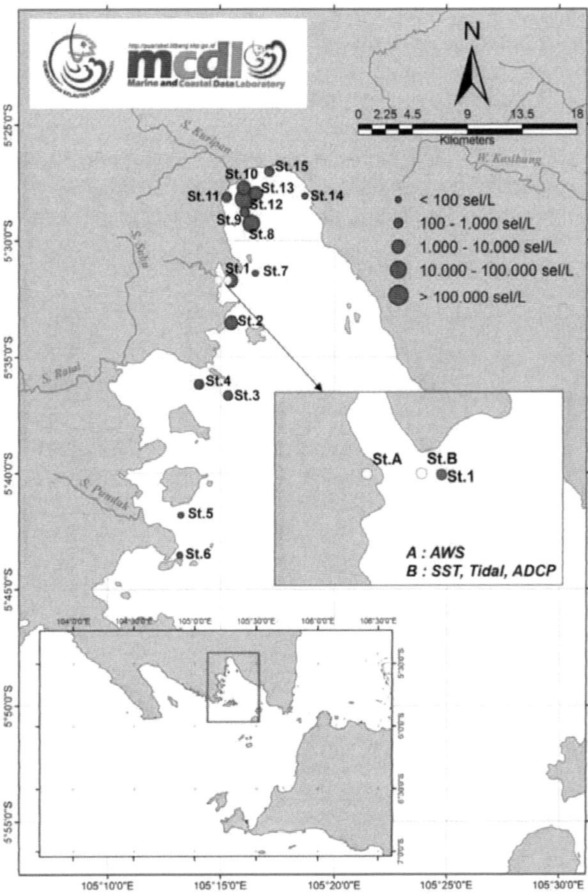

Figure 4. Spot distribution of phytoplankton bloom on 22 – 23 December 2012.

Furthermore, it is evident that *Choclodinium* sp. was highly abundant in the upper part of the bay, from Betung Bay to Ringgung Sub-district coastal areas. The highest abundance was found in Pasaran Bay (station 12) with the average abundance of around 1.1 x 10^5 cells/L. In this area, anthropogenic activities such as marine fish culture in floating net cages (Station1, 2 and 9), sea pearl culture (station 7), and fish landing (station 8 and 15) were found. Moreover, an estuary was also located in the area. [19] reported that eutrophication resulted from water run-off from the river, nutrient-rich water arisen from aquaculture, heavy rainfall, and water stirring in mixing layer could lead to the bloom event of this phytoplankton.

Interestingly, the abundance of *Cochlodinium* sp. was lower in the southern and middle parts of the bay than the upper part. The abundance of *Cochlodinium* sp. in Ratai Bay particularly in ST 3 and 4 was 2.38 x 10^2 and 2.00 x 10^2 cells/L, respectively. Moreover, a much lower abundance of *Cochlodinium* sp. (< 10^2 cells/L) was observed in station 5 and 6 (a more southern area). The southern part of the bay is a more open connecting directly to Sunda Strait.

Some clumps of *Cochlodinium* sp. with the average abundance of 7.51 x 10^2 cells/L were still found in the middle of the bay. Meanwhile, the abundance of *Cochlodinium* sp. was found to be lower (69 cells/L) in around Panjang Port (St. 14). This suggested that the distribution of *Cochlodinium* sp. abundance was significantly influenced by water stirring / distribution and tides.

3.4. Distribution of nutrients (nitrogen and phosphate)

Table 4 shows the dissolved nutrient conditions in Lampung Bay. It is evident that high concentrations of ammonia and nitrate were obtained indicating that the coastal water was in eutrophication. The results showed that the concentration of nitrate (NO_3) increased 4.5 fold compared to that in normal condition, and nitrite was 0.7 times lower than in the normal condition. Meanwhile, ammonia was significantly high during this bloom event, i.e., 6.5 times higher than that in the normal condition, whereas the concentration of nitrate obtained in the study was lower than that in the normal condition. Upon the bloom event, the concentration of phosphate in the water was lowest in seawater of the bay, i.e., 0.2 times lower than the common condition.

Table 4. Dissolved nutrient conditions in Lampung Bay.

Variable	Normal	Observation on 22 - 23 Dec 2012		
		Range	Average	Increment
NO_2	0.012	0.003 - 0.02	0.008	0.7
NO_3	0.036	0.032 - 0.413	0.162	4.5
NH_4	0.103	0.085 - 1.241	0.670	6.5
PO_4	0.119	0.080 - 0.075	0.024	0.2

According to nutrient data in the present study suggested that *Cochlodinium* sp. is a phytoplankton species that could grow well in seawater containing high N and low P. The present results showed that N/P ratio was 35.5, significantly higher than the normal N/P ratio [16]. In the Philippines, *Cochlodinium* bloom events occurred when N to P ratio was 14:1 [12]. Meanwhile, in Jejudo Korea, *Choclodinium polykrikoides* blooms occurred when the nitrogen content in seawater was higher than that in its normal condition [3].

High nitrogen content in seawater in the present study could be attributed to water run-off upon heavy rainfall occurred after a long dry season in early October 2012. Some outbreaks of *Choclodinium* sp. blooms could also be triggered by heavy rains happened before red tides, as the outbreaks found in Korea coastal areas [3] and Chesapeake Bay [5].

3.5. Conditions of physical oceanography

The results of temperature measurement using HOBO showed that seawater temperature of Hurun Bay was relatively warm (29.9-30.3°C) and had low T variability. Meanwhile, temperature data derived from World Ocean Database 2009 showed that the lowest seawater temperature was obtained in the season of December – February ranging from 28.25-28.50°C [20]. Compared to the results indicated that the temperature condition in December 2012 was in anomaly which was higher than the seasonal temperature. This warm temperature with low variability obtained in the study is favorable for red-tide organisms. The water mixing in Lampung Bay is affected by the tidal phenomenon.

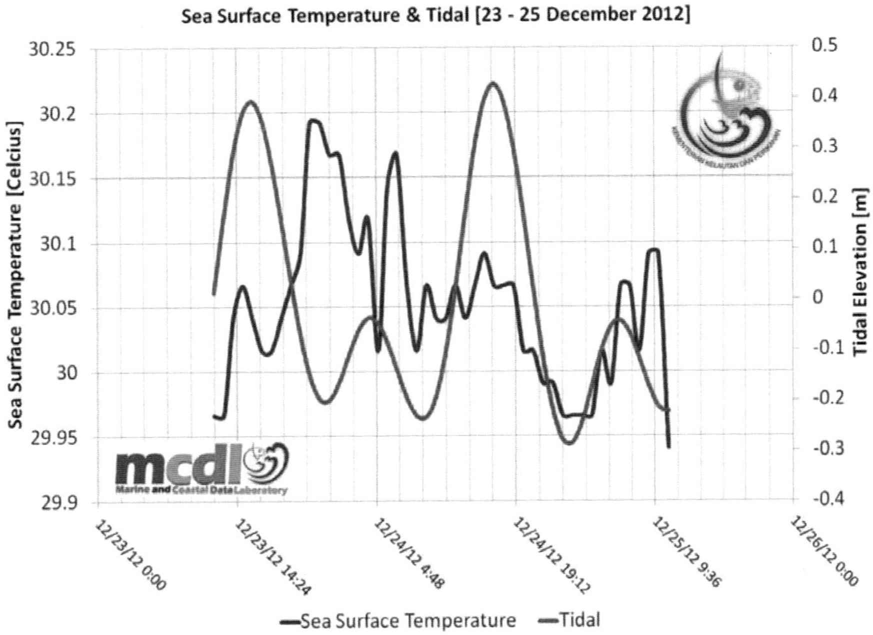

Figure 5. Water column temperature and tidal profile in Lampung Bay measured on 23 -25 December 2012.

The water mixing could be determined from vertical currents measured by ADCP in as shown in figure 6 [21]. Some outbreaks of *Choclodinium* sp. blooms in southern California coastal waters were initiated by the mixing of warm and cold water resulted from upwelling and freshwater run-off entering into the sea [22]. *Cochlodinium* sp. could grow well and rapidly in warm seawater [3]. In Korea coastal waters [3, 24] and Mexico coastal waters [17], the outbreaks of *Cochlodinium* blooms occurred in temperature of 29-30°C. Upwelling and downwelling upon monitoring occurred when no wind was present (figure 7). This suggested that wind was not the main factor inducing up and downwelling in the bay, tides were, instead [25]. During the high tide, water current moved into the bay, whereas during the low tide, the current would move out from the bay [26].

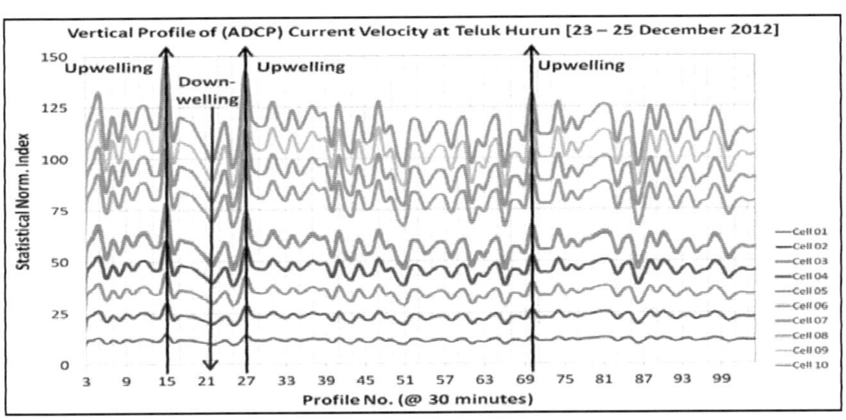

Figure 6. Vertical Velocity Profile Measured using ADCP at Teluk Hurun (23 – 25 December 2012).

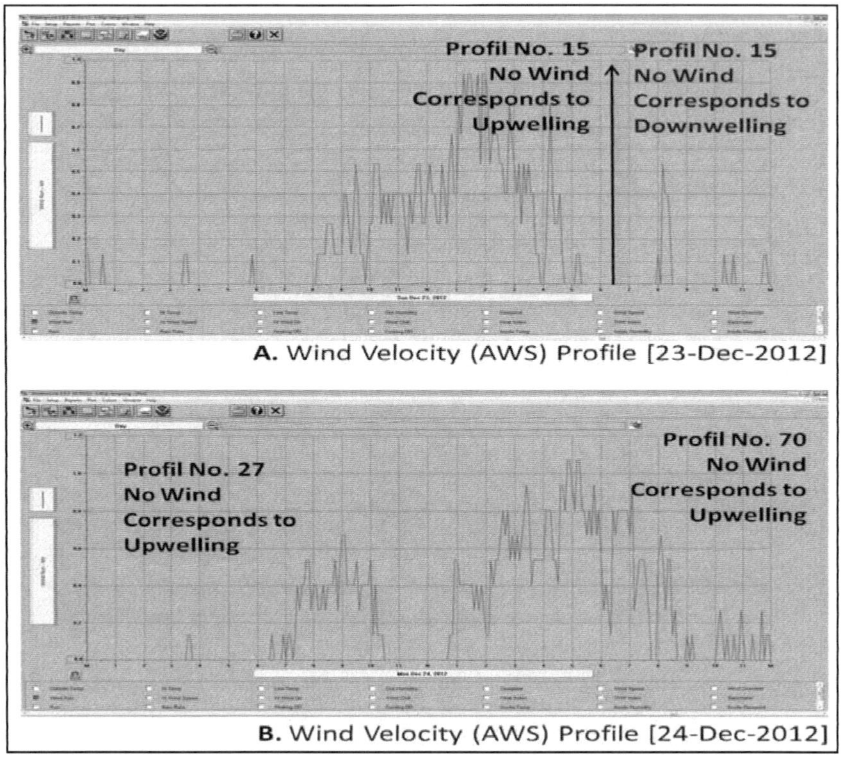

Figure 7. Wind profile in 23 (A) and 24 December 2012 (B).

Sea current movement driven by tides in Lampung Bay affected the *Cochlodinium* sp. distribution [27]. Tidal currents are moving into and out from the bay likely resulted in the water mass circulating within the gulf/bay [28]. The circulating water mass accumulated the water mass more in western coast areas of the bay where many small islands such as Puhawang, Kelagian, Maitem, Tegal islands were located. On the other hand, water mass running along the eastern coast areas was not obstructed by any islands or any neighboring corals, leading to no water mass accumulation in the area [29]. This indicated that the water mass accumulation in the western coast areas of the bay influenced the *Cochlodinium* sp. distribution, providing clumped propagation found only in the western bay.

4. Conclusions

The outbreak of *Choclodinium* sp. blooms in Lampung Bay was induced by the high content of dissolved N resulted from land intrusion and also organic decomposition from the sea bottom arising to the surface due to upwelling. Moreover, the bloom event was also triggered by the anomaly temperature in December 2012, in which the sea surface temperature was warmer than the usual. Water mass circulation resulted in *Cochlodinium* sp. distribution more concentrated in the west coast areas of Lampung Bay.

Acknowledgments

The surveys were conducted in 2012 funded by APBN DIPA TA. 2012 for the Indonesian Ministry of Marine Affairs and Fisheries. Further analysis of algae is undertaken during 2014 -2015 in The Research Institute for Fish Recovery & Resources Conservation. Discussions and hydrodynamics data analysis has been done in the Marine and Coastal Data Laboratory (MCDL), at Marine Research Center. This publication is funded by APBN DIPA TA 2017 for the Fisheries Research Center. Sp.ecial thanks go to Army and Tussadiah (MCDL) for making the beautiful maps (Fig. 1 & Fig. 3).

References

[1] Praseno D P 2000 *Red Tide in Indonesia Coastal Waters* (Jakarta: LIPI) pp 82
[2] Muawannah, Wahyu W and Tri H 2012 Evaluation of red tide monitoring results: *Cochlodinium* in Hurun Bay coastal water and its neighboring areas Presentation Material Development Institute for Mariculture, Lampung (in Bahasa Indonesia)
[3] Lee Y S and Lee S Y 2006 Factors affecting outbreaks of *Cochlodinium polykrikoides* blooms in coastal areas of Korea *Marine Pollution Bulletin* **52**(6) 626-634
[4] Onitsuka G, Miyahara K, Hirose N, Watanabe S, Semura H, Hori R, Nishikawa T, Miyaji K and Yamaguchi M 2010 Large-scale transport of *Cochlodinium polykrikoides* blooms by the Tsushima warm current in the southwest Sea of Japan *Harmful Algae* **9**(4) 390-397
[5] Mulholland M R, Ryan E M, George E B, Peter W B, Katherine C F, Leo A P, Jose L B G, Harold G M, Todd A E, William S H, Kenneth A M, Dianna L B and Christopher J G 2009 Understanding causes and impacts of the Dinoflagellate *Cochlodinium polykrikoides*, blooms in the Chesapeake Bay *Springer: Estuaries and Coasts* doi:10.1007/s12237-009-9169-5
[6] Iwataki M, Kawami H and Matsuoka K 2007 *Cochlodinium fulvescens* sp. nov. (Gymnodiniales, Dinophyceae), a new chain-forming unarmored dinoflagellate from Asian Coasts *Phycological Research* **55** 231-239
[7] Tanto T A, Wisha U J, Kusumah G, Pranowo W S, Husrin S, Ilham I and Putra A 2017 Karakteristik arus laut perairan Teluk Benoa, Bali *J. Ilmiah Geomatika* **23**(1) 37-48
[8] Arifiyanto, Pranowo W S, Kuswardani A R T D and Fatoni K I 2016 Pengolahan dan penyajian data arus pasang surut hasil pengukuran acoustic doppler current profiler (ADCP) SonTek Argonout-XR menggunakan perangkat lunak T_TIDE_V1.3beta *J. Hidropilar* **1**(2) 56-67
[9] Pranowo WS, Adi R A and Puspita C D 2013 Analisis daya dukung sumberdaya laut dan pesisir Sumba Timur untuk pembukaan ladang produksi garam. *Prosiding Seminar Hasil Penelitian Terbaik Tahun 2013*. Badan Penelitian & Pengembangan Kelautan & Perikanan p 336-342
[10] [APHA] American Public Health Association 2005 *Standard Methods for the Examination of Water and Waste Water*, 17th *ed* (Washington DC: APHA) pp 1193
[11] Yamaji I 1984 *Illustration of the Marine Plankton of Japan* (Osaka: Hoikusho)
[12] Vicente H J, Gaid R D and Dejarme H E 2002 Harmful algal bloom in Iligan Bay, Southern Philippines *Science Diliman* **14**(2) 59 -65
[13] Anton A, Teoh P L, Mohd-Shaleh S R and Mohammad-Noor N 2008 First occurrence of *Cochlodinium* blooms in Sabah, Malaysia *Harful Algae* **7**(3) 331-336
[14] Munir S, Naz T, Burhan Z U, Siddiqui P J A and Morton S L 2012 First report of the athecate chain forming Dinoflagellate *Cochlodinium fulvescens* (Gymnodiniales) from Pakistan *Pak. J. Bot.* **44**(6) 2129-2134
[15] Gobler C J, Berry D L, Anderson O R, Burson A, Koch F, Rodgers B S, Moore L K, Goleski J A, Allam B, Bowser P, Tang Y and Nuzzi R 2008 Characterization, dynamic, and ecological impact of harmful *Cochlodinium polykrikoides* blooms on eastern Long Island, NY, USA *Harmful Algae* 7 293-307
[16] Steidinger K A and Tangen K 1997 *Identifying Marine Phytoplankton. Editor: Carmelo R. Thomas* (USA: Academic Press)
[17] Lizarraga G I, Lopez C D J, Bustillos G J J and Hernandez S F 2004 Blooms of *Cochlodinium polykrikoides* (Gymnodiniaceae) in the Gulf of California, Mexico *Revista de Biologia Tropical* **52**(1) 51-58
[18] Jahromi S T, Ahmad S O, Fereshteh S, Eisa A, Maryam M, Kiumars R, Samad H and Mohammad RS 2011 Identification and molecular phylogeny of the dinoflagellate (*Cochlodinium polykrikoides*) from the Persian Gulf *International Journal of Review in Life Science* **1**(4) 193-200
[19] Lizzaraga I G 2008 Occurrence of *Cochlodinium fulvescens* (Gymondiniales: Dinophyceae) in The Southwestern Gulf of California *Rev. Biol. Mar. Oceanogr.* **49**(1) doi.org/10.4067/S0718-19572014000100013

[20] Pranowo W S, Adi R A, Permana H and Hananto N D 2012 Sirkulasi arus permukaan pasang surut di Muara Pegah, Delta Mahakam, Kalimantan Timur *J. Segara* **8**(1) 56
[21] Kudela R M, Jhon P R, Melissa D B, Jenny Q L and Tawnya D P 2008 Linking the physiology and ecology of *Cochlodinium* to better understand harmful algal bloom events: a comparative approach *Harmful Algae* **7** 278-292
[22] Lee M O, Choi J H and Park I H 2010 Outbreak conditions for *Cochlodinium polykrikoides* bloom in the southern coastal waters of Korea *Marine Environmental Research* **70**(3) 227-238
[23] Pranowo WS, Phillips H and Wijffels S 2005 Upwelling event 2003 along South Java Sea and the Sea of Lesser Sunda Islands *J. Segara* **1**(3) 116-123
[24] Pranowo W S, Puspita C D, Bramawanto R, Adi R A and Kuswardani A R T D 2014 Dinamika arus dalam mendukung perikanan budidaya laut di Teluk Bone *J. Harpodon Borneo* **7**(2) 135-152
[25] Pranowo W S, Herdiani Y and Radjawane I M 2004 *Barotropic tidal and wind-driven larval transport on Saleh Bay, Sumbawa, Indonesia APEC/MRC/OMISAR* The Twelfth Workshop on Ocean Models (WOM-12) 7-10 September, 2004, Dalian, P. R. China, p 12
[26] Pranowo W S 2012 Dinamika upwelling dan downwelling di Laut Arafura dan Timor. *J. Widyariset* **15**(2) 415-424
[27] Munasik D N, Sugianto, Pranowo W S, Suharsono, Situmorang J and Kamiso H N 2006 Pola arus dan kelimpahan karang *Pocillopora damicornis* di Pulau Panjang, Jawa Tengah *Indonesian Journal of Marine Science* **11**(1) 11-18

Biological characteristics of Indo-Pacific King Mackerel (*Scomberomorus guttatus*, Bloch and Schneider 1801) in Moro Waters part of Kepulauan Riau, Indonesia

T Noegroho[1*], **M Boer**[2], **L Adrianto**[2] **and Sulistiono**[2]

[1] Graduate Student of Aquatic Resources and Fisheries Management Program
[2] Lecturer at Department of Aquatic Resources Management,
Faculty of Fisheries and Marine Sciences, Bogor Agricultural University, Bogor-Indonesia
*e-mail: teguhnug80@gmail.com

Abstract. Research related to biological characteristics of Indo-Pacific King Mackerel in Indonesia is still rare, whereas fishing activity has been carried out. The study was conducted from 2014-2017 in Moro Waters, Riau Islands, Indonesia. The purpose of this study was to analyze the biological characteristics including length frequency, length weight relationship, Gonado Maturity Stages, Gonado Somatic Index (GSI), sex ratio, length at first maturity, fecundity and ova diameter. From length frequency were obtained mode size at 37-39 cm for male, 52-54 cm for female and 43-45 cm for transition sex. The growth pattern of king mackerel was isometric. The gonad fish were dominated of mature gonads by 52%, and the immature gonads by 48%, with level V was almost found every month. GSI shows that spawning season in January-August and peak season occurs in April-May. The results of chi-square test showed that there were significant differences number of males and females fish. The length at first maturity was at 47,4 cm, in the range between 46.3 to 48.5 cm. The fecundity ranged on 28.082-1.506.075 eggs, on a length of 45-55 cm. Based on the development of the ova diameter and Gonad Maturity Stage, the King Mackerel spawning pattern is partial spawner.

1. Introduction

Indo-Pacific King Mackerel (*Scomberomorus guttatus*) is one of the species of mackerel fish found in Indonesian Waters. Spread almost in all coastal waters and islands throughout Indonesia. The mackerel is a neritic species and migrates no further than Spanish mackerel (*Scomberomorus commerson*), has a habitat in less clear waters with lower salinity [1].

Indo-Pacific King Mackerel in the waters of the South China Sea (SCS) caught by drifting gillnet. Most of the ships used are traditional <10 GT boats. Some locations in the SCS which became the landing of Indo-Pacific King Mackerels, among others, are in Bangka, Belitung, Moro Waters, Tanjung Pinang and several other islands. The composition of gillnet catches in SCS is dominated by King Mackerel. Therefore, a research related to biological characteristics of the Indo-Pacific King Mackerel is needed.

Research on the biology of Indo-Pacific King Mackerel has been done by RIMF (Research Institute for Marine Fisheries 2013) in Indonesia Fisheries Management Area (IFMA) 573 Indian Ocean South Java with the location in Cilacap Waters. While some others research related to Indo-Pacific King Mackerel was done by [2] in the Mandapam Water India, [3, 4, 5] in the Gulf of Mannar and Palk Bay of India, [6] in Japanese Waters, [7] in the Gulf of Mexico, [8] in FAO area 51 Indian Ocean (Oman Waters), [9] in Veraval India, [10] in the Bay of Bengal Bangladesh, and [11] in India.

The purpose of this research was to analyze the biological characteristics of King Mackerel which includes length frequency, length weight relationship, gonad maturity level, sex ratio, gonad maturity index, length at first mature, food habits, fecundity and ova diameter. This research was expected to provide information for sustainable fisheries management.

2. Methods

2.1. Time and location of research

The biological sample was collected from 2014-2017. Indo-Pacific King Mackerel (*Scomberomorus guttatus*) was the object in this research. The survey location on the Moro Waters, part of Riau Island (South China Sea). The location of survey and sampling can be seen in figure 1.

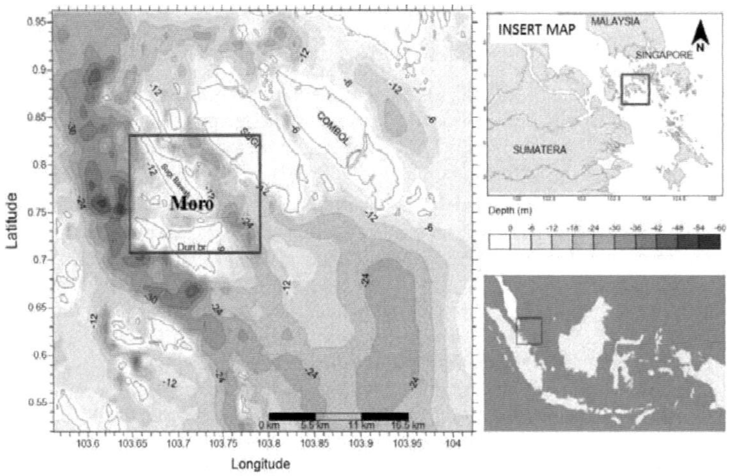

Figure 1. Study site and sampling of King Mackerel in Moro Waters.

2.2. Length frequency and length weight relationship

Length frequency and length weight relationship analysis is done by separating male and female fish. Length data and more weight are used to see changes between male and female fish related to the reproductive condition of the fish. Fork length data of fish and weight were collected by writer assisted by enumerator.

According to [12], length and weight relationship of fish of almost follow the cubic law, although not always follow the cubic law expressed by the formula $W = aL^b$, because during the live fish will grow and evolve mean length and weight of the fish is always changing, then the formula written in the general form:

$$W = aL^b \tag{1}$$

where W is the weight of fish in grams, L is the length of fish in cm, while a and b are constants. Knowledge regarding the length and weight of fish is we can guess the weight of the length of the fish or otherwise, regarding the description of fish growth, plumpness, which is influenced by the environment.

2.3. Level of gonad maturity

Observation of the gonad maturity level is done by splitting the sample by the dissecting set carefully so that the gonad does not participate in split until it can simplify the process of observation. After that it was then to observe directly to the color and size of the gonad. Thus, the identification of the gonad with reference to the order of the maturity level of gonads corresponding by [13] which divides the maturity stage male and female gonads into 5 stages (table 1).

Table 1. Criteria of gonad maturity stage by [13].

Female

Stage	Category	Description
I	Immature	Gonads elongated and slender, about one-third the length of the abdominal cavity, sex can be seen with a magnifying glass, ovarian clear gray to reddish.
II	Maturing	Gonads enlarged but the eggs cannot be viewed one by one with the naked eye, ovary ovoid-shaped, reddish color with capillary vessels, ovaries fill about half the space below.
III	Mature	Gonads enlarged and swollen, egg easily seen with the naked eye, ovaries reddish orange, ovaries filling 2/3 of the body cavity.
IV	Ripe	Gonads filled the basement, have enlarged ovaries, eggs and ripe crisp, easy to get out of the lumen of the stomach of fish ovaries when pressed.
V	Spawned/ Spent	Eggs already released (post-spawning). Gonads shrin and loose wall. Ovari very large and soft

Male

Stage	Category	Description
I	Immature	Testis very smooth, flat like a ribbon but the sexes can be distinguished with a magnifying glass. Most sperm contained in the center channel.
II	Maturing	Testes enlarged, triangular cross sections, colored reddish or white starting with capillaries, small and dense.
III	Ripe	Testes enlarged and swollen, slightly opaque white color, a little sperm out freely when in press / gonads cut, can be seen with the naked eye.
IV	Partly spent	Testes greatly enlarged, opaque white color no red spots, sperm easy out if the stomach is pressed a little fish or cut when the gonads.
V	Spent	Testis spent soft, dark reddish, mushy and gonad wall of blood vessels.

2.4. Sex Ratio

The sex ratio is analyzed by comparing the number of male fish with female fish, for example:

$$X = M/F \tag{2}$$

where: X = sex ratio; M = number of male fish; F = number of female fish.

The sex ratio was made in a bar graph from several months to see the difference in the percentage ratio. The sexs ratio test follows the way suggested by [14] with a Chi - Square test formula, as follows:

$$X^2 = \sum_{i=1}^{k} \frac{(f0 - fn)^2}{fn} \tag{3}$$

where : X^2 = Chi – Square; fo = Observed frequency; fn = Hoped frequency; Test table in real level 95% (n-1) with hypothesis as follows; H_0 : no real difference between male and female fishes.; H_1 : real difference between male and female fishes; if, χ^2 counts <χ^2 table, H_0 accepted, H_1 denied; iIf, χ^2 counts >χ^2 table, H_1 accepted, H_0 denied [12].

2.5. Gonad Maturity Index (GSI)

Gonad Maturity Index (GSI) was analyzed with the formula put forward by [12] as follows:

$$GSI = Wg/W \times 100\% \tag{4}$$

where: *GSI* = Gonad Maturity Index; *Wg* = weight of gonad (g); *W* = Somatic weight = fish weight minus gonad weight (g).

2.6. Length at first maturity (L_m)

The value of L_m was analyzed by Spearman-Karber method in [16] as follows:

$$m = Xk + X/2 - \left(X \times \sum pi\right) \quad (5)$$

where: *m* = size logarithm of first matured gonad; *Xk* = logarithm of the last middle grade where it 100% mature gonad occurs; *X* = difference of mean value of logarithm; *pi* = the proportion of gonad ripe fish in the i-th class.

$$CL = Anti\log\left[m \pm 1{,}96\sqrt{X^2\left(\frac{\sum pi \times qi}{ni+1}\right)}\right] \quad (6)$$

CL = *Confident limit* (maximum and minimum); *m* = fish length of firstly matured gonad; *ni* = the number of fish in the i-th class; *qi* = *1 – pi*.

2.7. Food habits

Food habits were analyzed using the frequency occurrence method. Each organism contained in the stomach is expressed in the percentage of all the stomach studied, except the empty stomach. Thus, we can see the frequency of occurrence of an organism that fish eats [15].

2.8. Fecundity

Fecundity was analyzed by the formula put forward by [15] as follows:

$$F = G/g \times N \quad (7)$$

where: *F* = fecundity (egg); *G* = total gonad weight (g); *g* = sub sample gonad weight (g); *N* = number of gonad sample (items).

The number of gonad sample is calculated with the gravimetric method, by calculating some parts of gonad samples given the Gilson solution. The samples were then weighed. After weighing the sample, it was then dissolved with water, 30 ml. From 30 ml of water containing the gonad sample, 1 ml was taken to be observed under a microscope.

2.9. Ova diameter

The ova diameter was measured with a scaled micrometer. The division of the ova diameter group was made at the 0.05 mm class interval [17]. Devaraj has divided the ova diameter of King Mackerel into 12 levels, as seen in table 2.

Table 2. Ova diameter category of King Mackerel [17].

Gonad Level	Category	Diameter (mm)	Gonad Level	Category	Diameter (mm)
Immature	A	0.03-0.10	Ripe	H	0.56-065
	B	0.10-0.15		I	0.65-0.70
Maturing	C	0.15-0.20		J	0.75-0.80
	D	0.20-0.30	Spent	K	0.80-0.88
	E	0.30-0.36		L	0.90-0.98
	F	0.46-0.53	Partly spent	S	0.23-0.35
	G	0.53-0.56			

3. Results

3.1. Length Frequency

From the range of length obtained were 10-75 cm, the length of male fish by 25% and female fish by 50% and transitional fish by 25%. From length frequency were obtained mode size at length 37-39 cm for male and 52-54 cm for female and 43-45 cm for transition on a sex change. Visually by comparing the morphological conditions of the gonads and the length of the fish can be separated the length of the fish as shown in figure 2.

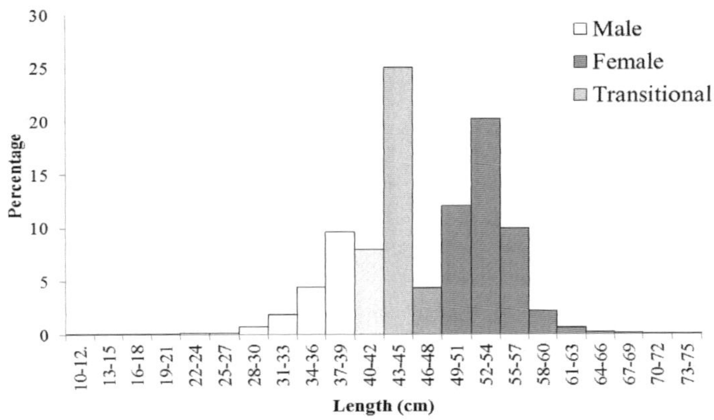

Figure 2. Length frequency of Indo-Pacific King Mackerel and the length of the transition on a sex change.

3.2. Length weight relationship

From the length weight relationship of King Mackerel fish was obtained a value = 0.00007; the value of b = 2.425 and the value of R^2 = 0.752, while the female fish value a = 0.00002, b = 2.782 and R^2 = 0.664. Length weight relationship of the male female was obtained values a- 0.00002, b = 2.738 and R^2 = 0.905. The average weight of captured male fish is 0.2 kg (200gr), while female fish is 1.04 kg (1040 gr). From the value of b value shows the growth pattern of King Mackerel both of male and females is negative allometric, while the b value from both sexes after t test was isometric. The length weight relationship of King Mackerel is presented in figure 3.

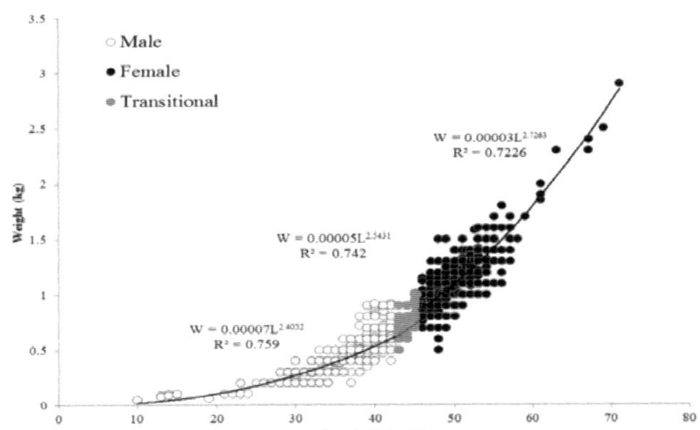

Figure 3. Length weight relationship of Indo-Pacific King Mackerel, for male and female fish.

3.3. Gonad maturity level

The number of gonad samples were 3307, consisting of 15% male gonads, 70% female gonad and 15 % from transition gonads of fish. Gonad maturity level obtained showed that gonad level III and IV were more dominant from month to month (2014-2017). Gonad level I-II could mostly be found in male fish, whereas gonad level V, which was an indication of spawning fish, was found almost every month. Especially for male fish sperm cells in gonads consist of several levels, some are ready for fertilization and some are still developing. Gonad maturity level is presented figure 4.

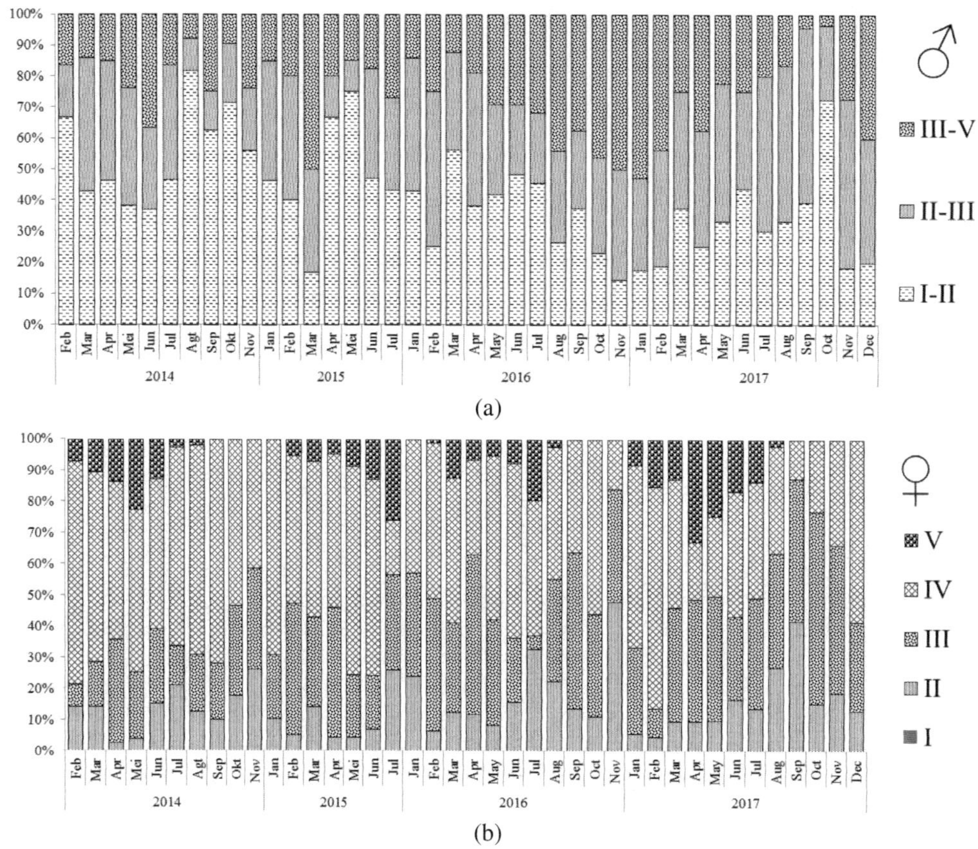

Figure 4. Gonado maturity stages of King Mackerel for (a) male and (b) female.

3.4. Gonado Somatic Index (GSI)

The King Mackerel spawning season was suspected in January-August. Spawning season usually occurs one or several months after the highest peak of GSI. Peak spawning season occurs there April-May. GSI value is presented in figure 5.

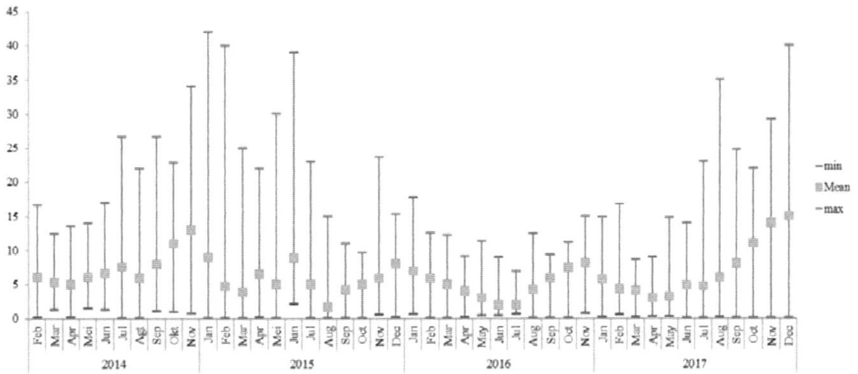

Figure 5. Gonado Somatic Index of Indo-Pacific King Mackerel.

In figure 6, the parabolic line indicates that the GSI value rises up to a length of 48 cm, then down at length > 55 cm. At the same length of fish, the magnitude of GSI shows varying values as well.

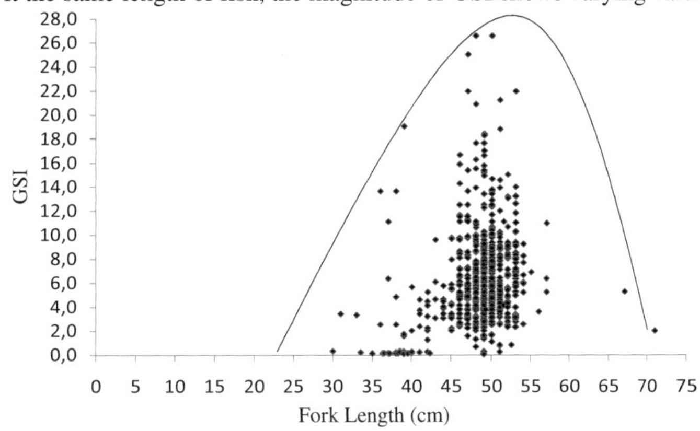

Figure 6. Relationship of fork length and Gonado Somatic Index.

The GSI value of Indo-Pacific King Mackerel also changes with the increasing of Gonad Maturity (table 3). In gonad level V, there was a decrease in GSI value for female fish.

Table 3. The relationship of Gonado Maturity Stages and Gonado Somatic Index of Indo-Pacific King Mackerel.

Sex	Gonad Level	GSI	
		Range	Average
Male	I	0.105-2.564	0.441
	II	1.350-2.040	1.695
Female	III	1.652-10	3.893
	IV	2.354-16.625	6.467
	V	0.836-12.5	4.907

3.5. Length at First Maturity (L_m)

The calculation of length at first maturity value of King Mackerel which was 47.4 cm, in the range of 46.3-48.5 cm. Based on the L_m value, the percentage of Indo-Pacific King Mackerel catch by gillnet that its immature fish reached 48% and the mature fish reached 52%. The percentage of monthly gonad maturity can be seen in table 4.

Table 4. Monthly gonad maturity of Indo-Pacific King Mackerel.

Month	N	Gonad Maturity (%)	
		Immature	Mature
Feb	229	43	57
Mar	272	40	60
Apr	542	38	62
May	383	42	58
Jun	183	24	76
Jul	275	43	57
Aug	196	52	48
Average		48	52

3.6. Sex ratio

From 3307 samples from February 2014 to June 2017, it could be obtained that the number of male and female fish was 0.4: 1. The sex ratio is shown in figure 7. The chi square test showed that there was a significant difference of the number of male and female fish (table 5).

.

Table 5. Chi square test of sex ratio Indo-Pacific King Mackerel.

Month	Obs. Frequency		Hoped Frequency	X2 Hitung	
	Male	Female	(fn)	Male	Female
Feb	6	12	9	1	1
Mar	7	24	15.5	4.7	4.7
Apr	13	93	53	30.2	30.2
May	21	64	42.5	10.9	10.9
Jun	19	39	29	3.4	3.4
Jul	25	56	40.5	5.9	5.9
Aug	38	49	43.5	0.7	0.7
Sep	15	36	25.5	4.3	4.3
Oct	20	46	33	5.1	5.1
Nov	21	41	31	3.2	3.2
Total				69.5	69.5
Average				138.9	

X^2 calculated = 138.9
X^2 table at 0.05 (n-1) = 9
X^2 table = 16.9
X^2 calculated> X^2 table, h1 is rejected; h0 accepted
There was significant difference of sex ratio between male and female of King Mackerel

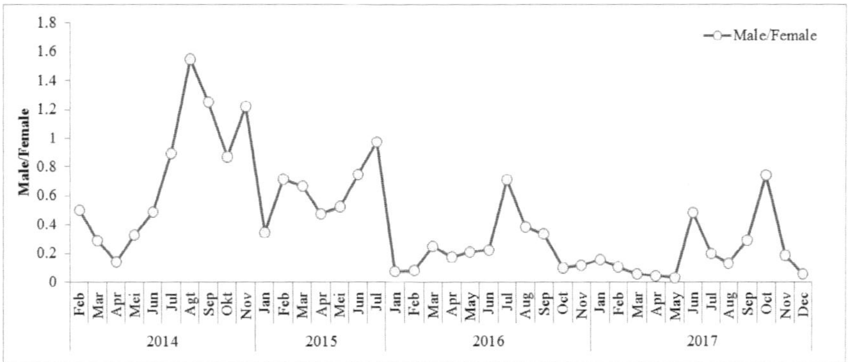

Figure 7. Sex ratio of Indo-Pacific King Mackerel

3.7. Food habits
Types of stomach contents showing that the King Mackerel diet cannot be obtained because all samples indicate empty stomach condition and contain crushed fish only, so that were difficult to identify. Some fish stomach samples contain juvenile fishes that were still intact, but the morphological form was difficult to identify.

3.8. Fecundity
Fecundity of King Mackerel ranged from 28,082 - 1,506,075 eggs, at 45-55 cm length. Fecundity and fork length relationship of King Mackerel showed a strong relationship, with an R^2 value of 0.8457 (figure 8). This indicates that each increase in length greatly affects the number of eggs, up to a certain size.

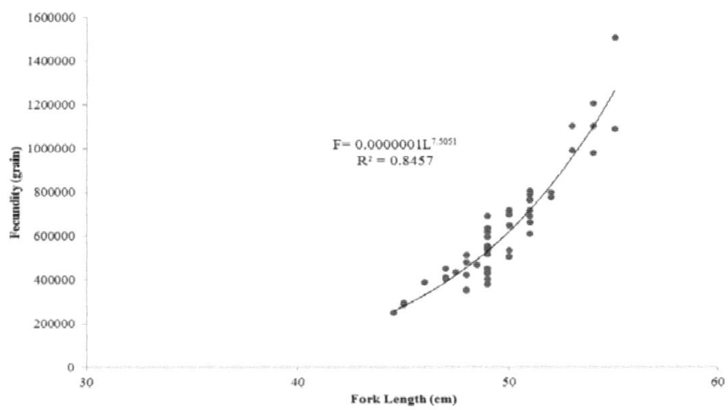

Figure 8. Fecundity and fork length relationship of King Mackerel.

3.9. Ova Diameter
The ova diameter at gonad level IV ranged from 0.03 to 1.98 mm, with an average of 0.64 mm. From all sizes of ova diameters, the range from 0.03-1.97 mm with interval 0.05 was obtained. In this study, it could be found that the condition of immature eggs with < 0.53 mm diameter has matured diameter of > 0.53 mm and < 0.78 mm and spawn egg conditions (diameter > 0.80 mm). In February, the condition of immature eggs of 1%, has matured by 13% and ready to spawn by 86%. In August, the percentage of immature eggs by 42%, has matured by 54% and ready to spawn at 4%, while in

October the condition of immature eggs by 19%, has matured by 80% and ready to spawn by 2%, as seen in figure 9.

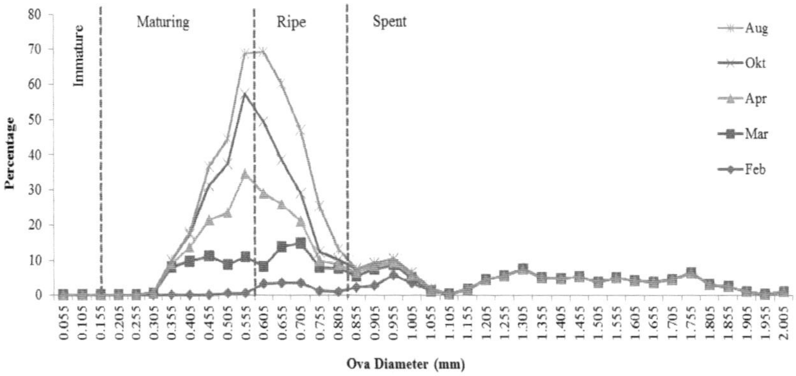

Figure 9. Ova diameter of Indo-Pacific King Mackerel.

4. Discussion

The length of the male fish is dominated by the mode range at 37-39 cm, while the female fish mode values at length 49-51 cm. The average length of male fish caught is 38 cm, while the female fish at length 49 cm. Transition of sex change from male to female at length 43-45 cm. The obtained sample shows the different growth of King Mackerel individual fish. This can be proved by the presence of fish samples of the same length but different conditions of gonadal development. Sometimes the length of 44 cm fish has turned into a female, and sometimes until the new length 45 seen there is a change. The average length of King Mackerel fishing with gillnet were 49-51 cm. This shows that the fish caught predominant mature fish. In hermaphrodite fish the use of fishing gear is very influential on the proportion of the size of the fish caught. Gillnet operated in Moro Waters and its surroundings is good because it dominantly catches the size of the adult fish, so that young fish can grow. In the protandrous hermaphrodite species the number of male fish remaining in the waters is primary factor to the development of the testes to the ovaries [18].

Length weight relationship in both male and female fish shows an allometric pattern. Changes in the value of b from young fish to mature fish indicate the growth of fish, one of which is the development of fish reproduction organs. Negative allometric growth can mean weight gain of fish slower than its length increase. Similarly, the length, weight of the fish caught should also be in equal proportion between males and females in order to the sustainability of fish populations to be maintained.

The maturity level of gonad was dominated by gonad level IV, the maturity of male fish gonads was only found from level I-II. The research has not found male fish gonads in level IV or V. While female dominant fish was found in level III-V. The length of the male fish on level I-III was found in the length of fish < 45 cm, whereas the female fish began to be seen at length > 45 cm. This leads to the notion that Indo-Pacific King Mackerel (*Scomberomorus guttatus*) allegedly strong is a protandri fish (hermaprodit protandrous), a fish whose body has a gonad that performs a process of differentiation from the male phase to the female phase.

This research has been going on for four years. From all samples that have been obtained showed female fish can be seen visually on the length of fish > 45 cm. Male fish found at length < 45 cm, and not found at length > 45 cm. Females are widely found starting at 45 cm length, and are not found at lengths <45 cm. The question is why no male fish are found in larger fish sizes, and vice versa for female fish. Judging from the number of samples that have been obtained (2674 gonad samples) in four years, it is thought there is something strange about this Indo-Pacific King Mackerel. Very little literature supports to decide that the fish is hermaphrodite. Finally, I proceed to histology test of gonad samples for both visually visible male and female. Temporary results indicate strongly that the species is protandrous hermaphrodite. How details of the pattern of changes that occur cannot I submit because the process is being done in the laboratory. This certainly needs further research to prove the truth.

The value of GSI increases with the increase of fish length to a certain size. GSI indo-Pacific King Mackerel value continued to increase from level I to IV, then began to fall on level V because there were fish spawning. The peak spawning season of King Mackerel in Moro Waters is thought to take place in January-August, so the indo-Pacific King Mackerel has partial spawner type. Spawning season takes place at a relatively long time. Gonad level V as a fish spawn has been found almost every month. This spawning season is different from the one reported by [2] in Indian Waters, such as in May-August with a spawning peak in July. Whereas [5] reported indo-Pacific King Mackerels in Mannar Gulf Waters and Palk Bay spawning from January to August, with spawning peaks in April-May.

King Mackerel gonad was dominated by ripe fish gonad, it was suspected that gillnet drift has a high selectivity in terms of the length at first capture. The average length at first captured should be greater than the length at first maturity (Lm), i.e. 47.4 cm, in the range of 46,3-48,5 cm. The mesh size should be made larger so that young fish can escape and multiply (until spawning).

The Lm value of King Mackerel in Moro waters is almost the same as in Indian waters, but bigger than the one reported by [19] in Cilacap waters. There are several things that affect the value of Lm, including: genetic differences, environment, and food habits. The value of Lm from the previous study is presented in table 6.

Table 6. Length at first maturity of Indo-Pacific King Mackerel.

No	Location	Lm (cm)	Ref
1	India	48-52	[2]
2	Teluk Mannar and Palk Bay	46.5 (45.1-48)	[5]
3	Cilacap water, Indian Ocean	42.3	[19]
4	Moro water, Kepulauan Riau	47.4 (46.3-48.5)	Present study (2017)

From the Chi square test to the sex ratio of Indo-Pacific King Mackerels showed a significant difference between the number of male and female fish. This shows that the number of male fish is less than that of female fish. This condition is still good for the sustainability of the resources of indo-Pacific King Mackerel in Moro waters.

In the hermaphrodite population, sex ratios are also strongly influenced by the high fish catch pressure on species [20]. If the operated gillnet has a small meshsize then the sex ratio will be dominated by the male fish, if the large meshsize (>4 inch) then the sex ratio in female fish dominance. The sustainability of the hermaphrodite species depends greatly on the balance of the sex ratio between the captured male and female.

Food habits of the indo-Pacific King Mackerel cannot be identified because the stomach is dominantly dissected containing the destruction of fish and in a state of emptiness. [11] from Indian waters reported that types of King Mackerel diet, such as, *Sardinella* spp., *Dusumerei* sp., *Rastrelliger* spp., *Decapterus* spp., *Trichiurus* spp., and *Stolephorus* sp. (anchovy). Many stomach samples were dominated by empty conditions. This condition was occurred because the fishermen mostly catch fish at night when feeding period of King Mackerel has been completed or were not eating, so the prey fish have been destroyed digested. Pelagic fish such as King Mackerel is supposed have a feeding period in the morning until late afternoon. [21] reported that indo-Pacific King Mackerel is a surface feeder, eating some of the pelagic fish on the surface. Active foraging in the morning at 6 to 10 a.m, and late in the afternoon until 10 p.m. The young fish (1-3 years) consume more food than the old fish. Indo-Pacific King Mackerel in Moro Waters allegedly different feeding times than those reported by [21]. Traditional fishermen operating gillnet (setting) between 7 or 8 p.m, and take the result (hauling) at 2 or 3 a.m. Indo-Pacific King Mackerel caught dominant stomach is empty, it can be concluded that the

King Mackerel is more active during the day (diurnal), including in foraging (prey) from the early morning until late afternoon or early evening.

Fecundity of King Mackerel in the Moro Waters almost identical to that reported by [8] in the Indian Ocean (FAO area 51), Siddeek report that fecundity increased from 385,000 eggs at age 2 years to 1,100,000 eggs at age 4 years. The relation of the number of eggs to the length indicates a strong R^2 value, meaning the strong relationship between the length and number of mackerel eggs or the variation of the number of eggs in the same length [15]. From the ova diameter data of three months, it showed that spawning period of King Mackerel takes place over a long period. In this study it was also found that the ova diameter was more than 1 mm, unlike the one reported by [17] that did not find ova diameter of >1 mm in the waters of Mannar Bay and India Palk Bay. The ova diameter of King Mackerel in Moro Waters is greater than in Indian Waters. This could be due to genetic and environmental differences. Ova diameter and fecundity of mackerel vary in each waters influenced by age composition, environmental factors such as food availability, population density, water temperature, dissolved oxygen and racial differences [12].

In figure 9, it is seen that in February, March, and April there is an egg diameter that is in size ready to spawn. In February and March the percentage of mature eggs has been reduced by spawning, while in April the number of mature eggs still has a larger percentage. This indicates that spawning in April is still longer than in February and March. In August and October there are also egg sizes ready for breeding but the percentage is smaller than February-April. From figure 9 the development of egg diameter per month shows the type of Indo Pacific King Mackerel is asynchronous, the oocytes from each stage of development of egg diameter is not indicated by the dominant group [22]. The spawning pattern of Indo-Pacific King Mackerel based on egg diameter is partial spawner, that is fish spawning his eggs at once, but several times in one spawning season [23].

5. Conclusion

The Gonad Maturity Level of King Mackerel was dominated by matured gonads rather than immature gonads, with gonad level V was found almost every month. The spawning season of mackerel fish takes place in January-Agustus. The Gonad Maturity Index continues to rise as the fish size increases to a certain length, then it will drop to a fish length of > 55 cm. There was a significant difference of sex ratio between male and female of King Mackerel. The length at first maturity of Indo-Pacific King Mackerel was 47.4 cm, in the range of 46.3-48.5 cm. The fecundity of King Mackerel ranges from 28,082-1,506,075 eggs, with 45-55 cm length. Relationship of fork length and fecundity showed strong correlation with the R^2 value of 0.728. The ova diameter of King Mackerel caught in Moro water has a larger than that reported in Indian Waters. The spawning pattern of mackerel fish is partial spawner.

6. Suggestion

It is suggested to clarify the observation of indo-pacific king mackerel gonads, need to be done histological analysis on gonads. The results of the histological analysis can later be used to validate visual observations of the gonads, since gonad morphological assessment is highly subjective

Acknowledgments

This paper is a part of the first author's dissertation. We would like to thank all researchers from Research Institute for Marine Fisheries, Prof. Dr. Ali Suman, and Dr. Fayakun Satria, M. App. Sc who help the authors. Our gratitude goes to all enumerators and to all lecturers and staffs in Faculty of Fisheries and Marine Sciences, Bogor Agricultural University, as well as to all classmates of the Aquatic Resources Management class of 2015 (Rudi, Hatta, Malik, Meuthia, and Ana).

References

[1] Collette B B and Nauen C E 1983 An annotated and illustrated catalogue of tunas, mackerels, bonitos, united nations development programme food and agriculture organization of the united nations and related species known to date Rome FAO Fisheries Synopsis *FAO Species Catalogue Scombrids of The World* **2**(125) 1-137

[2] Krishnamoorthy B 1958 Obesrvation on the spawning season and the fisheries of the spotted seer, *Scomberomorus guttatus* (Bloch & Schneider 1801) *Indian J. Fish* **5** 270-281
[3] Devaraj M 1976 Discovery of the Scombrids *Scomberomorus koreanus* (Kishinouye) in India, with taxonomic discussion on the species *Japanese Journal of Ichthyology* **23** (2) 1-9
[4] Devaraj M 1981 Age and growth of three species of seerfishes *Scomberomorus commerson, S. guttatus* and *S. lineolatus Indian Journal Fisheries* **28** 104 -127
[5] Devaraj M 1987 Maturity, spawning and fecundity of the spotted seer, *Scomberomorus guttatus,* in the Gulf of Mannar and Palk Bay *Indian Journal of Fisheries* **34** 48-77
[6] Nakamura I and Reiko N 1982 New records of two species of *Scomberomorus* from Japan *Japanese journal of ichthyology* **28**(4) 1-5
[7] Collette B B and Russo J L 1984 Morphology, systematics, and biology of the spanish mackerels (*Scomberomorus,* Scombridae) *Fishery Bulletin* **82**(4) 1-148
[8] Siddeek M S M 1995 Review of fisheries biology of *Scomberomorus* and *Acanthocybium* species in the Western Indian Ocean (FAO Area 51). Working Group on Pelagics, Gulfs Committee for Fisheries Management and Development *Working Paper WGP* **95**/2. 32 p.
[9] Ghosh S, Pillai N G K, and Dhokia K K 2009 Fishery, population dynamics and stock assessment of the spotted seer in the gillnet fishery of Veraval *Indian J. Fish* **56** (3):157-161
[10] Rashid H, Mustafa M G and Dewan S 2010 Population dynamics and the management of the indo-Pacific King Mackerel *Scomberomorus guttatus* from the upper Bay of Bengal of Bangladesh coast *The Bangladesh Veterinarian* **27**(2) 82-90
[11] Rohit P and Ghosh S 2012 Fishery, biology and population dynamics of the Indo-Pacific King Mackerel, *Scomberomorus guttatus* (Bloch & Schneider 1801) exploited in India *IOTC-WPNT02* 1-9
[12] Effendie M I 2002 *Fisheries Biology* (Bogor: Dewi Sri Foundation) p 116
[13] Schaefer M B and Orange C J 1956 Studies on sexual development and spawning of yellowfin tuna (*Thunnus albacares*) and skipjack (*Katsuwonus pelamis*) in three areas of the Eastern Pacific Ocean by axamination of gonads *Bulletin I-ATTC* **1** (6) 282-349
[14] Sugiyono 2004 *Nonparametric Statistical for Research* (Bandung: CV Alfabeta) p 306
[15] Effendie M I 1979 *Fisheries Biology Method* (Bogor: Dewi Sri Foundation) p 112
[16] Udupa K S 1986 statistical method of estimating the size at first maturity in fishes ICLARM Metro Manila *Fishbyte* **4**(2) 8–10
[17] Devaraj M 1983 Maturity, spawning and fecundity of the king seer, *Scomberomorus commerson* (Lacepede), in the seas around the Indian Peninsula *Indian Journal Fisheries* **30** (2) 203-230
[18] Wu G C, Hau Wen Li, Huang H H, Lin H J, Lin C J, and Ching-Fong C 2016 The testis is a primary factor that contributes to epigenetic modifications in the ovaries of the protandrous black porgy, *Acanthopagrus schlegelii Biology of Reproduction* **94**(6) 1-13
[19] Restiangsih Y H, Noegroho T, Wagiyo K 2016 Biological aspects of indo-Pacific King Mackerel (*Scomberomorus guttatus*) in Cilacap and adjacent waters *Bawal Fishing Research Journal* **8**(3) 191-198
[20] Provost M M and Olaf P Jensen 2012 Management and assessment of sex changing fishes. PMAFS protogynous hermaphrodite modeling workshop (Institute of Marine and Coastal Sciences: Rutgers University) p 1-30
[21] Devaraj M 1998 Food and Feeding habits of the spotted seer, *Scomberomorus guttatus* (Bloch and Schneider), in the Gulf Of Mannar and Palk Bay *J. Mar. Biol. Ass. India* **40**(1 and 2) 105-124
[22] Murua H and Kraus G 2003 Procedur to estimate fecundity of marine species in relation to their reproductive strategy *Journal North. Atl. Fisherie Science* **33** 23-32
[23] Lowe-McConnell R H 1991 *Ecological Studies in Tropical Fish Communities* (Australia: Cambridge University Press) p 400

Ecological risks from zinc contamination to preys of the Irrawaddy dolphin (*Orcaella brevirostris*) in Banten Bay

M A Khalifa[1*], M M Kamal[2], E M Adiwilaga[2] and A Sunuddin[3]

[1] Fisheries Department, University of Sultan Ageng Tirtayasa, Indonesia,
[2] Aquatic Resources Management Department, Bogor Agricultural University, Indonesia
[3] Marine Science and Technology Department, Bogor Agricultural University, Indonesia
*e-mail: ma.khalifa@untirta.ac.id

Abstract. Banten Bay is an estuarine ecosystem which serves as important habitat for Irrawaddy dolphin (Orcaella brevirostris), also functions as a fishing ground for artisanal fishers, maritime port, shipyard, and surrounded by industrial zones. One major impact from land-based industrial development is the load of heavy metal pollution (Zinc –Zn), which cause a potential threat to Irrawaddy dolphin population as a predator in the ecosystem through bioaccumulation process from its prey. This research aimed to measure ecological risks in Irrawaddy dolphin due to Zn contamination in its potential prey, i.e., sardines (Sardinella fimbriata) and squids (Loligo sp.). Survey on dolphin population was conducted in Banten Bay from Jan 2013 to Feb 2014 with participatory mapping and visual observation, while Zn concentration in dolphin's preys was measured using Atomic Absorption Spectrophotometer (AAS) and analyzed by Risk Quotient (RQ) with base assumption that Maximum Allowable Concentration (MAC) of Zn is 6.7 mg/kg wet weight. Results showed that concentration of Zn in squids (7.67±1.02 mg/kg WW) was higher than that of sardines (3.46±0.29 mg/kg WW). Thus, squids have a moderate risk, while sardines have low risk to bioaccumulation of Zn in Irrawaddy dolphin.

1. Introduction

Irrawady dolphin (*Orcaella brevirostris*) is one of protected species where found in Banten Bay [1]. Local name of Irrawaddy dolphin in Indonesia is Pesut Mahakam. Mahakam is one of the rivers in Kalimantan Island where was Irrawaddy dolphin's habitat [2]. Irrawaddy dolphin also found in other coastal waters, river, and lake in Indonesia [3], [4], [5]. This member of Delphinidae has "Vulnerable" status, and some sub-population has "Critically endangered" category in International Union for Conservation of Nature (IUCN) [6].

Banten Bay is a shallow estuarine ecosystem, with a maximum depth is 30 meters. Banten Bay has rapid economic development, like fisheries activity, shipyard, and surrounded by industrial area [7]. Cilegon and Bojonegara was an industrial area in the west of the bay and produced the highest volume pollution to Banten Bay [8]. There are 16 metal industries [9], which could generate heavy metals contamination to environment. Previous Research [10] indicates high zinc contamination existed in Banten Bay.

Heavy metals pollution in Banten Bay will threaten Irrawady dolphin's life. Ecological impact from pollution like habitat destruction and bioaccumulation in Irrawaddy dolphin's body. Heavy metals contamination in cetacean could lead organs and function damage until causing death [11]. This research aims to evaluate ecological risk from zinc contamination to prey of the Irrawaddy dolphin (*Orcaella brevirostris*) in Banten Bay.

2. Method

This research is part of Irrawaddy dolphin research in Banten Bay at 2013-2014. Irrawaddy dolphin main prey is small pelagic fish, cephalopods, and some species of macrozoobenthos [12]. This research sample is Sardines (*Sardinella fimbriata*) and Squid (*Loligo* sp.). Local fishing port reported during the year, sardines and squid are existed in Banten Bay. Prey's sample analyzed in Environmental Analysis Laboratory, Faculty of Agricultural Technology, Bogor Agricultural University. Samples analyzed by Atomic Absorption Spectrophotometer (AAS) with five replication.

Before ecological risk from metals, contamination analyzed, Maximum Allowable Concentration (MAC) metals information is required. MAC value (mg/ kg WW) was estimate refer to [13]:

$$MAC = \frac{RfD \times BW \times AT}{IR \times FI \times EF \times ED} \quad (1)$$

where: *RfD* = Refrences Dose Zn (mg/ kg ww per day); *BW* = Body Weight (kg); *AT* = Average Time (days); *IR* = Ingestion Rate (kg per day); *FI* = Fraction Ingested; *EF* = Exposure Frequency (days/ year); *ED* = Exposure Duration (year).

Some of the variables for calculated MAC was not available for Irrawaddy dolphin. Therefore, some variable information applied from other species with the same ecological niche. This paper refers to Finless porpoises (*Neophocaena phocaenoides*) [14], shown in Table 1.

Table 1. Variable for estimate MAC.

Variables	Values
References Dose Zn	0.3 mg/ kg WW
Body Weight	60 kg
Average Time	10.220 days
Ingestion Rate	3 kg/ day
Fraction Ingested	0.9
Exposure Duration	28 years
Exposure Frequency	365 days/ year

Ecological risk evaluates with Risk Quotient (RQ) value from metals concentration in prey compared with MAC value [13]:

$$RQ = \frac{Zn\ Concentration}{MAC} \quad (2)$$

Risk classification: RQ < 1 : Low Risk, 1 ≤ RQ < 10 : Moderate Risk, RQ ≥ 10: High Risk.

3. Result and discussion

Last research reported about heavy metals pollution in Banten Bay's water and sediment, the highest concentration was Zinc [10]. These case predicted because of some metal industry near Banten Bay used Zinc for material in the industrial process [9]. Zinc is one of essential element for organism, with particular function for metabolism. However, an excess of zinc can cause organism health disorders. Zinc concentration from sardines and squid samples are shown in figure 1

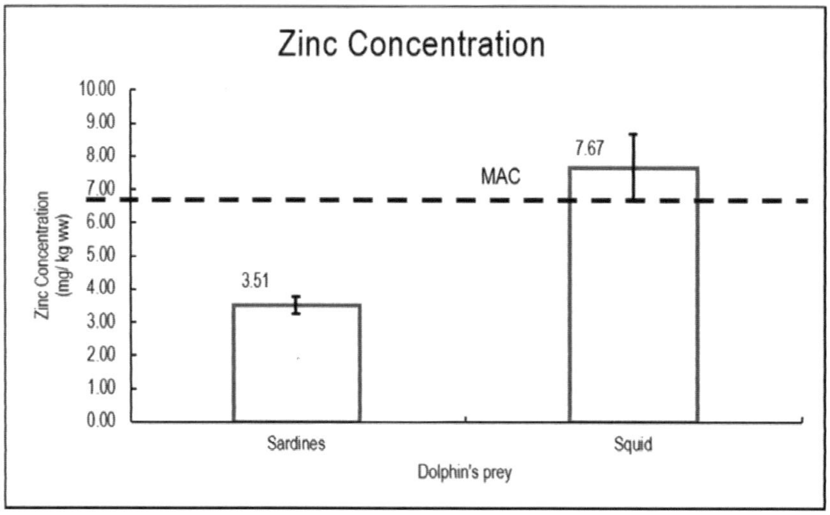

Figure 1. Zinc concentration at Irrawaddy dolphin's prey.

Figure 1 shown Zn concentration in sardines is 3.51 ± 0.52 mg/ kg WW and while in squid is 7.67 ± 1.03 mg/ kg WW. Zinc concentration in squid higher than in sardines due to the squid has bioaccumulation capacity greater than small pelagic fish [15]. Bioaccumulation capacity related to the ecological trophic food chain, squid was on the higher trophic level than sardines.

The result of zinc MAC value calculation is 6.7 mg/kg WW. Less information about Irrawaddy dolphin became an entire constrain for this research. Therefore, Irrawaddy dolphin research enhancement was required for making conservation effort. MAC value higher than zinc concentration in sardines, inversely in squid. Comparison of MAC value and Zinc concentration define as RQ value. RQ value for sardines and squid shown in figure 2.

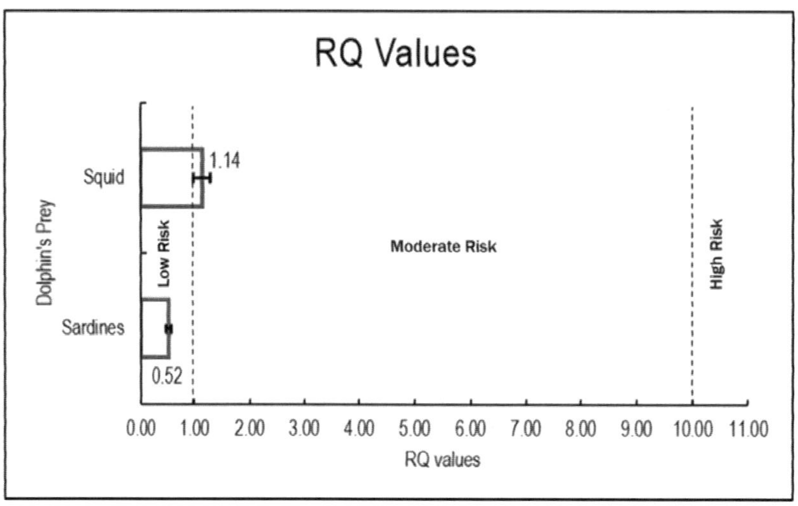

Figure 2. RQ values of squid and sardines Zn concentration.

Based on RQ value in figure 2, zinc contamination to Irrawaddy dolphin from sardines has low risk. Meanwhile, zinc concentration from squid has moderate risk contaminated to Irrawaddy dolphin. Zinc contamination could cause trouble for Irrawaddy dolphin's health.

High concentration of mercury and zinc founded in harbor porpoise died from infectious disease than physical trauma [16], [17]. Those indicate that excess mercury and zinc in the cetacean body are decreasing bodies resistance capacity from infectious disease (from bacteria, fungus, virus, parasite-like pneumonia disease).

Potential risk from zinc contamination to Irrawaddy dolphin is the devastation of the nervous system, degrade immunity, may even cause death. Previous research reported necropsy from several Irrawaddy dolphin accumulated low concentration of heavy metal in liver and kidney but was not mortality factors [18], [19]. In Mekong river, a low concentration of zinc in liver and kidney of Irrawaddy dolphin were found [20].

Irrawaddy dolphin's habitat is often overlapping with an anthropogenic activity that produces pollutant. This case requires deep research for resulting detail information about the effect of metal pollution to Irrawaddy dolphin, contaminant accumulation in Irrawaddy dolphin and many more. That information will be principle information for holistic conservation effort.

4. Conclusion

Banten Bay is already contaminated by zinc. Ecological risk of zinc contamination to Irrawaddy dolphins from sardines is low, while squid has moderates risk contaminated to Irrawaddy dolphins in Banten Bay.

Acknowledgments

This research funded by Education Fund Management Institution, Ministry of Monetary and Excellence High Education Scholarship, Ministry of Research, Technology and Higher Education, Indonesia. This research supported by Karangantu National Fishing Port, Indonesia.

References

[1] Khalifa M A, Kamal M M, Adiwilaga E M and Sunuddin A 2014 Preliminary study on the distribution of Irrawaddy Dolphin, *Orcaella brevirostris,* in Banten Bay *Open Journal of Marine Science* **4**(4) 338-343

[2] Kreb D and Noor I Y 2012 *Abundance and Threats Monitoring Surveys During Low Water Levels July and September 2012-Pesut Mahakam Conservation Program Technical Reports* (Samarinda: Yayasan Konsservasi RASI)

[3] Tas'an and Leatherwood S 1984 Cetaceans Live-Captured for Jaya Ancol Oceanarium, Djakarta 1974-1982 *International Whaling Commission* Document SC/35/SM2: 485-489

[4] Rudolph P, Smeenk C and Leatherwood S 1997 *Preliminary checklist of Cetacea in the Indonesian Archipelago and adjacent waters* (Leiden: Nationaal Natuurhistorisch Museum)

[5] Kreb D and Budiono 2005 Cetacean Diversity and Habitat Preferences in Tropical Waters of East Kalimantan, Indonesia *The Raffles Bulletin of Zoology* **53**(1) 149-155

[6] Reeves R R, Jefferson T A, Karczmarski L, Laidre K, O'Corry-Crowe G, Rojas-Bracho L, Secchi E R, Slooten E, Smith B D, Wang J Y and Zhou K 2008 *Orcaella brevirostris* The IUCN Red List of Threatened Species 2008: e.T15419A4579987 http://dx.doi.org/10.2305/ IUCN.UK. 2008.RLTS. T15419A4579987.en

[7] Indonesia Institute of Sciences 2001 *Laporan Penelitian Wilayah Pesisir Teluk Banten Tahap Kedua* (Jakarta: Indonesian Institute of Sciences)

[8] Sjaifuddin 2007 *Pengelolaan Lingkungan Wilayah Pesisir dan Laut Teluk Banten Berkelanjutan* [Disertation] (Bogor: Bogor Agricultural University)

[9] Salim J 2009 *Model Pengelolaan Limbah Industri Baja Sebagai Upaya Untuk Mempertahankan Kelestarian Wilayah Pesisir Kawasan Industri Krakatau Cilegon* [Disertation] (Bogor: Bogor Agricultural University)

[10] Rochyatun E, Lestari and Abdul R 2005 Kualitas Lingkungan Perairan Banten dan Sekitarnya Ditinjau dari Kondisi Logam Berat *Oseanologi dan Limnology di Indonesia* **38** 23 – 46

[11] Das K, Debacker V, Pillet S and Bouquegneau J M 2003 *Toxicology of Marine Mammals,* ed Vos J G, Bossart G D, Fournier M, and O'Shea T J (London: Taylor&Francis)

[12] Baros N B and Clarke M R 2009 *Encyclopedia of Marine Mammals* (*2nd edition*) ed WF Perrin B Würsig and JGM Thewissen (Oxford: Elsevier Inc)

[13] Hung C L H, So M K, Connel D W, Fung C N, Lam M H W, Nicholson S, Richardson B J and Lam P K S 2004 A preliminary risk assessment of trace elements accumulated in fish to the Indo-Pacific Humpback Dolphin (*Sousa chinensis*) in the Northwestern Waters of Hongkong *Chemosphere* **54** 643-651

[14] Hung C L H, Lau R K, Lam J C W, Jefferson T A, Hung K S, Lam M H W and Lam P K S 2007 Risk assessment of trace elements in the stomach contents of Indo-Pacific Humpback Dolphins and Finless Porpoises in Hongkong Waters *Chemosphere* **66** 1175-182

[15] Wahyuni H, Sasongko S B and Sasongko D P 2013 Konsentrasi logam berat di perairan, sediment dan biota dengan faktor biokonsentrasinya di Perairan Batu Belubang, Kab Bangka Tengah *METANA* **9**(2) 8-18

[16] Bennet P M, Jepson P D, Law R J, Jones B R, Kuiken T, Baker J R, Rogan E and Kirkwood J K 2001 Exposure to heavy metals and infectious disease mortality in Harbour Porpoise from England and Wales *Environmental Pollution* **112** 33-40

[17] Kaskuschke A and Prange A 2007 The Influence of metal pollution on the immune system a potential stressor for marine mammals in the North Sea *International Journal of Comparative Psychology* **20** 179-193

[18] Beasley I, Chooruk S and Piwpong N 2002 The status of the Irrawaddy Dolphin, *Orcaella brevirostris*, in Songkhla Lake, Southern Thailand *The Raffles Bulletin of Zoology* **10** 75-83

[19] Beasley I 2007 *Conservation of the Irrawaddy Dolphin, Orcaella brevirostris* (*Owen in Gray, 1866*) *in the Mekong River: Biological and Social Considerations Influencing Management* [Ph.D. Thesis] (Queensland: James Cook University)

[20] Siebert U and Das K 2011 *Evaluation of The Ecotoxicological Effects of POPs and Heavy Metals, Reflecting Pathological, Microbiological and Genetic Analyses, on The Mekong River Population of Irrawaddy Dolphins* (*Orcaella brevirostris*) (Washington: World Wildlife Fund)

IOP Conf. Series: Earth and Environmental Science **176** (2018) 012024 doi:10.1088/1755-1315/176/1/012024

Typology of Olive Ridley turtle (*Lepidochelys olivacea*, Linn 1958) nesting habitat in Kuta Beach, Serangan Beach and Saba Beach, Bali Province

R Septiadi[1*], D G Bengen[1], N M N Natih[1]

[1]Departement of Marine Science and Technology, Faculty of Fisheries and Marine Science, Bogor Agricultural University
*e-mail: rivalniseptiadi@apps.ipb.ac.id

Abstract. In Indonesia, the population sizes of Olive Ridley turtles in Kuta Beach, Serangan Beach and Saba Beach, Bali, tend to decrease due to habitat degradation. This study aims to describe the typology of Olive Ridley turtle nesting habitat in Kuta Beach, Serangan Beach and Saba Beach, Bali Province. Observation and in-situ measurements were conducted in November 2016 and February 2017 at nine stations using systematic sampling methods. The parameters measured were the number of nests, shore elevation, sand fractions, coastal vegetation, and shore width and length, which were analyzed using Principal Component Analysis (PCA) in order to see the correlation between the amount of nesting and nesting habitat typology. Our results show that the nesting habitat is characterized by low elevation of shore (0.89°-3.71°), shore width between 39.9-95.4 m, finest sand fraction (28%), and coastal vegetation dominated by *Hibiscus tiliaceus* (68.98%). PCA results indicated that fine sand fraction (grain size of 0.1-0.25 mm) and shore width are the main factors which characterizing the Olive Ridley turtle nesting habitat.

1. Introduction

1.1 Background

The Olive Ridley turtle (*Lepidochelys olivacea*) is one of the six sea turtle species found in Indonesia, and one of seven globally. Besides the Olive Ridley turtle, the green turtle (*Chelonia mydas*), loggerhead turtle (*Caretta caretta*), hawksbill turtle (*Eretmochelys imbricata),* leatherback turtle (*Dermochelys coriacea*), and flatback turtle (*Natator depressus*) are also found in Indonesia. Olive Ridley turtles are distributed all over the Pacific Ocean and the Indian Ocean [1]. In Indonesia, the Olive Ridley turtle are known to lay eggs on the beaches of large islands such as East Java, West Kalimantan, and Papua [2].

Olive Ridley turtle population in Indonesia tends to continue to decline, which cause is thought to be mainly anthropogenic rather than natural factors and predators [3]. Explicitly, the main threat of sea turtle population are anthropogenic activities, such as coastal and marine pollution, nesting habitat and feeding areas destruction, migration path disturbance, sea turtle consumption and trades, and the conversion of habitat as a place of tourism. As a result of the population decline, Olive Ridley turtle have been on the red list of the International Union for Conservation of Nature (IUCN) Vulnerable category [4].

The nesting habitat of Olive Ridley turtle are found to be located in tropical regions, in areas generally not overgrown by trees in a wide areas and bright with a lot of light [5]. Sea turtles generally choose coastal areas to nest, which is wide-ranging with an average slope of 30°, and inside the tidal range between 30-80 m. The eggs are placed in nests located between 8-41 m from the highest tide point in order not to be submerged in water. Olive Ridley turtle lay eggs 4 to 5 times a year with intervals between 10 to 14 days. Olive Ridley turtles are able to migrate as far as 2300 km from the nesting area [5].

Kuta Beach, Serangan Beach and Saba Beach in Bali Province are known to be the Olive Ridley turtle's nesting habitat. The Kuta Beachis a sloping white sandy beach. The Serangan Beach has a sloping coastal topography with shore elevations ranging between 0%-2 %, and the highest point 3 m above sea level [6]. The Saba Beach has a length beach of 5 km, bordered the Pering beach to the east and Purnama beach to the west. Saba Beach black sandy and coastal vegetation such as a spruce shrimp. The topographies of the Kuta, Serangan, and Saba Beaches are thus ideal for Olive Ridley turtles to land and nest.

1.2 Research Purposes

Olive Ridley turtles are one of the many species of sea turtle landed and laying in Bali Province. Olive Ridley turtle nests are found in areas of Kuta Beach, Serangan Beach and Saba Beach. Despite the ecological characteristic of Olive Ridley turtle nesting habitat in these three beaches are still not yet unknown. Therefore, in order to understand the ecological characteristics of Olive Ridley turtle nesting habitat in these three beaches, research is needed. We thus conducted this research, which aims are to describe the typology of Olive Ridley turtle nesting habitat in Kuta Beach, Serangan Beach and Saba Beach, Bali Province. From this study, we expect to be able to obtain know the relationship between the typology of turtle nesting habitat and the number of turtles nest, especially those spread in Kuta Beach, Serangan Beach and Saba Beach, Bali Province, which will be useful for providing inputs for the management of Kuta, Serangan, and Saba Beaches as tourism spots, while considering the Olive Ridley turtles nesting habitats.

2. Method

2.1. Location and time

The research was conducted in Kuta Beach (Station Kt1, Kt2, Kt3), Serangan Beach (Station Sr1, Sr2, Sr3) and Saba Beach (Station Sb1, Sb2, Sb3), Bali Province (figure 1). Research is divided into three steps: (1) field observations in November 2016, (2) observations and measurements of the characteristics of nesting habitat in February 2017, (3) sediment analysis and data processing in March 2017 at the Environment Laboratory, Department of Aquaculture and Marine Bioprospecting Laboratory, Department of Marine Science and Technology, Faculty of Fisheries and Marine Science, Bogor Agricultural University.

2.2. Tools and materials

The tools used in this research is GPSmap Garmin 60CSx, stationery, laptop, digital camera, sample bottle, roll meter 100 m, rope scale 20 m, rope scale 10 m, sewing meter, waterpass, rapia strap, small shovel. The materials used are data of the number and type of sea turtles that landed and sand samples

2.3. Data collection

We used both primary data and secondary data. Primary data were obtained from observation and measurement of Olive Ridley turtle nesting habitat in Kuta Beach, Serangan Beach, and Saba Beach, all three in Bali Province. Measurements include the amount of nest parameter, shore elevation, shore width and length, sand fractions, type of closures and vegetation density. Secondary data includes the number of data and sea turtles species, sea turtle eggs findings from 2016 to March 2017.

2.3.1. Station determination. Determination of research station for data collection was based on preliminary observations by myself and preliminary records by tracing all shore areas previously reported to be often used as a place for sea turtles to land, either for laying eggs or not, in Kuta Beach, Serangan Beach and Saba Beach. Sampling based on systematic sampling. The distance between research stations is 500 m from the first point set at Kuta Beach, Serangan Beach and Saba Beach.

IOP Conf. Series: Earth and Environmental Science **176** (2018) 012024 doi:10.1088/1755-1315/176/1/012024

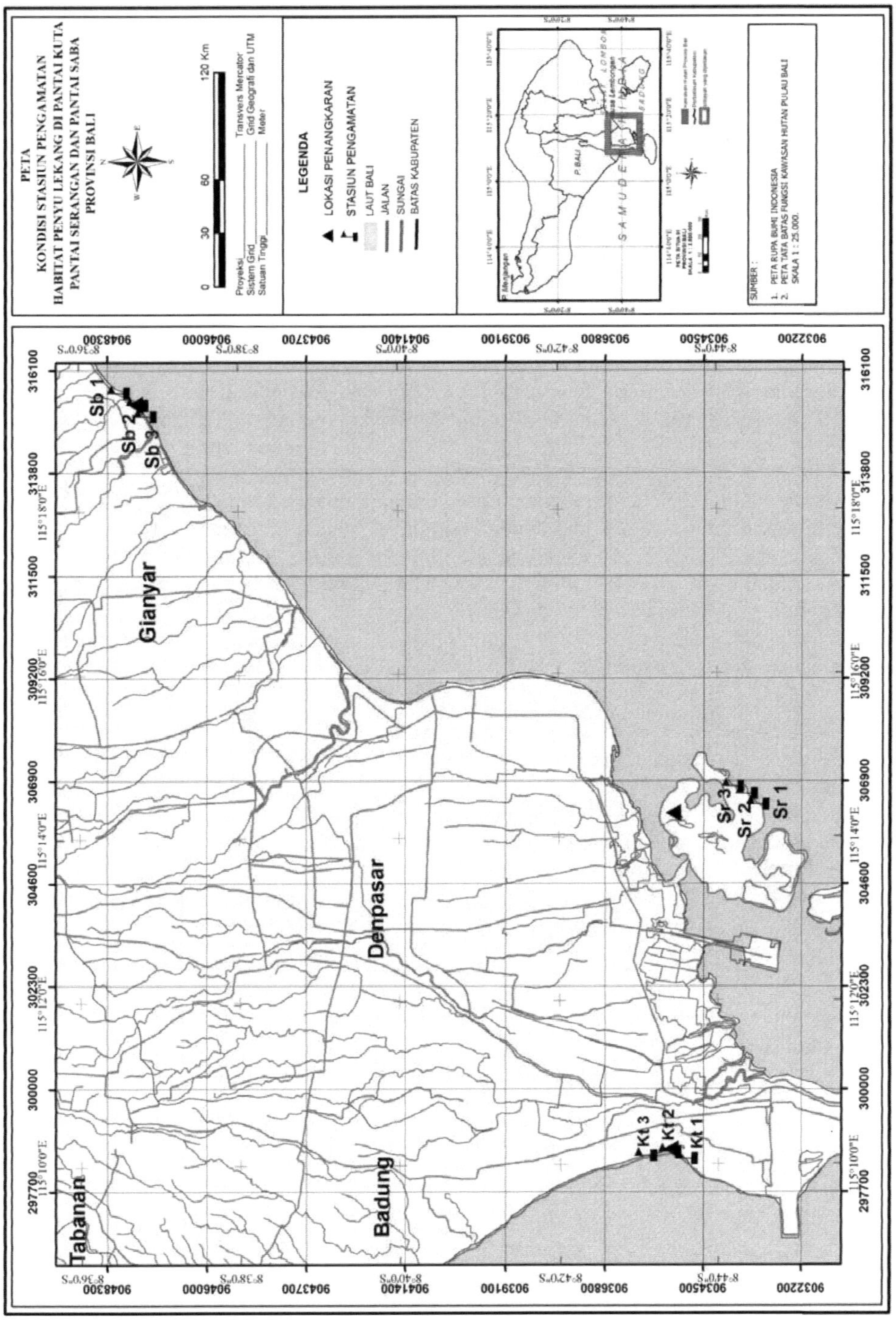

Figure 1. Location and research station in Kuta Beach, Serangan Beach and Saba Beach, Bali Province.

IOP Conf. Series: Earth and Environmental Science **176** (2018) 012024 doi:10.1088/1755-1315/176/1/012024

2.3.2. Shore elevation. The shore elevation is measured using a 50 m and 3 m scale meter for the long and altitude from the outermost vegetation to the lowest ebb. Waterpass used to maintain the straightness of the scale. Measurements start from boundary of outer vegetation to the lowest ebb.

2.3.3. The amount of nesting distribution. Sea turtle nesting distribution used belt transect method. Calculation the amount of nesting is done by draw a perpendicular line from the shore to the boundary of coastal vegetation, then the amount of nesting is computed visually in belt transect area as wide as 100 m.

2.3.4. Substrate fraction. Substrate sampling is performed on the nesting from the surface up to a depth 5-20 cm using a small shovel. Samples of sand are stored in 200 ml sample bottles, for subsequent analysis of substrate fractions.

2.3.5. Coastal vegetation. The method used in the measurement and inventory of vegetation is the multilevel sample plot method performed at each station. A sample plot of 20x20 m^2 is used to calculate mature trees (height > 1.5 m and diameter > 20 cm). The area of 10x10 m^2 to calculate pole trees (height > 1.5 m and diameter < 20 cm), while the area of 5x5 m^2 to calculate the saplings (height > 1.5 m and diameter < 10 cm), and the area of 2x2 m^2 to calculate the seedlings (height < 1,5 m and diameter < 10 cm).

2.3.6. Shore long and width. The measurement of shore long used meter and based on benchmark location. The measurement of shore width divided into three, is supratidal width measured from the outmost vegetation to the highest tidal boundary, the intertidal width measured from the highest tidal boundary and the total width calculated from the sum of the supratidal width and the intertidal width. Measurements were made at each station with a distance of 500 m.

2.4. Data analysis

Shore elevation measured by the formula:

$$\tan \alpha = \frac{y}{x} \quad (1)$$

Information: α = elevation, x = total height (m), y = total width (m)

Analysis of the size of the sediment fraction is intended to determine the composition of sediments. The process of sediment fraction analysis is:

a. The sample was dried by oven 70 °C for 24 hours.
b. The sample was filtered using a multilevel filter.
c. Weight the samples that have been filtered from the size 2-0,063 mm.
d. Subtrates that have been known percentage are then analysed and determined the type of substrate.

$$T_oP = \frac{a-b}{a} x100 \quad (2)$$

Information: *ToP* = Total particles (%), *a* = Dry weight of sample (mg), *b* = Weight after spying (mg)

The measured coastal vegetation stucture includes the following parameters [7].

1) Vegetation Density

$$\text{Density}\left(\frac{\text{ind}}{\text{m}^2}\right)(\text{K}): \frac{\text{Number of individuals (ind)}}{\text{Plot area (m}^2)} \quad (3)$$

Relative Density

$$\text{Relative Density(\%)(KR)}: \frac{\text{Relative density (\%)(KR)}}{\text{Density of all species}} x100\% \quad (4)$$

Frequency

$$\text{Frequency (F)}: \frac{\Sigma \text{ Plot found a species}}{\Sigma \text{ Plot of all species}} \quad (5)$$

Relative Frequency

$$\text{Relative Frequency (\%)(FR)}: \frac{\text{Frequency of a species}}{\text{Frequency of all species}} x100\% \quad (6)$$

Domination

$$\text{Domination (D)}: \frac{\text{Area of vegetation area}}{\text{Plot area}} \quad (7)$$

$$\text{Area of vegetation area}: \frac{\pi \text{DBH2}}{4} \quad (8)$$

$$\text{DBH}: \frac{\text{The length of the tree circumference}}{\pi} \quad (9)$$

2) Relative domination

$$\text{Relative Domination (\%)(DR)}: \frac{\text{Domination of a species}}{\text{Domination of all species}} x100\% \quad (10)$$

3) Important value index

$$INP = KR+DR+FR \quad (11)$$

Information: *INP* = Important Value Index, *KR* = Relative Density, *DR* = Relative Domination, *FR* = Relative Frequnecy

2.5. The amount of nesting and habitat characteristic
Principal component analysis is a method of factorial analysis that allows a more readable representation or interpretation of the data structuure by drawing essential information [8].

Centralization is the difference between the parameter value and the average value of the parameter [8].

$$C = N_i - \bar{x} \quad (12)$$

Information: C = central value, N_i = initial parameter value, $\bar{x}$ = the average value of the parameter

Reduction is the result between the value of the centralized parameter, with the standard deviation value parameter [8].

$$\text{R}=\frac{C}{Sd} \quad (13)$$

Information: R = Reduction value, C = Initialized parameter value, S_d = Standard deviation

3. Results and discussion

3.1. Olive Ridley turtle nesting distribution
Olive Ridley turtle is the most abundant species of sea turtles nesting on the beaches of Bali Province. Olive Ridley turtle nesting season spawned from February to October. In the breeding season, the sea turtle mothers will ride to the beach then dig the sand to lay the eggs. Olive Ridley turtle spawn an average of 100 grains, with Olive Ridley turtle eggs shaped elips or round white, and skin that is chewy [5]. When compared to other sea turtles, the Olive Ridley turtle is because Olive Ridley turtle are not afraid of interference from outside like light and sound, although Olive Ridley turtles rose from the open seas and not to laying eggs. The incubation period of Olive Ridley turtle eggs is between 45 to 65 days [1].

The temperature of the nest affects the successful incubation of sea turtle eggs, which require a sufficient temperature range so that embryos do not rot and can thrive. A decent temperature for the development of turtle egg embryos is between 25-32 °C. Temperature will also determine the sex ratio of juveniles, where turtles are born from nests whose incubation temperature is greater than 28°C will result in female genitals [1]. [5] states that the temperature fluctuations occur at a depth of 15 cm below the surface, the more into the fluctuation diminishing and even gradually achieves stability.

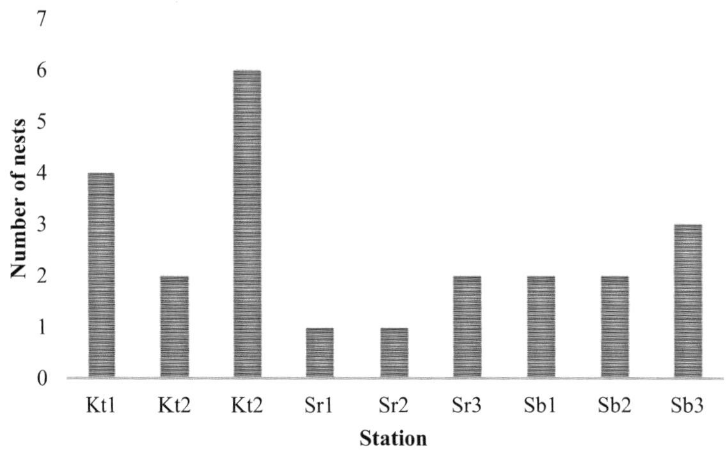

Figure 2. Findings of Olive Ridley turtle nesting at research station in Kuta Beach, Serangan Beach and Saba Beach, Bali Province.

During the research, there were 23 Olive Ridley turtle nests from nine stations (figure 2). The largest number of nests are found at Kt3 station located in Kuta Beach, while the smallest number of nests is located at Sr1 and Sr2 stations located in Serangan Beach. During the research from February to March 2017, the number of Olive Ridley turtle eggs found in Kuta Beach is 1677 eggs, the number of Olive Ridley turtle eggs foun in Serangan Beach is 634 eggs, the number of Olive Ridley turtle eggs found in Saba Beach is 1822 eggs. The findings of Olive Ridley turtle eggs are dominated in nesting areas in Kuta Beach, Serangan Beach and Saba Beach. The findings of Olive Ridley turtle nesting were found in each research station.

3.2. Shore elevation

Shore elevation affect the amount of sea turtles that will land and make a nest, because the beach conditions are sloping and has fine sand can facilitate turtles to the mainland to find the location and make the nest as a nesting place. The steeper the beach will make it more difficult for sea turtles to see more distant objects because the sea turtles eyes is only able to see well at an angles of less than 150° [9]. In addition, sea turtles lay their nests 30 to 80 m above the highest tide [5].

Shore elevation in Kuta Beach, Serangan Beach and Saba Beach categorized slope slightly with a slope between 0.89°-3.71° and average slope 2.05°. The biggest slope located in Serangan Beach which has strong currents compared to Kuta Beach and Saba Beach. The number of coral fractions in Serangan Beach is more and has coral reefs in the intertidal zone and there is a coral dike nearby. Kt1 station located in Kuta Beach has the smallest slope and Sr1 station located in Serangan Beach has the biggest slope (figure 3)

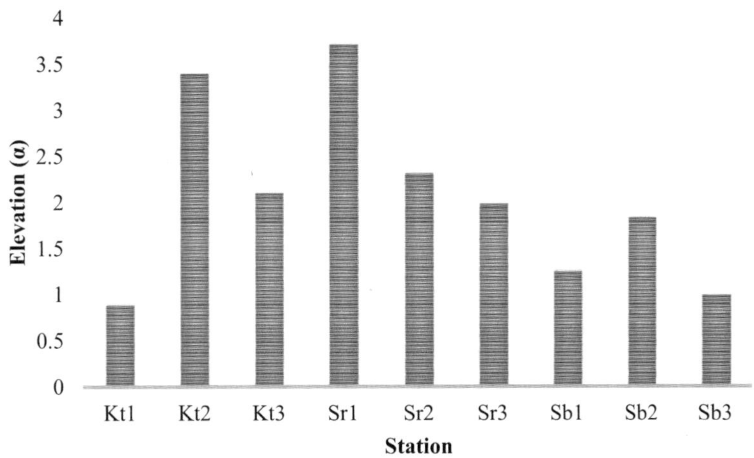

Figure 3. Shore elevation in research station.

3.3. Long and shore width

Kuta Beach has a length of about 2 km consisting of 3 observation stations. Serangan Beach has a length of about 1.5 km consisting of 3 observation stations. Saba Beach has a length of about 5 km consisting of 3 observation stations. The distance between observation stations on each coast is 500 m. The largest beach length is Saba Beach and the lowest beach length is Serangan Beach.

Kuta Beach has a shore width that ranges from 78-80.97 m to average 79.69 m. Serangan Beach has a shore width that ranges 58.7-78.86 m to average 71.02 m. Saba Beach has a shore width that ranges 39.90-95.40 m to average 63.60 m. The longest shore width is located at Sb3 station located in Saba Beach, while the shortest shore width is located at Sb2 station in Saba Beach. The intertidal width of each station ranges from 27.60-64.76 m with an average width of 50.23 m. The supratidal width ranges from 10.10-46.10 m with an average width 21.43 m. Kuta Beach, Serangan Beach and Saba Beach have different beach widths (figure 4).

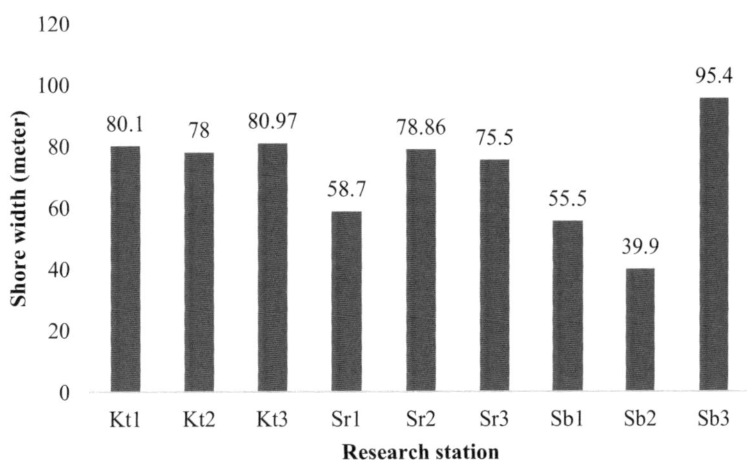

Figure 4. Shore width in Kuta Beach, Serangan Beach and Saba Beach, Bali Province.

Shore width in Kuta Beach, Serangan Beach and Saba Beach has a significant difference. Sr1 station located in Serangan Beach, Sb1 station and Sb2 station located in Saba Beach has a narrower shore width compared to other stations. Sr1, Sr2, Sr3, Sb1, and Sb2 station has a narrow width in the supratidal zone and the amount of Olive Ridley nests found is small. The size of nesting beacheas greatly affects the accessibility of Olive Ridley turtle reaching suitable areas for making nests. Supratidal area is a dry area and is not exposed to sea tides, making it easier for sea turtles to make nests. The shore width which are used as sea turtles nesting ranged between 20-80 m [10]. The results showed the average width of the beach was 71.44 m which was in the wide range of beaches used as sea turtle nesting places.

3.4. Substrate fraction

Sand texture is related to the level of ease of digging the nest. Olive Ridley turtle generally prefer large and sloping beaches located on the top of the beach or above the highest tide line for nesting, with sediment arrangements of not less than 90% in the form of sand and the remainder being dust and clay, with fine grain diameters of sand and medium [5].

Kuta Beach and Serangan Beach generally have characteristic of medium and fine white and gray sand substrate which can be found at Kt1, Kt2, Kt3, Sr1, Sr2 and Sr3 station. Saba Beach generally has the characteristic of fine dark sand substrate found in Sb1, Sb2, and Sb3 station. Classification of granular diameter sand are grouped into 5 categories, which is very fine sand with a diameter of 0.05-010 mm, fine sand with a diameter of 0,10-0,25 mm, medium sand with a diameter of 0.25-0.50 mm, coarse sand with a diameter of 0.50-1.00 mm, very coarse sand with a diameter of 1.00-2.00 mm.

The texture of Olive Ridley turtle sand substrate was composed by very fine sand component 28.04%, 23.98% fine sand, 41.28% medium sand, 5.77% coarse sand, and 0.93% very coarse sand. Very fine sand and fine sand dominate the nesting beach more than 50% (figure 5). The percentage of very fine sand that has a value of more than 50% is located at Kt1, Sb1, and Sb2 station, and the lowest is in Kt2 station of 0.78%. The greatest percentage of fine sand is found at Sb3 station with value 35.90%, medium sand is in Kt2 station with value 87.84%, coarse sand found in Kt1, Kt2, Kt3, Sr1, Sr2 and Sr3 station, very coarse sand found in Kt2, Kt3, Sr1, Sr2 and Sr3 station (figure 5).

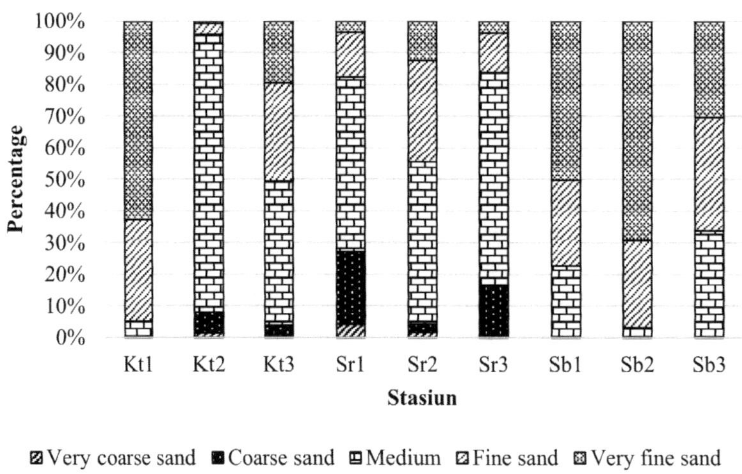

Figure 5. Substrate fractions in nesting habitat Kuta Beach, Serangan Beach and Saba Beach, Bali Province.

3.5. Coastal Vegetation

According to [11] coastal vegetation in the form of dense trees for sea turtle nesting gives a good influence on the stability of sea turtle nesting population. If the trees grow dense, the fallen leaves gradually decompose into mineral particles and drift into the sea. The process takes place continuously, so the fertility of the waters can still be guaranteed. Fertility of waters become the needs of biota living in the area, such as the growth of seaweed and the availability of marine invertebrates in the form of zooplankton. Marine invertebrates are food needed by sea turtle populations that are still juvenile (hatchling). Vegetation along the nesting beach is dominated by coconut tree (*Cocos nucifera*), ketapang tree (*Terminallia catappa*), and waru tree (*Hibiscus tiliaceus*).

Tabel 1. Important value index (INP) vegetation in Kuta Beach, Serangan Beach and Saba Beach as an Olive Ridley turtle nesting places.

No	Species	Important Value Index
1	*Hibiscus tiliaceus* (Waru)	68.98
2	*Cocos nucifera* (Kelapa)	60.56
3	*Terminallia catappa* (Ketapang)	56.24
4	*Pandanus tectorius* (Pandan)	21.63
5	*Barringtonia asiatica* (Butun)	20.38
6	*Casuarina equisetifolia* (Cemara)	16.07
7	*Morinda citrifolia* (Mengkudu)	12.79
8	*Ziziphus mauritiana* (Bidara)	11.74
9	*Corypha utan* (Gebang)	7.11
10	*Kibatalia arborea* (Kayu Santan)	5.72
11	*Pongamia pinnata* (Kacang Kayu Laut)	5.65
12	*Leucaena glauca* (Lamtoro)	3.52
13	*Acacia auriculiformis* (Akasia)	3.16
14	*Diospyros maritima* (Ki Geseng)	3.11
15	*Calophyllum inophyllum* (Nyamplung)	2.34

Coastal vegetation found from 9 stations covering 15 species of trees consisting of coconut (*Cocos nucifera*), waru (*Hibiscus tiliaceus*), akasia (*Acacia auriculiformis*), butun (*Barringtonia asiatica*), kayu santan (*Kibatalia arborea*), ketapang (*Terminallia catappa*), pandan (*Pandanus tectorius*), bidara (*Ziziphus mauritiana*), cemara (*Casuarina equisetifolia*), lamtoro (*Leucaena glauca*), mengkudu (*Morinda citrifolia*), kacang kayu laut (*Pongamia pinnata*), gebang (*Corypha utan*), ki geseng (*Diospyros maritima*), nyamplung (*Calophyllum inophyllum*). Waru (*Hisbuscus tiliaceus*) is the species of vegetation that has the highest INP 68.98 %. Nyamplung (*Calophyllum inophyllum*) is the species of vegetation that has the lower INP 2.34 %.

The species of coconut is found in Kuta Beach. Serangan Beach and Saba Beach. At Kt1 station there are 6 types of trees. namely coconut (*Cocos nucifera*). waru (*Hibiscus tiliaceus*). akasia (*Acacia auriculiformis*). butun (*Barringtonia asiatica*). kayu santan (*Kibatalia arborea*). ketapang (*Terminallia catappa*). At Kt1 station grow the plant creep type of katang-katang (*Ipomoea pescaprae*). Types of coastal vegetation that have morphological stature (stems. leaves. crowns) such as coconut trees (*Cocos nucifera*). waru (*Hibiscus tiliaceus*). dan ketapang (*Terminallia catappa*) causing the area to become more naturally protected from predatory disturbances [12]. This is one of the factors that is appropriate for sea turtles to nest and lay eggs.

3.6. The amount of nesting and habitat characteristic correlation

The parameter of Olive Ridley turtle nesting habitat used in the principal component analysis to analyse the association of nest numbers and habitat characteristics. the amount of nest (JS). shore elevation (KP). long and shore width (LP). coastal vegetation (VP). and sand fractions which grouped into 5 categories. which is very fine sand with a diameter of 0.05-010 mm. fine sand with a diameter of 0.10-0.25 mm. medium sand with a diameter of 0.25-0.50 mm. coarse sand with a diameter of 0.50-1.00 mm. very coarse sand with a diameter of 1.00-2.00 mm.

The results of PCA centered on three major axes where each axis (F) contributes to the relationship between nest numbers and nesting habitat characteristics in Kuta Beach. Serangan Beach and Saba Beach. The first axis (F1) contributes 52.40% (the root value of feature 4.72). the axis 2 (F2) contributes

19.34% (the characteristic root value 1.74). and the third axis (F3) contributes 13.58% (value of root characteristic of 1.22).

Parameters that have a strong enough correlation with the first axis (F1) are shore elevation (KP) with a correlation value of 0.92 and medium sand fraction (PS) with a correlation value of 0.85. Shore width (LP) with a correlation value of 0.98 and the amount of nest (JS) with a correlation value of 0.45 has a fairly strong correlation with axis 2 (F2). Coastal vegetation (VP) with a correlation value of 0.89 and the amount of nest (JS) with a correlation value of 0.59 has a fairly strong correlation with axis 3 (F3).

Graph of relationship of the two major axes (F1 and F2) on figure 6 (a) shows the existence of two clusters that illustrate the relationship between the distribution of observation stations and habitat variables. Group 1 consists of stations Kt1. Kt3 and Sb3 which are characterized by the amount of nest (JS) ≥ 3 on a fine sand beach (PH) with a percentage 31.05-35.9 % and shore width (LP) between 39.9-95.4 m. At Sb1 station has a very fine sand fraction (PSH) with a percentage 50.25 %. Group 2 includes Sr1 station which has shore elevation (KP) with an angle of 3.71°. coarse sand fraction (Pkr) with a percentage 22.97 %. shore width tends to be narrow with pandan vegetation (*Pandanus tectorius*) with INP 55.56. Graph of F1 and F3 axis relation on figure 6 (b) shows Sb3 station has a shore elevation of a sloping beach with an angle of 0.99° there is fine sand fraction (PH) and very fine sand fraction (PSH) the numerous and wide beaches are long. In addition to this station found the amount of nest (JS) more eggs than Sb1 and Sb2 station.

Graph in figure 6 shows a close correlation between variables the amount of nest (JS). fine sand (PH) and shore width (LP). Fine sand makes it easy for Olive Ridley turtles to dig up the nest. and the width of the unbroken beaches makes it easy for Olive Ridley turtles to make nests.

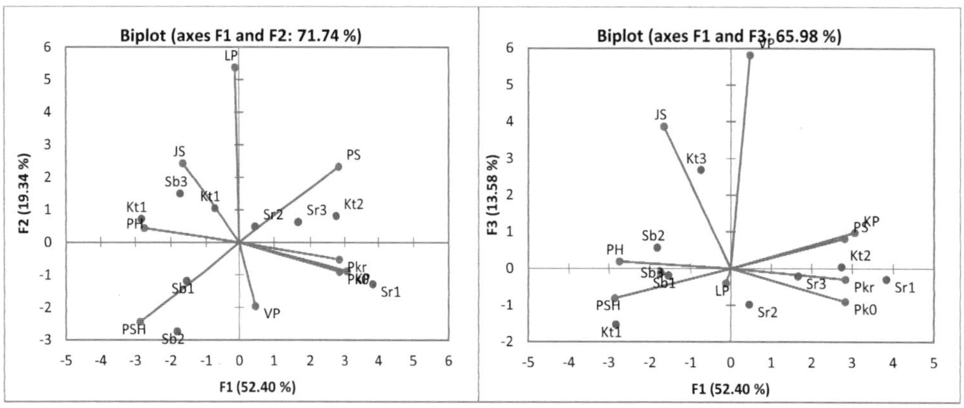

(a) (b)

Information:

St: Station JS: The amount of nest; KP: Shore elevation; LP: Shore width; VP: Coastal vegetation; PSH: Very fine sand; PH: Fine sand; PS: Medium; Pkr: Coarse sand; Pk0: Very coarse sand.

Figure 6. Results of PCA habitat parameter and research station on axis F1. F2. dan F3; (a) Axis F1 and F2. (b) Axis F1 and F3.

4. Conclusions

Olive Ridley turtle is one of sea turtle species that many nesting Kuta Beach. Serangan Beach and Saba Beach. Olive Ridley turtle nesting habitat typology in this three beaches a gently sloping beach with a slope 0.89°-3.71°. shore width between 39.90-95.40 m. very fine sand fraction (diameter 0.1-0.25 mm) dominant and coastal vegetation dominated by Waru trees (*Hibiscus tiliaceus*) with the average INP 68.98 %. Of the rhree nesting habitat locations. Kuta Beach (Kt1. Kt2. Kt3 stations) has the highest percentage of nesting followed by Saba Beach. Habitat in Kuta Beach and Saba Beach high potential as a sea turtle protection area. due to the slope of its shore elevation. shore width with vast expanses. Olive

Ridley turtle nesting habitat at Kuta Beach. Serangan Beach and Saba Beach can be an ideal nesting beach for Olive Ridley turtles.

Recommendation

Need to do research on other species of sea turtle nesting habitat profile that ever nesting at several tourist beaches in Bali Province.

References

[1] Lutz P L. Musick J A and Wynekken J 2003 *The Biology of Sea Turtles Volume II* (Florida: CRC Press)
[2] Yusuf A 2000 *Mengenal Penyu* (Jakarta: Yayasan Alam Lestari)
[3] Adnyana I B W and Creusa H 2009 *Panduan Melakukan Pemantauan Populasi Penyu di Pantai Peneluran di Indonesia* (Jakarta: WWF-Indonesia dan Universitas Udayana)
[4] IUCN 2016 Red List of Threatened Spesies IUCN 2016. Version 3.1
[5] Nuitja I N S 1992 *Biologi dan Ekologi Penyu Laut* (Bogor: IPB Press)
[6] Darmawan I G S 2013 *Pemanfaatan Lahan Pra dan Pascareklamasi di Pulau Serangan* [tesis] (Denpasar: Universitas Udayana)
[7] Kusuma C 1997 *Metode Survey Vegetasi* (Bogor: IPB Press)
[8] Bengen D G 2000. *Teknik Pengambilan Contoh dan Analisis Data Biofisik Sumberdaya Pesisir* (Bogor: Pusat Kajian Sumberdaya Pesisir dan Lautan. Institut Pertanian Bogor)
[9] Dharmadi N and Wiadnyana N 2008 Kondisi habitat dan kaitannya dengan jumlah Penyu Hijau (*Chelonia mydas*) yang bersarang di Pulau Derawan Berau Kalimantan Timur *Pusat Riset Perikanan Tangkap* **14** 19-204
[10] Mathenge S M. Mwasi B N. Mwasi S M 2012 Effects of anthropogenic activities on sea turtle nesting beaches along the Mombasa-Kilifi Shoreline. Kenya *Marine Turtle Newsletter* **135** 14-18
[11] Rubio M 2009 *Nesting Beach Characteristics of Endangered Sea Turtles in Las Perlas Archipelago. Panama* [Dissertation] (Edinburgh: Heriot-Watt University)
[12] Roemantyo. Nastiti A S and Wiadnyana N N 2012 Struktur dan komposisi vegetasi sekitar sarang penyu hijau (*Chelonia mydas* Linnaeus) Pantai Pangumbahan. Sukabumi Selatan. Jawa Barat. *Berita Biologi* **11**(3) 373-387

IOP Conf. Series: Earth and Environmental Science **176** (2018) 012025 doi:10.1088/1755-1315/176/1/012025

Discrete EquiSpaced Unshaded Line Array method for target identification using side scan sonar imagery

M Z Lubis[1*], W Anurogo[1], H Kausarian[2], T Choanji[2], S Antoni[3], S Pujiyati[4]

[1] Geomatics Engineering, Politeknik Negeri Batam, Batam 29461, Indonesia
[2] Faculty of Engineering, Universitas Islam Riau, Pekanbaru, 28284, Indonesia
[3] Marine Geology Department, King Abdulaziz University, Jeddah. Saudi Arabia
[4] Department of Marine Science and Technology, Institut Pertanian Bogor, Indonesia
*e-mail: zainuddinlubis@polibatam.ac.id

Abstract. Beam pattern resulting from Discrete Equispaced Unshaded Line Array method is to analyze the pattern of two-dimensional beam pivot at the angle of sound waves coming from the direction of origin on the axis of the received array, will affect the sound beam angle. This research was conducted in December 2016 in the Punggur Sea, Batam, Riau Islands-Indonesia, and its coordinate system is 104° 08.7102 E and 1° 03.2448 N until 1° 03.3977 N and 104° 08.8133 E, DEULA method in target 4 has the highest value in the directivity pattern is 21.08 dB. The results of the beam pattern model show that neither the central value at the incidence angle (°) of the directivity pattern (dB) were not at the 0 (zero) or the beam pattern central have been generated by the target 6 with incident angle -1.5° and 1.5°. In addition, it has declined by 40 dB. Highest result of line trace is target 1 with 191, 88 cm on target 1, and highest of time result is 13568 cm/second on target 4. Seismic figure of side scan sonar imagery have total line trace is 4479, time: 77.9547 cm/s, and gain: 0.00271091.

1. Introduction

Punggur Sea is the part of the Riau Islands in Indonesia. Generally Punggur Sea still rarely done research on the identification of seabed using the acoustic wave technology. Worldwide, these marine environments have not been studied using high-resolution hydro-acoustic methods (eg, bathymetry, and base profiles), useful for investigation of sediment processes, benthic habitats, mineral resources, artificial artifacts, and other features. Side-scan sonar imagery and underwater photography, coupled with bathymetric and sedimentology data, revealed noteworthy features and the interplay of multiple processes on the Riau Island and Punggur Sea, Indonesia [1]. The seafloor imagery, target identification, and image processing with two dimensional was conducted on the Punggur Sea in 2017 [1, 2].

The side scan sonar (SSS) instrument is a sonar development that able to observe and show the two-dimensional surface pictures by the counter of the sea floor, topography, and the target simultaneously. This instrument is able to distinguish the small particles of the seabed structure such as rocks, muds, sand, gravel, or basic types other waters [3]. Side-scan sonar is an active sonar system which implemented the characteristics of sideways look, two channels, narrow beam, and towed body [4, 5]. The passive acoustic monitoring method with species-specific transmission beam pattern influences the facility with which animals can be localized by a hydrophone array [6].

The principle of a side scan sonar can be seen in figure 1, and the system resolution: governed by the shape of the acoustic beam and the length of the transmitted pulse [7, 8]. So it depends on the three-dimensional distribution of the acoustic energy of the system that affects the size of the footprint. If the side-scan sonar has low frequency sources (10-30 kHz), the sound pulse will be transmitted and received at long range (covering a large area in a short time) [9, 10]. More particularly, the application of the system is wide-ranging in marine geology, marine biology, hydrography, underwater archaeology, oceanic engineering, and surveys for military purposes [11]. Regarding the monitoring technologies of benthic artificial habitats, the side-scan sonar system has been proposed as the most practical [12]. However, due to the cost of the system, the complexities of its operation procedures, the necessities of a precise track chart for the ship, and extensive period of time and experience for data processing.

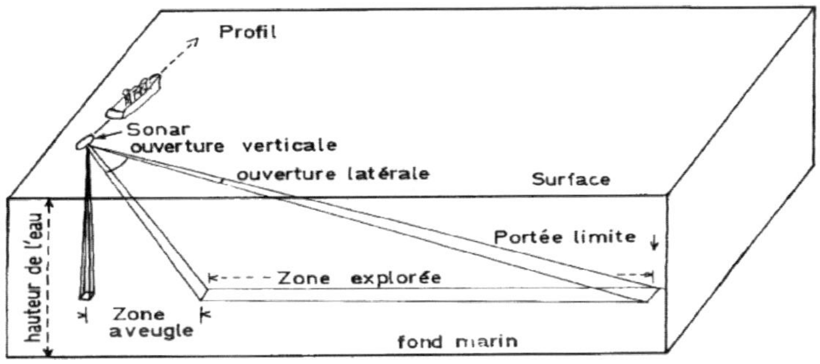

Figure 1. Principle of a side scan sonar [1].

2. Methods

2.1. Study area

This research was conducted in December 2016 at the Punggur Sea, Batam, Riau Islands, Indonesia at the coordinate of 104°08.7102 E and 1° to 1° N 03.2448 03.3977 08.8133 N and 104° E . This research has 3 tracks (figure 2). Acoustic data acquisition was done using a Side Scan C-Max CM2 sonar tow fish instrument also it was set at a frequency of 325 kHz with performs a seabed mapping process that will get features down with two frequency (Port and starboard) display appearance, and a maximum distance of 200 m and 25 m cable length, and a maximum distance of 200 m and 25 m cable length, this study location can be seen in figure 2.

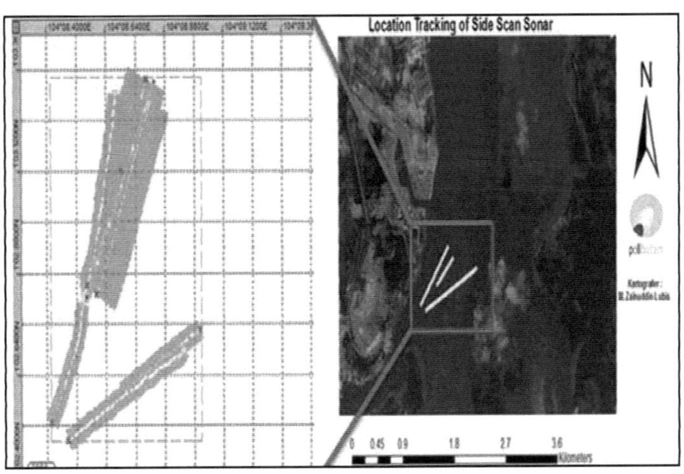

Figure 2.The location of the side scan sonar recording and its observation (3 line of the survey to collecting data in Punggur Sea with the seen on yellow line).

2.2. *Two-dimensional beam pattern*

The characteristics of beam pattern C-Max CM2 of Side Scan Sonar Tow Fish can be seen in figure 3.

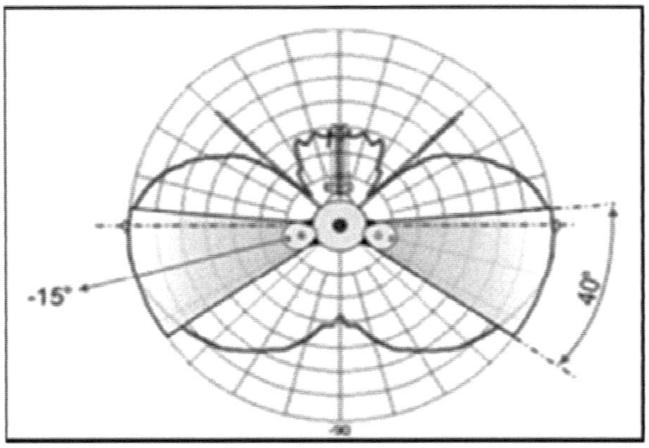

Figure 3. The beam pattern C-Max CM2 of Side Scan Sonar Tow Fish characteristic [5].

Calculating the two-dimensional beam pattern depends on the angle of the incoming sound waves from the array axis, the received power depends on the angle at which light sound incident to array. We can describe this angular dependence with an equation to relate the actual strength that was accepted for the average time power to the shaft (where, θ = 0° and the maximum power is a). The Beam Pattern Discrete-Equispaced Unshaded Line Array can be seen in (figure 4). This ratio is a function of beam pattern two-dimensional array, b (θ) [4] where:

$$b(\theta) = \frac{(P(\theta))}{(P(\theta=0°))} = \frac{\frac{(MP_{max})^2}{R}(1+\cos\delta)}{\frac{(MP_{max})^2}{R}(1+\cos 0°)}$$

$$b(\theta) = \frac{\frac{(MP_{max})^2}{R}(1+\cos(kd\sin\theta))}{2\frac{(MP_{max})^2}{R}} = \frac{(1+\cos(kd\sin\theta))}{2} \quad (1)$$

Using a geometric identification in the sonar system:

$$1+\cos\theta = 2\left(\cos^2(\frac{\theta}{2})\right)$$

$$b(\theta) = \left[\cos^2(\frac{kd\sin\theta}{2})\right]$$

or

$$b(\theta) = \left[\cos^2(\frac{\pi d\sin\theta}{\lambda})\right] \quad (2)$$

$$b(\theta) = \left(\frac{V}{2}\right)^2 \quad (3)$$

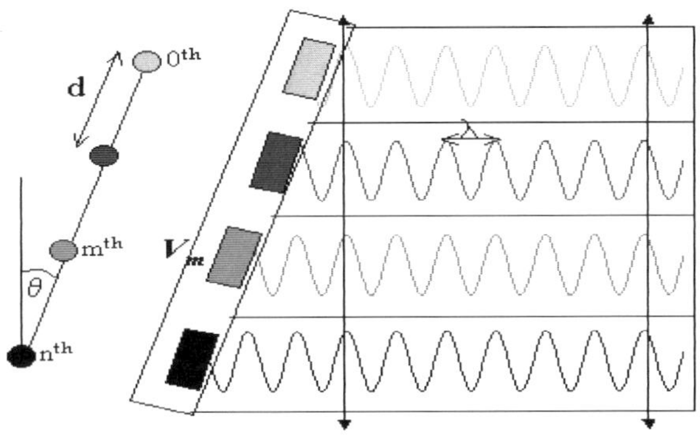

Figure 4. Model the beam pattern, using simple mathematical modeling of Discrete-Equispaced Unshaded Line Array.

3. Results and discussion

Generally, the upper slope of the seafloor will be filled by fine-grained sediments and the bottom will be filled by coarse-grained sediments. Variations of dark and bright colours on the side scan sonar image show the backscattered energy generated from the seafloor. Bright colours will show a high backscatter value whereas dark colours indicate a low backscatter value [22, 23]. It was caused by the gravity force. However, based on the side scan sonar image can be seen the differences between the texture and roughness clearly on the sediments of sand, biogenic sand and mud. There are 7 targets were detected by recording the image of Side Scan Sonar used C-MAX with an estimation of target detection could be observed the large object by the visual images on the target 6 at coordinates 1° 03.3977 N and 104° 08.8133 (figure5). These objects can be indicated was a sunken wreck on the seabed of Punggur Sea, Batam, Riau Islands, Indonesia. The result of the time acquisition, ping position, gain and elevation of the SSS recording can be seen in figure 1. In addition, figure 5 shows the image result of the seabed.

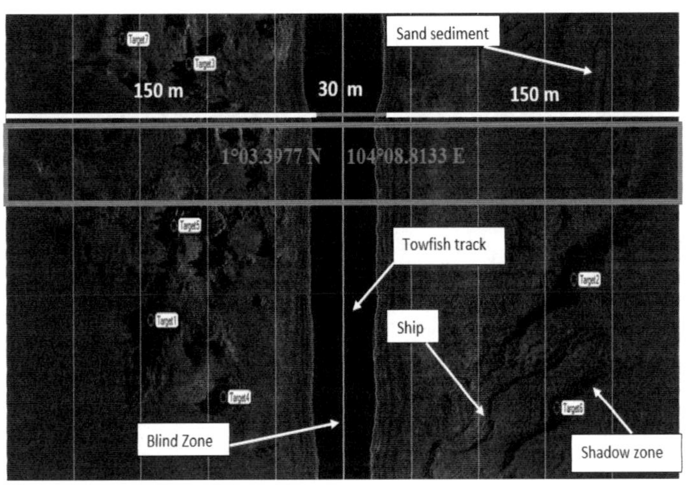

Figure 5. The classification of the seafloor sediments in the Punggur Sea.

There are several lines on the sediment image. Its indication was caused by the influence of the boat and tow vehicle movements so that disrupt the appearance of the image (figure 7). The physical characteristic of the Punggur Sea was having the dynamical current sea. This causes movement of the tow vehicle more difficult to control. Based on data from the movement of side scan sonar was founded the sensor pitch movements until 15 degrees from a standstill 0 degrees roll movements occur up to 10 degrees from its position as well as 0 degrees.

Image of seafloor sediments with the 7 targets. The target of 6 was the object of the sunken ship on the seafloor, while sand object can be seen clearly (figure 6), while the other targets (target 1-5 and target 7) do not indicate the object of the sinking vessel, but the seabed in the form of sand and rough sediment on the seabed. The ship object was founded at the coordinate of 1°03.3101 N and 104°08.7362 E. On the port (right side) was seen the lighter entrenchment. Their excavation led to differences in texture, roughness and slope of the seafloor sediments. Furthermore, the excavation that occurred on the seabed was indicated causing the larger particle sediments was lifted up.

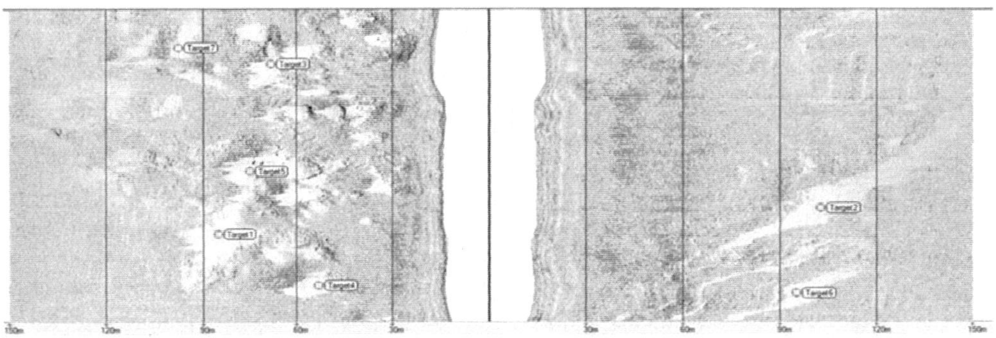

Figure 6. Image of seafloor sediments with the 7 targets.

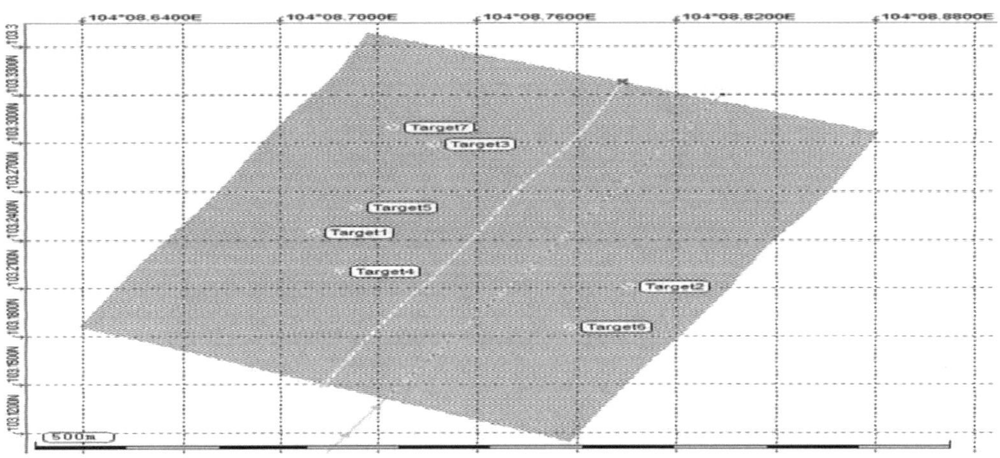

Figure 7. The target positions and the seafloor image by SSS tracking.

According to the results of visual detection [13, 14] grain size, sediment surface roughness scale and significant slope variations could be important role in the acoustic response. The swift currents would precipitate coarse grain sediments and the weak current would precipitate the fine-grained sediments. However, the seafloor condition will affect the location of the sediment. According to [15, 16] a sweep of side scan sonar can produce mosaics, geological and sedimentology features that are easily to observe and interpreted qualitatively so could provide the information about the dynamics of the ocean floor.

IOP Conf. Series: Earth and Environmental Science **176** (2018) 012025 doi:10.1088/1755-1315/176/1/012025

The beam pattern incidence angle (°) of the directivity pattern (dB) and beam pattern can be seen in figure 8.

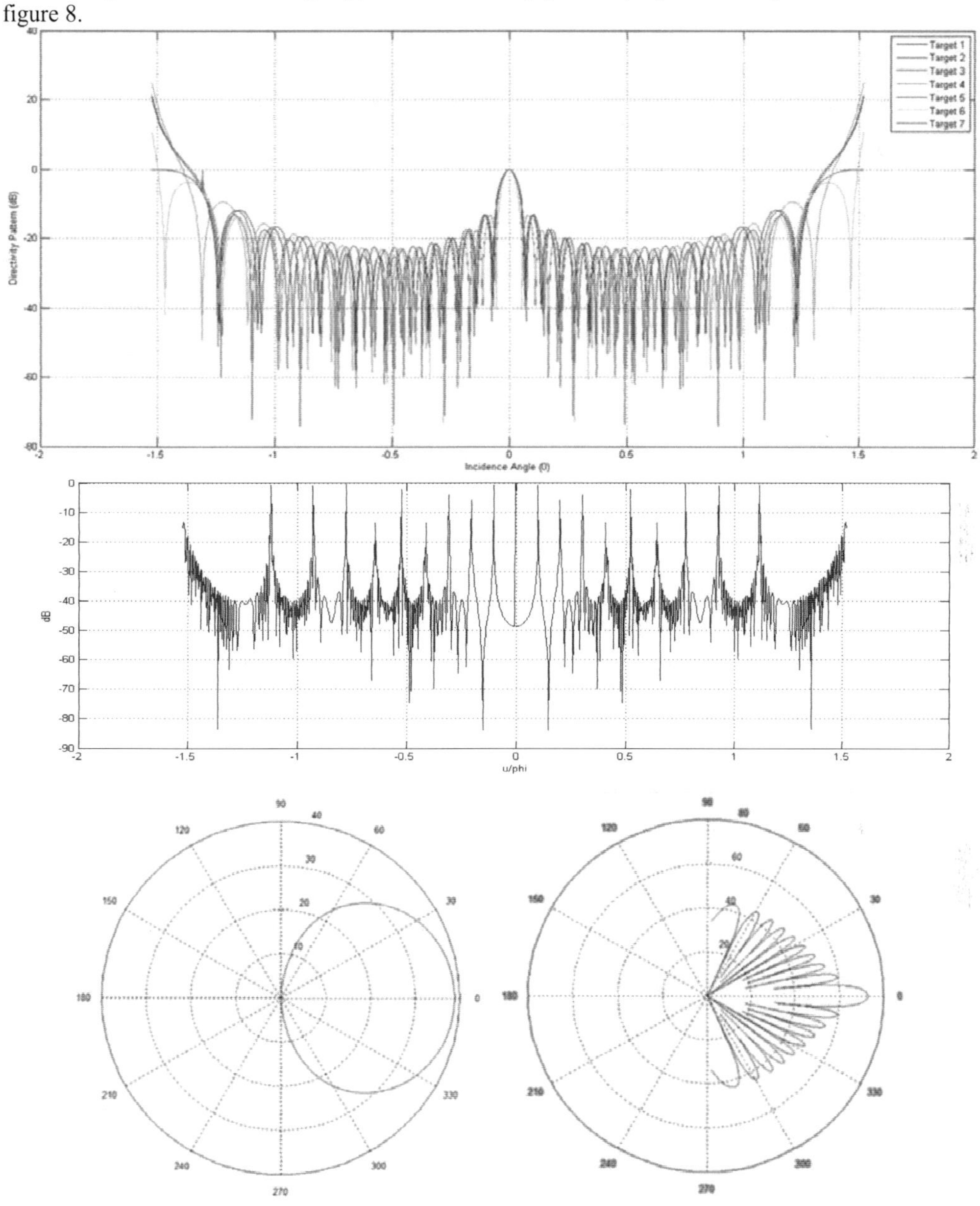

Figure 8. The beam pattern incidence angle (°) of the directivity pattern (dB) and Beam pattern (the results are obtained from the processing of data frequency, altitude and power generated on the side scan sonar (SSS) data on MATLAB software.

The target amount as well as 7 targets on Beam pattern of Discrete Equispaced Line Array unshaded method, the target of 4 has the highest value at 21.08 dB of its directivity Pattern was shown in light blue. By looking at the centre of the model beam pattern result, the target of 6 has a central value of the

IOP Conf. Series: Earth and Environmental Science **176** (2018) 012025 doi:10.1088/1755-1315/176/1/012025

incidence angle (°) of the directivity pattern (dB) are not in a value of 0 or at the centre of the resulting beam pattern. On target of 6 there is a value of incident angle -1.5° and 1.5° has decreased to -40 dB. This relation can be seen to the value of altitude (m) which is 18.1m and a gain of 6 dB which allegedly was the object of the shipwreck on the seabed. On the target of 4 has the values of incident angle -1.32° and 1.32° and also decreased up to -50 dB. The de-convolution of this method result shows the beam pattern which was still concentrated by touching the second quadrant of the existing signal, these results provide a similar pattern the beam and a beam generated by looking at the value of the wavelength and frequency is generated [17-18]. This can be seen in relation to the value of altitude (m) which is worth 17.8 and gain value of 5 dB which is the object of the shipwreck on the seabed allegedly.

Seismic line trace of target detection have 41 number of data collection from side scan sonar imagery after processing (is the result of extraction of side scan feature (.xtf into SEG-Y file). Highest of seismic line trace of target detection is target 3 with 1664 (figure 9).

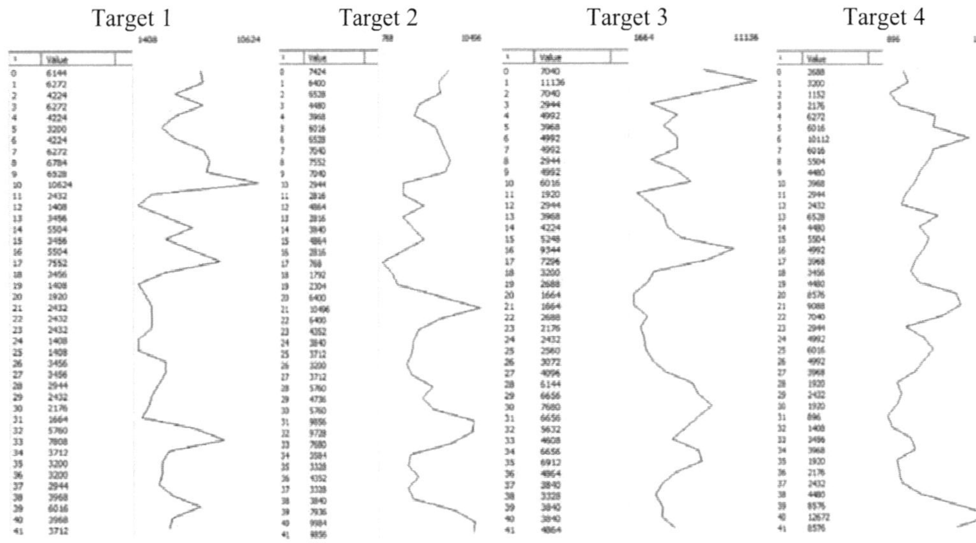

Figure 9. Seismic line trace of target detection.

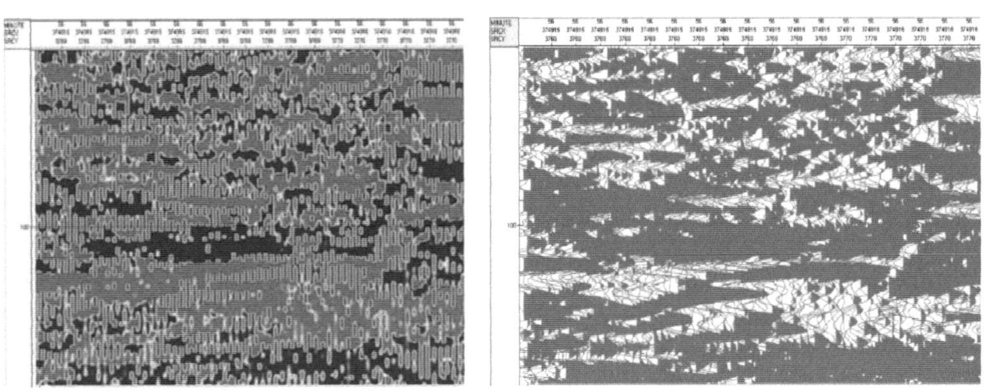

Figure 10. 2D seismic cross-section from side scan sonar (SSS) data extraction results.

IOP Conf. Series: Earth and Environmental Science **176** (2018) 012025 doi:10.1088/1755-1315/176/1/012025

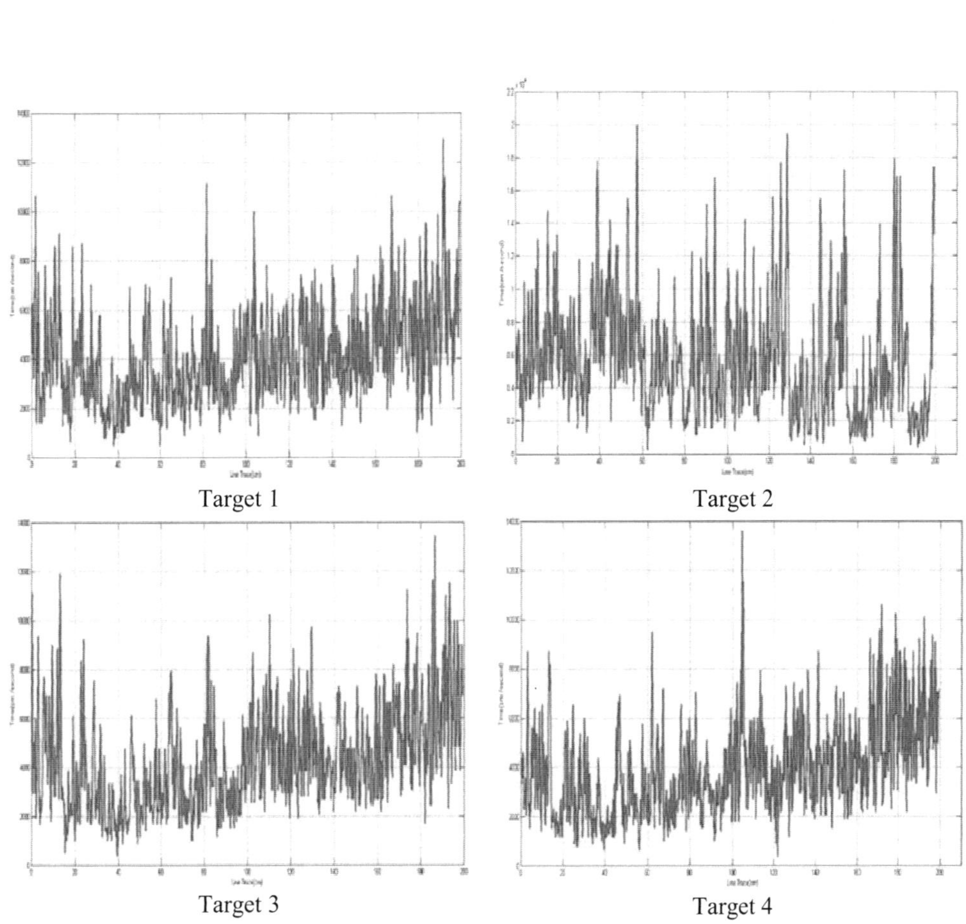

Figure 11. Line trace (cm) vs time (cm/second) target 1-4.

Identification of seismic line trace with identification target must be important for seismic investigation and analysis [19, 21]. Line trace (cm) vs time (cm/second) target 1-4 has obtained from the data extraction using MaxView software that is exported to .txt, by generating time data. While the data line trace is the process of data processing resulting from 2D cross-sectional data (figure 10) Figure of line trace vs time have max data is 200 on line trace and 220 x 103 time in target 2. Highest result of the time in figure 9 is 12928 cm/second and 191, 88 cm in line trace target 1 of side scan sonar imagery. The highest result in figure 7 with the value 9968 cm/second and 57, 525 cm in line trace target 2 of side scan sonar imagery. The highest result in figure 10 with the value 13440 cm/second and 186, 615 cm in line trace target 3 of side scan sonar imagery. Highest result of the time in (figure 11) is 13568 cm/second and 104, 325 cm in line trace target 4 of side scan sonar imagery. Highest result of line trace is target 1 with 191, 88 cm on target 1, and highest of time result is 13568 cm/second on target 4. Seismic figure of side scan sonar imagery have total line trace is 4479, time: 77.9547 cm/s, and gain: 0.00271091 (figure 11).

4. Conclusions

The research in Punggur Sea, Batam using the C-Max CM2 Side Scan Sonar Tow fish at a frequency of 325 kHz, the result of the seabed image is that the more seabed sediment is sand with the target gain and altitude values. The result of the side scan sonar (SSS) image recorded have identified seven targets, while the target of 4 and 6 based on the Discrete-Equispaced Unshaded Line Array method was discovered the sunken wreck. There is a relationship that has a real relationship between the incident angle (°), directivity pattern (dB), Altitude (m), and a gain value (dB). The Discrete Equispaced unshaded Line Array was a method which can be used to identify the targets on the seabed. The highest result of the line trace was the target 1, and highest of time result on target 4. Target 1 have a relationship with results with highest target detection.

Acknowledgments
We would like to thank the P3M Politeknik Negeri Batam and Geomatics Engineering D3 Programs Politeknik Negeri Batam who helped and guided us so that we have completed our research.

References

[1] Lubis M Z, Anurogo W, Khoirunnisa H, Irawan S, Gustin O and Roziqin A 2017 Using Side-Scan Sonar instrument to characterize and map of seabed identification target in punggur sea of the Riau Islands, Indonesia *Journal of Geoscience, Engineering, Environment, and Technology* **2**(1) 1-8

[2] Lubis M Z, Lubis R A and Lubis R U A 2017 Two-dimensional wavelet transform de-noising and combining with side scan sonar image *Journal of Applied Geospatial Information* **1**(1) 1-4

[3] Bartholomä A 2006 Acoustic bottom detection and seabed classification in the German Bight, Southern North Sea *Geo-Marine Letters* **26**(3) 177

[4] Hamilton L J and Parnum I 2011 Acoustic seabed segmentation from direct statistical clustering of entire multibeam sonar backscatter curves *Continental Shelf Research* **31**(2) 138-148

[5] Bryant R S 2015 Side scan sonar for hydrography-an evaluation by the Canadian Hydrographic Service *The International Hydrographic Review* **52**(1) 43-56

[6] Frey-Martínez J, Cartwright J and James D 2006 Frontally confined versus frontally emergent submarine landslides: a 3D seismic characterisation *Marine and Petroleum Geology* **23**(5) 585-604

[7] Alves T M 2010 3D seismic examples of differential compaction in mass-transport deposits and their effect on post-failure strata *Marine Geology* **271**(3) 212-224

[8] Grothues T M, Newhall A E, Lynch J F, Vogel K S and Gawarkiewicz G G 2016 High-frequency side-scan sonar fish reconnaissance by autonomous underwater vehicles *Canadian Journal of Fisheries and Aquatic Sciences* **74**(2) 240-255

[9] Ruffell A (2014) Lacustrine flow (divers, side scan sonar, hydrogeology, water penetrating radar) used to understand the location of a drowned person *Journal of Hydrology* **513** 164-168

[10] Zhang J, Tao B, Liu H, Jiang W, Gou Z and Wen F 2016 A mosaic method based on feature matching for side scan sonar images *In Ocean Acoustics* (*COA*) IEEE/OES China pp 1-6

[11] Lubis M Z, Taki H M, Anurogo W, Pamungkas D S, Wicaksono P and Aprilliyanti T 2017 Mapping the Distribution of Potential Land Drought in Batam Island Using the Integration of Remote Sensing and Geographic Information Systems (GIS) *IOP Conf. Series: Earth and Environmental Science* IOP Publishing **98**(1) p 012012

[12] Garner J T, Buntin M L, Fobian T B, Holifield J T, Tarpley T A and Johnson P D 2016 Use of side-scan sonar to locate *Tulotoma magnifica* (Conrad, 1834) (Gastropoda: Viviparidae) in the Alabama River *Freshwater Mollusca Biology and Conservation* **19** 51-55

[13] Bond L J, Kepler W F and Frangopol D M 2000 Improved assessment of mass concrete dams using acoustic travel time tomography Part I—theory *Construction and Building Materials* **14**(3) 133-146

[14] Kenny A J, Cato I, Desprez M, Fader G, Schüttenhelm R T E and Side J 2003 An overview of seabed-mapping technologies in the context of marine habitat classification *ICES Journal of Marine Science* **60**(2) 411-418

[15] Taki H M and Lubis M Z 2018 Signal detection for identification energy and behaviour of male dolphin bottle nose (Tursiopsaduncus) using NTD Model *IOP Conf. Series: Earth and Environmental Science* IOP Publishing January **105**(1) p 012029

[16] Owen M J, Maslin M A, Day S J and Long D 2015 Testing the reliability of paper seismic record to SEGY conversion on the surface and shallow sub-surface geology of The Barra Fan (NE Atlantic Ocean) *Marine and Petroleum Geology* **61** 69-81

[17] Singh A, Modi M H, Sinha A K, Dhawan R and Lodha G S 2015 Study of structural and optical properties of zirconium carbide (ZrC) thin-films deposited by ion beam sputtering for soft x-ray optical applications *Surface and Coatings Technology* **272** 409-414

[18] Zhang Y, Wang Y, Li W, Huang Y and Yang J 2015 Scanning radar angular superresolution with fast standard Capon beamformer *In Geoscience and Remote Sensing Symposium* (*IGARSS*) *IEEE International* 2015 pp 3111-3114

[19] Pauselli C, Barchi M R, Federico C, Magnani M B and Minelli G 2006 The crustal structure of The Northern Apennines (central Italy): an insight by the CROP03 seismic line *American Journal of Science* **306**(6) 428-450

[20] Russell B, Hampson D, Schuelke J and Quirein J 1997 Multiattribute seismic analysis *The Leading Edge* **16**(10) 1439-1444

[21] Lagomarsino S and Giovinazzi S 2006 Macroseismic and mechanical models for the vulnerability and damage assessment of current buildings *Bulletin of Earthquake Engineering* **4**(4) 415-443

[22] Chavez P S, Isbrecht J, Galanis P, Gabel G L, Sides S C, Soltesz D L, *et al.* 2002 Processing, mosaicking and management of the Monterey Bay digital sidescan-sonar images *Marine Geology* **181**(1) 305-315

[23] Chang Y C, Hsu S K and Tsai C H 2010 Sidescan sonar image processing: correcting brightness variation and patching gaps *Journal of Marine Science and Technology* **18**(6) 785-789

Mapping of mangrove coverage and canopy height using LiDAR data at Sangkulirang District, East Kutai, East Borneo

A T Mahadi[1*], V P Siregar[1], Nursugi[2]

[1] Marine Science and Technology Department, Faculty of Fisheries and Marine Science, Bogor Agricultural University
[2] Marine and Coastal Environment Mapping Research Centre, Geospatial Information Agency, Indonesia
*e-mail: atmahadi2@gmail.com

Abstract. Light Detection and Ranging (LiDAR) is remote sensing technology using transmitted properties of scattered light to detect the intended target. This technology potentially used for spatial planning and management, including mangrove monitoring. The purpose of this research was to map the mangrove coverage and canopy height using airborne LiDAR data at mangrove areas of Sangkulirang district, East Kutai, East Borneo. The corrected point cloud LiDAR data inputed into 10 blocks boundary. This data is classified into 7 classes (ground, mangroves, non mangroves, water, vehicle, low point and isolated point). Ground class used as the data source for digital terrain model (DTM) while mangrove class used as the data source for digital surface model (DSM). The substraction from overlaying DTM - DSM will produce canopy height model (CHM) that represent the height of mangroves canopy from land. The result of this research showed that mangrove distribution had an area of 64.07 km^2. The height of mangrove canopy was dominated by 10-30 meters height while the maximum height reached 54.04 meters.

1. Introduction

Mangrove forest is one component of vital enviromental ecosystem in coastal area. Indonesia had the largest mangrove area (3,112,989 Ha or 22.6% from global mangrove distribution), even larger than mangrove area in Australia (7.1%) and Brazil (7.0%) [1]. Mangrove forest known as an unique and complex ecosystem, high ecological and economical value. Mangrove has ecological function as spawning grounds, nursery grounds, feeding grounds for marine biota and coastal protection from abration [2]. Losing this ecosystem as protector from abration can lead to unstable activity of abration and sedimentation in coastal area [3]. This condition led to the need of mangrove mapping and spatial planning to reach the suitable management of environment in coastal area.

The rapid development of remote sensing technology nowadays is so helpfull for the researcher to monitor mangrove forest. One of remote sensing technology that developed and proved to give high precision and high resolution data is Light Detection and Ranging (LiDAR) technology. LiDAR is an active remote sensing technology that emits intense-focused beams of light and measures the time it takes for the reflection to be detected by the sensor, using transmitted scattered light to measure range (height) and information from an object on the earth. The three-dimensional coordinates (x,y,z or latitude, longitude, and elevation) from an object on the earth computed using interval time information of transmitted pulse to reflected pulse [4].

LiDAR system consist of 3 components, namely global positioning system (GPS), Inertial Navigation System (INS) and laser scanner. These components work independently and integrated to produce LiDAR data. GPS component is used to define three-dimensional orientation of the acquitition vehicle. The positioning of the vehicle refers to the geodetic GPS mark on land used as a control point.

GPS and INS accuracy will determine the accuracy of the coordinate and elevation position that result in the accuracy of the targeted object [5].

LiDAR technology has the same principle with Radio Detecting and Ranging (RADAR) which uses active remote sensing sensor. The differences between these two technologies is the use of electromagnetic energy source. LiDAR uses infrared (IR) wave tansmission while RADAR uses microwave transmission. These two technologies are very often used to produce Digital Elevation Model (DEM). Mahmudi had analyzed the different accuracy of DEM using several data source, namely the Advanced Spaceborne Thermal and Reflection Radiometer (ASTER), Shuttle Radar Topography Mission (SRTM) and LiDAR. The result showed that LiDAR data had the highest DEM density accuracy (± 0.3674 m) than SRTM (± 8.0916 m) and ASTER (± 9.8854 m) [6]. Vertical height accuracy was also performed on the control point. Lidar data had the smallest altitude difference with ± 0.078 m standard deviation than ASTER (± 0.297 m) and SRTM (± 0.422 m) [6].

LiDAR technology has great potential for spatial planning and management, including monitoring of mangrove ecosystem. The advantage of this technology is can be operated during day or night, operational cost is more effective than conventional survey method, can get elevation value with high accuracy and the ability to retrive large amount of data in short time. The height detection from LiDAR data can give accuracy up to 14 cm [7]. This technology is also capable of displaying three-dimensional (3D) vegetation data, measuring vegetation canopy height, canopy cover distribution, canopy volume, subcanopy topography, biomass, large tree density, trees diversity, foliage area index, and physiographic change [8].

Research on mangrove using LiDAR data is still rare in Indonesia, but has been widely practiced in other countries [9, 10, 11]. Development of study on mangrove ecosystem areas in Indonesia is still needed for suitable planning and management. Therefore, this study intends to map the mangrove canopy cover and canopy height in Sangkulirang District, East Kutai, East Borneo using the potential of airborne LiDAR data.

2. Methods

2.1. Time and resesarch location

LiDAR data processing was done on September 2016 to January 2017 at Earth and Toponymous Mapping Centre, Geospatial Information Agency and Laboratory of Mapping and Spatial Modeling, Faculty of Fisheries and Marine Science – Bogor Agricultural University. Area of interest in this study located in the mangrove area, sorrounding Tanjung Manis Village, Sangkulirang District.

2.2. Materials and tools

The tool used in this study is a set of computers equipped with software of spatial data processing, such as Microstation v8 + Terrasolid, ArcGIS 10.2.2 and GlobalMapper v15. The materials used in this study are *Airborne Laser Terrain Mapper* (ALTM) LiDAR data using infrared wavelength (IR) (λ=1,064 nm) that aquired on August 2016. Aerial imagery that acquired in the same time as the LiDAR data acquisition and land-use map from Ministry of Forestry and Environment (KLHK) are required in this study.

2.3. Data procesing procedure

2.3.1. Acquitition and pre-processing LiDAR data. The acquitition and pre-processing LiDAR data step were carry out by the company that colaborated with Geospatial Information Agency (BIG). At the acquitition step, LiDAR data is recorded using Optech Orion M200 ALTM instrument supported with Optech CS-10000 (80 MP) camera system on July to August 2016. All of these instrument integrated with airborne double engine fixed wing, *Cessna-402B*.

LiDAR acquisition conducted at an altitude of 700 m-800 m above sea level with a speed of 59.2 knots (30,192 m/s). The ffrequency of pulse rate used is 150.000 Hz with 40° viewing angle and 500 m-600 m sweep area. The use of ALTM Optech Orion M200 has the ability to detect up to 4 reflections with a density of 4 points per square meter with a vertical accuracy of 15 cm. The acquisition step produce aerial imagery raw data, GPS / INS raw data, exterior orientation parameter data (EO) GPS and INS, LiDAR raw data and flight path data (trajectory) controlled according to BIG LiDAR acquisition procedure.

At the pre-processing step, raw data of GPS, INS and *laser scanner* are required to obtain dataset with good quality control. This step consist of sevral processes, including trajectory processing and LiDAR data matching. Trajectory processing used to correcting the LiDAR point cloud features along to the flight paths. Point clouds that overlayed between flight paths will be set aside (cut overlap). In addition, the outliers feature is also set aside. After this step was completed, the LiDAR data was ready for further processing step. Commonly, output from this step was *.las extention format.

2.3.2. Block boundaries. The block boundaries are conducted to minimize the scope of area to be studied. In addition, because of the LiDAR data big size, these block boundaries will make an ease data processing. Block boundaries in this study were made based on the delineation area of mangrove vegetation. The blocks size refers to BIG procedure. In Borneo index map, these block boundaries located on map with index number of 1916-63. Block boundaries creation adjust with large scale map format (1:10,000). The use of large scale map was in line with the increasing need for geospatial information in large scale map to support spatial planning and disaster management [12].

This process was done by making blocks in the study area that covered mangrove vegetation. Mangrove vegetation in the study area can be determined by visual interpretation of aerial imagery and KLHK land use map of the latest version, which is the 2012 version. This land use map is used to see the general description of mangrove location at the study area then followed by delineation of mangrove vegetation using aerial imagery reference.

After delineation of the mangrove area and the creation of boundary blocks, corrected LiDAR data is filled into each block that had been created. Furthermore, data processing will be done on each block to ease the processing data and improve the accuracy of data processing. In this study, 10 blocks were made in the study area using TerraSolid software (Tscan module). The number of blocks created depends on the ability of the computer to process the large-size of LiDAR data.

2.3.3. LiDAR data classification. LiDAR data classification was done using TerraSolid software (Tscan /Tphoto/Tmodel module). Classification using this software can be done automatically and manually method. In Tscan and Tmodel modules, the point features are classified using Macros (automatic) and Model Tool Box (manual). Additional data such as aerial imagery or high resolution satellite images are required in the LiDAR data classification process. The data is used as a reference to classify LiDAR point cloud in accordance with the appearance of the earth's surface. In this study, LiDAR point clouds are overlayed with aerial imagery (using Tphoto modules) so that they classified in line with visual interpretations of the earth appearance. Data flow diagram presented in figure 1.

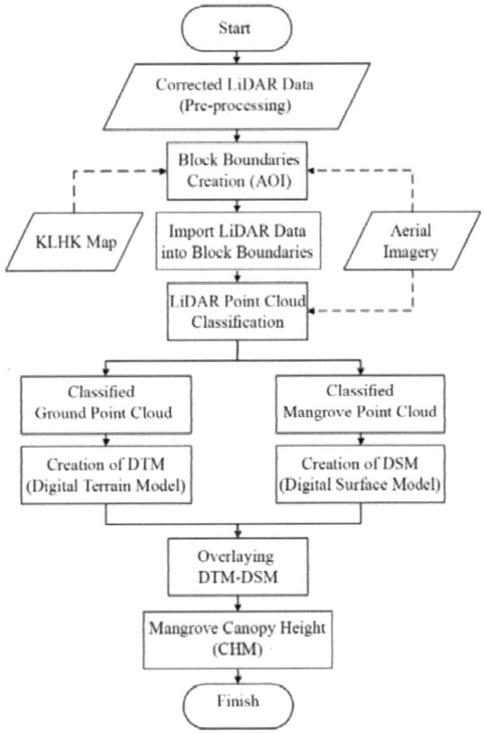

Figure 1. Flow chart of data processing.

LiDAR data classification was done to categorize the point cloud features in LiDAR data into a particular class. In this study, the points feature is categorized into 7 classes, consist of 2 main classes (ground class and mangrove class) and 5 supporting classes (non-mangrove class, waters, low point, isolated point, and vehicle). The classes are made according to the needs and appearance of the objects in the study site.

Generally, LiDAR point cloud classification was made into two classes (ground class and non ground class). In this study, the categorization of 7 classes consisting of main classes and support classes needs to be done because the classification of point cloud directly into the ground class and non ground class (mangrove class) still contain the noise features of the point cloud [13]. Therefore, additional classes for noise filtration are required in order to obtain good classification results in the main class. LiDAR data in ground class was required as a data source to produce digital terrain model (DTM) models that represent bare earth. LiDAR data of the mangrove class is needed as a data source to produce a mangrove surface model or digital surface model (DSM) that represents the surface shape of mangrove vegetation.

2.3.4. Digital Terrain Model (DTM). DTM is a model representation of bare-earth that obtained from ground class LiDAR data point cloud. Ground class LiDAR data that used to obtain DTM had to pass editing step to clear up point cloud that intersected with water area and corrected proportionally. The editing process of ground class was conducted by setting a minimum height of ground (0 meter) and filtering point cloud that intersected with water and vegetation. In addition, noise data such as low point (point cloud under minimum height of ground class) have to classify into other class in order to give good data source for DTM processing. This step was conducted in TerraSolid software (Tscan and Tmodel module) and the output in *.las extension format.

The result from TerraSolid was continued in GlobalMapper software. This sofware will determine data projection of LiDAR data that had been proceed (zone UTM 50 North). LiDAR data from ground class classification was needed as an input in order to conduct DTM. Elevation model from input data was made by binning (minimum value - DTM) method. By this method, gridding of point cloud data will use minimum value (lowest height) to produce DTM.

2.3.5. Digital Surface Model (DSM). DSM is a model representation of object surface above the earth (mangrove canopy) that obtained from mangrove class LiDAR data point cloud. Vertical value from this model was computed from mean sea level. Mangrove class LiDAR data that used to obtain DSM had to pass editing step. The editing process of mangrove class was conducted by filtering point cloud that intersected with water and ground. In addition, noise data such as isolated point (point cloud that have extremely height of mangrove class) have to classify into other class in order to give good data source for DTM processing. This step was conducted in TerraSolid software (Tscan and Tmodel module) and the output in *.las extension format.

The result from TerraSolid was continued in GlobalMapper software. This sofware will determine data projection of LiDAR data that had been proceed (zone UTM 50 North). LiDAR data from ground class classification was needed as an input in order to conduct DSM. Elevation model from input data was made by binning (maximum value - DSM) method. By this method, gridding of point cloud data will use maximum value (highest height) to produce Digital Surface Model (DSM). The differences between DTM and DSM was illustrated in figure 2.

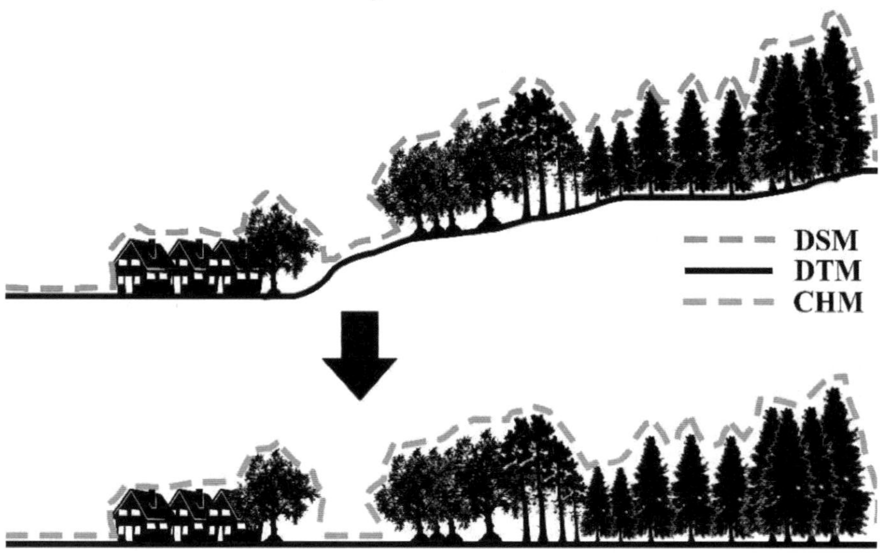

Figure 2. Canopy Height Model/ CHM (orange) conducted by substracting DTM (black) and DSM (green).

2.3.6. Canopy Height Model (CHM). Digital Surface Model (DSM) still not represent the real height of mangrove because the mangrove height was conducted from mean sea level (geoid). So, it is necessary to obtain Canopy Height Model (CHM) that represent the real mangrove canopy conducted from the ground (figure 2). CHM was proceed and produced in GlobalMapper software. DTM and DSM grid data that had been obtained are the source to proceed this model. Substraction analysis of DTM and DSM was done to observe the vertical differences from the ground to mangrove canopy.

3. Results and discussion

3.1. Block boundary

The blocks boundary that created adjusted to the ability of the computer processing large-size LiDAR data. The total area covering the block is 213.90 Km^2. Figure 3 shows the result of block boundaries.

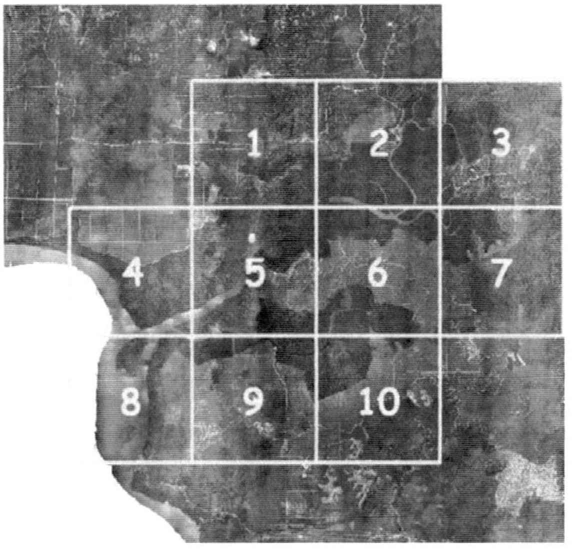

Figure 3. Block boundary in study area.

The total amount of cloud points that filled the entire block boundary was 535,044,728 point clouds. The block that has the fewest number of point cloud was block number 8 with the amount of 30,815,485 points (figure 4). Block number 8 had the fewest point cloud distribution because it contain large water body area inside the block. LiDAR acquisition using ALTM is sensitive to water area. ALTM using scattered light with λ=1,064 nm (Near Infrared/ NIR). This wavelength had high reflectance to vegetation and ground but this wavelength had low reflectance to water area. Because of this condition, LiDAR acquitition using this instrument will not well-detected area of water [4].

3.2. Point cloud LiDAR classification

In this study, point cloud LiDAR classified into 7 classes that consist of 2 main class (ground class and mangrove class) and 5 supporting class (non-mangrove class, water, low point, isolated point, and vehicle). Point cloud distribution in each block boundary shown in figure 4.

Supporting class was made for unnecessary point cloud (noise) so it was not included as the main class that could distract the process of DTM and DSM. Direct classification of point cloud into ground points still produce noise point cloud while classification into several class could minimize noise point cloud in ground class (main class) [13]. It is necessary to produce good quality of DTM and DSM in order to give better result.

IOP Conf. Series: Earth and Environmental Science 176 (2018) 012026 doi:10.1088/1755-1315/176/1/012026

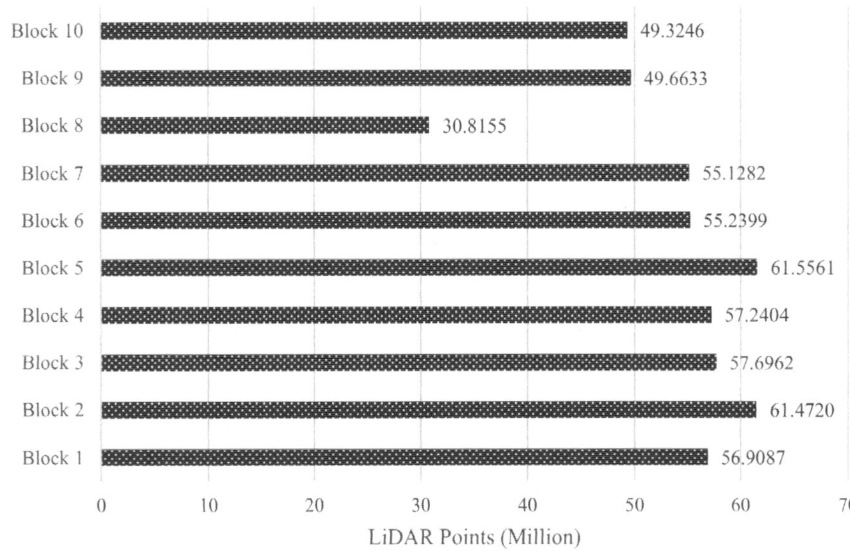

Figure 4. Point cloud distribution in each block boundary.

Generally, non-mangrove class had the most amount of point cloud distribution. It is because non-mangrove class in this study define as all point feature in vegetation area that not belong to mangrove vegetation. In this study, non-mangrove had an area about 70% of all block boundary and that condition made this class had the most amount of point cloud (figure 5).

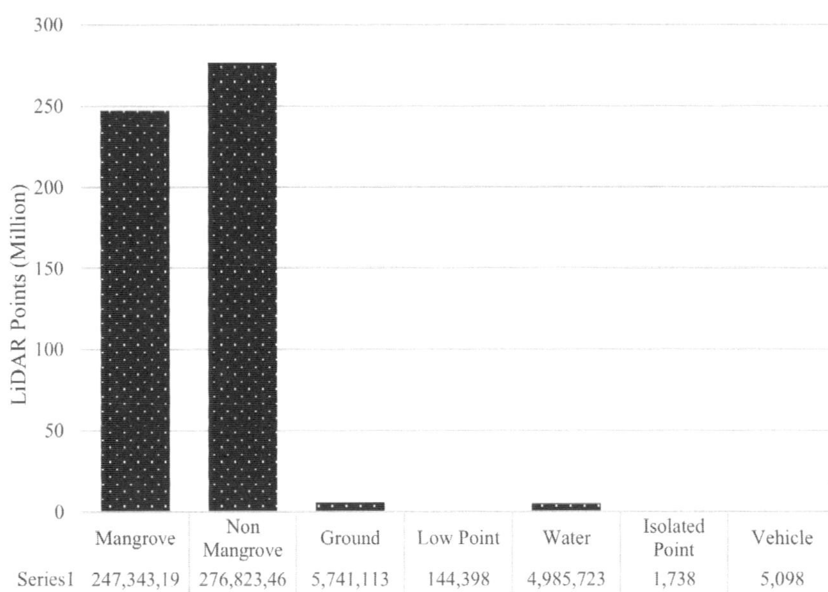

Figure 5. Point cloud distribution in each class.

Based on the main class clasification, point cloud distribution in mangrove class had more point cloud than ground class. Generally, there are two kind of reflectantion of LiDAR data, last reflectance and others reflectance (first reflectance). The last reflectance is the reflectance from ground while the the first reflectance is the reflectance of vegetation canopy [14]. ALTM LiDAR has an advantage to detect up to 4th reflectance in single transmisson of scattered light. So, point cloud from first to third reflectance are reflectance from object above the ground, including mangrove vegetation. This situation caused mangrove class had more point cloud distribution than ground class.

3.3. Digital Terrestrial Model (DTM)

DTM consist of point cloud that classified in order to represent bare earth of the ground [15]. In the result of grid interpolation from ground class, elevation of ground had a height of 0 meter to 39.76 meter, dominated by blue colour that indicated domination of 0 meter to 2 meter height. This condition show that the ground topography in studied area is located at coastal area that adjoin with water area. It is in line with visualisation of earth from aerial imagery that show the river body, estuary, and open water area. The result of DTM that proceed with GlobalMapper shown in figure 6.

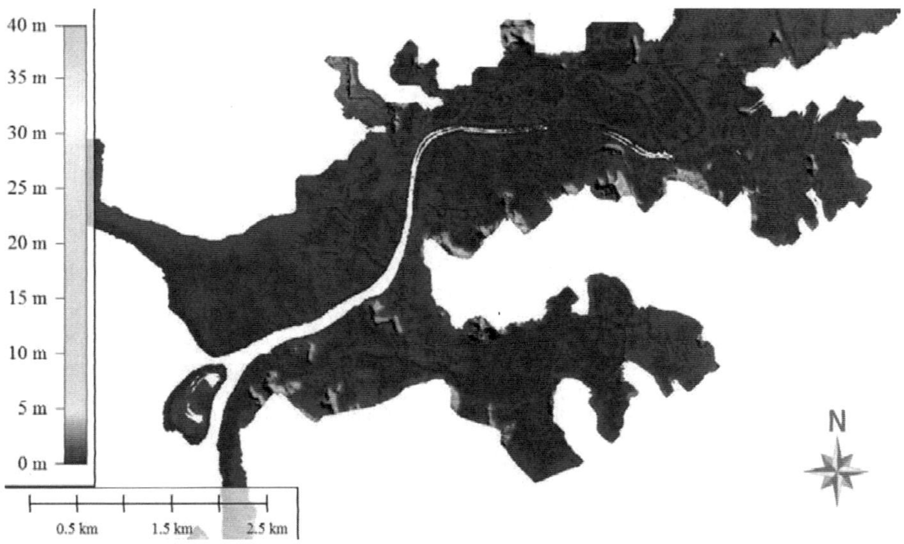

Figure 6. Digital terrain model (DTM) in study area.

3.4. Digital Surface Model (DSM)

DSM can be generally defined than DTM, DSM is a model production that represent surface (ground surface, building surface, or canopy surface) [4]. In the result of grid interpolation from mangrove class, the elevation of mangrove canopy had a height up to 68.09 meter. Mangrove canopy dominated by green colour that indicated height of 10-30 meter. The result of DSM that proceed with GlobalMapper shown in figure 7.

DSM using LiDAR data is very relevant for telecommunication management, vegetation management, air traffic safety, and 3D model simulation [16]. Elevation of surface can be used to produce canopy height model (CHM). DSM value can not directly used to represent the height of canopy from the earth because the height of LiDAR data corrected by geoid layer (SRGI 2013 datum) that assumed as the mean sea level. So, DSM value had to corrected by DTM value to produce the height value that count from the ground.

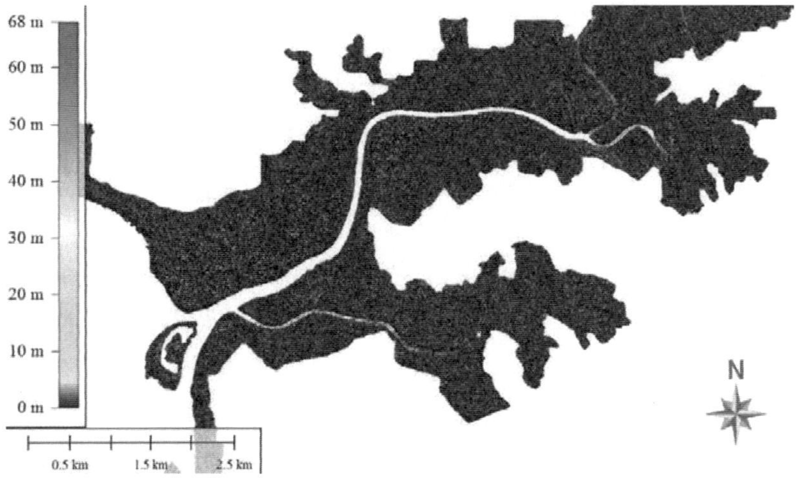

Figure 7. Digital surface model (DSM) in study area.

3.5. Canopy Height Model (CHM)

Generally, CHM is a representation of vegetation canopy [17]. CHM produced by point cloud DSM and point cloud DTM [9]. CHM can be applied to map the important caracteristic of vegetation such as biomass for understanding carbon contentin vegetation [18].

The result of CHM that proceed with GlobalMapper shown in figure 8. CHM show the vegetation height that dominated by green colour. That colour indicated a heoght of 10 – 30 meter. The maximum height of mangrove canopy reache a height of 54.04 meter. From this model, we can get the mangrove coverage in study area had an area of 64.07 km^2.

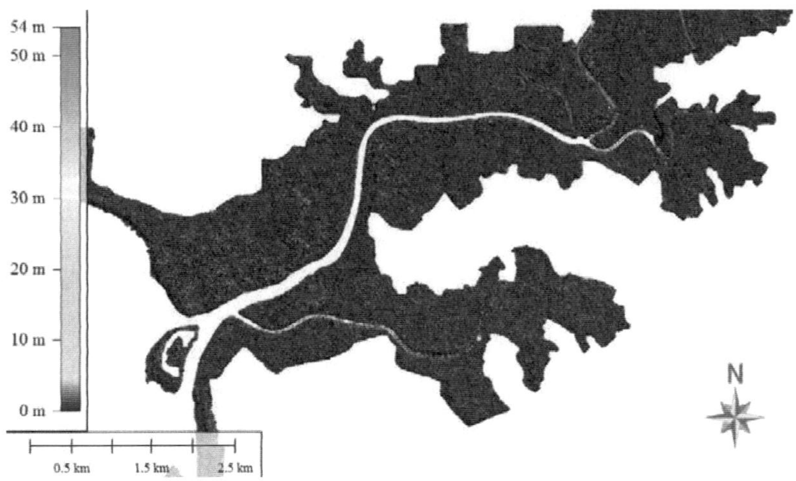

Figure 8. Canopy Height Model (CHM) in study area.

Indonesia has the largest mangrove area distribution (22.6%) and has many unexplored area that not affected by human activity (such as Papua amd Borneo). These condition had a potential for mangrove vegetation to growth naturally and optimally. Similiar study of mangrove height at the area of Mimika, Papua using SRTM elevation data. Mangrove canopy height in that area reach up to 40 meter height [19]. Borneo island known as the island that have a lot of river and estuary at coastal area. A lot of estuary that protected from strong oceanography dinamics, especially at Sangkulirang Bay, provide well natural condition for mangrove vegetation to grow optimally. It is proved by 54.04 meter height of mangrove canopy gained in this study area.

The main natural parameters that affected mangrove growth are fresh and salinity water supply, nutrient, and substrate stability [20]. Mangrove vegetation in study area lived around Sangkulirang Bay, at Kerayan River estuary. This estuary is protected the mangrove habitat because the Northern area has peninsula at Berau District and Southern area has peninsula at Kaliorang Sub-district. These peninsulas will slow down current water flow from Makassar Strait into the estuary. Moreover, the estuary also protected by Hantu Island. Geographic position of this island caused low penetration of current waterflow into mangrove habitat so that the fresh water and salinity supply stability are controlled. Geographic condition and mangrove existing will get the stability of subtrat condition. Unique mangrove root system will slowing down the fresh water supply and banding particle of sediment so that particle with high orgic level will precipitate rapidly and form sediment layers [20]. These condition show that Sangkulirang District has good suitability environment parameters for optimally mangrove growth.

4. Conclusions

Point cloud LiDAR can be used to produce canopy height model that can be applied to analyse other important caracteristic of vegetation. The area that covered all of studied area was 213.90 km^2. In this study area, mangrove coverage at Sangkulirang District, East Kutai, East Borneo has an area of 64.07 km^2 mangrove distribution. Mangrove vegetation dominated by the height of 10 – 30 meter. The maximum height of mangrove in this area reached 54.04 meter. LiDAR data is potentially used as the data source of mangrove canopy mapping.

Acknowledgments

Best appreciation to Geospatial Information Agency (BIG) that provided the main materials of this research such as LiDAR data, aerial imagery and hardware for processing data. Thank you to all of people from Marine Science and Technology Department, FPIK-IPB that help this study to be done.

References

[1] Giri C, Ochieng E, Tieszen L L, Zhu Z, Singh A, Loveland T, Masck J and Duke N 2011 Status and distribution of mangrove forests of the world using earth observation satellite data *Global Ecol. Biogeogr.* **20** 154-159

[2] Dharmawan I W E and Pramudji 2014 *Panduan Monitoring Status Ekosistem Mangrove* ed. Pramudji and Nontji A (Jakarta: COREMAP CTI LIPI) p 1

[3] Astra A S, Etwin K S, Arief M H and M Bagus M 2014 *Keterlibatan Masyarakat Dalam Pengelolaan Pesisir dan Laut, Studi Kasus: Kawasan Perlindungan Pesisir Desa Timbulsoko, Kecamatan Sayung, Kabupaten Demak* (Bogor: Wetlands International Indonesia) p 2

[4] [NOAA] National Oceanic and Atmospheric Administration, Coastal Service Centre 2012 *Lidar 101: An Introduction to Lidar Technology, Data, and Applications* (Charleston SC: NOAA Coastal Service Centre) 3p

[5] [KNG] Karvak Nusa Geomatika. 2016 *Akuisisi Lidar dan Pemotretan Udara Digital KEK Maloy Batuta Trans Kalimantan (MBTK) dan Sekitarnya: Laporan Tahapan Pengukuran Titik Kontrol* (Jakarta: PT Karvak Nusa Geomatika) p 4

[6] Mahmudi, Subiyanto S and Yuwono B D 2015 Analisis ketelitian DEM ASTER GDEM, SRTM dan LIDAR untuk identifikasi area pertanian tebu berdasarkan parameter kelerengan (studi kasus: Distrik Tubang, Kabupaten Merauke, Provinsi Papua) *J. Geodesi Undip* **4**(1) 95-106
[7] Ballhorn U, Navratil P, Jubanski J and Siegert F 2014 LiDAR survey of the Kalimantan Forest and Climate Partnership (KFCP) project site and EMRP area in Central Kalimantan Indonesia, *Technical Working Paper* (Munchen: Remote Sensing Solution GmBH p 4
[8] Behera M D and Roy P S 2002 Lidar remote sensing for forestry applications: the Indian context *Current Science* **83**(11) 1320-1328
[9] Wannasiri W, Nagai M, Honda K, Santitamnont P and Milphokasap P 2013 Extraction of mangrove biophysical parameter using airborne LiDAR *Remote Sens* **5** 1787-1808
[10] Faelga R A G , Paringit E C, Perez G J P, Argamosa R J L, Ibanez C A G, Posilero maV, Tandoc F A M and Zaragosa G P 2016 Separability and variability of rhizophocraceae and avicenniaceae in a natural mangrove forest using point density distribution from LiDAR data. *J. of Phil. Geos. Rem. Sens. Soc* 18-25
[11] Olagoke A, Proisy C, Feret J P, Blanchard E, Fromard F, Mehlig U, Menezes M M, Santos V F and Berger U 2016 Extended biomass allometric equation for large mangrove trees from terrestrial LIDAR data *Trees* **30** 935-947
[12] [BIG] Badan Informasi Geospasial 2013 *Laporan Akuntabilitas Kinerja Institusi Pemerintah-BIG* (Cibinong: Badan Informasi Geospasial) p 30
[13] Ainiyah R 2013 Perbandingan Metode Klasifikasi Data Lidar dengan Fotogrametri [thesis]. Bandung: Institut Teknologi Bandung p 63
[14] Kato A, Schreuder G F, Calhoun D, Schiess P and Stuetzle W 2007 Digital surface model of tree canopy structure from LIDAR through implicit surface reconstruction *ASPRS 2007 Annual Conference* p 2
[15] GeoBC 2014 *Specification for LiDAR* (Victoria BC: Ministry of Forest, Lands and Natural Resources Operation) p 32
[16] Heideman H K 2014 *LiDAR Base Specification. U.S. Geological Survey Techniques and Method.* Version 1.2 book 11(B4) (November 2014) Virginia (US) p 20
[17] Leeuwen M V, Coops N C and Wulder M A 2010 Canopy surface reconstruction from a LiDAR point cloud using Hough Transform *Rem. Sens. Letters* **1**(3) 125-132
[18] Vega C and St Onge B 2008 Height growth reconstruction of a borel forest canopy cover a period of 58 years using a combination of photogrammetric and lidar model *Remote Sens. of Env.* **112** 1784-1794
[19] Aslan A, Rahman A F, Warren M W, and Robenson S M 2016 Mapping spatial distribution and biomass of coastal wetland vegetation in Indonesia Papua by combining active and passive remotely sensed data *Rem. Sens. of Env.* **183** 65-81
[20] Dahuri R 2003 *Keanekaragaman Hayati Laut: Aset Pembangunan Berkelanjutan Indonesia* (Jakarta: PT Gramedia Pustaka Utama) p 58

IOP Conf. Series: Earth and Environmental Science **176** (2018) 012027 doi:10.1088/1755-1315/176/1/012027

Empirical acoustic TS - Length relationship for Torpedo Scad (*Megalaspis cordyla*)

W Setiawan[1], I Jaya[1*], T Hestirianoto[1], S Pujiati[1], A Priatna[2], A Ma'mun[2]

[1]Department of Marine Science and Technology, FPIK IPB Bogor Indonesia
[2]Marine Fisheries Research Center, Ministry of Marine Affairs and Fisheries, Jakarta
*corresponding e-mail : indrajaya@apps.ipb.ac.id

Abstract. One of the main challenge in fish stock assessment with acoustic method is how to improve the accuracy of the estimated acoustic target strength (TS) of the fish under investigation. Various factors contribute to the bias in acoustic TS, namely the high diversity of fish species, body shape, orientation and size. It is therefore necessary to have a database of acoustics TS of fish on a particular species, size (length), and orientation as a reference to be used in grouping and in estimating the over all TS-length relationship of the species. In this paper we reported the result of in situ TS measurements of Torpedo Scad (*Megalaspiscordyla*). The TS data analysis is performed numerically through the best-fit technique to establish the relationship between length, object orientation, and TS. The TS-length relationship obtained for Torpedo Scad (*Megalaspis cordyla*) is TS = 20 log (FL)–53.68.This relationship could be used to eliminate bias in the fish stock assessment by acoustic method.

1. Introduction

In order to achieve sustainable fisheries management, where exploitation rates are balanced with the rate of recovery and growth of fish resources, adequate information is needed on the condition or status of fish stock resources. In addition, one of the main information that needs to be known in determining of fisheries management that is realistic, measurable and can be evaluated objectively the initial description of the condition of fish stocks to serve as the starting point of reference. The effectiveness of management actions that have been and will be taken can be determined with a high level of trust. The status of fish resources can be known through a study of fish stocks. There are various methods of assessing fish stocks that are developing today. One of them is the acoustic method, which has evolved since 50 years ago [1]. The main challenge in assessing fish stocks with acoustic surveys is how to improve the accuracy of the results. One of the factors of bias is the high diversity of fish species, both shape and size [2]. It is therefore necessary to have a database of acoustic values (target strength, TS) of fish on a particular type and size, as a reference used in grouping and estimating the length of the species [3].

Acoustical method is one of the standard method stock assessment, in addition to other methods such as from the catch of fish. Currently, the accuracy of fish stock estimation with acoustic method is still not satisfactory due to the high diversity of fish species, both shape and size. While very little target strength (TS) measurements from individuals of various dominant commercial fish species [4]. It is therefore necessary to measure the physical parameters of fish (length, weight, orientation) with acoustic parameters (TS, target strength).

The result of the measurement of the target value of each individual/fish species on each particular size (FL) structure will be a reference to verify the acoustic data of fish size (dB) to length (cm), in order to reduce the bias in further analysis towards estimation biomass and stock potential of Fish Resources based on hydro-acoustic method [5]. The objective that is expected to be achieved through

IOP Conf. Series: Earth and Environmental Science **176** (2018) 012027 doi:10.1088/1755-1315/176/1/012027

this research is the increasing accuracy of estimation fish stock assessment using acoustic method. Activities in the study of TS Characteristics (Target Strength) and TS-Length Formulation on Fish Key Indicators for Increasing Fishery Stock Estimation Accuracy, including calculating TS values of some dominant commercial fish species. Formulating empirical relationships between acoustic parameters and physical parameters. The formulation of the relationship between TS Acoustic values and the length or weight of fish for some commercial fish is important.

Formulation of the relationship between TS Acoustic value and length or weight of fish for some commercial fish is important for increasing accuracy of fish stock estimation for sustainable fisheries management [4]. The research results are used as input for the formulation of fishery management policy in the form of model as one component in estimation calculation fish stock. The more valid and measurable management steps required for sustainable and optimal fisheries management.

2. Methods

2.1. Time and study area
Acoustic measurements were conducted in Palabuhan Ratu Bay, with floating lift net (Figure 1), in two trips, namely August and November 2016.

Figure 1. Floating rides (lift net) vehicle as a place to do research with the transducer placed above the cage.

2.2. Tools and materials
The tools and materials used in this study include: Scientific Echosounder Simrad EK 80-200 kHz tranducer split beam, sphere ball (target standard) type tungsten carbide (WC) 38.1 mm diameter, cage where live fish are measured by their TS value and confinement of live fish sampling before measured, measuring instruments (meter), digital scales and CTD SBE 19 plus to measure temperature and salinity.

Before performing the TS measurement activity of each individual fish with an acoustic system, the acoustic instrument is first calibrated [6]. Measurement of profiles of temperature and salinity of waters to obtain the speed of sound value is done with CTD SBE 19 plus. Limited scope of the lift net the maximum sphere ball can only be placed at a depth of 4.5 m from the surface of the transducer. The parameter setting used in calibration is in table 1, and is used at the time of acquisition of TS value of each individual fish.

IOP Conf. Series: Earth and Environmental Science **176** (2018) 012027 doi:10.1088/1755-1315/176/1/012027

Table 1. Setting of acoustic parameters.

Parameter	Value
Frequency	200 KHz
Pulse Duration	0.512 ms
Power transmit	150 watt
Sound speed	1542 m/s
Absorption	82.9 dB/Km
SV threshold	-60 dB
TS threshold	-60 dB

2.3. Measurement of TS value
TS measurements were performed on each individual species of fish collected under living conditions at various lengths. For a fish that cannot be sustained alive then it is measured in a fresh die condition. Each individual live fish is placed inside the cage, then the individual fish is scanned with an acoustic system, with the transducer placed above the cage, so the target (fish) is in the beam acoustic beam area [2]. The echosounder parameter settings used at the time of acquisition must correspond to the setting of parameters at calibration time (table 1). The measurements of each individual fish are declared successful, until a normal distribution of TS values is generated.

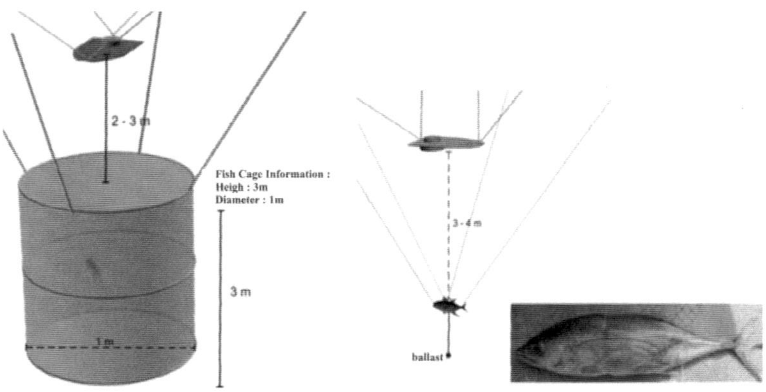

Figure 2. Measurement of TS fish values a) cage method (confinement) for live fish; b) tethered method for dead fish.

2.4. Data analysis
Acoustic data is processed using SONAR5-pro software, extraction results are target strength values (TS) in decibels (dB). The number of TS values per fish individual measured by acoustics consists of many variations of TS values that spread within a certain range, but generally spread normally and have a certain modus value.

2.4.1. The target strength value (TS) of each individual fish. The amount of data used to obtain the average TS value of one individual fish is the TS data in the range of ± 3dB from the middle peak value of its modus mode. According to [2] TS is a backscattering cross section (σbs) of the returned target signal, the relationship between TS and σbs is expressed as TS = 10 log (σbs). Then the average TS is TS = 10 Log (σbs). The mean TS score is assumed to be the TS value of each individual fish, and is used in the regression analysis to obtain TS-length formulations on similar species.

IOP Conf. Series: Earth and Environmental Science **176** (2018) 012027 doi:10.1088/1755-1315/176/1/012027

2.4.2. Empirical relationship of TS values with length of fish. A series of coupled data between length (L) and average TS for the same species is used to analyse the empirical relationship between the two so that TS-length formulation is formed. The empirical relationship between TS and L refers to TS = 20 log (L) + A, where A is the target strength value for 1 cm of normalized strength depending on the species of fish species [6]. Based on the equation, then to get TS-length formulation only use linear regression analysis between independent variable L and dependent variable TS. Previously a series of L and TS data of each individual in the same species tested the normality of the data as a condition of regression analysis. After the regression equation TS = a Log L + b is formed, then the partial test is the significance of variable L to TS, with the hypothesis: H_0 = L variable has no significant effect on TS; H_1 = L variable has significant effect on TS. Regression equation obtained, then compared with equation [6] that is TS = 20 Log L + A, so will be obtained equation to get value of A on certain species. If A has been obtained, then by entering the L value of each measured individual fish, it will get the average value of A or call C. So the TS-length formulation is TS = 20 Log L + C [7].

3. Results and discussion

3.1. Target strength values

A total of 100 live fish species Torpedo Scad fish (*Megalaspis cordyla*), which successfully collected at night successfully measured the TS value in living conditions only as many as 35 pieces of samples. Result of measurement with 200 kHz acoustic frequency of 35 fish of Torpedo Scad, its TS value is distributed starting from -61.67 ± 1.6 dB up to -49.22 ± 1.5 dB. Each fish size has a different modus class value. The value of the TS taught fish class modus tends to increase with the larger size (table 2). A number of fish are measured TS value, spread from the size 18.8-22.0 cm FL, with the average number of samples measuring 300 times for each fish, the average value of TS fish Torpedo Scad at size 18.8 - 22.0 cm FL is -58.04 dB (table 2).

Table 2. Composition of size and value of fish TS Torpedo Scad (*Megalaspis cordyla*).

No. Fish	TL (cm)	FL (cm)	W (g)	Swimblader (cm)	Average TS	Deviation Standard
1	22.0	20.1	103.8	5.0	-58.13	1.6
2	22.1	20.6	104.6	5.0	-58.28	1.7
3	20.2	19.0	94.5	5.0	-58.64	1.5
4	21.3	20.0	98.0	6.8	-60.94	1.8
5	20.9	19.6	100.2	5.0	-55.15	1.7
6	21.1	19.8	101.3	4.5	-58.27	1.7
7	20.5	19.4	92.2	4.1	-58.57	1.6
8	20.9	19.5	101.2	4.5	-57.99	1.7
9	21.9	19.7	105.5	4.5	-58.20	1.7
10	21.9	20.4	94.2	4.2	-58.20	1.6
11	21.3	20.9	108.0	5.0	-55.32	1.7
12	21.5	19.6	96.6	5.0	-55.85	1.5
13	21.6	19.6	96.2	5.0	-57.99	1.7
14	21.8	18.9	83.1	5.0	-58.11	1.7
15	21.5	19.6	103.4	4.5	-58.04	1.7
16	20.5	18.8	92.9	4.5	-58.08	1.6

No.	TL (cm)	FL (cm)	W (g)	Swimblader (cm)	Average TS	Deviation Standard
17	21.4	19.8	96.3	5.5	-55.34	1.7
18	21.2	20.0	101.6	5.5	-58.09	1.7
19	20.8	18.9	88.9	4.0	-58.12	1.6
20	21.2	19.3	93.5	6.0	-58.01	1.7
21	21.3	20.3	108.6	5.5	-55.23	1.7
22	22.2	20.6	94.6	5.0	-49.22	1.5
23	20.8	19.3	96.6	5.0	-58.09	1.7
24	23.8	20.9	102.2	5.5	-58.00	1.7
25	23.8	21.7	101.3	5.0	-61.67	1.6
26	21.9	20.4	86.3	5.0	-58.39	1.7
27	21.7	19.6	93.1	5.0	-58.35	1.5
28	22.6	20.4	105.3	5.5	-58.58	1.6
29	21.9	20.0	97.5	4.5	-60.91	1.7
30	21.2	19.8	96.4	6.0	-60.88	1.6
31	21.4	20.0	92.2	5.5	-58.29	1.7
32	23.3	22.0	91.1	5.5	-58.33	1.6
33	21.1	19.2	86.7	5.5	-58.11	1.6
34	21.2	19.3	87.6	5.0	-60.99	1.6
35	21.8	19.9	106.6	5.0	-60.91	1.7

3.2. Formula TS-Length Fish Torpedo Scad (Megalaspis cordyla)

The analysis of fish length relationship (FL) Torpedo Scad (*Megalaspis cordyla*) to TS value from each individual to get the empirical value in the form of TS = 20 Log (FL) + C was done based on the measurement data presented in table 2. Number of samples fish measured Torpedo Scad is 35 piece of samples.

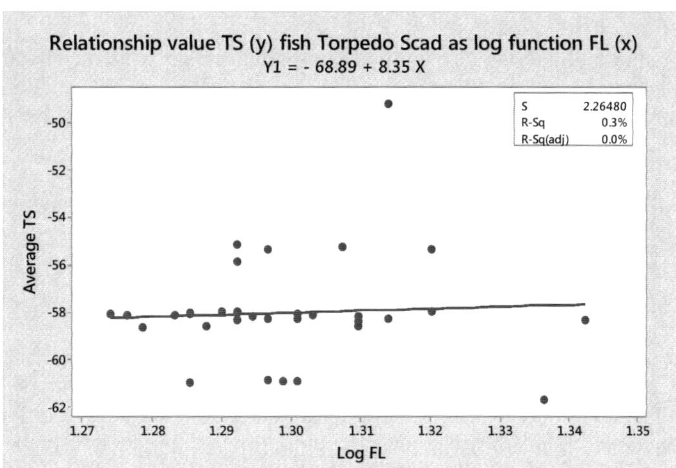

Figure 3. Relationship value TS (y) fish Torpedo Scad as log function FL (x).

Table 3 shows the result of the regression analysis for the significance of the independent variable (x) Log FL to average TS value. The result of t-test at 95% confidence level indicates that the length size (FL) has no significant effect on the TS value, where the absolute t_{count} (0.34) $<t_{\text{-table}}$, also insignificant ($P_{\text{-value}}$ (0.737)> 0.05). Regression equation model based on research data is not significant meaning, linear regression model does not meet the criteria of linearity. The regression equation for TS-length fish formulation TS = 8.35 Log (FL) - 68.9 with coefficient of determination (R^2 = 0.30%) and R^2 (adj) = 0% (figure 4).

Table 3. Independent variable coefficient (x) is the length of fish (FL) in the regression equation.

Variable	Coefficient	SE Coef	T	p	VIF
Constant	-68.9	32.2	-2.14	0.040	
FL	8.35	24.81	0.34	0.738	1.00

The result of $F_{\text{-test}}$ or ANOVA (table 4) shows the value of $F_{\text{-count}}$ (0.11) $<F_{\text{-table}}$ (1; 34; 0.05) (4.13), then reject H_0 which means linear model between fork length (FL) with its TS value is insignificant. Test the suitability of models in the ANOVA table shows that the significance (p-value = 0.000) $<\alpha$ = 0.05, then reject H_0). The independent variable that is long (FL) Torpedo Scad fish effect simultaneously to the dependent variable (TS value), so regression model is feasible to be used.

Table 4. ANOVA table for TS-length Torpedo Scad (*Megalaspis cordyla*) regression equation.

Source	Degree of freedom	SS	Mean Square	F value	P value
Regression	1	0.584	0.58412	0.11	0.000
Error	33	169.267	5.12931		
Total	34	169.51			

If the relation of the length of the fish to the TS value of each individual is based on the formula TS = 20 Log (FL) + C, then the following calculation applies:
The equation y = 8.35x-68.89
If TS = 20 log (FL) + C, then
20 log (FL) + C = 8.35log (FL) - 68.89
C = 8.35log (FL) - 68.89 - 20 log (FL)

By entering the values of FL in the range of fish measured Torpedo Scad, then obtained variation of value C (table 5). The average C value for fish Torpedo Scad with size 18.0 - 22.5 cm FL is C = -53.68. Foote equation for fish Torpedo Scad (*Megalaspis cordyla*) obtained is TS = 20 Log (FL) - 53.68.

Table 5. C value of FL value per length of fish Torpedo Scad (*Megalaspis cordyla*).

FL (cm)	18	18.5	19	19.5	20	20.5	21	21.5	22	22.5
C (dB)	-54.27	-54.1	-53.99	-53.86	-53.73	-53.61	-53.49	-53.37	-53.25	-53.14

3.3. TS measurement torpedo scad samples in the dead fresh condition

Paired T Test were conducted on fish samples to see if there were differences in TS values resulting from two different treatments. Paired T Test is also used as a comparative test or difference if the data scale of the two variables is quantitative (interval or ratio) [8]. The treatment performed in this research is the calculation of TS fish Torpedo Scad for the conditions in living conditions and the condition of fish in a dead fresh (table 6.). The measuring time range between the treatment of live and dead fish ranges from 2 hours to 9 hours after the measurement of the initial condition (life). Within the span of time, it is expected that fish morphology and elasticity of fish meat is not much different from the initial condition (life).

Table 6. Measurement of TS fish with Torpedo Scad two different treatments.

Variables	Torpedo Scad A	Torpedo Scad B	Torpedo Scad C	Torpedo Scad D	Torpedo Scad E
TL (cm)	22.2	21.3	20.2	23.3	23.8
FL (cm)	20.6	20.9	19	22	20.9
Log FL	1.313867	1.320146	1.313867	1.3424	1.3201
W (g)	94.6	108	94.5	91.1	102.2
L Swimblader (cm)	5	5	5	5.5	5.5
TSavg (dead)	-58.75	-57.98	-57.31	-53.15	-57.85
TSavg (live)	-49.22	-55.32	-58.64	-58.33	-58
Δ t	2:56:00	5:57:00	8:31:00	1:48:00	3:50:00

Δ t = time lapse difference in fish for living conditions and dead fresh condition (within hours)

The hypothesis of this case can be written:
H_0: μ_1-$\mu_0 = 0$
H_1: μ_1-$\mu_0 \neq 0$
$\alpha = 0.05$
H_1 means that the actual difference of the two averages is not equal to zero.

Table 7. Paired T test for fish TS Torpedo Scad Live and Dead Fresh conditions.

	N	Mean	StDev	SE Mean
TS Live	5	- 55.90	3.965	1.77
TS dead fresh	5	- 57.01	2.22	0.99
Perbedaan	5	1.1	5.49	2.45
95% CI for mean difference: (-5.71, 7.92)		T-Test of mean difference = 0 (vs ≠ 0): T-Value = 0.45 P-Value = 0.676		

The p-value of the paired t-test above is 0.676, which is greater than 0.05. Thus, the statistical conclusion we take is THRIVE H_0. This means that the TS measurement values for fish Torpedo Scad living and dead fresh conditions equal to zero. Thus, TS measurements for living conditions and fresh dead conditions have no significant difference.

4. Discussion

4.1. *Empirical relationship of TS value with length of fish*

Based on the results of measuring activities of Target Strength (TS) values for Torpedo Scad fish (*Megalaspis cordyla*) it is possible to perform a simple statistical TS-length analysis. The Central Limit Theorem says that the more observations (in this case the experimental sample unit) the data is getting closer to the normal distribution. The general rule, if the experimental unit is more than or equal to 30 then the data is considered to follow the normal distribution. In this study the experimental unit was 35 (above 30), so the normality test did not become necessary.

Based on table 2, regression equation for TS-length fish formulation is TS = 8.35 Log (FL) – 68.9 with coefficient of determination (R^2 = 0.30%) and R^2 (adj) = 0% (figure 3). The t-test result at the 95% confidence level (table 3) shows that the length (FL) does not significantly affect the TS value. The regression equation model based on the research data is not significant, the linear regression model does not meet the linearity criteria so that this equation cannot be used for prediction or forecasting. Test the suitability of models in the ANOVA table (table 4.) shows that the significance (P_{value} = 0.000) <α = 0.05, then reject H_0. The independent variable that is long (FL) fish Torpedo Scad effect simultaneously to the dependent variable that is TS value, so regression model is feasible to be used.

If the relation of the length of the fish to the TS value of each individual is based on the formula TS = 20 Log (FL) + C, then 20 Log (FL) + C = 8.35 Log (FL) - 68.89. So C = -11, 65 Log (FL) - 68.89. By entering the values of FL in the range of fish measured Torpedo Scad, the variation of the value of C (table 5) is obtained. Thus, the average C value for Torpedo Scad fish with size 18.0 - 22.5 cm FL is C = - 53.68. So we get the equation Foote for Torpedo Scad fish (*Megalaspis cordyla*) is TS = 20 Log (FL) - 53.68.

4.2. Paired T test Measurement TS Torpedo Scad fish

Paired T Test is performed on fish samples to see if there are any differences in TS values resulting from two different treatments. The treatment conducted in this research is the calculation of TS Torpedo Scad fish for living conditions and the condition of fish in a dead fresh (table 6.). The $p_{\text{-value}}$ of the paired $t_{\text{-test}}$ is 0.676, which is greater than 0.05 (α). Thus, the statistical conclusion we take is THRIVE H_0. This means that the TS measurement values for Torpedo scad fish living and dead fresh conditions equal to zero. Thus, TS measurements for living conditions and fresh dead conditions have no significant difference. The condition of the morphology of fish samples measured at live and dead fresh is not much different is very influential on the paired t test results.

5. Conclusion

The measurements of TS values of each individual fish in the same species in this study have shown evidence and reinforce the hypothesis that TS values are a function of the length of the fish. TS value has a linear relationship with the length of the fish, the larger the size of the fish the TS value is greater. The empirical relationship in the form of TS-length for Torpedo Scad fish (*Megalaspis cordyla*) is TS = 20 Log L - 53.68. TS measurements for living conditions and fresh dead conditions have no significant difference.

Acknowledgement

The data provided in this paper is the results from join research collaboration between BPPL-KKP and ITK-IPB under the research project "Measurement of Target Strength Characteristics and TS-Length Formulation on Fish Key Indicators for Improving Accuracy of Fish Stock Estimation TA 2016.

References

[1] Johanesson K A and Mitson R B 1983 *Fisheries acoustic a practical manual for acoustic biomass estimation* (Roma: FAO Fisheries Technical Paper)
[2] Simmonds E J and MacLennan D N 2005 *Fisheries Acoustic: Theory and Practice 2nd ed* (United Kingdom: Blackwell Science Ltd)
[3] Lee D J, Shin H I and Shin H H 1995 Fish stock assessment by hydroacoustic methods and its applications I Estimation of fish school target strength *J. Kor. Soc. Fish Techno.* **31** 142-15
[4] Mun J H, Lee D J, Shin H I and Lee Y W 2006 Fish length dependence of target strength for black rockfish, goldeye rockfish at 70 kHz and 120 kHz *J. Kor. Soc. Fish. Techno.* **42** 30-37
[5] Korneliussen R J, Heggelund Y, Eliassen I K and Johansen G O 2009 Acoustic species identification of schooling fish *ICES J. Mar. Sci.* **66** 1111-8
[6] Foote K G 1982 Optimizing copper spheres for precision calibration of hydroacoustic equipment *J. Acoust. Soc. Am.* **71** 742–747
[7] Hannachi M S, Abdallah L B and Marrakchi O 2004 Acoustic identification of small-pelagic fish species: target strength analysis and school descriptor classification *MedSudMed Technical Documents No.5*
[8] Walpole R E 1995 *Pengantar Statistika Edisi Ke-3* (Jakarta: Gramedia Pustaka Utama).

IOP Conf. Series: Earth and Environmental Science **176** (2018) 012028 doi:10.1088/1755-1315/176/1/012028

The effect of atmospheric correction on object based image classification using SPOT-7 imagery: a case study in the Harapan and Kelapa Islands

V P Siregar[1], N W Prabowo[2*], S B Agus[2], T Subarno[2]

[1]Department of Marine Science and Technology, Faculty of Fisheries and Marine Sciences, Bogor Agricultural University, Bogor 16680, Indonesia
[2]Graduate School of Marine Technology, IPB Dramaga Campus, Bogor Agricultural University, Bogor 16680, Indonesia
*e-mail: prabowonico95@gmail.com

Abstract. Object-based classification (OBIA) has become a new paradigm in remote sensing data analysis. Errors in interpretation of benthic habitat classification image can be caused by deviation of values in reflectance. The absorption, scattering, and atmospheric reflections caused differences between the reflectance value of satellite images and real objects. The author have tried to compare two different approaches of OBIA classification with FLAASH atmospheric correction and OBIA classification without atmospheric correction. In this research, atmospheric correction with FLAASH method on SPOT-7 imagery conducted to eliminate atmospheric effect on the image. The Support Vactor Machines (SVM) algorithm is selected for the classification of benthic habitat in Harapan and Kelapa Island. Atmospheric-corrected image was analyzed visually then compared to ortho image in reflectance pattern, object segmentation and classification result. Results showed an improvement of visualization and reflectance pattern in shallow water objects. This study concluded that the FLAASH method as one of the appropriate methods to be used in atmospheric correction in SPOT-7 images.

1. Introduction

The research of marine ecosystem mapping, with spatial scaling and object feature labels mapped in accordance with the geomorphic structure and benthic communities are still a major challenge in the development of remote sensing applications. Benthic habitat mapping using satellite imagery along with the accuracy assessment has been widely performed [1,2,3,4,5,6,7,8,9]. This research aims to compare the object-based image classification result with corrected and uncorrected image. These studies focused on the differences between accuracy derived from the use of hyperspectral and multispectral images; between high and low resolution images; between uncorrected images and corrected image with water coloumn correction [1,2].

Errors in interpretation of benthic habitat classification results can be caused by deviation values of reflectance values. This is due to absorption, scattering, and atmospheric reflections that cause differences in the reflectance value of satellite images and actual objects [10]. The reflectance value is very important in the process of object classification on the image digitally. The atmospheric correction able to eliminate the influence from the atmosphere and return the reflectance value according to the actual object's reflectance value on the earth's surface.

The method of atmospheric correction consists of three methods: radiative transfer, relative correction based on image characteristics and surface linear regression. Among these methods, radiative transfer models are more widely used in satellite imagery with higher accuracy of reflectivity calculations [11]. Methods with radiative transfer algorithms such as the 6S (Second Simulation of the Satellite Signal in the Solar Spectrum) and MODTRAN (Moderate Spectral Resolution Atmospheric Transmittance Algorithm and Computer Model) methods [12].

The MODTRAN4 method is available on the ENVI software with the FLAASH (Fast Line-of-sight Atmospheric Analysis of Spectral Hypercubes) modul. FLAASH is an atmospheric correction tool using the MODTRAN4 method which has been able to extract visible, NIR (Near Infrared) and SWIR (Short-wave Infrared) wavelength. FLAASH can eliminate the effect of atmospheric disturbances by obtaining more accurate parameters of reflectivity, emissivity, surface temperature and physical surface. FLAASH has an aerosol and mean value retrieval method using a dark pixel reflectance ratio based on Kaufman's research [13]. The MODTRAN4 model contained in FLAASH reduces the effect of the atmosphere effectively on SPOT-7 thereby enhancing the image information with better accuracy than the QUAC model [14]. The author will try to compare two different approaches of OBIA classification with FLAASH atmospheric correction and OBIA classification without atmospheric correction.

2. Method

2.1. Survey area

The study area is located in shallow water area of Harapan and Kelapa Island, Kep. Seribu, Jakarta (Fig 1.). Collection of benthic habitat data was done by systematic random sampling method with distance of each sampling point about 20 m.

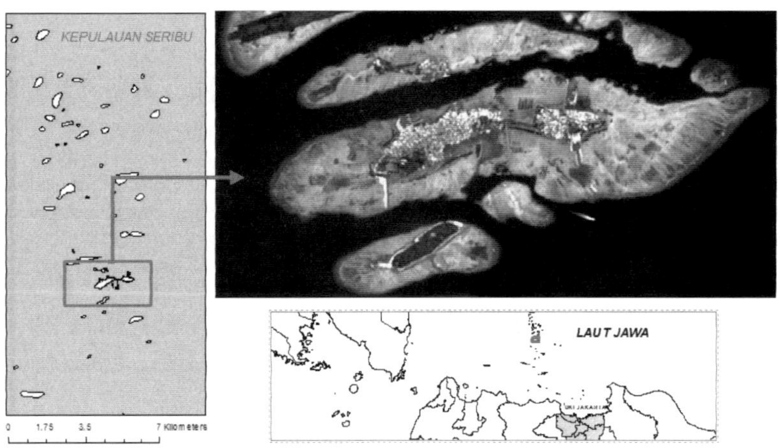

Figure 1. Field survey location (data source: SPOT 7 acquisition image dated June 12, 2016) (5°32'50.64"S - 5°47'21.48"S and 106°30'59.4"E - 106°37'53.4"E).

2.2. Data acquisition

The data sources used are multispectral images (SPOT-7 Kepulauan Seribu), and benthic habitat data sampling. SPOT-7 multispectral image is acquired on June 12, 2016 with a projection system of UTM coordinate zone 48S WGS84. The SPOT-7 image characteristic consists of 4 multispectral bands (Blue, Green, Red, NIR) with a spatial resolution of 6 meters (multispectral). Field data were collected in April 2017 using photo transect technique (1 x 1 meters) with 343 observation point. Benthic habitat data is integrated (spatial join) with polygon of image segmentation result, integration result then used as region of

IOP Conf. Series: Earth and Environmental Science **176** (2018) 012028 doi:10.1088/1755-1315/176/1/012028

interest (ROI) in process of classification. The description of the habitat classes is based on the percent of the benthic component cover of the 1x1 m squared photo field observation. The description of the classification scheme is derived from the percentage value of eight benthic components based on the Brail-Curtis coefficient insignificant value of 40% (AHC). The value of dissimiliarity (40%) indicates that each habitat class constructed has a resemblance to the benthic habitat component of at least 60%.The classification scheme was then developed using the percentage of benthic habitat closure data by eliminating components having presence frequency based on centroid values of less than 4% [14].

2.3. Pre-processing image

Pre-processing image consisted of atmospheric correction process was performed using the FLAASH correction module contained in ENVI 5.1 software [15]. FLAASH is a first-principles atmospheric correction tool that corrects wavelengths in the visible through near-infrared and a shortwave infrared region, up to 3 mm. It incorporates the MODTRAN4 radiation transfer code [16, 17]. The input image for FLAASH must be a radiometrically calibrated radiance image. Other information such as flight date, start time, time in GMT, scene center location, sensor altitude, and ground elevation is also required for this correction. Atmospheric correction aims to eliminate the influence of atmospheric particles such as dust and water vapor. This atmospheric correction stage generates new images that have been corrected from atmospheric disturbances.

$$L = \left(\frac{A\rho}{1-\rho_e S}\right) + \left(\frac{B\rho_e}{1-\rho_e S}\right) + L_a \qquad (1)$$

Where is: L = the spectral radiance from sensor; P = the surface reflectance from pixel; S = the albedo in the atmosphere; A, B = the coefficient from atmospheric and geometric condition; ρ_e = the average of surface reflectance; L_a = the radiance of atmospheric reflectance; $\left(\frac{B\rho_e}{1-\rho_e S}\right)$ = the radiance energy from object scatter; $\left(\frac{A\rho}{1-\rho_e S}\right)$ = the radiance energy from object reflectance

$$L_e = \left(\frac{(A+B)\rho_e}{1-\rho_e S}\right) + L_a \qquad (2)$$

L_e = the average of image radiance

The calculation to convert Digital Number to reflectance for Blue, Green, Red and NIR bands in SPOT 7 imagery:

Reflectance = DN x Gain – Bias

Band	Gain	Bias
Band-1/Blue	1.66345	0
Band-2/Green	1.559910	0
Band-3/Red	1.085514	0
Band-4 NIR	10.468422	0

The image used in the study belongs to latitude (-5.5474, -5.7893), longitude (106.5165, 106.6315), recording date of SPOT-7 (June 12, 2016), recording time (08:42 : 38.81), height of sensor (695 km), ground level (0.1 km), pixel size (6 meters), tropical model, aerosol model (maritime), visibility (40 km), solar azimuth angle (121.38 °) and the sun elevation angle (52.84°). Visual analysis is done by taking the same area in the image before and after the atmospheric correction, observed the visible color changes of the image. The reflectance pattern analysis is performed by taking the same pixel on shallow water objects, deep water, and terrestrial areas (vegetation) on images before and after atmospheric correction. Then the FLAASH corrected image is used as the input image layer of object-based classification (OBIA) and then compared the resulting classification accuracy. The reflectance pattern image (before and after correction) compared with reference to USGS's reflectance pattern.

2.4. Object based image analysis
Object-based classification consists of two stages: segmentation and classification. Segmentation is the concept of building objects / segments from pixels into segments or objects that have the same properties [18]. The segmentation algorithm used in this research is multiresolution segmentation (MRS). This algorithm starts with a single pixel segment and combines neighboring segments until the threshold of heterogeneity is reached. The image segmentation process depends on three components, shape, compactness, and scale [19]. Segmentation is done using shape and compactness parameters of 0.1 and 0.5, respectively.

2.5. Image classification
The image used as the input layer in the classification process is an image that has been corrected and uncorrected image. The rule set classification was created in tree process component. The classification algorithm used is the support vector machines (SVM) algorithm. SVM is a supervised classification algorithm that can search a vector or line that serves as a two-class separator by maximizing margins between classes [19]. This classification algorithm is based on line factor classifiers classified machine learning classification [20].

2.6. Accuracy assessment
The results of image classification accuracy testing was conducted to determine the accuracy of each thematic maps. Accuracy was obtained by comparing the results of image classification with field observations. The accuracy test is performed using an error matrix (confusion matrix). Testing accuracy produce overall accuracy (OA), producer accuracy (PA), user accuracy (UA), and the value of kappa statistics [14]. Kappa value statistics and the data variance was used to compare the accuracy of the test results of two thematic maps in order to know the significance of changes in thematic map accuracy after sunglint correction (Z statistic).

3. Results and discussion

3.1. Atmospheric correction image
The images show visually different images before and after atmospheric correction (Fig. 2). The results of atmospheric correction produce a new image that has been free from atmospheric disturbances. The atmospheric corrected image is seen more clearly, especially on shallow water areas. The application of atmospheric correction has proven to increase the contrast of the image.

Based on Figure 2, there is a change in the reflectance pattern in the corrected image at the object in the shallow water (on reef flat zone) area. There was a significant increase reflectance pattern in blue band (450-520 nm) and green (530-590 nm), while the red band reflectance pattern (625-695 nm) and near infrared (760-890 nm) did not change. As for the deep water area, the reflectance pattern shown did not change significantly. The lagoon area also significant increases reflectance patterns in blue bands (450-520 nm) and green (530-590 nm). While on the vegetation object visible there is a significant decrease of reflectance pattern on blue band (450-520 nm) and green (530-590 nm), while the red band (625-695 nm) and near infrared (760-890 nm) reflectance pattern are changed (table 1).

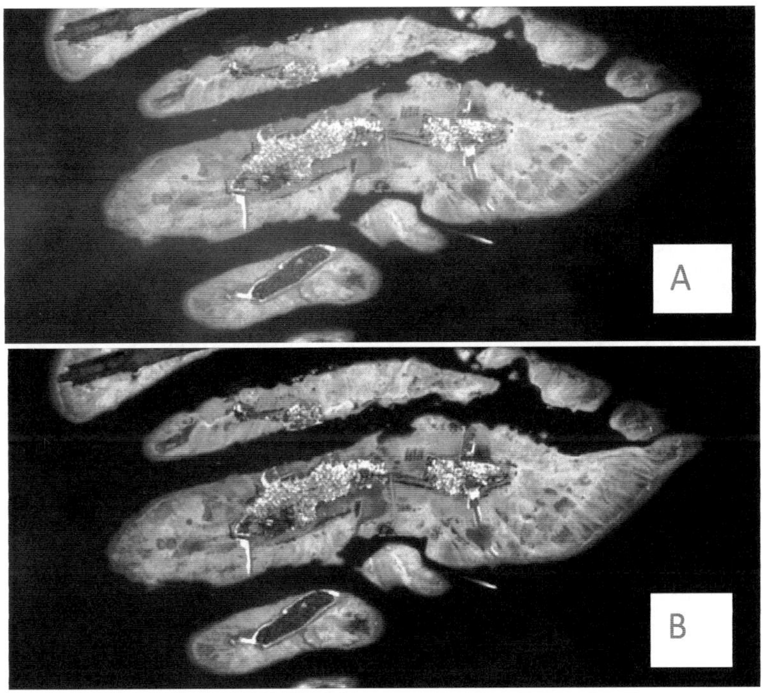

Figure 2. Comparison image before correction (a) and after atmospheric correction (b) (composite band 321).

Shallow water areas have almost the same spectral response for each objects included. Here is a picture of the spectral pattern from each object found in the field survey based on direct measurements using a spectrometer. The image shows that the spectral pattern of benthic habitat object has the same pattern as the spectral increase in the blue band and the green band. However, the spectral measurements from survey are highly dependent on the sun, the turbidity of the waters and the weather conditions [11].

IOP Conf. Series: Earth and Environmental Science **176** (2018) 012028 doi:10.1088/1755-1315/176/1/012028

Table 1. Comparison of spectral patterns between uncorrected and corrected image.

<table>
<tr><th>Object</th><th>Spectral pattern (uncorrected image)</th><th>Spectral pattern (corrected image)</th><th>References image (USGS)</th></tr>
<tr><td>Shallow water
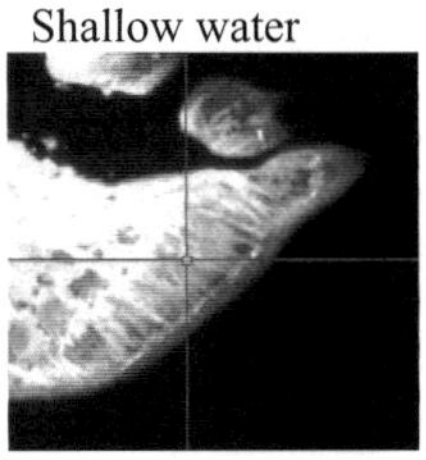</td><td>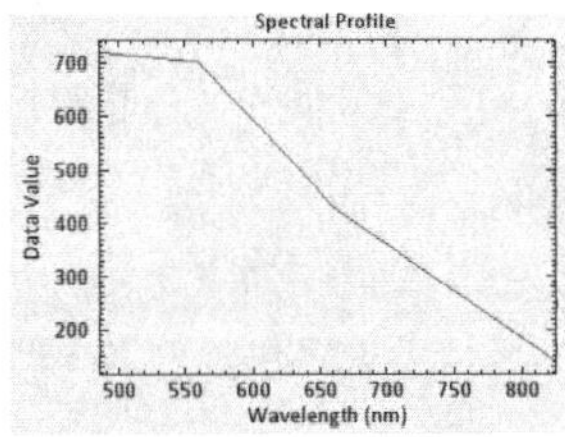</td><td>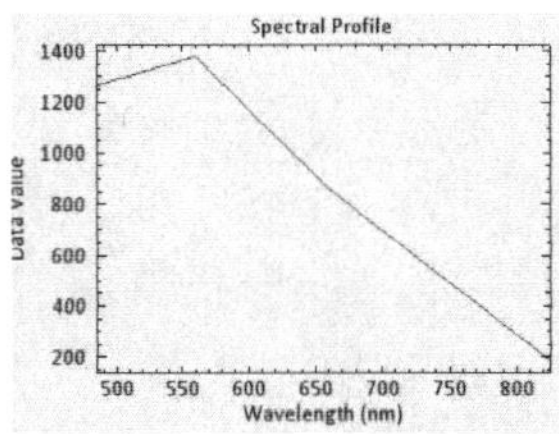</td><td>Seawater_Coast_SW1[W1R1Ba AREF]
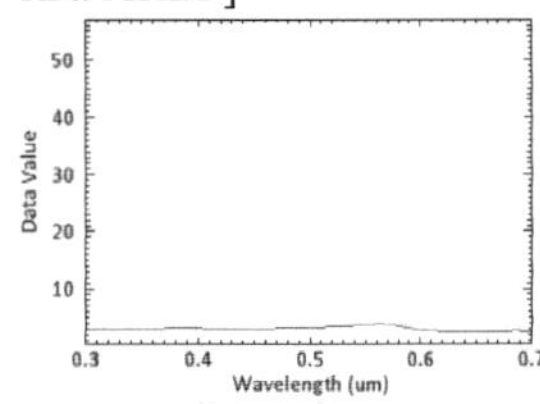</td></tr>
<tr><td>Karang Mati Alga
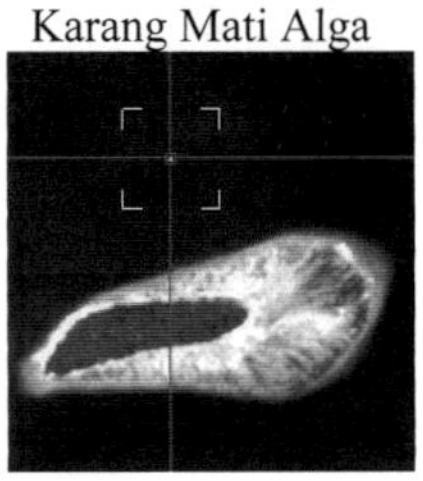</td><td>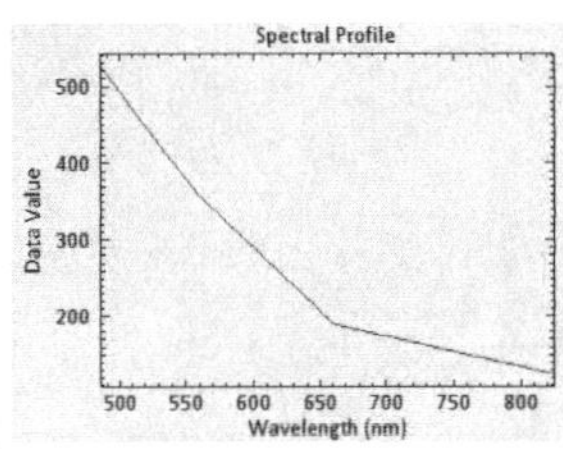</td><td>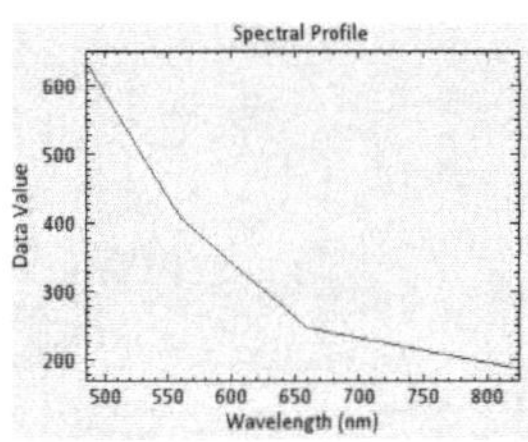</td><td>Ocean_SW2[W1R1Ba AREF]
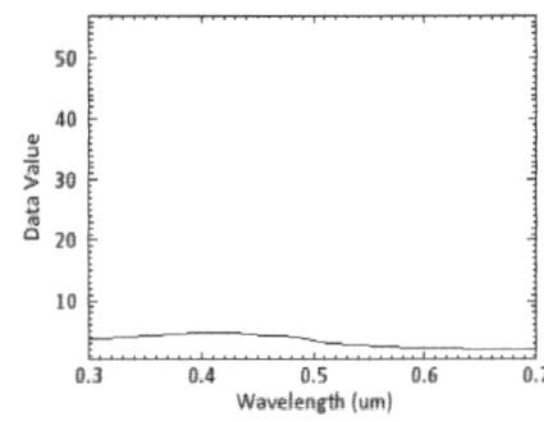</td></tr>
<tr><td>Pasir Rubble Alga
</td><td>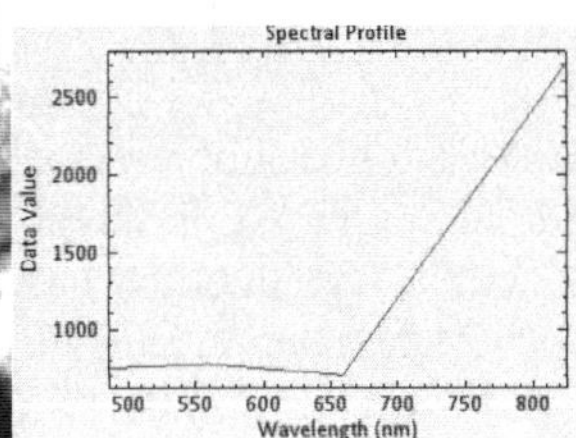</td><td>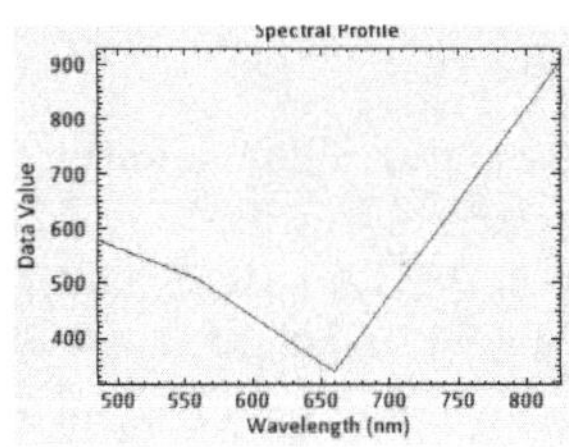</td><td>Vegetation_Dry Grass [W1R1Ba AREF]
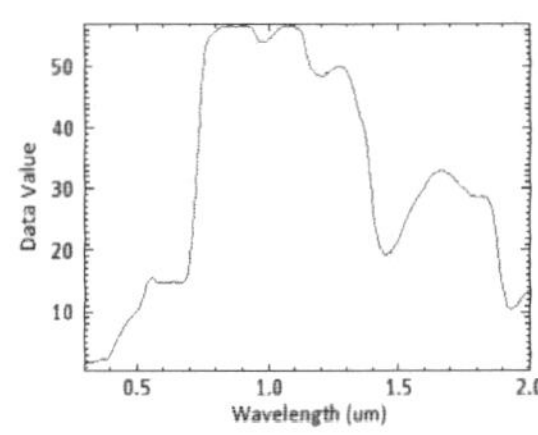</td></tr>
</table>

3.2. Object Based Image Classification (OBIA)

3.2.1. Scale optimization

The optimization of object-based classification segmentation refers to the results of previous research [21]. Image segmentation on each applied scale produces a different number of polygons. Scale is the average size of the object generated in the segmentation process [19]. The optimal scale parameter value is 15, and at this value the number of generated classes is quite detailed i.e. 11 classes with the highest OA (76.85%). The image shows the difference of the object segment formed from the image before and after the atmospheric correction with the same segmentation parameter values. The corrected image has a segment shape that more closely resembles with the actual object pattern in the field (Fig. 3). This is can caused by the atmospheric correction can increase the contrast of objects in the image so that differences between objects are more visible. The uncorrected image show that each segment relatively homogen and the corrected image show that the variability in each segment rather high.

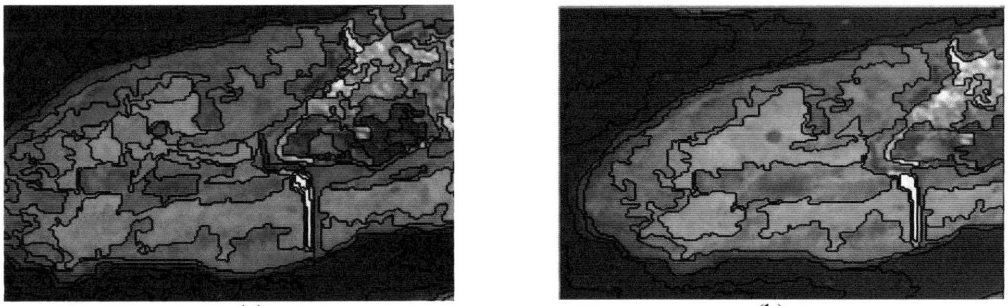

(a) (b)

Figure 3. Visual comparison of image of segmentation result before (a) and after (b) atmospheric correction (image composite 321).

3.2.2 Classification of benthic habitat

The classification method used is the SVM classification with scale factor 15 [21]. The object-based classification shows the benthic habitat distributed in shallow waters of Harapan and Kelapa Island (Fig. 4, Fig. 5). The characteristics benthic habitat components were identified as 8 habitats, live coral (C), sponge (SP), softcoral (SC), macroalgae (MA), dead coral algae (DCA), sand (P), rubblel R), and seagrass (L). The description of the classification scheme is further derived from the percentage value from 8 benthic components based on the Bray-Curtis coefficient insignificant value of 40%. There is no provision for using similarity values to define a classification scheme based on clustering analysis. It is caused by different conditions and variations of observation site and is adapted to the satellite imagery platform used. Then the classification scheme was developed using the percentage of benthic habitat cover by eliminating centroid value components less than 4% [14]. The 11 benthic habitat classes were obtained based on cluster analysis results.

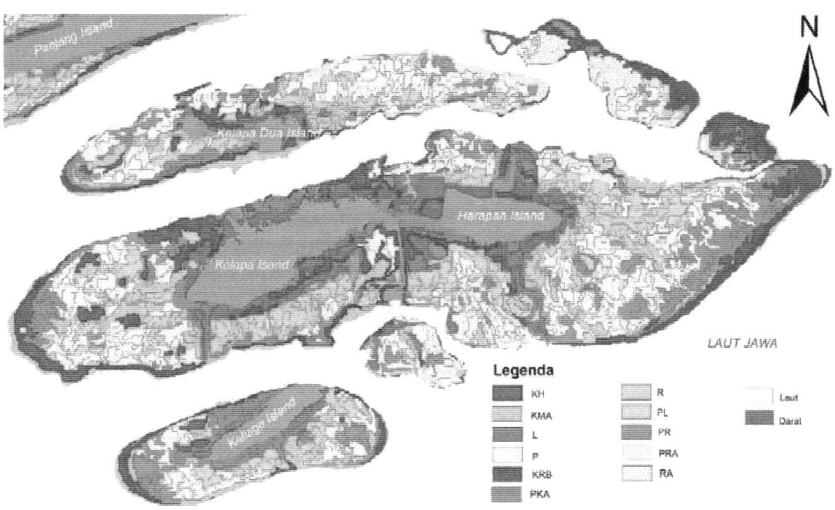

Figure 4. Classification of benthic habitat mapping using uncorrected image with SVM algorithm.

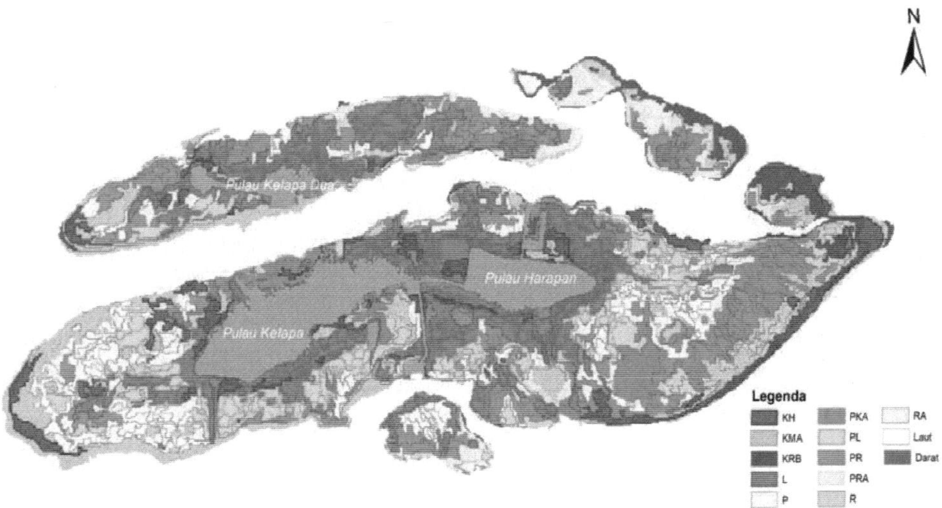

Figure 5. Classification of benthic habitat mapping using corrected image with SVM algorithm.

The 11 benthic habitat classes formed are sand + rubble + algae (PRA) dead coral + algae (KMA), live coral + rubble (KRB), seagrass (L), sand (P), sand + seagrass (PL), sand + live coral (PKA), sand + rubble (PR), live coral (KH), rubble R) and rubble + algae (RA). The naming of these classes is based on a dominant benthic closure type entirely constructed by benthic components. The application of the classification scheme is influenced by the complexity of the constituent components of coral reefs and the limitation of remote sensing data [22]. The object of 11 classes that have been identified are further used as input data for the classification process.

The classification result shows the uncorrected, the sand habitat (66 Ha) shown by the yellow and mixed class of sand and seagrass (53 Ha) shown by light green dominates in the north, west and east of the Harapan Island. Meanwhile, in the corrected image, the benthic habitat dominated by seagrass habitat shown by dark green in the north and sand habitat dominates in the west and then mixed rubble and sand class shown by pink color dominates the east area.

3.3. Accuracy assessment

Comparison of image classification before and after atmospheric correction looks different from comparison of its accuracy. The confusion matrix using 237 sampling point to produce the accuracy assessment. Image classification accuracy test results before and after atmospheric correction produce an overall accuracy with about 69.43% and 76.68% (Table 2). While at UA and PA between 18-77%. The overall accuracy of the images after atmospheric correction increased with about 7% compared to the uncorrected image. The spectral similarity between coral reef habitat classes is unavoidable. The result of atmospheric correction is able to increase the accuracy of the mapping when compared with the image without correction, although there are two possibilities on the reflectance value of underestimate and overestimate caused by aerosol parameters that are less suitable with field conditions during image recording [17].

Table 2. Comparison of accuracy assessment (PA, UA, OA) before and after correction.

Class	Before		After	
	PA	UA	PA	UA
Seagrass	54.84	54.84	69.70	77.19
Dead coral with algae	37.50	28.57	58.82	47.62
Sands	70.59	73.47	69.39	69.39
Live coral	66.67	69.57	70.83	73.91
Coral with rubble	58.82	58.82	72.22	76.47
Sand with rubble	42.86	60.00	43.48	66.67
Sand with seagrass	64.10	71.43	62.50	57.14
Sand-coral-algae	45.45	50.00	50.00	30.00
Rubbles	40.00	37.50	47.37	56.25
Sand-rubble-algae	25.00	22.22	45.45	55.56
Rubble with algae	50.00	18.18	80.00	36.36
OA	69.43		76.68	

Significance test of the image accuracy before and after atmospheric correction was about 1.73. The value of the Z statistic between -1.96 to 1.96 mean that the comparison was not significant [15]. It means that the application of atmospheric correction on the images before and after correction of this study did not produce significant differences. The OBIA classification on corrected and uncorrected image was not significantly different based on Z statistic value of 1.73.

4. Conclusion

The atmospheric correction process produce a new image that has been almost free from atmospheric disturbances. The atmospheric correction by FLAASH method can be used to improve the shallow water object reflectance pattern on SPOT-7 images according to reference data. There is a visual difference in OBIA classification results from corrected and uncorrected image with the FLAASH Module. This research shows that the atmospheric correction made in pre-processing image able to the accuracy although the results were not significantly different.

Acknowledgments
The author would like to thank the Ministry of Reaserch, Technology and Higher Education (KEMENRISTEKDIKTI), for financially supporting this research through BOPTN (PUPT) research scheme 2016. This research is also supported by Mapping and Spatial Modelling Laboratory Department of Marine Science and Technology, IPB.

IOP Conf. Series: Earth and Environmental Science **176** (2018) 012028 doi:10.1088/1755-1315/176/1/012028

References

[1] Mumby P J and Edwards AJ 2002 Mapping marine environments with IKONOS imagery: enhanced spatial resolution can deliver greater thematic accuracy *Remote Sensing Environmental* **82** 248-257

[2] Andrefouet S *et al.* 2003 Multi-site evaluation of ikonos data for classification of tropical coral reef environments *Remote Sensing of Environment* **88** 128-143

[3] Andrefouet S and Guzman H 2005 Coral reef distribution, status a geomorphology biodiversity relationship in Kuna Yala (San Blas) archipelago, Caribbean Panama *Coral Reefs* **24** 31-42

[4] Siregar VP 2010 Pemetaan substrat dasar perairan dangkal karang congkak dan lebar kepulauan seribu menggunakan citra satelit quick bird *Jurnal Ilmu dan Teknologi Kelautan Tropis* **2** 19-30

[5] Congalton RG and Green K 2009 Assessing the accuracy of remotely sensed data principles and practices (second edition)

[6] Phinn S R, Roelfsema C M and Mumby PJ 2011 Multi-scale object-based image analysis for mapping geomorphic and ecological zones on coral reefs *International Journal of Remote Sensing* **33** 3768-3797

[7] Siregar V P *et al.* 2013 Pemetaan habitat dasar dan estimasi stok ikan terumbu dengan citra satelit resolusi tinggi *Jurnal Ilmu dan Teknologi Kelautan Tropis* **5** 453-463

[8] Wahiddin N, Siregar V P, Nababan B, Jaya I and Wouthuyzen S 2015 Object-based image analysis for coral reef benthic habitat mapping with several classification algorithms *Procedia Environmental Sciences* **24** 222-227

[9] Anggoro A, Siregar V P and Agus A 2015 Pemetaan zona geomorfologi ekosistem terumbu karang menggunakan merode obia, studi kasus pulau pari *Jurnal Pengindraan Jauh* **12** 1-12

[10] Fibriawati L 2016 Koreksi atmosfer citra SPOT-6 menggunakan metode MODTRAN4. Posiding Seminar Nasional Penginderaan Jauh

[11] Gao B C *et al.* 2006 A review of atmospheric correction techniques for hyperspectral remote sensing of land surfaces and ocean colour In: Geoscience and Remote Sensing Symposium p 1979e1981

[12] Guo Y and Zeng F 2012 Atmospheric correction comparison of SPOT-5 image based on model FLAASH and model QUAC In: International Archives of the Photogrammetry, Remote Sensing and Spatial Information Sciences **39** ISPRS Congress, 25th August - 1st September 2012 Melbourne Australia

[13] FLAASH 2009 Atmospheric correction module quac and flaash user's guide version 4.7 ITT Visual Information Solution Inc

[14] Green EP *et al.* 2000 Remote sensing handbook for tropical coastal management: UNESCO

[15] Yunkai and Fan Z 2012 Atmospheric correction comparison of SPOT-5 image based on model flaash and model quac, international archives of the photogrammetry *Remote Sensing and Spatial Information Sciences* **39** 7-14

[16] Agrawal G, Sarup J and Bhopal M 2011 Comparison of QUAC and FLAASH atmospheric correction modules on EO-1 Hyperion data of Sanchi *International Journal of Advanced Engineering Sciences and Technologies* **4** 178e-186

[17] Rani N, Singh T and Mandla RV 2015 Comparison of atmospheric correction of Hyperion image using FLAASH and QUAC methods In: Rao, Madhava V, Rao, Kesava P, Kumar Phanindra, T, Murthy, DSR (Eds), Geoinformatics Applications in Rural Development Professional Books Publisher, Hyderabad, ISBN 978-81-909728-9-5,452e259

[19] Trimble 2014 Ecognition developer: User guide Munich, Germany Trimble

[18] Navulur K 2007 Multispectral image analysis using the object-oriented paradigm Taylor & Francis Group LLC

[20] Nugroho AS, Witarto AB and Handoko D 2003 Support vector machine teori dan aplikasinya dalam bioinformatika *Ilmu Komputer* **20** 120-126

[21] Siregar V P *et al.* 2016 Mapping of shallow water benthic habitat on reef flat zone using object based and pixel based image classification in Harapan-Kelapa Islands, Kepulauan Seribu. *Proceeding of 3rd International Symposium LISAT*
[22] Mumby P J *et al.* 1999 The cost-effectiveness of remote sensing for tropical coastal resources assessment and management. Journal of Environmental management **55** 157-166

Formulating marine sanctuary designation for sustainable management of blue swimming crab in Lancang Island, Kepulauan Seribu

Syamsul Bahri Agus*, **Tarlan Subarno, Adriani Sunuddin**

Department of Marine Science and Technology, Faculty of Fisheries and Marine Science, Bogor Agricultural University, Indonesia
*e-mail: allanawani@gmail.com

Abstract. Lancang Island is the center of fisheries of blue swimming crab in Kepulauan Seribu, Jakarta. Most of the inhabitants in this island are fishermen, and blue swimming crab is the main target without considering the stocks. One of the management effort of blue swimming crab by the government is the restriction of catchment size. To support the government policy and to maintain the population stocks and the sustainability of blue swimming crab around Lancang Island, we need another effort such us proposing area of Marine Sanctuary. The analysis of possibility area of the marine sanctuary was performed with simulated annealing using Marxan. The datasets were used in analysis included geomorphic model of habitat (sea floor), coastal ecosystems, blue swimming crab catches (sample size, location, and fishing pressure). The results obtained three possibility locations of the marine sanctuary, which is around of Laki Island, Southern, and Southeastern of Lancang Island. From the possibility locations, we recommend the first location as the Marine Sanctuary with considering the more diverse of features represented.

1. Introduction

Kepulauan Seribu is a seascape that forming up a group of islands and consist of more than 100 islands. Geographically, Kepulauan Seribu is located in the north of Java Island and administratively included within DKI Jakarta Province. Some of the area in northern of Kepulauan Seribu is a conservation area, known as Taman Nasional Kepulauan Seribu (TNKpS) and managed by Ministry of Environment and Forestry. Most of the islands are uninhabited, and only 11 islands are occupied, one of which is the Lancang Island located in southern of Kepulauan Seribu.

Most of the islands outside of core zone of the conservation area in Kepulauan Seribu have become major tourist destinations. Sebira Island in the most north and Lancang Island in the southern of Kepulauan Seribu are the two inhabitant island that remains to be the fisheries center in Kepulauan Seribu besides as the tourist destination. Meanwhile, most of the other inhabitant islands have relied on tourism as the main economy drives. In Lancang Island, the catching of fishery commodity which is currently the mainstay is tini fish, squid, and blue swimming crab.

The blue swimming crab (*Portunus pelagicus*) is one of the fishery resources with high economic value. High demand coupled with high price became one of the driving factors of exploitation intensively, especially around of Lancang Island. On the other hand, the government has issued regulations by restricting the conditions and captured sizes (PERMEN-KP/No.1/2015). Among the objectives of the regulation is to restrict population decline and to maintain the availability of blue swimming crab resources in nature. One of those regulated in the policy is the limitation of catchments sizes, where the allowed sizes of blue swimming crab are at least 10 cm.

The concept of sustainable of blue swimming crab fisheries management needs to be implemented in the Lancang Island. The management concept what is meant is the proposed the designation of a marine sanctuary in around of Lancang Island. Some others inhabitant islands in Kepulauan Seribu have established marine sanctuaries in 6 sites and managed by the local community. The marine sanctuaries were established to preserve the ecosystems especially coral reefs from the massive pressure of tourism in Kepulauan Seribu. Similar to the six existing marine sanctuaries in other inhabitant islands, in Lancang Island is necessary to propose and establish some of the environment area as the marine sanctuary to preserve the marine resources. One of important benefit if the marine sanctuary can be established, is expected to maintain the preservation of blue swimming crab fisheries, so that it can remain a major economic driver in Lancang Island. To determine the suitable location of the designation of the marine sanctuary, it is necessary to integrate geospatial aspects and systematic analysis. In this study, we integrate some geospatial information covering coastal ecosystems, geomorphological features of the seafloor, and geospatial aspects of blue swimming crabs.

The utilization of remote sensing technology to obtain information in coastal and marine is one way out to gather the required spatial information, including as one of the inputs in feasibility assessment of marine sanctuary. Spatial information on the distribution of mangrove, seagrass, and coral reefs can be derived from satellite imagery with several approaches [1, 5]. The characteristics of seafloor such as the geomorphological zones can be obtained from the modeling of the benthic habitat [6, 8]. The identifying of potential sites for a marine sanctuary, there is a well-known model or software and has been widely used, Marxan. Marxan is a tool that helps in decision making, one of its functions is to identify conservation sites, including marine conservation [9, 10].

2. Materials and Method

2.1. Study site

The research was conducted in the around of Lancang Island; there are three small islands are included in the study site, Lancang Island, Laki Island, and Bokor Island. Administratively the study location is included in District Administration of Kepulauan Seribu – Jakarta, and geographically located in the northern of Tangerang – Banten. The Lancang Island is one of the inhabited islands in Kepulauan Seribu, while the other two islands in this study site are uninhabited islands.

The total area of study site is about 154.88 Km^2 from western of Laki Island to the eastern of Bokor Island (figure 1). The study location was then divided into 4.84 km^2 of the spatial grid, and resulting 32 square grids (A1 – D8). The construction of this square grids was done to facilitate the collection of spatial information through participatory mapping and converting of non-spatial to spatial datasets.

2.2. Data collections

The datasets used in this study include primary and secondary data collected from several sources. The kinds of datasets used are presented in Table 1.

Table 1. The kinds and sources of datasets.

The kinds datasets	Data Sources
Fishing ground	Participatory mapping
Fishing pressure	Participatory mapping
The catchment size	Sample of catchment measurement
Water depth	Survey sounding
Bathymetric map	Indonesian Naval
Tidal	In-situ measurement (using moritide)
WorldView-2 imagery	Digital Globe
Distribution of coastal ecosystems (mangrove, seagrass, and reefs)	Imagery analysis

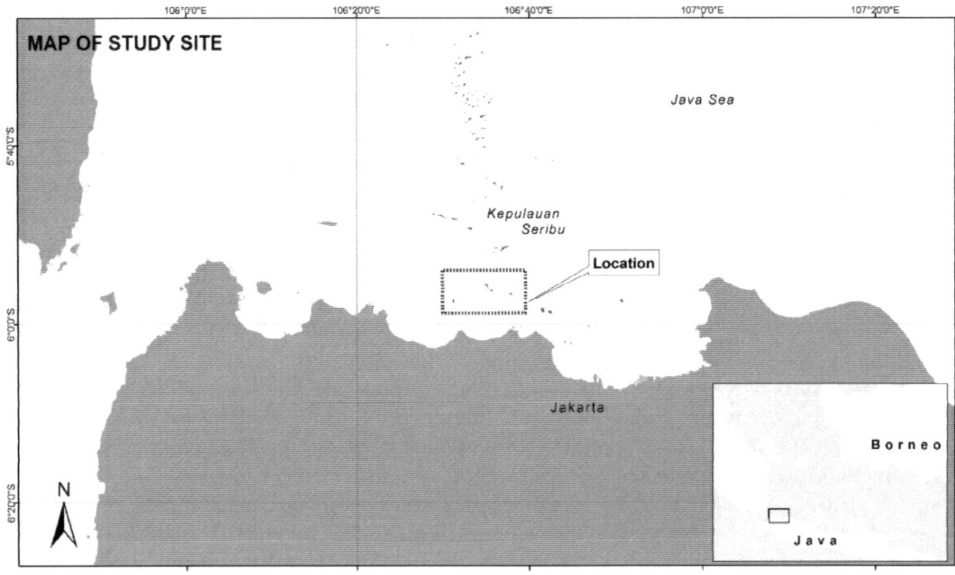

Figure 1. The study site located in southern of Kepulauan Seribu.

Data collection was conducted from 2015 to 2016, covering the entire fishing/monsoon season. Data collected are catchment and environmental data of fishing ground. The catchment data includes fishing ground, fishing intensity (fishing pressure), catchment sizes, catchment weight, and berried female. Identification of fishing ground was done through participatory mapping with fishermen when landing their catchment and conducted by ground check of the fishing ground using Global Position System (GPS). To facilitate the identification of fishing ground, the questionnaire was completed with a map of study site that has been divided into 32 grids with 4.84 km^2. The sampling of the catchments was done during the interview to measure the carapace, weigh of samples, and to identify the berried female.

Environmental data include water depth and coastal ecosystems. The depth data was obtained from a combination of the results from the survey and the secondary data of nautical chart from the Indonesian Navy Hydrographic. Tidal measurements for the depth correction of survey results were conducted using marine automatic tidal detection (Moritide).

2.3. Data analysis

Data analysis includes processing of satellite imagery, geomorphological modeling of benthic habitats, analysis of catchment data, and analysis the potential area of the marine sanctuary. Preparation and data processing in this research are illustrated as presented in figure 2.

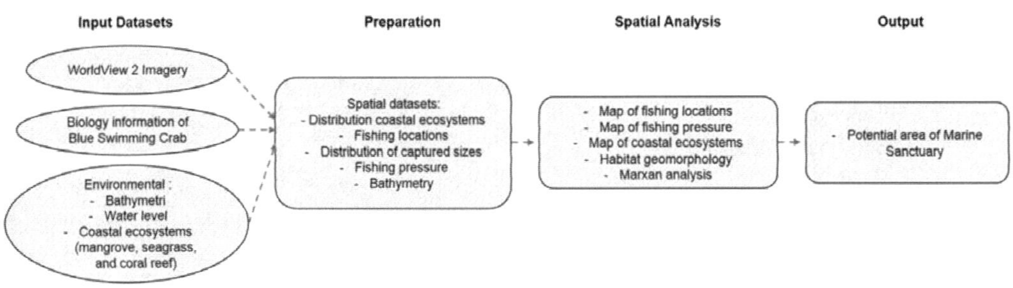

Figure 2. Workflow of preparation and data processing.

2.3.1. Spatial distribution of Blue Swimming Crab. Tabular data, i.e., catchment data then mapped using a quadratic grid, the characteristics of grid map have been described in the data collection section. Tabular data results then converted into spatial data and then obtained the spatial distribution of fishing ground of blue swimming crab for four monsoons. The location of fishing ground obtained within a year period, then grouped into four monsoons. The classification of the fishing season based on the terms by local fishermen, namely the southeast monsoon, alternate one monsoon, northwest monsoon, and alternate two monsoons. The nonspatial data obtained based on the measurements of the catchment samples was converted to attribute data to complement the fishing ground information of the identification resulted from participatory mapping and field checks. As such, from the map of the distribution of fishing ground then obtained information on the distribution of catchment sizes, distribution of the berried female, and the difference of fishing intensity in each grid.

2.3.2. Image processing. Image processing was conducted to derive spatial distribution of coastal ecosystems, including mangroves, seagrasses, and coral reefs on three small islands in the study site. The imagery used was WorldView-2, with the acquisition in June 2015. The classification method was supervised classification using maximum likelihood (MLC) algorithm. The MLC algorithm is one of the common algorithms used in remote sensing data processing [11]. The interpretation of imagery was done by referring to the information gathered from the field survey.

2.3.3. Modeling of benthic geomorphology. The seafloor around of Lancang Island was analyzed to obtain its geomorphological characteristics. Geomorphological characteristics were used as one of the inputs in determining the priority location as the marine sanctuary. The benthic geomorphological characteristics were then grouped into slope, depression, crest, and flat. The modeling of geomorphological characteristics was performed using Benthic Terrain Modeler (BTM) [6, 8, 12]. BTM analysis used bathymetry data as input in the analysis process, in this study the input analysis was the integration between the survey data and the data extracted from the nautical map.

2.3.4. Planning scenario to identify area for marine sanctuary. Marxan has been widely used as spatial analysis tools to identify potential sites for conservation areas, including marine protected area [10, 13, 14]. In this study, Marxan was used to identify potential sites as a marine sanctuary, to maintain the preservation of fishery resources around Lancang Island, especially blue swimming crab. The first step in analyzing to identify location was to divide the study area into planning unit (pu). The study area was divided into the hexagonal shape of pu with the size of each pu was 25980.76 m^2, the number of pu produced by that size was 6187 units. The size of the pu was adjusted to the study location and the representation of input parameters into each pu.

The analysis was done by simulating annealing with interactive improvement, as its marxan work, through 1000 repetitions [9, 15]. Target features and cost are factors that greatly affect the results of the marxan analysis. The target features used were ecosystems distribution (mangrove, seagrass, and coral reefs), a map of fishing ground (size and condition of berried females), and habitat geomorphology. One of the purposes of using Marxan analysis is to minimize the possibility of conflict occurring from the use of marine space. For that purpose, it was developed scenario analysis to represent the use of existing marine spaces used as cost. Scenario cost used were the fishing intensity, distance from the mainland, and the fishing boat traffics.

3. Results

3.1. Coastal ecosystems and benthic habitat

Spatial distribution of coastal ecosystems in three islands obtained from the results of analysis of Worldview-2 imagery. The spatial information of this coastal ecosystems based on the result of image processing using field data as a reference. Based on imagery analysis and validated by field observation, it is known that mangroves exist only on Lancang Island and Laki Island, and are not found on Bokor Island. The Lancang island consists of Lancang Kecil and Lancang Besar islands, and there are mangroves in both of them. Seagrass and coral reefs were found in shallow waters on three islands.

Figure 3 shows the distribution of coastal ecosystems on the islands of Lancang, Laki, and Bokor. Generally, Lancang Island dominates shallow water areas, as well as the distribution of coastal ecosystems.

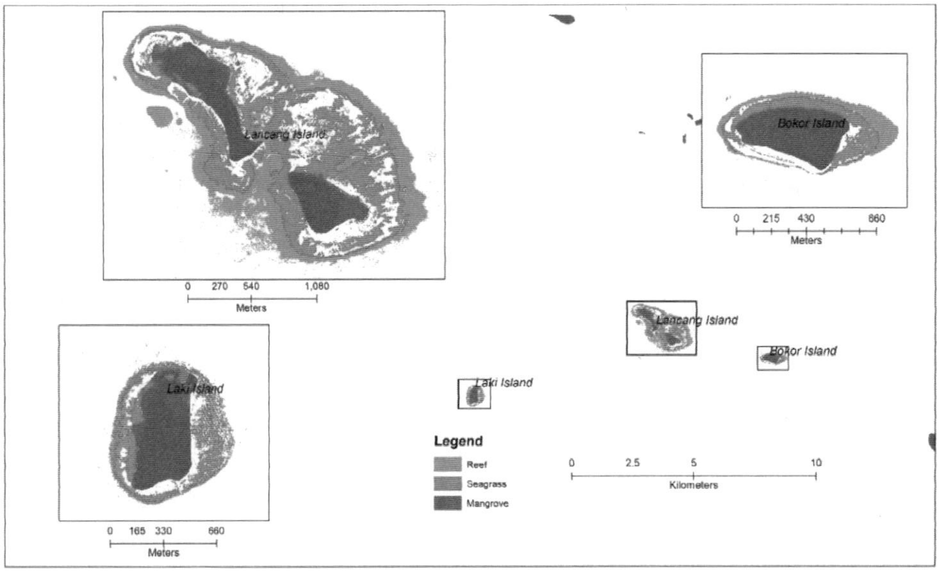

Figure 3. The distribution of coastal ecosystems in the study area.

Table 2 shows the area of coastal ecosystems of mangroves, seagrasses and coral reefs derived from satellite imagery. Each coastal ecosystem with the largest area found on the Lancang Island. Based on the area of both land and shallow waters, Lancang Island is bigger than the other two islands (figure 3, so that the area of the coastal ecosystem is wider on Lancang Island. The smallest area of seagrass and coral reefs were found in Laki Island. Based on field observations, in the shallow waters of that location, there are many rubbles that cover some of the benthic areas of the island.

Table 2. The area of each coastal ecosystems derived from imagery.

Islands / Ekosystem	Lancang Island (ha)	Bokor Island (ha)	Laki Island (ha)
Mangrove	9.38	-	3.93
Seagrass	105.34	10.29	15.27
Reefs	72.19	6.62	6.87

Geomorphological characteristics of benthic habitat obtained from modeling using BTM were grouped by geomorphological zone. There were four classes of geomorphological zone used in the classification of benthic habitats, i.e. Crest, Slope, Depression and Flat [8, 16]. The results of the analysis indicate that the benthic geomorphology in the study sites tends to be flat, the crest zone is found in the area close to the mainland, and some of the crests are found in areas close to Laki Island. Generally, the results of the analysis indicate that the geomorphological characteristics around of the Laki Island are more complex than others.

3.2. Spatial distribution of fishing ground

The location of the crab fishing by the Lancang Island fishermen has varied distribution based on the spatial grid. At each fishing season, the distribution of fishing ground was generally concentrated in the

South and West of Lancang Island. As can be seen in figure 4 on the southern Lancang Island be fishing every monsoon, which means that the area is almost a center for the fishing ground of blue swimming crab throughout the years. Certain locations are only be fishing in some seasons, for example on grid A4 and B2 are only fishing in the northwest monsoon, at other locations, ie, on grid A8 and D2 only fishing during alternate two monsoons.

The information obtained from fishermen that among their considerations in selecting fishing sites is a high chance of catching in certain areas. When they get a catch at a location, it is likely they will revisit the same location. As can be seen in figure 4, several grids location have fished in 3 - 4 seasons, i.e., in grids A5 and C5 while in other locations it is only in 1 - 2 fishing seasons, i.e., A8 and B8.

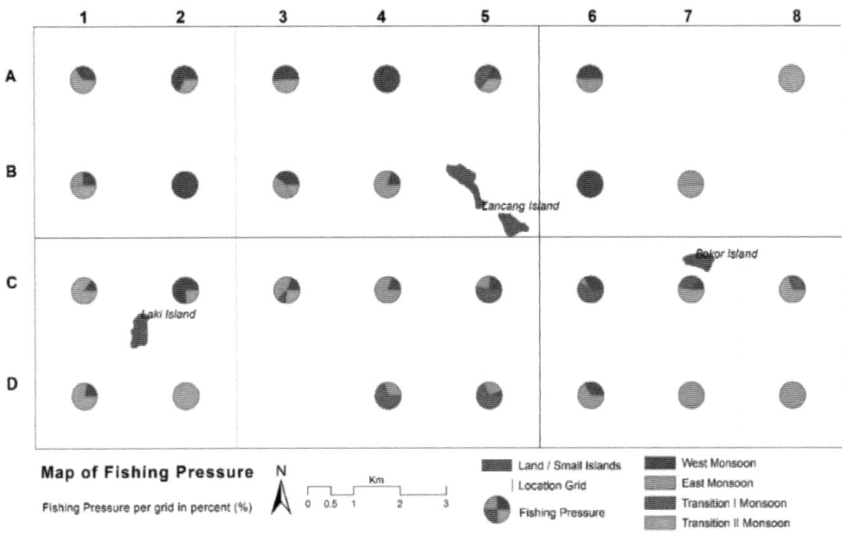

Figure 4. The distribution fishing intensity based-on seasonal fishing monsoon.

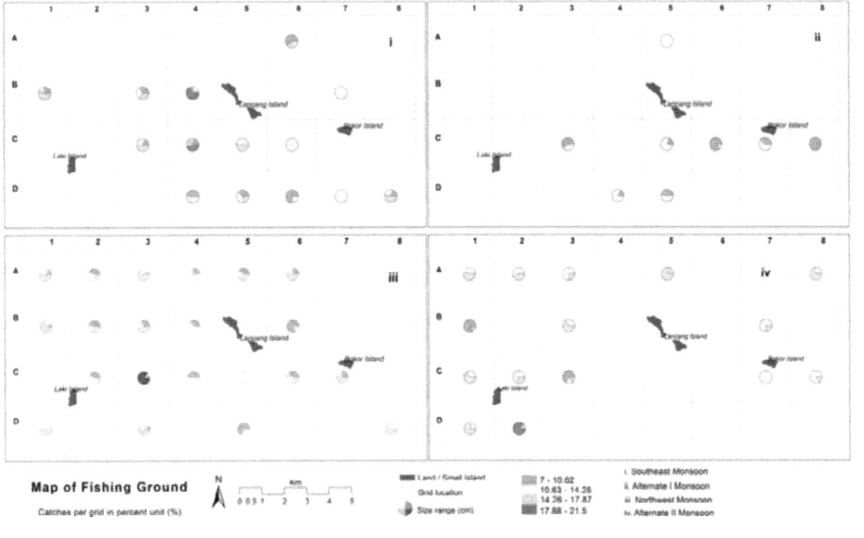

Figure 5. Map of distribution of fishing ground in every monsoon.

The blue swimming crab caught by fishermen have a size range of 7 - 21.5 cm distributed varying in each grid. Based on fishermen information, their fishing sites are located at 10-26m of water depth. In Lampung waters, Zairion, et al. [17] obtained crab sampling catch with a size of 5.1 cm at the area with depth less than 5m. To facilitate the analysis process, the range of sizes were grouped into 4 classes: C1 (7 - 10.62 cm), C2 (10,63 - 14,25 cm), C3 (14,26 - 17,87 cm), and C4 (17.68 - 21.5 cm). Based on the fishing season, the C1 and C2 classes almost dominate in the alternate two monsoons, while in the other three seasons vary with the highest catchment variation in the northwest monsoon (figure 5). In the northwest seasons, besides the variations in different size classes, it also appears to be the fishing season with the highest intensity than other seasons.

3.3. Marxan results

The analysis for determining potential sites using Marxan was done by applying several scenarios. Scenarios in the analysis were more emphasized on the representation of conservation features at each potential location. There were three scenarios values of the proportion of conservation features applied, i.e. 30%, 40%, and 50%. After applying some experimental parameter values in the analysis, the final selected parameters are the proportion of 30% (0.3), boundary length modifier (BLM) 2, species penalty factor (SPF) 1, and 1000 repetitions.

Potential areas identified from spatial analysis results, obtained in three locations (figure 6). The first location is around of Laki Island, which is concentrated in the East to the North of Laki Island. The second location is South of Lancang Island, and the third location is in the Southeast of Lancang Kecil Island, especially between Lancang Kecil Island and Bokor Island. From the three potential sites, based on the researchers' judgment, such as the ease of monitoring and possible conflicts of marine spatial use, the proposed sites for marine sanctuary are in the potential site 1 (around of Laki Island). In this area represents the distribution of coastal ecosystems as well as the uniqueness of geomorphological characteristics of the benthic habitat. In this region, a significant number of crabs with less than 10 cm of sizes, besides that this area is not a main of fishing area of blue swimming crab by local fishermen.

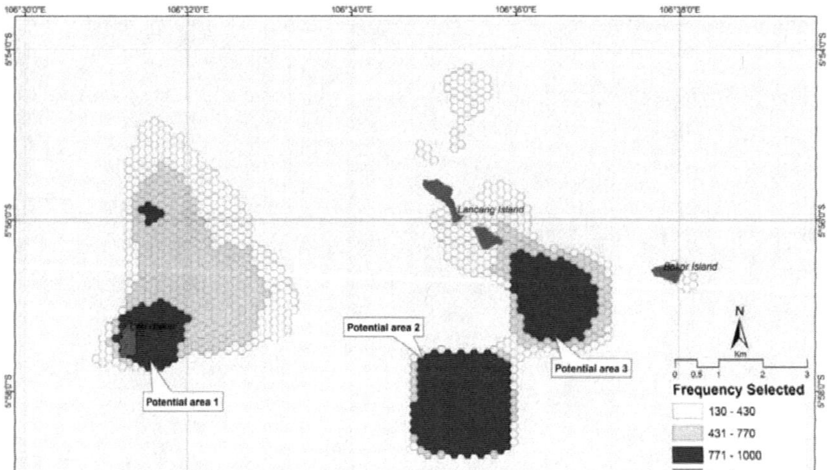

Figure 6. Potential locations for a marine sanctuary.

4. Discussion

Marxan was able to find potential sites for proposing as marine sanctuary around Lancang Island. Considering coastal ecosystems, habitat geomorphological characteristics, and spatial distribution of blue swimming crab, especially the captured sizes. The cost parameters represented by the intensity of fishing ground and the vessel traffic are considered to avoid conflicts in the utilization of marine spaces. In this case marxan is quite reliable in determining the planning decisions to minimize the possibility of marine space utilization conflicts [9, 10]. Marxan has been widely known and widely used as a tool in designing conservation areas, including marine protected areas. Even other studies use Marxan to design a watershed as a development area of economic centers as marine aquaculture areas [15].

Most of the analysis conducted in designing the marine conservation has not considered the complex parameters. This study tries to include one of the parameters in the analysis process that unnoticed, namely the habitat complexity of the study location. The habitat complexity describes the complexity of seafloor, in this case, is the variety of benthic features present in the study site. The habitat complexity can be measured through several approaches, the approach taken in this research was based on geomorphological characteristics of the seafloor. In shallow waters where electromagnetic waves can penetrate to the bottom of the water column, geomorphological characteristics can be derived from satellite images [18,21]. In deeper waters, the benthic geomorphological characteristics of can be analyzed by using bathymetric information obtained from several sources [6,22].

Most of the region in northern of Kepulauan Seribu are included as conservation area known as Taman Nasional Kepulaun Seribu (TNKpS). For the areas that do not include in TNKpS, i.e. some inhabited islands, some of the waters have been designated as a marine sanctuary (Daerah Perlindungan Laut-DPL) and managed by local communities. Marine sanctuaries on the other islands, i.e. as many as six sites in Kepulauan Seribu, the threat to some of these areas comes from the ever-increasing marine tourism activities. Lancang Island or Laki Island is proposed as a marine sanctuary to minimize exploitation and pressure of fisheries resources, especially blue swimming crab that does not pay attention to the sustainable stock in marine environment.

Lancang Island is currently the only one fisheries center in the South of Kepulauan Seribu, so it needs a sustainable fisheries management model. One of the management models is to set part of the marine area as the marine protection, that is conservation of marine in small scale. The expected benefits to be achieved if the marine sanctuary around the Lancang Island established, among them is the protected of shallow water ecosystem (seagrass and coral reefs) so that it retained its function as a supplier of ecological processes in the marine environment. By applying some locations as a marine sanctuary, it is also expected to realize the utilization of fishery resources, especially the wiser in exploitation of blue swimming crab to achieve sustainable fisheries.

5. Conclusion

From this research can be made some conclusions, first by integrating some aspects geospatially, marxan able to identify the potential locations for designation of the marine sanctuary. Second, there are three potential locations found for marine sanctuary in the study site, and the recommended location is around of Lancang Island. We hope the results of this study can be considered by decision-makers to determine some areas in the around of the Lancang Island to designate as a marine sanctuary.

Acknowledgement
Special thanks to Ministry of Research, Technology and Higher Education who has funded this research through BOPTN in 2015 – 2016. And the community of Lancang Island for the cooperation when conducting this research.

References

[1] Traganos D and Reinartz P 2017 Mapping Mediterranean seagrasses with Sentinel-2 imagery *Mar. Pollut Bul.* Jul 01
[2] Pu R and Bell S 2017 Mapping seagrass coverage and spatial patterns with high spatial resolution IKONOS imagery *International Journal of Applied Earth Observation and Geoinformation* **54** 145-158
[3] Heumann B W 2011 An object-based classification of mangroves using a hybrid decision tree-support vector machine approach *Remote Sensing* **3** 2440-2460
[4] Roelfsema C, Phinn S, Jupiter S, Comley J, and Albert S 2013 Mapping coral reefs at reef to reef-system scales, 10s–1000s km^2, using object-based image analysis *International Journal of Remote Sensing* **34** 6367-6388

[5] Haya L O M Y and Fujii M 2017 Mapping the change of coral reefs using remote sensing and in-situ measurements: a case study in Pangkajene and Kepulauan Regency, Spermonde Archipelago, Indonesia *Journal of Oceanography* **2017**

[6] Lundblad E R, Wright D J, Miller J, Larkin E M, Rinehart R, Naar D F, *et al.* 2006 A Benthic Terrain Classification Scheme for American Samoa *Marine Geodesy* **29** 89-111

[7] Wedding L M, Friedlander A M, McGranaghan M, Yost R S and Monaco M E 2008 Using bathymetric lidar to define nearshore benthic habitat complexity: Implications for management of reef fish assemblages in Hawaii *Remote Sensing of Environment* **112** 4159-65

[8] Subarno T, Siregar V P, Agus S B, and Sunuddin A 2016 Modelling complex terrain of reef geomorphological structures in Harapan-Kelapa Island, Kepulauan Seribu *Procedia Environmental Sciences* **33** 478-486

[9] Ball I R, Possingham H P and Watts M 2009 Marxan and relatives: software for spatial conservation prioritization *Spatial conservation prioritization: quantitative methods and computational tools Oxford University Press Oxford* pp 185-195

[10] Ardron J A, Possingham H P, and Klein C J 2010 *Marxan Good Practices Handbook*

[11] Otukei J R and Blaschke T 2010 Land cover change assessment using decision trees, support vector machines and maximum likelihood classification algorithms *International Journal of Applied Earth Observation and Geoinformation* **12** S27-S31

[12] Wright D J, Lundblad E R, Larkin E M, Rinehart R W, Murphy J, Cary-Kothera L, *et al.* 2005 ArcGIS Benthic Terrain Modeler

[13] Pasnin O, Attwood C and Klaus R 2016 Marine systematic conservation planning for Rodrigues Island, western Indian Ocean *Ocean and Coastal Management* **130** 213-220

[14] Watts M E, Ball I R, Stewart R S, Klein C J, Wilson K, Steinback C, *et al.* 2009 Marxan with Zones: Software for optimal conservation based land- and sea-use zoning *Environmental Modelling and Software* **24** 1513-1521

[15] Henriques N S, Monteiro P, Bentes L, Oliveira F, Afonso C M L and Gonçalves J M S 2017 Marxan as a zoning tool for development and economic purposed areas - Aquaculture Management Areas (AMAs) *Ocean and Coastal Management* **141** 90-97

[16] Erdey-Heydorn M D 2008 An ArcGIS Seabed Characterization Toolbox Developed for Investigating Benthic Habitats *Marine Geodes* **31** 318-358

[17] Zairion Z, Boer M, Wardiatno Y and Fahrudin A 2016 Composition and size of the blue swimming crab (*Portunus pelagicus*) caught at several bathymetric stratifications in East Lampung Waters *Jurnal Penelitian Perikanan Indonesia* **20** 199-206

[18] Phinn S R, Roelfsema C M and Mumby P J 2012 Multi-scale, object-based image analysis for mapping geomorphic and ecological zones on coral reefs *International Journal of Remote Sensing* **33** 3768-3797

[19] Hamilton S M and Spencer T 2011 Geomorphological modeling of tropical marine landscapes: Optical remote sensing, patches and spatial statistics *Continental Shelf Research* **31** S151-S161

[20] Anggoro A, Siregar V P, and Agus S B Geomorphic zones mapping of coral reef ecosystem with OBIA method, case study in Pari Island *Jurnal Penginderaan Jauh* vol **12** pp 1-12 2015

[21] Selamat M B, Jaya I, Siregar V P and Hestirianoto T 2012 Geomorphology zonation and column correction for bottom substrate mapping using Quickbird Image *Jurnal Teknologi Perikanan dan Kelautan* **3**

[22] Wright D J and Heyman W D 2008 Introduction to the special issue: marine and coastal GIS for geomorphology, habitat mapping, and marine reserves *Marine Geodesy* **31** 223-230

IOP Conf. Series: Earth and Environmental Science **176** (2018) 012030 doi:10.1088/1755-1315/176/1/012030

Comparison between acoustic Target Strength of Yellowfin Tuna (*Thunnus albacares*) and Longtail Tuna (*Thunnus tonggol*)

Fatimah Nur Islaminingdiah[*], **Indra Jaya**[*], **Totok Hestirianoto**

Department of Marine Science and Technology, Faculty of Fisheries and Marine Science, Bogor Agricultural University, Bogor 16680, Indonesia
* e-mail : indrajaya@apps.ipb.ac.id; +62 85736251040

Abstract. Yellowfin Tuna (*Thunnus albacares*) and Longtail Tuna (*Thunnus tonggol*) are among the two commercially important fishes and highly migratory species, subject to be conserved by the convention on the law of the sea in Annex I of the 1982 Convention on the Law of the Sea (FAO, Fisheries Department 1994). Morphologically, *T. albacares* and *T. tonggol* have a striking difference, where the former has swim bladder while the latter has not. This paper comparing the result of acoustic target strength (TS) measurement of the two species using SIMRAD EY 60 scientific acoustic instrument. The in-situ measurement was conducted on each species for two different fish length. The fish is tethered using a monofilament string and the distance between the surface of the transducer to fish is ± 4 m. The result showed that the $\overline{TS}$ value of *T. Albacares*$_1$ (51.70 cm) is -40.46 dB and *T. Albacares*$_2$ (29.20 cm) is -43.70 dB, whereas value of *T. tonggol*$_1$ (19.4 cm) is -46.53 dB and *T. tonggol*$_2$ (17.1 cm) is -49.29 dB. Therefore, teh presence of swim bladder made a difference on the measured mean $\overline{TS}$ between T. albacares and T. tonggol. Whilw for the length of fish, is not always the case.

1. Introduction

T. albacares and *T. tonggol* are economically important Scombrid. *T. albacares* has an expensive price that makes fishermen very happy to catch this fish. Although the price is expensive the fish is still much to love it. This makes *T. albacares* has a high selling competitiveness [1]. While *T. tonggol* is a small tuna fish that is not less popular with *T. albacares*. This fish has a cheaper price compared to *T. albacares*, because of its smaller size. They are highly migratory species and widely distributed in tropical and subtropical waters around the world. The impact that fish are overfishing in various places. Overfishing is an adverse action for ecological sustainability and harmful to fishermen. Fishermen are losing money because they are depleting their catch. Consequently, fishermen have to sail farther to get more catches, but this also affects the cost that they have to spend to sail further. Therefore, it is necessary to study that fish. Hydroacoustic technology has one of the solutions, the solution is to know the TS (Target Strength) value of the fish. TS value is one of the important parameters to identify the fish density.

According to [2] TS is the ability of the target to reflect the sounds that hit it. Based on the domain used, TS is defined as the logarithm of the output between the intensity of the incoming sound about the target and multiplied by the number ten (10), as the following equation:

IOP Conf. Series: Earth and Environmental Science **176** (2018) 012030 doi:10.1088/1755-1315/176/1/012030

$$TSi = 10\log\frac{Ir}{Ii} \quad (1)$$

$$TSe = 10\log\frac{Er}{Ei} \quad (2)$$

where *TSi* is an intensity of target strength, *Ir* is the sound intensity of the target, *Ii* is the intensity of the target's reflected sound. *TSe* is the energy of target strength, *Er* is the sound energy of the target and *Ei* is the reflected energy at a distance of 1 m from the target.

T. albacares has a swim bladder, but *T. tonggol* has no swim blader. The swim bladder is an important organ that is squeezed by right and left ribs in the middle banging between the head and tail. Not all fish have swim bladder just like pelagic fish. Acoustically fish can be divided into two categories, with swim bladder fish and fish without swim bladder. Fish with swim bladder have a maximum TS value is not right on the dorsal aspect, while fish without swim bladder has maximum TS value right on the dorsal aspect.

The purpose of this research is to know the $\overline{TS}$ value of fish with swim bladder (Yellowfin Tuna, *T. albacares*) and without swim bladder (Longtail Tuna, *T. tonggol*) using Simrad EY 60 hydro acoustic instrument in Bagan Apung of Palabuhanratu Bay, Sukabumi, West Java, ID.

2. Method

This research was conducted on August 29, 2016 until September 2, 2016 using the tethered and in-situ method in Bagan Apung, Palabuhanratu Bay, Sukabumi, West Java. Processing of value data is done TS at Marine Acoustic and Instrumentation Laboratory, Department of Marine Science and Technology, Bogor Agricultural University.

The instrument used in this research is Simrad EY 60 hydro acoustic instrument with 38 kHz frequency, sphere ball (Tungsten 38.1 mm) with TS of -42.11 dB, GoPro Hero 4 underwater camera, Canon camera, rope, cold box, fish suture equipment, sewing meter, digital scales, analog scales, and monofilament. The software used to process data are Echoview, StatSoft Statistica 6, MS Office Word 2013 and MS Office Excel 2013. The materials used are Yellowfin Tuna (*T. albacares*) and Longtail Tuna (*T. tonggol*). The fish length is measured in terms of its Fork Length (FL) using the standard ruler. The same also applied for swim bladder (SB), it was measured in length.

TS measurements in the field were conducted using the tethered method. The first thing to do is wrapped-fish around with net using thin monofilament to contain the fish so that easy to handle. Then using the same monofilament, the fish was tethered at the end of the mouth (front), the fish's upper body (top), and the rear. Next put the ballast in the middle (bottom) as a ballast. Lower the target fish slowly so that the fish is not damaged. After reaching a depth of ± 4 meters below sea level, keep the angle of the fish so as not to carry the current by pulling and stretching the rope that has been installed at three points on the target fish before it is lowered. If the target fish angle is stable and visible on the echo sounder monitor, start recording what echosounder reads. The recording process was carried out for 5 to 10 minutes. If the target fish is unstable due to the current, pause the tape and pull the strap until the target fish reenter the main lobe echo sounder. After the recording is enough, stop the recording and raise the target fish back to the chart. Remove the sarong from the target fish. Repeat the process to the next target fish. After all the fish have been recorded, the field data measurement process has been completed.

Data processing is done using echoview software. The setting parameter used were: operating frequency 38.00 kHz, sound speed 1544.24 m/s, draft transducer 1.0 m, sample interval 0.000064 s, transmit power 300.0 W, pulse length 0.256 ms, transducer gain 23.250 dB, absorption coefficient 0.005304 dB/m, Sa correction -0.850 dB. *.raw files obtained using minor-axis beam width 8,680 °, major-axis beam width 8.730 °, two-way beam angle -18.00. The color display used is the color min - 70.00 dB, color max -30.00 dB and color range 40.00 dB. Then integrate the cells that have been set, then record the target value on MS Office Excel. The upper and lower ranges was set 1.5 dB, for example, the value of TS -45 dB has a range of -43.5 dB up to -46.5 dB. The standard deviation value was calculated using MS Office Excel and for normal distribution and mean value was obtained using Statistica software. The mean value of $\overline{TS}$ obtained were then compare between *T. albacares* and *T. Tonggol*.

3. Results

3.1. Yellowfin Tuna (T. albacares)

The normal distribution of $\overline{TS}$ values *T. albacares* is shown in the graph (figure 1). Graphic of normal distribution have bar chart and line chart. The bar chart explained the number of occurrence of data. The line chart showed a pictures of the data when it is spreading normally. The appearance of the first fish data more than the second fish, because the first fish was detected longer than the second fish. The first fish is detected longer because the water conditions are not supporrtive (fast current), consequently the fish is unstable and not in the main lobe.

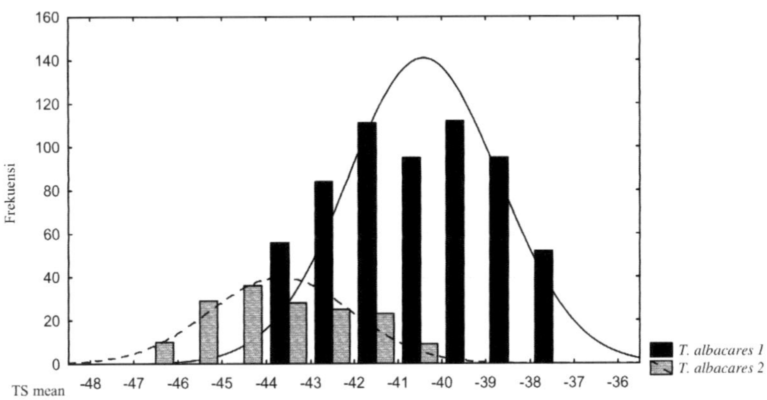

Figure 1. Normal distribution display assumption of the *T. albacares*.

The TS value of *T. albacares*$_1$ (51.70 cm) (figure 2a) in the range -43.48 dB to -37.52 dB. The $\overline{TS}$ value of *T. albacares*$_1$ based on that range is -40.46 dB ± 1.71. The *T. albacares*$_2$ (29.2 cm) (figure 2b) have TS values between -40.68 dB to -46.43 dB. The $\overline{TS}$ value of the *T. albacares*$_2$ is -43.70 dB ± 1.60. This indicates that the larger fish has a longer $\overline{TS}$ value.

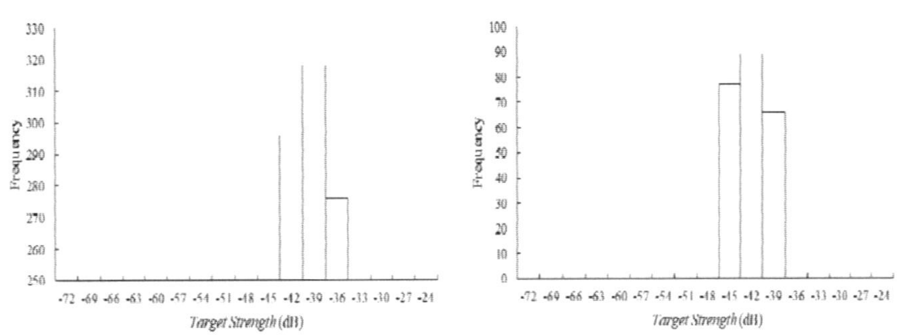

(a) (b)

Figure 2. TS-frequency data distribution of *T. albacares*.

3.2. Longtail Tuna (T. tonggol)

The $\overline{TS}$ values of T. tonggol shows in the graph (figure 3). The graph be compose of the $\overline{TS}$ values and the frequency of the occurence of the data. The frequency of accurrence of the data is indicated by the bar chart. The line chart is normal distribution assumption of the *T. tonggol*. Line chart is shapes of the Beta Distribution who shows the normal distribution data of TS values. That figure shows frequency of *T. tonggol*$_2$ have larger data be compared frequency of *T. tonggol*$_1$.

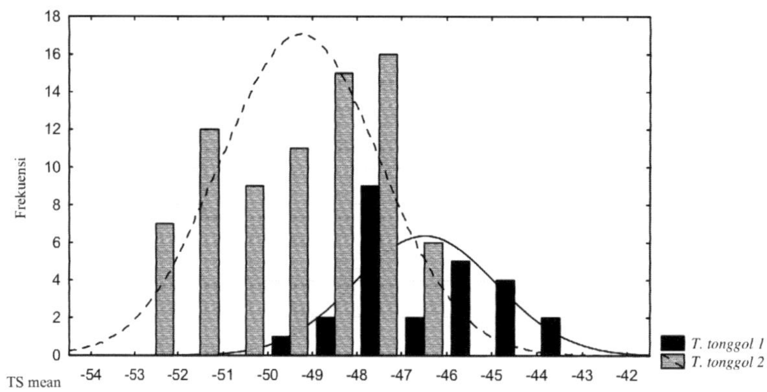

Figure 3. Normal distribution display assumption of the *T. tonggol*.

The *T. tonggol* $\overline{TS}$ values shows in figure 4a and 4b. Figure 4a is graph of frequency distribution of *T. tonggol*$_1$ and figure 4b is graph of *T. tonggol*$_2$. The graph in figure 4a has a climax bar chart. Figure 4b have two climax bar chart. The two climax bar chart in figure 5b is not good, because the current so fast when measured data and position of fish is unstable. The position of fish are no on-axis and out from main lobe. On-axis is the position of fish to center of beam are at an angle of 180° (straight). On-axis position has a maximal beam, so the result of data is good. On-axis position on the fish with bladderless are precise on the dorsal aspect.

The *T. tonggol* have a range of TS values between -52.45 dB to -43.77 dB. In the range, $\overline{TS}$ value of *T. tonggol*$_1$ is -46.57 dB ± 1.56 and $\overline{TS}$ value of *T. tonggol*$_2$ is -49.29 dB ± 1.78. The results indicate that the length of the fish affects the TS value.

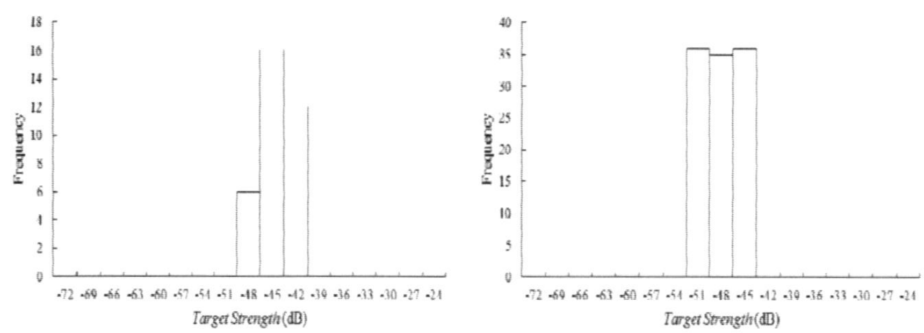

(a) (b)

Figure 4. TS-frequency data distribution of *T. tonggol*.

3.3. Relations of the TS and Fork Length

Data of *T. albacares* and *T. tonggol* shows in table 1. The data ploted at 3rd quadrant. This data is result of measurement in the field and data processing.The relationship graph for TS values and Fork Length (figure 5), TS values in x axis an Fork Length in y axis. First *T. albacares* point at 51.70 cm, -40.46 d). Second *T. albacares* point at 29.20 cm, -46.70 dB. The two points are connected to the linear line. The *T. alabacares* line named α_1. Then, First *T. tonggol* point at 19.40 cm, -46.53 dB. Second *T. tonggol* point at 17.10 cm, -49.29 dB. The first and second points of *T. tonggol* are connected to the linear line. The *T. tonggol* line named α_2. The lines of T. albacares and T. tonggol are drawn lines to intersect. The intersection of the two lines is the equilibrium point. The equilibrium of TS values is 20.80 cm, -44.93 dB. Estimated at 20.80 cm fork length, both fish species have the same TS value, as an ecological strategy. Ecological strategy is the ability of the fish to cluster.

Table 1. Data of *T. albacares* and *T. Tonggol.*

	FL (cm)	SB (cm)	TS *min* (dB)	TS *max* (dB)	TS (*mean*) dB	Stdev
T. albacares 1	51.70	8.40	-43.48	-37.52	-40.46	± 1.70
T. albacares 2	29.20	4.00	-46.43	-40.68	-43.70	± 1.60
T. tonggol 1	19.40	-	-49.11	-37.77	-46.57	±1.56
T. tonggol 2	17.10	-	-52.45	-46.43	-49.29	±1.78

Note : FL = Fork Length, SB = Swim Bladder, TS = Target Strength, Stdev = Standar Deviation

4. Discussion

Figure 4 shows the difference in $\overline{TS}$ values occurring between two *T. albacares* due to FL differences and the presence of swim bladder. $\overline{TS}$ values in fish with swim bladder were greater than $\overline{TS}$ values without swim bladder [2]. The larger the size of the fish, the greater the value of $\overline{TS}$ obtained. In accordance with the research [3] obtained data indicating that the longer the size of fish $\overline{TS}$ value obtained greater. This shows that the size of the fish and the value of $\overline{TS}$ are directly proportional. The presence of swim bladder in the target fish (Yellowfin Tuna) seems to affect the sensitivity to the $\overline{TS}$ value. In the study, [4] obtained the results that the value of $\overline{TS}$ fish with swim bladder is greater than fish without swim bladder (Catfish).

Swim bladder serves as a regulator of hydrodynamics, to regulate the pressure generated by the surrounding waters. However, it also serves as a reflector and a resonator of the incoming sound waves. The swim bladder, which contained gas/membrane and meat around it, creates a strong contrast or high impedance that play important factor in the reflection of the incoming sound. It almost incompressible liquid. Gas is easier to move than liquid. If there is no central organ, external pressure waves will reflect back into the air, following the smallest path of resistance. In other words, the incoming sound energy will be reflected back from the rigid membrane at the entrance of the internal organ. The function of the swim bladder is the floating organ/regulates the gas circulation, the ballance regulator when swimming vertically, oscillating organ, and as a reflector of hidroacoustic force [5]; [6]; [7]; [8]. The working mechanism of swim bladder applies the Law of Newton II, the Law of Archimedes and the Law of Bernoulli simultaneously [8, 9, 10]. The action given due to contraction/relaxation of the swim bladder membrane causes a floating/drowning reaction with a distance certain (vertical), proportional to the increasing/decreasing air volume (Newton and Archimedes Laws sinergies) [8].

For *T. tonggol* which does not have swim bladder make target fish is not too sensitive to $\overline{TS}$ value. According to [11], the main factors affecting the value and characteristics of the target strength of fish are the frequency of sound used, the shape and length of the fish body, the fish body structure and the physical parameters and the orientation and behavior of the fish. This factor is a general trend. According to [12] that for a particular fish species (with a particular anatomy) and wavelength (frequency) there is a close relationship between the target strength (equivalent cross section) and the size of the fish ie larger fish have greater target strength.

5. Conclusion

In situ TS measurement for Yellowfin Tuna (*T. albacares*) and Longtail Tuna (*T. Tonggol*) has been conducted with different FL. The presence of swim bladder made a difference on the measured mean $\overline{TS}$ between *T. Albacares* and *T. Tonggol*. While for the length of fish, is not always the case.

References

[1] Kantun W, Malik A A and Harianti 2015 Kelayakan limbah padat Tuna Lion Madidihang *Thunnus albacares* untuk bahan baku produk diversifikasi *JPHPI* **8**(3): 303-314
[2] Johanesson K A and Mitson R B 1983 *Fisheries Acoustic A Pracical Manual For Acoustic Biomass Estimation* FAO Fisheries Tech www.fao.org/docrep/x5818e/x5818e01 htm
[3] Manik H M 2009 Measurement of acoustic reflection of tuna fish using echosounder instrument *Ilmu Kelautan* **14**(2): 84-88
[4] Hamim M 2011 Pengukuran Target Strength Ikan Mas dan Ikan Lele pada Kondisi Terkontrol Menggunakan *Quantified Fish Finder* [Skripsi] (ID)
[5] Kleckner R C 1980 Swimbladder wall guanine enhancement related to migratory depth in silver phase *Agulilla rostrata.Comparative Biochemisrty and Physiology Part A: Physiology.* **65**(3): 351-354
[6] Kobayashi H, Palster B and Scheid P 1990 CO_2 back-diffusion in the rete aids O_2 secretion in the swim bladder of the eel *Respiration Physiology* **79**(3): 231-242
[7] Taylor G K, Holbrook R I and de Perera T B 2010 Fractional rate of change of swim bladder volume is reliably related to absolute depth during vertical displancements in teleost fish *Journal of The Royal Society Interface* **7**(50): 1379-1382
[8] Kartika I W D 2017 Karakterisasi dan Biokompatibilitas Kolagen Gelembung Renang Ikan Cunang (*Muraenesox talabon*) sebagai Biomaterial *Scaffold* Kultur Sel [Thesis] (ID)
[9] Henshawn J M 2014 *An Equation for Every Occasion: Fifty-two Formulas and why They Matter Maryland* (*EN*): JHU Press
[10] Safiullin R K, Tayurskii D A, Skvortsov A I and Aganov A V 2014 *Physics around us: Qualitative Problems in Physics*
[11] Furusawa M 1988 Prolate spheroidal models for predicting general trends of fish target strength *Journal of the Acoustic Society of Japan* **9** 169-180
[12] Burczynsky 1982 *Introduction to the Use of Sonar System for Estimating Fish Biomass* FAO of United Nation Rome (IT) p 89

IOP Conf. Series: Earth and Environmental Science **176** (2018) 012031 doi:10.1088/1755-1315/176/1/012031

In situ TS measurement of Mackerel Scad (*Decapterus macarellus*), Leopard Coral Trout (*Plectropomus leopardus*), and Largehead Hairtail (*Trichiurus lepturus*)

W Setiawan[1], I Jaya[1*], T Hestirianoto[1], S Pujiati[1], A Priatna[2], A Ma'mun[2]

[1] Dept. Marine Science & Technology, Bogor Agricultural University, Indonesia
[2] Marine Fisheries Research Center, Ministry of Marine and Fisheries Affairs, Indonesia
*corresponding e-mail: indrajaya@apps.ipb.ac.id

Abstract. Acoustic target strength (TS) measurements were performed on one small pelagic fishes, and two demersal fishes, which are very important for Indonesia fisheries, namely: mackerel scad (*Decapterus macarellus*), coral trout (*Plectropomus leopardus)*, and largehead hairtail (*Trichiurus* sp.) on various fish sizes (fork length, FL) and orientation. The fish was successfully measured using tethered technique, where each individual fish was places and hang in web using monofilament string. The TS data analysis is performed numerically using the best-fit technique to establish the relationship between length, fish orientation, and TS. The TS-length relationship was obtained as follow: for mackerel scad, TS = TS = 20log - 84,38; for coral trout, TS = 20Log - 76,68, and for large headhairtail, TS = 20Log -91.88.

1. Introduction

The existence of small pelagic fish in Indonesian waters is very diverse and abundance continues to change due to migration and environmental changes. The calculation of potential existence of fish has been done by the Ministry of Marine Affairs and Fisheries in the last several years. Various efforts have been made by conducting surveys of fish stocks using acoustic technology and catch sampling as well as recording of fishermen catch data at ports throughout Indonesia [1]. The results of the axis analytical measurements are strongly influenced by the accuracy of measured values of fish target strength (TS). Therefore it is necessary to accurately measure the target strength value (TS) of small pelagic fish for verification of the results of an axis survey for fish stock estimation [2]. Fisheries management in the territorial waters of Indonesia, especially in coastal areas need to be done. One of the first actions to begin is to know precisely the stock or abundance of fish in fishing ground [3]. This can be done by utilizing acoustic technology to obtain the value of abundance, fish biomass based on the reflected sound of the object/target (fish). Before making the determination of fish abundance and biomass must be known exactly the actual value of the target strength of the target fish. This is especially important because target strength is the main parameter for determining fish density, abundance and biomass in acoustic technology applications [4]. The results of this study are expected to be useful to assist decision makers for the management of fishing so as to have a controlled stock, which in the long term will help in the preservation of coastal fisheries communities.

2. Methods

2.1. Time and location

In situ TS measurement was performed from the floating lift net [Figure 1], in Pelabuhan Ratu Bay waters during November 2016. The live fish was placed inside of the cage, while for dead fish was tethered using monofilament line.

Figure 1. Floating rides (lift net) vehicle as a place to do research.

2.2. Tools and materials

The tools and materials used in this study include: Scientific Echosounder Simrad EK 80-200 kHz transducer split beam, sphere ball (target standard) type tungsten carbide (WC) 38.1 mm diameter, cage where live fish are measured by their TS value and confinement of live fish sample before measured, measuring instruments (meter), digital scales and CTD SBE 19 plus to measure temperature and salinity.

Before performing the TS measurement activity of each individual fish with an acoustic system, the acoustic instrument is first calibrated [4]. Measurement of temperature and salinity of waters to obtain the speed of sound value is done with CTD SBE 19 plus. Limited scope of the lift net and strong current conditions, so the maximum sphere ball can only be placed at a depth of 4.5 m from the surface of the transducer. The parameter setting used in calibration is in table 1, and is used at the time of acquisition of TS value of each individual fish.

Table 1. Setting of acoustic parameters.

Parameter	Value
Frequency	200 KHz
Pulse Duration	0.512 ms
Power transmit	150 watt
Sound speed	1542 m/s
Absorption	82,9 dB/Km
SV threshold	-60 dB
TS threshold	-60 dB

2.3. Measurement of TS value

In the previous TS measurements, a test was conducted to examine the differences in TS values between live fish and dead fresh fish for the same fish samples. The result of the TS measurements after paired t test on the 5 fish were obtained conclusion that the measurement value of TS for the fish (*Megalaspis cordyla*) living and dead conditions is equal to zero. Thus, TS measurements for living conditions and fresh dead conditions have no significant difference. The results obtained are the basis for TS measurement testing for other small pelagic fish species conducted on dead-fresh fish samples. Measurements of fish were measured done using tethered techniques, where each fish was located and hung on a monofilament strings net [5]. Each individual fish is scanned with an acoustic system, with the transducer placed over a hanging fish sample, so the target (fish) is in the beam acoustic beam area. The echosounder parameter settings used at the time of acquisition must correspond to the setting of parameters at calibration time (table 1). The measurements of each individual fish are declared successful, until a normal distribution of TS values is generated.

2.4.Data analysis
Acoustic data is processed using SONAR5-pro software, extraction results are target strength values (TS) in decibels (dB). The number of TS values per fish individual measured by acoustics consists of many variations of TS values that distributed within a certain range, but generally distributed normally and have a certain modus value [6].

2.4.1.The target strength value (TS) of each individual fish. The amount of data used to obtain the average TS value of one individual fish is the TS data in the range of ± 3dB from the middle peak value of its modus mode. According to [5] TS is a backscattering cross section (σbs) of the returned target signal, and the relationship between TS and σbs is expressed as TS = 10 log (σbs). Then the average TS is TS = 10 Log (σbs). The mean TS score is assumed to be the TS value of each individual fish, and is used in the regression analysis to obtain TS-length formulations on similar species.

2.4.2.Empirical relationship of TS values with length of fish. A series of coupled data between length (L) and average TS for the same species is used to analyze the empirical relationship between the two so that TS-length formulation is formed. The empirical relationship between TS and L refers to TS = 20 log (L) + A, where A is the target strength value for 1 cm of normalized strength depending on the species of fish species [4]. Based on the equation, then to get TS-length formulation we only use linear regression analysis between independent variable L and dependent variable TS. Previously a series of L and TS data of each individual in the same species tested the normality of the data as a condition of regression analysis. After the regression equation TS = a Log L + b is formed, then the partial test is the significance of variable L to TS, with the hypothesis: H_0 = L variable has no significant effect on TS; H_1 = L variable has significant effect on TS.Regression equation obtained, then compared with equation [4] that is TS = 20 Log L + A, so will be obtained equation to get value of A on certain species. If A has been obtained, then by entering the L value of each measured individual fish, it will get the average value of A or call C. So the TS-length formulation is TS = 20 Log L + C [7] .Measurements of TS values for dead fresh fish conditions in this study used Mackerel Scad (*Decapterus macarellus*), Leopard Coral Trout (*Plectropomus leopardus*), and Largehead Hairtail (*Trichiurus lepturus*).

3. Results

3.1 Target Strength value
The measurement result with 200 kHz acoustic frequency of the 17 samples Mackerel Scad (*Decapterus macarellus*) showed that the TS value was distributed from -66 s / d -42dB. Each fish size has a different modus class value. The value of the TS Mackerel Scad modus class tends to increase with the larger size (table 2). A number of Mackerel Scad measured for their TS values, scattered from size 20.0-33.5 cmbFL, with an average number of samples measuring 300 times for each fish. Measurement results showed that,TS values ranging from -61.14 ± 1.7 dB to -46.06 ± 1.7 dB. The average value of TS Mackerel Scad in the 20.0-33.5 cm FL size is -55.63 ± 1.7 dB (table 7). The results of measurements of Mackerel Scad with acoustic frequency have the same diversity values.

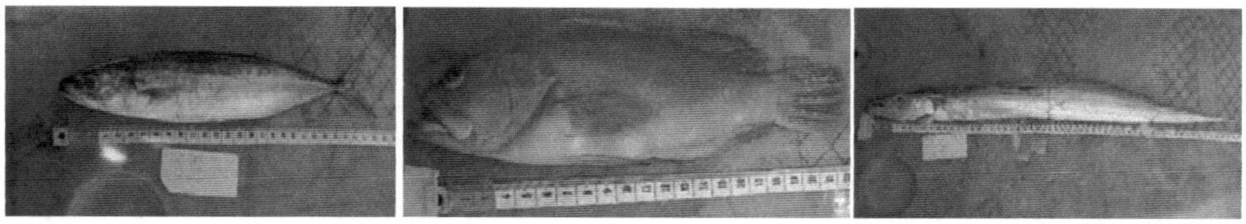

Figure 2. Samples of fish Mackerel Scad (*Decapterus macarellus*), Leopard Coral Trout (*Plectropomusleopardus*), and Largehead Hairtail (*Trichiurus lepturus*) from left to right.

The TS value from 6 samples Leopard Coral Trout (*Plectropomus leopardus*) was distributed from -56.04 s/d -40.67 dB. Each size of fish has a different modus class value. The value of the TS Leopard Coral Trout modus class tends to increase with the larger size (table 2). A number of Leopard Coral Trout measured their TS values, scattered from size 21 – 28 cm TL, The average value of TS Leopard Coral Trout in the 21 – 28 cm TL size is -49.63 dB (table 2), and average TS value from 5 samples of Largehead Hairtail (*Trichiurus lepturus)* is -56.73 dB distributed from -59.02 sd – 40.67 dB with scattered from size 43 – 50 cm TL. Both the results of measurements of Coral Trout and Largehead Hairtail with acoustic frequency have the same diversity values.

3.2.Formula TS-Length Fish

Based on the results of measuring activities of Target Strength (TS) from three fishes (Mackerel Scad, Leopard Coral Trout, and Largehead Hairtail) it is possible to perform a simple statistical TS-length analysis. The Central Limit Theorem says that the more observations (in this case the experimental sample unit) the data is getting closer to the normal distribution. The general rule is, if the experimental unit is more than or equal to 30 then the data is considered to follow the normal distribution. In this study the experimental unit was below 30, so the normality test did become necessary [8].

The analysis of fish length relationship three fishes to TS value from each individual to get the empirical value in the form of TS = 20 Log (FL) + C was done based on the measurement data presented in table 2. In sequence number of samples fish measured Mackerel Scad, Leopard Coral Trout, and Largehead Hairtail is 17, 6 and 5 piece of samples.

Based on the results of treatment of data to meet the assumptions required in the regression analysis, then after the exploration of data, from 17 fish samples measured TS value there were values of deviants (outliers). The normality test (figure 3-5. left) showed that plot results in residual values are in the vicinity of the normality line or approximately on the diagonal line, so the residual normality assumption is fulfilled. Homoskedasticity test (figure 3-5. right) to see the homogeneity of variance, showed fitted values vs. residual values randomly distributed, meaning that the residual variation is constant, so that residual homoskedasticity assumption is fulfilled. So the data of measurement result is feasible for regression analysis.

Table 2. Structure of size and value of fish TS Mackerel Scad (*Decapterus macarellus*), Leopard Coral Trout (*Plectropomus leopardus*), and Largehead Hairtail (*Trichiurus lepturus*).

No. DeadFish	TL (cm)	FL (cm)	W (g)	Swimblader (cm)	TS rata-rata (dB)
MS1	24.5	22.5	147.3	7	-52.11
MS2	31.5	28.5	305	10	-58.21
MS3	26	24	120	7.5	-61.02
MS4	37	33.5	440	15	-58.2
MS5	28.2	25.8	225	7	-61.14
MS6	27.2	24.8	180	6.5	-55.1
MS7	24.5	23	123	8	-55.78
MS8	28	25.8	193.7	8	-58.22

No. Dead Fish	TL (cm)	FL (cm)	W (g)	Swimblader (cm)	TS rata-rata ((dB
MS9	24.7	22.9	164	7	-55.15
MS10	28.5	26	146	8	-59.08
MS11	25.4	23.6	176	7	-46.06
MS12	25.5	23.5	155	6	-49.37
MS13	24.5	22	170	6	-58.26
MS14	24.7	22.5	123	not visible	-52.36
MS15	27	24	173	10	-52.11
MS16	21.5	20	103	5	-58.33
MS17	28	23.5	178	7	-55.27
LC1	25.5	25.5	220	5.5	-51.12
LC2	21	21	138	5	-51.52
LC3	23	23	264	6	-40.67
LC4	28	28	407	6.5	-46.51
LC5	22	22	209	5.4	-51.91
LC6	21.9	21.9	187	5	-56.04
LH1	80.5	80.5	170	-	-54.69
LH2	49	49	115	-	-59.02
LH3	43	43	59	-	-58.07
LH4	60	60	180	-	-55.12
LH5	55	55	124	-	-56.76

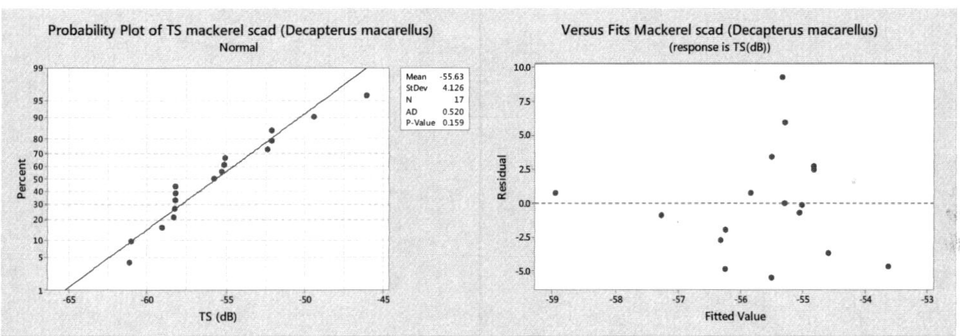

Figure 3. Normality plot (left) and Interdependent between error (right) for value TS Mackerel Scad fish (*Decapterus macarellus*).

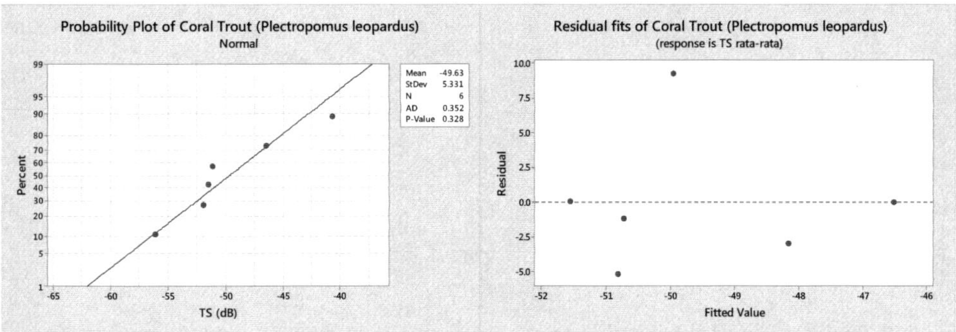

Figure 4. Normality plot (left) and Interdependent between error (right)for value TS Leopard CoralTrout (*Plectropomus leopardus*).

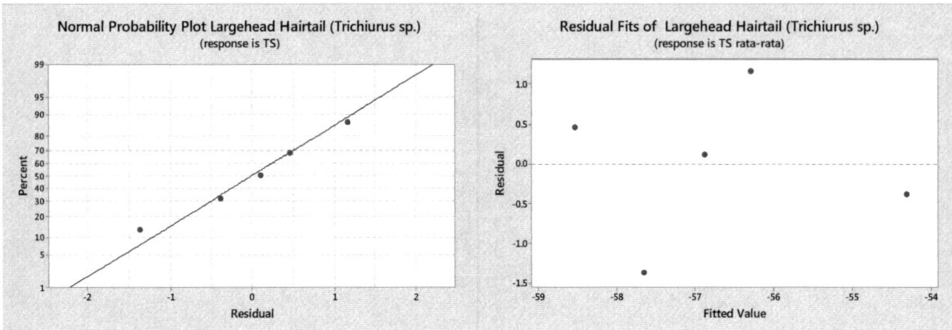

Figure 5. Normality plot (left) and Interdependent between error (right) for value TS Largehead Hairtail (*Trichiurus lepturus*).

After normality and homoscedasticity tests for three small pelagic fish, both tests were performed to look at the homogeneity of the data, whether the data were spreading randomly so that the residual homoskedasticity assumptions were met. So the data of the measurement result is feasible for the regression analysis for the three fish species.

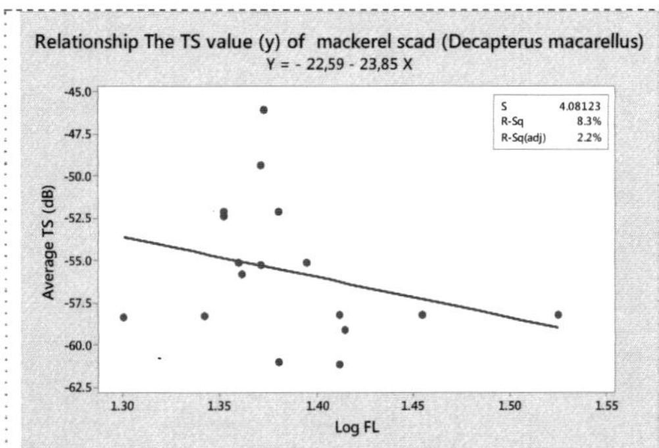

Figure 6. Relationship value TS (y) Mackerel Scad fish (*Decapterus macarellus*) as log function FL (x).

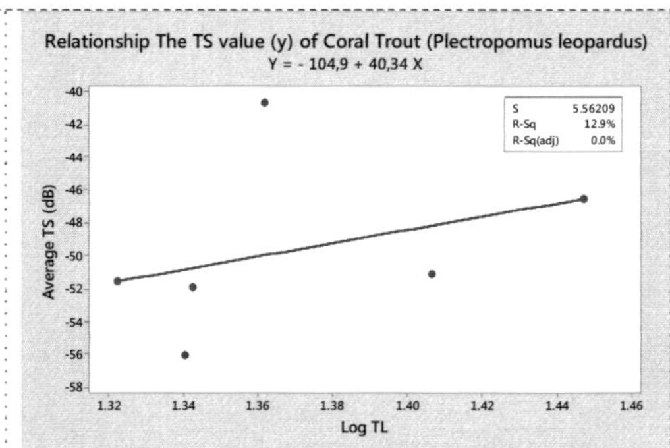

Figure 7. Relationship value TS (y) Coral Trout fish (*Plectropomus leopardus)* as log function FL (x).

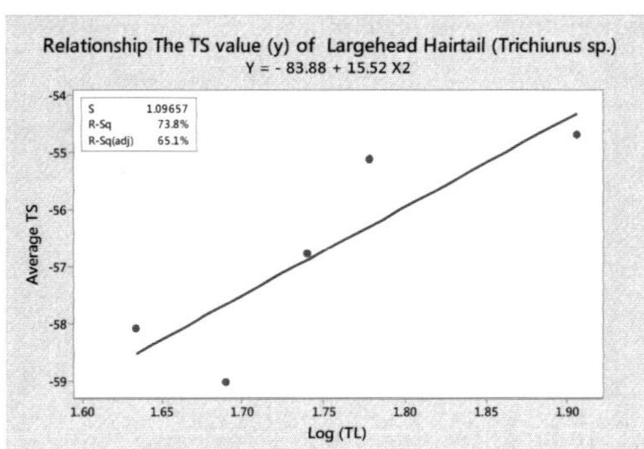

Figure 8. Relationship value TS (y) Largehead Hairtail fish (*Trichiurus sp.*) as log function FL(x).

The regression equation for TS-length Mackerel Scad fish (*Decapterus macarellus*) fish formulation is TS = -23.85 Log (FL) – 22.59 with coefficient of determination (R^2 = 8.3%) and R^2 (adj) = 2.2% (figure 6). The regression equation for TS-length Leopard Coral Trout fish formulation is TS = 40.34 Log (FL) – 104.9 with coefficient of determination (R^2 = 12.9%) and R^2 (adj) = 0 % (figure 7). and the regression equation for TS-length Largehead Hairtail fish formulation is TS = 15.52 Log (FL) – 83.88 with coefficient of determination (R^2 = 73.8%) and R^2 (adj) = 65.1 % (figure 8).

3.3.Formula TS-Length

Mackerel Scad (*Decapterus macarellus*)

If the relation of the length of the fish to the TS value of each individual is based on the formula

$$TS = 20 \text{ Log (FL)} + C \quad (1)$$

then the following calculation applies:

Equation $y = -23.85x-22.59$ (2)

if $TS = 20 \log(FL) + C$

then $20 \log(FL) + C = -23.85 \log(FL) -22,59$

$C = -23.85 \log(FL) – 22.59 – 20 \log (FL)$

$C = -43.85 \log(FL) – 22.59$

By entering the values of FL in the range of fish measured Mackerel Scad, then obtained variation of value C (table 5). The average C value for fish Mackerel Scad with size 20.00 – 33.5 cm FL is C = -84,38. Foote equation for fish Mackerel Scad (*Decapterus macarellus*) obtained is TS = 20 Log (FL) -84,38.

Table 3. C value of FL value per length of fish Mackerel Scad.

FL	19	19.5	20	20.5	21	21.5	22	22.5	23	23.5	24
C	-78.66	-79.16	-79.64	-80.11	-80.57	-81.02	-81.46	-81.88	-82.30	-82.71	-83.11

24.5	25	25.5	26	26.5	27	27.5	28	28.5	29	29.5	30
-83.50	-83.89	-84.27	-84.64	-85.00	-85.36	-85.70	-86.05	-86.38	-86.72	-87.04	-87.36

30.5	31	31.5	32	32.5	33
-87.68	-87.99	-88.29	-88.59	-88.89	-89.18

Final Foote Equation for Mackerel Scad fish: TS = 20log – 84.38

If we repeat the procedure above for two other fish it will get the value of C = 20.34 log(FL) – 104.90 for Leopard Coral Trout (*Plectropomus leopardus)* and C = -4.48 log(TL) – 83.88 for Largehead Hairtail fish (*Trichiurus lepturus*), then entering the values of TL in the range of fish measured Coral Trout fish (21 – 28 cm TL) we get final foote equation Coral Trout fish : TS = 20 Log (TL) - 76,68. Same for Largehead Hairtail fish (43-50 cm TL), final foote equation: TS = 20 Log (TL) - 91.88.

4. Conclusion

1. In case of Torpedo Scad fish (*Megalaspis cordyla*), TS measurements for living condition and fresh dead condition treatments have no significant difference using T Paired test.
2. The measurements of TS values of each individual fish in the same species in this study have shown evidence and reinforce the hypothesis that TS values are a function of the length of the fish. TS

value has a linear relationship with the length of the fish, the larger the size of the fish the TS value is greater.

3. The empirical relationship in the form of The TS-length relationship was obtained as follow: for mackerel scad, TS = 20log-84.38; for coral trout, TS = 20Log-76.68, and for large headhairtail, TS = 20Log-91.88.

Acknowledgements
The data provided in this paper is the results from join research collaboration between BPPL-KKP and ITK-IPB under the research project "Measurement of Target Strength Characteristics and TS-Length Formulation on Fish Key Indicators for Improving Accuracy of Fish Stock Estimation TA 2016.

References

[1] Utama A A and Wudianto 2016 Dinamika spatial ikan mesopelagis (*Ceratoscopeluswarmingii,* Lütken 1892) di Samudera Hindia *J.Lit.Perik.Ind* **22**(4) 263-270
[2] Johanesson K A, Mitson R B 1983 *Fisheries Acoustic a Practical Manual for Acoustic Biomass Estimation* (Roma: FAO Fisheries Technical Paper)
[3] Sprintall J, Gordon A L, Murtugudde R and Susanto R D 2000 Semi annual Indian Ocean forced Kelvin wave observed in the Indonesian seas in May 1997 *J. Geophys. Res.* **105** 17217 17230
[4] Lee D J, Shin H I and Shin H H 1995 Fish stock assessment by hydroacoustic methods and its applications - I - Estimation of fish school target strength *J. Kor. Soc. Fish. Techno.***31**142-15
[5] Foote K G 1982 Optimizing copper spheres for precision calibration of hydroacoustic equipment *J. Acoust. Soc. Am.* **71** 742–747
[6] Simmonds E J and MacLennan D N 2005 *Fisheries Acoustic: Theory and Practice 2nd ed* (UK:Blackwell Science Ltd.)
[7] Hannachi M S, Abdallah L B and Marrakchi O 2004 Acoustic identification of small-pelagic fish species: target strength analysis and school descriptor classification *MedSudMed Technical Documents No.5*
[8] Walpole, Ronald E 1995 *Pengantar Statistika Edisi Ke-3* (Indonesia : Gramedia Pustaka Utama)

IOP Conf. Series: Earth and Environmental Science **176** (2018) 012032 doi:10.1088/1755-1315/176/1/012032

Feature extraction of coastal surface inundation via water index algorithms using multispectral satellite on North Jakarta

Asmadin[1*], V P Siregar[2], I Sofian[3], I Jaya[2], A B Wijanarto[3]

[1)] Faculty of Fisheries and Marine Science, Halu Oleo University, Kendari-Indonesia
[2]Departement of Marine Science and Technology, Bogor Agricultural University Bogor-Indonesia
[3]Geospatial Information Agency, Cibinong-Indonesia
*e-mail: asmadin234@gmail.com

Abstract. The study aims to characterize water index algorithms using Sentinel 2A-MSI and Landsat 8-OLI sensors to generate the performance of CSI and its proper uses in science. Feature extraction of the 35 combinations of CSI specifically uses information deriving from a number of water index spectral bands. Supervised classification is used to categorize water and non-water features using Spectral Angle Mapper (SAM). Furthermore, Dynamic Surface Water Extent (DSWE) algorithm was taken to mapping inundation and non-inundation. Result of this research shows that the best water index algorithms is exhibited by the MNDWI algorithm where it can enhance the water reflectance to be positive value. Re-classification on all water index algorithms indicates a similar pattern to the feature extraction of water and non-water. It is determined by overall accuracy (99.56-99.99%), Kappa coefficient (0.97-1.00), producer accuracy (99.05-99.99%) and user accuracy (93.43-99.94%). Substituting all water index algorithms into DSWE algorithm points out that the MNDWI is still the good performance and other substituted algorithms generate low performances. Overall feature extractions of CSI in North Jakarta at this research manifests that the MNDWI algorithm using multi-spectral and satellites is the best algorithm.

1. Introduction

Global climate change causes North Jakarta included in 10 major cities in Southeast Asia, which is vulnerable to environmental pressure [1]. Its location in the coastal area makes it the most vulnerable coastal city to environmental pressure in Indonesia. This area is potentially vulnerable to coastal surface inundation (CSI), because of its lowest topography, severe impact of land subsidence, and the impact of local sea level rise due to ongoing climate change. Over the last five years, the flood-prone points have increased rapidly and spread evenly across all regions. Information about CSI is important to take into account, because the causes of inundating in this city by natural and anthropogenic factors affect the environmental and ecological aspects [2], as well as efforts to protect the regional environment, health and economic activities.

The main approach to obtain information about CSI changes in recent years can use remote sensing satellite [3-6]. *U.S. Geological Survey* (USGS) specifically developed Landsat 8, to assess aspects of Dynamic Surface Water Extent (DSWE), which focused on ground surface inundation [7]. Development of Landsat 8 has more bands and better bandwidth design as a form of evolutionary technical improvement [8]. Likewise with Sentinel 2A developed by European Space Agency (ESA) since it was launched in mid-2015. Sentinel 2A data has a better special resolution than Landsat 8. The study of CSI feature extraction refers to the approach of water body extraction. The water body extraction approach itself initially used a spectral band-based approach. It is then developed into a Water Index Algorithm to map the body of water. The method of feature extraction of CSI specifically is still evolving, but in general the extracting process refers to two major Water Index Algorithm approaches; first is the Traditional Supervised and Unsupervised Classifications by using one band or multiple bands [9-11], and another approach is the Spectral Index approach related to water and threshold [12-16]. This approach has been widely used, due to the unique spectral characteristics of water bodies in Visible and Infrared bands. Generally, spectral analysis can effectively identify and detect bodies of water based on supervised classification techniques. The Water Index Algorithm remains an important option to be implemented in rapid water mapping in large-scale areas [17], although this approach has limitations [18].

Implementation of Water Index Algorithm itself to separate the object of water and non-water has experienced a fairly rapid development and has lasted for more than three decades. Recently, a Water Index Algorithm implementation was developed to map surface inundation using the DSWE algorithm [7]. History records that in the beginning, Crist and Cicone [19] and Crist [20] suggested Tasselep Cap Wetness (TCW) index derived from six-band surface reflection data and setting a threshold of 0 to separate the water and non-water objects. Furthermore, McFeeters [21] proposed Normalized Difference Water Index (NDWI) by using the green band value minus the near-infrared band (NIR) and divided by total of both bands, and the water body has a positive value while the non-water has a negative value. NDWI can suppress and eliminate non-water to a large extent, but NDWI cannot efficiently suppress ground signals. After nearly a decade, then developed various modifications of Water Index Algorithms. Successively developing Modified Normalized Difference Water Index (MNDWI) by Xu [22], Sum457 index (Al-Khudhairy et al. [23], combined by Sum457, ND5723, and ND571 indices by Beeri and Phillips [24], New Water Index (NWI) by Haibo et al. [25], Automatic Water Extraction Index (AWEI) by Feyisa et al. [26], Land Surface Water Index (LSWI) by Dong et al. [27]. Combinations of each of the three water indices (LSWI, MNDWI, and NDWI) with EVI and NDVI by Menarguez [28]. The latter index is more sensitive to water bodies, especially mixed water pixels and vegetation pixels.

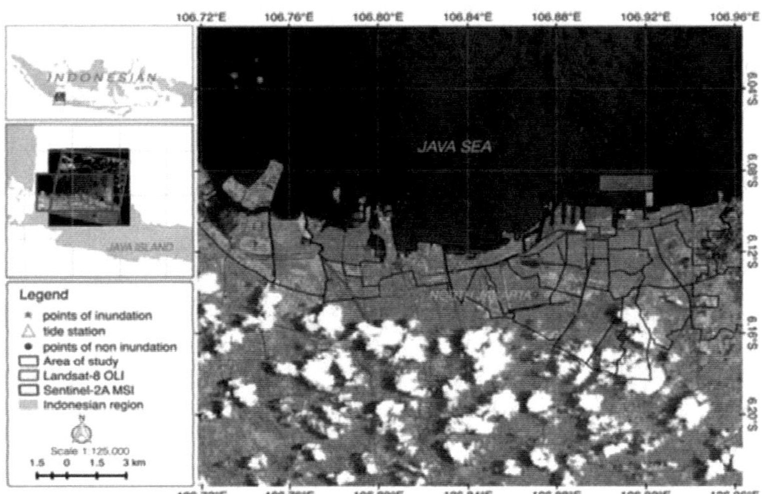

Figure 1. Photo of Satellite imagery using composite GREEN, NIR and MIR bands of Sentinel 2A (Right view). In Left view show Sentinel 2A (BLUE insert) and Landsat 8 (GREEN insert), ground truth survey area inundation (RED) and non-inundation (BLUE). Photos of cloud/shadow coverage also appear on Sentinel 2A.

Apart from the various implements of Water Index Algorithm above using multispectral satellites, especially for extracting the performance of CSI in North Jakarta that has not been conducted. It is the main objectives of this research.

2. Method

2.1. Time and description of study area

This study was conducted in the area of North Jakarta City, DKI Jakarta Province, Indonesia (figure 1) on June 2016 until July 2017. This city is a coastal city located in the lowlands of DKI Jakarta (6°7-22'S and 106°40-55'E).

2.2. Data acquisition and processing

The acquisition of satellite data is listed in table 1. The data are corrected products that have surface reflection data from Level 1 Terrain (L1T) Landsat 8-OLI from USGS, and Level 2A products Sentinel 2A from the European Space Agency (ESA). All such data are in the format of the Archive Standard European Format (SAFE) [29].

Table 1. Landsat 8-OLI and Sentinel 2A-MSI used in this study.

Sensors	Data type take	Acquisition Date/ Time at Scene	Sun Zenith Angle (°)	Sun Azimuth Angle (°)
Landsat 8 OLI	Path/Row 122/64	13 May 2016/02:59:3	35.98	47.04
	Path/Row 122/64	30 June 2016/02:59:55	49.72	43.06
	Path/Row 122/64	17 August 2016/03:00:10	55.68	56.02
Sentinel 2A MSI	Orbit number 32	19 July 2016)/03:14:23	35.83	41.16
	Orbit number 32	07 October 2016)/ 03:18:51	20.53	90.25

The Landsat 8 OLI satellite is the new generation of Landsat launched since February 11, 2013, bringing two earth observation sensors, Operational Land Imager (OLI) and Thermal Infrared Sensor (TIRS). OLI collects imagery data for 9 spectral bands with 30 m spatial resolution (15 m panchromatic band) and TIRS collects image data for 2 thermal infrared spectral bands with a 100 m spatial resolution. Both sensors have an area of view coverage of 15° with a spatial resolution of 185 km across a height of 705 km [29]. Sentinel 2A was launched on June 23, 2015, equipped with 13 multispectral bands producing high resolution imagery and a wide sweeping capacity for a new perspective of science. The combination of high resolution, spectral capabilities, and 290 km of sweep width will provide a view of the earth with an unprecedented 20.6° field of view coverage. Characteristics of the Landsat 8 and Sentinel 2A sensor bands are shown in table 2 and figure 2.

Table 2. Characteristics of Landsat 8 and Sentinel 2A used in this study.

Landsat 8-OLI			Sentinel 2A-MSI		
Bands	Wavelength (μm)	Resolution (m)	Bands	Wavelength (μm)	Resolution (m)
1 Coastal/Aerosol	0.43-0.44	30	1 Coastal/Aerosol	0.43-0.46	60
2 Blue (B)	0.45-0.51	30	2 Blue (B)	0.44-0.54	10
3 Green (G)	0.53-0.59	30	3 Green (G)	0.55-0.58	10
4 Red (R)	0.63-0.67	30	4 Red (R)	0.65-0.68	10
			5 Red edge 1 (RE1)	0.70-0.72	20
			6 Red edge 2 (RE2)	0.73-0.75	20
			7 Red edge 3 (RE3)	0.77-0.79	20
5 Near infrared (NIR)	0.85-0.88	30	8 NIR	0.76-0.90	10
			8a NIR narrow (NIRn)	0.86-0.88	20
			9 Water vapour	0.94-0.96	60
9 SWIR/cirrus	1.36-1.39	30	10 Shortwave infrared/cirrus	1.36-1.39	60
6 Shortwave infrared 1 (SWIR-1)	1.57-1.65	30	11 SWIR-1	1.54-1.68	20
7 Shortwave infrared 1 (SWIR-1)	2.10-2.29	30	12 SWIR-2	2.08-2.32	20
8 Panchromatic	0.50-0.68	15			

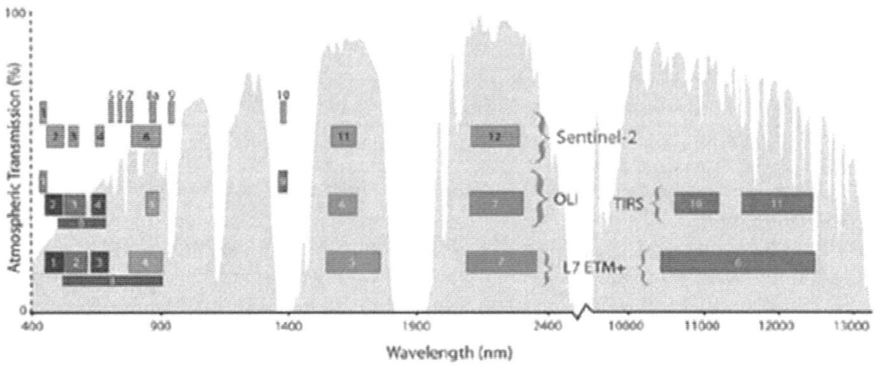

Figure 2. Sentinel-2 and Landsat Spectral Bands (Source: https://earth.esa.int/).

Data processing of Landsat 8 and Sentinel 2A to map water bodies and inundation can divide into 2 stages, pre-processing and post-processing (figure 3). Data pre-processing used the Semi-Automatic Classification Plugin (SCP) developed by Congedo [30] to perform atmospheric correction [31,32] to convert the images into product 2A, which is the Bottom Of Atmosphere (BOA) reflectance data. QGIS has an interface with Free Open Source Plugin (FOSS). One of them was used in this research process that is Semi-Automatic Classification Plugin (SCP). It included assisting in the downloading of satellite data, determining ROI signatures, classification, error matrix test, masking, and etc. Multispectral data processing refers to the equation in table 3.

Table 3. A summary of open surface water body mapping algorithms based on different Water Indices Algorithms using Landsat 8 and Sentinel 2A.

Water Index Algorithms	Formulas	References
NDVI	(NIR – Red)/(NIR+Red)	[33]
NDWI	(Green – NIR)/(Green+NIR)	[21]
NDMI	(NIR – MIR)/(NIR+MIR)	[35]
MNDWI	(Green – MIR)/(Green+MIR)	[22]
WRI	(Green + Red)/(NIR+MIR)	[34]
NWI	Blue – (NIR+MIR+SWIR) / Blue + (NIR+MIR+SWIR)	[25]
AWEI	4 x (Green – MIR) – (0.25 x NIR + 2.75 x MIR)	[26]
DSWE	(MNDWI > 0.123) or MNDWI>-0.5 and NIR<0.2 and SWIR<0.1	[7]

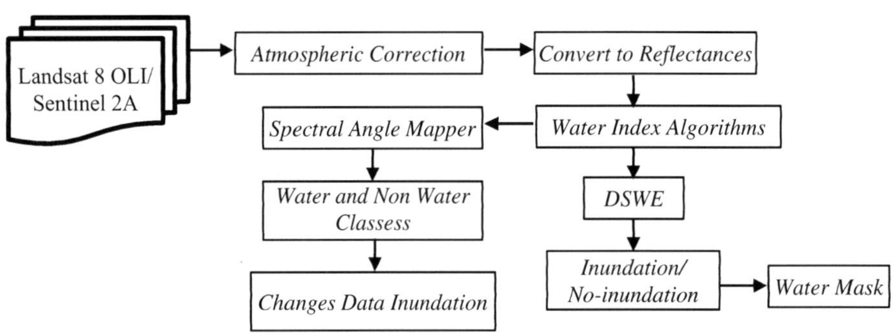

Figure 3. Data processing of Landsat 8 and Sentinel 2A to map water bodies and inundation.

2.3. Data validation

The result of CSI mapping based on the visible extraction of the Landsat 8-OLI and Sentinel 2A-MSI satellites needs to be validated. Data validation, such as by direct measurement using a tide gauge MOTIWALI (Mobile Tide Water Level Instrument). MOTIWALI has a good precise. This instrument was first developed by the Laboratory of Marine Instrumentation of Department Marine Science and Technology, Bogor Agricultural University since 2007. It is designed to measure real tidal data in real time to predict the Sea Level Rise during May-June 2016, even water levels during peak flooding (3-6 June 2016). In addition, tracing information from various online mass media sources and direct surveys (figure 1), including checking of some reservoir dikes and coastal dikes were destroyed (table 4).

Table 4. The Location of coastal walls were destroyed, estimated the height of inundation, and predict of tide in North Jakarta during early period June, 3 until 6, 2016.

Date/Day/Time	Locations	Height of Inundation (m)	Height of Tide (m)*
6/3/2016 (Friday) 20.55 PM	Mutiara Beach	0.30-0.50	0.84 (2:12 AM)
6/4/2016 (Saturday)	Muara Angke Penjaringan	0.40	0.79 (1:12 AM)
6/5/2016 (Sunday)	East side of coastall wall of Muara Baru Penjaringan (near of the TPI)	0.50	0.30 (15:24 AM)
6/6/2016 (Monday) 22.00 PM	Reservoir wall of Muara Angke Penjaringan	0.80	0.35 (15:36 AM)

Source: Primer data processed from trace study (2016)
*Data analysis of real time of tide gauge station was designed by AIK Laboratory IPB

3. Results and discussion

3.1. Inundation extraction of coastal surfaces based on water index algorithm

This result study shows that there are similarities of feature extraction results of similar spectral characteristics and different spatial resolution between Sentinel 2A and Landsat 8 sensors (figure 4). A contrasting difference exists in the extraction of water and non-water feature of each Water Index Algorithm classification. Sentinel 2A can be partially attributed to higher spatial resolution. All classifications of Water Index Algorithms indicate any abilities to increase water reflectance, and the MNDWI is more capable than other water index.

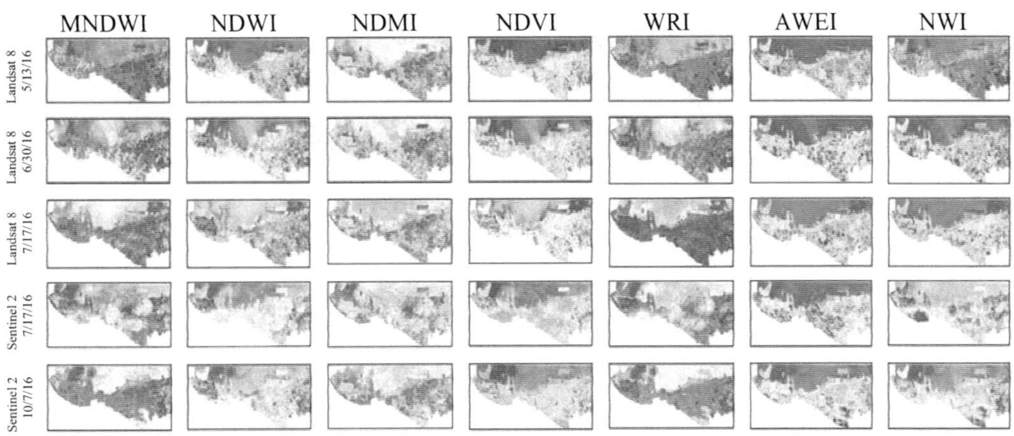

Figure 4. Feature extractions of different combination of the seven Water Index Algorithms.

All results of the CSI feature of 35 combinations generating from seven algorithms (except DSWE algorithm) of both satellite sensors in figure 4 above, show similar patterns, whereas Sentinel 2A-based results show resolution against higher water areas than Landsat 8. Nevertheless, the feature extraction of CSI's is showing that Landsat 8 sensor on condition some days after inundating and dikes were destroyed (30th of June 2016) shows a better performance than Sentinel 2A (July 17, 2016). Unlike the case of CSI's feature on both sensors prior to inundation of Landsat 8 (May 31, 2016) data and after inundation of Landsat 8 (August 17, 2016) and Sentinel 2A (October 7, 2016) sensors. Clouds/shadows cause the water's feature to be reduced on both sensors. It was clearly the worse by the Sentinel 2A sensor (July 17, 2016). In addition, the degree of clouds/shadows of varying optimum thresholds at different locations and times of all indices cannot effectively eliminate shadow and cloud noise in some places.

Of the seven Water Index Algorithms above, the extraction of MNDWI results over other algorithms, the quality is different in yielding three things: (1) water has a positive value greater than any other index, since water absorbs more MIR than NIR; (2) the wake land has a negative value; and (3) the soil and vegetation have a negative value, because the soil is more reflective of MIR than the NIR and its vegetation reflects MIR still more than GREEN. Consequently, compared to other Water Index Algorithms, the contrast between water and artificial landmass MNDWI is greatly enhanced due to an increase in the value of water reflectance and a decrease in the value of land built-up from positive to negative. A reflectance water increase in the MNDWI image will result in a more accurate extraction of open water, such as water bodies on land. While the mixture of soil and vegetation built up throughout the negative value and therefore can be suppressed, even thrown away, this condition is also described equally by Xu [22]. Unlike the case with the NWI, although the ability to extraction the appearance of water bodies includes small bodies of water both, but it is not good to separate the residents and water bodies. This is reflected from the water and non-water reflectance values, which are all negative. While the water and non-water reflectance values of WRI are all positive and larger than all other water indexes, making it difficult to separate water with other objects.

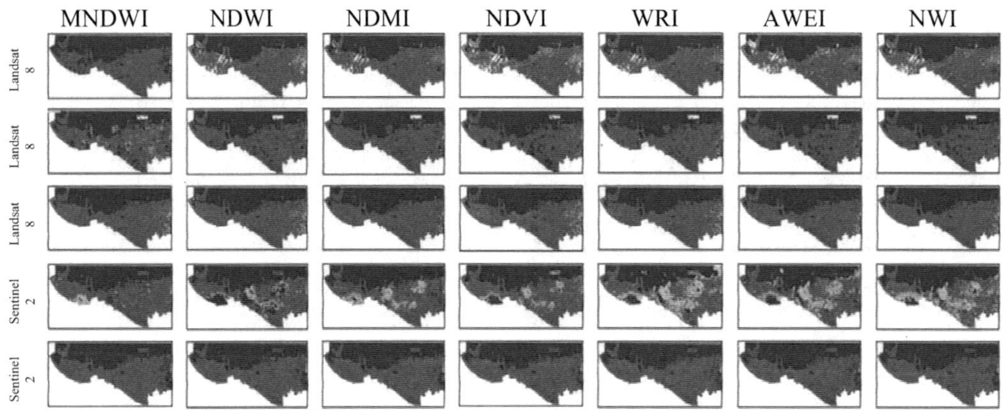

Figure 5. Feature extractions of different combination of the seven algorithms using SAM.

The result study show also similar performance and pattern on both sensors and all Water Index Algorithms for water and non-water classes using supervised classification Spectral Angle Mapper (SAM) (figure 5). The feature is more clearly seen. SAM is a physical-based spectral classification that uses n-D angles to match pixels with the reference spectrum. The spectrum membership is used by SAM coming from the ROI file extracted directly from the image of the water index algorithm classification results. The overall performance of CSI based on the classification of Water Index Algorithms and SAM (figures 2 and 3) of Landsat 8 and Sentinel 2A sensors using MNDWI was more accurate to identify CSI compared to other algorithms, since it is supported by four things: (1) water has a larger positive absorbing value more MIR than NIR; (2) the land built-up has a negative value more reflecting the MIR than the NIR; (3) land and vegetation have negative values, reflecting more MIR than GREEN; and (4) the contrast between water and the land build-up use MNDWI is clear and more accurate than other algorithms based on an increase in the value of water reflectance and a decrease in the value of land built-up from positive to negative.

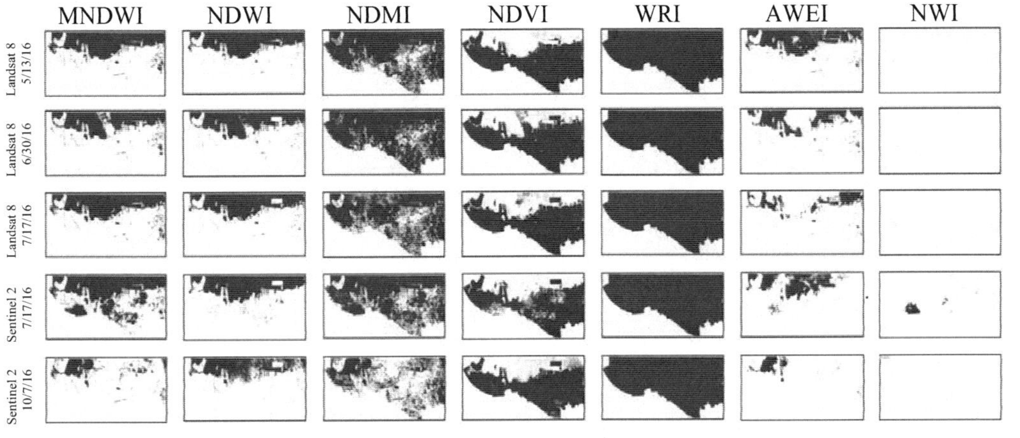

Figure 6. Contrast of the surface inundation from 35 combinations of the seven algorithms refers to Jones [7] using $DSWE_1$ in table 4 and the two sensors.

The positive or zero water reflectance value allows to extract the inundation and non-inundation, not only water bodies or shallow waters use the $DSWE_1$ algorithm [7] that not only applies MNDWI to the Landsat sensor, but also Sentinel 2A and the other Water Index Algorithms. Based on figure 6 and 7 shows that the contrast of inundation and non-inundation appears clearly, for example using the $DSWE_1$ and $DSWE_2$ algorithms [7]. $DSWE_1$ = MNDWI> 0.123 applied to Landsat sensor [7,36] and developed for Sentinel 2A sensor. The inundation feature is still clearly from the $DSWE_1$ algorithm using NDWI> 0.123 substitution, although not optimal than MNDWI. While $DSWE_1$ is substituted with other Water Index Algorithms (NDMI, NDVI, WRI, AWEI, NWI) shows the ability to extract the feature to be reduced even not seen at all.

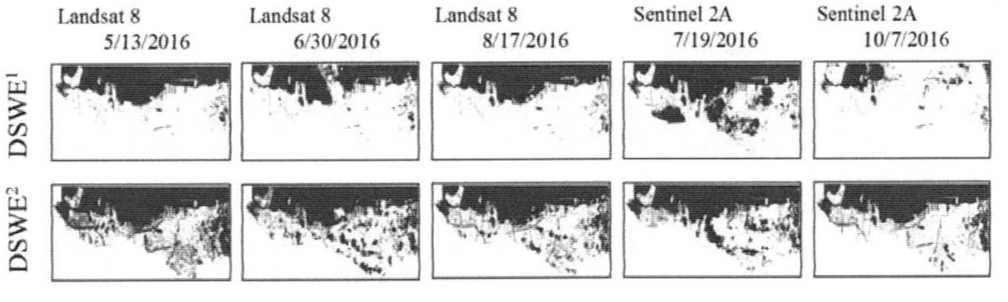

Figure 7. Contrast of the surface inundation from DSWE algorithms refers to Jones [7] in table 4 and the two sensors.

The changes of CSI are also apparent (figure 7) from Landsat 8 and Sentinel 2A data using $DSWE_2$ [7], where $DSWE_2$ = MNDWI> -0.5 AND NIR <0.2 AND SWIR <0.1. $DSWE_2$ uses a combination of GREEN, MIR, NIR and SWIR bands. The feature of CSI processed by $DSWE_1$ is more contrast than $DSWE_2$. Both of the DSWEs are not able to effectively eliminate shadow and cloud noise in different locations. Nevertheless, the feature of CSI is still clearly from Landsat 8 sensors prior to inundation (May 31, 2016), after subsequent inundating in the next few months (August 17, 2016), and even the results of sedimentation in shallow waters in the days after the dykes destroyed (June 30, 2016). The application of DSWE using Sentinel 2A sensor shows the performance of CSI's appearance was not clearly visible compared to Landsat 8. Unlike the case with the Sentinel 2A sensor (October 7, 2016)

IOP Conf. Series: Earth and Environmental Science **176** (2018) 012032 doi:10.1088/1755-1315/176/1/012032

after subsequent inundating in the next few months, the appearance of CSI is clearly indicated with better spatial resolution. Overall it was formulated that the application of DSWE using the threshold of MNDWI compared to other water indices greatly affects the inundation mapping, since the MNDWI water threshold value is not the same on other water indices.

3.2. Assessment of sensor accuracy and water index algorithms for water and non-water mapping

The results of the study found that the overall accuracy averaged of 95.56-99.99% and Kappa coefficient 0.97-1.00 (table 5). Both Landsat 8 and Sentinel 2A sensors perform a better performance in mapping the CSI. The study area is an urban area dominated by land built-up, that it has own advantages for satellites to extract CSI visibility by separating water and non-water objects.

Table 5. (a) Overall accuracies and (b) Kappa coefficients of different combination of the seven algorithms and the two sensors.

Sensors	Data acquitions	Overall accuracy (%)							Kappa Coefficients						
		MNDWI	NDWI	NDMI	NDVI	WRI	AWEI	NWI	MNDWI	NDWI	NDMI	NDVI	WRI	AWEI	NWI
Landsat 8	13-May-16	100.00	100.00	100.00	100.00	100.00	100.00	100.00	0.99	1.00	1.00	1.00	1.00	1.00	1.00
	30-Jun-16	100.00	100.00	100.00	100.00	100.00	100.00	100.00	1.00	1.00	1.00	1.00	1.00	1.00	1.00
	17-Aug-16	100.00	100.00	100.00	100.00	100.00	100.00	99.98	1.00	1.00	1.00	1.00	1.00	1.00	0.99
Sentinel 2A	17-Jul-16	99.90	99.40	94.34	99.96	99.82	99.91	97.80	0.99	0.97	0.84	0.99	0.99	0.99	0.96
	7-Oct-16	99.97	100.00	100.00	100.00	100.00	100.00	100.00	0.99	1.00	1.00	1.00	1.00	1.00	1.00
Average		99.97	99.88	98.87	99.99	99.96	99.98	99.56	0.99	0.99	0.97	1.00	1.00	1.00	0.99

The results of Landsat 8 and Sentinel 2A image accuracy based on producer accuracy reaching 99.05-99.99% and user accuracy is about 93.43-99.94% (table 6). The achievement of such accuracy test results, as the result of spectral classification of water bodies/inundation and non-water/non-inundation on land built-up is easily detected in the study area. Even the detection of inundations along the highway at the bottom along the highway, the appearance of water is still clearly through the classification. This result is verily clear performing water in residential areas, mangrove ecosystems, reservoirs and others.

Table 6. Producer accuracies (a) and User accuracies (b) of different combination of the seven algorithms and the two sensors.

Sensors	Data acquitions	Producer accuracy (%)							User accuracy (%)						
		MNDWI	NDWI	NDMI	NDVI	WRI	AWEI	NWI	MNDWI	NDWI	NDMI	NDVI	WRI	AWEI	NWI
Landsat 8	13-May-16	99.99	100.00	100.00	100.00	100.00	100.00	100.00	100.00	100.00	100.00	100.00	100.00	100.00	100.00
	30-Jun-16	100.00	100.00	100.00	100.00	100.00	100.00	100.00	100.00	100.00	100.00	100.00	100.00	100.00	100.00
	17-Aug-16	100.00	100.00	100.00	100.00	100.00	100.00	99.98	100.00	100.00	100.00	100.00	100.00	100.00	100.00
Sentinel 2A	17-Jul-16	99.96	96.07	96.90	99.96	99.82	95.89	95.26	99.25	67.15	74.30	99.50	99.72	79.23	92.96
	7-Oct-16	99.98	100.00	100.00	100.00	100.00	100.00	100.00	99.96	100.00	100.00	100.00	100.00	100.00	100.00
Average		99.99	99.21	99.38	99.99	99.96	99.18	99.05	99.84	93.43	94.86	99.90	99.94	95.85	98.59

4. Conclusion

Feature extraction of 35 combined CSI can be generated from water index algorithms (MNDWI, NDWI, NDMI, NDVI, WRI, AWEI, NWI, and DSWE and also two sensors form Landsat 8 OLI and Sentinel 2A MSI in North Jakarta area. To enhance the characteristics of those both sensors, atmospheric correction is conducted in best performance. Supervised classification using SAM shows a good performance in water and non-water where indicating by overall accuracy (99.56-99.99%), coefficient Kappa (0.97-1.00), producer accuracy (99.05-99.99%) and user accuracy (93.43-99.94%). Both sensors and all various Water Index Algorithms have good performance in open surface water mapping. MNDWI has different quality for water body, inundation and non-inundation mapping accurately in study area. Performance of all Water Index Algorithms using both sensors will be optimal if the capture has low cloud coverage. The results of DSWE Algorithm are not good enough to be applied in conjunction with other water indices, except MNDWI. DSWE algorithm applied with MNDWI can map the inundation and non-inundation, but cannot effectively eliminate the shadow and cloud noise in different areas and locations for Landsat 8 and Sentinel 2A sensors. It is needed a caution and wisely decision in implementing those algorithms above to mapping CSI.

References

[1] Yusuf A A and Francisco H A 2009 *Climate Change Vulnerability Mapping for Southeast Asia* http://www.eepsea.org/pub/tr/12324196651Mapping_Report.pdf

[2] Sutrisno D, Windiastuti R, Sofyan I and Ramdhani D 2011 Spatial modeling of sea level rise projections for health vulnerability estimation *Globë* **13**(2) 102-111

[3] Ji L, Zhang L and Wylie B 2009 Analysis of dynamic thresholds for the normalized difference water index *Photogramm. Eng. Remote Sens.* **75** 1307–1317

[4] Chandrasekar K, Sesha Sai M V R, Roy P S and Dwevedi R S 2010 Land surface water index (LSWI) response to rainfall and NDVI using the MODIS Vegetation Index product *Int. J. Remote Sens.* **31** 3987–4005

[5] Barrett D and Frazier A 2016 Automated method for monitoring water quality using landsat imagery *Water* **8** 257

[6] Torres-Rua A, Ticlavilca A, Bachour R and McKee M 2016 Estimation of surface soil moisture in irrigated lands by assimilation of landsat vegetation indices, surface energy balance products, and relevance vector machines *Water* **8** 167

[7] Jones J W 2015 Efficient Wetland surface water detection and monitoring via landsat: comparison with in situ data from everglades depth estimation network *Remote Sens.* **7** 12503-538

[8] Loveland T R and Irons J R 2016 Landsat 8: The plans, the reality, and the legacy *Remote Sens. Environ.* **185** 1–6

[9] Sheng Y, Shah C A and Smith L C 2008 Automated image registration for hydrologic change detection in the Lake-Rich Arctic *IEEE Geosci. Remote Sens. Lett.* **5** 414–418

[10] Sivanpillai R and Miller S N 2010 Improvements in mapping water bodies using ASTER data *Ecol. Inf.* **5** 73–78

[11] Huang C, Chen Y and Wu J 2014 Mapping spatio-temporal flood inundation dynamics at large river basin scale using time-series flow data and MODIS imagery *Int. J. Appl. Earth Obs. Geoinf.* **26** 350–362

[12] Li W, Du Z, Ling F, Zhou D, Wang H, Gui Y, Sun B and Zhang X 2013 A Comparison of land surface water mapping using the normalized difference water index from TM, ETM+ and ALI *Remote Sens.* **5** 5530–49

[13] Li W, Qin Y, Sun Y, Huang H, Ling F, Tian L and Ding Y 2015 Estimating the relationship between dam water level and surface water area for the Danjiangkou Reservoir using Landsat remote sensing images *Remote Sens. Lett.* **7** 121–130

[14] Xie H, Luo X, Xu X, Tong X, Jin Y, Pan H and Zhou B 2014 New hyperspectral difference water index for the extraction of urban water bodies by the use of airborne hyperspectral images *J. Appl. Remote Sens* **8** 085098

[15] Hui F, Xu B, Huang H, Yu Q and Gong P 2008 Modelling spatial-temporal change of Poyang Lake using multitemporal Landsat imagery *Int. J. Remote Sens.* **29** 5767–84

[16] Ji L, Geng X, Sun K, Zhao Y and Gong P 2015 Target detection method for water mapping using Landsat 8 OLI/TIRS Imagery *Water* **7** 794–817

[17] Ouma Y O and Tateishi R 2006 A water index for rapid mapping of shoreline changes of five East African Rift Valley lakes: an empirical analysis using Landsat TM and ETM+ data *Int. J. Remote Sens.* **27** 3153–81

[18] Du Z, Li W, Zhou D, Tian L, Ling F, Wang H, Gui Y and Sun B 2014 Analysis of Landsat-8 OLI imagery for land surface water mapping *Remote Sens. Lett.* **5** 672–681

[19] Crist E P and Cicone R C 1984 A physically-based transformation of thematic mapper data—The TM tasseled cap *IEEE Transactions on Geoscience and Remote Sensing* **22** 256-263

[20] Crist E P 1985 A TM Tasseled cap equivalent transformation for reflectance factor data *Remote Sens. Environ.* **17** 301–306
[21] McFeeters S K 1996 The use of the Normalized Difference Water Index (NDWI) in the delineation of open water features *Int. J. Remote Sens* **17** 1425–32
[22] Xu H 2006 Modification of normalised difference water index (NDWI) to enhance open water features in remotely sensed imagery *Int. J. Remote Sens.* **27** 3025–3033
[23] Al-Khudhairy D, Leemhuis C, Hofhnann V, Shepherd I, Calaon R, Thompson J, Gavin H, Gasca-Tucker D, Zalldls Q and Bilas G 2002 Monitoring Wetland Ditch Water levels using Landsat TM and Ground-Based Measurements *Photogramm. Eng. Remote Sens.* **608** 809–818
[24] Beeri O and Phillips R L 2007 Tracking palustrine water seasonal and annual variability in agricultural wetland landscapes using Landsat from 1997 to 2005 *Glob. Chang. Biol.* **13** 897–912
[25] Haibo Y, Zongmin W, Hongling Z and Yu G 2011 Water body extraction methods study based on RS and GIS *Procedia Environmental Sciences* **10** 2619-262
[26] Feyisa G, Meilby H, Fensholt R and Proud S R 2014 Automated Water Extraction Index: a new technique for surface water mapping using Landsat imagery *Remote Sens. Environ.* **140** 23–35
[27] Dong J, Xiao X, Menarguez M A, Zhang G, Qin Y, Thau D, Biradar C and Moore B 2016 Mapping paddy rice planting area in Northeastern Asia with Landsat 8 images, phenology-based algorithm and Google Earth Engine *Remote Sens. Environ.* **185** 142–154
[28] Menarguez M A 2015 *Global Water Body Mapping from 1984 to 2015 Using Global High Resolution Multispectral Satellite Imagery* (Norman: University of Oklahoma)
[29] Irons J R, Dwyer J L and Barsi J A 2012 The next Landsat satellite: The Landsat Data Continuity Mission *Remote Sens. Environ.* **122** 11–21
[30] Congedo L 2017 *Semi-Automatic Classification Plugin Documentation Release 5.3.6.1.* https://www.researchgate.net/publication/307593091_SemiAutomatic_Classification_Plugin_Documentation_Release_5361 [accessed Aug 29, 2017]
[31] Novelli A, Aguilar M A, Nemmaoui A, Aguilar F J and Tarantino E 2016 Performance evaluation of object based greenhouse detection from Sentinel-2 MSI and Landsat 8 OLI data: a case study from Almería (Spain) *Int. J. Appl. Earth Obs. Geoinf.* **52** 403–411
[32] Vermote E, Justice C, Claverie M and Franch B 2016 Preliminary analysis of the performance of the Landsat 8/OLI land surface reflectance product *Remote Sens. Environ.* **185** 46–56
[33] Rouse J W, Haas R H, Schell J A and Deering D W 1973 Monitoring vegetation systems in the great plains with ERTS (Earth Resources Technology Satellite) In Proceedings of Third Earth Resources Technology Satellite Symposium, Greenbelt, ON, Canada, 10–14 December 1973; Volume SP-351, pp. 309–317
[34] Shen L and Li C 2010 Water body extraction from Landsat ETM+ Imagery using adaboost algorithm In Proceedings of 18th International Conference on Geoinformatics, 18–20 June 2010, Beijing, China; pp. 10.1109/geoinformatics.2010.5567762
[35] Wilson E H and Sader S A 2002 Detection of forest harvest type using multiple dates of Landsat TM imagery *Remote Sens. Environ.* **80** 385–396
[36] Carroll M L and Loboda T V 2017 Multi-decadal surface water dynamics in North American Tundra *Remote Sens* **9** 497

IOP Conf. Series: Earth and Environmental Science **176** (2018) 012033 doi:10.1088/1755-1315/176/1/012033

Acoustic approach for estimation of Skipjack (*Katsuwonus pelamis*) abundance in Bone Bay

R S Putri[1], I Jaya[2*], S Pujiyati[2], A Priatna[3], A Makmun[3] and A Suman[3]

[1] Department of Marine Technology, Graduate School of Bogor Agricultural University, Indonesia
[2] Faculty of Marine Science and Technology, Bogor Agricultural University, Indonesia
[3]Research Institute for Marine Fisheries, Ministry of Marine and Fisheries Affairs, Indonesia
*e-mail: indrajaya@ipb.ac.id

Abstract. Skipjack is one of the most important economical fish. As a result, skipjack stocks are vulnerable to over-exploitation. Therefore, to anticipate and to set sound management plan, it is necessary to monitor from time to time the abundance of these fish to ensure the sustainability of its resources. The purpose of this study was to determine the abundance distribution of skipjack fish based on acoustic method in Bone Bay waters. The abundance estimate of the skipjack was conducted using Split Beam Echosounder Simrad EY 60 with frequency 120 kHz. The results of this study indicated that the range of skipjack target strength (TS) was -67.0 dB to -55.3 dB, or equivalent to fish length of 8 cm to 31 cm, detected from surface to a depth of 40 m. Temporally, skipjack density during the day was more towards the surface with an average of 43 ind/1000m^3 and at night in deeper waters with an average density of 36 ind/1000m^3. Vertically, the highest density was detected at a depth of 5 m - 15 m with a density of 82 ind/1000m^3. While horizontally, the highest skipjack density of 188 ind/1000m^3 was found around Kolaka waters.

1. Introduction

Skipjack is one of fishery commodities in Indonesia which has important economic value. This makes this fish as one of the main target in fishing activities. One of Skipjack producing areas in Indonesia is Bone Bay waters. Bone Bay is a potential waters for Skipjack fishing in Eastern Indonesia. The potential abundance of Skipjack is related by the spatial and temporal pattern of biophysical distribution of the oceanographical condition in Bone Bay. The oceanography condition near the mouth of the bay (Flores Sea) is certainly different from the other area [1]. It is suspected as a trigger of the high abundance of Skipjack in this waters.

Skipjack fishing mostly use pole and line, purse seine, gillnet and handline. The skipjack behaviour that form a schooling gives the opportunity to catch Skipjack in large quantities. As the result, Skipjack's resource is vulnerable to over-exploitation. Although fishery resources are a renewable resource, the rate of recovery may be unbalanced with the rate of utilization.

In 2005-2014, Skipjack production in WPP 713 experienced an annual increase of 11.40%, with production in 2014 was 56.299 tons. Skipjack production in Indonesia in 2014 was 496.682 tons, meaning that 11% of it came from WPP 713 [2]. With the increasing of skipjack production, it is necessary to have information about the status of Skipjack's stock. The estimation of Skipjack stock is important so that the potential of Skipjack resources can be utilized sustainably and to avoid the overfishing. One of the most effective methods for estimating fish stocks is by using the acoustic method. The principle of acoustic method is based on the echo system, that is the reflection of the emitted sound by something that prevents it [3]. The acoustic method has several advantages in estimating schooling of fish and its abundance, including information that can be generated faster and covering larger areas, stock estimation can be done in real time and in situ regardless of data fishery statistics, have high accuracy, can be used when other methods are unusable, and are not harmful or destructive because the frequency of sound used is not harmful to the users or the target of the survey [4].

The existence of Skipjack's schooling can facilitate the detection of Skipjack using acoustic method. The estimation of abundance of fish resources through acoustic survey generally has the same condition with other fishery survey, that is to gather data about a fisheries resource in an area. This analysis can produce information about the conditions of a fisheries resources [5]. One approach to estimate the fish abundance with acoustic methods is by measuring the target strength (TS). Measurement of fish target strength is the main parameter in estimating fish stock with acoustic method [6]. Generally, the higher the target strength of fish, the larger the size of the fish, but not always so. In addition to size, TS values are also influenced by the swim bladder in the fish, fish body shape or type of fish itself.

The purpose of this research was to estimate the density of Skipjack temporally (day and night), vertically (depth) and horizontally (spatial) using the acoustic method in Bone Bay waters, and to validate the results with the results from fishery survey. It was expected that the results of this study could provide information about the existence and abundance of Skipjack. This information could be used to support the sustainable management of Skipjack fisheries resources in the waters of Bone Bay.

2. Methods

The acoustic survey was conducted in April 2014 in Bone Bay waters and was validated using a survey on landing site in September - October 2016. The survey on landing site was conducted to collect data about the size of Skipjack in two fishing port with high Skipjack production, namely Sinjai and Luwu. The acoustic survey track that was done in Bone Bay waters was shown in figure 1.

Acoustic data was collected using Split Beam Echosounder Type Simrad EY 60 with frequency 120 kHz. Software Echoview, R, and ArcGIS 10.3 were used for acoustic data analysis. Fork Length (FL) data were obtained by following the Skipjack fishing operation by using pole and line in Bone Bay waters. In addition, FL measurement was also conducted at the fishing port in Sinjai and Luwu. FL measured using roll meter.

Acoustic data processing followed the method in [7] with the following processing steps :
Target Strength analysis and average TS are calculated by equations (1) and (2).

$$TS = 10 \log ts \quad (1)$$

$$\overline{ts} = \frac{\sum ts}{n} \quad (2)$$

where TS is acoustic reflection strength of fish as a single target (dB).

Scattering volume (SV) and average SV are calculated by equations (3) and (4).

$$SV = 10 \log sv \quad (3)$$

$$\overline{sv} = \frac{\sum sv}{n} \quad (4)$$

where SV is *volume backscattering strength* (dB).

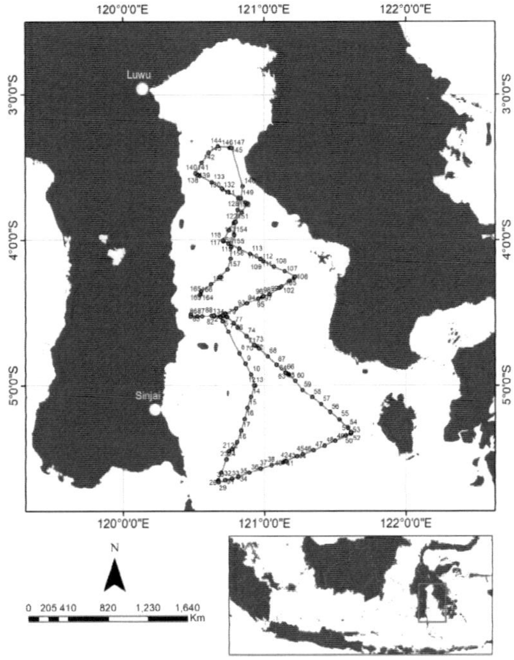

Figure 1. Figure 1 Research location and track of acoustic survey.

To obtain fish density value (ind / 1000m^3), equation (5) is used.

$$SV = 10\log\rho + \overline{TS} \tag{5}$$

So that:

$$\rho = 10^{\frac{SV-\overline{TS}}{10}} \tag{6}$$

where ρ is density of fish (ind/1000m^3).

The prediction of TS value can be known based on the length of fish caught by using target strength of Skipjack in [8]:

$$\mathrm{TS} = 20\ log\mathrm{FL} - 85{,}08 \tag{7}$$

3. Results and discussion

3.1. Fisheries survey

The fishery survey was conducted to obtain the fork length data which was then was used in the estimation of the target strength (TS) of Skipjack. The total of samples used were 626 fishes, which were obtained from two different locations, Sinjai (406 fishes) and Luwu (220 fishes). Figure 2 showed that the dominant of fish length was ranged from 40 to 49 cm, with the total of sample was 188 and the lowest size was ≥70 cm. The catchable size of Skipjack in the waters of Bone Bay is > 47 cm in TKG IV [9]. Based on this result, it could be concluded that from the total samples, 65% of them was not eligible to catch. Skipjack production in South Sulawesi in 2015 was 21038.8 tons, 32% comes from Sinjai and Luwu [10].

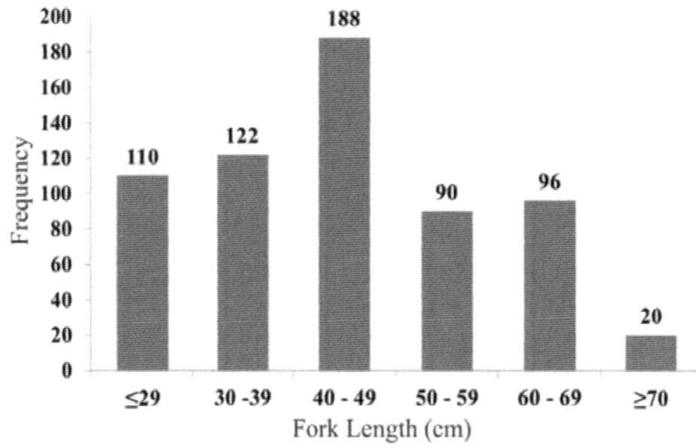

Figure 2. Fork Length composition of Skipjack from the fisheries survey.

3.2. Estimation of target strength (TS)

The Acoustic method was not able to provide definitive information on detected fish species, so the TS were predicted. Target Strength (TS) describes the ability of a target to reflect an incoming sound wave. TS is expressed in decibels (dB) [5]. TS was a function of size, shape, and presence or absence of swim bladder [11]. Based on the size of Skipjack catch in fishing port, it could be known that the range of Skipjack TS was -59 dB to -47 dB (figure 3).

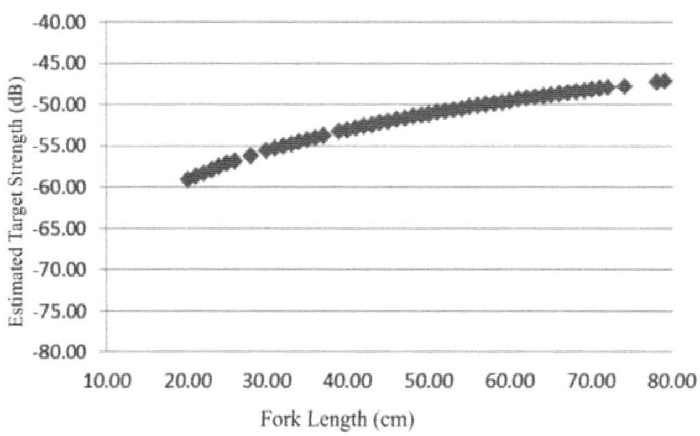

Figure 3. The relationship between FL (Fork Length) and estimated TS of Skipjack using the equation (TS=20 log FL – 85.08) [8].

The graphic relationship between FL (Fork Length) and estimated TS of Skipjack in this study indicated that the longer size of the fish, the higher its target strength (figure 3). Measuring the target strength of the same species, the target strength had a linear relationship with the length of the fish [12-13].

The length of fish is one factor that was very influential on the TS. In general, the relationship between TS and the fork length of the fish is positive, which means that the longer the fish, the higher the TS. In addition, TS values are also influenced by swim bladder, behavior and the instrument factors. Skipjack tuna is a fish that does not have a swim bladder, unlike other types of tuna fish, so the TS value of skipjack is lower than other tuna species [14].

3.3. Acoustic survey

Based on the results of acoustic survey, the obtained acoustic data was in the form of echogram. Figure 4 was an example of an echogram that indicated the existence of a fish schooling and vice versa. This reference was used in performing acoustic data analysis.

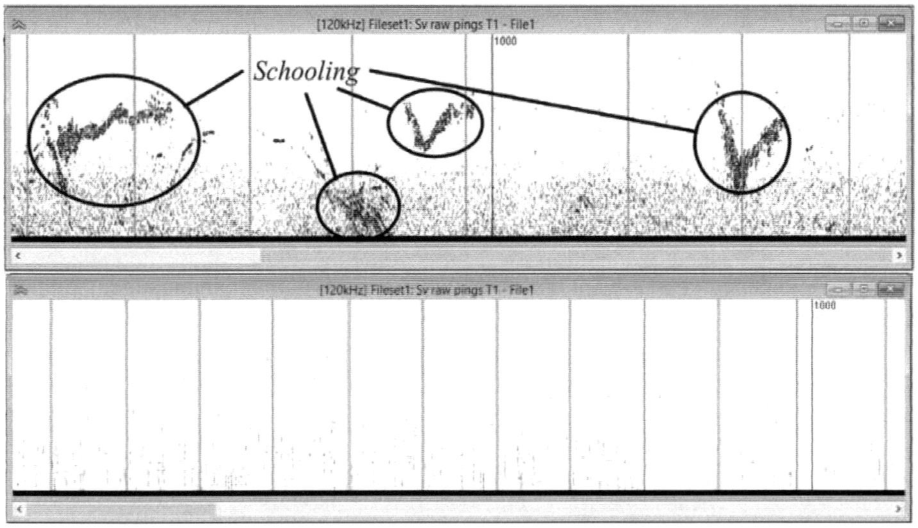

Figure 4. Echogram with an indication and no indication of fish schooling.

3.3.1. Temporal distribution. Distribution of fish in night and day tend to be different. Pelagic fish during the day is more often to form schooling as an effort to facilitate foraging, mating and spawning, to avoid or defend themselves from predators while at night is more distributed in the water column [15]. Figure 5 showed that the temporal distribution of Skipjack during the daytime was in the water column with a depth range of 7.1 m to 38.9 m, density 1 to 188 ind/1000m^3 and estimated FL was ranged from 4 cm to 76 cm. The central of Skipjack distribution was at 21.4 m depth, with a density of 43 individual and an estimated FL of 29 cm.

In the evening, Skipjack was distributed in the depth range of 11.8 m to 39.3 m with a range density 4 to 141 ind/1000m^3 and estimated FL range of 4 to 73 cm. The central of Skipjack distribution at night time was at a depth of 29.1 m with a density of 36 ind/1000m^3 and an estimated FL of 27 cm. In general, the distribution of Skipjack at night time located in deeper water while in the day was located closer to the surface of the waters. Skipjack is more active during the day time than at night because they are looking for food, its peak occurence is in the morning and mid of noon and afternoon [3].

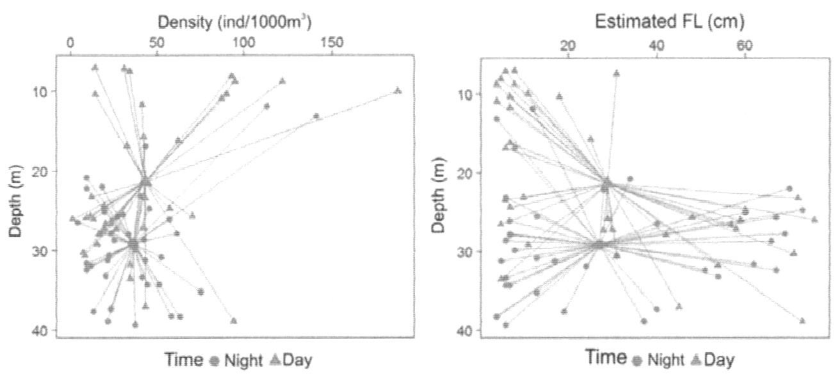

Figure 5. Skipjack distribution in the day and night.

3.3.2. Vertical distribution. Skipjack tuna is often referred as surface tuna [16], because Skipjack generally distributes in surface waters. Skipjack performs repetitive ups and downs movement in the waters, but the main distribution is in the upper layer where the water temperature is greater than 15°C [17]. Skipjack was distributed in the depth range up to 40 m, so the echo analysis was limited to a depth of 40 m. Skipjack was indicated high concentrated in the depth range up to 40 m, and used as a reference in the analysis of acoustic data conducted in this study [18].

The average target strength and density of Skipjack was then plotted in a graphic to evaluate their differences in the depth of the waters. Generally, the deeper the water, the higher the target strength due to the longer of fish size. However, in Figure 6, the target strength values was fluctuated to a depth of 40 m, this might be due to the calculated target strength was come from the average of the target strength of schooling fish that might had varying sizes. Skipjack is a tuna that has no swim bladder, so that their target strength was highly influenced by the size their body. The highest average of target strength was -55.3 dB and was occurred in the depth range of 20 m - 25 m, with the estimated FL was 31 cm.

The highest density of skipjack was 82 fish/1000m^3, and was occurred in the depth of 5 m - 15 m or closer to the surface of the water (figure 6). It could be assumed that these fishes were generally small fish because their average of target strength was small (TS was -66.8 dB and estimated FL was 8 cm). Generally, the deeper the waters, the lower density of fish [19]. The highest average of target strength (-55.3 dB) as previously mentioned was estimated to be the longest fish that was detected by acoustics, with density of 29 ind/1000 m^3.

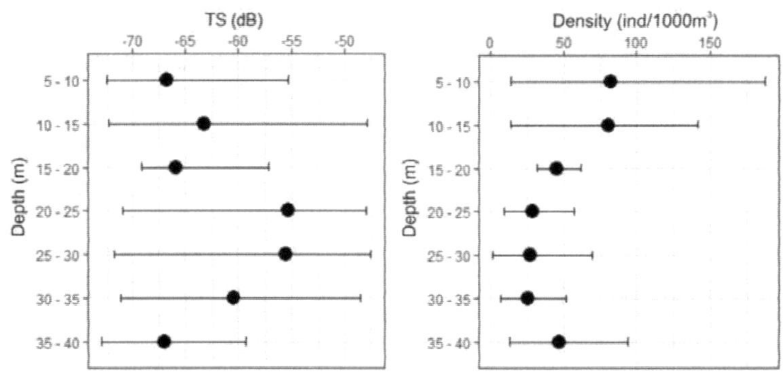

Figure 6. The relationship of depth with target strength and Skipjack density.

The estimated FL of Skipjack based on TS was then presented in table 1 to give an information about the estimated of Skipjack size vertically. Based on TS average from acoustic survey, an average of estimated FL was ranged from 8 cm to 31 cm. The overall data from estimated FL showed that the catchable fishes according to their size were distributed in the depth range of 10 m - 35 m, with size of 48.1 cm - 76.0 cm.

Table 1. Depth, $\overline{TS}$, estimated $\overline{FL}$ and density of Skipjack.

Depth (m)	$\overline{TS}$ (dB)	Estimated $\overline{FL}$ (cm)	Density (ind/1000m^3)
5 - 10	-66.8	8	82
10 - 15	-63.3	12	81
15 - 20	-65.9	9	45
20 - 25	-55.3	31	29
25 - 30	-55.5	30	27
30 - 35	-60.4	17	26
35 - 40	-67.0	8	47

3.3.3. Horizontal distribution. The horizontal distribution of fish in aquatic ecosystems is complex. There are several factors that influence the distribution of fish populations, such as migration, feeding behaviour, patterns of predation, reproduction and habitat selection [20]. The skipjack density was showed horizontally (figure 7) to determine the position and the abundance of Skipjack in Bone Bay waters. Figure 7 showed that the highest density of Skipjack in Bone Bay waters was located in Kolaka waters precisely at 121°10'84,00 "E and 4°55'11,56"S, with the total density of 188 ind/ 1000m^3. This might be due to high disribuition of small pelagic fish in these waters . Small pelagic fish is food source for Skipjack, thus inviting Skipjack to concentratein that waters. Skipjack generally distributed in areas with high productivity. This is due to the presence of small pelagic fish that is high in these water condition for feeding [21].

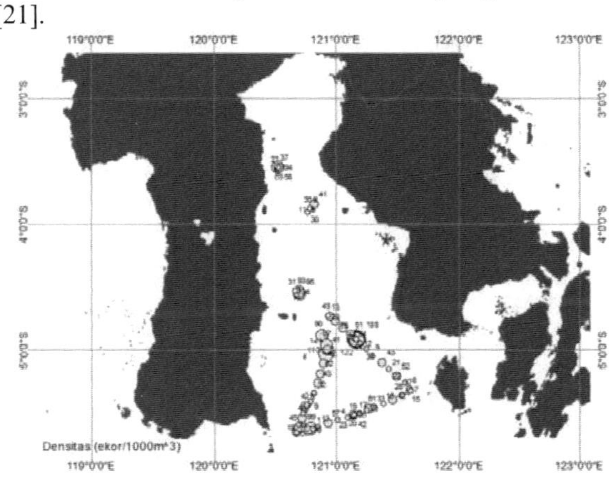

Figure 7. Map of skipjack density in Bone Bay.

4. Conclusion

The range of skipjack target strength was -67.0 dB to -55.3 dB, or equivalent to fork length of fish of 8 cm to 31 cm that was detected from the surface to a depth of 40 m. Temporally, the Skipjack density during the day was more toward the surface at 21.4 m depth with average density of 43 ind/1000m^3 and at night in deeper waters at 29.1 m depth with average density of 36 ind/1000m^3. Vertically, the largest density at the time of the acoustic survey was conducted was at a depth of 5 m - 15 m with a density of 82 ind/1000m^3. While horizontally, the highest Skipjack density, with the total density was 188 ind/1000m^3, was located in Kolaka waters or in the southern part of Bone Bay.

Acknowledgments

We acknowledge the support of DITJEN DIKTI for the scholarship of PMDSU under the contract number: 330SP2H / LT / DRPM / IX / 2016 dated September 8, 2016 on behalf of Prof. Indra Jaya.

References

[1] Jamal M, Sondita M F A, Haluan J and Wiryawan B 2011 Pemanfaatan data biologi ikan Skipjack (*Katsuwonus pelamis*) dalam rangka pengelolaan perikanan bertanggung jawab di Perairan Teluk Bone *Jurnal Natur Indonesia* **14**(1) 107-113
[2] [KKP] Kementerian dan Kelautan dan Perikanan 2015 *Statistik Perikanan Tangkap di Laut menurut Wilayah Pengelolaan Perikanan Negara Republik Indonesia (WPP-NRI), 2005-2014* (Jakarta: Kementerian Kelautan dan Perikanan)
[3] Baskoro M S, Taurusman A A and Sudirman. 2011 *Tingkah Laku Ikan Hubungannya dengan Ilmu dan Teknologi Perikanan Tangkap* (Bandung: CV. Lubuk Agung)
[4] MacLennan D N and Simmonds E J 1992 *Fisheries Acoustic* (London: Chapman and Hall) p 325
[5] MacLennan D N and Simmonds E J 2005 *Fisheries Acoustic : Theory and Practice, 2nd ed.* (London: Blackwell Science Ltd.) p 437
[6] Sweierzowski A and Doroszczyk L 2003 *Seasonal Differences in situ Measurement of the Target*

Strength of Vandace (Coregonus albuta L.) in Lake pluszne M. Oczapowskiego (Poland : Inland Fisheries Institute)
[7] Pujiyati S, Wijopriono, Mahiswara, Pasaribu B P, Jaya I and Manurung D 2007 Estimasi hambur balik dasar perairan dan sumber daya ikan demersal menggunakan metode hidroakustik. *J. Lit. Perikanan. Ind.* **13** 145-155
[8] Jun Z, Zuo-zhi C, Peng Z, Guo-bao C and Yong-Song Q 2014 In situ target strength measurements of Skipjack tuna *Katsuwonus pelamis* and Yellowfin tuna *Thunnus albacares* in the South China Sea. South China Sea Fisheries Research Institute, Chinese Academy of Fishery Sciences. Guangzhou, China
[9] Mallawa A 2012 Aspek Perikanan dan Prediksi Tangkapan Per Unit Upaya Ikan Skipjack (*Katsuwonus pelamis*) di Perairan Luwu Teluk Bone, Sulawesi Selatan. [Internet]. [Waktu dan tempat pertemuan tidak diketahui]. Makassar [ID] : Universitas Hasanuddin. [diunduh 2017 Feb 14]. Tersedia pada http://repository.unhas.ac.id/bitstream/handle/123456789/1768/ASPEK%20PERIKANAN%20SKIPJACK%20%28PAPER%20UGM.pdf?sequence=1
[10] [DKP] Dinas Kelautan dan Perikanan 2016 *Statistik Perikanan Laut Sulawesi Selatan* (Makassar: Dinas Kelautan dan Perikanan Sulawesi Selatan)
[11] Chu D, Stanton T K and Wiebe P H 1992 Frequency dependence of sound backsacttering from live individual zooplankton *ICES Journal of Marine Science* **49** 97-106
[12] Foote K G 1987 Fish target strength for use in echo integrator surveys *J. Acoustic Soe of America* **82**
[13] Johannesson K and Mitson R B 1983 *Fisheries Acoustic, A Practical Manual for Acoustic Biomass Estimation* (Rome: FAO Fisheries Tech.) p 240
[14] Okamoto H, Sato K, Imaizumi T, Abe K, Takao Y, Akamatsu T, Matsuo I, Ito M, Nishimori Y, Wang Y and Ogawa S 2010 Application of broadband dolphin mimetic sonar for discriminating target fish species *Scientific Committee Sixth Regular Session* [10 – 19 Agustus 2010]
[15] Fauziyah 2005 *Identifikasi, Klasifikasi dan Analisis Struktur Spesies Schooling Ikan Pelagis berdasarkan Metode Deskriptor Akustik* (Bogor: Institut Pertanian Bogor)
[16] Collette B B and Nauen C E 1983 FAO Species Catalogue. Vol. 2. Scombrids of the world. An annotated and illustrated catalogue of tunas, mackerels, bonitos and related species known to date. (Rome: FAO). *FAO Fish. Synop.* **125**(2) 137 p (Ref 168)
[17] Schaefer K M and Fuller D W 2007 Vertical movement patterns of skipjack tuna (*Katsuwonus pelamis*) in the eastern equatorial Pacific Ocean, as revealed witharchival tags. *Fish. Bull.* **105** 379–389
[18] Blackburn M and Williams F 1975 Distribution and ecology of skipjack tuna, *Katsuwonus pelamis*, in an offshore area of the eastern tropical Pacific Ocean. *Fish. Bull. U.S.* **73** 382-411
[19] Pujiyati S 2008 *Pendekatan Metode Hidrokustik untuk Analisis Keterkaitan Antara Tipe Substrat Dasar Perairan dengan Komunitas Ikan Demersal* (Bogor: Institut Pertanian Bogor)
[20] Jurevics P, Skute A, Brakovska A and Stepanova M 2012 Spatio-temporal dstribution of fish in the Northern Part of Lake Svente *Acta Biol. Univ. Daugavp. Suppl.* **3** 50-61
[21] Trigueros-Salmeron J A, Ortega-Garcia S 2001 Spatial and seasonal variation of relative abundance of the skipjack tuna *Katsuwonus pelamis* (Linnaeus, 1758) in the Eastern Pasific Ocean (EPO) during 1970-1995 *Fisheries Research* 227-232

IOP Conf. Series: Earth and Environmental Science **176** (2018) 012034 doi:10.1088/1755-1315/176/1/012034

An assessment of a coastal altimetry data product in the Indonesian Waters

Jonson Lumban-Gaol[1]*, Danu Adrian[1], Stefano Vignudelli[2], Robert. R. Leben[3], I Wayan Nurjaya[1], Takahiro Osawa[4], Parluhutan Manurung[5], Risti Endriani Arhatin[1]

[1]Department of Marine Science and Technology, Faculty of Fisheries and Marine Science, Bogor Agricultural University, Indonesia.
[2]Consiglio Nazionaledelle Ricerche, Istituto di Biofisica, Area Ricerca CNR San Cataldo, 56127 Pisa, Italy.
[3]Colorado Center for Astrodynamics Research, Colorado University, Boulder, CO 80309-0431, USA.
[4]Center for Remote Sensing and Ocean Sciences, Jalan Jenderal Sudirman, Udayana University, Bali Indonesia.
[5]The Indonesian Geospatial Information Agency (BIG), JalanRaya Jakarta - Bogor Km 46, Cibinong 16911, Indonesia.
*e-mail: jonson_lumbangaol@yahoo.com

Abstract. We analyzed the percentage of valid coastal altimetry Jason-2 X-TRACK-SLA data by Center for Topographic Studies in Sea and Hydrosphere (CTOH) and the waveforms from the Sensor Geophysical Data Record (SDGR) that are distributed by NOAA National Ocean Data Center around the Indonesian waters. In general, the percentage of valid data after the first point of the shoreline is greater than 90%. The percentage of valid data in steeper waters (86%) is higher than sloping waters (34%). The waveform types formed in coastal waters are peaky and Brown. However, spatially there is a difference where in a steep coast at a distance of > 5 km from the coastline the type of waveform is identified Brown, while on the sloping coast the brown type is found at a distance > 10 km. The SLA time series indicate seasonal variations in which the SLA is negative during the Southeast Monsoon (May-October) and positive during Northwest Monsoon (November-April).

1. Introduction

Indonesia is an archipelagic nation consisting of nearly 14,000 islands with a total coastline length exceeding 90 thousand kilometers. The Indonesian coastal and deep seas area highly dynamic environment with many physical processes, such as Indonesian Throughflow [1], equatorial origin Rossby and Kelvin waves [2;3], seasonal upwelling [4;5], sea level rise (SLR) and their variability [6]. Altimetry satellite data can be used for this related study; however, as it has been mentioned before, monitoring coastal waters circulation with altimetry is still a challenging issue because of the numerous constraints [7].

Global SLR is one of the impacts of global climate change and causes inundation of many small islands and coastal area in Indonesia. The scientific research indicates that global sea levels have been rising at a rate of 3.5 millimeters per year since the early 1990s, and the rate of increase varies considerably in different locations. As sea level is expected to rise further in the future (0.5 to 1 meter by 2100 and possibly more) [8]. The Sea level rise trend in South East Asia Seas including Indonesia seas has increased by 38 mm per year [6]. SLR may contribute to a loss of arable land through inundation, affecting crop growth and yield [9]. Since brackish water fish pond and rice fields in Indonesia are mostly located in coastal zones, it is estimated that by year 2050 the area of paddy rice fields could be reduced by 182,556 ha in Java and Bali, 78,701 ha in Sulawesi, 25,372 ha in Kalimantan, 3,170 ha in Sumatra, and 2,123 ha in Lombok [10]. In South Kalimantan, tidal swamps have been reclaimed for various agricultural activities. This changing can be used to reduce coastal vulnerability due to SLR [11]. Therefore, monitoring of SLR and their impact on coastal area is very important to study in the Indonesian coastal waters.

Traditionally, sea level change has been estimated from tide gauge measurements, collected over the last century. Recently, approximately 100 units of tide gauges were installed in the Indonesia waters. This amount is certainly not adequate to cover the length of the Indonesian coastline of about 90,000 km. Thus, it is really required to have sea level information that is complete and accurate both spatially and temporally to collect data of sea level. To monitor the dynamics of sea level along the coastlines will be needed more than 400 units of tide gauges. The satellite altimeter data can be used as a complementary tool for covering the lack of sea-level data from tide gauge in Indonesian coastal waters [12].

Sea level measurement by satellite altimetry has been done since the early 1990s: ERS-1, Topex/Poseidon, ERS-2 GFO Jason-1 and ENVISAT, Jason-2, etc. [13, 14]. Radar altimetry has been developed to monitor the sea level with high accuracy [15, 16]. However, despite the many satellite instruments showing some ability to measure sea level, it is still hard to provide accurate altimeterderived-measurement of sea level in the coastal area.

Most altimetry datasets are designed for the deep oceans because land contaminates the return signal of the altimeter. To resolve these issues, there was an effort in the last decade to improve the processing using a set of techniques (e.g., re-tracking, new corrections). One example is the coastal datasets that were corrected regionally through the use of X-TRACK, which is developed by Center for Topographic Studies in the Sea and Hydrosphere (CTOH). Before utilizing the dataset widely, it is necessary to assess the potential of this CTOH data, especially in the Indonesian coastal waters. This study attempts to analyze the potential altimetry Jason-2 data along the coast of Indonesian waters.

2. Methods

2.1. Coastal Altimetry Data

In this study we used the along-track (X-TRACK) satellite altimeter data product, developed by CTOH (http://ctoh.legos.obs-mip.fr/products/coastal-products/), and Sensor Geophysical Data Record (SDGR) that are distributed by NOAA National Ocean Data Center (https://podaac.jpl.nasa.gov/dataset/OSTM_L2_SGDR_D).

SLA projected onto reference tracks with a spatial interval of about 6-7 km between points (1 second) are computed using the X-TRACK software [7], developed at LEGOS France. The CTOH computes regional along-track sea level anomaly (SLA) products for altimeter Jason-2 missions from 2008-2013. We used Jason-2 data along-track (1 Hz) of Indonesian coastal tracks (figure 1). Sea level data were derived from level 2 of GRD in NetCDF format.

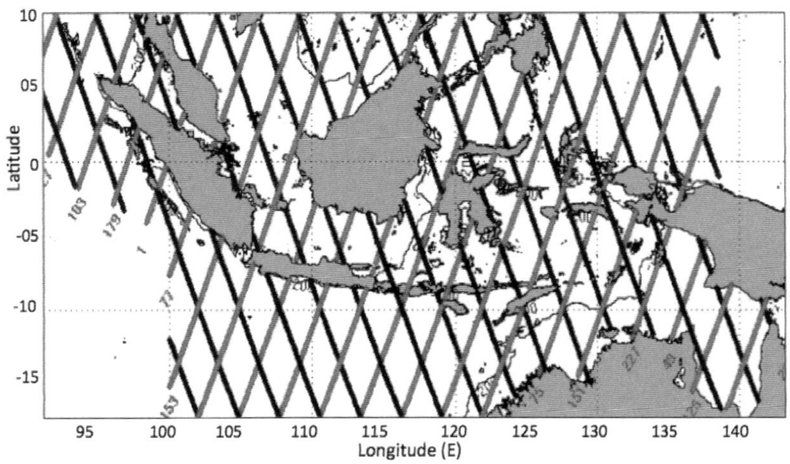

Figure 1. Jason-2 satellite altimetry of track of 2008-2013 in Indonesian waters. The red lines are ascending pass and black lines are descending pas.

2.2. Sea Surface Height (SSH) measurement

Satellite observations of sea surface height are available along the ground-tracks of Jason-2 satellite altimetry missions. Measurement of SSH was performed with various corrections in order to obtain actual distances between a satellite and sea level, as the following [17]:

$$R_{cor} = R_{obs} - \Delta R_{dry} - \Delta R_{wet} - \Delta R_{iono} - \Delta R_{ssb} \quad (1)$$

where (R_{cor}) is the corrected range, (R_{obs}) = $c\ t/2$is the computed range from travel time t observed and c is the speed of the radar pulse neglecting refraction.

Altimeters measure the distance between the sea surface and the satellite. The height, h, of the sea surface above references ellipsoid is given as:

$$h = H - R_{cor} = H - (R_{obs} - \Delta R_{dry} - \Delta R_{wet} - \Delta R_{iono} - \Delta R_{ssb}) \quad (2)$$

where H is the height of the spacecraft determined through orbit determination. Atmospheric disturbances are corrected, among others: the effects of the ionosphere (Δh_{iono}), wet troposphere (Δh_{wet}), dry troposphere (Δh_{dry}), the bias condition of the sea (Δh_{ssb}), tidal correction value (h_{tides}), and dynamic atmospheric corrected value (h_{atm}).
The actual sea surface height is a superposition of these geophysical signals and the dynamic sea surface height, $h_D$such that [10]:

$$SSH(h) = h_D + h_{geoid} + h_{tides} + h_{atm} \quad (3)$$

The sea surface high anomaly (h_{sla}) determined as:

$$h_{sla} = H - R_{obs} - \Delta h_{dry} - \Delta h_{wet} - \Delta h_{iono} - \Delta h_{ssb} - h_{tides} - h_{atm} - h_{MSS} \quad (4)$$

2.3. Data processing and analysis
The altimeter satellites were originally designed to observe offshore waters so that the raw data produced around the coast was very limited in both quality and quantity. Therefore, various studies have been conducted to improve the quality and quantity of altimeter data around the coast. One of

them is the Altimeter-Based Investigations in Corsica, Capraia and Contiguous Areas (ALBICOCCA) research project in the Northwest Mediterranean Sea. Through this project developed Additional developments lead to the X-TRAC tool. With this tool, new classical GDR processing is done to produce coastal altimeter data [19].

The approach is to use more detailed models for responding to the high frequency of tidal and atmospheric responses, reducing satellite orbit errors and increasing mean sea surface resolution. The application of x-track tool is expected to improve the quality of altimeter products i.e.: SSH, MSSH, SLA in coastal areas [19]. A more complete explanation of data processing and analysis can be found in the User Handbook of CTOH Along-Track Sea Level Anomalies regional products (X-TRACK).

3. Results and discussion

3.1. Potential coastal altimeter data

Figure 2 shows the percentage of along-track available valid data from the Jason-2 missions 2008 - 2013 period in Indonesian coastal waters and adjacent seas. The along-track valid data indicate that the availability are greater than 90%, except one near shore points from the coastal line. The average percentage of valid data on the first footprint from the coastline in shallow water such as the Java Sea (34%) is lower than the coastal in the deep sea (86%) such as the eastern Indian Ocean (figure 2b).

Generally, the percentage of data availability of the X-TRACK-SLA product after the first point from the coastline is greater than 90%. However, the percentage of valid data in deep waters is higher than in coastal waters (figure 2c). The results of this study are consistent with those shown in the Northwestern Mediterranean Sea, the X-TRACK-SLA data significantly increase the number of valid data [18]. The altimeter satellite data around the coast is usually low quality due to land contamination of the satellite footprints, the high frequency ocean response to tidal and atmospheric loading [19].

3.2. Types of waveform

The radar altimeter waveforms provide the range between the satellite and the surface at nadir via two-way travel time of the transmitted pulse, the Significant Wave Height (SWH) via the slope of the waveform leading edge, and the backscattering coefficients which represent the surface roughness and characteristics via the returned power [20]. Figure 3 shows the Jason-2 track 246 waveforms at a distance of 0-5 km, 5-10, and 50-100 km from the southern coast of Java Island.

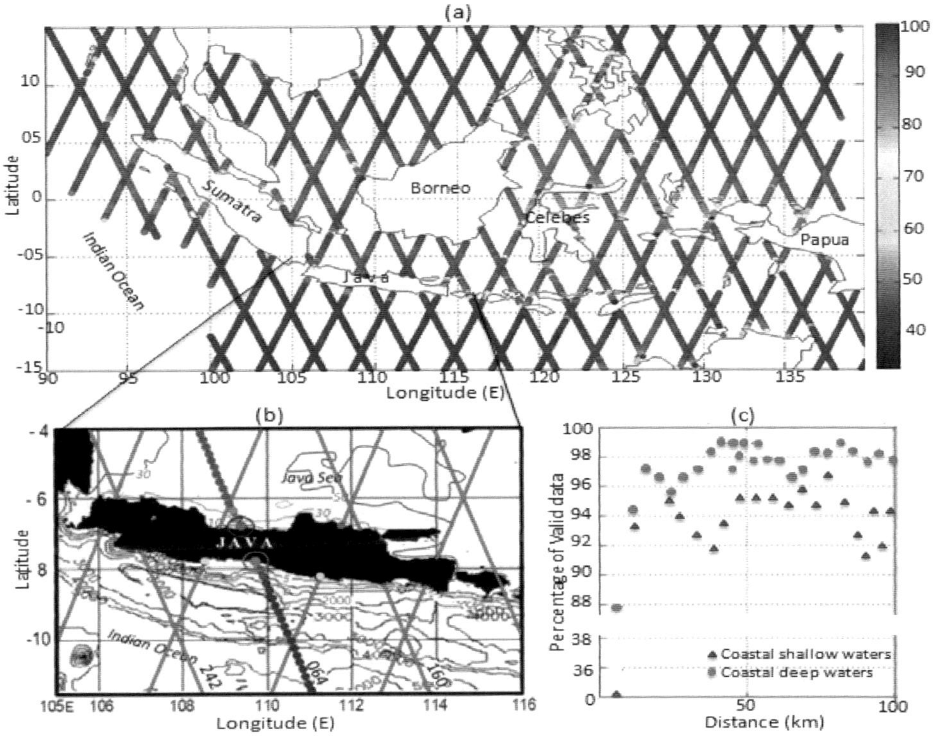

Figure 2. *Top*: (a) percentage of data availability for along-tracks data from the Jason-2 missions. Bottom: (b) the blue circles highlight region where the percentage of availability data in coastal shallow waters and the red circles highlight region the coastal deep waters, and (c) Comparison of the percentage of valid data in shallow water coastal (the Java Sea) and deep coastal waters (the eastern Indian Ocean off Southern Java) calculated from the first point up to 100 km from coast line.

The waveform formed near the coast (0-5 km) on track 246 is a peaky type (figure 3a) and more than 10 km is Brown type. Waveform near the coast is strongly influenced by the noise generated by terrestrial and shallow waters [19]. The ideal waveform is the Brown waveform generally formed in the open seas far away from the main land. An average 94% Brown waveform type can be found at a distance of more than 15 km from the shore [21]. However Waveform in the southern coast of Java island at a distance of 5-10 km from the coast is a Brown type (figure 4b). Meanwhile the waveform types in the same distance in the north coast of Java island is peaky (figure 4a) [22], where the depth is shallow with an average depth of 30 meters [23].

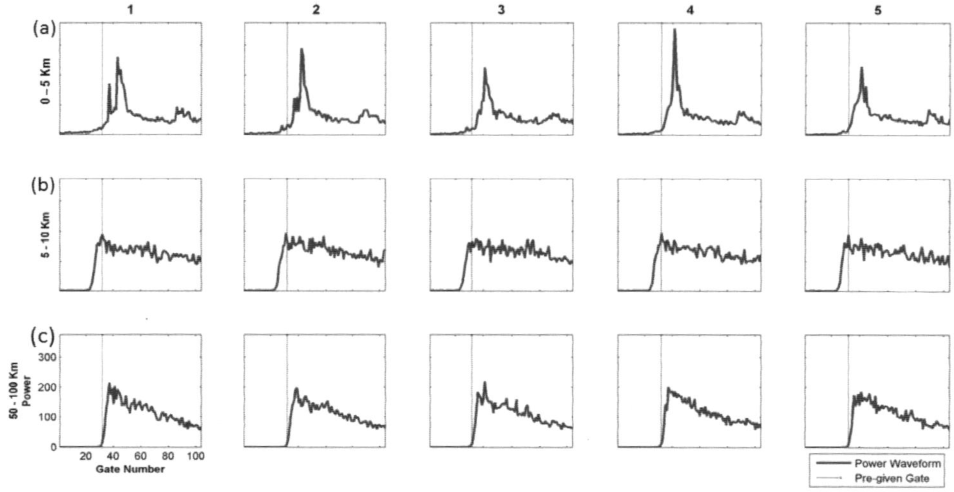

Figure 3. Types of Waveform Jason-2 track 242 of cycle 104.

The bathymetry in southern coast of Java Island shows that the depth of coastal waters at a distance of 5 km and more from the coast > 500 m. The south coast of Java Island is also directly related to the Indian Ocean so that the water characteritics near the coast it is classified as an deep sea. The ideal Brown waveform type is generally formed in the high seas (figure 4b) [24]. The other factors that may affect signals received by satellite sensors and affect the waveforms are the depth and shape of the water surface, coastal environmental conditions, atmospheric aerosols, buildings (e.g., lighthouses or vessels), and others [25].

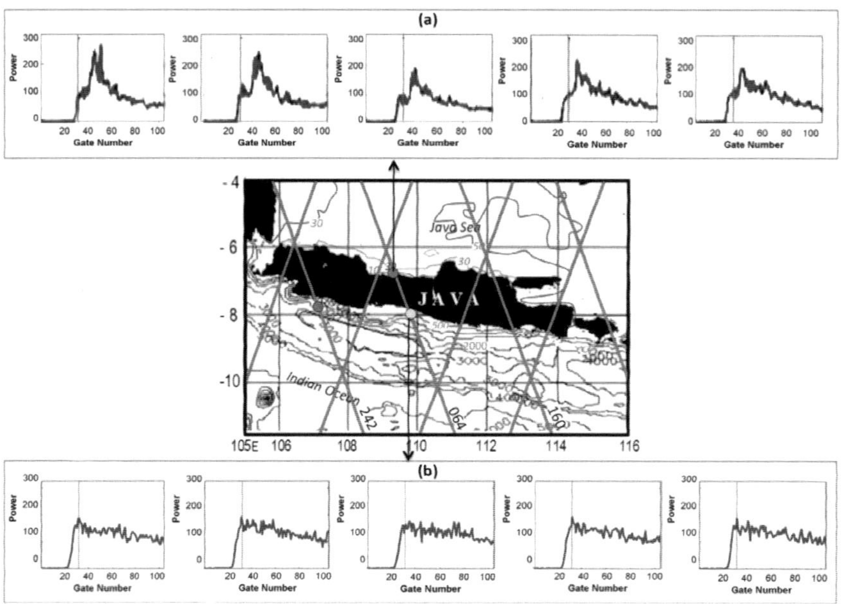

Figure 4. Types of waveform at a distance of 5-10 Km from the coast (a) North coast of Java Island (shallow waters) [21] and (b) South coast of Java Island (deep waters).

3.3. Mean Sea Surface Height (MSSH)

The MSSH is substantially obtained by averaging all valid SSHs at each point along track. It accounts for geoid and Mean Dynamic Topography (MDT). MSSH is estimated at each of the altimeter Jason-2 foot points. Generally, the MSSH increases from offshore to coast. The MSSH in Indonesian waters is shown in figure 5. The MSSH located at each foot points covers in the Indonesian water ununiformly. And it indicates well-known patterns, such as the MSSH is much higher in eastern Indonesia water (western Pacific) than that in eastern Indian Ocean off of south Java.

The highest of MSSH is 80 cm occur in Eastern Indonesia waters off North Papua, and the lowest is -20 cm in Eastern Indian Ocean off Sumatera and Java. This sea-level difference leads to establish the Indonesian Throughflow from the Pacific to the Indian Ocean [24].

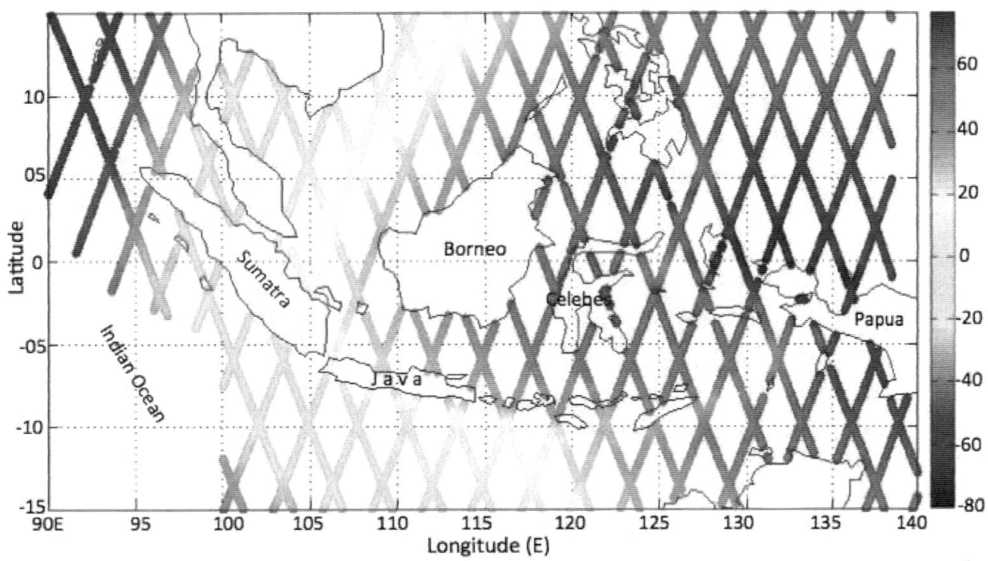

Figure 5. Mean Sea Surface Height [in cm] along the track of Jason-2 missions (2008-2013).

3.4. Sea-Level Anomaly (SLA)

Sea-level anomaly is the difference between the sea-level and the mean sea level for this time of year. For example, we analyzed the SLA time series on the Jason-2 track 064, and cycle-latitude plot (Hovmoller diagram) of SLA along selected track in the southern cost of Java Island (figure 6). The sea-level displays the number of months/cycles each year experiencing positive or negative anomalies. The SLA is negative during the Southeast Monsoon (SEM) period from June to October, meanwhile positive SLA occurs during the Northwest Monsoon (NWM) period from November to March. This seasonal change of SLA expresses a response of sea surface to reversal monsoon winds system. During the SEM period, the wind-driven Ekman transport in the southern cost of Java is toward offshore causing sea-level decreases. In contrary, during the NWM near the coastal area of southern Java Ekman transporttoward the coast that pills up surface water near the coastal region.

The difference of sea-level between the SEM and NWM period in the southern of Java Island is approximately 80 cm. Generally during NWM period, monthly mean of sea-level height in Indonesian waters increases such as in south Java waters, Makassar Strait, and Banda Sea [5, 26-27]. The SLA changes in the southern cost of Java Island are strongly influenced by the monsoon winds in the Indonesian waters [28-29]. During the SEM period the easterly monsoon winds blow along southern coast of Java and produces Ekman transport toward offshore, which then generate coastal upwelling. This causes sea level to drop due to an offshore Ekman transport [4,5].

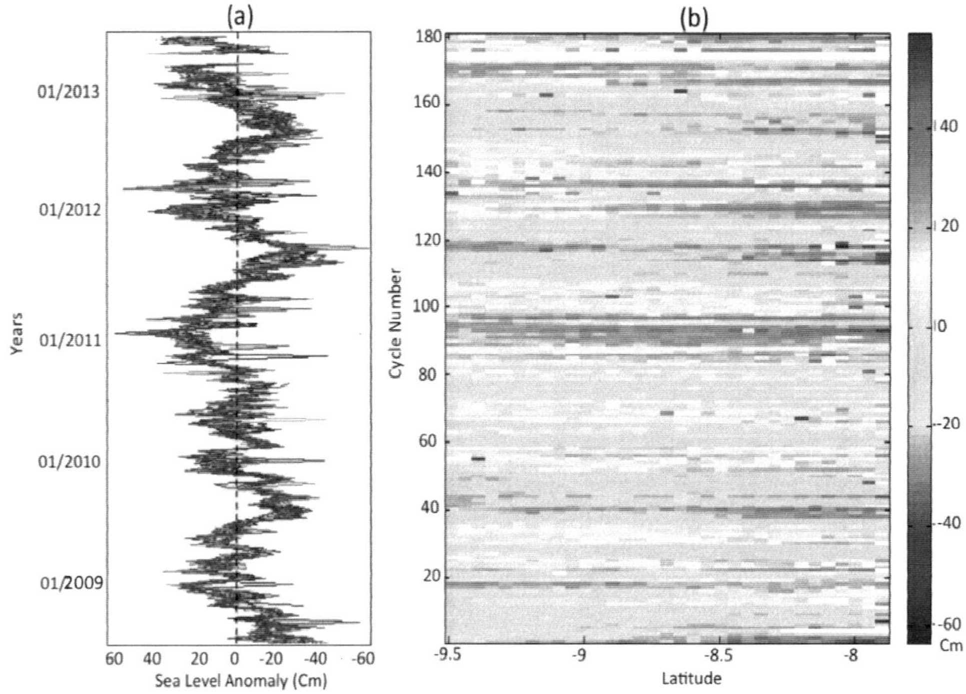

Figure 6. (a) Time series of SLA and (b) cycle-latitude plot (Hovmoller diagram) of SLA for all ground point along selected Jason-2 track 064 in the cost of southern Java Island.

4. Conclusion

The altimetry data of X-TRACK-SLA from CTOH product in Indonesian coastal waters showed that more than 90% of recorded SLA data are valid for 5 years of the Jason-2 missions. In general, the percentage of valid data in steeper coastal waters is higher than the sloping coastal waters. It shows that coastal altimetry data are potential to be used for coastal dynamics and sea level variability studies in the Indonesian coastal waters. The type of waveform formed in Indonesian coastal waters is a type of peaky and Brown type. The waveform type in Indonesian coastal waters is not only affected by distance from a coast but also influenced by water depth. The variability of SLA anomaly in Indonesian coastal waters exhibits large seasonal change that is influenced by monsoon winds system. Negative SLA occurs during the SEM period, while positive SLA is during the NWM period.

Acknowledgments

I wish to extend my appreciation to the Directorate General of High Education and Research and Technology for the funding support with grant No. 1667/IT3.1/PN/2018, and IPB staff mobility program. Altimetry data used in this study were developed, validated, and distributed by the CTOH/LEGOS, France. Anonymous reviewers provided useful suggestions.

References

[1] Gordon A L, Susanto R D and Vranes K 2003 Cool Indonesian Throughflow as a consequence of restricted surface layer flow *Nature* **425**(824)
[2] Meyers G, Bailey R J and Worby A P 1995 Geostrophic transport of Indonesian Throughflow *Deep Sea Research Part I: Oceanographic Research Papers* 1163-1174.
[3] Sprintall J, Gordon A L, Murtugudde R and Susanto R D 1997 A semiannual Indian Ocean forced Kelvin wave observed in the Indonesian seas in May *Journal of Geophysical Research: Oceans* 2000 **105** 17217-17230

[4] Wyrtki K 1962 The upwelling in the region between Java and Australia during the south-east monsoon *Marine and Freshwater Research* 217-25

[5] Susanto R D, Gordon A L and Zheng Q 2001 Upwelling along the coasts of Java and Sumatra and its relation to ENSO *Geophysical Research Letters* **28** 599-1602

[6] Strassburg, M.W., Hamlington, B.D., Leben, R.R., Manurung, P., LumbanGaol, J., Nababan, B., Vignudelli, S. and Kim, K.Y. 2014. Sea level trends in South East Asian Seas (SEAS). *Climate of the Past Discussions*, *10*:4129-4148.

[7] Roblou L, Lamouroux J, Bouffard J, Lyard, F, Le Hénaff M, Lombard A, Marsaleix P, De Mey, P and Birol F 2011 *Post-processing altimeter data towards coastal applications and integration into coastal models In Coastal altimetry* Springer Berlin Heidelberg pp 217-246

[8] Church J A W and N J W 2011 Sea-level rise from the late 19th to the early 21st century *Surv. Geophys.* **32**(4–5) 585–602

[9] Wassmann R, Hien N X, Hoanh C T and Tuong T P 2004. Sea level rise affecting the Vietnamese Mekong Delta: water elevation in the flood season and implications for rice production *Climatic Change* **66**(1-2) 89-107

[10] Förster H, Sterzel T, Pape C A, Moneo-Lain M, Niemeyer I, Boer R and Kropp J P 2011 Sea-level rise in Indonesia: on adaptation priorities in the agricultural sector *Regional Environmental Change* **11** 893-904

[11] Saidy A R and Azis Y 2009 *Sea level rise in South Kalimantan, Indonesia: an economic analysis of adaptation strategies in agriculture* EEPSEA, IDRC Regional Office for Southeast and East Asia, Singapore, SG

[12] Manurung, P, Leben, R R, Vignudelli S, and Lumban-Gaol J 2014 Reconstruction of sea level change in Southeast Asia waters using combined coastal sea level data and satellite altimetry data *APN science bulletin: global environmental change* **4** 23-29

[13] Leben R R, George H B, and Benjamin R E 2002 Operational altimeter data processing for mesoscale monitoring *Mar. Geod.* **25** 3-18

[14] Ablain M, Legeais J F, Prandi P, Marcos M, Fenoglio-Marc L, Dieng H B, Benveniste J and Cazenave A 2017 Satellite altimetry-based sea level at global and regional scales *Surveys in Geophysics* **38**(1) 7-31

[15] Carton J A, and Chao Y 1999 Caribbean Sea eddies inferred from TOPEX/POSEIDON altimetry and a 1/6° Atlantic Oceanmodel simulation *J.Geophys. Res.* **104**: 7743–7752.

[16] Andersen and Scharroo 2011 Range and geophysical correction in coastal region: and implication for mean sea surface determination *Coastal Altimetry* S Vignudelli et al Editor Springer Berlin

[17] Vignudelli, S Kostianoy A G, Cipollini P, and Benveniste J 2011 (Eds) *Coastal Altimetry* Springer Berlin/Heidelberg Germany

[18] Bouffard J, Roblou L, Birol F, Pascual A, Fenoglio-Marc L, Cancet M, Morrow R and Menard Y 2011 Introduction and assessment of improved coastal altimetry strategies: case study over the Northwestern Mediterranean Sea In *Coastal altimetry* (297-330) Springer, Berlin, Heidelberg.

[19] Roblou L, Lyard F, Le Henaff M, and Maraldi C 2007 X-TRACK, a new processing tool for altimetry in coastal oceans *Geoscience and Remote Sensing Symposium* IGARSS 2007 IEEE International 5129-5133

[20] Lee H, Shum C K, Emery W, Calmant S, Deng X, Kuo C Y, Roesler C and Yi Y 2010 Validation of Jason-2 altimeter data by waveform retracking over California coastal ocean *Marine Geodesy* **33**(S1) 304-316

[21] Deng X. 2004. Improvement of Geodetic *Parameter Estimation in Coastal Regions from Satellite Radar Altimetry* [thesis] Perth (AU) Curtin Univesity of Technology

[22] Hakim M R, Nababan B and Panjaitan J P 2016 Accuracy improvement on sea surface height estimation based on waveform retracking analyses of Jason-2 satellite in Java Sea *Jurnal Ilmu dan Teknologi Kelautan Tropis* **7** 771-790

[23] Ningsih N S, Yamashita T, and Aouf L 2000 Three-dimensional simulation of water circulation in the Java Sea: influence of wind waves on surface and bottom stresses *Natural Hazards*

145-171

[24] Deng X and Featherstone W E 2006 A coastal retracking system for satellite radar altimeter waveforms: application to ERS-2 around Australia *Journal of Geophysical Research: Oceans* **111**(C6)

[25] Chelton D B, Ries J C, Haines B J, Fu L L and Callahan P S 2001 Satellite altimetry *International geophysics* **69**

[26] Radjawane I M and Azminuddin F 2016 Seasonal and semi-annual variability of sea surface height in Makassar Strait *Journal of Physics: Conference Series* **739**(1) p012067)

[27] Gordon A L and Susanto R D 2001 Banda Sea surface-layer divergence *Ocean Dynamics* **52**(1) p2-10

[28] Bray N A, Hautala S, Chong J and Pariwono J 1996 Largescale sea level, thermocline, and wind variations in the Indonesian Throughflow region *Journal of Geophysical Research: Oceans* **101**(C5) 12239-12254

[29] Lumban-Gaol J, Leben R R, Vignudelli S, Mahapatra K, Okada Y, Nababan B, Mei-Ling M, Amri K, Arhatin R E, and Syahdan M 2015 Variability of satellite-derived sea surface height anomaly, and its relationship with Bigeye tuna (*Thunnus obesus*) catch in the Eastern Indian Ocean *European Journal of Remote Sensing* **48** 465-477

Review management policy marine conservation area of Wakatobi National Park

H E Adimu[1,2*], M Boer[3], F Yulianda[3], A Damar[3,4]

[1]Program Management of Coastal and Marine Resource Management
[2]Halu Oleo University, Indonesia
[3]Bogor Agricultural University, Indonesia
[4]Center for Coastal and Marine Resources Studies (CCMRS), Bogor Agricultural University
*e-mail: eldin.kahedupa@gmail.com

Abstract. Marine conservation areas have an important role in the protection of species and biodiversity. This paper aims to review policy issues in the management of marine conservation areas in Wakatobi National Park (WNP). The research is located in WNP Southeast Sulawesi Province, Indonesia. The method used in this research is qualitative method, Technique of data collecting done through, interview and observation beside study of policy document. Policies are reviewed based on text analysis. Marine conservation areas have an important role in the protection of species and biodiversity. The results of the Wakatobi national park management policy review, there are two authorities that have a role in the management of the area. Coordination and cooperation are required in the management so that the sustainability of conservation areas can run well.

1. Introduction

Conservation areas have an irreplaceable role as bastions of species protection and for conservation of biodiversity. Currently there are more than 150,000 conservation areas in the world with an area of about 19 million km² or 12.7% of the earth's surface [1]. Law of the Republic of Indonesia No.5/1990 on Conservation of Natural Resources and Ecosystems, national parks are defined as natural conservation areas that have native ecosystems, managed by zoning systems, and utilized for research, science, education, aquaculture, nature tourism, and recreation. Wakatobi National Park (WNP) is seven of the national marine parks managed by the zonation system. The area of WNP at 13,900 km^2 including the marine area and terrestrial islands of the region comprising 97% of the sea and 3% of the land [2]. Conservation policies as an integral part of sustainable development at present still face very tough challenges as well as complex and interrelated issues [3]. Marine Protected Areas (MPAs) are promoted as an important marine ecosystem management tool. However, they are complex systems that, from a governance perspective, raise serious challenges with regard to their effectiveness [4]. The stakeholders WNP are divided into several groups: the government consists of central government and local government, private groups, local communities, and academics [5]. The purpose of this paper is to review policy issues in the management of marine conservation areas.

Rules and regulations governing the use of resource within Indonesia's protected areas are formulated centrally by the Ministry of Forestry in Jakarta and apply nationally [6]. However, management of protected areas is devolved to the provincial offices of the Conservation of Natural Resources (CNR) a sub-department of the Ministry of Forestry. Conventional management schemes for areas such as WNP assume that it is possible to manage resource and environmental use through rules and regulations [8]. Marine protected areas, specifically no take areas (NTAs) designed to restrict all extractive uses, are now widely-used to promote sustainable fisheries and protect marine biodiversity [9].

2. Methods

The study is located in WNP South East Sulawesi in eastern Indonesia, and is centrally located within the Coral Triangle, a region with exceptional marine biodiversity [7] (figure 1). The study is qualitative method, technique of data collecting done through, interview and observation beside study of policy document [10]. Policy analysis is done descriptively qualitative by using text analysis [11]. The analysis is conducted on policies that are both vertical and horizontal in relation to the content of policy texts and policy contexts on regional management, coordination among policy actors, and the interrelationship between policy content. Technically, content analysis includes the efforts, the classification of symbols used in communication, using criteria in classification, and using certain analytical techniques in making production. This analysis is often used in verification analyses. The workings or logic of this analysis are in fact the same as most qualitative data analysis. Researchers begin the analysis by using certain symbols, classify the data with certain criteria as well as predict with certain analytical techniques as well. Conservation area management in WNP.

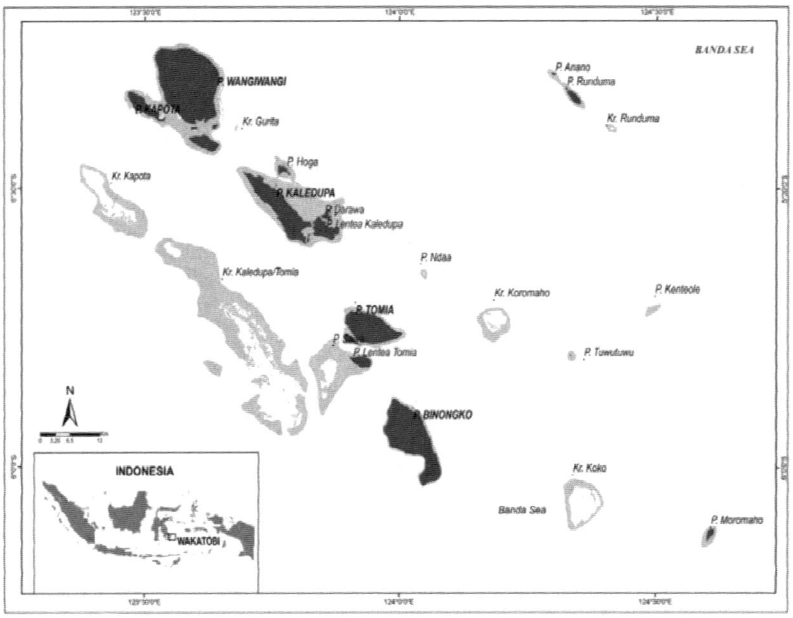

Figure 1. Location map of Wakatobi National Park.

3. Results

3.1. General condition of WNP area

WNP is one of the seven marine national parks in Indonesia, with the second largest area after Teluk Cendrawasih Bay National Park. Appointment through Decree of the Minister of Forestry No. 393/KPTS-VI/1996 dated 30 July 1996, and its stipulation is based on the Decree of the Minister of Forestry. 7651/Kpts-II/2002 dated 19 August 2002 covering 1.39 million hectares including the territorial waters and the entire islands of the region [13]. Wakatobi is an acronym for the four maind islands of Wangi-Wangi, Kaledupa, Tomia, and Binongko with 97% of the sea and 3% of the island's land [13]. The population is 94,985 people by 2015, up 0.29% from 2014, while the population density is 223/ km^2 [15]. The purpose of the establishment of this national park is to ensure the life buffer system for biodiversity conservation as representative of ecosystem of Banda-Flores marine ecology area (Banda Flores Marine Eco-region), ensuring the realization of sustainable development of the regional economy, especially from fisheries and tourism sectors, and ensuring the availability of sustainable livelihood for local communities. UNESCO has designated Wakatobi as a world biosphere reserve at the 18th International Conference on the Committee for the UNESCO MAB Biosphere Reserve Program in Paris on 2-4 April 2012 [19]. Included in the world's triangle coral reef marine biodiversity is one of the highest in the world, there are 396 species of corals scleractanian, 942 fish species, 31 species of fungal corals, 31 species of foraminifera and 34 species of stomatopods [17].

The focus of resource management on 8 important resources, coral reefs, mangroves, turtle habitats, seagrass ecosystems, coastal bird habitats, marine mammal habitats, and important economic fish. The National Park management partnership has intensive cooperation with Wakatobi District Government, Ministry of Marine Affairs and Fisheries, Indonesian Institute of Sciences (LIPI), TNC-WWF, customary institutions, local communities and other institutions. Results of monitoring of resources include: (1) coral reefs; to date there are 396 hard coral species, 28 soft coral genus and 31 species of mushroom corals. Coral reef types in WNP consist of: fringing reef, barrier reef, atoll reef, and reef patches. The area of coral reefs in Wakatobi is ± 54,500 ha, of which Kaledupa Atoll is the longest atoll in WNPA [15]. Results of coral reef monitoring 2012-2016 by WNPA show coral health conditions 2012 to 2016 in WNP is significantly higher in the no-take zone than in the utilization zone, with the best conditions in main island habitats. Bleaching coral reef health monitoring results from 2012 to 2016 tend to be low (<2%).

The highest percentage occurred in 2010. Even the results of monitoring by quadratic method, showed more surprising results, where nearly 60% of coral colonies affected bleaching in Wakatobi in 2010. (2) reef fish; as many as 590 species of fish found in Wakatobi, even extrapolated using coral fish diversity index, estimated reef fish in Wakatobi reach 942 species [13]. The number of fish from 2012 to 2016 on monitoring of SPAGs in WNP is fluctuating, the highest number of fish found in 2013 in December, 2014 in November and 2016 in March. The number is >1000 tail during observation in that year. (3) mangroves; the results of mangrove monitoring by WNPA in 2008-2012 were conducted on 14 sites around three major islands namely Wangi-Wangi, Kaledupa and Tomia in WNP encountered 13 genera and 28 species, 24 species found in Kaledupa Island, 12 species in Wangi- fragrance and Tomia 12 types. The Rhizophoraceae genus has the highest density of up to 30 individuals per m^2 in the growth stages of seedlings, 1 individual per m^2 at the stake, and 1,400 ind per ha at the tree level. The distribution of mangrove species in the community in the last four years is relatively stable with the highest frequency of encounters in the Rhizophoraceae genus. (4) Seagrass; the results of seagrass monitoring by WNPA 2008-2012, 11 of 12 types of seagrasses in Indonesia are found in Wakatobi. The seagrass cover in the Wangi-Wangi area is generally higher than Kaledupa and Tomia, seagrass cover at all locations fluctuates throughout the year. (5) turtle habitat; the type of turtle that is in the WNP is, hawksbill eretmochelys and green turtle chelonian mydas while turtle nesting beaches spread on the island of Runduma, Anano, Kentiole, Cowo-cowo and Moromaho. The peak of turtles that rise and lay eggs in WNP occurs in March-May and September-December. (6) Habitat of marine mammals; type of marine mammals (whales and dolphins), monitoring in 2004-2009 by WNPA there are 6 types of whales from a total of 1181 tails that are found, beaked whale, whale pilot, blue whale, whale sperm, bryde's whale and melonhead whale. The most common findings were of the whale 890 whale melonhead, the highest whale population in 2004-2005. There are still 89 species of unidentified whales. (7) coastal bird habitat: monitoring results of coastal / marine birds in WNP 2005-2012 identified 74 species of coastal / marine birds and 14 species (19%) protected status. Wakatobi endemic species are found, the type (gray pelvic chili) which is only found on the island of Hoga, Kaledupa throughout the year of observation, (8) economically important fish; economical fish consists of a type of herbivore that has a high selling value [18].

3.2. Management of WNP regions

The management of the area is carried out by Wakatobi National Park Authority (WNPA), consisting of three (3) National Park Management Section (SPTNW) SPTNW 1 located at Numana Wangi-Wangi Island, SPTNW II is located in Ambeua Kaledupa Island and SPTNW 3 on Waha Island

IOP Conf. Series: Earth and Environmental Science **176** (2018) 012035 doi:10.1088/1755-1315/176/1/012035

Tomia. The basis of management law refers to Law No. 5/1990 on the Conservation of Biological Natural Resources and Ecosystems, which regulates efforts to conserve biological natural resources and their ecosystems through the following activities: a) protection of life support systems, b) preservation of diversity of plant and animal species and their ecosystems c) sustainable use of biological natural resources and their ecosystems.

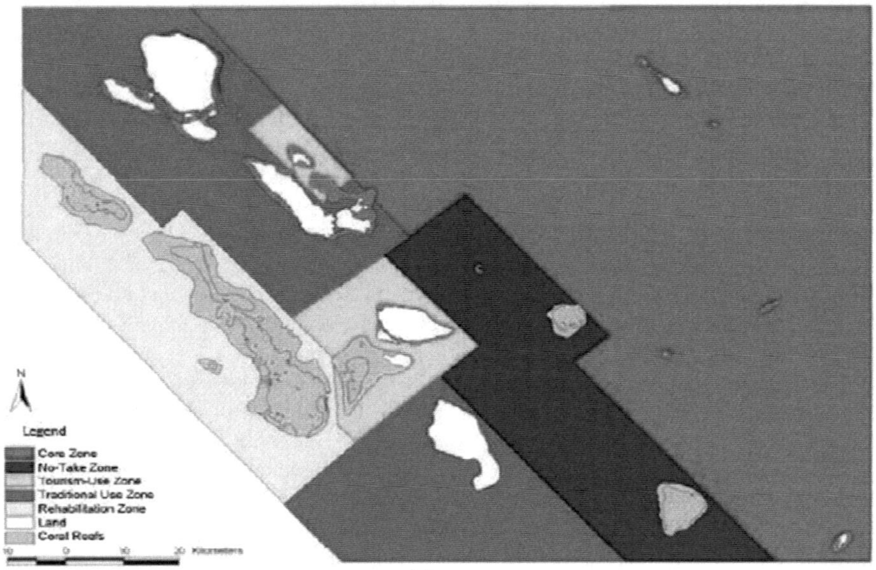

(a)

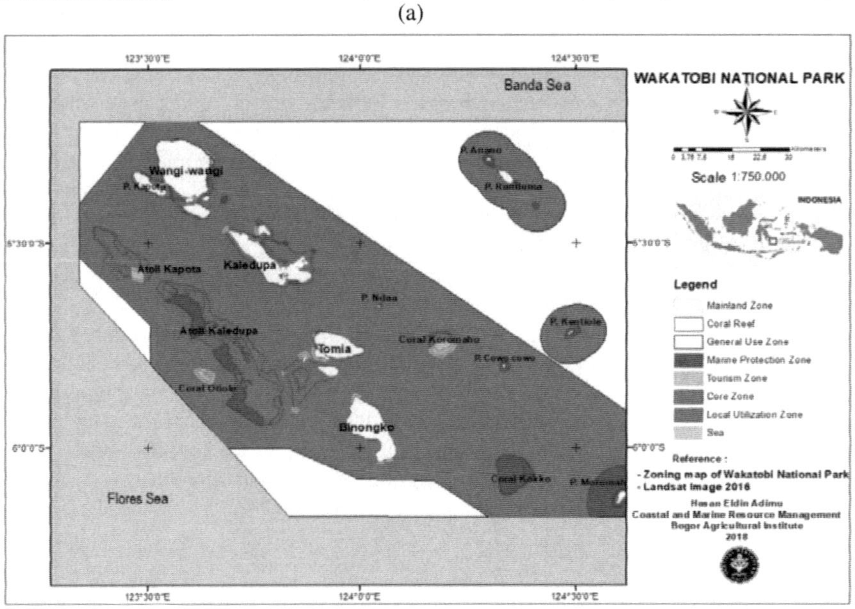

(b)

Figure 2. (a) Initial zonation of 1999-2006 (before revision) (b) zoning 2007-current (after revision) Source: WNPA (2008).

Management carried out by a zoning system (figure 2) consists of: (1) a core zone of ± 1,300 ha, (located on the island of Moromaho Island which is considered a pristine area intended for preservation and preservation of biological resources); (2) marine protected zones covering an area of ± 36,450 ha, (resource protection and functioning as a "bank" for fish and other resources); (3) tourism zones covering ± 6,180 ha (intended for tourism activities that also function as zones without extractive use); (4) local exploitation zone of ± 804,000 ha (the most extensive zone devoted to the benefit of Wakatobi's community use) (5) general utilization zone of ± 495,700 ha (located in deep sea 4 miles from coral or island and intended for interests of developing large-scale pelagic fisheries); and (5) special zone / land area of ± 46,370 ha (intended for development of community infrastructure and Wakatobi regency government).

Forms of protection and security of the area, WNPA by conducting routine patrol activities, combined patrol and species monitoring, reef check, and resource inventory and coastal ecosystems. Focus on 8 (eight) natural resources targeted for conservation (corals, fish spawning areas, mangroves, seagrass beds, cetaceans, turtles, coastal bird habitats and other marine species). Utilization is only done for the benefit of research, science, education, support cultivation, nature tourism and recreation. Approach to the community is done by counselling, training, environmental campaign, assistance and guidance to the fishermen community in the form of alternative livelihoods as well as the formation of tourist villages.

The issuance of Law No. 29 of 2003 (establishment of Wakatobi District) provides authority in the management and utilization of the area. Based on Law No 23 of 2014 (regional autonomy) utilization activities that can be done in the sea area consists of; business of aquaculture, fishing affairs and fisheries empowerment affairs small (coastal). In addition development activities carried out in coastal areas such as port infrastructure, roads and others. Wakatobi Regency which is the division of Buton Regency, which is located and the same extent with the location and the area of Wakatobi National Park. So that in the waters of Wakatobi there are two status of the area, one as the area of Wakatobi National Park and as the district area [13]. Harmonized legislation related to the management of natural resources and environment in Wakatobi area. Conflicts of interest and authority take place in the tourism licensing function; marine and fisheries; as well as the utilization of WNP zonation [14].

3.3. WNP Management Policy

The management of WNP aims is to preserve the source the biological nature and its ecosystem, to fulfill its function as a protection area of life support systems, preservation of the diversity of flora and fauna species, as well as sustainable use of biological resources and ecosystem, to be used for research purposes, development of science, education, support cultivation, nature tourism and recreation.

Management is at the level of central government policy (Ministry of Environment and Forestry) under the Director General of Natural Resources and Ecosystem Conservation, responsible for managing the national park and implemented by the WNP. Based on Law No. 5 of 1990, mentioned that the Park National managed by zoning system, with management system which still refers to regional management policies conservation. But along with the development of regional autonomy, with the formation of Wakatobi Regency, then the system existing management must be synergized with policies that exist in legislation regional autonomy. One of the management models considered appropriate is a collaborative management model [16]. Management of WNP, then the park management program described in the form of activities based on priority scale every 5 (five) years which are elaborated as follows:

1. Natural Resources Conservation and Conservation Program
   - The consolidation of the WNP area is implemented in order management effectiveness aimed at obtaining clear legal certainty with structuring and granting boundaries, distribution of management spaces based on the designation function embodied in forms of activities, outer boundaries and zonation, monitoring and evaluation of zoning, and socialization and zoning security.

- The management of biodiversity and ecosystem aims to conserve biological natural resources and their ecosystem according to the purpose of designation and determination of WNP, to be able to fulfill the protection function of buffer system life, preservation and diversity of plant and animal species and their ecosystems, and the sustainable utilization of natural resources and their ecosystem. This will be realized in the form of survey activities (identification and inventory) and monitoring of natural resources, rehabilitation, restocking, marine aquaculture development and wildlife breeding.
- Protection activities areas are focused on crime prevention and eradication theft or illegal fishing of sea products (illegal fishing).
- The development of the utilization of environmental services and natural tourism shall be carried out in order to develop the products of environmental services and nature tourism that are capable of generating economic benefits from natural resources within the area to provide business space for communities, local governments and businesses with creating a competitive business climate.
- Other research activities that examine the potential of fisheries that can be developed as a sustainable livelihood alternative. One of the triggers pressure on the region is the high dependence of the community on natural resources (sea), therefore need to develop research research development of alternative livelihoods sustainable livelihood (alternative sustainable livelihood) that does not interfere with the sustainability of natural resources.

2. Source Management Capacity Development Program Natural Resources and the Environment
   - Institutional capacity building of WNPA is implemented in order to establish institutional management steadily supported by human resources qualified based on appropriate qualifications and competencies with management needs.
   - Planning and control of WNP area management implemented with the purpose of management of the area is based and refers to systematic planning based on priority scale supported by the mechanism of control and coaching and accountability.
   - The development of a collaborative management system is implementation of an activity or handling a problem in order to help improve management effectiveness jointly and synergistically by the parties on a basis mutual understanding and mutual agreement prevailing laws and regulations.
   - Development of regulations that support the management is intended as a form of integrating regulation laws on conservation of natural resources and natural resources its ecosystem into the draft local regulations so direction of development and development of Wakatobi Regency always in line with the goal management by encouraging the drafting of local regulations supporting the management, such as the draft law on collaborative management, arrangement of fishing gear, district and district spatial planning entrance fee and tariff for tourism activities within the region, etc.
3. Awareness Program, Increased Participation and Community Empowerment in and around WNP
   - Increased public awareness and natural love development intended to further enhance understanding and public awareness of the importance of resource conservation biological nature and its ecosystem.
   - The development of community participation is intended for encouraging the active role of society is increasing in field, so the management becomes more effective and efficient and full support from the community as well all parties.

- Community empowerment is meant to encourage enhancement of potential utilization contained in community, to accelerate the improvement of welfare communities around the region as well as its support for areas in the conservation of natural resources.

Wakatobi Regency which is an expansion of the Regency Buton, which is the same location and width as the location and area WNP area. So in the water area Wakatobi has two regional statuses, one as a region WNPA and as a district area. With the condition of the region having two statuses, then there is some things that could be the pressure and threat ie inter-agency policies that are not synergistic, either on the policy order at the center as well as in the regional order. Furthermore, the problem of different perceptions about legislation. These things if not immediately communicated and sought solutions would potentially be a deep conflict management area in Wakatobi area. Therefore to in the future, has become a liability, the WNPA must build synergies and work together to create mechanisms collaborative management adaptive to the local situation.

With the condition of the region having two statuses, then there is some things that could be the pressure and threat ie inter-agency policies that are not synergistic, either on the policy order at the center as well as in the regional order. Furthermore, the problem of different perceptions about legislation. These things if not immediately communicated and sought solutions would potentially be a deep conflict management area in Wakatobi area. Determination of national parks and districts will have an impact on the management in the conservation area let alone the area is located exactly or overlap. The analytical tool used to project the plan management of the national park is the Conservation Action Plan (Conservation Action Plan / CAP) [13]. The condition of the region that has two statuses, then there are some things that can be the pressure and the threat of inter-agency policy is not synergistic chance, both in the policy setting at the central as well as in the framework in the region. [12]. Furthermore, the problem of different perceptions about legislation. These matters if not immediately communicated and sought the solution will potentially be a conflict in the management of the area in the Wakatobi region. Therefore, in the future, it has become an obligation, the Hall of WNP should build synergies and work together to create an adaptive collaborative management mechanism for the local situation [12]. This has led to the limitation of the drafting agencies, parliaments and other decision-makers to know and make the groundwork for the drafting of natural resource conservation policies and regulations or to recognize them. Disharmony also often occurs because of the conflict between the law and its implementing regulations. Often laws and regulations with government agency policies are not aligned. Policies between government agencies are often conflicting.

3.4. Authority of management WNPA and district government Wakatobi

Based on the results of the policy analysis in table 1 it is seen that there is a fundamental difference in the basis of the determination of the area between Wakatobi national park and Wakatobi District that have an impact on policy making. Law No.5 of 1990 (on the conservation of biological natural resources and its ecosystem) focuses on the protection of natural resources, utilization accommodated but sustainably and measurably. While law No. 23 of 2014 (on regional autonomy) and Law no. 29 of 2003 (on the establishment of Bombana District, Wakatobi District, and North Kolaka District) focus on regional development and development. The basis of operational reference of the area between Wakatobi RPTN and Wakatobi RPJMD has not yet been seen the synergistic mechanisms that are developed jointly in the management of the area. If this is not done synergy immediately, will trigger the impact of disharmonic policies that adversely affect the sustainability of regional management. Utilization of zonation-based zones; for supervision in the utilization of natural resources in each zone ie core zone, marine protection zone, tourism zone, local utilization zone, general utilization zone and special zone/land while demography-based Wakatobi District; for economic activities such as capture/empowerment of small-scale fishermen, cultivation, tourism, infrastructure development and inter-island sea transportation services.

Table 1. Regional management policies by WNPA and district governments are seen from various aspects.

Objects	Wakatobi National Park	Wakatobi District
Legal basis	- Law No.5 of 1990 (on conservation of biological natural resources and its ecosystem) - Determination of Wakatobi Islands and surrounding waters of 1.390.000 Ha as National Park. (SK Menhut RI No. 7651 / Kpts-II / 2002 dated August 19, 2002) - The change of name of Wakatobi Islands National Park (WINP) into Wakatobi National Park (WNP) (Permenhut No. P.29 / Menhut-II / 2006 on Organization and Administration of Technical Implementation Unit of National Park	- Law No. 23 of 2014 (about change) Law No. 32 of 2004 (on regional autonomy) - Law No. 29 of 2003 (about the establishment of Bombana District, Wakatobi District, and North Kolaka District)
Basic operational reference region	- Regulation of director general of conservation of natural resources and ecosystem (about strategic plan of directorate general of conservation of natural resources and ecosystem year 2015-2019) - Wakatobi National Park Management Plan 1998-2023	- Regional Regulation No. 1 of 2012 (on long-term development plan of the region (RPJPD) of Wakatobi Regency 2005-2025) - Wakatobi District Medium Term Management Plan 2016-2021 is a description of the vision, mission and program of Regent and Vice Regent of Wakatobi period 2017-2021
Utilization of the area	Research, science, education, cultivation, nature tourism and recreation. zone and special zone / land	Demographic bases; for economic activities such as catching / empowering small-scale fishermen, cultivation, tourism, infrastructure development and inter-island sea transportation services
Natural resource management	Focus on eight important resources Coastal or marine bird habitats (Aves), mangrove ecosystems, turtle nesting beaches, seagrass beds, coral reefs, economically important fish, Spawning Agregation Site (SPAGs) and marine mammal trails (cetacean).	Management of coral reef and mangrove ecosystem as well as other protected resources

Source: interviews, policy documents, and observations

4. Conclusion

Based on the results of a review of policy review of Wakatobi National Park management, it was found that there are two authorities that have a role in formulating the area policy of Wakatobi National Park as the manager of the National Park Office and Wakatobi District government. Such conditions affect the configuration of policy makers that impact on the management of the area.

Acknowledgements

The first author is grateful to the Ministry of Research, Technology and Higher Education for funding research through doctoral dissertation research scheme, Wakatobi National Park Authority for granting permission to conduct research in WNP and Halu Oleo University.

References

[1] [CNRE] conservation of Natural Resources and Ecosystems 2015 *Statistics of Directorate General of Resources and Ecosystem Conservation 2015* Ministry of Environment and Forestry Indonesia

[2] [WNPA] Wakatobi National Park Authority 2015 *Compilation Report Monitoring, Spawning Aggregation Site (SPAGS) 2012-2016* Baubau p 47
[3] Santosa A 2008 *Conservation of Indonesia A Portrait of Management and Policy* p 50
[4] Jentoft S, Thijs C, Son V and Bjorkan 2007 Marine protected areas: a governance system analysis *Hum. Ecol.* **35** 611–622
[5] Adimu H E, Boer M, Yulianda F and Damar A 2017 The role of stakeholders in marine conservation areas in Wakatobi National Park, Indonesia *AACL Bioflux* **10**(6) 1483-1491
[6] Clifton J 2003 Prospects for co-management in Indonesia's marine protected areas *Marine Policy* **27** 389–395
[7] Clifton J 2013 Refocusing conservation through aculturallens: improving governance in the Wakatobi National Park, Indonesia Marine Policy **41** 80-86
[8] Elliott G, Mitchell B, Wiltshire B, Manan A and Wismer S 2001 Community Participation in Marine Protected Area Management: Wakatobi National Park, Sulawesi, Indonesia *Coastal Management* **29** (4) 295-316
[9] Boonzaier L and Pauly D 2016 Marine protection targets: an updated assessment of global progress *Oryx* **50**(1) 27–35
[10] Sugiono 2013 *Combination Research Methods (Mixed Methods)* (Bandung: Alfabeta CV)
[11] Krippendorff K (1989) *Content Analysis.* In Barnouw E, Gerbner G, Schramm W, Worth T L and Gross L (Eds.) Internationalen cyclopedia of communication (Vol. 1, pp. 403-407) (New York: Oxford University Press)
[12] [WNP] Wakatobi National Park 2017 *Wakatobi National Park Management* [Online] Bogor available : http://wakatobinationalpark.com/
[13] [WNPA] Wakatobi National Park Authority 2008 Wakatobi National Park Management Plan 1998 2023 (Revision 2008) Baubau
[14] Awang C L and Marsono I H D 2010 A participatory policy study on the management of Wakatobi National Park after the establishment of Wakatobi Regency [thesis] (Yogyakarta: Gajamada University)
[15] [CBS] Central Bureau of Statistics Wakatobi 2016 *Statistics of Wakatobi District* Wangiwangi
[16] [MEAF] Ministry of Environment and Forestry 2015 *Strategic Plan for 2015-2019* Jakarta pp 132
[17] Pet-Soede L and Erdmann M V 2003 Rapid Ecological Assessment Wakatobi National Park WWF/TNC Joint Publication Denpasar Bali pp 187
[18] TNC-WWF 2003 Rapid Ecological Assessment Wakatobi Natioal Park pp 187
[19] Purwanto, Wijonarno A, Putu S, Orba L D and Hardin 2012 Monitoring of Fish Breeding Locations in Wakatobi National Park, Sulawesi, Indonesia (2005-2009) The Nature Conservancy- Indonesia Marine Program Report 1B/12

IOP Conf. Series: Earth and Environmental Science **176** (2018) 012036 doi:10.1088/1755-1315/176/1/012036

Fishery sustainability study with sustainability window (SuWi) analysis in the South China Sea (Indonesia Fisheries Management Area 711)

R Perangin-angin[1*], Sulistiono[1], R Kurnia[1], A Fahrudin[1], A Suman[2]

[1]Department of Aquatic Resources Management, Faculty of Fisheries and Marine Science, Bogor Agricultural Institute, Bogor, West Java, Indonesia
[2] Agency for Marine and Fisheries Research and Development, Ministry of Marine Affairs and Fisheries, DKI Jakarta, Indonesia
*e-mail: robert.peranginangin@yahoo.com

Abstract. Sustainable development can be defined in several different ways, but usually, the term refers to the definition given by the Brundtland Commission in the publication of Our Common Future. Sustainable development is often understood by relying on three pillars, namely environmental/ecology, economic and social. The demersal fishery in the South China Sea (Indonesia Fisheries Management Area / IFMA 711) became one of the supporting sectors of the economy of the fishermen communities around of these waters. This study aims to analyze the sustainability of demersal fisheries with the Sustainability Window (SuWi) approach. Demersal fishery sustainability condition in the South China Sea (IFMA 711) waters tend to be better from year to year. However, the width of demersal sustainability window tends to be low and indicate the status of sustainability of demersal fish is not maximal in the South China Sea (IFMA 711). Therefore, care should be taken to manage demersal fishery in the South China Sea (IFMA 711).

1. Introduction

Fisheries management begin to grow, after a growing awareness on the importance of ecological sustainability based on the condition of the stock of fishery resources that tend to decline, such as that occurred in the decline of the stock of Pacific salmon, California sardine, and Peruvian anchovy [1]. FAO released an analysis related to the decline in global annual fish landing growth in the 1980s, and in 1990 a decline of 3% of global annual catches compared to the previous year [2].

This trend continued in the next few years and between 1990 and 1992, global landings fell by an average of 1.5% per year [3]. Overfishing has caused a total collapse of many fisheries including important species [4,5]. In addition, excessive exploitation of fisheries has led to the reduction of biodiversity, and ecosystem function [6]. On the other hand, most coastal communities in the world depend on marine resources. Marine products, such as fish, are an important source of protein for about a billion people worldwide [7,8,9,10]. Management efforts to reduce overfishing have been widely pursued, but very little progress has been made due to the general inability to bear the short-term economic and social costs [11].

In an economic perspective, fisheries exist to meet social and economic demands and one would hope to find that impacts on fishery resources have resulted in measurable social and/or economic benefits [12]. However, in terms of sustainability issues, the economic crisis symptoms in some of the above-described fisheries should be considered to produce solutions for economic sustainability [13].

From a social point of view, fisheries are seen as a means to generate economic benefits. The role of fisheries as a source of employment, especially in rural or remote areas, has also been given high priority

[12]. In this case, the social aspect of fisheries will be very important to understand. One of the main features of fisheries in recent decades, related to social crises, has been the introduction of modern fishing technologies and also the increasing trade globalization affecting fishing communities. Modernization of the fishery has two faces that actually contribute to the welfare of the fishing communities, while others have caused social problems associated with the depletion of fishery resources [14]. According to [15], in the North Atlantic and the Mediterranean, there has been a decline in the quality of life and living standards among many fishing communities, as has been experienced in some areas of Southeast Asia [16]. From this, the science of fisheries itself progressively diverts their attention from single species to ecosystem approaches, from a micro to a macro perspective, increasing the need to quantify the impact of fishing on natural and man-made systems [11,17].

Ensuring sustainable fisheries and community development requires the integration and harmonization of environmental, social and economic issues that can, in turn, lead to an effective management plan [18]. The purpose of this research is to analyze demersal fishery sustainability condition using environmental or ecological integration through estimation of demersal, social, and economic resource condition with SuWi (Sustainability Window) analysis. Measuring different dimensions of sustainability is not easy because there is no general method that has been developed for that purpose.

2. Tools and methods

2.1. Data collection

Data for the analysis of demersal dominance and demersal fisheries sustainability level in the South China Sea (IFMA 711), obtained from secondary data in the form of production and production value of marine catch fishery, and fisherman's welfare level presented by the price indices received by fisherman for period 2007 to 2015.

2.2. Data analysis

2.2.1. Resource Domain Analysis. Production data and production value of marine capture fisheries, both demersal resources and other fishery resources located in the South China Sea (IFMA 711) and national are expressed in the form of drawings and graphs and described descriptively. This quantitative descriptive analysis is used to determine the level of dominance in terms of production and economic value of demersal fish resources in the South China Sea (IFMA 711). This information is important to understand the priorities of demersal fisheries management in the South China Sea (IFMA 711).

2.2.2. Sustainability windows (SuWi) analysis. Simultaneous analysis of three dimensions of sustainability can be done using a new analysis tool, Sustainability Window (SuWi), developed by Jyrki Luukkanen [19]. Suwi's approach does not postulate the type of relationship between the development of different sustainability dimensions that will be tested with empirical data. Suwi's goal is to analyze the three-dimensional developments in the case study area and to see if they simultaneously meet the Brundtland Commission type sustainability criteria [20]. Sustainability Window is a new analytical tool for assessing the sustainability of development simultaneously in three dimensions (environmental, economic, and social). This analytical method provides information on the maximum and minimum economic development required to keep the direction of social and environmental development toward a more sustainable goal. With the Sustainability Window (SuWi) method, it is easier to analyze sustainability using different indicators and different time periods so that comparative analysis becomes easy. This new method is also possible to analyze the dynamics of sustainability and changes over time from the thickness (figure 1).

IOP Conf. Series: Earth and Environmental Science **176** (2018) 012036 doi:10.1088/1755-1315/176/1/012036

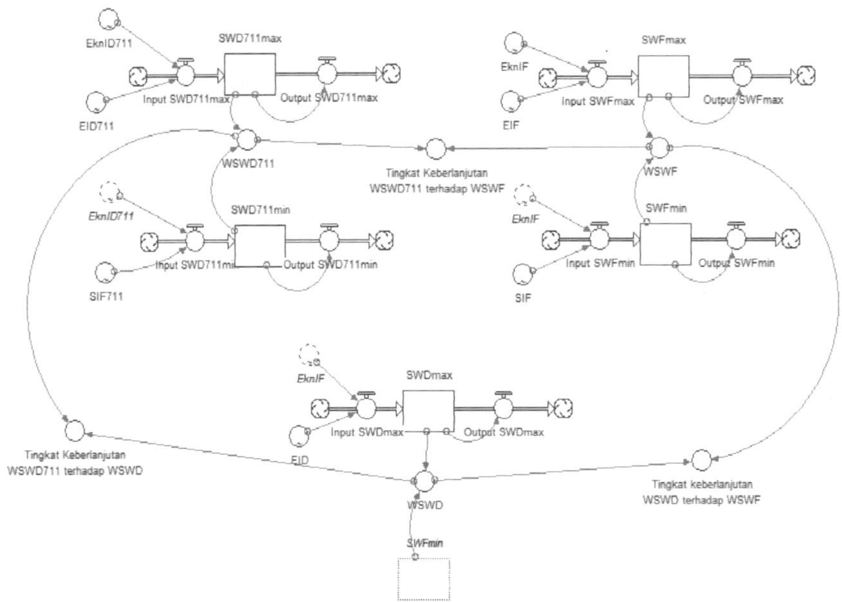

Figure 1. Dynamic Model : (a) Sustainability Level of Demersal Fisheries in IFMA 711; (b) Sustainability Level of National Capture Fisheries; (c) Sustainability Level of National Demersal Fisheries.

It provides a new perspective to analyze the sustainability trends and impacts of the underlying sustainability policy. The width of Sustainability Window (SuWi) for demersal fish resources in the South China Sea (IFMA 711) can be obtained from the difference between Sustainability Window maximal demersal fishery (SWD711max) and Sustainability Window minimum demersal fishery (SWF711min), calculated by using the following equation with indexed data:

$$SWD711_{max} = \frac{EknID711_{t1}}{EID711_{t1}} EID711_{t0} \qquad (1)$$

and

$$SWF711_{min} = \frac{EknID711_{t1}}{SI711_{t1}} SIF711_{t0} \qquad (2)$$

Which EknID711 is an index of demersal fishery production value in the South China Sea (IFMA 711), EID711 is a demersal fish capture index in the South China Sea (IFMA 711), and SIF711 is an acceptance index of fishermen who conduct fishing activities in the South China Sea (IFMA 711).

The thickness of Sustainability Window (SuWi) of demersal fish resources in the South China Sea (IFMA 711) compared with the thickness of Suwi of national demersal fish resources and national capture fisheries to determine the demersal fish resources sustainability position in the South China Sea (IFMA 711). The thickness of the national demersal fishery SuWi is calculated using the following equation with indexed data:

$$SWD_{max} = \frac{EknID_{t1}}{EID_{t1}} EID_{t0} \qquad (3)$$

and

$$SWF_{min} = \frac{EknID_{t1}}{SIF_{t1}} SI711_{t0} \qquad (4)$$

Which EknID is a national demersal fishery production value index, EID is a national demersal fishery capture index, and SIF is an accepted index of fishermen conducting fishing activities in Indonesian waters. Meanwhile, the thickness of SuWi capture fisheries at the sea nationally is calculated using the following equation with indexed data:

$$SWF_{max} = \frac{EknIF_{t1}}{EIF_{t1}} EIF_{t0} \quad (5)$$

and

$$SWF_{min} = \frac{EknIF_{t1}}{SIF_{t1}} SIF_{t0} \quad (6)$$

Which EknIF is a production value index of national capture fisheries at the sea, EIF is national capture fishery index and SIF is an accepted index of fishermen who conduct fishing activities in Indonesian waters.

By applying the Sustainability Window (SuWi) analysis into the dynamic model (figure 2), a demersal fisheries sustainability position in the South China Sea (IFMA 711) can be obtained compared to the national demersal fisheries and capture fisheries in the Indonesian Ocean.

3. Results

3.1. Analysis of production priority

Fishery Management Areas in Indonesia have enormous potential to be managed and utilized. The potential of this great fishery resources can be seen from the production of marine capture fisheries. There are 3 groups of species that dominate biomass from the sea, which are small pelagic fish resources, large pelagic fish resources, and demersal fish resources, which are small pelagic fish resource has the largest dominance proportion. The South China Sea (IFMA 711) is one of the waters that have the enormous potential of capture fisheries resources, especially demersal fishery. Figure 2 presents the quantities of demersal fish resource production that dominate capture fisheries in the South China Sea (IFMA 711).

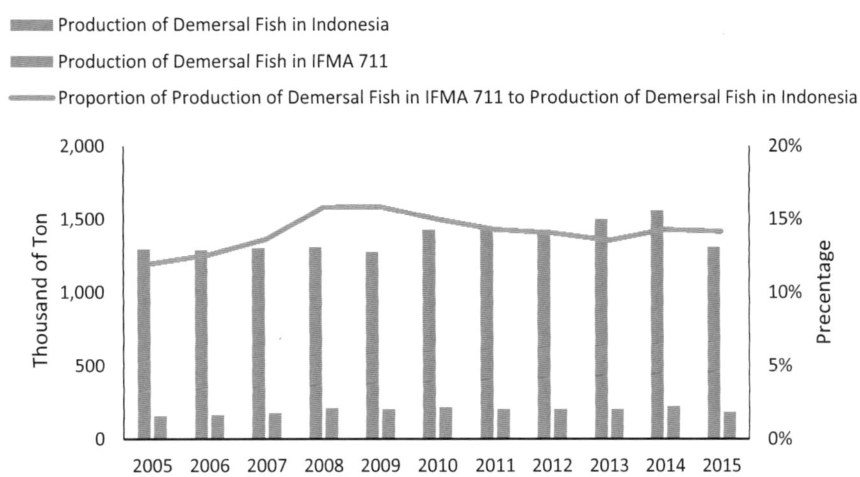

Figure 2. Demersal fish production inSouth China Sea to the average of national production of demersal fish 2005-2014.

Based on figure 2, this condition makes demersal fish resources as a resource that has an important and decisive role in terms of biomass on the sustainability of capture fisheries in the South China Sea (IFMA 711). Nationally, demersal fisheries production in the South China Sea (IFMA 711) is above the average of demersal fisheries production of all the Indonesian Fisheries Management Areas (IFMA) which currently amounts to 11 of IFMA.

3.2. Analysis of economic priority

The capture fisheries sector in the South China Sea (IFMA 711) has a significant economic contribution to the Indonesian economy through the GDP of the marine capture fisheries sub-sector. Figure 3a presents the average percentage of the value of capture fisheries production in the South China Sea (IFMA 711) with 11.3% of the total value of national capture fisheries production for the period 2005 to 2015. Percentage of production value in capture fishery produced by the South China Sea (IFMA 711) is quite significant by looking at the comparison of economic contribution to 10 other Indonesian Fisheries Management Areas (IFMA). Much of the value of capture fisheries production in the South China Sea (IFMA 711) was contributed by demersal fisheries, with a percentage of 32.6% to 41% for the period 2005 to 2015 (figure 3b).

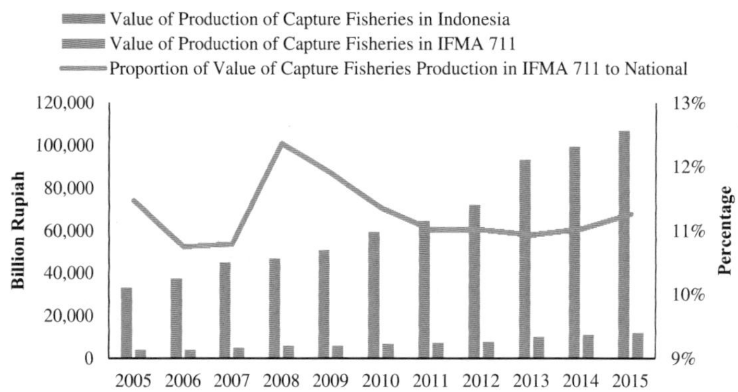

(a)

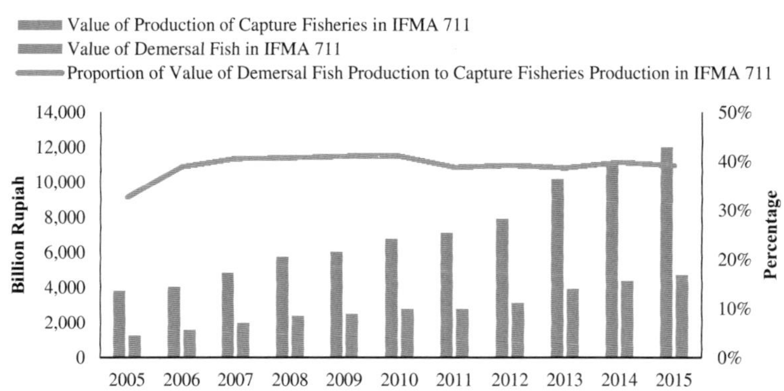

(b)

Figure 3. The proportion of economic value : (a). The Economic value of catchment fisheries in the South China Sea (IFMA 711) to the Economic value of National catchment fisheries; (b). Economic value of demersal catchment fisheries in the South China Sea (IFMA 711) to Economic value of catchment fisheries in the South China Sea (IFMA 711).

3.3. Sustainability evaluation

Sustainability window (SuWi) change analysis can be done during the time available information on trends related to sustainable development. Figure 4 illustrates the change in the minimum and maximum values of the sustainability window (SuWi). The maximum value of Sustainability Demersal fishery window in the South China Sea (IFMA 711) is obtained by comparing the value of demersal fishery production to the magnitude of its production. Since 2007, the maximum demersal fisheries sustainability window (SuWi) value in the South China Sea (IFMA 711) has increased rapidly until 2013. The 2014-2015 period has experienced a significant upward trend, possibly due to a decrease in demersal fisheries production in the South China Sea IFMA 711), but on the other hand, there is an increase in the production value of demersal fish resources.

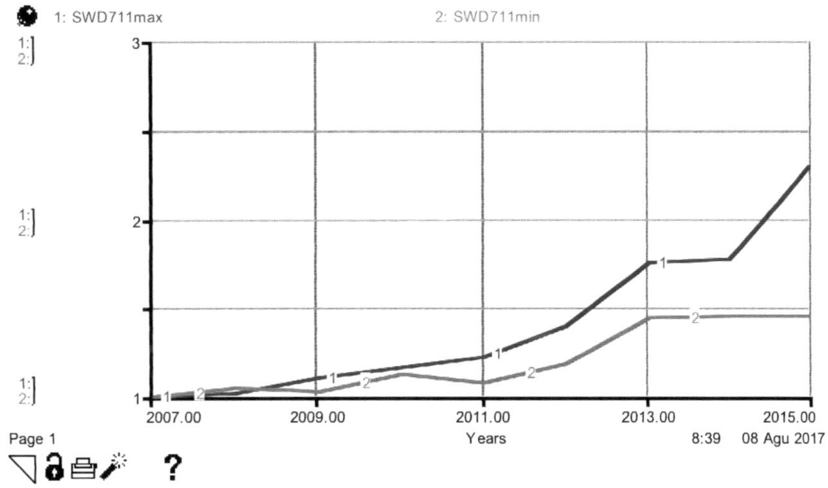

Figure 4. Change of upper limit (SWD711max) and lower limit (SWD711min) of Sustainability Window (SuWi) of demersal fish in the South China Sea (IFMA 711).

The lower limit of the demersal fisheries demersal in the South China Sea (IFMA 711) tended to be constant until 2009, indicating an increase in production values and indices received by fishermen but not significantly changed. The 2009 to 2010 period increased and then again declined until 2011, indicating that in 2009 to 2010 there was an increase in the value of demersal fishery production in the South China Sea but not accompanied by a significant increase in the index received by fishermen, but this is the opposite for 2010 until 2011. In the period of 2011 to 2013, tends to rise due to temporal increase in the intensity of demersal fishery production value in the South China Sea (IFMA 711) on the index received by fishermen living around the South China Sea (IFMA 711). The time period 2013 to 2015 tends to be constant, which is an implication of the change or increase in the value of the index received by fishermen which are comparable to the change or increase in the value of demersal fisheries production in the South China Sea.

The maximum value of sustainability of national demersal fisheries and national capture fisheries tends to be the same until 2013. However, in 2014 to 2015, national demersal fisheries tend to rise significantly while national catch fisheries tend to be constant. This condition is caused by the intensity of national capture fishery production per production value tends to rise, while the production of national demersal fishery per production value tends to decrease (figure 5).

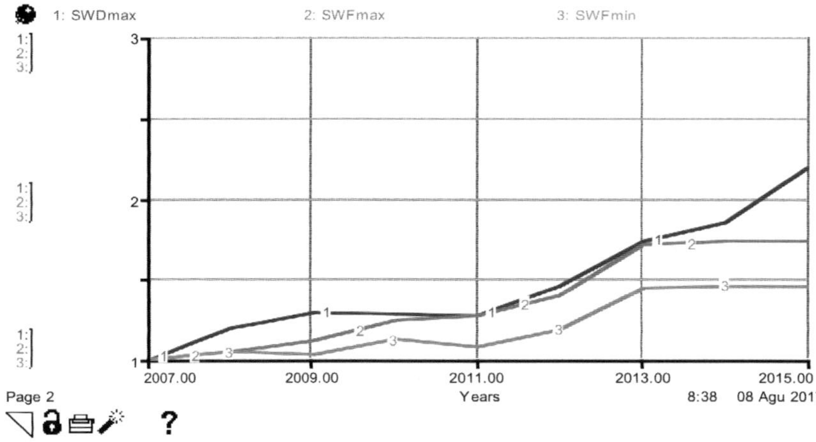

Figure 5. Change of upper limit of demersal fish (SWDmax) and upper capture fisheries limit (SWFmax) to lower limit (SWFmin) of national capture fisheries Sustainability Window (SuWi).

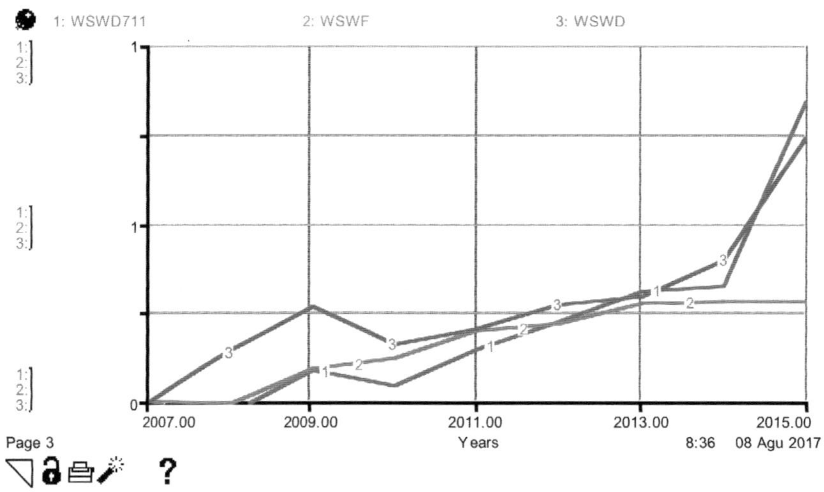

Figure 6. The thickness of Sustainability Window (SuWi) from demersal fish in the South China Sea (WSWD711), national demersal fishery (WSWD), and national capture fisheries (WSWF).

Figure 6 illustrates the comparative trends related to sustainable development of national marine fisheries, national demersal fisheries, and special demersal fisheries in the South China Sea (IFMA 711). The maximum value of sustainable economic growth appears to be quite critical, in 2008 showing the thickness of demersal fishery Sustainability Window (SuWi) in the South China Sea (IFMA 711) tended to rise, but then declined since 2009. Then, 2010 showed significant thickness, with a sharp rise taking place in 2014. The demersal thickness trend of demersal fisheries in the South China Sea (IFMA 711) tends to be similar to the national demersal fisheries. This condition indicates that national policies on demersal fisheries will have a significant impact on demersal fisheries in the South China Sea (IFMA 711). From 2007 to 2015, national fisheries Sustainability Window (SuWi) thickness tends to increase and higher than demersal fisheries in the South China Sea (IFMA 711). However, since 2014, the opposite is true with the demersal Sustainability Window (SuWi) in the South China Sea higher than national fisheries and national demersal fisheries.

4. Discussions

National capture fisheries production is dominated by three groups of fish species, namely small pelagic, large pelagic, and demersal. Demersal fish resources dominate the production and economic value of capture fisheries production in the South China Sea (IFMA 711). Capture fisheries, especially demersal fishery, is one of the most dependable fishing venues for fishing in the South China Sea (WPP NR 711), so it needs to be managed in such a way, including managing resources that are natural inputs for the sustainability of the fishery business. [21] state that capture fisheries are managed within a wide range of institutional structures (policies) and include the granting of rights to fish resources that take into account the moral-social aspects.

In 2010 to 2015, the distance range of thickness SWD711min and SWD711max tended to increase significantly thepercentage of the thickness level. This indicates that policies enforced during that period support the improvement of demersal fish resource sustainability in the South China Sea (IFMA 711). Similarly, for SWDmax (demersal) and SWFmax (capture fisheries) thickness ranges, each compared to SWFmin, which tends to widen as the year increases. The fishery resource utilization policy of both sea-catching fishery in general and demersal fishery shows improvement in sustainability as shown by the increasing of SuWi thickness level for each fishery.

[19] in his research on sustainable development in China with the SuWi method, obtained a sustainable window thickness level between 0 and 1.2. Based on the existing graph with 2007 base year, the level of thickness of window of national fisheries sustainability between 0 - 0.28; while the level of demersal fishery sustainability window width at IFMA 711 in 2007 s / d 2010 between 0 - 0,04 with increase of thickness level SuWi happened in 2011 until 2014 that is equal to 0.14 – 0.32 and significant increase of thickness of SuWi to 0.84 occurs in 2015. With the level of thickness SuWi tend to increase from 2011 to 2015, it can be said the condition of demersal fishery sustainability, especially in the South China Sea gradually recovered. However, the level of thickness of SuWi that tends to have a value below 1, indicates the need to be careful in managing demersal fisheries in the South China Sea (IFMA 711). So the management policy made should be more consider the ecological condition of its fishery resources. The thinness of this sustainability window, can be caused by several things, such as the high pressure of fishing by the rampant illegal fishing activities in this waters (according to the research results in chapter 5), which puts pressure on the condition of ecological and social indicators of demersal fisheries in the South China Sea (IFMA 711).

Law no. 45 of 2009 on anamendment to Law no. 31 of 2004 concerning fisheries, granting additional authority to fishery supervisors to perform special actions in the form of burning and/or sinking of foreign flagged fishing vessels with sufficient initial evidence [22]. This policy has succeeded in suppressing the number of foreign-flagged illegal fishing actors doing fishing especially in South China Sea (IFMA 711). The combination of punishment methods had a significant impact on the number of fisheries violations. The decline in the number of illegal fishing actors has resulted in the scarcity of fishery production in the flag state of illegal fishing vessels, causing high prices of fishery commodities in the country. This condition also affects the rising prices of fish resources in Indonesia as a consequence of the increasing demand for Indonesian capture fisheries from other countries. So that has an impact on improving the welfare of fishermen, especially fishermen who do activities in the South China Sea (IFMA 711). To this end, this policy has had a positive effect on the sustainability of demersal fisheries in the South China Sea (IFMA 711).

In an effort to maintain the sustainability of fishery resources, the Indonesian government through the Ministry of Marine Affairs and Fisheries issued a regulation on the moratorium on fishing activities on fishing vessels whose construction is done abroad [23]. This policy has been effective since its enactment in 2014 and has had a very significant impact, especially on demersal fisheries activities in Indonesia. The rate of growth and life time of demersal fish resources that tend to be faster and shorter than other resources, leads to demersal fisheries being more responsive to existing policies. The adopted fisheries management policy has an impact on the significant increase of sustainability level in demersal fisheries especially in South China Sea (IFMA 711) compared to national fisheries.

5. Conclusions

Based on the SuWi analysis, the condition of demersal fishery sustainability tend to be better from year to year in the South China Sea (IFMA 711). However, the width of demersal sustainability window tends to be low and indicate the status of sustainability of demersal fish is not maximal in the South China Sea (IFMA 711). Therefore, care should be taken to manage demersal fishery in the South China Sea (IFMA 711).

References

[1] Ludwig D E, Hilborn R, Walters C 1993 Uncertainty, resources exploitation, and conservation: lessons from history *Science* **260** 17-36

[2] FAO 1993 *Review of the State of World Marine Fishery Resources* FAO Fisheries Technical Paper 335 p 136

[3] Garcia S M and Newton C 1994 Current Situation, Trends and Prospects in World Capture Fisheries. In: Pikitch EK, Huppert DD, Sissenwine MP, editors. Global Trends: Fisheries Management Proceedings of the American Fisheries Society Symposium 14–16 June 1994;1997

[4] Chu C 2008 Thirty years later: The global growth of ITQs and their influence on stock status in marine fisheries *Fish Fish* **10** 1-14

[5] Milich L 1999 Resource mismanagement *versus* sustainable livelihoods: the collapse of the Newfoundland Cod fishery. *Soc. Natur. Resour.* **12** 625-642

[6] Worm B, Barbier E B, Beaumont N, Duffy J E, Folke C, Halpern B S, Jackson J B C, Lotze H K, Micheli F, Palumbi S R, Sala E, Selkoe K A, Stachowicz J J and Watson R 2006 Impacts of biodiversity loss on ocean ecosystem services *Science* **314** 787-790

[7] Dalzell P and Adams T J H 1997 Sustainability and management of reef fisheries in the Pacific Islands. In: 8th Intl Coral Reef Symp. Proceedings of the Eight International Coral Reef Symposium, Panama City, 24 June- 29 June, 1996, vol. 2 University of Panama and the Smithsonian Tropical Research Institute, Panama City pp 2027e2032

[8] FAO 2005a *Increasing the Contribution of Small-scale Fisheries to Poverty Alleviation and Food Security* FAO Technical Guidelines for Responsible Fisheries p 98

[9] Ram-Bidesi V 2008 Development of marine resources, fisheries policies and women's rights in the Pacific Islands *SPC Women Fish. Inf. Bull.* **18** 3e10

[10] FAO 2012 *The State of World Fisheries and Aquaculture Food and Agriculture Organization of the United State* (Rome: FAO Fisheries and Aquaculture Department) p 241

[11] Charles A T 1998 Living with uncertainty in fisheries: analytical methods, management priorities and the Canadian groundfishery experience *Fisheries Research* **1998** 37-50

[12] Cochrane K L 2000 Reconciling sustainability, economic efficiency and equity in fisheries: the one that got away? *Fish and Fisheries* **2000** 3–21

[13] Whitmarsh D, Pipitone C, Badalamenti F and D'Anna G The economic sustainability of artisanal Fisheries: The case of the trawl ban in the Gulf of Castellamare, NW Sicily *Marine Policy* **2003** 489-97

[14] Maiolo J R and Orbach M K 1982 Modernization and marine fisheries policy (Michigan: Ann Arbor Science) p 330

[15] Crean K and Symes D 1996 Sailing into calmer waters? In: Crean K, Symes D, editors. Fisheries Management in Crisis (Oxford: Fishing News Books)

[16] Bailey C 1994 Employment, labor productivity, and income in smallscale fisheries of thesouth and southeast Asia. In: Socioeconomics issues in coastal fisheries management. Bangkok, Thailand: RAPA Publication, Indo-Pacific Fishery Commission

[17] Pauly D, Christensen V, Guenette S, Pitcher T J, Rashid Sumaila U, Walters C J, Watson R and Zeller D Towards sustainability in World fisheries *Nature* **2002** 689-95

[18] Germain N, Hartmann H J, Melo F J F and Reyes-Bonilla H 2015 Ornamental reef fish fisheries: New indicators of sustainability and human development at a coastal community level *Ocean and Coastal Management* **104**(2015) 136-148

[19] Luukkanen J, Kaivo-oja J, Vehmas J, Panula-Ontto J and Hayha L 2015 Dynamic sustainability:

Sustainability window analysis of Chinese poverty-environment nexus development *Sustainability* **6**(2015): 14488-14500

[20] [WCED] World Commission on Environment and Development 1987 *Our Common Future* (Oxford: University Press)

[21] Hilborn R, Orensanz, Lobo J M and Parma A M 2005 Institutions, incentives and the future of fisheries *Philosophical Transaction of the Royal Society B.* **360** 47–57

[22] KKP 2009 Undang-Undang No. 45 tahun 2009 tentang Perubahan Atas Undang-Undang No. 31 Tahun 2004 tentang Perikanan (Jakarta: KKP)

[23] KKP 2014 PerMenKP No. 56 Tahun 2014 tentang Penghentian Sementara (*Moratorium*) Perizinan Usaha Perikanan Tangkap di Wilayah Pengelolaan Perikanan Negara Republik Indonesia (Jakarta: KKP)

Predator-prey model of exploited fish population Arowana (*Scleropages* spp.)

S Triharyuni*, **Aisyah**

Fisheries Research Center
*e-mail: setiya_triharyuni@yahoo.co.id

Abstract. Interactions in the ecosystem are inseparable from predator-prey interactions. Population model of two species that the predator interacts with its prey for its survival has been developed. The growth rate of a predator depends on predation on its prey. The model used to describe this predation is the Holling Tanner II function. This model also includes a catch variables of prey and predator populations. This formulation provides more realistic picture of exploited fisheries, such as *Scleropages* spp. (arowana) that prey on small fish and shrimp. The purpose of this study is to analyze the model of predator-prey and to interpret on the model based on simulation. High of exploitation rate will lead to rapid population decline. Exploitation rate ranges from $0 \leq h \leq h\ max$, where $h\ max$ is the maximum exploitation rate. Results of these simulation show that population of the predator fish will become extinct if exploitation rate exceeds the maximum exploitation rate.

1. Introduction

There are two types of fisheries in Indonesia, namely marine fisheries and inland fisheries. Exploitation activities are carried out on resource populations in those two ecosystems. Exploitation carried out continuously regardless of sustainability, can lead to the extinction of fish resources. To overcome this, the management steps need to be applied to maintain the sustainability of fishery resources. Management that has been done so far is to set the maximum capture or often called the maximum sustainable yield (MSY). However, on MSY based management it is based solely on a single species without considering the interaction of species within it [1].

Several previous studies related to prey-predation models by including MSY in them include research on MSY in fisheries with prey-predator structure [2,3,4,5]. [6] conducted research on Michaelis-Menten prey-predator models with constant harvesting in the prey population. Then research on stability analysis of prey-predator models with constant harvesting on prey fish [7,8]. The studies discussed the stability of the prey-predator models with harvesting in the form of function in the population of prey fish and constant harvesting. Effects of harvesting on various types of prey-predator models have been considered by many researchers [9,10,11,12,13], and [14] focused on the harvesting in predator prey interaction of seals, steelhead trout and herring MSY formulation shows that the highest level of harvesting of predator fish occurs when the growth of prey fish is on its maximum level [15]. Continues harvesting until it passes the limit causes the extinction of these resources [16]. Therefore, it is necessary to restrict the catch to maintain the sustainability of these resources.

Arowana fish is one of economical fishery commodities, usually exported to foreign countries such as Singapore, Hongkong, Taiwan, Japan, Brunei Darussalam and Malaysia [17]. This fish is classified as a predator in inland waters. Population density of arowana fish is very low, and the number of adult population become declined [18]. Some that threaten the life of arowana fish are caused by low fecundity, the habit in accommodating tillers in the mouth and spawning in open water [19]. The intensive level of exploitation shown by an increase in the number of exports from 1995-2001, that also become a factor that threatens the arowana population [17]. This paper aims to analyze the stability and make simulations of the prey-predator model by harvesting on arowana fish as the development of existing models. The results of the study are expected to provide an overview and information in arowana resource management so that it is sustained and maintained.

2. Mathematical models

The model formed in this study is the interaction of prey-predation with harvesting on predator species, namely the interaction between arowana fish with small fish. Arowana fish is a type of predatory fish and the main prey is small fish [20]. The other assumptions used in this model include:

a. Small fish in one population are prey with logistic growth.
b. Arowana fish are predatory fish that depend on the existence of small fish. Small (juvenile stage) arowana feeds insect & invertebrate; while the larger one (adult) feeds on small fishes and invertebrate. To simplify the model, the study focused on the predatory prey interaction between arowana as a predator and small fish as a prey.
c. In conducting predation there is a limit of saturation prey so that the function of predator follows Holling Tanner II Models.
d. Only Arowana is exploited, and henceforth, exploitation is called harvesting.
e. No environmental changes/predator or prey responding independently to the environmental factors.

The predator-prey interaction is shown in figure 1.

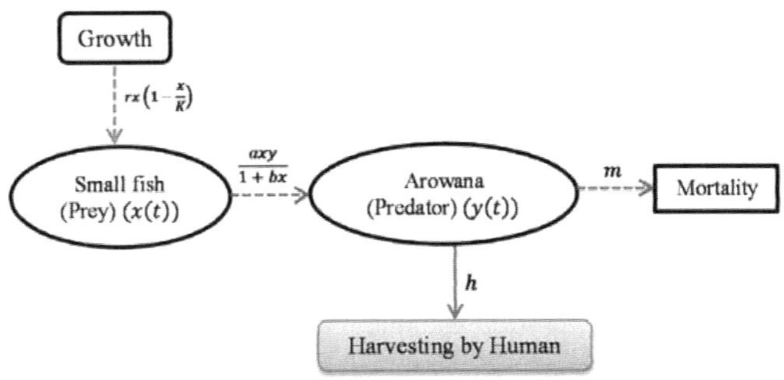

Figure 1. Interaction diagram between arowana and small fish with harvesting rate.

From the interaction and assumptions of variables above, the model is formulated as below:

$$\frac{dx(t)}{dt} = rx(t)\left(1 - \frac{x(t)}{K}\right) - \frac{\alpha x(t)y(t)}{1+bx(t)} \qquad (1)$$

$$\frac{dy(t)}{dt} = \frac{c\alpha x(t)y(t)}{1+bx(t)} - my(t) - hy(t)$$

r is the intrinsic growth rate of small fish, **K** is carrying capacities, $\boldsymbol{m}$ is mortality rate coefficient of arowana, $\boldsymbol{a}$ coefficient small fish predation by arowana, $\boldsymbol{c}$ conversion factor from the number of calories required by the new arowana for each prey small fish caught, $\boldsymbol{b}$ is the half saturation of predation on small fish co-efficient and $\boldsymbol{h}$ are harvesting rate of arowana.

IOP Conf. Series: Earth and Environmental Science **176** (2018) 012037 doi:10.1088/1755-1315/176/1/012037

3. Equilibrium and stability analysis

To obtain the equilibrium point, let the system (1) if $\frac{dx(t)}{dt} = 0$ and $\frac{dy(t)}{dt} = 0$. Meanwhile, to know the stability of the system (1), is used Jacobi Matrix and finding the negative real parts of the eigenvalues evaluated at each equilibrium point. The Jacobi Matrix of system (1) is:

$$J_E = \begin{bmatrix} \frac{-2rx(1+bx)^2+K(r(1+bx)^2-ay}{K(1+bx)^2} & -\frac{ax}{1+bx} \\ \frac{acy}{(1+bx)^2} & -h-m+\frac{acx}{1+bx} \end{bmatrix} = \begin{bmatrix} A & B \\ C & D \end{bmatrix} \quad (2)$$

The result of equilibrium point analysis shows there are three points. Point $E_1(0,0)$, which means that in the ecosystem between arowana fish and its prey are extinct. The Jacobian matrix evaluated at E_0 gives, $J_{E_1} = \begin{bmatrix} -r & 0 \\ 0 & -h-m \end{bmatrix}$ and the eigenvalues of J_{E_0} are $\lambda = r > 0$ and $\lambda = -h - m < 0$. The next equilibrium point is $E_2(K, 0)$, which means that arowana fish is extinct so that small fish as arowana prey grow until reaching the carrying capacity. The Jacobi Matrix of these point is

$J_{E_2} = \begin{bmatrix} -r & -\frac{ak}{1+bk} \\ 0 & -h-m+\frac{ack}{1+bk} \end{bmatrix}$ and the eigenvalues is $\lambda = -r < 0$ and $\lambda = ack - (1 + bk)(h + m)$.

The third equilibrium point is $E_3\left(\frac{h+m}{\alpha c-bh-bm}, \frac{rc(Kac-Kbm-Kbh-m-h)}{(ac-bm-bh)^2K}\right)$, which is the expected condition, because with those condition the ecosystem will remain maintained. This condition implies that between prey and predator populations live together. Existence of this will be valuable biogically if it has a non-negative value, so the requirement for point E_3 is biologically significant if $\frac{b(h_y+m)}{\alpha c} < 1$ and $\frac{(m+h_y)(Kb+1)}{Kac} < 1$.

4. Maximum harvesting

Maximum harvesting is the maximum allowable harvest to maintain the sustainability of a population. In the fisheries is usually called the maximum sustainable yield (MSY). Model (1) implements harvesting of arowana population in the form of parameter h. It is known that the maximum growth value of the prey fish population is half of its capacity $\left(\frac{K}{2}\right)$ [8]. Based on the assumptions given for harvesting rate ie $0 \leq h \leq h_{maks}$, where h_{maks} is the maximum harvest rate. Then the maximum harvest rate in model (1) is :

$$h_{max} = \frac{Kac-Kbm-2m}{Kb+2} \quad (3)$$

5. Simulation and discussions

This paper studied the dynamical behaviors of one prey one predator system. The interaction between the prey and the predator is assumed to be established by a Holling type II response function because the predator needs sufficient handling time for other prey and there is saturation. A good example of a prey and predator system is arowana (predator) and small fish as a prey. Existing problem is the real world data are not available for this model. Therefore, we using parameters from [7][8]) to illustrate the results numerically, e.i. r = 0.8; K = 100; b = 0.1; c = 0.75; m = 0.001 and the initial value $x(0) = 80$ and $y(0) = 20$, whereas a = 0.01, these value are small because arowana feeds insect & invertebrate. By using those parameters and harvesting rate at $h = 0.02$, system (1) has three equilibrium points, i.e $E_1(0,0)$; E_2 (100.0) and $E_3(4.107)$. The result show that all of equilibrium points are unstable (figure 2).

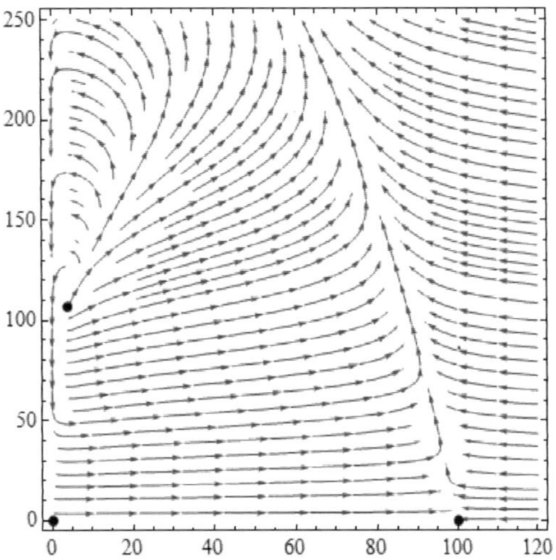

Figure 2. Phase potrait of prey x and predator y, with $h = 0.02$.

Figure 2 describe the points of $E_1(0,0)$ indicated that arowana and small fish are extinct, E_2 $(100,0)$ indicated that arowana extinct and the growth of small fish reach to the carring capacity on the ecosystem. The points of $E_3(4,107)$ means that arowana and small fish can coexist in an ecosystem with a conserved individual population of 4 for small fish and 107 for arowana.

One of the causes of population extinction is the high predation rate of prey and the low level of prey growth or the low initial population of the prey population [21] [22]. Harvesting the predator population may control those populations that can lead to the extinction of prey populations. Conversely, by controlling the predator populations to avoid those extinction, harvesting restrictions will be imposed on predator populations. The restriction of harvesting on predatory fish in this case is arowana fish, because of the exploitation that has been done since 1970 and the number of exports of these fish continues to increase [17].

Based on (3), the maximum harvesting limit is 0.0615. The study of [23] states that the exploitation of arowana fish does not exceed of 50%. To determine the effect of harvesting on the population of prey and predator is done simulation with three treatments, when $h < h_{max}$ (harvesting less than maximum harvesting), $h = h_{max}$ (harvesting and maximum harvesting are equal) and $h > h_{max}$ (harvesting more than maximum harvesting). In prey–predator system, maximum harvesting that only apply on the prey population will tend to extinction of the predator population [4] and/or other species [3]. The fact is both predator and prey has the opposite condition, when predator is decreased conversely with prey, or vice versa. In this condition, standing stock of the predator are still high when the condition of the effort is small, and the yield becomes again lower when the effort is larger [3]. In order to keep the sustainability of prey and predator populations, a maximum harvesting is required, in accordance with the statement [3] that in the Holling–Tanner prey–predator system with any type of response function, fishing of the predator species at the MSY level is a safe harvesting policy for the coexistence of the species. Harvest restrictions need to be done because of the population declining. On the other hand, these species is listed as threatened under CITES Appendix II, which mean that this fish must be protected [17].

5.1. Conditions when $h < h_{max}$

First simulation is done if harvesting less than maximum harvesting, by taking the value of h = 0.02. The simulation results show that when $h < h_{max}$, population of arowana will increase and population of small fish tend to decrease. This is due to the increasing predation of the small fish by arowana. This condition can be interpreted with less harvesting rate than maximum harvesting, arowana fish population and its prey can coexist and mutual source of both stay maintained (figure 3). In fact this condition does not occured in arowana fishery, because arowana has an economic value. The high demand for arowana fish is a threat to those population [17].

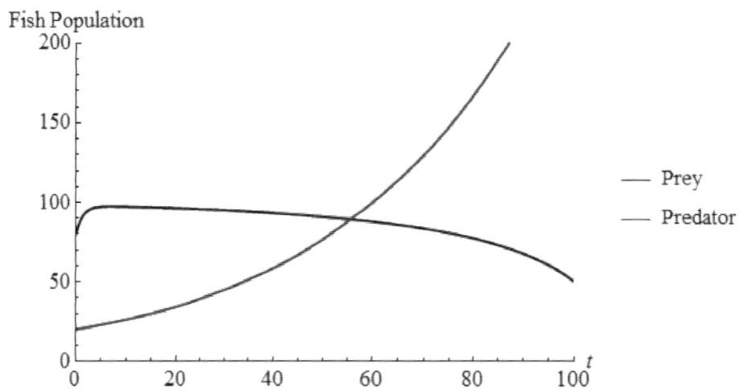

Figure 3. Effect of harvesting of arowana when $h < h_{max}$ on ecosystem.

5.2. Conditions when $h = h_{max} = 0.0615$

The second simulation, harvesting rate is equal to the maximum harvest level. In this case, the harvesting rate of 0.0615 is the maximum level of harvesting that can be done to maintain resource sustainability and give the maximum catch based on model (1). The result shows that small fish as a prey with initial value of individual is 80 tend to unchange. As for arowana fish population, with initial value of 20 tend to increase but the increase was not as significant in the first simulation ($h < h_{max}$). This can be interpreted that, arowana fish population and small fish can still coexist and its existence is maintained in ecosystem if harvesting done maximally (figure 4).

Maximum harvesting is the concept of Maximum Sustainable Yield (MSY). The concept of MSY aims to maintain the size of the fish population at its maximum point and provide economic benefits to society [24]. Management of arowana fisheries that is already done is based on number of capture quotas. For example, the capture of Arowana in Kumbe River reach to 50% of the total production of arowana in these region [25]. Maximum fishing on this predator is one of the strategy in maintaining the sustainability of resources within an ecosystem, however both prey and predator can coexist under this harvesting policy. Maximum harvesting that only apply on the prey population will tend to extinction of the predator population, while maximum harvesting on both prey and predator will cause a single isolated population (composed of prey only, i.e., the predator population goes to extinct) [26].

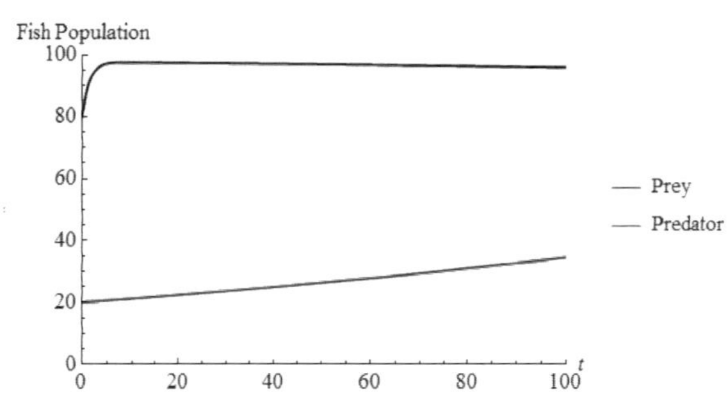

Figure 4. Effect of harvesting of arowana when $h = h_{max}$ on ecosystem.

5.3. Conditions when $h > h_{max}$

Harvesting is done continuously and exceeds the limit will cause the extinction of a resource [16]. High level of harvesting will lead to a decrease in the number of fish populations in an ecosystem. Those condition is reflected in figure 5, when the harvesting rate of arowana is added to be $h = 0.1$, the arowana fish population will decrease, even at a time this fish population will be extinction. The declined of predatory fish population shows that the predation rate of small fish is reduced so after that the population of small fish will increase it reaches carrying capacity condition. Different condition will occurred when harvesting is done on prey fish, which is high harvesting of prey fish populations will decrease even to the extinction condition. This condition will be followed by the extinction of predatory fish populations respectively [8]. Harvesting rate that exceeds the maximum limit is similar to over exploitation. over exploitation of fish resources causes unsustainability [27]. Legality is applied to avoid those condition. Therefore, the Government has issued a regulation No.7/1999, Decree of the Minister of Forestry No.209/kpts-II/2001 renewed by Regulation of the Minister of Forestry Number P.12/Menhut-II/2005 to protect arowana fish population.

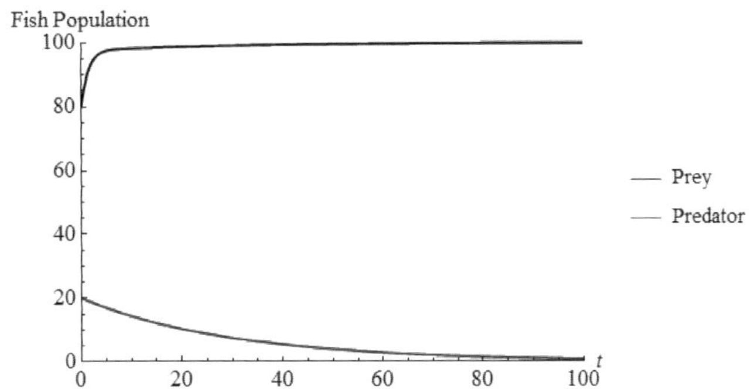

Figure 5. Graphs of stability of prey x and predator y, with $h > h_{max}$.

6. Conclusions and suggestion

6.1. Conclusion

The results of the analysis conducted on the prey-predator model with harvesting on predators, obtained a maximum harvest rate h_{maks} is 0.0615. The model produces three equilibrium points, namely the condition when all populations become extinct, in the ecosystem only the remaining fish are prey and both populations between prey and predator coexist. The simulation indicates that it is necessary to restrict the exploitation of predatory fish to maintain its sustainability. High exploitation coefficient will cause the decrease or extinct of predator population.

6.2. Suggestion

In the next study, the prey-predator model will conduct harvesting treatment of both species, also by giving a treatment of existed conservation area.

References

[1] Kar T K and B Ghosh 2013 Maximum sustainable yield and species extinction in a prey–predator system: some new results *J. Biol. Phys.* **39** 453–467

[2] Hertini E and Gusriani N 2013 Maximum Sustainable Yield (MSY) structure of fishing with prey predator *Proceedings of National Seminar in Sains and Nuclear Technology* 307-311

[3] Ghosh B and Kar T K 2013 Maximum sustainable yield and species extinction in a prey–predator system: some new results. *J. Biol. Phys.* **39** 453–467

[4] Paul S, Bhattacharya P and Choudhury K S 2013 Maximum sustainable yield policy in prey-predator system-a study *Proceedings published by International Journal of Computer Applications (IJCA)* 46-49

[5] Kinfe H H and Zenebe T M 2016 Mathematical modelling of Fish Resources Harvesting with Predator at Maximum Sustainable Yield, *International Journal of integrative Sciences, Innovation and Technology (IJIIT)* **5**(4) 1-11

[6] Dwaradi H 2011 Prey-Predator Model Analysis of Michaelis-Menten with harvesting on prey population (Bogor: Bogor Agricultural University)

[7] Kar T and Cakraborty K 2010 Effort dynamics in a prey-predator model with harvesting **6**(3) 318-332

[8] Afifah L I and Pagalay U 2015 Stability analysis of prey-predator model with constant harvesting on prey fish *CAUCHY* **3**(4) 14-22

[9] Xiao D, Li W and Han M 2006 Dynamics in a ratio-dependent predator–prey model with predator harvesting. J. *Math. Anal. Appl.* **324** 14–29

[10] Kar T K 2005 Conservation of a fishery through optimal taxation: a dynamic reaction model *Communications in Nonlinear Science and Numerical Simulation* **10** 121–131

[11] Leard B, Lewis C and Rebeza J 2008 Dynamics of ratio-dependent predator-prey models with nonconstant harvesting D*iscrete and Continuous Dynamical Systems Series* **1**(2) 303-315

[12] Lv Y, Yuan R and Pei Y 2013 A prey-predator model with harvesting for fishery resource with reserve area. Applied Mathematical Modelling 37 (2013) 3048–3062

[13] Belkhoja K, Moussaoui A and Alaoui M A A 2018 Optimal harvesting and stability for a prey–predator model. Nonlinear Analysis: Real World Applications **39** 321–336

[14] Triharyuni S and Aldila D A 2016 Mathematical model of predator-prey interaction between seal-herring and steelhead trout. *International Symposium on Current Progress in Mathematics and Sciences 2016 (ISCPMS 2016) AIP Conf. Proc.* 1862 doi:10.1063/1.4991256

[15] Schaefer M B 1957 A Study of the dynamic of fishery for yellowfin tuna in the Eastern Tropical Pacific Ocean *Bulletin of the Inter-American Tropical Tuna Commission* **2** 247-285

[16] Eaton P 2005 *Land tenure, conservation and development in Southeast Asia.* First published by Routledge Curzon p 191

[17] Kartamihardja E S K, Umar C and Aisyah 2014 Lesson learned in community based management and conservation of red arowana (*Scleropages formosus*, Muller and Schlegel, 1844) at Empangau Lake, West Kalimantan *J. Kebijak. Perikan. Ind.* **6**(2) 65-75

[18] Kottelat M 2011 *Scleropages formosus*. The IUCN Red List of Threatened Species 011:e.T20034A9137739.doi.org/10.2305/IUCN.UK.2011 1.RLTS.T20034A9137739.en.

[19] Rahman S, Ismail M Z, Tang P Y, Idid M R and Muniandy S 2010 Genetic variability and estimation of effective population sizes of the natural populations of green arowana, *Scleropages formosus* in Peninsular Malaysia *Advances in Environmental Biology* **4**(1) 14-23

[20] http://www.fishbase.org/summary/6357. *Scleropages formosus* (Müller & Schlegel, 1840) Asian bonytongue. Downloaded on Agustus 28, 2017

[21] Campbell N A, Reece J B and Mitchell L G 2004 *Biologi* Translated by Wasmen Manalu; Edited by Amalia Safitri, p 501

[23] Marom S 2013 Establishment of models of a predator prey with threshold harvesting on the predator *δELTΔ* **1**(2) 115-199

[23] Satria H and Kartamihardja E S 2010 Stok abundance and conservation area development of Saratoga, *Scleropages jardinii* (Saville-Kent, 1892) at Maro River, Merauke Regency, Papua Province. *J. Lit. Perikan. Ind.* **16**(1) 49-62

[24] Conrad J M and Clark C W 1987 *Natural Resource Economics, Notes and Problem* (New York: Cambridge University Press)

[25] Kartamihardja E S, Purnomo K, Tjahyo D W H and Koeshendrajana S 2013 Ecosystem approach to management of saratoga, *Scleropages jardinii* resources at Maro River, Merauke-Papua *J. Kebijak. Perikan. Ind.* **5** (2) 87-96

[26] Legovic T, Klanjscek J and Gecek S 2010 Maximum sustainable yield and species extinction in ecosystems *Ecol Mode* **221** 1569–74

[27] Eaton P 2005 *Land Tenure, Conservation and Development in Southeast Asia* (First published by RoutledgeCurzon) p 191

IOP Conf. Series: Earth and Environmental Science **176** (2018) 012038 doi:10.1088/1755-1315/176/1/012038

Modeling of sea surface temperature using linear models with autocorrelation Indian Ocean

M Miftahuddin[1*], Y Ilhamsyah[1,2]

[1]Faculty of Mathematics and Science, Syiah Kuala University,
Jln. Teuku Nyak Arief Darussalam, Banda Aceh, Aceh, 23111, Indonesia
[2]Applied Climatology, Graduate School of Bogor Agriculture University, Bogor, Indonesia
*e-mail: miftah@unsyiah.ac.id

Abstract. Sea surface temperature (SST) is one of several indicators of climate system of the Earth. We used model SST to observe SST dataset from buoy arrays in the eastern equatorial Indian Ocean. Relationship between SST and other climate parameters can be represented in linearity approach. This approach shows that temporal variability of the SST as a dominant effect. Linear model fitting (LMF) has been examined with four treatments, with and without: covariate transformation, interaction, centering, and addition time covariate in the model. The LMF chosen as basic construction in the model with covariate interaction combination and transformation, which increases magnitude of multiple-R^2 (56.62%) and adjusted-R^2 (56.13%), i.e. 0.31% and 0.43% respectively. This shows that time covariates have a strong significance effect in the model, compared to continuous covariates. However, the model has autocorrelation, which has large Akaike Information Criterion (AIC) value then this deletion of effects can be done through the autoregressive moving average. Moreover we obtained that LMF which suitable to SST is model with AIC value 403.2987 by using three climate features include two time covariates. Furthermore, we observed that using GAM model fitting showed an increase in explained deviance to 65.90%, a significant decrease in AIC from 678.24 to 634.99 and significant increase in adjusted-R^2 from 51.20% to 64.40% by using sixteen climate features include two times covariates without interaction and transformation.

1. Introduction

Sea surface temperature (SST) is one of many important indicators that indicates/marks the climate system of the Earth. The SST data are useful to early detection climate change and global warming. In fact, the ocean is the particular region of the earth where 70.9% areas and 29.1% is mainland. The largest ocean is the Pacific Ocean (165,759,239.06 km^2) and the second largest is the Atlantic Ocean (106,448,511.33 km^2), and third largest is the Indian Ocean (72,519,667.09 km^2). SST data is essential indicator to recognize climate variability in the earth [1-3], such as El Nino and La Nina phenomena in the Pacific Ocean that affects on dry and wet seasons in Indonesian and its adjacent region.

There are several reasons we modeled the SST data in the eastern equatorial Indian Ocean. First, the observed SST data show complex structure like missing values that represented as gaps which vary among buoys array. Second, in addition Aceh province is located in the western most Indonesia region and it directly connect with Indian Ocean on southern and western waterbody. Third, in 2012 Magnus et all proposed a climate model to investigate the effects of solar radiation and the greenhouse effect on global warming [4]. Their analysis is based on the data from land stations only and does not consider the relationship between sea and land dataset. Hence, the objectives our study, the SST data is used to reveal the relationship among variables in both by using linear and generalized additive model (GAM). Further, firstly, we reviewed the basic construction regarding linear model fitting without and with covariate transformation. Secondly, we used through linear combination of covariates interaction and without/with transformation in the linear model. Thirdly, we applied the second steps with centering approach. Fourth, we developed model by using linear model with consider autocorrelation aspect. Finally, we developed model by using GAM model with the number of extended climate features.

2. Material and Methods

2.1. Materials and data

In this study we used one of moored buoy arrays in the Indian Ocean at 1.5^{0}N,90^{0}E, operated by JAMSTEC Japan as part of the Tropical Atmosphere Ocean (TAO) program [5]. The buoy collect real-time SST daily data at 1 m depth from 1 January 2006 to 8 June 2012. The SST dataset consist of 2,066 daily observations with the response variable SST (^{0}C) from 00.00 - 12.00 pm GMT [6].

To show essential time covariates in the linear model, we used *f*(.) function as represented by model with three continuous covariates and two time covariates, i.e. month and year. We assume that three continuous covariates which correspond linearly with SST, i.e. the relationship will be the same for all levels of the time covariate and without interaction between the covariates, as follows [6],

$$SST_i = \beta_0 + \beta_1 Temperature_i + \beta_2 Humidity_i + \beta_3 Rainfall_i + \eta_k Month_i + \gamma_l Year_i + \varepsilon_i \quad (1)$$

for $k = 1, \ldots, 12$; $l = 1, 2, \ldots, 6$, and $i = 1,\ldots,n$, where η and γ are vector parameter of time covariate for month and year respectively. We construct unrestricted model where the model has seasonal effect η_m and annual effect γ_l are restricted to 0. We evaluate several models with different order include time covariate in the model fitting.

Previously, modeled SST data is as linear combination of three parameters of climate features and two time covariates [2]. These parameters are air temperature (^{0}C), relative humidity (%) covariates. They have mean with the same time records on 07.00 WIB, 13.00 WIB and 18.00 WIB, whereas rainfall (mm) during 3-hours period, seasons and annual factors. We extend to SST dataset to buoy position at 4^{0}N; 90^{0}E, depth 1 m, on period 2011 to 2015 with 16 climatic parameters, such as wind speed, subsurface temperature, shortwave radiation, salinity, humadity, precipitation, dynamic height, density, air temperature, isoterm, conductivity, zonal and merid current velocity, and time covariates (month and year). The complexity of the linking between the SST and the parameters becomes challenging in construction a dynamic model. Treatments with respect to the model constructed are without and with covariate transformation, without and with covariate interaction, and mixed both mentioned treatment.

2.2. Linear and Interaction Model Fitting

In preliminary modelling, daily SST data are used with linear in parameters assumption that denoted as (x_i, y_i), for $i = 1, 2,\ldots,n$ where x_i as covariates and y_i as response variables. The relationship between the random variables X and Y can be written in the matrix form as [6,7]:

$$\boldsymbol{Y} = \boldsymbol{X\beta} + \boldsymbol{\varepsilon} \quad (2)$$

where $\boldsymbol{Y}$ represents a vector of observations $(y_1,\ldots, y_n)^T \epsilon\ R^n$, $\boldsymbol{\beta}$ represents a vector of parameters $(\beta_0,\beta_1,\ldots,\beta_p)^T \epsilon\ R^{p+1}$, $\boldsymbol{X} \epsilon\ R^{nx(p+1)}$ represents a matrix with rows n and columns $p + 1$ of a set p covariates $X_0, X_1,\ldots, X_p$ of length n including an *intercept* and *errors* ε are assumed *independent and identically distributed*, i.e. normal random variable $\boldsymbol{\varepsilon} \sim N(\mathbf{0_n}, \sigma^2\mathbf{I_n})$, where ε represents a column vector $(\varepsilon_1,\ldots,\varepsilon_n)^T$ of error and $\mathbf{I}_n$ as the identity matrix. In formulation, for the covariates we can describe as quantitative values, qualitative, transformation, interaction among covariates, and various data types.

Let random variable Y then the probability density function (pdf) of a continuous (or the probability mass function (pmf) for Y is discrete) is referred as a probability distribution and denoted as $f(y; \boldsymbol{\theta})$, where $\boldsymbol{\theta}$ represents the parameters of the distribution. The conditional expectation describes the linear or functional relationship of parameter in model,

$$E[Y|X] = \textstyle\sum_{j=0} \beta_j X_j \quad (3a)$$

or

$$\mu_i = E[Y|X_i] = f(X_i),\ i = 1,\ldots, n \quad (3b)$$

Assumption *additivity* with interaction model:

$$\mu_{ij} = \mu + \alpha_i + \beta_j + (\alpha\beta)_{ij} \quad (4)$$

where μ is constant, α_i and β_j are main effects, and $(\alpha\beta)_{ij}$ is interaction effect.

2.3. Generalized Additive Models Fitting

In a generalized additive models, suppose denoted Y is conditionally independent of the covariates x given the additive covariates [6,7,8]

$$Y = \beta + \sum_{j=1} f_j(xj)$$

where $f_j(xj)$ for some functions fj. Let Y and $\boldsymbol{X}$ are random variables that representing response (*output*) and covariate (*input*), respectively then conditional relationship between both the variables can be written as

$$Y = E[Y|X_1, ..., X_p] + \varepsilon, \quad (4a)$$

$$\text{where } E[Y|X_1, ..., X_p] = \beta_0 + \sum_{j=1}^{p} f_j(X_j). \quad (4b)$$

Thus, model aditive is defined by the following:

$$Y_i = \beta_0 + \sum_{j=1}^{p} f_j(X_{ij}) + \varepsilon_i, \quad i = 1, ..., n, \quad (5)$$

where the β_0 is an intercept, f_j is model types, such as linear, nonlinear, smooth function, spatial, interaction, etc, that combining covariate effects. Expectation, variance and covariance ε errors independently of X_j,

$$E[\varepsilon_i] = 0, \text{ and } \mathrm{var}(\varepsilon_i) = \sigma^2, \mathrm{cov}(\varepsilon) = \sigma^2 I_n. \quad (6)$$

GAM model is an extension of the Linear Models (LM) and Generalized Linear Models (GLM) through a link function g(.), with assume that variable response follows several exponential family distributions. In general GAM structure is given as follows:

$$g(\mu) = g(E(Y|X_1, X_2, ..., X_p)), \quad (7)$$

In other words, from equation (5) is,

$$f^*(\mathbf{X}) = \beta_0 + \sum_{j=1}^{p} f_j(\mathbf{X}_j). \quad (8)$$

where f^* is response expectation by estimating additive function. From equation (5) we assume $\varepsilon_i \sim N(0, \sigma^2)$, [7]. In GAM model, [7] used loess smoothers or smoothing splines functions, whereas GAM via mgcv package by Simon Wood [9] and the model for large data sets [10]. By using spline regression method, we used knots to minimalize penalize as in the following:

$$\|Y - X\beta\|^2 + \lambda \int f''(x)^2 dx \quad (9)$$

where given λ or through cross validation to select optimal penalty.

3. Results and discussion

3.1. Linear model fitting of the SST dataset

In this section we apply our methodology to SST dataset on several treatments in model fitting.
In preliminary, the linear model is developed by without and with covariate transformation, as follows

M1: SST = $\beta_0 + \beta_1$Temp + η_k Month + γ_lYear + β_2Humd + β_3Rain + ε

By analysis of its variance, it is revealed that humidity and rainfall have insignificant effects as shown in table 1. Temperature and time covariates (Month and Year) have p-value <2e-16. It means that three covariates have significant effects to sea surface temperature compared with humidity and rainfall effects by using M1 model. Through linier model (M1) reveals that air temperature around ocean given affect to sea surface temperature condition on the certain month and year, as seen in figures 1 and 2.

Table 1. ANOVA of the M1 model.

Source	df	SS	MS	F-value	Pr(>F)
Temperature	1	83.331	83.331	581.2248	<2e-16 ***
Month	11	236.830	21.530	150.1698	<2e-16 ***
Year	6	57.463	9.577	66.7999	<2e-16 ***
Humidity	1	0.183	0.183	1.2750	0.2590
Rainfall	1	0.002	0.002	0.0132	0.9087
Residuals	2045	293.193	0.143		

By the parameters estimation of the M1 model showed that time effects on September and October have insignificant, whereas for 2011 and 2012 are insignificant effects as well. The model fitting reaches 56.31% (R^2 value). The specific time effects on the SST data can be showed by M1 model as seen in figure 1. Season effect occurs on monthly unit, exclude on September and October. Probably on the both month occurs time transition of the change season for the sea surface temperature phenomenon in the investigated period.

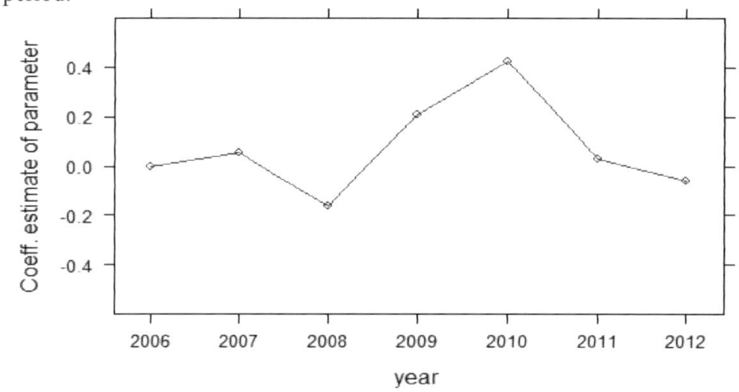

Figure 1. Annual effect of the M1 model.

In figure 1, the annual effect shows slightly increase of annual variation that happened between 2006-2007, rapid decreasing that occurred between 2007-2008, and rapid increasing between 2008-2010. Hereinafter, it decreased sharply to 2011 and slowly decreased between 2011-2012. In general, between 2007-2008 the SST data fitting by using M1 model showed non-positive effect. Likewise, it is between 2011-2012. However, between 2006-2007 and between 2008-2011, it gives positive effect.

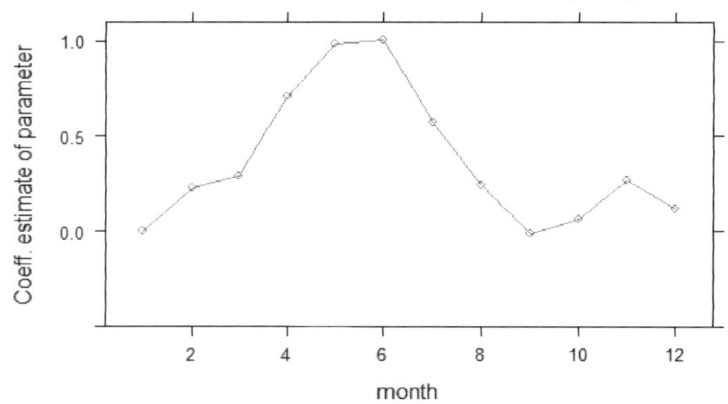

Figure 2. Seasonal effect of the M1 model.

The seasonal effect showed that increase trend that happened from January to June and a decrease trend from June to September (figure 2). Furthermore, a second feeble increase trend appeared again until November and subsequent decrease until December. A peak of the highest month effects occurs in June and lowest in September (figure 2).

Whereas the seasonal effect between January and February occurred rapid increase, and slightly increase from February to March. Further it occurs rapid increase return from March to May. The slightly increase the seasonal effect also occurs again from May to June. However, it rapid decrease happens from June to September. The slightly increase the seasonal effect go on September to October and rapidly increase go on October to November. Finally, it decreasing again occurs from November to December.

3.2. Linear model with treatment of the SST dataset

We know that generalized additive models (GAMs) are an extension function of linear models (LM) and generalized linear models (GLMs). Before our discussed regarding GAMs models fitting for the SST dataset, in this section we explore linear models fitting by using several treatments. Each treatment can be constructed as a M model and symbolized as M1, M2, M3, …, M15 models. The main treatments are transformation, interaction, and centering.

3.2.1. Linear model fitting with covariate transformation. We constructed M2 model as seen in M1 model with covariate transformation. It is used Rain = log(RAIN+0.01). By analysis of variance process it revealed that humidity and rainfall does not significant effects. The transformation rainfall covariate did not change fitting magnitude on model in the multiple R^2. However, it changed on adjusted R^2 i.e., 0.01%. This is caused by continuous covariate effect which is very small, compared to time covariate effect in the linear model fitting (see M1 and M2 models).

3.2.2. Linear model fitting with covariate interaction and without transformation. Effect of interaction modeling on the continuous covariate as in M1 and M2 models is developed to the M3 model. Result of ANOVA analysis for this model showed that humidity, rainfall and all interactions of continuous covariates did not have significant effect. Interaction effect in the model without transformation provided the multiple R^2 increasing on 0.06%. Although magnitude of adjusted R^2 is stable on 55.88% (see R^2 values pre and post interaction in the M2 and M3 models).

3.2.3. Linear model fitting with covariate interaction and transformation. In this section, we can be expressed again M3 model as M4 model with rainfall transformation. From ANOVA table knows that humidity, rainfall and interaction of continuous covariate does have insignificant effect, exclude interaction between humidity and rainfall covariates gives strong significant effect. It is also for temperature, Month and Year covariates. By parameter estimation of the M4 model shows the combination effect between interaction and transformation in the M4 model gives changing multiple R^2 and adjusted R^2 magnitudes on 0.31% and 0.43% respectively (see R^2 value on M1 and M4 models).

3.2.4. Linear model fitting with covariate interaction and without transformation and centering. Analog M3 model, we construct M5 model without centering. By analysis of variance as seen in the ANOVA table shows that humidity, rainfall, and all interaction of does have insignificant effect of treatments without transformation and centering. The centering effect in the interaction model does not changes R^2 value. This is caused by there is domination effect of time covariate bigger than continuous covariate effect (see R^2 value in the M3 and M5 models).

3.2.5. Linear model fitting with covariate interaction, transformation and centering. In this section, analog the construction of model M5 we developed the M6 model with transformation. Through ANOVA table knows that humidity, rainfall and several interaction of continuous covariate does not have significant effect, exclude interaction between humidity and rainfall covariate gives strong significant effect. It shows that treatment by interaction with transformation more effect than only with centering without transformation. Treatment for covariate interaction, transformation of continuous covariate and centering in the M6 model interaction gives effect with respect to R^2 values, multiple-R^2 and adjusted-R^2 respectively on 0.18% and 0.12% (see R^2 values on the M1 and M6).

3.2.6. Rainfall transformation and interaction of time covariates. The construction M7 model can be written, as follows SST = (Temp + Humd + Rain)*Month + (Temp + Humd + Rain)*Year. Furthermore, interaction between continuous covariate and factor covariate is combined in the linear model, where treatments with and without transformation. From this treatment knows that an increase R^2 and adjusted R^2 values, from 56.62% (M4) to 60.26% (M7) and from 56.13% (M4) to 58.85% (M7). However, by this treatment combination effect causes changing in annual affect patterns and monthly effects, so that this interaction combination can be ignored in the model fitting. Although with and without transformation as in M7 and M8 models provide significance between several factor interactions of time covariates and continuous covariates. It can be obtained by ANOVA table. It shows that significant effect, except for humidity and interaction between rainfall and year as time covariates.

3.2.7. Without rainfall transformation and with interaction of time covariates. The construction M8 model can be written as analog with the M7 model without transformation. In this treatment knows also that an increase R^2 and adjusted R^2 values, from 56.37% (M5) to 60.20% (M8) and from 55.88% (M5) to 58.78% (M8). Based on results of interaction combination effect between continuous covariate and factor covariate then further investigation with respect to autocorrelation in the model fitting.

Table 2. Summary of linear model fitting for SST data.

Model	Residual SE	df	Multiple R^2	Adjusted R^2	F-statistic	AIC
M1	0.3786	2045	56.31%	55.88%	131.8 (df=20 & 2045)	1864.176
M2	0.3786	2045	56.31%	55.89%	131.8 (df=20 & 2045)	1872.767
M3	0.3786	2042	56.37%	55.88%	114.7 (df=23 & 2042)	1876.128
M4	0.3776	2042	56.62%	56.13%	115.9 (df=23 & 2042)	1864.176
M5	0.3786	2042	56.37%	55.88%	114.7 (df=23 & 2042)	1876.128
M6	0.3781	2042	56.49%	56.00%	115.3 (df=23 & 2042)	1870.220
M7	0.3657	1994	60.26%	58.85%	42.59 (df=71 & 1994)	1779.054
M8	0.3660	1994	60.20%	58.78%	42.48 (df=71 & 1994)	1782.316

Table 2 shows that the M1-M2 models without interaction and the M3-M6 models with interaction have a similar residual standard error (SE) values, except for the M4 model. The model (M1- M8) has t he same p-value < 2.2e-16. However, all models mentioned has large AIC so that it needs fitting impr ovement.

Residual on within and between groups for time covariate become concerns in linear model fitting by autocorrelation structure. The M9-10 models are constructed through a correlation ARMA(p = 1, q =1) M11 model with ARMA(p = 2, q =2) produces AIC negative values (as seen in table 3).For this reason then we needs upgrading fit performance with respect to model fitting with the autocorrelation construction by using AR(1) model, in the M12 and M13 models without and with transformation respectively.

Table 3. AIC of Model Structure Constructed.

Model	Structure	AIC
M9	SST = Temp + Month + Year + Humd + Rain	-1357.3650
M10	SST= Temp + Month + Year + Humd + Temp*Humd + Temp*Rain + Rain + Humd*Rain	-1313.4470
M11	Analog M10 with transformation	-1318.3260
M12	Analog M10 with transformation	-755.2367
M13	Analog M10 with transformation	-771.0071
M14	Analog M10 with transformation	556.4355
M15	Analog M10 with transformation	403.2987

IOP Conf. Series: Earth and Environmental Science **176** (2018) 012038 doi:10.1088/1755-1315/176/1/012038

The result as in table 3 gives negative point of half AIC values on previous models. Further by using autocorrelation ARMA approach with ~1| Month formulation implemented on the M14 and M15 models gives AIC positive value. In this study, correlation matrix of the model is not shown. Through generalized least squares (GLS) fit by using Restricted Maximum Likelihood to construct model is as follows:

Table 4. GLS fit by REML for Constructed Model.

Model	AIC	BIC	Loglik	Parameter Est.	df (total)	Residual
M12	-755.2367	-609.0729	403.6184	0.88001	2066	2042
M13	-771.0071	-624.8433	411.5035	0.87993	2066	2042
M14	556.4355	702.5993	-252.2178	0.71406	2066	2042
M15	403.2987	549.4625	-175.6493	0.78685	2066	2042

Table 4 shows similar parameter estimation with different Akaike Information Criteria (AIC), Baye sian Information Criteria (BIC) and Logic values of the M12 to M15 models. All models have the same of total degrees of freedom (df) and residual. Based on ANOVA table, then the M15 model has smalles t value of residual SE compared with other model constructed.

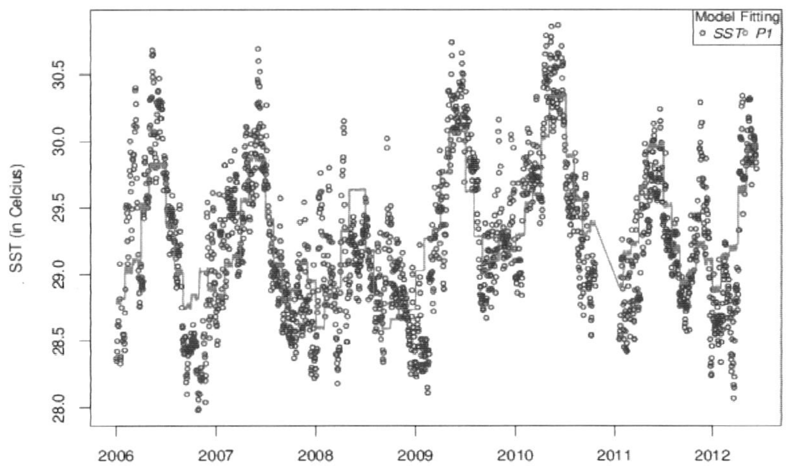

Figure 3. The M15 linear model fitting of the SST data.

Results of ANOVA analysis of M12 and M13 models with coefficients correlation in the Rho = 0.5 – 0.9 are discussed, as follows. The significant effect on all parameters, except for humidity, rainfall and interaction between continuous covariate terms in both models. Based on parameter estimation shows that rainfall transformation given effect pretty significant with respect to changing magnitude of parameter values, such as intercept value changes from 26.89 become 28.181. Although several magnitude of parameters shows similar value or category. If we observed the SST data fitting between the M12 and M13 models indicates that the M13 model is smoother than the M12 model.

Using the M14 and M15 models with the correlation coefficient in the Rho= 0.5-0.9 obtained ANO VA table shows that the significance effect for all parameters, except for humidity, rainfall and interact ion between continuous covariates in the M14 and M15 models. Further based on parameter estimation is known that the transformation gives a significant effect on the change of parameter values such as intercept value from 30.134 to 31.501. Although some parameters show quantities that it can be categ orized as similar or not statistically significant change as in the M14 and M15 models. figure 3 shows t hat the M15 model fitting smoother than M14 model. It has AIC value smaller than M14 model. The M 15 model fitting more appropriate compared in the M4 model due to it have autocorrelation effect.

3.3. Generalized additive models fitting of the SST dataset

By using generalized additive model (GAM) model, we have extended to use 16 climate features, where 14 continuous covariates and 2 times covariates. We obtained several significanse parameters as seen table 5 with AIC value is 634.99, GCV is 0.10, and the adjusted R-squared is 64,4%. In this model, we used wind speed and conductivity with large missing values, as seen in table 5.

In model fitting, we separates between continous covariates and time covariates to find time covariates contribution in the model. We obtain several parametric coeficients and smoothing parameter approximation for AIC and GCV is 665.82 and 0.10 respectively with adjusted R-squared value is 63,1%. Thus it shows that this approach (separate and compound) has the smallest AIC value and the largest R^2 adj. Both models show that time effects have a strong significance and the largest contribution in applied GAM models to the SST data.

Table 5. Analysis of variance for GAM model.

Predictor	edf	Red.df	F-value	P-value
Wind speed	4.951	6.094	4.371	0.000
Subsurface temperature	2.979	2.999	61.932	<2e-16
Shortwave radiation	1.896	2.381	0.639	0.555
Salinity	1.800	2.221	0.688	0.518
Humidity	1.000	1.000	12.112	0.000
Prec. rain	4.370	5.318	2.508	0.026
Dynamic height	2.850	2.978	21.432	2.21e-13
Density	2.845	2.976	6.774	0.000
Air temperature	1.956	2.215	3.882	0.017
Isoterm	1.000	1.000	0.035	0.852
Conductivity (SSS)	2.757	2.938	6.459	0.000
Current velocity	2.293	2.563	2.435	0.073
Zonal	1.000	1.000	30.385	4.36e-08
Merid	3.734	4.765	0.722	0.601
Year	8.000	8.000	97.955	<2e-16
Month	6.908	7.000	40.567	<2e-16

4. Conclusion

Linear model fitting of the SST data with interaction covariates in a combination and transformation gives the highest R^2 value among treatments given to the model i.e., with and without interaction, with and without transformation, with and without covariate centering, and a combination among given treatments. Although the effect value of interaction covariates in combination and transformation on R^2 value small classified (under 0.5%) gives magnitudes changing in multiple R^2 and adjusted R^2 on 0.31% and 0.43% respectively. This proof shows that time covariate does have strong significant effect in the linear model fitting for SST dataset.

In addition, for one buoy investigation shows that year effect does have significance highest peak happen on 2010 and lowest on 2008. Whereas for month effect shows increase trend occurs from January to June and decrease trend from June to September. Furthermore, trend continuous increase again till November and decrease on December. Thus, the highest peak of month effect occurred in June and the lowest in September. To obtain the optimal value of multiple R^2 and adjusted R^2 and the smallest AIC value, then the autocorrelation is a major concern in the fitting treatment to overcome the complexity of the SST dataset. Time effects have a strong significance and the largest contribution in the linear and GAM models in modeled the SST dataset, for with and without the number of climate features extended and different periods.

References

[1] Schott F A, Xie S P and McCreary J P 2009 Indian ocean circulation and climate variability *Reviews of Geophysics* **47** 1–46
[2] North G R and Stevens M J 1998 Detecting climate signals in the surface temperature record *Journal of climate* **11**(4) 563–577
[3] Deser C, Alexander M A, Xie S P and Phillips A S 2010 Sea surface temperature variability: Patterns and mechanisms *The Annual Review of Marine Science* **2** 115–143
[4] Magnus J R, Melenberg B and Muris C 2011 Global warming and local dimming: the statistical evidence *Journal of the American Statistical Association* **106**(494) 452–464
[5] McPhaden M J, Ando K, Bourlès B, Freitag H P, Lumpkin R, Masumoto Y, Murty V S N, Nobre P, Ravichandran M, Vialard J, Vousden D and Yu W 2010 The global tropical moored buoy array *Proceedings of Ocean Obs.* **9** 668-682
[6] Miftahuddin M 2016 Fundamental fitting of the SST data using linear regression models *IEEE* 128-133
[7] Hastie T, Tibshirani R and Friedman J 2008 *The Elements of Statistical Learning* (Stanford, California: Springer)
[8] James G, Witten D, Hastie T and Tibshirani R 2013 *An Introduction to Statistical Learning* (Stanford, California: Springer)
[9] Wood SN 2009 mgcv. R Package Version 1.6-0. (Available from http://CRAN.R-project.org/ package=mgcv)
[10] Wood S N, Goude Y and Shaw S 2015 Generalized additive models for large data sets *Applied Statistics* **64** 139-15

Study of fresh Frigate Tuna (*Auxis thazard*) quality sold by mobile fish retailer in Makassar

A A Rivai*, Metusalach, Kasmiati

Faculty of Marine Science and Fisheries, Department of Fishery Resources Utilization, Hasanuddin University
*e-mail: andi.alamsyah.03@gmail.com

Abstract. Frigate tuna (*Auxis thazard*) is widely marketed and consumed in Makassar. Most of the fresh fish for household needs are purchased from traditional mobile fish retailer (Pa'gandeng) in Makassar. However, data on the quality of fresh frigate tuna reaching the consumer of household level is still lacking. This research was conducted in three areas, namely east, south, and north of Makassar City, South Sulawesi. This study was aimed to describe the quality characteristics of fresh frigate tuna retailed by Pa'gandeng. The study was conducted by randomly taking four samples of fish from Pa'gandeng at three points of sale (sale time) in each part of the city. Physical and organoleptic parameters were measured directly at the sampling locations, while the chemical and microbiological parameters were analyzed in the laboratory. Effect of time on the fish freshness parameters were tested using ANOVA with a confidential level of 95% ($\alpha = 0.05$). When ANOVA indicated the presence of a significant effect, LSD (Least Significant Difference) test was applied to determine which points of observation that the corresponding parameter exhibited a significant difference. The relationship between the observed parameters was analysed using regression analysis. Results of organoleptic, microbiological, and chemical analysis showed that the fresh frigate tuna retailed by Pa'gandeng in Makassar was in good quality and, therefore, suitable for consumption. The ANOVA test indicated that the parameters that significantly different between the sale points were the temperature and organoleptic values.

1. Introduction

Makassar is one of the big cities in Indonesia with high level of fish consumption. Makassar stretches along a coastal line and has good infrastructures to support the distribution of fish products, such as fishing ports and fish markets. In addition, people in Makassar have high preference to consume fish. Frigate Tuna (*Auxis thazard*) is very popular because it is widely available in the market and its fishing ground is close to Makassar. Frigate tuna has a relatively cheap price, and is known to have a high nutritional value.

Being popular and nutritious fish, frigate tuna has some flaws. Frigate tuna, can be toxic for some reason. Poor handling of frigate tuna often results in high level of histamine (a biogenic amine compound) in the flesh and may cause poisoning when consumed. The red meat of frigate tuna contains high concentration of histidine which is converted to histamine during the deterioration of fish. These histamine compound can cause itching, poisoning, and even death.

The household need of fish in Makassar is mostly supplied by the traditional mobile fish retailers Pa'gandeng). The varieties of fish and the prices offered by Pa'gandeng are competitive with those of in traditional wet markets, so they can meet the consumer need. On the other hand, the fish sold by Pa'gandeng are generally lower in quality. This is due to the poor handling of the fish during the sales and distribution.

[1] reported that the pH and organoleptic parameters of fish caught with trawl and purse seine after being auctioned on the fishing port were categorized as good and marketable. However, data on the quality of fresh fish delivered to the household consumer are still lacking. This study aimed to describe the status of the quality of fresh frigate tuna sold by Pa'gandeng in Makassar based on organoleptic tests, physical (temperature), microbiological (total plate count and coliform), and chemical (total volatile bases and value of peroxide) parameters. The result could be beneficial for the government to take some policies to improve fish handling at Pa'gandeng and to provide safe fish for the community.

2. Materials and methods

This research was conducted from October 2013 until November 2013 in the northern, southern, and eastern part of Makassar City (figure 1). The materials used were fresh frigate tuna, distilled water, ice, tissue paper, styrofoam box, zipped-plastic bags, cutting board, digital balance, stomacher, pipettes, petri dishes, incubator, portable pH meter, portable thermometer, stopwatch, score sheet of organoleptic test, and laboratory equipment for analyzing of total plate count (TPC), coliform, total volatile bases (TVB), and peroxide value. The fish were collected from Pa'gandeng at fishing port in Paotere and in two selling location in 3 parts of Makassar.

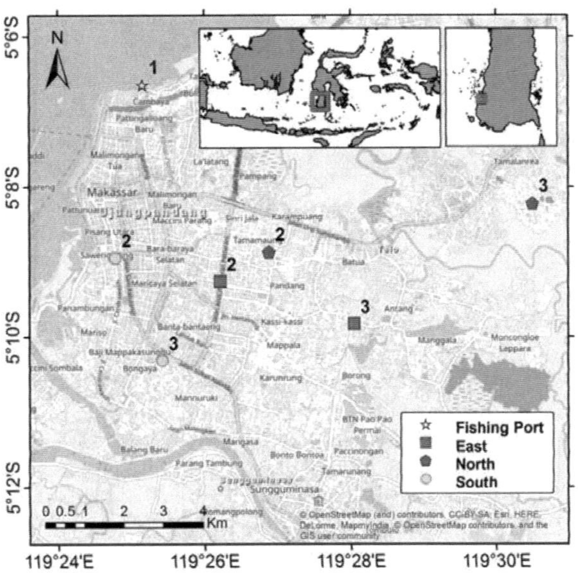

Figure 1. Research location.

This research was conducted using survey method. The selling activity of Pa'gandeng was directly observed. Temperature and pH of fish, as well as the organoleptic parameters of fish were evaluated and measured on site. The temperature and pH of fish were measured at the center of the fish body. The TPC, *coliform*, TVB, and peroxide value of the fish were determined in the laboratory. Sampling was done between 07:00 and 12:00 AM. At each sampling (sale) point, 4 fish samples were taken randomly for the analysis. Each part of the city was represented by one Pa'gandeng for three consecutive days of sampling. For the microbiological and chemical analysis, the frigate tuna samples were taken at each point of sales and then wrapped in zipped plastic bag. Samples were then stored in a Styrofoam box along with ice and taken to the laboratory for analysis.

The collected data for each parameter studied were subjected to ANOVA at a 95% of confidential level ($\alpha = 0.05$). When ANOVA indicated the presence of significant differences, LSD (least significant difference) test was employed to determine which points of sale differed from one another. The relationship between parameters and time (duration of sell) was analyzed using simple regression analysis. Data were processed SPSS 17.0 computer software and the results were presented in the form of graphs.

3. Results and discussion

3.1. Pa'gandeng

Pa'gandeng is a fish seller who buys fish from a fishing port and sells it again directly to the household consumers. Pa'gandeng generally sells the products or fish directly to consumers by traveling around the village or city using a bicycles or motorcycles. The vehicle used was equipped with fish containers mounted on the motorcycles (figure 2). The containers are covered with large plastic bags so that the fish are protected from direct sunlight and air pollution during the distribution to the consumers. Each Pa'gandeng has his own definitive route to sell the fish.

Figure 2. The vehicle (left) and the container (right) used by Pa'gandeng to distribute and sale the fish around Makassar.

3.1.1 Fish handling process. Fish purchased by Pa'gandeng at the fishing port were sorted and packed separately into plastic bags according to their type. According to [2], the advantages of sorting are maintaining the quality of the fish and preventing cross contamination so that the fish can be marketed longer. Further, the fish were arranged in the fish container as shown in figure 3. On top of the fish layer, small amount of crushed block ice was given to lower the temperature of the fish. For 10-15 kg of fish, Pa'gandeng used only 2-3 kg of ice, which was not proportional to the amount of the fish and was too small for cooling fish in the tropical area. The appropriate ratio of ice to fish in the tropics is 1:1 to ensure the freshness of the fish. Ice is the best cooling medium compared to other cooling medium because ice is able to lower the fish temperature quickly without altering its quality and the required cost is also relatively lower than the other cooling medium [3]. The insufficient amount of ice used by the Pa'gandeng in this study caused rapid increase in the fish temperature. Along the way to the consumer, Pa'gandeng never added ice to the fish containers, and therefore the process of fish handling became ineffective to preserve the quality of the fish.

Figure 3. Frigate tuna sold by Pa'gandeng in Makassar.

3.1.2. Sanitation and hygiene. Based on the observations, all the equipment used were non-hygienic although they were always washed every time after being used. The reason was because the container was used repeatedly and the way of cleaning was not good. According to [4], most of the bacteria causing the fish damage are contaminants that come from the surrounding environment. When the equipment used for the distribution of fishery products was clean, the deterioration process of fish became slow. This is due the less number of contaminants in the surrounding environment. [5] reported that fish prepared under hygienic conditions had a lower microbial value.

3.2. Organoleptic value

Organoleptic test is a subjective test that is based on the assessment from direct observation by the panelists. The assessment followed the score sheet value and standard. This test is the most common way to determine the freshness level of the fish because it is easier, faster, and does not require any laboratory test. The higher the organoleptic value, the better the freshness of the fish [4].

The standard of organoleptic value in this study was calculated based on the examination of condition of eye, gills, body surface slime, odor, and texture of fish using score ranging from 1 to 9. Figure 4 showed changes of the organoleptic value from the first point to the end point of frigate tuna sale in Makassar.

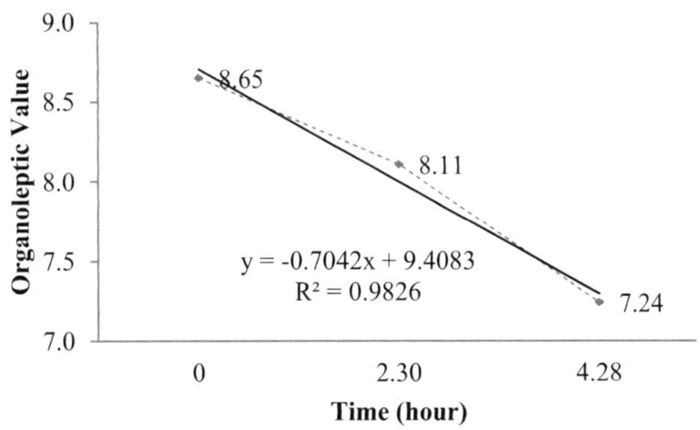

Figure 4. Changes of the organoleptic value of fresh frigate tuna *(Auxis thazard)* during the study.

The organoleptic value of fresh frigate tuna decreased from 8.65 to 7.24. At the first point of sale (0 hour), the fish were recently purchased by the Pa'gandeng so that the condition of eye, gills, body slime, odor, and texture of the fish were good. The organoleptic value at the midpoint (2.30 hours) and the end point of sale (4.28 hours) were 8.11 and 7.24. The greatest decrease of the fish quality occurred when the fish were distributed from midpoint to the end point of sales. This might be due to sharp increase of the fish temperature since the ice used by the Pa'gandeng has completely melted and no new ice was added into the fish container. In addition, the fish was also exposed to direct sunlight. Despite the lack of ice and direct exposure to sunlight, the organoleptic test indicated that the frigate tuna marketed by Pa'gandeng in Makassar was of good quality, although there was a decrease in the organoleptic value from the first point to the end point of sales. According to the standard of fresh fish quality, good quality of fish should have organoleptic value of at least 7 [6].

The organoleptic value at each point of sale had a significant difference based on ANOVA results. Least Significant Difference (LSD) test showed that each observation point had an organoleptic value that was significantly different from each other ($p < 0.05$). Based on the result of linear regression analysis, it was known that the duration of sale (time) and the organoleptic value of frigate tuna had a strong relationship ($R^2 = 0.9826$).

IOP Conf. Series: Earth and Environmental Science **176** (2018) 012039 doi:10.1088/1755-1315/176/1/012039

3.3. Physical parameter

One of the factors affecting the quality of the fish is temperature. High temperature accelerates the rate of fish deterioration. [7] reported that low temperature can extend the period of *rigor mortis* and so extending the period of fish quality in a good condition. The average temperature of fresh frigate tuna marketed by Pa'gandeng in Makassar was between 14.67 – 24.17 °C for 4.28 hours of sale duration. The change in temperature of the fish occurred from the first point to the end point of sale (figure 5). The temperature of fish at first point of sale was higher than at the midpoint of sale. This was because the fish at the first point of sale have been without ice for some times since unloaded from the boat, while the fish at the midpoint of sale were still with ice added prior to transport to the midpoint of sale. The increase of the fish temperature from the midpoint to the end point of sale was due to the ice added before the midpoint of sale has completely melted and also the fish were exposed to heat from direct sunlight during transport to the final point of sale.

Based on the results of ANOVA, it was known that there were differences in temperature values from each point of sale. The LSD test showed that the temperature at the end point of sale differed with temperature at the first point and the midpoint of the sale ($p < 0.05$). The duration of sale (time) and temperature of the frigate tuna had normal relationship ($R^2 = 0.6177$).

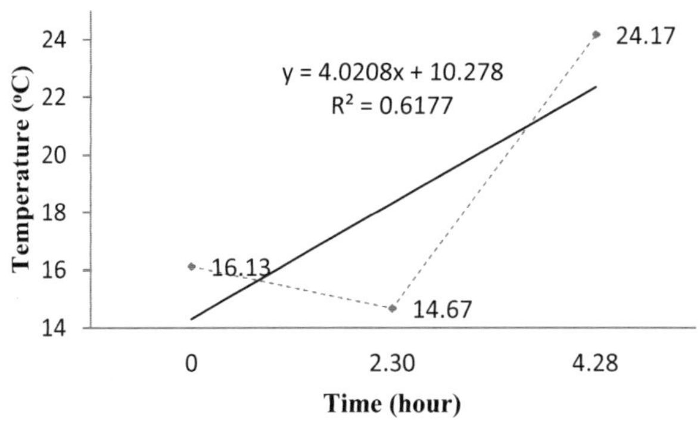

Figure 5. Changes of the temperature of fresh frigate tuna *(Auxis thazard)* retailed by Pa'gandeng in Makassar.

3.4. Microbiological parameters

3.4.1. Total plate count. Total plate count (TPC) is one the fish freshness parameters and is referring to the number of bacteria in one gram of the fish meat. The TPC method is based on a cup count method with assumption that every living cell of bacteria will develop into a colony. This test is used to detect all aerobically grown microorganisms and reflects the general cleanliness of the sample. Figure 6 showed that bacteria in the fish meat grew fast from the first point to the end point of sales. The sharp increase in the TPC was in line with the sharp increase of the fish temperature between the midpoint and the end point of the sale due to the absence of ice in the fish containers. In addition, poor handling applied by the Pa'gandeng may cause contamination, thus contributing to the addition of bacteria to the fish. Nonetheless, the frigate tuna in this study could be categorized as fresh fish. According to SNI 01-2729-1992, the standard of maximum limit of TPC for fresh fish is 5×10^5 cfu/g [4]. Based on the results of ANOVA, TPC value of frigate tuna at each point of sale had no significant difference ($p > 0.05$). The duration of sale (time) and TPC of the frigate tuna had a strong relationship ($R^2 = 0.9682$).

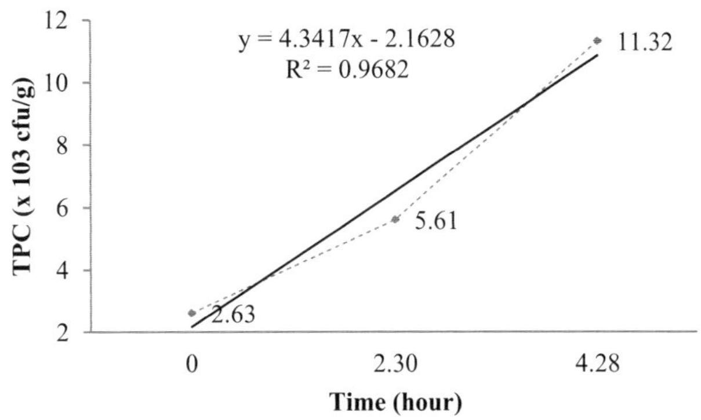

Figure 6. Changes of total plate count (TPC) of fresh frigate tuna *(Auxis thazard)* retailed by Pa'gandeng in Makassar.

3.4.2. Coliform. Coliform bacteria are indicator commonly used to measure the sanitary quality of food and water. When these bacteria are present in foods, they can cause diarrhea, stomach cramps, vomiting, fever, and interfere the kidney function. These bacteria can be found in the intestine of livestock and humans [8]. Figure 7 showed a decrease in the number of coliform from the midpoint to the end point of sales. This might be due to the effect of washing the fish and the replacement of water in the containers. The number of coliform in the frigate tuna samples were 25.83 MPN/g indicating that the fish were of good quality and suitable for consumption. Based on the results of ANOVA, coliform value of frigate tuna at each point of sale had no significant difference ($p > 0.05$). The duration of sale (time) and coliform of the frigate tuna had a normal relationship ($R^2 = 0.6118$). According ICMSF [9], the acceptable limit for total coliform and faecal coliform are 10^2 MPN/g and 10 MPN/g, respectively, and E. coli must not be present in fish. Coliform bacteria contained in the water and the fish's body is not pathogenic to humans, but may indicate the higher presence of pathogenic bacteria [10].

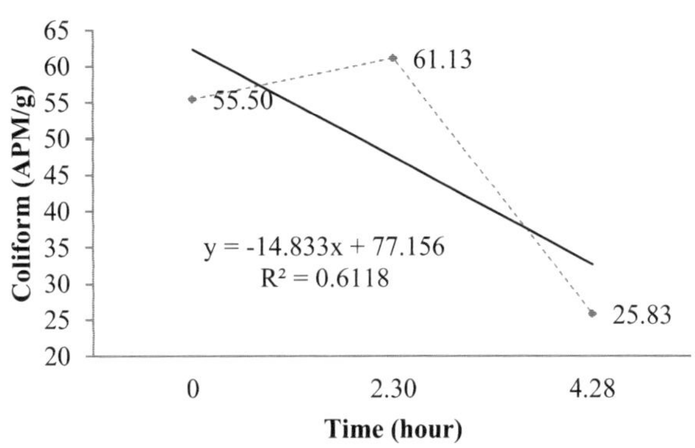

Figure 7. Changes of total coliform in fresh frigate tuna *(Auxis thazard)* retailed by Pa'gandeng in Makassar.

3.5. Chemical parameters

3.5.1. pH. The pH value expresses the level of acidity or alkalinity of a substance and may be used as an indicator of fish freshness. Prime quality or very fresh fish usually has a pH of about neutral (6.8-7.2) while rotten fish has a pH of over 9. Figure 8 showed that the average pH of fresh frigate tuna was between 5.55 and 5.71. The pH values decreased from the first point to the end point of sale. Based on the results of ANOVA, pH value of frigate tuna at each point of sale had no significant difference ($p > 0.05$). The duration of sale (time) and pH of the frigate tuna had a strong relationship ($R^2 = 0.9672$). The decreasing in pH was due to the glycolysis process which produced lactic acid. Lactic acid is a result of the glycogen breakdown by the enzymes in the fish meat. [11] reported that during the production and the auction, the fish pH decreased due to the process of glycogen breakdown into lactic acid.

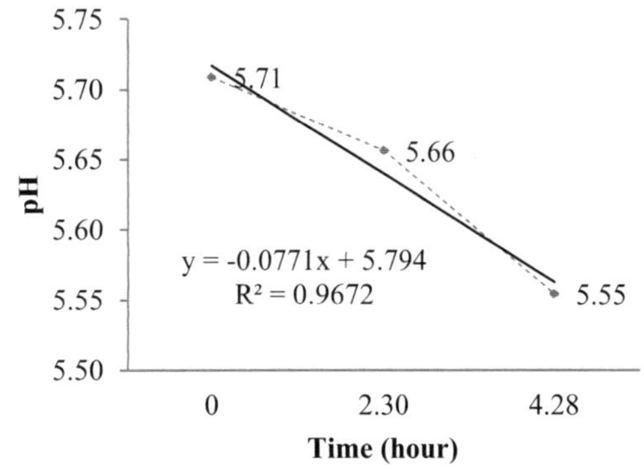

Figure 8. Changes of pH of fresh frigate tuna *(Auxis thazard)* retailed by Pa'gandeng in Makassar.

Bad quality fish will have a high pH. According to [1], the pH of fish will decrease only to a certain extent (about pH 5.5). This is related to the availability of glycogen in fish meat. If the glycogen has totally been exhausted, the formation of lactic acid will not take place any longer and the pH of the fish meat will stop to decrease. Lactic acid formed during the glycolysis process will be broken further to its simpler derivatives resulting in the decrease of the fish meat acidity, and therefore increasing the pH. The meat of the frigate tuna sold by the Pa'gandeng in Makassar was still acidic indicating that the fish was fresh and good for consumption.

3.5.2. Total volatile bases (TVB). TVB measurement is one of the methods to evaluate the freshness of fish chemically. The principle of TVB measurement is to vaporize the volatile compounds which are formed by the decomposition of amino acids in fish meat [4]. Figure 9 showed that the average TVB of fresh frigate tuna increased from the first point to the end point of sale. The average TVB at the final point of sale was 23.18 mg N/100 g.

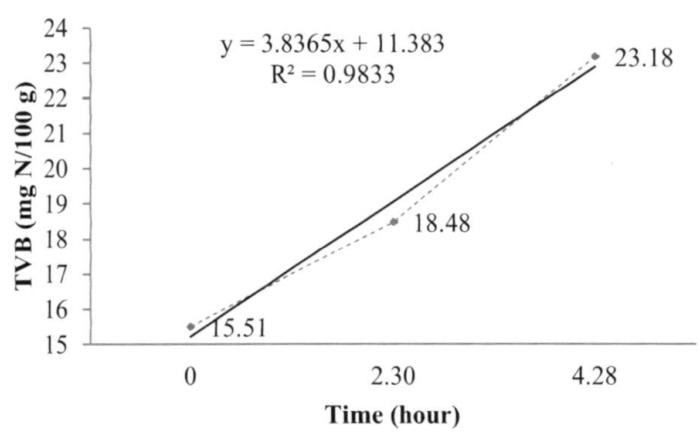

Figure 9. Changes of total volatile bases (TVB) of fresh frigate Tuna *(Auxis thazard)* sold by Pa'gandeng in Makassar.

[12] classified the freshness level of fish based on the TVB values as follow:
1. Very fresh fish (TVB <10 mg N/100 g);
2. Fresh fish (10 ≤ TVB ≤ 20 mg N/100 g);
3. Still feasible for consumption (20 ≤ TVB ≤ 30 mg N/100 g);
4. Not feasible for consumption (> 30 mg N/100 g).

Based on the above classification, the frigate tuna sold by Pa'gandeng in Makassar was still fresh at the first point and midpoint of sale, while at the end point of sales the fish was already deteriorate but was still suitable for consumption. ANOVA results showed that TVB value of frigate tuna at each point of sale had no significant difference ($p > 0.05$). The duration of sale (time) and TVB of the frigate tuna had a strong relationship ($R^2 = 0.9833$). An attention has to be paid to controlling the fish temperature since it is correlated positively with the bacterial growth and TVB level in fish. The higher the temperature of fish in handling process, the bacterial growth and the increasing of TVB will be faster [13]. High population of bacterial load in fish is in line with high formation of TVB. [14] reported a significant increase was occurred in TVB during salting of anchovy and this was due to the increasing of bacterial counts and enzyme activity. [5] reported that moisture content and pH could affect the microbial value in fish.

3.5.3. Peroxide value. The process of rancidity, according to [15], is caused by the oxidation of unsaturated fatty acids in fish. Oxidation is a major cause of quality deterioration in pelagic fish such as mackerel and herring. These fish have high levels of fat in their meat. Figure 10 showed that the peroxide value in fresh frigate tuna increased constantly to the end point of sale, with peroxide value at the end point if sale was 33.24 mEq/kg. This peroxide value indicated that the quality of the frigate tuna sold by Pa'gandeng was good and safe for consumption. Based on the results of ANOVA, peroxide value of frigate tuna at each point of sale had no significant difference ($p > 0,05$). The duration of sale (time) and peroxide value of the frigate tuna had a strong relationship ($R^2 = 0.9833$). [16] reported that fresh fish that started to decay typically has a peroxide value between 20-40 mEq/kg.

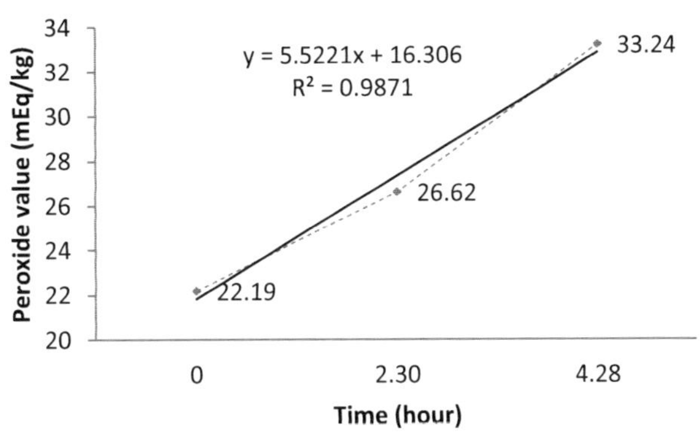

Figure 10. Changes of peroxide value of fresh frigate Tuna *(Auxis thazard)* retailed by Pa'gandeng in Makassar.

4. Conclusion

The frigate tuna *(Auxis thazard)* sold by Pa'gandeng in Makassar was not of prime quality, but was still fresh (good quality) and therefore suitable for consumption. All fish quality indicators (pH, TPC, coliform, TVB, peroxide value, and organoleptic values) were within the pre-determined standard of quality. However, the high fish temperature during selling and the presence of the coliform bacteria at a relatively high number necessitate more efforts to ensure a better quality of fish to be delivered to the consumer. Application of good handling process of the fish, application of good cold chain system, implementation of good sanitation and hygiene, prevention of contamination and use of more appropriate containers to store and transport the fish are among the key factors which require further improvements.

References

[1] Metusalach, Kasmiati, Fahrul and Jaya I 2012 *Analisis hubungan antara cara penangkapan dan cara penanganan dengan kualitas ikan yang dihasilkan* (Makassar: LP2M Hasanuddin University)
[2] Wardani D I 2002 *Kajian Sistem Penanganan Pasca Panen Ikan Laut Segar (Studi Kasus Tempat Pelelangan Ikan Muara Angke Jakarta Utara* (Bogor: Institut Pertanian Bogor)
[3] Afrianto E dan Liviawati E 1989 *Pengawetan dan Pengolahan Ikan* (Jakarta: Kanisius)
[4] Hadiwiyoto S 1993 *Teknologi Pengolahan Hasil Perikanan* (Jogjakarta: Penerbit Liberty)
[5] Udochukwu U, Inetianbor J, Akaba S O and Omorotionmwan F O 2016 *American Journal of Microbiological Research* **4** 37–40
[6] [BSN] Badan Standar Nasional 2013a *SNI 01-2728.1-2006: Udang Segar* (Jakarta: Badan Standardisasi Nasional)
[7] Murniyati AS dan Sunarman 2000 *Pendinginan, Pembekuan dan Pengawetan Ikan* (Yogyakarta: Penerbit Kanisius)
[8] [BSN] Badan Standar Nasional 2013b *SNI 01-2332-1-2006*: *Penentuan Coliform dan Escherichia coli pada Produk Perikanan* (Jakarta: Badan Standardisasi Nasional)
[9] ICMSF 1982 *Microorganisms in Food* (Canada: Univ. Toronto Press)
[10] Doyle M P and Ericson M C 2006 *Microbe* **1** 162-163
[11] Wangsadinata V 2008 *Sistem pengendalian mutu ikan swanggi (Priacantus macracanthus) (studi kasus di CV Bahari Express, Pelabuhan Ratu, Sukabumi)* (Bogor: IPB)
[12] Farber L 1965 *Fish as Food Vol IV (*New York: Academic Press)
[13] Denny H 2004 *Evaluasi dan identifikasi tingkat kemunduran mutu hasil perikanan tangkap ikan belanak (Mugil thazard) (Studi kasus Muara Angke, Kecamatan Penjaringan, Jakarta Utara)* (Bogor: Institut Pertanian Bogor)
[14] Hernández-Herrero M M, Roig-Sagués A X, López-Sabater E I, Rodrı´guez-Jerez J J and Mora-Ventura M T 1999 *J. Food Sci.* **64** 344-347
[15] Winarno F G 1997 *Kimia Pangan dan Gizi* (Jakrta: Gramedia)
[16] Eyo A A 2001 *Fish processing technology in the tropics* (New Bussa Nigeria: National Institute for Fresh Water Fisheries Research)

IOP Conf. Series: Earth and Environmental Science **176** (2018) 012040 doi:10.1088/1755-1315/176/1/012040

Volatile components of raw Patin Catfish (*Pangasius hypophthalmus*) and Nile Tilapia (*Oreochromis niloticus*)

Rusky I Pratama*, **I Rostini and E Lviawaty**

Fisheries and Marine Science Faculty, Padjadjaran University, West Java, Indonesia
*e-mail: rusky@unpad.ac.id

Abstract. Volatile components are groups of compounds which could affect the aroma and acceptance of commodities. The objective of this research was to identify volatile components of two most common freshwater commodities in Indonesia which are patin catfish (*Pangasius hypophthalmus*) and nile tilapia (*Oreochromis niloticus*). The method used in this study were to detect volatile compounds using Gas Chromatography/Mass Spectrometry (GC/MS) on both fresh fish samples with Solid Phase Micro Extraction (SPME) as an initial extraction method. The volatile components analysis successfully detects and identify as many as 46 volatiles compounds in raw nile tilapia and 59 compounds were detected in raw patin catfish. Most of the detected components came from hydrocarbons, aldehydes, alcohols and ketone groups. Most of those volatile compounds detected were derived from the results of enzymatic reactions, lipid auto oxidation and various environmental influences.

1. Introduction

West Java Province, Indonesia has great potential in freshwater fisheries resources and its production has steadily grown over the year. Aquaculture plays an important role in providing fish supply nationally. Among West Java main production in aquaculture freshwater fish commodities are patin catfish (*Pangasius hypophthalmus*) and nile tilapia (*Oreochromis niloticus*). Most of nile tilapia production in West Java came from Purwakarta, Cianjur and West Bandung Districts and mainly cultured in ponds and floating net cage. In 2010, national nile tilapia production had reached 912,613 tons with 29.97% average increase in 5 years. Patin catfish is considered as one of the important freshwater fish export commodity in Indonesia among others. Patin catfish production has also shown a significant growth in number over the year. Its national production was 403,133 tons in 2014 with average increase as much as 39.90% [1] and in West Java, patin catfish are mainly coming from the floating net cage in Cirata and Jatiluhur Reservoir.

Fresh seafood is more susceptible to deterioration than other foods. Fresh fisheries commodities are a valuable nutritional source for protein and lipids, especially essential amino acids and polyunsaturated fatty acids. Each fishery commodity would have differences in their chemical composition and flavour compounds and that included their volatile components. Flavor characteristics and thus its components are an important determinants of food quality. Identifying flavor components in various foodstuff including fish and seafood that characterize them is important to carry out because flavor in general could influence sensory characteristics, consumption level, overall acceptance and choices of the end product by its consumers. The flavor is formed as a result from a combination of experiences and sensation which we perceived on product characteristics [2]. The flavour is generally divided into two categories, one is volatile flavour which contributes to aromas and other is non-volatiles flavour which contributes to taste characteristics. Volatile components of the flavor normally originate from within the raw material composition and its compounds are significantly affecting product aroma. Volatile compounds in fishery commodities generally derive from various chemical compound groups such as hydrocarbon, aldehydes, alcohol, ketones, sulfurous and nitrogenous compounds, heterocyclic and esters [3,4].

Volatile compounds can be absorbed from the water, during fish respiration and then stored under the fish skin in the lipid layer. Lipid derived volatile compounds such as aldehydes and ketones are frequently generated by oxidative enzymatic reactions and autooxidation of lipids. Formation of volatile compounds derives from fatty acid oxidation leading to specific aromas could also be seen as an opportunity to develop and enhance the sensory quality of food product containing fish or seafood [5]. Volatiles flavors are easily evaporating therefore the precaution of handling and transporting the raw material or product are becoming utmost important. The production of volatiles in seafood is influenced by several factors such as species, food parts, environment, ripeness, biological factors, processing and other conditions. Some distinctive volatiles have been used to identify the freshness, deterioration and transformed seafood, and to evaluate the rearing water quality and the geographical origin [6].

In several countries such as China, Japan and Scandinavian countries, the study of fishery commodities volatile components were already carried out for more than a decade ago with various fish species such as carp, pond smelt, loach and silver carp [4,7], sea bream, black bream [8,9], rainbow trout [9], scallop [10], and salmon [11]. To our best knowledge, there was still no comprehensive study found in Indonesia that has been conducted in investigating the volatile components of raw patin catfish and nile tilapia.

The results of this research will be useful in the fisheries flavor research field for example, in conducting volatile components mapping to identify the key component which could influence aroma characteristic of certain fish species and also beneficial for various flavor studies application such as storage quality parameters and in producing flavor extract derived from fishery commodity. Therefore an attempt on volatile flavor analysis of two common fishery commodities is important to carry out. The objective of this research is to identify the volatile compounds of two main aquaculture in Indonesia which are raw patin catfish and nile tilapia.

2. Research method

Patin catfish and nile tilapia samples were taken from floating net cage complex in Cirata Reservoir, Purwakarta District, West Java. Sample preparation was carried out in the Fisheries Product Processing Laboratory, Fisheries and Marine Sciences Faculty, Padjadjaran University and volatile compounds analysis were carried out in Flavor Laboratory, Indonesian Centre for Rice Research, Sukamandi, Subang.

2.1. Samples preparation

After sampling, all samples were then promptly transported in a cool box which contained layers of bulk ice and transported to the laboratory. Samples were then washed, descaled, beheaded and weighed. Each fish samples were then filleted. A portion of fish meat (white meat) samples from the fillet were selected and weighed sufficiently for volatile analysis. All fish samples were then labelled and packaged in three different packaging layers. The first primary packaging was aluminium foil, the secondary packaging was cling wrap plastics and the tertiary packaging is a zip-lock plastic bags. The purpose of this layered packaging is to minimize the changes and degradation that could happen to the samples during transportation to the analysis laboratory which could be caused by various environmental factors such as air, light and temperatures [12]. Samples transported were placed inside a cool box which contain sufficient amount of packed ice to keep the samples dry and maintaining cool temperatures.

2.2. Volatiles compound analysis

Volatiles component from both fish species were analyzed according modification of [9] procedure. The analysis was carried out using waterbath for samples extraction and Gas Chromatography (GC) (Agilent Technologies 7890A GC System) and Mass Spectrometry (MS) apparatus (Agilent Technologies 5975C Inert XL EI CI/MSD) for detecting and identifying the volatile components.

Samples extraction method was done by Headspace Solid Phase Micro Extraction (HS/SPME) using DVB/Carboxen/Poly Dimethyl Siloxane fiber. Sample extraction time used on waterbath was 40°C for 45 minutes. GC column used was HP-5MS (30 m x 250 µm x 0.25 µm), helium carrier gas, initial temperature was 45°C (hold 2 minutes), temperature escalation as much as 6°C/minutes, final device temperatures 250°C (hold 5 minutes) with an overall running time 41.17 minutes.

2.3. Data analysis
Samples volatile components mass spectrums detected from GC/MS were then compared with the mass spectrum pattern which were available in computer database or NIST (National Institute of Standard and Technology) library 0.5a version. The data then were further analyzed with Automatic Mass Spectral Deconvolution and Identification System (AMDIS) software [13]. The resulting data from volatile compound analysis were discussed descriptively based on identification and the semi quantification intensity of the compounds detected from the analyzed samples.

3. Results and discussion
Volatile compounds analysis results showed that raw patin catfish has higher quantities of volatile compounds compared to the compounds that were identified from raw nile tilapia samples. Wide variation of compounds was also observed from both species. Volatile compounds analysis results from raw patin catfish samples successfully detected and identified as much as 59 volatile compounds and from raw nile tilapia as many as 46 compounds. The compounds detected were then categorized into several major groups such as hydrocarbon, aldehydes, alcohols, ketones and others which categorized as nitrogenous and furan compound groups.

Among 59 identified volatile components in raw patin catfish, the dominant five volatiles were hexadecane, hexanal, 1-octen-3-ol, 2,3-octadione and nonanal, whereas in nile tilapia were hexanal, 6-methyl-3-heptanone, nonanal,heptanal and 2-decanone. The compound with the highest proportion (%) in raw patin catfish originated from hydrocarbon groups (aliphatic, cyclic and aromatic) which is hexadecane (23.904%), however aldehyde group has the highest compounds variety (20 compounds) with hexanal as the most abundant compounds in this group (16.978%) followed by hydrocarbons (17 compounds). In addition to hydrocarbons and aldehydes, GC/MS have also detected alcohol group (14 compounds) with 1-octen-3-ol (14.317%) that has the highest proportions and ketones (6 compounds) with 2,3-octanedione (10.333%) that has the highest proportions. Volatile component analysis had also detected two furan compounds, 2-pentylfuran previously detected in steamed silver carp and smoked black bream [4,9] and 2-ethylfuran. Furan compounds are heterocyclic and usually derive from glucose dehydration (cellulose thermal degradation), but several of them could also derive from Maillard reaction [14,15].

Volatile compounds analysis result from raw nile tilapia samples successfully detected as many as 46 compounds which then were categorized into several major groups such as hydrocarbon, aldehydes, alcohols, ketones and others. Aldehydes in raw nile tilapia samples was found to be the highest in quantity (16 compounds) with hexanal as the most abundant compounds (43.066%). Similar results were found in [16] regarding study on raw tench fish which showed two major categories of volatile compounds identified and they were aldehydes (21 compounds) and alcohols (6 compounds). In addition to aldehydes, GC/MS have also detected hydrocarbon group (10 compounds) with pentadecane (2.395%) has the highest proportions, alcohols (10 compounds) with 1-octen-3-ol (2.684%) has the highest proportion, ketones (8 compounds) with 6-methyl-3-heptanone (23.527%) has the highest proportion. The volatile component analysis was also detected one nitrogenous compound (N,N-dimethyl-methylamine) and one furan group compound. Volatile compounds analysis results for raw patin catfish and for raw nile tilapia are shown in table 1 with their proportions (based on area percentage).

Table 1. Volatile compounds in raw patin catfish and nile tilapia samples.

Groups	Raw Patin catfish		Raw Nile tilapia	
	Area	Proportion (%)	Area	Proportion (%)
Hydrocarbons (aliphatic, cyclic, aromatic)				
Toluene	39992	0.016	-	-
1-Hexene, 3,5,5-trimethyl-	6765244	2.631	-	-
Benzene, 2-ethyl-1,4-dimethyl-	797887	0.310	-	-
Cyclohexene, 1-methyl-4-(1-methylethenyl)-, (S)-	348133	0.135	59013	0.070
1,3,6-Heptatriene, 5-methyl-	303507	0.118	33927	0.040
1,4-Cyclooctadiene, (Z,Z)-	97076	0.038	-	-
Undecane	812821	0.316	531167	0.632
Naphthalene	1076584	0.419	-	-
2-Decanone	1263020	0.491	-	-
Pentadecane	8274270	3.218	2013723	2.395
Hexadecane	61467667	23.904	-	-
Heptadecane	-	-	1209960	1.439
Nonadecane	-	-	106768	0.127
Tridecane	1999636	0.778	313326	0.373
Tetradecane	2060130	0.801	554156	0.659
8-Heptadecene	1339404	0.521	-	-
1-Nonadecene	70830	0.028	-	-
Hexadecane	1806020	0.702	290366	0.345
Dodecane	835765	0.325	340471	0.405
Aldehydes				
Butanal	-	-	625630	0.744
Pentanal	1539386	0.599	1681918	2.000
Hexanal	43657045	16.978	36214661	43.066
2-Hexenal, (E)-	71176	0.028	43815	0.052
4-Heptenal	3172954	1.234	5728	0.007
Heptanal	9021083	3.508	4505298	5.358
2-Heptenal, (E)-	151174	0.059	150091	0.178
Tridecanal	3283716	1.277	-	-
Octanal	45860	0.018	-	-
2-Octenal, (E)-	519663	0.202	278148	0.331
2,4-Heptadienal, (E,E)-	168136	0.065	12256	0.015
Nonanal	13993674	5.442	4746845	5.645
2,6-Nonadienal, (E,Z)-	409543	0.159	177678	0.211
Benzaldehyde, 4-ethyl-	282525	0.110	-	-
Decanal	722177	0.281	571619	0.680
2-Undecenal	412303	0.160	31930	0.038
2-Nonenal, (E)-	183830	0.071	300092	0.357
2,4-Hexadienal, (E,E)-	219346	0.085	-	-
Tetradecanal	630001	0.245	-	-
Hexadecanal	1319190	0.513	106734	0.127
Dodecanal	255578	0.099	60201	0.072
Alcohols				
2-Penten-1-ol, (E)-	950089	0.369	-	-
2-Penten-1-ol, (Z)-	351592	0.137	-	-
1-Hexanol	2193194	0.853	826277	0.983
1-Octanol	82325	0.032	1078548	1.283
1-Octen-3-ol	36814527	14.317	2257040	2.684
2-Octen-1-ol	3895611	1.515	140399	0.167

Groups	Raw Patin catfish		Raw Nile tilapia	
	Area	Proportion (%)	Area	Proportion (%)
1-Pentanol	9824	0.004	85491	0.102
1-Hexanol, 2-ethyl-	160342	0.062	271530	0.323
2-Octen-1-ol, (E)-	9275557	3.607	-	-
1-Heptanol	29756	0.012	210093	0.250
1-Penten-3-ol	2187	0.001	5768	0.007
9,12-Octadecadien-1-ol, (Z,Z)-	350136	0.136	-	-
2-Hexen-1-ol, (E)-	16338	0.006	5306	0.006
1-Nonanol	82560	0.032	52637	0.063
Ketones				
2,3-Octanedione	26571925	10.333	16396	0.019
2,3-Pentanedione	2889548	1.124	5916	0.007
2-Heptanone	52743	0.021	127909	0.152
5-Hepten-2-one, 6-methyl-	900410	0.350	155822	0.185
3,5-Octadien-2-one	2652161	1.031	11400	0.014
3-Heptanone, 6-methyl-	377390	0.147	19783813	23.527
3-Undecanone	1121205	1.333	-	-
2-Decanone	-	-	2442910	2.905
Others				
Furan, 2-ethyl-	22348	0.009	-	-
Methylamine, N,N-dimethyl-	-	-	460332	0.547
Furan, 2-pentyl-	47900	0.019	65869	0.078

In general, fresh fish are characterized by sweet, mild, green, plant-like, metallic and fishy aromas and volatile compounds contributing to these aromas are generated mainly by oxidative enzymatic reactions and auto oxidation of lipids [7]. According to [16], the flavor associated with freshwater fish is usually mild, delicate and pleasant and most fish have a common sweet and plant-like aroma. This fresh fish flavor is due to volatile aldehydes and alcohols which are mainly derived from the oxidative deterioration of polyunsaturated fatty acids. Based on panelists sensory evaluation, aroma description range of fresh patin catfish are fresh, neutral and has species specific aroma [17] while there are still no information found for raw nile tilapia sensory evaluation.

Wide variation of volatile compounds detected on raw patin catfish and nile tilapia in general were derived mainly from proteins and lipid content, thus the quantities and type of volatile compounds are related to chemical composition contained in both samples. Most of the volatile compounds were derived from the results of enzymatic reactions, microorganism activities, lipid auto oxidation, resulting substance from various thermal involved reactions and environmental impacts [8]. Several studies have been published regarding volatile components in fishery commodities, among them were [4,7,8,9,16,18,19,20,21]. The differences observed on volatile compounds might reflect the different volatile extraction methodologies used, the different species, age, part of the sample, biological factors and/or geographical origin of the samples [6].

In general, it can be assumed that raw patin catfish possessed a slightly higher number of volatile components compared to nile tilapia. One of the reasons is that the further increasing of such components was a result of biochemical pathways of protein and lipid of fish. Other factors that could affect the numbers and types of compounds detected in this volatile compounds analysis were extraction method (including time and temperature of extraction), type of samples and GC/MS column and its running parameters [20].

According to [4,10,14], hydrocarbon compounds groups detected in both samples could derive from decarboxylation reaction and the splitting process of fatty acids carbon chains, a secondary reaction from carotenoid (if present) and unsaturated fatty acids thermal oxidations. Aldehydes group compounds detected could derived from fatty acids carbon double bonds oxidation whether they were saturated or unsaturated [4,9,10,22,23,24]. The odor thresholds of aldehydes are generally lower than those of other volatile compounds, thus they have a great potential effect on total flavor [16].

Almost all reaction that could produce volatile compounds would involve saturated and unsaturated fatty acids, which present in most of fishery commodities. Patin catfish has lipid content range from 0.89-1.23%. These results were were not including the patin catfish belly parts. The belly parts of patin are known to contain high lipid content [17]. Lipid content in fish flesh directly affect odor and flavor intensity [25,26]. According to [27], nile tilapia has a broad fatty acid composition, including saturated, monounsaturated and polyunsaturated fatty acids. Its myristic acid and oleic acid content are relatively high with fatty acid profile consists of 45% saturated fatty acids, 13.9% monounsaturated fatty acids and 16.2% are polyunsaturated fatty acids.

Some of the hydrocarbons, aldehydes, alcohols and ketone groups which detected in samples such as hexadecane, nonadecane, tridecane, dodecane, 1-nonadecene, naphthalene, toluene, nonanal, hexadecanal, pentanal, heptanal, 1-octen-3-ol, 1-nonanol, 1-hexanol, 2-ethyl-1-hexanol, 2-heptanone, 2-decanone were also known being detected in raw, cooked and recooked silver carp [4], wild and cultured sea bream [8] and raw black bream [9]. Heptane, tridecane, pentadecane were previously detected in squid volatile components. Hydrocarbons are generally considered to have less impact on the overall flavor of foods because their high aroma thresholds. The presence of the aromatic compounds such as naphthalene in fresh seafood could be associated with fuel pollution in the ocean [6].

Aldehydes such as hexanal was detected in both fish species and are known to have green-like flavour characteristic, pentanal was detected in both fish and has green flavour, (E,E)-2,4-heptadienal has cinnamon-like/oxidized oil-like and detected in both fish, (E)-2-octenal has cinnamon like flavour and detected in both fish, octanal were detected in raw patin catfish and has citrus like flavour characteristics, (E,Z)-2,6-nonadienal has cucumber like/insect like flavour characteristics and detected in both fish. Short chain volatile compounds such as E-2-nonenal that have been found in both fish were previously detected in grey mullet fish and have been related to rancid and bitter taste in fish. Aldehydes, specifically Strecker degradation derives, might impart nutty/malty nuances to the product while heptanal, octanal or nonanal might impart a more characteristic fishy flavor [5], [28]. Linoleic acid is a fish oil component and a precursor of hexanal. Octanal and nonanal might possibly derive from the high amount of oleic acid. EPA and DHA are the leading precursor for the development of oxidation products such as 4-heptenal and 2,4-heptadienal, that derive mainly from the oxidation of linoleic and linolenic acid, both of the compounds were detected in raw patin catfish and nile tilapia. Aliphatic aldehydes are prone to produce fishy or rancid type odors, such as (Z)-4-heptenal, hexanal and (E,E)-2,4-heptadienal [5].

The odor description of volatile aldehydes formed by autooxidation of *n*-3 polyunsaturated fatty acid (PUFA), such as 2-pentenal, 2-hexenal and 2,6-nonadienal, are generally pleasant and associated with green, cucumber, apple, mushroom and grass and 2,4-heptadienal is associated with green, cucumber but also oily and fatty notes. The odors of volatile aldehydes derived from *n*-6 PUFA, such hexanal, 2-octenal, 2-decenal and 2,4-decadienal are described as tallowy, fatty, herbaceous, nutty, oily and associated with fat-fried, cod oil and oxidized oil. The differences in the fatty acid composition were responsible of different volatile compound profile and different sensory characteristics of the fish fillet [16]. Benzaldehyde which detected in raw patin catfish is known for its almond/fruity/creamy/nutty aroma [7]. Aldehydes could provide significant aromas, either pleasant or rancid, to foodstuffs, and their odor threshold values are usually lower than those of alcohols. Thus, even trace amounts of aldehydes might override the flavor effect of some other substances [6].

Alcohol compound 1-penten-3-ol that was detected in all samples is responsible for the flavor of fresh marine products and generated from polyunsaturated fatty acid [28]. Volatile alcohols that were detected in both fish samples such as 1-nonanol has little impact on food flavors because of their high odor threshold, unless they are unsaturated or present at high concentrations [6]. Branched chain alcohols are formed by the secondary decomposition of hydroperoxides of the *n*-3 and *n*-6 PUFA. Although alcohols have relatively high odor threshold values, unsaturated alcohols such as 1-octen-3-ol, with usually lower threshold values, are expected to have higher impact on the overall aroma [5], [29] and individually 1-octen-3-ol has mushroom odor description [30].

Ketone compound 2,3-pentanedione that was detected in raw patin catfish were previously detected in fresh sardine [18], turbot, mussel, clam are produced by oxidation of *n*-3 PUFA [30] and has a caramel like flavour characteristic [18]. Generally, ketone group that presents in samples are known to contribute to the sweet aroma of many crustaceans [7]. Ketones together with aldehydes are the main products of lipid autooxidation of fatty acids or the auto-oxidation of unsaturated fatty acids via hydroperoxides. In addition, they can also produce as a secondary product from the Strecker reaction in Maillard reaction systems. Similar to aldehydes and ketones, alcohols may be formed by secondary decomposition of hydroperoxides of fatty acids [5,29]. However, they have also been attributed to enzymic peroxidation of *n*-3 and *n*-6 polyunsaturated fatty acids [5]. Ketones are derived from enzymatic degradation of poly-unsaturated fatty acids, amino acids or microbial oxidation. 2,3-pentanedione and 3,5-octadien-2-one (detected in both fish) also detected in squid, are the intense odors in squid and the latter has been reported as a lipid oxidation product in seafood. Several ketones might contribute to the cheesy odor of fish sauce [6]. Furans compound that were detected in raw patin catfish and nile tilapia could contribute to the flavor and aroma of fish products, although they have been found in raw fish, they are mainly derived from Maillard reaction after drying or grilling [5]. Furans can also be produced by oxidation of fatty acids as 2-pentyl furan (detected in both fish samples) that impart reversion, beany, grassy, and licorice-like flavors in soybean oil [29].

4. Conclusion

The detected volatile compounds from samples could be categorized into several groups which are hydrocarbons, aldehydes, alcohols, ketones and others (nitrogenous compounds and furans). As many as 59 volatile compounds were detected and identified from raw patin catfish samples and 46 volatile compounds were from raw nile tilapia. It is shown from these results that raw patin catfish has a slightly higher quantity of volatile compounds compared to the compounds that were identified from raw nile tilapia samples. Hexadecane is being the most abundant proportion compound detected in raw patin catfish while in raw nile tilapia, hexanal has the highest proportion. The volatile component information reported in this research could provide important basic information and be further exploited in more advance flavor research and applications.

References

[1] Statistical Data and Information Centre 2014 *Marine and Fisheries in Figures 2014* (Jakarta: Ministry of Marine and Fisheries) pp 31-44

[2] Burdock A G 2002 *Fenaroli's Handbook of Flavor Ingredients, 4th Edition* (Boca Raton: CRC Press)

[3] Thanchotikul U and Hsieh T C Y 1989 Volatile Flavor Components in Crayfish Waste *Journal of Food Science* **54** 1515-20

[4] Liu J K, Zhao S M and Xiong S B 2009 Influence of recooking on volatile and non-volatile compounds found in silver carp (*Hypophthalmichthys molitrix Fish Sci*) **75** 1067-75

[5] Peinado I, Miles W and Koutsidis G 2016 Odour characteristics of seafood flavour formulations produced with fish by-products incorporating EPA, DHA and fish oil *Food Chemistry* **212** 612-19

[6] Jin Y, Deng Y, Yue J, Zhao Y, Yu W, Liu Z and Huang H 2015 Significant improvements in the characterization of volatile compound profiles in squid using simultaneous distilation-extraction and GCxGC-TOFMS *Journal of Food* **13** 434-44

[7] Morita K, Kubota K and Aishima T 2003 Comparison of aroma characteristics of 16 fish species by sensory evaluation and gas chromatographic analysis. *Journal of the Science of Food and Agriculture* **83** 289-297

[8] Alasalvar C, Taylor K D A and Shahidi F 2005 Comparison of volatiles of cultured and wild sea bream (*Sparus aurata*) during storage in ice by dynamic headspace analysis/gas chromatography-mass spectrometry *J. Agric. Food Chem.* **53** 2616-22

[9] Guillen M and Errecalde M 2002 Volatile components of raw and smoked black bream (*Brama raii*) and rainbow trout (*Onchorhynchus mykiss*) studied by means of solid phase micro extraction and gas chromatography/mass spectrometry *J. Sci. Food Agric.* **82** 945-52

[10] Linder M and Ackman R G 2002 Volatile compounds recovered by solid-phase micro extraction from fresh adductor muscle and total lipids of sea scallop (*Placopecten magellanicus*) from Georges Bank (Nova Scotia) *J. Food Sci.* **67** 2032-37
[11] Wierda L R, Fletcher G, Xu L and Dufour J P 2006 Analysis of volatile compounds as spoilage indicators in Fresh King Salmon (*Oncorhynchus tshawytscha*) during storage using SPME–GC–MS *Journal of Agricultural and Food Chemistry* **54** 8480-90
[12] Pratama R I 2011 *Flavor Characteristics of Several Indonesian Smoked Fish* Thesis (Bogor: School of Postgraduate Bogor Agricultural University)
[13] Mallard G W and Reed J 1997 *Automatic Mass Spectral Deconvolution and Identification System (AMDIS) User Guide* (Gaithersburg: U.S. Department of Commerce)
[14] Chung H Y, Yung I K S, Ma W C J and Kim J 2002. Analysis of volatile components in frozen and dried scallops (*Patinopecten yesseonensis*) by gas chromatography/mass spectrometry. *Food Research International* **35** 43-53
[15] Maga J A 1987 The flavor chemistry of wood smoke *Food Review International* **3** 139-83
[16] Turchini G M, Moretti V M, Mentasti T, Orban E and Valfre F 2007 Effects of dietary lipid source on fillet chemical composition flavour volatile compounds and sensory charactersitics in the freshwater fish tench (*Tinca tinca* L.) *Food Chemistry* **102** 1144-55
[17] Suryaningrum D T, Muljanah I, and Tahapari E 2010 Sensory profile and nutrition value of several types of patin catfish and nasutus hybrid *Jurnal Pascapanen dan Bioteknologi Kelautan dan Perikanan* **5** 153-64
[18] Ganeko N, Shoda M M, Hirohara I, Bhadra A, Ishida T, Matsuda H, Takamura H and Matoba T 2007 Analysis of volatile flavor compounds of sardine (*Sardinops melanostica*) by solid phase micro extraction *Journal of Food Science* **73** 83-8
[19] Miyasaki T, Hamaguchi M and Yokoyama S 2011 Change of volatile compounds in fresh fish meat during ice storage *Journal of Food Science* **76** 1319-25
[20] Pratama R I, Rostini I and Awaluddin M Y 2013 Flavor composition of fresh and steamed common carp (*Cyprinus carpio*) *Jurnal Akuatika* **4** 55-67
[21] Duflos G, Moine F, Coin V M and Malle P 2005 Determination of volatile compounds in whiting (*Merlangius merlagus*) using headspace-solid-phase microextraction-gas chromatography-mass spectrometry *Journal of Chromatographic Science* **43** 304-12
[22] Guillen M D, Errecalde M C, Salmeron J and Casas C 2006 Headspace volatile components of smoked swordfish (*Xiphias gladius*) and cod (*Gadus morhua*) detected by means of solid phase micro extraction and gas chromatography–mass spectrometry *Food Chem.* **94** 151-6
[23] Sakakibara H, Yanai T, Yajima I and Hayashi K 1988 Changes in volatile flavor compounds of powdered dried bonito (katsuobushi) during storage *Agric. Biol. Chem.* **52** 2731-39
[24] Cha Y J, Baek H H, Hsieh C Y 1992 Volatile components in flavour concentrates from crayfish processing waste *J. Sci. Food Agric.* **58** 239-48
[25] Puwastien P, Judprasong K, Kettwan E, Vasanachitt K, Nakngamanong Y and Bhattacharjee L 1999 Proximate composition of raw and cooked Thai freshwater and marine fish *Journal of Food Composition and Analysis* **12** 9-16
[26] Gokoglu N, Yerlikaya P and Cengiz E 2004 Effects of cooking methods on the proximate composition and mineral contents of rainbow trout (*Oncorhynchus mykiss*) *Food Chemistry* **84** 19-22
[27] Abelti L A 2017 Minerals content and fatty acids profile of nile tilapia (*Oreochromis niloticus*) fillet from Lake Zeway: effect of endogenous factors *Journal of Nutrition and Food Sciences* **7** 1-3
[28] Lazo O, Guerrero L, Alexi N, Grigorakis K, Claret A, Perez Z A, and Bou R 2017 Sensory characterization, physico-chemical properties and somatic yields of five emerging fish species http://dx.doi.org/10.1016/j.foodres.2017.07.023
[29] Girard B and Durance T 2000 Headspace volatiles of sockeye and pink salmon as affected by retort process *Journal of Food Science* **65** 34-9
[30] Prost C, Hallier A, Cardinal M, Serot T and Courcoux P 2004 Effect of storage time on raw sardine (*Sardina pilchardus*) flavor and aroma quality *Journal of Food Science* **69** 198-204

IOP Conf. Series: Earth and Environmental Science **176** (2018) 012041 doi:10.1088/1755-1315/176/1/012041

Design, construction, and stability test of aerial wireless coastal buoy

L O Wahidin[1], I Jaya[2*], A S Atmadipoera[2]

[1] Marine Technology Department, Bogor Agricultural University (IPB), Bogor, Indonesia;
[2] Department of Marine Science and Technology, IPB Bogor, Indonesia;
*e-mail : indrajaya.ipb@gmail.com; indrajaya@ipb.ac.id

Abstract. Due to its highly dynamic nature, coastal waters are necessary to be continuously monitor. Aerial wireless coastal buoy (AWCB) is one of the ocean instrument commonly deployed for observing oceanographic phenomena in coastal waters and around small islands. In this paper we describe the design, construction and the stability test result of new AWCB. The new AWCB is using water temperature sensors to collect data and transferring them through specific aerial wavelengths at real time and low cost in coastal areas. The AWCB is completed three steps. First, the buoy was designed using CAD software to determine its size, form and color. Next, the buoy was constructed using material from fiberglass. Last buoy was tested for its buoyancy and stability at the Acoustic and Instrument Laboratory, Faculty of Fisheries and Marine Sciences, Bogor Agricultural University. At sea-trial was also carried out around Pramuka Island, and during the trial it shows that the buoy is stable condition within waves swaying and surface water current. It is floating well on seawater surface and stable for six days installing duration.

1. Introduction

Coastal water only covers 8% of entire sea surface around the world, however it plays an important role dominantly for many process mainly relating with interchanging land and sea areas [1]. Coastal areas (shorelines and waters) function as a bridge system both ecological and social aspects to connecting many components. Its ecosystems are crucial for human being which provide huge complexity relationships especially in biological aspects differently in every area where human activities taking place. Coastal area and sea are facing some environmentally challenging pressures, such as increasing population and coastal developing pressures, followed by sea level raise and 21 century climate challenges [2]. Coastal water monitoring is a potentially effort to answer those challenges above. It provides further opportunities to understand developing geological hazards along the times. The monitoring also potential to seek solutions in increasing models and predicting the hazard's behaviors. However, coastal areas characteristics always exhibit obstacles for achieving those goals [3].

Coastal zones and sea monitoring are being necessary needs in serving data and information which are highly relating with coastal and ocean dynamics. These data availability are able to reveal hidden curtain phenomena. Providing these data and information *in situ* format especially for physical oceanography are prominently to observe quick temporal changes of coastal water. Therefore, it is needed an instrument that can be operated through times in order to watch the coastal phenomena in real time. [4] Coastal monitoring system using buoy can construct innovative solutions that intends to measure, such other activities, coastal water condition. Comparing other monitoring technologies, surface buoy supplies simple installing and autonomous. It is being a proper outlet to measure and to control coastal and sea environments.

Aerial wireless coastal buoy, a verily needed marine instrument for observing oceanographic phenomena in coastal areas which is completed with sensors to collect and transmit data through certain sound wavelengths. Developing this buoy, dimension of the buoy is an important indicator key to be bearing in mind. [5] Conducting research in aerial coastal monitoring expresses that a small instrument, low cost, and wear-resistant are being ideal solutions for a mooring buoy system which has resolution advantage, both spatial and temporal. Thereof, it is required a further study about the buoy design that can be proper deploying for coastal waters. The size of buoy is portable and does not need a heavy vehicle in installation are considered. Filling those objective above makes the research about it notably to be conducted. This paper focusses on exploring mechanical process of a set design and construction of Aerial Coastal Buoy. Explicitly, this paper aimed to develop design and construct aerial coastal buoy and to test its stability in laboratory scale and field testing.

2. Material and method

2.1. Description of the study

This research was conducted on July 2016 to November 2017 in Marine Instrumental Laboratory, Fisheries and Marine Science Faculty, Bogor Agricultural University (IPB) for mechanical process, field testing was conducted on May 2017 in around Pramuka Island, Kepulauan Seribu, DKI Jakarta Province. The research phase was divided into four research steps. First, designing the body parts of buoy using compatible three dimensions (3D) software that aimed to reflect the expected buoy. Second, detailing and transforming the 3D designed buoy into flatted 2D parts that were beneficial to construct the buoy according to the available materials, tools and equipment in the laboratory. The two last steps were to test the constructed buoy in freshwater media (fresh water body) to seeking buoyancy and stability of the buoy, and field testing such as coastal waters to find out its rill condition where it was swayed by sea surface currents and waves.

2.2. Designing buoy

Initializing step to contrive a design was to crystallize an idea about the expected buoy, including form and size (dimension). Designing started from rough sketch in a paper, then followed with sketching in a 3D software. In this research, the Sketchup 16.1.1449 trial version was using and it was downloaded from http://www.sketchup.com/. This software is a powerful in the design process. Form and size were considered in this design step. The buoy was expected to be cycle on the upper surface (discuss) and cambering arrowed down. The upper surface was hoped to give more buoyancy and the lower part to be attached weight. The buoy dimension was determined as no more than one meter in length and width, and no more than 30 kg in weight in order to be portable use. Figure 1 below is a new aerial coastal buoy (AWCB) design with two displays, top and side views.

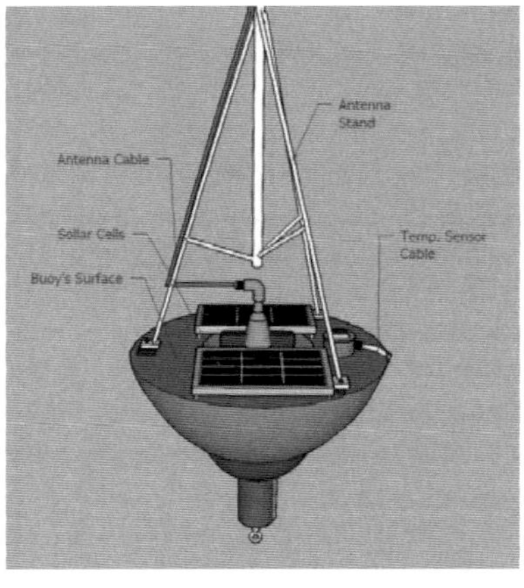

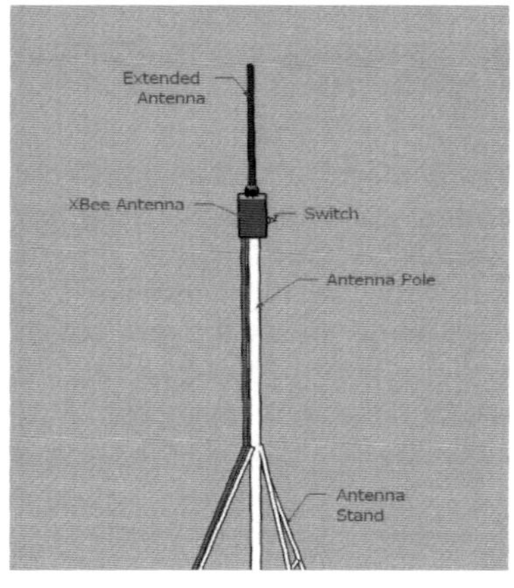

Figure 1. A new Aerial Wareless Coastal Buoy Design, side view (left) and top (right).

IOP Conf. Series: Earth and Environmental Science **176** (2018) 012041 doi:10.1088/1755-1315/176/1/012041

2.3. Construction

The buoy construction was a further step. The resulted 3D design was transformed into 2D forms (flatted forms). Tools and equipment availability was a crucial aspect in this step. Mechanical works conducted using woodworking equipment and materials which were available in the marine instrumental laboratory of IPB and some materials were ordered in close store building and online shopping. Materials and equipment were involved in all the buoy designing, construction and testing process are listed in the table 1 below.

Table 1. Materials and equipment of all Aerial Wireless Coastal Buoy (AWCB) making process in Marine Instrument Workshop of IPB.

Material/Tool	Type	Total/Function
A. Material		
Plywood 3 mm	3 mm thick	2 leaves
Matt		25 meters
Resin	157 BQTN	60 kg
Catalytic	Mepoxe	0.5 kg
Stainless steel 8 mm		3 meters
Nail 3 cm		½ kg
Plastic Pipe 4 inch	PVC	3 meters
Plastic Pipe ¾ inch	PVC	4 meters
Plastic Pipe ½ inch	PVC	4 meters
Set Dop 8 inch	PCV	4 units
B. Tool		
Laptop	ASUS S46C	Designing the buoy and data processing;
Software	Sketchup 16.1.1449 *trial version*	Software 3D;
Hand grinder machine	Krisbow	Cutting and cleaning up some parts of the buoy;
Hand drill	Krisbow KW07-1000	Perforating some parts of the master and mold of buoy;
Meter	Sumoku Eco	Measuring length of working process;
Screwdriver	Plus (+) and Minus (-)	Tightening screws to connect parts of the buoy;
Hummer		Helping to construct the buoy making process;
Paintbrush	Eterna 2'' dan 4 ''	Helping attaches resin and matts;
Stopwatch		Determining times for stability testing;
Camera		Documenting research process.

2.4. The buoy testing

Stability testing was carried out when all construction process over. This step aimed to seek the buoy ability on the water surface and how it could be stabilized its self from outside effects. The test assumed that when an object merged into the water column having some forces such as the object's weight itself (F_G) with it vertically downward direction and upward water pressure vertically. Weight force is centered in axis weight of the object. The upward force of merged object is called buoyancy (F_G). This force works on all surface areas of the object. All forces work on the object is weight force (F_G) and hydrostatic forces that happen to all the object' merged surface areas. Bear in mind that assuming the object is in static position (silent), therefore static force horizontal direction will be equal to be excluded. In this condition, the hydrostatic working force on surfaces of the object is a buoyancy. All forces working of the object is shown in figure 2 as below.

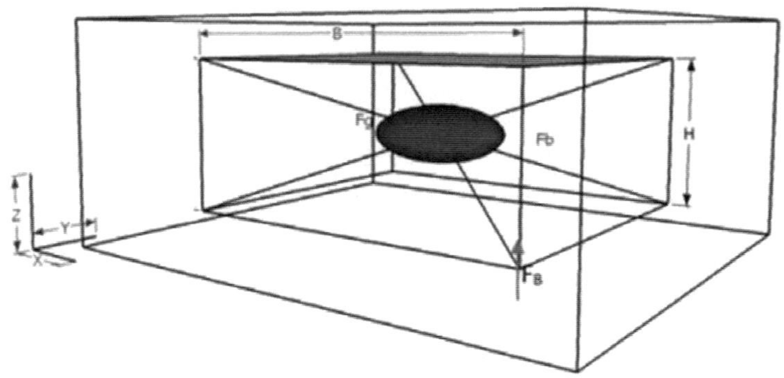

Figure 2. Working forces on a floating object.

In systematically, all those forces above working on a floating object are counted in wide unit measurement according to the formula [7] as follows:

$$F_G = \gamma bBH \quad (1)$$

$$F_b = \rho_o \,.\, B, \text{where } \rho_o = \rho_w \,.\, \mathrm{H} \quad (2)$$

where: F_G = Object weight force; F_b= Buoyancy; ρ_o= Object's density; B = Diameter of floating object; H = Depth; ρ_w = Water density; γb = Object mass.

In quit position, a resultant force either vertical or horizontal are equal to be zero, then by using formula [7]:

$$\sum F_x = 0 \quad (3)$$

$$\sum F_z = 0 \quad (4)$$

$$\rightarrow \quad F_B = F_G$$

$$\rho_o \,.\, B = F_G$$

$$F_G = \rho_w \,.\, h.B$$

$$F_G = \rho_w \,.\, A$$

where: F_{xy} = Weight force of object; F_z = Force on Z axis; F_B = Buoyancy; F_g = Weight force of object; ρ_o= Object's density; B = Diameter of floating object; H = Depth of merged object; ρ_w= Water density; A = Merged surface areas.

Stability of floating object is not affected by small disturbance (force) that tries to be un-stable. A floating object in stable position is happened when it's central weight point (*Bo*) is under the buoyancy point (*Ao*) and reversely, the unstable condition will be happened. When central weight point (*Bo*) is so close to the buoyancy point (*Ao*) causes the object will be in different position as seen in figure 3 below.

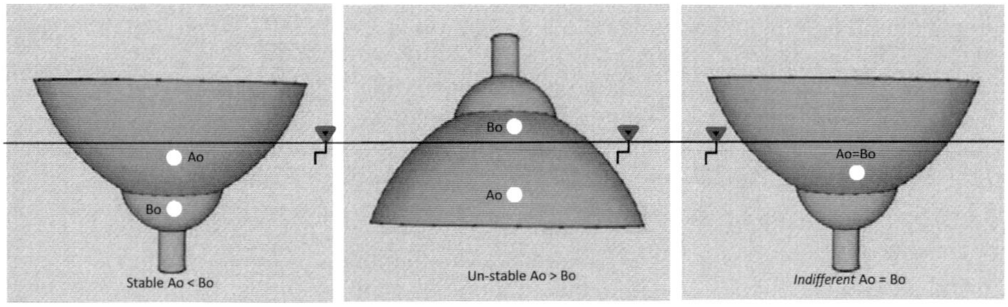

Figure 3. Floating stability of the buoy, stable (left), un-stable (center), and indifferent (right).

Metacentric high of all three body of buoys above is determined by [7] formula as below.

$$m = \frac{I_O}{V} - A_O B_O \tag{5}$$

This formula is consisted of *Io* is a moment inertia of cut object by surface of the liquid, *V* is a volume of transferred liquid of object, and *AoBo* is a distance between buoyancy point and an object point. According to the high value of metacentric (*m*) indicates that if $m>0$ showing the object is table position; $m = 0$ of the object reveals the neutral stability (indifferent), and if $m < 0$ is indicating that the object is not stable position.

3. Results and discussion

A new aerial wireless coastal buoy (AWCB) designing is a long process that has been passed among aspects of idea, mechanical process, data transmitting frameworks and field testing of it. Results of this paper are focusing on some of those process such as design, mechanicals process of construction and stability test of designed buoy and plotting transmitted temperature sensors data.

3.1. Buoy designing process

The three dimension (3D) designing of buoy was carried out using CAD software namely Sketchup 16.1.1449 *trial version*. This software is suitable for designing 3D which has many understandable features. The designed phase consists of outer body, electronic compartment holders and transmitting system (antenna). The body is expected to be a discuss form with 70 cm in diameter and it wanes toward down with an indention in 35 cm diameters that function as weight component from cements and moistened sands. The lower is made to be extending for 15 cm length and 10 cm in diameters, respectively. The last lower part is constructed by PVC pipe containing cement that functions as additional weight and stability of the buoy as seen in the following figure.

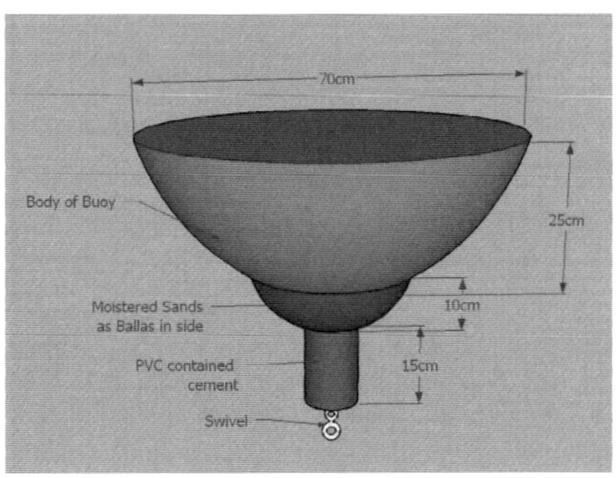

Figure 4. Body of aerial wireless coastal buoy and its size (side view).

Electronical compartment holder is made to be a tube which is a PVC DOP that has been completed with drat lock. The dop is set in center of the buoy with diameter and high are 20 cm and 22 cm, respectively. In addition, it is functioned as the foundation of antenna pole that is installed perpendicularly on body of the buoy. Solar panels were set in opposite on flattened cycle surface of the buoy. The set panel is looked for well operating the electronic components of the buoy that is about 10 watts power in each. Furthermore, they are posted slightly sidewise about 10^0 that is assuming the received sunshine in the solar panels is well distributed and no-obstacles. Parts of the design well shown in the following figure.

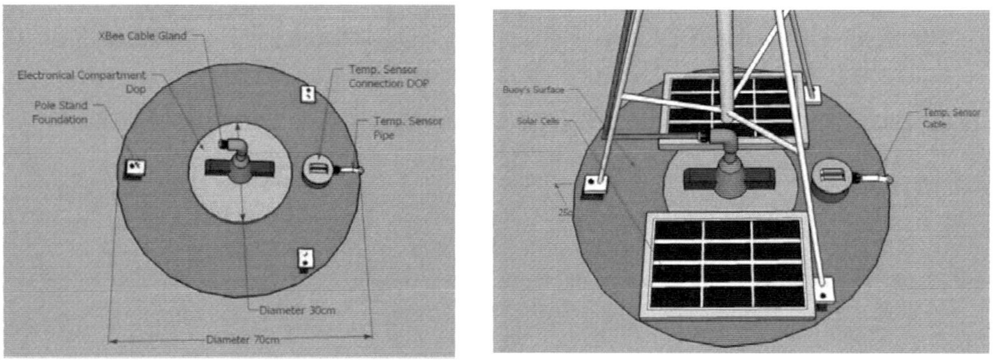

Figure 5. Surface part of the buoy, before setting solar cell panels (left) and after setting solar cell panels (right).

The last part of designing is antenna (transmission). Antenna high is referred on suggesting by [6] aimed that each instrument can transmit data as furthest. Hence, it will be consequence to the high of antenna. The designed antenna of this research is about 180 cm from the surface of buoy and about 200 cm from the seawater surface. This position is affected to data transmitting from nodes buoys in coastal water to the land-receiver in mainland. The antenna is slightly swinged by seawater waves and winds blowing. Therefore, it should be constructed from pipe iron. The top of antenna pole is set a very important transmitting component, an XBee, a transmission electronical shield and its extended antenna. This antenna and XBee are set together in order to be easy in controlling whenever any problems of transmitting. The antenna pole is also constructed to be portable and transportable. The whole body of design buoy is shown in the following figure.

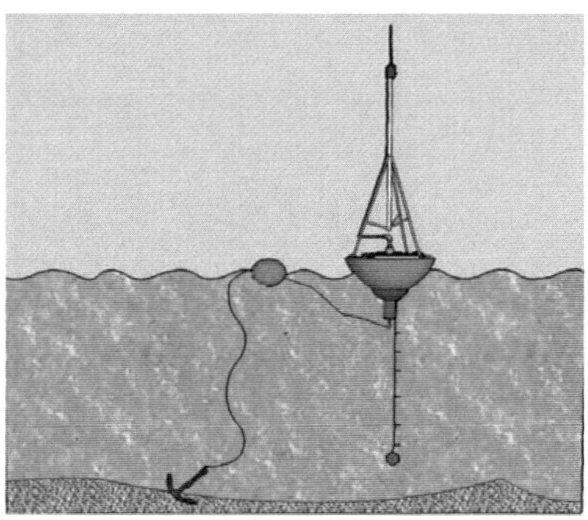

Figure 6. The whole view of designed buoy.

3.2. Construction

Constructing buoy is the second phase after the 3D designing over. The 3D design is transformed into 2D forms (flatted) as model to seek available way in constructing the buoy form at three dimensions in real condition. The construction process consists of three ways such as master making, moild making and constructing the desired buoy. The master is stringed up for being a half frame of buoy that are from plywood sheet which is cut to be pieces according to the printed 2D designed models. The half-framed model is covered with papers. Upper surface is smeared with mixed components which are resin, hardener, matt and talk. Those materials are compilers of fiberglass. The surface master is cleaned up using grinder machine and sandpapers until to be smoothly. Finally, the mirror glaze (wax) is stained on the master surface. Some process of master and mold making are shown below.

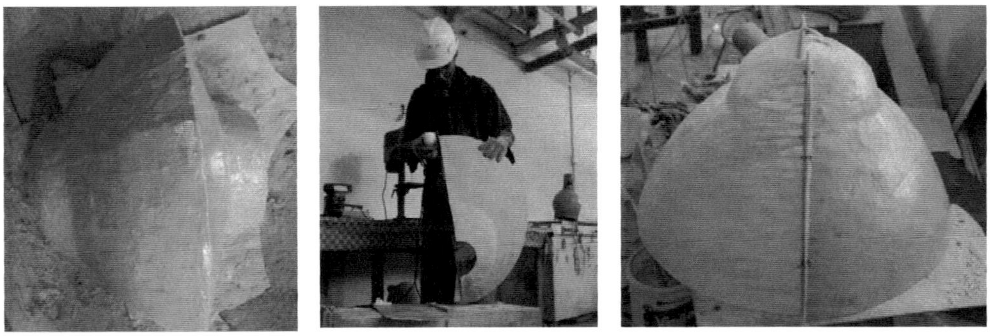

Figure 7. The buoy body's making process of Master (left), cleaning up the master (center) and set mold (right).

Next step is to make the buoy mold. The mold is made into two parts which is framed from fiberglass. Those parts then are gathered using small screws in both sides (Fig. 7 right above). Last step is the molding process. This process almost same with the process of mold making. However, the mixed fiberglass component is made thickly according to the expectation. The buoys are made three units according to the research scenario which consists of one end note units, two routers, and a land-receiver. All down parts of those buoys are combined with other components such as weights, electronic compartment DOPs, and antennas. Construction process of buoy making is displayed in the figure 8 below.

Figure 8. Process of buoy making. Uncovered buoy is floating in the freshwater (left), total molded buoy (center), coloring buoy (right).

3.3. Bouyancy and stability test

Buoyancy and stability tests are the last phase of aerial coastal buoy making process. Stability test is conducted to seek the buoy ability above the water surface in order to be deployed in the real condition, such as coastal waters and small islands. Hence, this test is a must to be conducted to all constructed buoys. Jordan and Beltran-Aguedo (2004) express that an important parameter of mooring buoy is stability, and the stability is expected to be returned on balancing condition. Measurement of some parameters which influences on constructed buoy are listed in the table 2 below.

Table 2. Test results of constructed coastal buoy.

Objects of Test	Results
FB (buoyancy)	9253.39 kg/m^3
FG (weight force)	264.6 kg/m^3
A (area)	0.850 m^2
P (object density)	561 kg/m^3
Psand (sand-weight density)	1400 kg/m^3
B (diameter of buoy)	0.70 m
H (merged buoy)	0.30 m
Ywater (water density)	1000 kg/m^3
Yb (buoy weight)	27 kg
M (Metasentrum)	27

Testing of buoyancy and stability of aerial coastal buoys as a resulted design is conducted on water surface in seeking its ability in real condition, before those buoys are deployed at the determined location, coastal areas and small islands. Displaying of test is shown as the following figure.

Figure 9. Stability and buoyancy test of the new AWCB in a pool at Fisheries and Marine Science Faculty, Bogor Agricultural University (IPB).

Deploying the aerial coastal buoy in a pool shows that the buoy is floating well on water surface through physical observation. An extreme action is done on the buoy by pressing one side of the buoy until its side reaches the water surface and releases it up. Time duration during the pressed side buoy and the buoy in stable position are recorded. The time shows that during the action is averaged about 28 seconds. Those testing shows that the buoy is stable on the water surface and suitable to face extreme situation at the pool scale. These tests are likely a physical simulation of the buoy for real situation. It is expected that the buoy can be deployed well in the real situation, even some seawater dynamic fluctuation. By those tests infer the buoy seems to be stable in deploying at the coastal areas.

3.4. Field testing
Field test of the new AWCB is the final step. This test is conducted in Pramuka Island sorrounding waters as long as six days of operation. Instalation of the buoy is carried out by mooring system. Three new AWCBs are installed in determining location and desired distance. Each buoy is moored to a small buoy using three meters long rope, and ten meters rope to an achor. Six temperature sensors are set as an array perpenducularly to the new AWCB. Distance of sensor is determined in a meter long. Physical

observation shows that the buoy is swinged well on seawater waves, winds, and seawater surface current. Moreover, buoyancy of those buoys are good in seawater. The land-receiver functions as a receiving node in main land.

Figure 10. Field testing of aerial coastal buoys in surrounding waters of Pramuka Island, Node Buoy (left) and A Land-receiver (right).

3.5. Transmitting of seawater temperatures

Measuring water quality parameter such as seawater temperature in this research used Dallas 1820 temperature sensors. Those sensors were mounted in distance one meter each in every buoy node vertically as long as six meters seawater surface. Sea water temperatures were collected by those sensors then transmitted using Xbee Module to the land-receiver node in the mainland. Those sensors worked during a full day long. The results of seawater temperature recording by three buoy nodes on May 22, 2017 are displayed in the folowing figures.

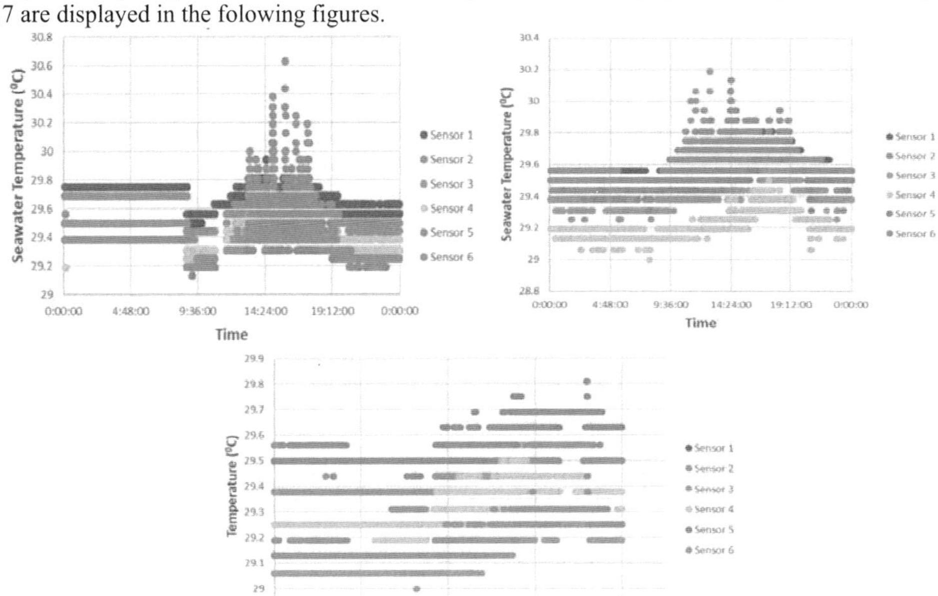

Figure 11. Seawater temperature fluctuation in three different node buoys such as buoy i (left), buoy ii (right), and buoy iii (center).

Those three figures above indicate that seawater tempeture in the serounding water are inranged of 29 to 30 ^{0}C. The water temperature are getting hot in midday at two o'clock as the peak of heat and then it getting down to the normal position. This trend is exhibited by almost all mounted sensors in the three nodes above. Seawater temperature fluctuation between day and night that is recorded in this research is small about 1 ^{0}C. It happens because of the buoy was deployed in shallow tropical coastal waters where variation of sea surface temperature is not so huge. More detail about seasurface temperature variation and its phenomena has been written by Kawai and Wada [7].

4. Conclusion

Steps of creating a new aerial wireless coastal buoy (New AWCB) in this research starts through some activities such design, construction, test of stability and buoyancy, and also field test. Detail works and availability of materials and tools are two crucial things along the working process. Understanding each work process and how to solve obstacles faced in construction of the buoy is a critical point. The accuracy between design process in computer and mechanical works is important to be beard in mind. Testing the buoy also important for making sure its stability and buoyancy. All process of buoy making and testing above infers that the buoy is already to be deployed in determined location. Field testing of those three new AWCB and its land-receiver points out those buoys are working well both recording and transmitting seawater temperature data.

Acknowledgements

Authors wishing to acknowledge assistance from staffs of Instrumentation, Acoustic and Robotics Laboratory, Marine Technology, Bogor Agricultural University (IPB). Special thanks to Balai Taman Nasional Kepulauan Seribu, DKI Jakarta that had been provided accomodation and transportation in the field work of this research.

References

[1] Stanev E V, Beckers J M, Lancelot C, Staneva J V, Le Traon P Y, Peneva E L and Gregoire M 2002. Coastal-open ocean exchange in the Black Sea: observations and modelling. estuarine *Coastal and Shelf Science* **54** 601-620

[2] Massaua J M, Thomas C W and Klinger T 2016 The use of science in collaborative management of marine environments *Costal Management* **44** (6) 606-627

[3] Miller P, Mills J, Edwards S, Bryan P, Marsh S, Hobbs P and Mitchell H 2007 A robust surface matching technique for integrated monitoring of coastal geohazards *Marine Geodesy* **30** 109-123

[4] Albaladejo C, Navarro H, Soto F, Iborra A and Torres R 2015 Low Cost Buoy for Monitoring Recreational Areas *Intrumentation Viewpoint* pp 61

[5] Withamana A 2013 Desing and Testing of Buoy Sistem Using A-WSN Protocol Zigbee in Coastal Areas Thesis Graduate School Bogor Agricultural University Bogor (In Bahasa)

[6] Hidayat R R 2015 Multi Hoop Transmitting Data System on Moored Buoy Using Wireless Sensor Networks Thesis IPB Graduate School. Bogor Agricultural University Bogor (in Bahasa)

[7] Munson B R, Young D F and Okiishi T H 2003 Fluid Mechanics Erlangga Jakarta (in Bahasa)

[8] Kawai Yand Wada A 2007 Diurnal sea surface temperature variation and its impact on the atmosphere and ocean: a review *Journal of Oceanography* **63** 721-744

The abundance and spatial distribution of plankton communities in Perancak Estuary, Bali

A W Hastuti[1*], Y Pancawati[2] and I N Surana[1]

[1] Institute for Marine Research and Observation – Perancak, Jembrana, Bali 82251, Indonesia
[2] Fish Quarantine and Inspection Agency, Juanda, Surabaya
*e-mail: amandangi.wahyuning@gmail.com

Abstract. Perancak estuary is a productive ecosystem and has important role in the balance of nature and the life of local communities. However, research about water quality and organisms in Perancak estuary is still low, especially on plankton abundance and distribution. The aim of this research was to analyze the abundance and spatial distribution of plankton communities. The research was conducted by field surveys in November 2015, samples were taken from 3 stations representing upstream (S1), stream (S2) and downstream (S3). Plankton sampling was carried out by filtering water samples as much as 100 liters of plankton net with mesh size 25μm, diameter 31 cm, and length 80 cm. Filtered samples were stored in the sample bottle and the preserved with 1 ml of 4% formaldehyde. The result showed, an abundance of phytoplankton ranged between 33,600 – 3,002,400 cell/m^3 and zooplankton between 6,000-57,000 ind/m^3. Total of 6 genera of phytoplankton and 14 genera of zooplankton was recorded during the research period. The ecological index showed that species diversity of plankton in Perancak estuary was categorized as low diversity. Plankton abundance in S1 which located in Loloan River is higher than others. Its caused by anthropogenic waste from agriculture, households and industrial waste.

1. Introduction

Perancak estuary is located in District Negara, Jembrana District, Bali has a total area of 876 ha, in the form of the pond either active or inactive with an area of 390 ha and mangrove forests with an area of 78.6 ha. Perancak estuary has distinctive characteristics which the water bodies affected by the intake of salty water of the Strait of Bali and fresh water from a river that flows in it, one of which is the River Loloan. Region Perancak estuary habitats for producers, consumers and consumer peaks in the food chain.

Plankton is one of the biological parameters that is influenced by other parameters and is a very important link in supporting the life of other organisms [1, 2]. Plankton is microscopic organisms floating in the water column or has a very weak swimming ability; the movement is always affected by the mass movement of water [2, 3]. The abundance of phytoplankton in the waters affected by several environmental parameters and physiological characteristics. The composition and abundance of phytoplankton will change at various levels in response to changing environmental conditions, whether physical, chemical and biological [4].

The existence of plankton can be used as one of the indicators of water quality. Phytoplankton is a biological indicator for evaluating the quality and fertility of aquatic ecology [5, 6]. Waters with nutrient content, both nitrate, phosphate, and silicate will enrich the estuary area because the nutrients are needed for plankton growth. Information on the plankton community structure in the Perancak still rare. Therefore, this

IOP Conf. Series: Earth and Environmental Science **176** (2018) 012042 doi:10.1088/1755-1315/176/1/012042

study aims to analyze the diversity and structure of the plankton community to describe the condition of the water condition in the Perancak Estuary.

2. Materials and Methods

The study was conducted in November 2015 representing the transitional season in the Perancak Estuary. Perancak Estuary is located in Jembrana district, Bali province. The choice of sampling stations was conducted according to the anthropogenic activities (S1), pond areas (S2), and gradient of salinity (S3) (figure 1). Data collection when high tide is assumed to be a mixing between sea water and freshwater that will affect the abundance of plankton species [7].

Plankton samples were taken at the surface layer at three observation stations using net plankton with mesh size 25 µm, diameter 31 cm, and length 80 cm, based on SNI 06-3963-1995 and Greenberg method [8]. Plankton samples were obtained from 100 liters of filtered water and concentrated to 30 ml. The water sample was placed in a white or dark film bottle, and 4% formalin added. Measurements of water quality parameters were carried out by handheld Multi-Parameter Water Quality Checker (WQC) -24. Parameters taken include temperature, salinity, pH, DO, phosphate, silicate, and nitrate referring to SNI 6989.57-2008.

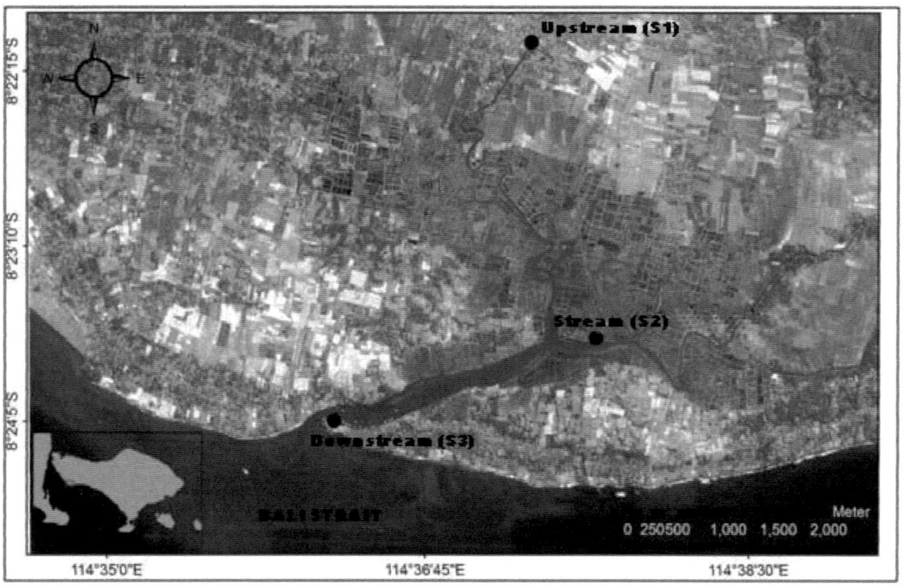

Figure 1. The study area.

Plankton identified by using Optica B310 binoculars microscope with 100 times magnification. 1 ml of the homogenized samples, dripped on Sedgewick Rafter and covered with cover glass, then identified and calculated the abundance. Identification of plankton refers to [9] plankton identification book. The abundance of plankton calculated referring to APHA (2005):

$$N = \frac{1}{Vd} \times \frac{Vt}{Vs} \times n \tag{1}$$

Description: N = Plankton abundance (cell/m^3 or ind/m^3); Vd = volume of filtered water sample (m^3); Vt = volume of filtered sample (ml); Vs = concentrate volume of Sedgwick Rafter Counting Cell (ml); n = number of observed plankton.

Diversity index was based on Shannon-Wiener index [10] with the following formula:

$$H' = -\sum_{i=1}^{n} pi \ln pi \quad (2)$$

Description: H' = Shannon-Wiener diversity index; $pi = ni/N$; ni = number of individual species-ith; N = total number of individuals

Evenness index was determined using the following formula [11]:

$$E = \frac{H\prime}{Hmax} \quad (3)$$

Description: E = Evenness index; H' = Shannon-Wiener diversity index; H_{max} = ln S; S = number of genera.

Dominance index was determined using the following formula [11]:

$$D = \sum_{i=1}^{s} \left(\frac{ni}{N}\right)^2 \quad (4)$$

Description: D = Simpson dominance index; ni = number of individual species-ith; N = total number of individuals; S = number of genera

3. Result and Discussion

3.1 The water quality of Perancak Estuary

The physic-chemical water quality of Perancak Estuary shows on table 1.

Table 1. The water quality of perancak estuary.

Parameter	Station		
	Upstream (S1)	Stream (S2)	Downstream (S3)
Temperature (°C)	35.03±0.12	32.70±0.00	30.13±0.20
Salinity (ppt)	16.27±0.15	33.70±0.00	34.07±0.12
pH	7.56±0.08	7.85±0.01	8.05±0.02
DO (mg/l)	5.13±0.10	3.42±0.07	4.07±0.05
Phosphate (mg/l)	0.43±0.00	0.25±0.00	<0.001±0.00
Silica (mg/l)	2.78±0.05	0.90±0.02	<0.001±0.00
Nitrate (mg/l)	<0.001±0.00	<0.001±0.00	<0.001±0.00

Based on table 1. the range of surface temperature from upstream to downstream is 30.13 - 35.03°C. In general. the salinity at the Perancak Estuary measured in November ranges from 16.27 - 34.07‰. where the lowest salinity is at upstream which Loloan River (16.27 ‰) and highest in downstream (34.07 ‰). Salinity fluctuations in the estuary have a profound effect on plankton distribution [12. 13. 14. 15]. The lowest average pH was found in the upstream (7.56). and the highest was in downstream (8.05). According to [16]. the ideal pH range for plankton life ranges from 6.8 to 8.0. Dissolved Oxygen (DO) at stream is lower (3.42 mg/l) than in the upstream (5.13 mg/l) and downstream (4.07 mg/l). Phosphate is often regarded as a limiting factor of phytoplankton growth in natural waters [17. 18. 19]. The measured phosphate values range from 0.25 to 0.43 mg/l and still within the optimum range for plankton growth. The phosphate values in the upstream are higher than the stream and downstream; it caused by the upstream is near from settlements that allow industrial waste disposal. household waste. and agricultural waste. The content of silicate in upstream is higher than two other stations that are 2.78 mg/l. Silica nutrients are needed and have an effect on the growth process and the development of the life of several kinds of phytoplankton such as diatoms and silicoflagellates for the formation of their cell. The nitrate content in the three research sites is almost undetectable; this is probably due to the absence of nutrient input from the land.

3.2 Phytoplankton abundance and distribution

The ranges of phytoplankton abundance in the Perancak Estuary is 42,000 to 3,002,400 cells/m^3. The highest abundance is found in the upstream and the lowest in downstream. The abundance of phytoplankton in aquatic can fluctuate in type and amount due to differences in nutrient content. Predation by zooplankton or herbivorous fish and tidal occurrence of sea water. Tides can affect plankton abundance and distribution. In high tide conditions. The flow of sea water to the river carrying the phytoplankton of sea water (Diatom) in it mixed with river water in which there is freshwater phytoplankton and vice versa. This is what causes Diatoms are found in the upstream which adds to the abundance of phytoplankton on the upstream and the discovery of Cyanophyceae at the stream and downstream.

The phytoplankton abundance in Perancak Estuary varied between 33,600 – 3,002,400 cell/m^3. Table 2 shows *Oscillatoria* sp. was dominance and distributes equally in all stations. However. the abundance of *Oscillatoria* sp. Has not reached endangers for the other lives organisms because according to [20] *blooming* will occurs when the abundance reaches $5x10^5$ to $2x10^6$ cells/l. Differences in the abundance of phytoplankton in Estuary Perancak are closely related to the physics-chemical conditions of the waters. According to [21]. Nutrients that get into the aquatic environment such as phosphate, nitrate, silica and ammonia will affect the development of phytoplankton and zooplankton.

The abundance of *Oscillatoria* sp. Which is high in the upstream can be caused by the high nutrients as a result of the runoff the organic materials coming from households. Ponds and farming activities. Consumption of fertilizer on agricultural supplies nitrate and phosphate to the river through washing and soil erosion [22].

Table 2. The abundance of phytoplankton in Perancak Estuary.

No.	Class of Phytoplankton	Density (cell/m^3)		
		Upstream (S1)	Stream (S2)	Downstream (S3)
	Cyanophyceae			
1	*Oscillatoria* sp.	2,995,200	33,000	42,000
	Diatom			
2	*Biddulphia* sp.	900	300	-
3	*Cocconeis* sp.	1,500	-	-
4	*Coscinodiscus* sp.	600	-	-
5	*Pleurosigma* sp.	2,100	-	-
6	*Thalassiosira* sp.	2,100	300	-
	Density (cell/m^3)	3,002,400	33,600	42,000

The phytoplankton composition in Perancak Estuary consists of 6 species classified into two classes: Diatoms and Cyanophyceae (figure 2). Judging from the physical factors the waters Estuary Perancak is a suitable habitat for the growth of Oscillatoria sp. Oscillatoria is an important member of Cyanophyceae because it can perform photosynthesis activities. Cyanophyceae live in neutral or alkaline waters and are not found in waters with a pH less than 4 [23]. This corresponds to a pH value in the Perancak Estuary that ranges from 7.56 to 8.05. Optimal temperature conditions also support the growth of Oscillatoria because the species can adapt to a certain temperature.

In contrast to Cyanophyceae, the phytoplankton of the Diatoms class in the Perancak Estuary has more species than Cyanophyceae. The presence of Diatoms is not found in any station. This is probably due to predation by zooplankton and herbivore fish, that Diatoms are a group favored by fish and shrimp larvae.

The diversity index phytoplankton (H') is in the range of 0.00 to 0.10. This value indicates that the phytoplankton diversity in Perancak Estuary is low (figure 3). The value of diversity index is inversely proportional to the index of dominance (D) when the index of diversity and evenness is low. The value of the dominant index is high and vice versa. The evenness index (E) between 0.00 - 0.09 indicates a low uniformity which means the number of species is not evenly distributed or it can be said there is dominance. This can be seen from the high index value of dominance between 0.96 - 1.00 which means there is a type of phytoplankton dominating, namely *Oscillatoria* sp. which can withstand the conditions of the waters of the Perancak Estuary.

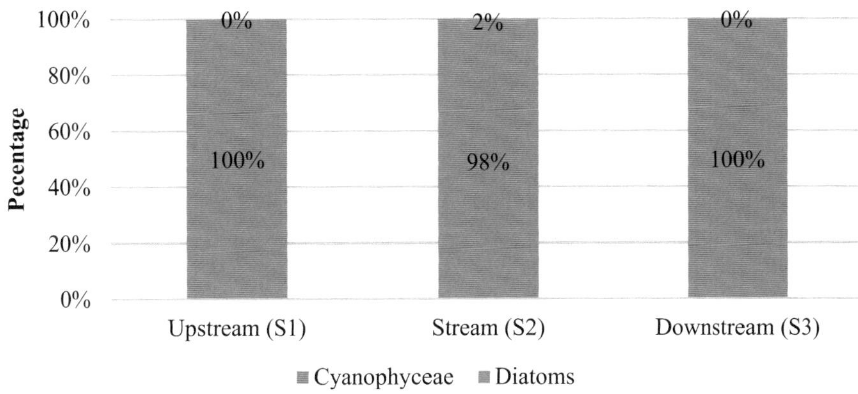

Figure 2. Phytoplankton composition in Perancak Estuary.

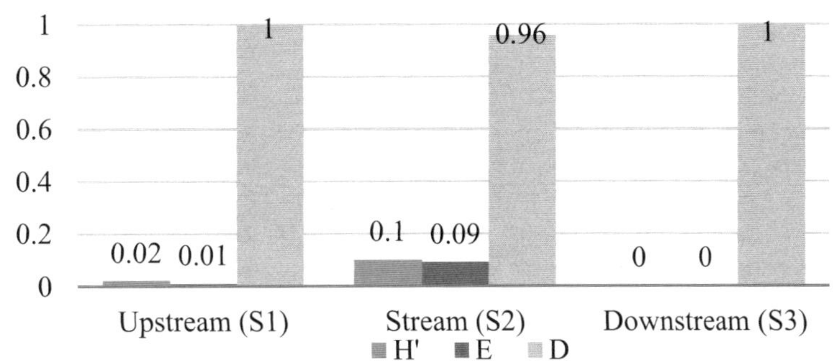

Figure 3. Ecological index of phytoplankton in Perancak Estuary.

3.3 Zooplankton abundance and distribution

The upstream (S1) has the highest composition and abundance of zooplankton in the Perancak Estuary. The abundance of zooplankton in the upstream reached 57.300 and/m3. The high abundance of zooplankton in the upstream is due to the presence of Phylodina sp. (Rotifera) is 33,900 and/m3. According to [24]. Phylodina sp. found in waters with high organic matter and mossy. This is by the condition of upstream containing organic material from household waste. farm activities and agriculture of the surrounding population. Analogous to phytoplankton abundance. the highest abundance of zooplankton in the upstream. The abundance of zooplankton depends on the abundance of phytoplankton.

Zooplankton affects the potential for the emergence and distribution of pelagic fish, in which almost all pelagic fish are high market demand. Fish communities typically develop well in areas where planktonic organisms are in high density because the grown fish has to make sure its juvenile can get enough food. Also. Zooplankton can be used as an indicator of water pollution (table 3).

Table 3. The abundance of zooplankton in Perancak Estuary.

No.	Class of Zooplankton	Density (ind/m^3)		
		Upstream (S1)	Stream (S2)	Downstream (S3)
	Bivalvia			
1	*Mytilus* sp.	9,300	-	-
	Ciliata			
2	*Undella* sp.	300	-	-
	Crustacea			
3	*Acartia* sp.	300	2,700	-
4	*Calanus* sp.	300	300	-
5	*Eurytemora* sp.	-	600	-
6	*Temora* sp.	300	-	-
	Cephalochordata			
7	*Amphioxides* sp.	7,800	300	-
	Gastropoda			
8	*Limacina* sp.	900	-	-
	Holothuroidea			
9	*Pelagothuria* sp.	300	-	-
	Rotifera			
10	*Brachionus* sp.	2,100	600	-
11	*Keratella* sp.	-	300	-
12	*Phylodina* sp.	33,900	300	-
	Polychaeta			
13	*Harmathoa* sp.	1.500	900	-
	Scaphopoda			
14	*Littorina* sp.	300	-	-
	Density (ind/m^3)	57,300	6,000	-

Rotifera is commonly found in freshwater ecosystems and plays an important role in the food chain as a natural food of shrimp and fish larvae. This high abundance of Rotifera can be attributed to the high abundance of phytoplankton in the upstream. Rotifers are food for larger animals such as worms and crustaceans. The crustacean was dominated in the stream (S2). The role of Crustaceans in a pelagic ecosystem is very important from a trophic point of view as a bridge between primary and secondary marine producers because the crustacean growth rate is relatively fast so that its abundance is higher than other zooplankton classes [25]. Figure 4 shows that crustacean abundance at station S2 is 60%.

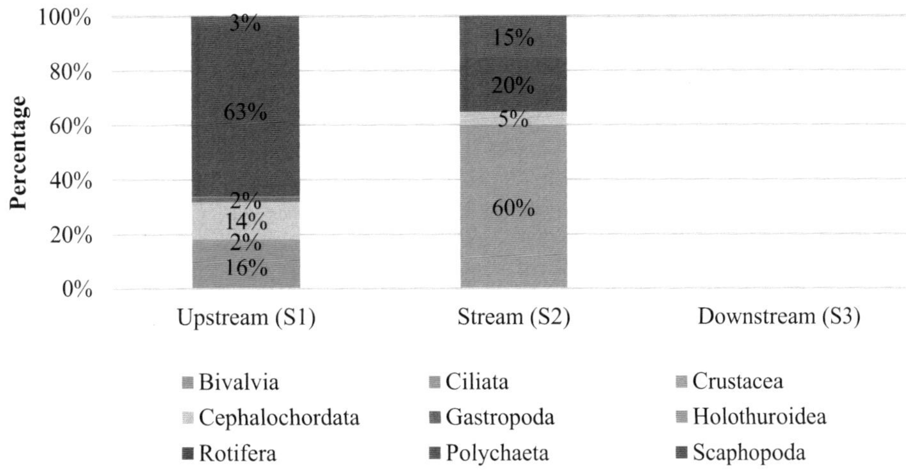

Figure 4. Zooplankton composition in Perancak Estuary.

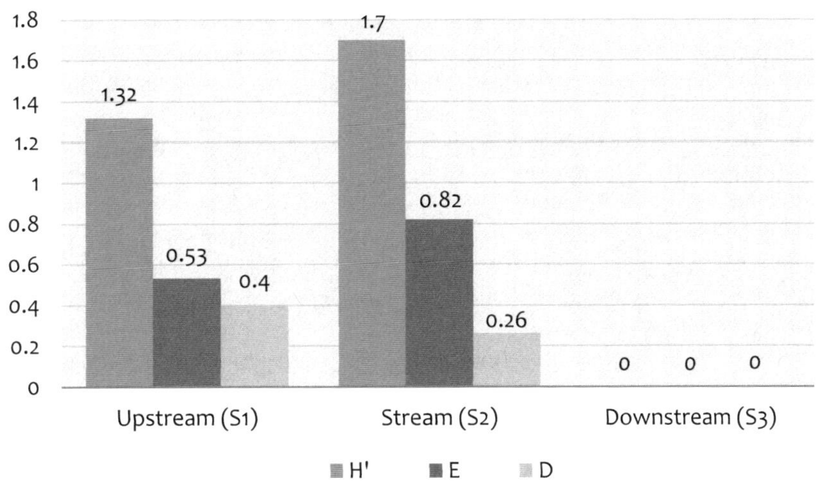

Figure 5. Ecological index of zooplankton in Perancak Estuary.

The diversity index of zooplankton (H') in the Perancak Estuary belongs to the medium diversity class in the range of 1.32 to 1.7. While for the evenness index (E) obtained range 0.53-0.82 shows high uniformity which means the number of inter-species is evenly or can be said there is no dominance. This can be seen from the low dominance index (D) value between 0.26 - 0.4 which means there are no species of zooplankton dominating in Perancak Estuary (figure 5).

4. Conclusion

The results showed that the type and abundance of plankton were quite varied with the number of species found as many as six types of phytoplankton and 14 types of zooplankton. The ecological index shows that the diversity of phytoplankton and zooplankton in the Perancak Estuary is low and there are no dominating species with stable uniformity and tend to be evenly distributed. Plankton abundance in Perancak area can be used as an indicator of aquatic ecology. Physico-chemical water quality in Perancak Perancak Estuary shows the normal range for the life of plankton and other aquatic biotas.

References

[1] Nybakken J W and Eidman H M 1992 *Biologi Laut: Suatu Pendekatan Ekologis* PT Gramedia Pustaka Utama
[2] Yuliana 2009 Komposisi dan Kelimpahan Plankton di Kepulauan Guraici Kabupaten Halmahera Selatan Maluku Utara *Lutjanus* **14**(1): 49–53
[3] Newell G E and Newell R C 1977 Marine Plankton Hutchinson
[4] Reynolds C S 2006 *The Ecology of Phytoplankton* Cambridge University Press
[5] Follows M J. S Dutkiewicz Montoya J M. Cermeno P. Loreau M. and Vallina S M 2014 Global Relationship between Phytoplankton *Nature Communications* 5 Nature Publishing Group: 1–10 doi:10.1038/ncomms5299
[6] Rochelle-Newall E J. Chu V T. Pringault O. Amouroux D. Arfi R. Bettarel Y and T Bouvier 2011 Phytoplankton Distribution and Productivity in a Highly Turbid. Tropical Coastal System (Bach Dang Estuary. Vietnam) *Marine Pollution Bulletin* **62**(11): 2317–29 doi:10.1016/ j.marpolbul.2011.08.044
[7] Cloern J E 1991 Tidal Stirring and Phytoplankton Bloom Dynamics in an Estuary *Journal of Marine Research* **49**(1) Sears Foundation for Marine Research 203–21
[8] Eaton A D. Lenore S C. Eugene W R. Greenberg A E and MAH Franson 2005 APHA: Standard Methods for the Examination of Water and Wastewater *Centennial Edition APHA. AWWA. WEF. Washington. DC*
[9] Yamaji I 1976 *Illustration Of The Marine Plankton Of Japan* Hoikusha Publishing Co. LTD Japan
[10] Shannon C E and Weaver W 1963 *The mathematical theory of communication* University of Illinois Press p 125
[11] Odum E P 1971 Fundamental of Ecology. WB Saunders *Company Toronto*
[12] Cervetto G. Raymond G and Marc P 1999 Influence of Salinity on the Distribution of Acartia tonsa (Copepoda. Calanoida) *Journal of Experimental Marine Biology and Ecology* **239**(1): 33–45 doi:10.1016/S0022-0981(99)00023-4
[13] Lawrence D. Valiela I and Tomasky G 2004 Estuarine Calanoid Copepod Abundance about Season. Salinity. and Land-Derived Nitrogen Loading. Waquoit Bay. MA *Estuarine. Coastal and Shelf Science* **61**(3): 547–57 doi:10.1016/j.ecss.2004.06.018
[14] Devreker D. Sami S. Gesche W. Joëlle Forget-Leray and François L 2009 Effects of Salinity. Temperature and Individual Variability on the Reproduction of Eurytemora Affinis (Copepoda; Calanoida) from the Seine Estuary A Laboratory Study *Journal of Experimental Marine Biology and Ecology* **368**(2) Elsevier B V: 113–23. doi:10.1016/j.jembe.2008.10.015
[15] Paturej E and Gutkowska A 2015 The Effect of Salinity Levels on the Structure of Zooplankton Communities *Arch. Biol. Sci.* Belgrade **67**(2) 483-492 DOI: 10.2298/ABS140910012P
[16] Boyd C E and Craig S T 2012 *Pond Aquaculture Water Quality Management* Springer Science & Business Media
[17] Schindler D W 1974 Eutrophication and Recovery in Experimental Lakes: Implications for Lake Management *Science* **184**(4139): 897–99 doi:10.1126/science.184.4139.897
[18] Henderson S B.. and Markland H R 1987 *Decaying Lakes The Origins and Control of Cultural Eutrophication* New York: John Willey & Son
[19] Jin L. Wu H.. and Chen M 2011 Effect of Nitrogen and Phosphorus on Phytoplankton Composition and Biomass in 15 Subtropical. Urban Shallow Lakes in Wuhan. China *Limnologica* **41**(2001) 48-56
[20] Kennish M J 1990 Ecology of Estuaries Vol **II** Biological Aspect CRC Press Boston

[21] Soedibjo B S 2007 *Pengaruh Faktor Lingkungan Terhadap Distribusi Spasial Komunitas Zooplankton Di Teluk Klabat. Perairan Bangka Belitung*. Oseanologi dan Limnologi di Indonesia Volume **33**(1): 47-63

[22] Mason C F 1991 Biology of Freshwater Pollution. Longman Scientific and Technical New York. John Wiley & Sons Inc.

[23] Bold 1985 Introduction to the Algae Structure and Reproduction 2nd Ed. Prentice Hall. Inc. Englewood Cliffs *In*: Prihantini 2008 *Biodiversitas Cyanobacteria Dari Beberapa Situ/Danau di Kawasan Jakarta-Depok-Bogor Indonesia* Makara Sains Volume **12**(1): 44-54

[24] Hochberg 2000. *Functional morphology of the muscle in Phylodina sp. (Rotifera: Bdelloidea)* Hydrobiologia Volume **432**: 57-64

[25] Razouls C. De Bovée F. Kouwenberg J and Desreumaux N 2005 Diversity and Geographic Distribution of Marine Planktonic Copepods *Available from WWW:< Http://copepodes. Obs-Banyuls. Fr/en>[cited 2008-08-10]*

IOP Conf. Series: Earth and Environmental Science **176** (2018) 012043 doi:10.1088/1755-1315/176/1/012043

Economic valuation and lost value of mangroves ecosystem due to oil spill in Peleng Strait, Banggai and Banggai Islands Regency Central Sulawesi

S Putranto[1*], N P Zamani[2*], H S Sanusi[2], E Riani[3], A Fahrudin[3]

[1]Graduate Student of Marine Science, Bogor of Agriculture University, Bogor
[2]Marine Science and Technology Department, FPIK-IPB Bogor
[3]Management of Aquatic Resources, FPIK-IPB Bogor
*e-mail: sugeng.putranto@yahoo.com; np_zamani@yahoo.com

Abstract. The increase of sea transportation in the Strait Peleng, Banggai and Banggai Islands Regency, Central Sulawesi such as tankers are potentially causing environmental degradation, which in case of a tanker collision that may cause an oil spill. The event of oil spills in the marine environment when entry into the mangrove ecosystem can cause damage to mangrove ecosystems. The research aims to determine how much the loss value of mangrove ecosystem when the oil spill contaminates into the mangrove ecosystem, so it is necessary to do an economic assessment for the damage of mangrove ecosystems. Furthermore, some data is needed, there :is important value index of mangrove, economic valuation of mangrove ecosystems, oil distribution patterns and the loos value of mangrove that caused by the oil spill. The research was conducted in nine districts in the Peleng Strait of Banggai and Banggai Islands, among others: South Batui, Batui, East Luwuk, Lamala, Masama, Balantak, Bualemo (located in the administrative area of Banggai) and the District Bulagi and Buko (located in the administrative area Banggai Islands). This research was conducted from August to November 2016 for taking the sample and interviews to the community and local government. Based on the results that showed, the mangrove ecosystem in Peleng Strait Coastal Water has a total area of ± 107 ha with a total value of economic value is US$ 1,702,605.73 /year. The biggest contributor comes from indirect value, that is as a buffer abrasion or as break water of US$ 1,237,277.78 /year. Base total economic value, the biggest effect of the oil spill could be losses per year is US$ 695,521.97 /year that occurred in the second transitional season (July – September).

1. Introduction

Banggai Regency and Banggai Islands Regency both are regions in Center Sulawesi Province which have several coastal ecosystems such as mangrove, coral reefs, and seagrass. The other site there are other potential resources in the mining sector, that is nickel mine producer (in an exploitation phase) and the natural gas in 3 locations, the Matindok Block, Senoro Block and Donggi Senoro LNG. This condition will have an impact on sea transportation thus increasing the traffic level. In case, oil tanker will provide a lot of coming into Banggai coastal area. It has the potential to cause some environmental degradation such as disposal of liquid waste, ballast water, and other incidents such as a tanker collision or leakage/ pipe cracking on loading arm.

Indonesia has encountered several oil spill incident, one of which occurred in Montara Well Head Platform in the Timor Sea in 2009. Although the oil spill location was in Australia's ZEE territory, however, the impact felt to the Indonesian region in 8 districts in East Nusa Tenggara province [1]. The Government of Indonesia claimed for environmental damage caused by the oil spill in the US $ 2,4 billion, or Rp 21,6 trillion. Unfortunately, it was rejected by PT TEP Australasia because of the claim was not supported by enough data.

The impact of an oil spill can cause the degradation of sea water quality in short-term effect or more harmful impact over a long term. Physically, the contamination of oil will be obvious because the sea surface will be covered with oil or tar balls. Chemically, the petroleum is classified as a hydrocarbon aromatic compound, it can be toxic to marine organisms and causes the damage to coastal ecosystems. Biologically, the liquid waste and oil spill incident affect to marine organisms living and humans both directly or indirectly. Because of it, the management of coastal areas is a very important thing. Referring to the concept of sustainable development, the focus is balancing between economic growth and natural resource quality.

Based on this background, the purpose of this research is to determine how much the loss value of mangrove ecosystem when the oil spill contaminates into the mangrove ecosystem, so it is necessary to do an economic assessment for the damage of mangrove ecosystems. Furthermore, some data is needed there eis important value index of mangrove, economic valuation of mangrove ecosystems, oil distribution patterns and the loss value of mangrove that caused by the oil spill.

2. Method

2.1. Location and time

The location of data observation is mangrove ecosystem area in Banggai Regency and Banggai Islands Regency, Central Sulawesi. The research was held in August - November 2016. Data were collected from 9 District; South Batui, Batui, East Luwuk, Lamala, Masama, Balantak and Bualemo (Banggai Regency), Bulagi and Buko (Banggai Island Regency).

2.2. The technique of data collection

2.2.1. Mangrove. Mangrove density was measured by Line Transect Plot method with a size of transect 10x10 m2 on each vertical/horizontal stratification zone (3 plots to data retrieval and repetition). The distance between each plot is about 20 m and coordinate points recorded by GPS locations. The measurement of diameter for individual mangrove refers to Ministry of Environment No. 201 the year 2004 regarding Standard and Guideline of Mangrove Degradation. The species identification refers to [2].

2.2.2. Valuation of economic. The sampling method is an accidental sampling, that will be taken from the people who live in around the mangrove area as the respondents. Base on pre-research, about 81 people as the respondent are working as: fishermen, Woodman and some people who related to mangrove resource. Primary data obtained by deep interview and questioners (directly collected the data from the local community) and secondary data collected from a local government office and other resources.

2.2.3. Oil spill. Oil spill model uses *Oilmap* software. This model can predict the spread of oil spills. This software uses a geographic information system which combines the environmental conditions and the oil characteristics (Table 1). Secondary data used for the distribution of oil spill, there are daily/monthly tide and annual tide, currents, waves, wind, and humidity. These data were obtained from the Luwuk Climatology and the *North Oceanic Atmospheric Administration (NOAA)* database.

Table 1. Characteristic of marine fuel oil [3].

Parameter	Value/Units
Specific gravity at 60/60 °F	0,87
Melting point	30-50 °C
Boiling point	150 °C
Flash point	70-230 °C
Solubility in water	No soluble in water

2.3. Data analysis

2.3.1. Economic Valuation of Mangrove Ecosystem. Economic valuation of mangrove ecosystem resources or *Total Economic Value (TEV)* is calculated from the sum of economic value based on utilization, which includes direct value, indirect value and option value [4]. Following is analysis method for economic valuation of mangrove ecosystems bellows:

A. Total Economic Value (TEV)

Total economic value of mangrove forest estimated from the sum of direct value, indirect value and option value can be written in the mathematical equation as follows:

$$TEV = DUV + IUV + OV \quad (1)$$

where;

Total Economic Value (TEV) = Measured in terminology as a willingness to pay (WTP)

Direct Use Value (DUV) = Goods and services provided by a resource that can be directly Utilized.

Indirect Use Value (IUV) = the value is derived from the goods and services which provided by a natural resources that used indirectly.

Option Value (OV) = Potential direct or indirect value of a natural resource which can be utilized in the future with assuming that the resource is not destroyed or permanently damaged.

B. Direct use value

Direct use value (DUV) is the economic value that derived from the direct utilization of an ecosystem.In the context of mangrove ecosystems, these economic values include the mangrove utilization as a material, cordwood and the value of fishery resources. Mangrove wood as a material (par value of wood) is estimated by the number of mangrove volume/ha/year multiplied by the price of mangrove wood [5]. The species which has potential wood value is *Rhizophora*, thus the calculation only involve this species. This type of wood usually uses as building material and Fish Aggregate Devices (FADs) by the local community.

$$\textit{par value of wood} = (V \times quantity) \times \textit{selling price /m3} \quad (2)$$

$$V = \pi r^2 t \quad (3)$$

where:

V	= volume (m^3)	r	= radius of a tree (m)
π	= 3.14	t	= height (m)

Mangrove value as cordwood (par value of wood fuel) is estimated by the amount of wood produced multiplied by the selling price [6].

$$par\ value\ of\ wood\ fuel = (V \times quantity) \times selling\ price\ /m^3 \quad ......(4)$$

Mangrove crab catching value (par value of crab)is estimated by the number of catches/year multiplied by the selling price.

$$par\ value\ of\ crab = (T \times B) - B\ (US\$/\ ha/year) \quad ......(5)$$

where:

T	= Total of crab catches (kg /ha /year)	B	= Operational price (US$)
H	= Selling price (US$ /kg)		

C. Indirect use value

Indirect use value is the economic value that derived from the indirect use of an ecosystem both biologically and physically. The biological function of mangrove indirectly as a spawning ground for fish. Biological value (biological par value) is estimated by the number of fish catches around mangrove area by fish price/kg, then reduced the investment and operational cost. The numbers of fish caught at offshore area are not included, which is not considered the function of mangrove forest.

$$Biological\ par\ value = (T \times H) - B\ (US\$\ /\ ha\ /\ year) \quad ......(6)$$

Physical function, the mangrove ecosystem also serves as coastal protection from the ocean waves. The initial par value is calculated through the wesel approach by making the concrete which equivalent to mangrove function of resisting the coastal from abrasion. According to Minister of General Services No. 28 the year 2016 which mention, to build a breakwater with 150 m length, 20 m width, and 5 m height and 20 years withstand that cost is 2,921,147,000 IDR or approximately 19,474,313 IDR/meter.

$$Physical\ par\ value = \frac{B \times PGP}{10\ (US4/m/year)} \quad ......(7)$$

where:

B = Wesel standard beton (US$)

PGP = long coastline (m)

D. Option value

Option value (par value of option) is the value calculated by the possible use of ecosystem goods or services in the future by yourself and others too; it's also called inheritance value [7]. This optional benefit value can be discovered by using direct method. The results of [8] research in Bintuni Bay, US $ 1,500 / km2 / year can be used as biodiversity benefit value on this research too; assuming that the mangrove forests ecologically are important and well maintained.

$$Par\ value\ of\ option = US\ \$\frac{ha}{year} \times Area\ of\ mangrove\ forest \quad ......(8)$$

2.3.2. The spread of an oil spill. Calculation of oil spill area in mangrove ecosystem is determined by using stochastic model. Stochastic model estimates the magnitude of marine fuel oil which entered or stranded to the mangrove then predicts the travel time of oil can reach the shoreline and amount the percentage of stranded oil on to the shore.

2.3.3. Determination of losses value. To calculate the losses value refers to the research result from [9]:

1) Par value of mangrove tree is calculated by the oil spill which impacting the mangrove plant with less than 3 m height
2) Par value of mangrove crab is calculated by considering the death impact on 0-15 day so that the annual percentage of death is 100% per year
3) Par biological value is calculated by the number of biota death which living around mangrove forest with the death impact on 0-15 day so that the annual percentage of death is 100% per year.
4) Par physical value is calculated by the existence of mangrove forest, so this is equal to the par value of mangrove tree calculation
5) Par value of option calculation similar to the par value of mangrove tree and par physical value

2.3.4. Spatial simulation model of losses value. This simulation will integrate the Oil Spill Trajectory and Economic Values that will be displayed in the form of a thematic map containing the value of the damage. This thematic map uses the GIS and Oilmap Arcs that combined with the formulation of economic valuations. Oil trajectory data was made simulated by incorporating oceanographic and climatologically parameters. The view of this spatial simulation will contain information on the Cost of Mangrove Ecosystem Damage at four seasons.

3. Result and discussion

3.1. The condition of a mangrove ecosystem

Referring to [10] regarding Standard and Guideline of Mangrove Degradation, the result shows that Banggai Regency and Banggai Island Regency have mangrove from good until bad criteria. Mangrove with good criteria located in Balantak with density is 3,183 ind/ha, Bualemo with 1,639 ind/ha, Bulagi 1,450 ind/ha, South Batui 1.222 ind/ha, Masama 1,160 ind/ha, and Lamala with 1,117 ind/ha.Based on the result of research, Balantak has the highest mangrove density among others due to many rivers that supply richness nutrient and sedimentation. Mangrove with bad criteria located in Batui with density is 750 ind/ha, Buko 675 ind/ha, and East Luwuk 550 ind/ha.

Degradation of mangrove area in East Luwuk caused by the area has been acquitted by the local community. Population growth is not accompanied by land available for residence area, thus becoming a trend that people to cut down the mangrove to build the house. The database mentions that East Luwuk is the district with the highest population density than other research area. In 1990, there was a permitting for the company from Taiwan to do logging and export the mangrove and no rehabilitation program until now [11]. The consequence is mangrove ecosystem becomes almost extinct in Banggai Coastal.

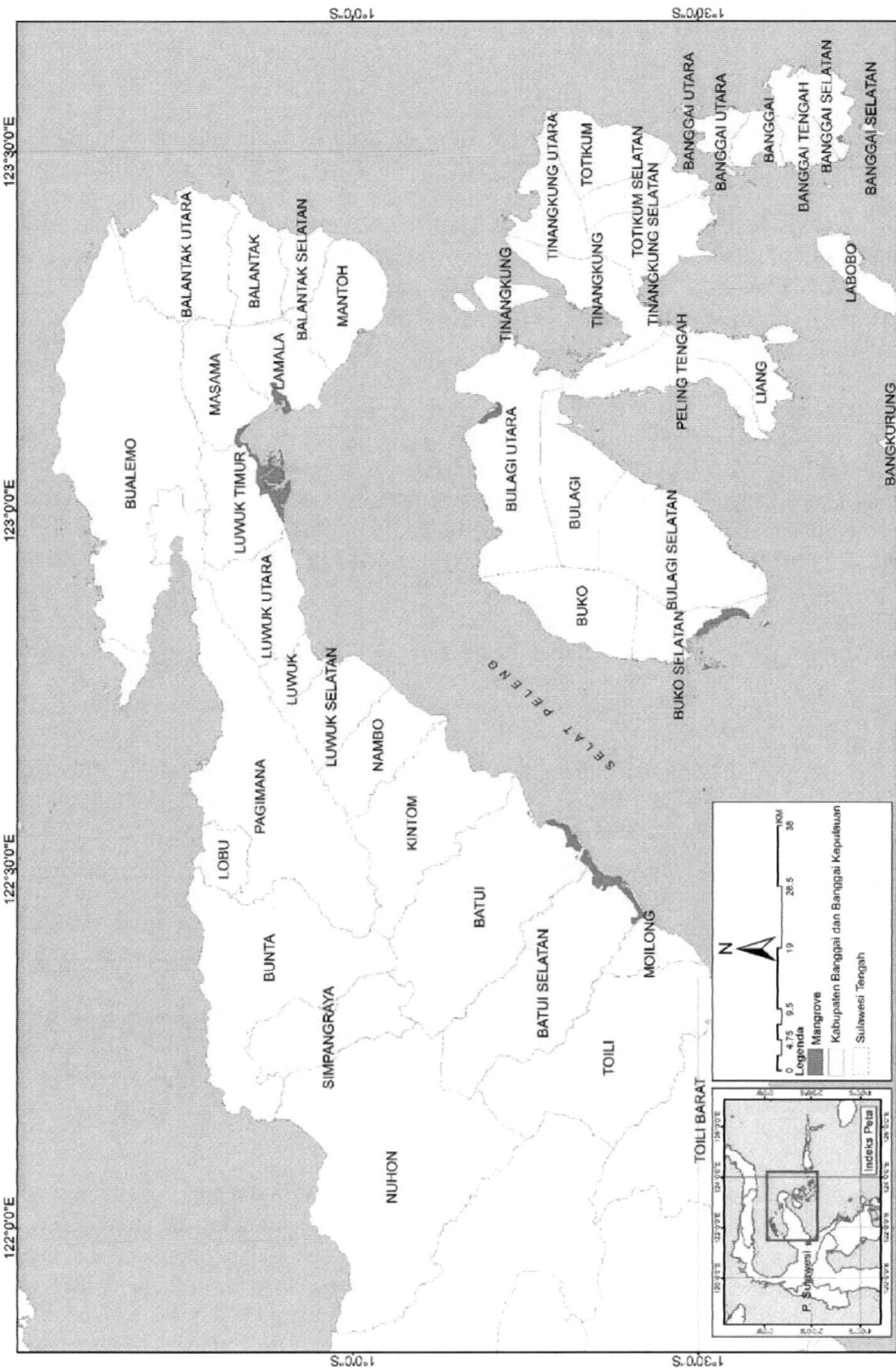

Figure 1. Map of mangrove distribution at the research location.

3.2. Economic Value

3.2.1. Direct use value. Direct use value on this research is estimated from the value of mangrove wood utilization for material (par value of wood), and the value of mangrove utilization as crab habitat (par value of crab). The value of mangrove as cordwood is not included in the calculation, because the people use it only for household needs and not for sale. Par value of wood is only calculated for Rhizophora sp. due to only its wood which utilized by local people. The value is obtained by multiplying the volume of tree log with a selling price of US$ 26.32/m^3. Balantak has the highest par value of wood US$ 101,512.98 /year while contradictive par value of wood in East Luwuk because there is no found Rhizophora tree in this location. The total log of the par value of wood from nine District is 438,137.4/year or 25.73% of the total economic valuation number.

Par value of crab is obtained by multiplying the number of crabs catches in one year with the selling price of US$ 3/kilogram then subtracted with the operational cost in one year. The number of crab catches and the amount of operational costs is obtained from the interviews with crab seekers. The highest par value of crab is in Bualemo (US$ 2,442.11 /year), while the lowest result is in Bulagi (US$ 649.62/year). The total par value of crab from nine District is US$ 10,641.94/year or 0.63% of the total economic value. The small value of crab is due to not optimal by people for utilizing mangrove ecosystem for crab fishing or aquaculture area.

3.2.2. Indirect use value. The indirect use value of mangrove ecosystem is estimated from biological and physical par values. The biological par value is obtained by multiplying the number of fish catches in one year with the selling price of US$ 2,26/kilogram then subtracted with the operational cost in one year. Fish catches value which included into the calculation is fishing around the mangrove forest area by using a fishing rod and a net (*sero*). The highest biological par value is located in Buko with US$ 2,426.92/year while the lowest is in Masama with US$ 970.83/year. The total biological par value in the 9 District is 14,943.64/year or 0.88% of the total utilization value.

The mangrove ecosystem has indirect value as a natural barrier, stabilizes the delicate sediment and prevents the shoreline from erosion. Mangroves also reduce the effects of storms and floods, maintain water quality and support the lives of various wildlife [12]. Indirect use of mangrove forests physically to be retention of abrasion thus a value estimated through replacement cost by building the breakwater. According to Minister of General Services No. 28 the year 2016 which mention that to build a breakwater with 150 m length, 20 m width, and 5 m height and 20 years with stand, the cost is US$ 219,635.10 or approximately 1,464.23/meter. Banggai Regency and Banggai Islands Regency have a total coastline which protected by mangrove forest are 16,900 m length, then the cost required to replace the physical function of mangrove ecosystem as a physical par value for 20 years withstanding is US$ 24,745,556 or equivalent to US$ 1,237,278/year.

3.2.3. Option value. The option value in the existing mangrove ecosystems in Banggai and Banggai Islands coastal areas can be approached using the benefits transfer method, by assessing the estimated value from other places (where the resources are available) and then the value is transferred to obtain a rough estimate of the value of the environment. The method is approached by calculating the value of existing biodiversity in this mangrove area. According to [8], Indonesia has biodiversity value of US$ 1,500 US$/km^2 or about US$ 15/ha/year. This value can be used in all mangrove forests throughout the Indonesia if the ecosystem of mangrove forest is ecologically important and naturally protected.

3.2.4. Total Economic Value. Total Economic Value (TEV) in this research is obtained by summing of direct value, indirect value and option value so that the result is amount US$ 1.702.606/year or approximately US$ 15.912/ha/year. The highest TEV is located in East Luwuk with US$ 368,775.45/year while the lowest is in Batui with US$ 33,610.09/year. Although East Luwuk has degradation of mangrove ecosystem however it has a high utilization value. Moreover, East Luwuk also has a long coast line which covered by mangrove forest to protect from abrasion.

Table 2. The result of calculation of a value of exploiting of mangrove ecosystem at the research location (in US$).

District	DUV		IUV		OV	TEV
	par value of wood	par value of crab	biological par value	physical par value		
South Batui	76,692.37	855.19	1,550.83	73,211.70	150.00	152,460.09
Batui	2,370.53	799.25	1,125.64	29,284.68	0.03	33,610.09
East Luwuk	0.00	770.18	1,721.76	366,058.52	225.00	368,775.45
Masama	5,552.89	1,257.14	970.83	146,423.41	150.00	154,354.27
Lamala	70,203.68	1,515.79	1,506.54	109,817.55	225.00	183,268.57
Balantak	101,512.89	933.87	1,691.20	109,817.55	225.00	214,180.52
Bualemo	88,312.63	2,442.11	1,970.90	183,029.26	375.00	276,129.90
Bulagi	46,293.16	649.62	1,979.02	146,423.41	150.00	195,495.21
Buko	47,199.21	1,418.80	2,426.92	73,211.70	0.75	124,331.63
Total (US$/thn)	438,137.37	10,641.94	14,943.64	1,237,277.78	1,605.00	1,702,605.73
Percentage (%)	25,73	0,63	0,88	72,67	0,09	100,00

The indirect use value takes the highest portion of TEV of 74.55% (indirect use value), while the lowest is the option value of 0.09% (option value) and the rest is from direct option value of 26.36% (direct use value). The highest of indirect use value is derived from the mangrove value as a breakwater. This below the comparison of TEV with other location (Table **3**3). Generally, the higher contribution of TEV is come from IUV (Indirect Use Value), due to the have long coast line which covered by mangrove forest to protect from abrasion.

Table 3. The economic value of mangrove ecosystem some regions in Indonesia (US $)

	Banggai dan Banggai Islands		North Minahasa[1]		Tanjung Pinang[2]		Kendal[3]	
	(US$)	(%)	(US$)	(%)	(US$)	(%)	(US$)	(%)
DUV	448,779	26,36	13,180	1,61	3,994,846	60,20	46,796	42,94
IUV	1,252,221	73,55	802,378	98,01	2,634,586	39,70	61,518	56,45
OP	1,605	0,09	3,105	0,38	6,833	0,10	668	0,61
TEV	1,702,606	100	818,663	100	6,636,265	100	108,983	100

[1][6], [2][13], [3][14]

Based on tabulation in hectare, the TEV of mangrove in this research is 15.912 US$ per hectare or 1.59 US$ per meter square. This research is much bigger than [16] research in Blanakan Bay West Java, Indonesia with 782,34 ha of plots mangrove sampling and 6,800 m length of coastline which the result of ecosystem

IOP Conf. Series: Earth and Environmental Science **176** (2018) 012043 doi:10.1088/1755-1315/176/1/012043

services is amount US$ 286,901.50/year or approximately US$ 366.72/year/ha or US$ 0,04/m2. The difference of both research due to the Indriyani research is not to calculate the direct use value such as the wood utilization and fishery and the costal line.

3.3. Oil spill spreading

This model is made with simulation of oil spill location in DSLNG Port (coordinate of 1° 15,104' S; 122° 35,630' E). The location is LNG loading and unloading activities area. The results of model simulation provide data information about the percentage of oil spills which are stranded on the shore. The results are displayed in the geographic information system (map) in each month in one year and for impact analyze in one year. The monthly data is grouped into four seasons; there is the west season (Dec-Feb), transitional season 1 (Mar-Apr), eastern seasons (May-Sep) and transition season 2 (Oct-Nov). This model simulation used MFO type of oil. The amount of spill is 700 ton or 4375 barrel with API index (American Petroleum Institute) is 33.7 with density is 0.87 gr/cm3. According to [9] this oil is categorized to the second group which has the characteristic of (1) volatile, (2) have residue after the end of evaporation, (3) spread rapidly, (4) does not have stable form emulsion and (5) more bioavailable than light oils, hence it more likely will affect animals in water and sediment.

Oil spill spreading in the west season is mostly influenced by the west season, and the oil spill moves towards to the east as shown in figure 2 (a). Oil will be stranded in the coastal of along Batui, Kintom, and Nambo. Mangrove in Batui Coastal is in bad criteria (with 750 ind /ha), while there is no mangrove habitat in Kintom and Nambo which is a sandy beach and settlement area. Oil spill also spread to the north of Peleng Island, precisely in Bulagi, which has mangrove forest with good criteria (1,450 ind/ha).

In the transition season 1, the wind tends to fluctuate not only from one direction, but it is from the north and northwest. Figure 2 (b) shows the oil spread to the southeast or towards Peleng Island. The oil spreads east ward along the coast of Banggai towards to the north, but its limited spreading and not reaches to Luwuk City. Oil will be stranded in the coastal of Batui, Kintom, and Nambo which is only Batui coastal has mangrove area with bad criteria. Oil is spreading towards to Peleng Island, precisely in Buko which has no mangrove area, it's rockya sand beach without settlement area.

Oil spreading in the east season is mostly influenced by the east season, and oil spill moves towards to the southwest as shown in figure 2 (c). The mangrove areas affected by the oil spill are located along Batui and South Batui. In May and June oil spills around Batui Beach then extends to South Batui in July and finally began to decrease in August and September.

In the transition season 2, the influence of the western season is beginning and added with pressure from the west wind. The result of oil spreads to the east as shown in figure 2 (d). The thin layer of oil began to spread eastwards along the coast of Banggai towards to the north. However, its finite spreading and not reaches to Luwuk city and will be stranded along the coast of Batui and Kintom.

3.4. The lost value due to the oil spill

[9] mention that mangroves are particularly vulnerable to oil exposure and then can cause the death within weeks to months. This level of sensitivity depends on the type of mangrove and the type of oil. Light oils are more toxic compared to heavy oils. Oil with slow decomposition will be more toxic than fast decomposed. Poly Aromatic Hydrocarbon (PAH) is a toxic compound which causes the death of mangrove. The death begins with yellowing leaves then defoliation and ultimately cause to death of the tree. The accumulation of hydrocarbons in the substrate disrupts mangrove function. The results of research in Panama mention that the effects to mangrove growth are different in diverse concentrations of hydrocarbons [16].

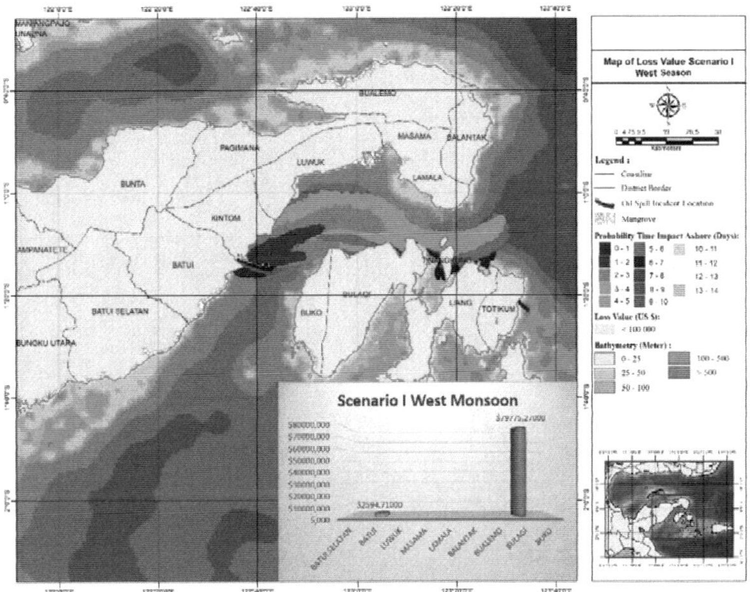
Map of Loss Value Scenario I
West Season
Scenario I West Monsoon
BUALEMO
MASAMA
BALANTAK
PAGIMANA
LUWUK
LAMALA
BUNTA
KINTOM
BATUI
BULAGI
BUKO
LIANG
TOTIKUM
BATUI SELATAN
BUNGKU UTARA

(a)

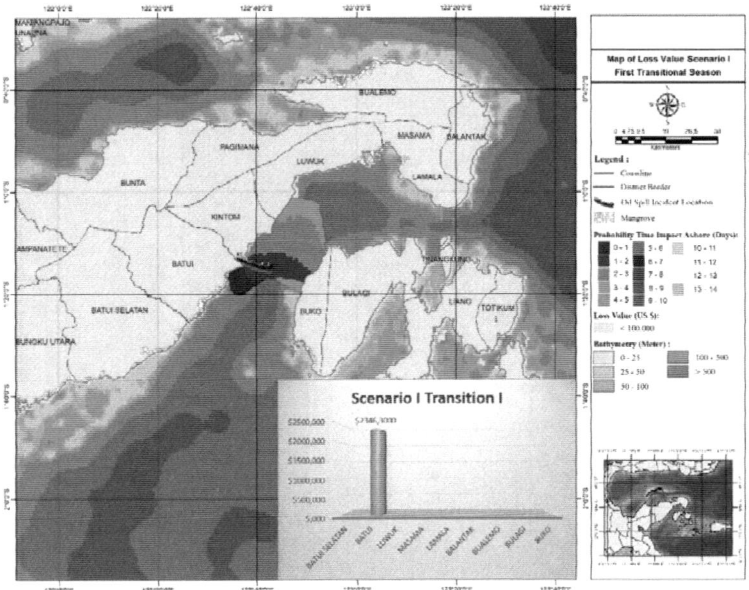
Map of Loss Value Scenario I
First Transitional Season
Scenario I Transition I
BUALEMO
MASAMA
BALANTAK
PAGIMANA
LUWUK
LAMALA
BUNTA
KINTOM
BATUI
BULAGI
BUKO
LIANG
TOTIKUM
BATUI SELATAN
BUNGKU UTARA

(b)

IOP Conf. Series: Earth and Environmental Science **176** (2018) 012043 doi:10.1088/1755-1315/176/1/012043

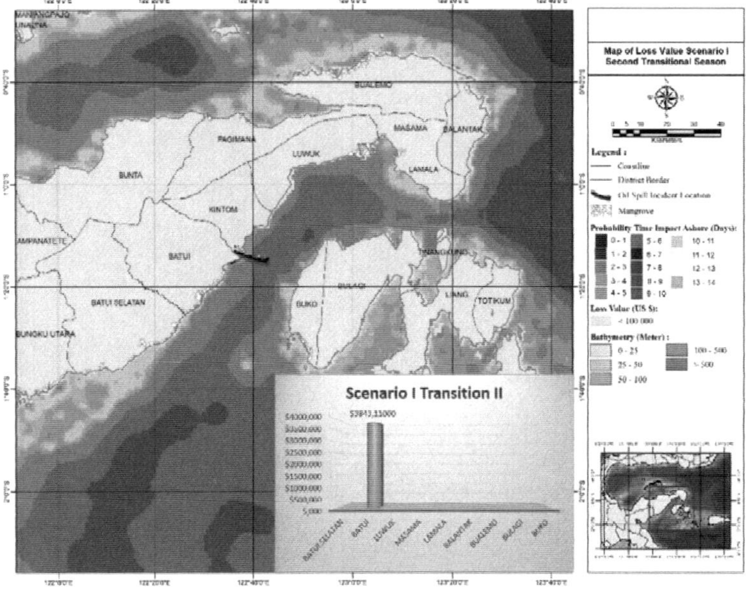

(c)

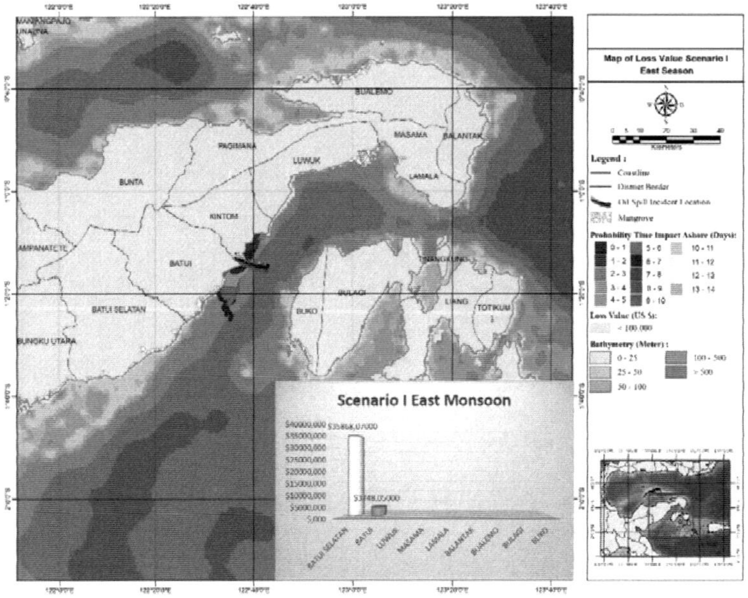

(d)

Figure 2. The oil spreading and lost value of the simulation in the four seasons (a) west season, (b) transitional season 1, (c) east season, (d) transitional season 2.

Table 4. The value of oil spill loss.

District	Loss Value by Season (US$)			
	West Season	Transition 1	East Season	Transition 2
South Batui	-	-	35,868.07	-
Batui	2,594.71	2,346.30	3,748.05	3,843.11
Bulagi	79,775.27	-	-	-
TOTAL	82,369.98	2,346.30	39,616.13	3,843.11

The total lost value due to the oil spill can be calculated = TEV (in affected areas) x affected mangrove area (%) x Response to damage/death (within one year). The simulation results (table 5): the biggest cost of an oil spill if occurs in the west season is US$ 79,775.27 in Bulagi while the lowest is in transition season 1 with US$ 2,346.30 in Batui.

4. Conclusion

There are 13 mangrove species found in Banggai and Banggai Islands Regency which *Rhizophora mucronata* is dominated almost found in all locations. Based on Ministry of Environment No. 201 the year 2004 regarding Standard and Guideline of Mangrove Degradation, the 6 locations which have a good category there are Balantak, Bualemo, Bulagi, South Batui and Lamala, while there is 3 location which has a bad category: Batui, Buko, and East Luwuk. Total Economic Value (TEV) result of mangrove ecosystem utilization is US$ 1,702,606 /year. The highest of economic valuation is derived from the mangrove benefit as breakwater which takes 72.67% of TEV, followed by the value of mangrove benefit as building materials is 25.73% and the lowest TEV as fishery products. It means that mangrove utilization as fishery resources has not been optimal by local community such as aquaculture and fishing. Oil spill simulation results by loss in scenario 1 is only US$ 82,369.98 which occurring in the west season.

References

[1] Meinarni N P S 2016 Dampak pencemaran lingkungan laut terhadap Indonesia akibat tumpahan minyak Montara di Laut Timor *Jurnal Komunikasi Hukum* **2**(2) 228-235
[2] Giesen, W, Wulffraat S, Zieren M and Scholten L 2006 Mangrove guidebook for Southeast Asia (Bangkok: FAO and Wetlands International) p 781
[3] Chopra S, Lines L, Schmitt D R and Batzle M 2010 Heavy-oil reservoirs: their characterization and production. Heavy Oils pp 1-69
[4] Sambu A H, Rahmi A and Khaeriyah 2014 Analysis of Characteristics of and Use Value of Mangrove Ecosystem (Case Study in Samataring and Tongketongke Sub-Districts, Sinjai Regency) *Journal of Environment and Ecology* **5**(2) 222-233
[5] Mangkay S D, Harahab N, Polii B and Soemarno 2013 Economic valuation of mangrove ecosystem in Tatapaan, South Minahasa, Indonesia *Journal of Environmental Science, Toxicology and Food Technology* **5**(6) 51-57
[6] Suzana B O L, Timbang J, Kaunang R and Ahmad F 2011 Valuasi ekonomi sumberdaya hutan mangrove di Desa Palaes Kecamatan Likupang Barat Kabupaten Minahasa Utara *ASE* **7**(2) 29-38
[7] Platon V, Frone S and Constantinescu A 2015 New developments in assessing forest ecosystem services in Romania *Journal Procedia Economics and Finance* **22** 45-54
[8] Ruitenbeek H J 1992 Mangrove management: an economic analysis of management options with a focus on Bintuni Bay, Irian Jaya. Jakarta and Halifax; Environmental Management Development in Indonesia Project (EMDI) p51
[9] Hoff R and Michel J 2014 *Oil Spills in Mangroves* (NOAA) p 97

[10] Minister of Environment Number. 201. 2004. Regarding Standard and Guideline of Mangrove Degradation
[11] SI 2012 Studi Indeks Kepekaan Lingkungan PT. Donggi Senoro LNG Jakarta pp 176
[12] Vo Q T, Kuenzer C, Vo Q M, Moder F and Oppelt N 2012 Review of valuation methods for mangrove ecosystem services *Journal Ecological Indicators* **23** 431-446
[13] Zen L W and Ulfah F 2014 Valuasi ekonomi hutan mangrove di Pulau Dompak Kota Tanjungpinang Provinsi Kepulauan Riau *Jurnal Dinamika Laut* **4**(1) 45-52
[14] Fadhila H, Saputra S W and Wijayanto D 2015 Nilai manfaat ekonomi ekosistem mangrove di Desa Kartika Jaya Kecamatan Patebon Kabupaten Kendal Jawa Tengah *Diponegoro Journal of Maquares* **4**(3) 180-187
[15] Indrayanti M D, Fahrudin A and Setiobudiandi I 2015 Penilaian jasa ekosistem mangrove di Teluk Blanakan Kabupaten Subang *Jurnal Ilmu Pertanian Indonesia* **20**(2) 91-96
[16] Duke N C, Pinzon Z S and Prada M S 1997 Large-scale damage to mangrove forests following two large oil spills in Panama *Biotropica* **29** 2-14

IOP Conf. Series: Earth and Environmental Science **176** (2018) 012044 doi:10.1088/1755-1315/176/1/012044

Application of multichannel seismic reflection method to measure temperature in Sulawesi Sea

R Fajaryanti[1]*, H M Manik[1]*, C Purwanto[2]

[1]Department of Marine and Science Techology, Bogor Agricultural University
Gedung Marine Center Jl. Agatis, No.1, FPIK, IPB Bogor 16680
[2]Marine Geological Institute of Indonesia
Jl. Dr. Djunjunan No.236 Pasteur- Bandung 40174
*e-mail : rinafajaryanti@gmail.com; henrymanik@ipb.ac.id

Abstract. Seismic oceanography is integrated discipline between seismology and physical oceanography that can be used to investigate phenomenon in water column. Seismic oceanography displays fine structure of water column seismically and in wide range. The aim of this research was to measure temperature in Sulawesi Sea with CTD and seismic data. The benefits of this research were to create integrity and effectiveness in seismic and oceanographic surveys as well as to help develop OTEC (Ocean Thermal Energy Conversion) studies in Indonesia. The acquisition of seismic and CTD data in line 17 Sulawesi Sea conducted by P3GL on Mei-Juni 2016. Data processing was conducted in Februari-April 2017 and divided into three processes: seismic data processing using ProMax 2D software, synthetic seismogram processing using Hampson-Russell software, and mapping temperature using Ocean Data View software. The results showed that temperature in Sulawesi Sea based on CTD data ranged 4.7-30.3 °C, and temperature based on the sesimic data ranged 4.49-30.29 °C. This difference occured due to difference in vertical sampling between CTD and sound velocity *semblance*. Acoustic impedance in Sulawesi Sea ranged 1.529-1.578 MRy. Vertical and horizontal resolution of seismic data in this research were 6.3 m and 93.5 m, respectively. It's well established that sound velocity the main contribution to acoustic impedance. Correlation between seismic field with synthetic seismogram from CTD data is 0.7.

1. Introduction

The information about physical oceanography can be obtained by satellite imagery, ground check, and acoustic survey. Thermohaline circulation is the primary forcing mechanism that keeps the ocean conveyor belt in perpetual motion, regarding ocean processes helps build a comprehensive understanding of the ocean's role in the distribution of heat around the planet and thereby affecting climate. Seismic oceanographer investigate oceanic thermohaline finestructure based on temperature and salinity profiles [1].

Seismic method is one of sea exploration method based on measurement of sound waves propagating on earth layer medium then reflected and refracted [2]. Seismic method has three application with different requirements for band-width and depth-width; engineering seismology, exploration seismology, and earthquake seismology. Engineering seismology is used to derive a velocity-depth model for near surface with bandwidth 1000 Hz and depth down to 1 km. *Exploration seismology* is used to derive an image of subsurface with bandwidth 100 Hz and depth of interest down to 10 km. Earthquake seismology is used to investigate oceanic and continental crust with bandwith 10 Hz and depth of interest down to 100 km [3].

Seismic oceanography is recently developed approach for investigating water column using seismic reflection method. This method utilizies vibration source from *air gun* and fired to water, the waves are exposed to the medium and energy. This propagation will be influenced by many factors, such as offset, attenuation, and other disturbances. Seismic waves are reflected and refracted by the boundary layer of the rock and received by the receiver. The recorded seismic data is data based time domain that is consisting of reflection waves, direct waves, and noise [2].

Over the past few years, multichannel seismic reflection method is used in petroleum industry to describe subsurface geological structure [4]. This method can be used to image oceanic termohaline finestructure as a reflection process in underwater that called seismic oceanography. [5] simulated seismic reflection and observation of physical oceanography in Norwegia Sea and established that this method have temperature sensitivity 0.03°C. In the same region, [6] demonstrated seismic reflection to express internal wave spectrum. This method provide great potential in ocean imaging and be able to observe ocean termohaline structure in great detail [7].

Seismic oceanography utilizies low-frequency sound waves (1-100 Hz) emitted by air gun to the water column then reflected by water column sturcture and received by the hydrophone to image internal oceanic structure. This work illustrates the great potential of multichannel finestructure of meddies with exceptional vertical resolution 10 m and horizontal resolution 100 m [8]. Several studies of seismic oceanoprahy have demonstrated the ability of seismic imaging to detect and map major features in the ocean, including thermohaline finestructure and water mass front [9], *internal waves* [5], currents [10], *eddies* in Cadiz Bay [10], and *boundary mixing* [6]. In Indonesia, this study was conducted by Wirda (2016) in detecting structure of water column in East Waigeo, Papua.

The aim of this research is to measure temperature in Sulawesi Sea using seismic data and Conductivity Temperature Depth (CTD) data. Seismic oceanography displays fine structure of water column seismically and in wide range. The application of seismic reflection is possible to create integrity and effectiveness in seismic and oceanographic surveys as well as help to develop OTEC (Ocean Thermal Energy Conversion) studies in Indonesia. Seismic oceanography in Indonesia has not been widely known. Through this research, it is expected to introduce and encourage the development of seismic oceanography in Indonesia.

2. Material and methods

The research area was located in Sulawesi Sea at line 17 (figure 1). Multichannel seismic and CTD data were acquired during May 2nd-June 3rd 2016 by Marine Geological Institute Indonesia in Sulawesi Sea. Air gun of 880 Cu.Inch were used for seismic source. A 900 m long streamter with 72 channels at an interval 12.5 m was used to receive the seismic signal. The near offset distance is 160 m and the data was recorded with a 2 ms sampling rate. The CTD data was collected in seicmic line is CTD 17 and CTD 20. The observed temperature and salinity profiles from CTD data were used to estimate the sound speed and density profiles which in turn are used for reflectivity modelling to generate synthetic seismogram.

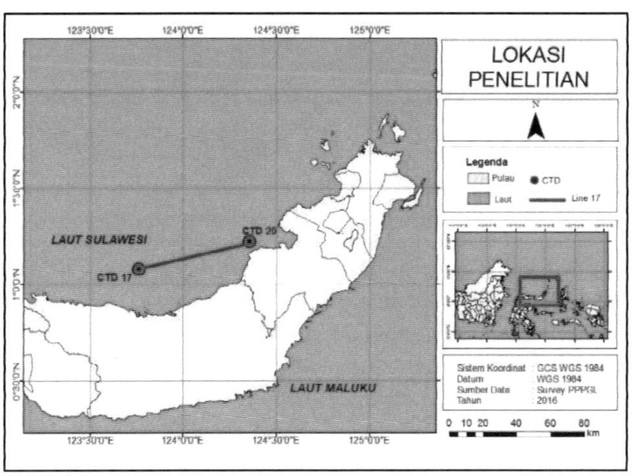

Figure 1. Location of seismic line 17 and CTD data in Sulawesi Sea.

Seismic data was processed wiith *ProMax 2D Version 5000.0.0.0 ©Landmark Graphic Corporation 1989-2008 All Right Reserved.* The process include input seismic data line 17 *.SEG-D into ProMax database. Subsequently, geometry assignment is performed to fit the field geometry into digital data. Information on geometry is the identity of seismic trace and becomes a vital attribute for the next process. Bandpass filter was used in this data using frequency 3-10-80-100 Hz to increase SNR. Trace muting is used to maximize observation of water column structure, by removing noise from direct wave and seafloor. In seismic oceanographic study, basin reflectors must be removed to improve the visibility of the water column reflector. Noise signals that cannot be seen periodically is separated using F-K filters, and convert time-distance domains (T-X) into frequency-waves number (F-K). The result is water column reflector in the form of a hyperbolic curve under direct arrival. Velocity analysis was processed to eliminate time lags due to separation of source and receiver on the surface and correct the speed of sound. Velocity analysis was performed every 100 CDP (CDP space is 6.25 m) with semblance method in each 8 points of picking velocity and extracted every 10 m depth. The resulting rms velocity is used for NMO (normal moveout) correction, ie the correction of travel time to eliminate the distance factor between the source and the receiver in one CDP and carry reflection waves from a dip reflection to a normal reflection. The next process is the stack by combining multiple traces into one trace. The result is a seismic cross-section of water column in the time domain. Seismic profiles are converted to depth using velocity estimated from CTD. Resolution of seismic profile can be estimated as :

$$L = \frac{c}{4\,f_o} \quad \text{and } R = \sqrt{\lambda h/2} \qquad (1)$$

where L represent vertical resolution (m), R represent horizontal resolution (m), c represent sound velocity (m/s), fo represent dominant frequency in seismic data (Hz), h represent depth of reflector (m), and λ represent wave length (m).

Synthetic seismogram was processed with *Hampson-Russel Software* (HRS). Generating synthetic seismogram from sonic and density log helps understanding the observed reflections in seismic data. The established empirical relationship are used to compute soundspeed from CTD data which are turn used to generate impedance (reflectivity) log. The impedance log is convolved with ricker wavelet to produce synthetic seismogram. Sound velocity varies with temperature, salinity, and depth and usually it ranges between 1490-1540 m/s (Sinha *et al.*, 2016). The empirical relationship by Wilson (1960) was used to calculate sound velocity from observed salinity and temperature profiles.

$$c = 1492.9\ + 3\,(T - 10) - 0.006\,(-10)^2\ 0.04\ \ (T - 18)^2 +\ 1.2\,(S - 35)\ 0.01\,(T - 18)\ (S - 35)\ \ + d/61 \qquad (2)$$

where c represent sound velocity (m/s), T represent temperature (°C), S represent salinity (psu), and d represent depth (m).

As sound wave travels through water column, it refracts and reflects from layers of different acoustic impedances. This layer property (i.e. acoustic impedance, Z) is calculated by multiplying density with soundspeed, $Z = \rho\, c$. The amplitude of reflected sound waves from different impedance layers is directly proportional to the reflection coefficient or reflectivity, determined by the impedance contrast between the two layers. For an incident normal sound wave, reflection coefficient is expressed as:

$$R = \frac{Z_2 - Z_1}{Z_2 + Z_1} \qquad (3)$$

Correlation was processed with *eLog* program in HRS using seismic stack data *.SEG Y to adjust CTD location (CTD 17 in CDP 8751 and CTD 20 in CDP 1751). The right wavelet phase in CTD data was processed by extracting statistical wavelet so that zero phase wavelet has the same amplitud with seismic data. A good correlation between synthetic seismogram and seismic reflection was 0.7.

Water mass in Sulawesi Sea can be observed by T-S diagram. T-S diagram was processed with *Ocean Data View* (ODV) 4. The spreadsheet file consist of ocean parameters like depth, temperature, salinity, latitude, and longitude was imported to ODV database then source variable is associated with meta variable. Salinity and Potential Temperatur profile is used to generate T-S diagram.

IOP Conf. Series: Earth and Environmental Science **176** (2018) 012044 doi:10.1088/1755-1315/176/1/012044

3. Result and discussion

Ocean imaging using reflection seismic method produce seismic profile whose quality depends on data processing, particularly sound velocity. Seismic oceanography is optimally obtained from sound velocity data approaching the actual conditions at sea. Reflection in water column is a manifestation of the difference in acoustic impedance generated by thermohaline structure. Accuration of sound velocity data means easily to interpret seismic profile.

According to [10], sound velocity in the sea ranged from 1490-1540 m/s and it decreases to depth relatively. Velocity analysis show that the sound velocity in Sulawesi Sea ranges from 1470-1540 m/s. Sound velocity profile shows the water column structure stratified into 3 layers (figure 2). The upper layer is red at 0-300 m depth has sound velocity of 1515-1540 m/s, the green middle layer at depth 310-550 m has sound velocity of 1495-1515 m/s, and the bottom layer is blue at depth 560-1050 m has sound velocity of 1470-1485 m/s. The sound velocity is used for NMO correction until the seismic profile is obtained in figure 3.

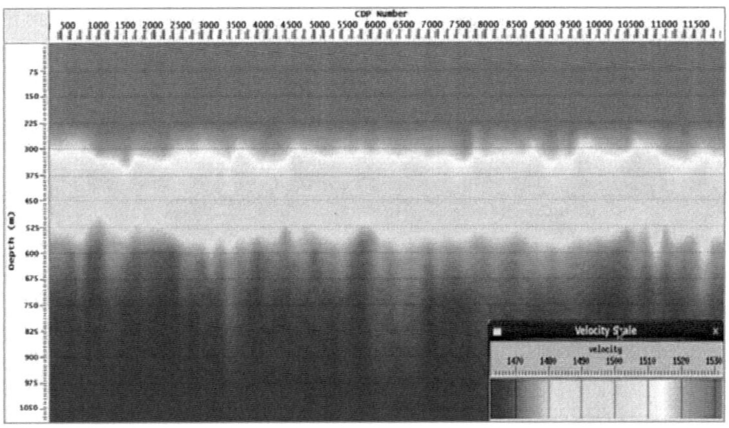

Figure 2. Sound velocity profile derived from semblance analysis shows stratification in Sulawesi Sea.

The maximum depth in Sulawesi Sea was varied from 1200-1400. Seafloor reflector is removed by trace muting process. figure 3 shows that stratification in water column formed by presence reflected seismic waves. The seismic waves are reflected by water column layers which is formed by difference in acoustic impedance. Those layers were recorded as reflectors depicted in red and blue seismic profil. The red reflector represent negative amplitude while the blue one represent positive amplitude. The lighter color represent the greater reflector amplitude value.

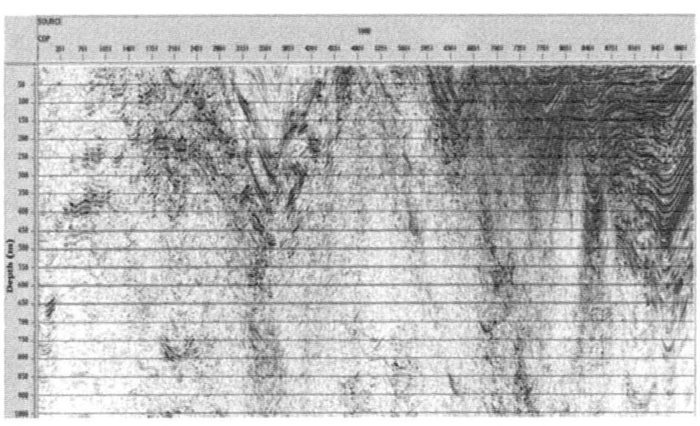

Figure 3. Seismic profile derived from semblance velocity analysis.

Acoustic reflector was clearly observed at 50-700 m depth. Most of the strong and continuous reflectors are confined in CDP 1401-4551 at 100-450 m depth and CDP 5951-9801 depth 50-550 m. Blur reflectors were confined in CDP 1-6301 at 500-700 m depth, CDP 1-6301 at 750-1000 m depth along the seismic line. Strong reflectors represent a relatively large acoustic impedance contrast. This difference is assumed due to significant changes in temperature and salinity [10].

Seismic profile can also be obtained from other sound velocity data. [11] made the seismic profile of Norwegia Sea derived from constant sound velocity (1500 m/s), sound velocity form Levitus database, sound velocity of CTD, and sound velocity from semblance analysis. [10] made seismic profile of Teluk Bengal using sound velocity obtained from CTD. In this research, seismic profile is derived from semblance sound velocity and CTD data. Seismic profile derived from sound velocity of CTD is shown in figure 4.

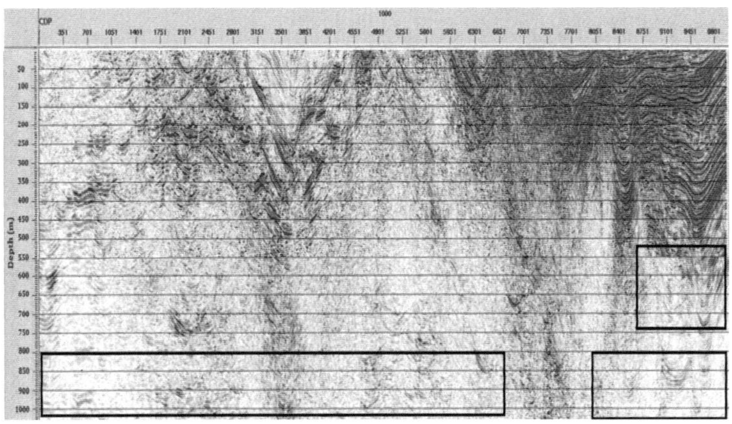

Figure 4. Seismic profile derived from sound velocity of CTD.

Seismic profile derived from sound velocity of CTD data is able to show the stratifictaion in Sulawesi Sea similar to the seismic profile in figure 3. However, in figure 4 shows blur reflector (black polygon area) so that the number of observed reflector decreases, ie CDP 8401-9801 at 500-750 m depth, CDP 1-6851and CDP 8051-9801 at 800-1000 m depth. According to [11], the decrease number reflectors are influenced by the lateral sampling of sound velocity profil. Sound velocity picked line wth lateral spacing ~8 km. Therefore, the seismic profile derived from semblance sound velocity is better and more clear in describing lateral variations more subtly that sound velocity of CTD data.

Seismic profile generated in this study has a horizontal resolution of 93.5 m and a vertical resolution of 6.3 m. The dominant frequency of seimic data is 60 Hz and the reflector is still clearly observed at 700 m depth. At a depth more that 700 m, the reflector blurred.

Seismic stratification in Sulawesi Sea is related to the temperature profile [1]. The sudden change in temperature and sound velocity (spike) has correlation to the layers in water column. According to [5], reflector arise when temepratture gradien is 4-8 4-8 °C. In Sulawesi Sea, spike at 100-200 m depth represent reflector occured in the same depth of seismic profile (figure 6).

Figure 5 shows the internal wave phenomenon at 100-450 m depth (black polygon area) with a flat reflector turns slightly wider with a high vertical displacement. Internal wave is ocean interior wave due to changes in hydrostatic balance [13]. Internal waves are generated when interfaces between layers of different water densities are disrupted, usually due to differences in temperature and/or salinity [12]. In addition, tilted reflector is observed at 150-500 m depth (black ellips area). Tilted reflectors arise from water *mass front*, *eddy*, *upwelling*, *downwelling*, etc. The tilted reflector that found in Sulawesi Sea is thought to be a water mass front that denotes two water masses distinct meet clearly.

Based on figure 7, salinity value in Sulawesi Sea from CTD ranged from 34.3-34.6 psu. Salinity profile shows that Sulawesi Sea divided into 3 layers, ie mixed layer, halocline layer, and deep layer. In the mixed layer at 0-40 m depth is 34.3 psu. In halocline at 50-450 m depth is 34.6 psu. Halocline is area within a body of water that marks a drastic change in salinity. Maximum salinity water from *North Pacific Subtropical Water* (NPSW) and minimum salinity water from *North Pacific Intermediate Water* (NPIW) meet within the body water of Sulawesi Sea [13]. In the deep layer, salinity ranged 34.5 psu and tends to increase slowly to the bottom of the water.

Based on figure 7, temperature in Sulawesi Sea from CTD ranged from 4.7-30.3 °C. Temperature profile shows that Sulawesi Sea divided into 3 layers, ie mixed layer, thermocline layer, and deep layer. Temperature in mixed layer at 0-50 m depth is 30 °C and tends to be homogeneous due to the mixing of water mass influenced by winds, currents, and tides. Salinity in thermocline at 50-250 m depth is 18 °C. In the deep layer, temperature ranged from 4.7 °C and tends to decrease slowly to the bottom of the water. Figure 8 shows the temperature stratification occured to 700 m depth and the rest relatively homogeneous. Reflector in seismic profile can be observed to a maximum depth of 700 m. Small gradient of temperature and relatively uniform temperature distribution at depths >700 m resulted blur reflector on seismic profile.

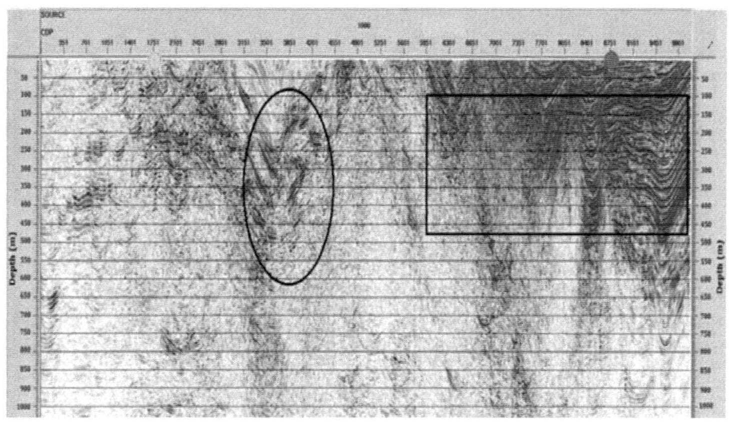

Figure 5. Seismic profil of Sulawesi Sea shows internal wave (black polygon area) and water mass *front* (black ellips area). Green symbol is CTD 20 site and the red one is CTD 17 site.

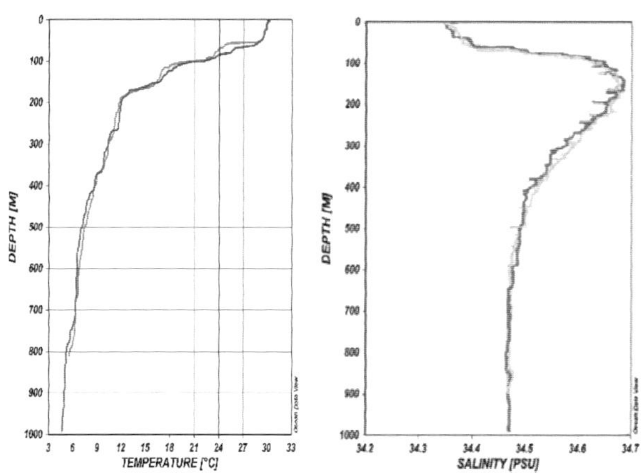

Figure 6. Temperature and salinity profile derived from CTD 17 (red) and CTD 20 (green).

IOP Conf. Series: Earth and Environmental Science **176** (2018) 012044 doi:10.1088/1755-1315/176/1/012044

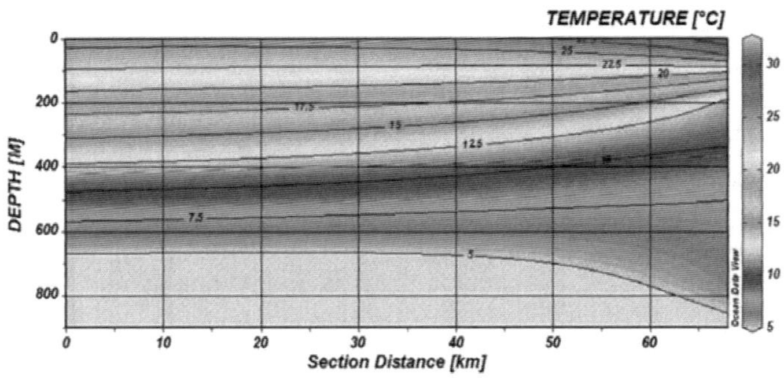

Figure 7. Temperature lateral profile of Sulawesi Sea derived from CTD data.

The properties of sea water can be plotted in one diagram to describe the characteristic of water mass. According to figure 8a, water mass of Sulawesi Sea derives from North Pacific Ocean and fills the upper thermocline layer, ie NPSW and NPIW. This water mass carried by Northern Equatorial Flow to eastern waters of Indonesia. The water current moves toward Phillipines and branches north into Kurushio current and south into Minandao stream. Some Minandao current move towards Sulawesi Sea and Makassar Strait, carrying NPSW with salinity 34.6-34.8 psu and temperature 15.79-23.83 °C at 120-200 m depth. At the bottom layer, NPIW with salinity 34.4-34.5 psu and temperaturr 5-7 °C at 250-400 m depth is transported [13]. Sulawesi sea is the part of Indonesia Through Flow (ITF), which is carry warm water from Pacific Ocean to cooler water from Indian Ocean. [13] states that during eastern monsoon Souther Equatorial Flow carried water mass along the northern Papua to the western and join the water mass that carried by Northern Equatorial Flow from north coastal of Sulawesi. Water mass characteristic in Sulawesi Sea during eastern monsoon is salinity of 33.8–34.4 psu and temperature 30 °C, while during western monsoon the salinity is 33 psu and temperature is lower. This changes is due to changes of surface current. Some of Sulawesi Sea current flowed into Sulu Sea and partly into the Makassar Strait.

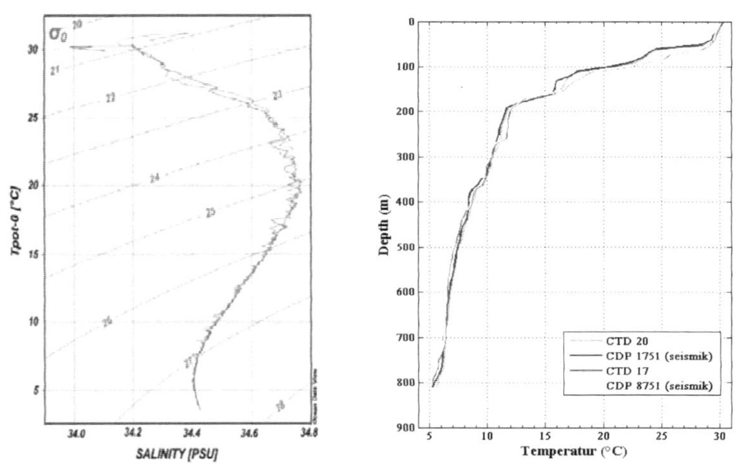

(a) (b)

Figure 8. (a) T-S diagram Sulawesi Sea from CTD data; (b) temperature profile for CTD and seismic data.

Temperature data can also be obtained from seismic sound velocity data using empirical relationship stated by Wilson that is calculated with Matlab *software*. Temperature was derived from seismic data ranged from 4.49-30.29 °C. Temperature profile derived from CTD and seismic data can be seen at figure 8b. According to figure 8b, temperature profile derived from seismic data is relatively similar to the temperature profile based on CTD data. Temperature from seismic data is smaller than CTD with the difference is 0.19 °C. This difference was caused by vertical sampling, CTD data was sampled each 1 m depth, while the sound velocity by seismic data is extracted by 10 m. According to [14], reflection seismic method is sensitive to the vertical sound velocity rather that the value of sound velocity itself. The propagation time is sensitive to average sound velocity but reflection amplitude is sensitive to vertical changes. Synthetic seismogram is used to see the correlation between CTD and seismic data. Density and sound velocity form the function of reflection coefficient to be convolved with wavelet. Density of Sulawesi Sea was 1021-1031 kg/m^3 and does not vary much to depth. Acoustic impedance in Sulawesi Sea is more influenced by sound velocity, while the sound velocity is more influenced by temperature. According to figure 10, salinity changes affect the change of synthetic seismogram character at 100-200 m depth, while temperature changes affect the change of synthetic seismogram character at 100-600 m depth. [15] estimates that reflection seismic derived from 90-95% sound velocity and 5-10% density and the contribution of temperature is 80% while salinity is 20%. [8] is also mention that the contribution of temperature is 83% and salinity is 17%.

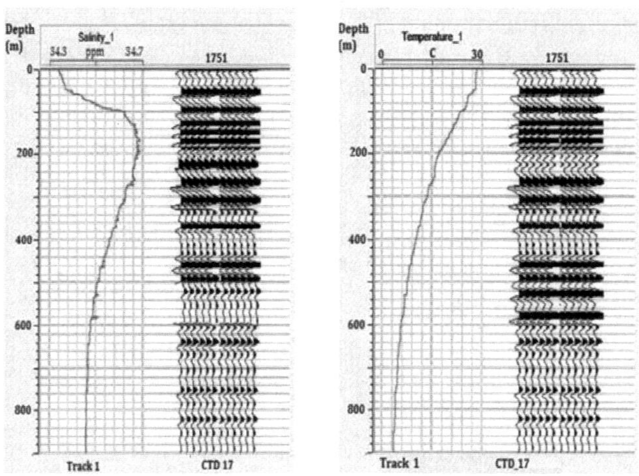

Figure 9. Synthetic trace of CTD data based on temperature and salinity profile.

Acoustic impedance profile is obtained by seismic inversion using seismic data as an input and log CTD data as control to invert into reflectivity data [16]. The impedance layers are essentially describe the thermohaline structures. Acoustic impedance is Sulawesi Sea ranged from 1.529-1.578 MRy. In figure 11, acoustic impedance at 0-400 m depth is relatively high that is characterized by blue-purple impedance color, while in the deeper layer (>500 m depth) the acoustic impedance weakens that is characterized by yellow impedance color. The layers with weak acoustic impedance is caused by a decrease in temperature and salinity [10].

The function of the reflection coefficient is convolved with the wavelet to form a synthetic seismogram. According to [3], wavelet consists of many components including source signature, recording filter, surface reflection, and geophone response. In this research, wavelet is formed from zero phase wavelet corrected their amplitude against seismic data. The resulting wavelet has length of 200 ms, taper length of 25 ms, sample rate of 2 ms, phase rotation 00 and constant phase. This synthetic seismogram is matched wth seismic data or composite trace to obtain good correlation. The final synthetic seismogram has length wavelet of 150 ms, taper length of 20 ms, and sample rate of 2 ms (figure 12). Synthetic seismogram is represent by red trace over seismic trace around the CDT location. The correlation between field seismic and synthetic seismic using CTD 17 and CTD 20 respectively is 0.731 and 0.72. This value indicateds a good or close relationship.

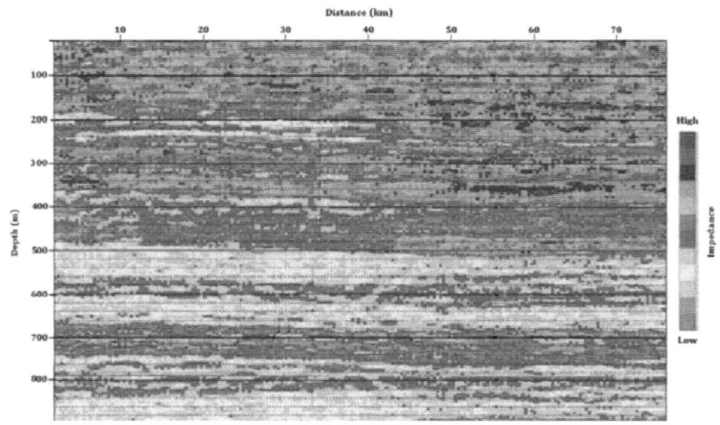

Figure 10. Acoustic impedance in Sulawesi Sea.

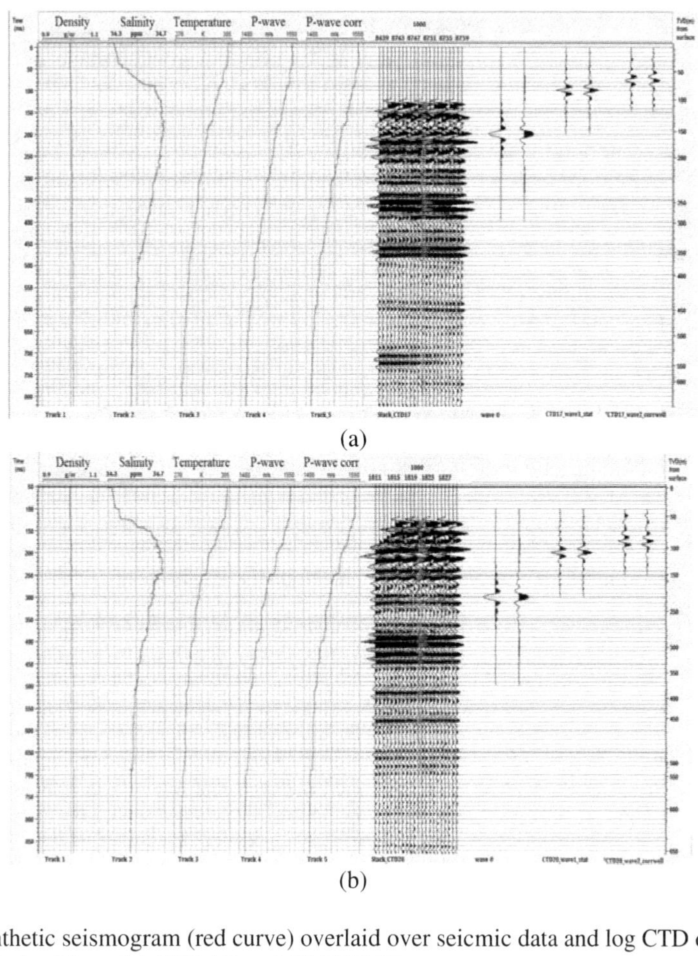

(a)

(b)

Figure 11. Synthetic seismogram (red curve) overlaid over seicmic data and log CTD data in Sulawesi Sea derived from (a) CTD 17; and (b) CTD 20.

4. Conclusion

Seismic reflection method was used to identify water column stratification. The seismic profile indicate the presence of internal wave and water mass front caused by encountering of two water mass with temperature and salinity distinc. Temperatur Sulawesi based on CTD data is ranged 4.7-30.3 °C while the temperature derived from seismic data is 4.49-30.29 °C. Correlation between syntetic seismogram of CTD data and seismic data is 0.7.

Acknowledgement

Acknowledgement delivered to Marine Geological Institute of Indonesia in Bandung for the data access and data processing permission in Seismic Laboratory.

References

[1] Nakamura Y T, Noguchi T, Tsuji S, Itoh H, Niino and Matsuoka 2006 Simultaneous seismic reflection and physical oceanographic observations of oceanic fine structure in the Kuroshio extension front *Geophys. Res. Let.* **33** L23605

[2] Hasanudin M 2005 Teknologi seismik refleksi untuk eksplorasi minyak dan gas *Oceana* **30** 1-11

[3] Yilmaz Q 2001. *Seismic Data Analysis: Processing, Inversion, and Interpretation of Seismik Data Vol I, Invest. Geophys* 2nd *ed.* (Tulsa: Society of Exploration Geophysis)

[4] Song H, Pinheiro L M, Ruddick B and Huang X 2012 Seismic oceanography: a new geophysical tool to investigate the thermohaline structure of the oceans, In: Marcelli, M., editor. *Oceanography*; 2012 Maret 23; Rijeka, Croatia (Rijeka: Intech) p 113-128

[5] Nandi P, Holbrook W S, Piramo S, Pearse and Schmitt R W 2004 Seismic reflection imaging of water as boundaries in the Norwegian Sea *Geophys. Res. Lett.* **31** L23311

[6] Holbrook W S, Fer I and Schmitt R W 2009 Images of internal tides near the Norwegian continental slope *Geophys. Res. Let.* **36** L00D10 doi: 10.1029/2009GL038909

[7] Buffett G G 2011 Seismic Oceanography: a new tool to characterize physical oceanographic structures and processes [tesis] (Barcelona: Universitat de Barcelona)

[8] Ruddick B, Song H B, Dong C Z and Pinheiro L 2009 Water column seismic image as maps of temperature gradient *Oceanography* **22**(1) 192-205

[9] Holbrook W S, Parametero P, Pearse S and Schmitt R W 2003 Thermohaline fine structure in an oceanographic front from seismic reflection profiling *Science* **301** 821-824

[10] Sinha S K, Dewangan P and Sain K 2016 Acoustic reflections in the water column of Krishna-Godavari offshore basin, Bay of Bengal *J. Acoust. Soc. Am.* **139** 2424-2431

[11] Fortin F J W and Holbrook W S 2009 Sound speed requirements for optimal imaging of seismic oceanography data *Geophys. Res. Let.* **36** 1-6 doi: 10.1029/2009GL038991

[12] Holbrook W S, and Fer I 2005 Ocean internal wave spectra inferred from seismic reflection transects. *Geophysical Research Letters* **32** L023733

[13] Wyrtki K 1961 *Physical Oceanography of Southeast Asian Waters. Naga Report 2* (La Jolla: University of California)

[14] Fer I, and Holbrook W S 2008 Seismic reflection methods for study of the water column. *Encyclopedia of Ocean Sciences* **00799** 1-11

[15] Biescas B, Sallares V, Pelegri J L, Machin F, Carbonell R, Buffet G, Andobeitia J J and Calahorrano A 2008 Imagine meddy finestructure using multichannel seismic reflection data *Geophys. Res. Lett.* **35** L11609 doi:10.1029/ 2008GL033971

[16] Sukmono S 2007 *Advance Seismic Methods for Field Exploration & Developments.* (Bandung: Institut Teknologi Bandung)

IOP Conf. Series: Earth and Environmental Science **176** (2018) 012045 doi:10.1088/1755-1315/176/1/012045

Diversity of meiofauna and its association to seagrass beds characteristics in Pramuka Island, Seribu Islands

M Rizqydiani*, **M S Ismet**, **D G Bengen**

Department of Marine Science and Technology
Faculty of Fisheries and Marine Science, Bogor Agricultural University
*e-mail: mutiarizqydiani@gmail.com

Abstract. Seagrass beds is one of the coastal ecosystem that has many biological benefits for other marine biota, like fishes, shrimps, and shells. Those marine biotas take advantages from seagrass beds as habitat for their communities. One example of marine biota that associated with seagrass beds is meiofauna. Meiofauna is an organism with size between 0.063 to 1 mm and live interstitially between substrate cavities. Ecologically, meiofauna plays important roles as mineral decomposer and food for bigger organisms. This research aims to study the relation between the diversity of meiofauna associated with seagrass beds in Pramuka Island, Seribu Islands. The research was conducted in 3 stations, each consists of 15 meiofauna sampling points. The samples were collected by using PVC core that embedded in the substrate. Meiofauna that has been found in research stations was consisting of 6 phyla, 8 classes, 17 orders, 31 families, and 35 genera. The most abundant phyla was Nematodes. Meiofauna has higher abundance in finer substrate which contain more organic material. Aside from that, the abundance of meiofauna was also affected by the seagrass density.

1. Introduction

Meiofauna is a sedimentarial organisms that can be associated in the seagrass ecosystem. It has individual size ranged between 0.044 to 1 mm and live in bottom layer of soft substrate. It is known to be the most abundant benthic group in marine ecosystem [1]. The characteristic of meiofauna is that they live interstitially on substrate cavities. The life cycle of meiofauna, specifically Foraminifera, plays an important role in the term of biogeochemical cycle and energy flow. Other than that, meiofaunal groups Foraminifera and Nematode are known as potential bioindicators of environmental quality [1, 3].

One of the factors that affect the availability of meiofauna in seagrass beds is seagrass density and other environment variables such as current, substrate types, pH and Eh of substrate, oxygen availability, and organic material in the substrate [4]. Pramuka Island has a community of seagrass that grows on sand substrate. The type of substrate is one of most important factors because it is related to oxygen and organic material concentration. This study aims to learn about the diversity of meiofauna and its association to seagrass beds' characteristics in Pramuka Island, Seribu Islands.

2. Methods

The data sampling was conducted on March 13–17, 2017 in Pramuka Island, Seribu Islands. There were three sampling stations that was set based on representation of seagrass ecosystem (figure 1). Laboratorium analysis was conducted in Marine Biology Laboratory, Bogor Agricultural University during April–June 2017.

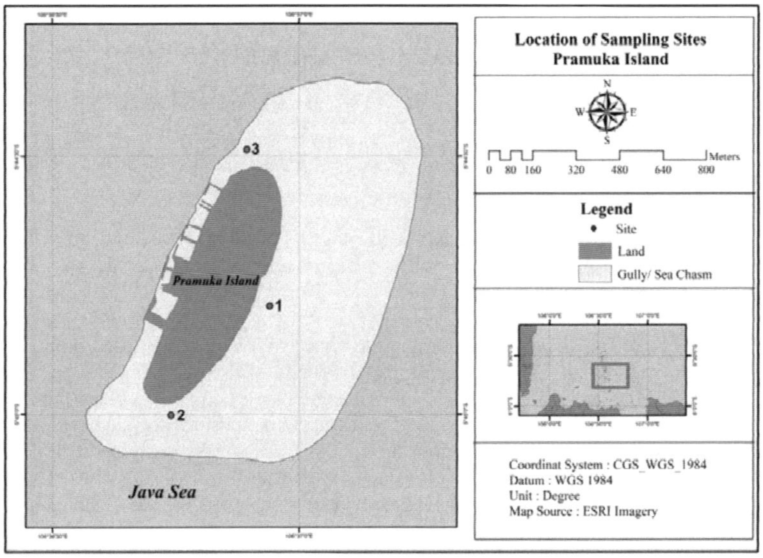

Figure 1. Location of sampling stations.

2.1. Sampling procedure

2.1.1. Seagrass density. The observation of seagrass density was done using *Seagrass Watch* method [5]. In each sampling stations there were three 50 m line transects with 10 m distance between them. In each line transect was placed 1 m x 1 m quadratic transect to observe the seagrass density (figure 2).

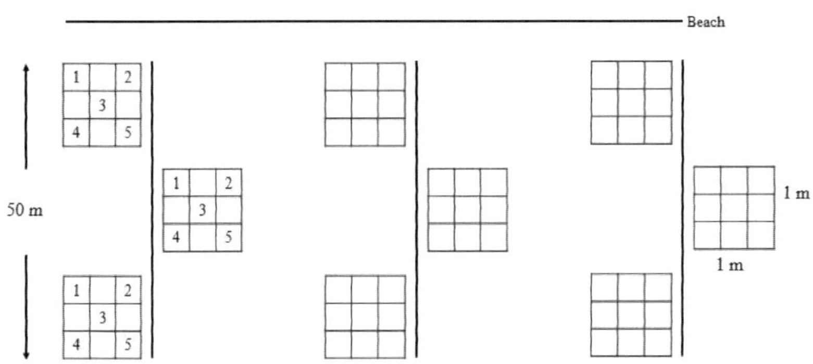

Figure 2. The layout of sampling station.

2.1.2. Meiofauna. Sampling of meiofauna was conducted using PVC core (ø: 5 cm; height: 30 cm) that embedded in the substrate. Then the substrate was strained using 1 mm mesh size strainer. After that, the sample was given solution of rose bengal and 4% formaldehyde to stain and preserve it so it would be easier to identify. The identification process was performed using a light microscope and book *Introduction to the Study of Meiofauna* [6].

IOP Conf. Series: Earth and Environmental Science **176** (2018) 012045 doi:10.1088/1755-1315/176/1/012045

2.2. *Analysis Procedure*

2.2.1. Classification of bottom substrate. The type of bottom substrate was classified based on Wenworth Scale [7] into three classes of sand : coarse, medium, and fine. The determination of substrate classification was done by using sieve shaker. Dried sample weighed as much as 20 grams, then sieved for about 10 minutes.

2.2.2. Seagrass density. The seagrass density in each sampling stations was calculated using formula below [8] :

$$\text{Density}_i = \frac{\text{Total individual}_i}{A} \quad (1)$$

2.2.3. Abundance of meiofauna. The abundance of meiofauna in each sampling station was calculated using formula below [8]:

$$\text{Abundance} = \frac{10000 \text{ x total counted individual}}{\text{area}} \quad (2)$$

2.2.4. Diversity index. The diversity index was calculated using formula below [8]:

$$H' = -\sum \text{pilog}_2 \text{ pi} \quad (3)$$

2.2.5. Evenness index. The evenness index was calculated using formula below [8]:

$$E = \frac{H'}{H'\text{max}} \quad (4)$$

2.2.6. Environmental data. Environmental data collected in this research including TOC, total phosphate, and nitrate concentration. The Walkley and Black method [9] was used to determine total organic carbon. While to determine nitrate and total phosphate concentration, Kjeldahl destillation and HCl extraction method were used respectively [10, 11].

2.2.7. Association of meiofauna with seagrass beds' characteristics. The association of meiofauna with seagrass density and substrate characteristis was analyzed using principal component analysis and correspondence analysis on *XLStat 2016* software. Principal component analysis is factorial analysis method that enable easier data interpretation by using essential information. Displayed matrix consists of sampling stations as individual (row), then seagrass density and substrate types as quantitative variable (column). This analysis was conducted to know the difference of seagrass distribution and substrate type in each sampling stations.
Correspondence analysis is factorial analysis that groups statistical units into homogenous groups from some variables or characters. Variables tested in correspondence analysis was the abundance of meiofauna in each sampling stations.

3. Results and discussion

3.1. *Classification of bottom substrate*

The bottom substrate was classified in to three classes : coarse sand, medium sand, and fine sand. Station 1 consists of rough sand, medium sand, and fine sand with similar percentages. Station 2 is dominated by rough sand (47.35%), while Station 3 is dominated by fine sand (51.88%) (figure 3). Coarse grained sand substrate has higher oxygen concentration compared to fine grained substrate, but it contains less organic material [4].

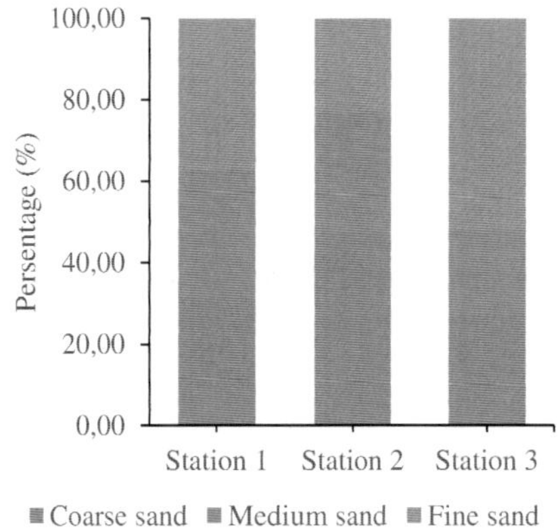

Figure 3. Classification of bottom substrate.

3.2. Seagrass composition in Pramuka Island

There were 6 seagrass species that found during the sampling in Pramuka Island namely *Thalassia hemprichii, Cymodocea serrulata, Cymodocea rotundata, Halophila ovalis, Enhalus acoroides,* and *Syringodium isoetifolium.* Nevertheless, *Thalassia hemprichii* was the most abundant species in all three sampling stations, with composition value 75.82% (Station 1), 56.20% (Station 2), and 61.84% (Station 3). Followed by *C. serrulata* with value 18.67% (Station 1), 32.46% (Station 2), and 29.80% (Station 3) (figure 4).

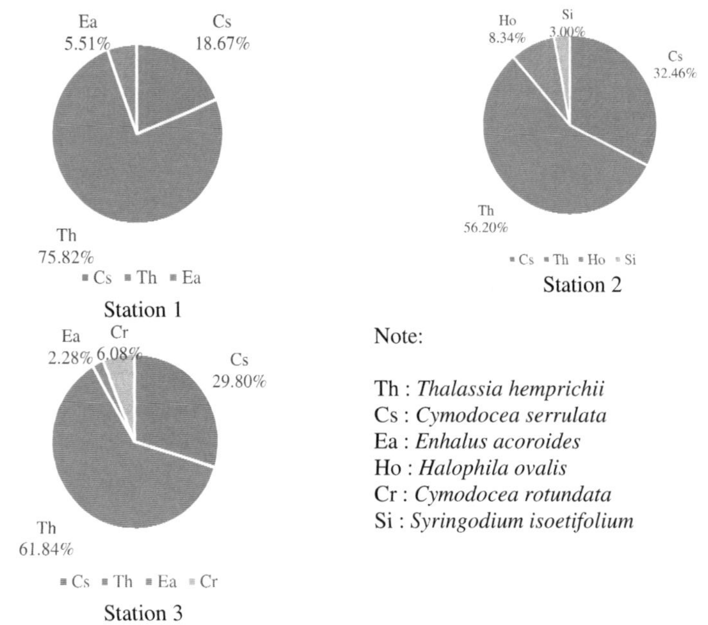

Figure 4. Seagrass composition in Pramuka Island.

3.3. Seagrass density in Pramuka Island

Seagrass *Thalassia hemprichii* has highest density value in all three sampling stations (figure 5). Thus, it indicates that *Thalassia hemprichii* is more adaptive to the aquatic environment compared to the other seagrass species. Density values of *Thalassia hemprichii* respectively are 94 ind/m^2, 243 ind/m^2, and 166 ind/m^2.

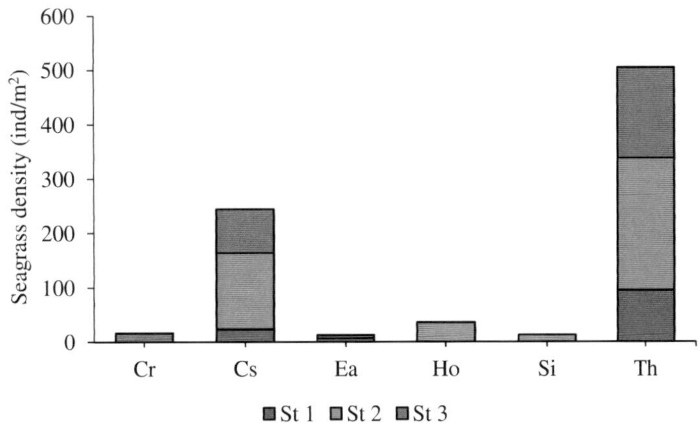

Figure 5. Seagrass density in each sampling stations.

3.4. The abundance of meiofauna

Meiofauna found in the seagrass beds in Pramuka Island consists of 6 phyla, 8 classes, 17 orders, 31 families, and 35 genera. The highest abundance of meiofauna was found in Station 3 (figure 6). It is considered that the high abundance in Station 3 was due to fine sand domination, which also contain more organic material. Nematode was the most abundant phyla in each sampling station, while Ciliophora was the most scarce phyla. According to [12], Nematode and Arthropod are the most common meiofauna to be found in sand substrate, although Nematode is actually able to adapt to different substrate types. [13] also explained that Nematode is the most viable meiofaunal group in low oxygen and high organic condition.

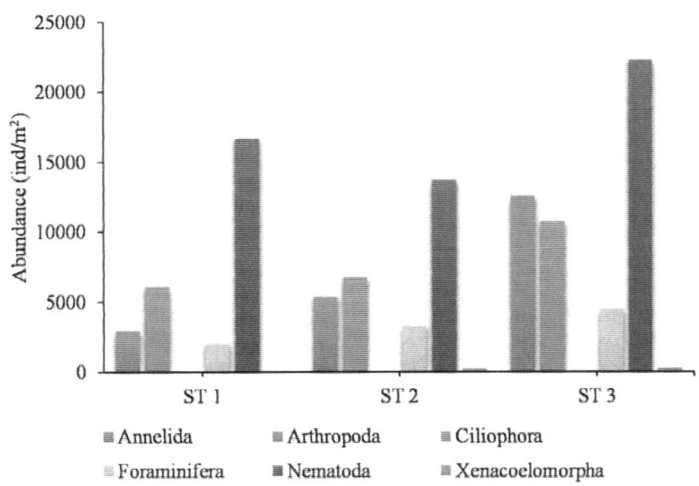

Figure 6. The abundance of meiofauna in each sampling station.

3.5. Environmental Data

Table 1. TOC and nutrients in each sampling station.

No	Variables	Sampling Stations		
		Station 1	Station 2	Station 3
1.	TOC (%)	0.33	0.23	0.48
2.	P-total (ppm)	37.62	55.73	49.19
3.	Nitrate (mg/l)	2.16	3.37	2.80

The result of analysis in table 1 showed that substrate from Station 3 contained the highest total organic carbon, which is 0.48%. This is linear with the highest abundance of meiofauna in Station 3. The result for nutrient analysis showed different result, as the highest nutrient content was in Station 2. This is also linear with the result of seagrass density calculation, which showed that Station 2 has the most dense seagrass vegetation. Nitrate is considered as one of the primary nutrient for seagrass growth. Nitrate is resulted from decomposition of organic, which involved the reaction of ammonia and bacteria. This process produces nitrate that is used to support seagrass growth [14].

3.6. Diversity and evenness index of meiofauna

Diversity index in Station 1, 2, and 3 respectively are 3.91, 3.95, and 4.55, thus the diversity of meiofauna in Pramuka Island can be categorized as high category (table 2). It can be said that meiofauna in Pramuka Island is evenly distributed and there is no dominance of any species. All three sampling stations have evenness index respectively 0.83, 0.81, and 0.89. The value can be interpreted that the meiofauna community in Pramuka Island is tend to be stable [15].

Table 2. Diversity and evenness index of meiofauna.

Station	H'	E
1	3.91	0.83
2	3.95	0.81
3	4.55	0.89

3.7. Association of Meiofauna with Seagrass Beds' Characteristics

The result of PCA analysis showed that the distribution of substrate types and seagrass in all sampling stations can be explained through axis 1 (F1) and axis 2 (F2) with total variance 70.65% (figure 7). At axis 1 (50.68%), it showed that Station 2 is dominated by *Thalassia hemprichii, Syringodium isoetifolium,* and *Halophila ovalis* that distribute in mixed substrate consists of medium and fine sand. [16] explained that *Thalassia hemprichii* is able to grow well in sand substrate. Otherwise, in axis 1 *Enhalus acoroides* is found in Station 3 which has fine sand dominated substrate. *Enhalus acoroides* tends to grow well on fine or muddy substrate [17]. It is due to the ability of fine or muddy substrate to bind more organic material than coarser substrate [18]. 2013). Axis 2 (F2) showed that *Cymodocea serrulata* and *Cymodocea rotundata* commonly found in Station 3 which has fine sand dominated substrate. [16] explained that fine sand with few mud can supports the growth of *Cymodocea serrulata.*

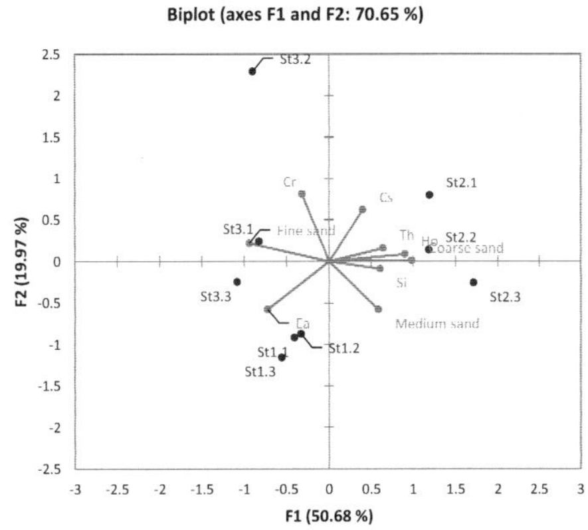

Figure 7. Results of principal component analysis.

The result of the correspondence analysis showed the distribution of meiofauna at sampling stations centered on axis 1 and 2 with total variance of 87.13% (figure 8). Nematode, Arthropod, Foraminifera, and Xenacoelomorpha are commonly found in Station 1 and 2 which have mixed substrate of coarse and medium sand. Nematode and Arthropod are common meiofauna that usually found on sandy beaches [12]. Nematode has highest abundance at the study station, followed by Arthropod and Foraminifera.

Annelid is commonly found in Station 1 and 3. Similarly with Nematode, members of the Annelid phylum generally found on finer substrate. Types commonly found in marine waters are from Polychaeta class, which inhabit various types of substrates such as mud, sand, or corals [19]. Ciliophora is found only in Station 3 with very small number, so it is not characterized on any station (figure 8).

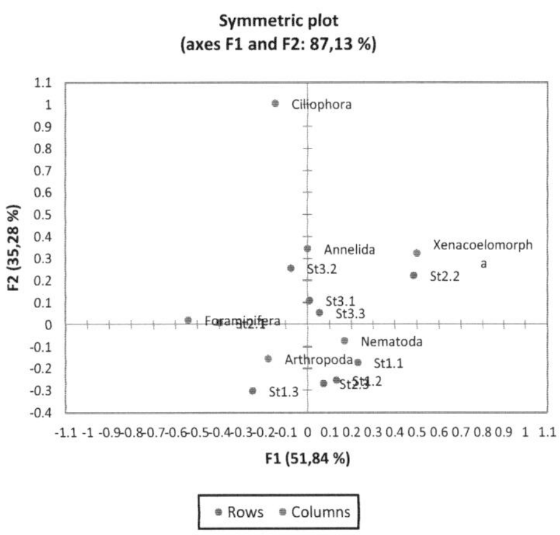

Figure 8. Results of correspondence analysis.

4. Conclusion

Meiofauna associated with seagrass beds in Pramuka Island consists of 6 phyla, 8 classes, 17 orders, 31 families, and 35 genera. The abundance of meiofauna is higher in seagrass beds with finer substrate. It is due to the ability of finer substrate to bind more organic material that become one of the factors that support the increase of meiofauna in seagrass beds.

Acknowledgment
This research was fully funded by Ministry of Research, Technology, and Higher Education through BOPTN 2016 funding scheme, with contract number 079/SP2H/LT/DRPM/II/2016.

References

[1] Balsamo M, Semprucci F, Frontalini F and Coccioni R 2012 Meiofauna as A Tool for Marine Ecosystem Biomonitoring In: *Marine Ecosystems* Cruzado A (Editor) London: INTECHOPEN Ltd.
[2] Yanko V, Arnold A J and Parker W C 1999 The effects of marine pollution on benthic foraminifera. In: *Modern Foraminifera.* Sen Gupta BK (Editor) Dordrecht: Kluwer Academic Publisher
[3] Heip C, Vincx M and Vranken G 1985 The ecology of marine nematodes *Oceanography and Marine Biology Annual Review* **23** 399-489
[4] Zulkifli, Soedharma D, Wardiatno Y and Sanusi H S 2007 Keterkaitan struktur komunitas meiofauna dengan kerapatan lamun dan parameter lingkungan di Selat Domak, Kepulauan Riau *J. Perikanan dan Kelautan* **12**(1) 15-22
[5] McKenzie L J, Campbell S J and Roder C A 2003 Seagrass-*Watch: Manual for Mapping and Monitoring Seagrass Resources by Community (Citizen) Volunteers 2nd Edition* Townsville: Northern Fisheries Center
[6] Higgins R P and Thiel H 1988 *Introduction to the Study of Meiofauna* London: Smithsonian Institution Press
[7] Blott S J and Pye K 2001 Gradistat: a grain size distribution and statistics package for the analysis of unconsolidated substrates *Earth Surf. Process Landforms* **26** 1237-1248
[8] Bengen D G 2011 *Sinopsis Teknik Pengambilan Contoh dan Analisis Data Biofisik Sumberdaya Pesisir* Bogor: Bogor Agricultural University
[9] (9) Nelson D W and Sommers L E 1982 Total carbon, organic carbon, and organic matter. In: *Methods of Soil Analysis Part 2, Second Edition.* Page AL, Miller RH, Keeney DR. Editor Madison: Soil Science Society of America, Inc. Publisher
[10] [APHA] American Public Health Association 2012 *Standard Methods for the Examination of Water and Wastewater 21th ed.* New York: New York Health Association
[11] Olsen S R and Sommers L E 1982 Phosporus. In: *Methods of Soil Analysis Part 2, Second Edition.* Page AL, Miller RH, Keeney DR. Editor Madison: Soil Science Society of America, Inc. Publisher
[12] Neves R C, Brand J, Bin Q G and Reichert H 2016 First time surveying meiofauna in Singapore *Raffles Bulletin of Zoology* **34** 8-12
[13] Soto E, Caballero W and Quiroga E 2015 Composition and vertical distribution of metazoan meiofauna assemblages on the continental shelf off central Chile *Lat. Am. J. Aquat. Res.* **43**(5) 922-935
[14] Conklin A R Jr 2014 *Introduction to Soil Chemistry* New Jersey: John Wiley and Sons Inc.
[15] Odum E P 1971 *Fundamentals of Ecology Third Edition* Philadelphia: WB Saunders Co.
[16] Phillips R C and Meñez E G 1988 *Seagrass* (Washington DC: Smithsonian Institution Press)
[17] den Hartog C 1970 *Seagrasses of The World* Amsterdam: North-Holland Pub. Co.
[18] Hongayo M C, Abad L A A, Acuesta T R, Ayeng T M A, Cano V N V, Guioguio E A, Lingas J M and Vito M P 2013 Marine meiofauna in Songculan Lagoon, Songculan Dauis, Bohol Philippines *J. Entomol. Zool. Stud.* **1**(3) 47-51
[19] Glasby C J, Hutchings P A, Fauchald K, Paxton H, Rouse G W, Russell C W and Wilson R S 2001 *Class Polychaeta.* In: *Polychaetes & Allies The Southern Syntesis. Fauna of Australia Vol 4A Polychaeta, Myoztomida, Pogonophora, Echiura, Sipuncula.* Beesley PL, Ross GJB, Glasby PJ. (Editor) Melbourne: CSIRO Publishing

IOP Conf. Series: Earth and Environmental Science **176** (2018) 012046 doi:10.1088/1755-1315/176/1/012046

Effect of steaming on physical and chemical characteristics White Shrimp (*Litopenaeus vannamei*) from Indramayu Waters

I Rostini* and R I Pratama

Department of Fisheries Product Processing, Faculty of Fisheries and Marine Science, University of Padjadjaran, Bandung Sumedang Highway Km.21 Jatinangor, Indonesia 45363
*e-mail: iis.rostini@unpad.ac.id

Abstract. White shrimp (*Litopenaeus vannamei*) was a fishery product that has potential to be developed, can be used as ready to eat food or as raw material for various processed products. This study aimed to determine the effect of steaming process on the physical and chemical characteristics of white shrimp collected from Indramayu waters. The research method used was experimental consisting of two treatments (fresh shrimp and steamed shrimp), each treatment was repeated 4 times consisting of three levels (headless, peeled tail on, and peeled undevined). The parameters observed included physical test (yield, cooking loss, color) and chemical test (moisture, protein, fat, and ash content). The results showed that fresh white shrimp yield after peeled completely of 61.92%, whole steamed shrimp 92.76%, and peeled tail on steamed shrimp was 36.02%. Yield of peeled undevined steamed shrimp has the lowest value compared to other that was 54.78%. Cooking loss during shrimp steamed process was 7.24%. The color values of steamed shrimp include L* 68.57, a* 12.17, and b* 47.41. Chemical composition of white shrimp after steaming process decreased water content, while protein, fat, and ash levels increased.

1. Introduction

Shrimp is one of the fishery products favored by all the people in the world, because it tastes good. Shrimps are caught, processed, and sold by several countries more than other fishery products. Shrimp flavor varies depending on the shape and species [1]. White shrimp (*Litopenaeus vannamei*) was a product of high value, for local consumption as well as export [2]. In waters of Indramayu produced lots of *Litopenaeus vannamei*. Shrimp acts as a source of animal protein. Protein content of vannamei shrimp (*Litopenaeus vannamei*) cooked by boiling process was 23.25% [3].

Processing fresh shrimp are various, such as steamed or boiled. Fresh shrimp is generally in the form of frozen whole fresh, head-off tail on, and peeled. Based on the variation of shrimp processing, it is necessary to know the yield obtained from each of the shrimp processing conditions.

Shrimp processing industry generally produce or sell shrimp in the form of frozen. When the shrimp will be used as raw material of steamed shrimp keep in mind how much meat can be obtained from the results of the product. This research aimed to determine the effect of steaming process towards the physical and chemical characteristics of *Litopenaeus vannamei* collected from the waters of Indramayu.

IOP Conf. Series: Earth and Environmental Science **176** (2018) 012046 doi:10.1088/1755-1315/176/1/012046

2. Material and method

2.1. Materials
Materials used in the study were vannamei shrimp (*Litopenaeus vannamei*) obtained from Indramayu waters of 40-50 (in 1 kg of 40-50 shrimps), water, ice, packing materials, and chemicals used for chemical analysis.

2.2. Method
The research method used was experimental consisting of two treatments (fresh shrimp and steamed shrimp), each treatment was repeated 4 times consisting of three levels (headless, peeled tail on, and peeled undevined). The physical test (yield, cooking loss, and colour test) and chemistry (moisture, crude protein, crude fat, and ash content) was analyzed on shrimps.

2.2.1. Preparation of steamed shrimp. Fresh shrimp washed by cold water, weighted for each treatment then made whole fresh, headless, peeled tail on, and peeled undevine. Steam treatment, after washed process shrimp steamed for 30 minutes. Steamed shrimp drained for 5 minutes, 100 °C temperature. Then the sample determined the yield, cooking loss, colour test, and proximate.

2.2.2. Determination of cooking loss. Cooking loss was determined based on the different weight of sample before and after steaming. The equation as follows [4]:

$$Cooking\ loss\ (\%) = \frac{(A-B)}{A}\ x\ 100 \qquad (1)$$

where: A = weight before steaming, B = weight after steaming and cooling.

2.2.3. Colour test. Colour measurement objectively using Chromometer CR200 with Hunter notation system (L*a*b*).

L*: the brightness parameter, represents the reflected light that produces white, grey, and black achromatic colours. L* value range from 0 (black) to 100 (white).

a*: red green gradation chromatic colour with plus (+) a value from 0 to 100 for red, and minus (-) a from 0 to -80 for green.

b*: yellow blue gradation chromatic colour with plus (+) b value of 0 up to 70 for yellow, and minus (-) b values from 0 to -80 for the blue colour.

2.2.4. Proximate analysis. Water content was measured by drying a sample of shrimp meat in an oven at 105 °C for 16 hours until the weight was constant. Crude protein content was analyzed by Kjeldahl method. Crude fat content was extracted with a 40-60°C boiling point petroleum ether. Ash content is measured by heating at 550°C for 4-5 hours using a muffle furnace [5].

3. Results
The vannamei shrimp have lost during the cooking process. The percentage of cooking loss during the steaming process is 7.24%. To know the part of shrimp vannamei that can be used as raw material for processed products then calculated the yield of the shrimp. Yield is the most important parameter to know the economic value and effectiveness of a product of material or material. Yield used to estimate how many body parts can be used as food ingredients [6]. Various forms of processed fresh shrimp and steamed are presented in figures 1 and 2.

The value of steamed vannamei shrimp was 92.76%. The heating process causes the denaturation of muscle protein denaturation and degree vary at each species [4]. Processing can destroy the protein or protein formation can produce a new one as a result of changes in protein conformation [7].

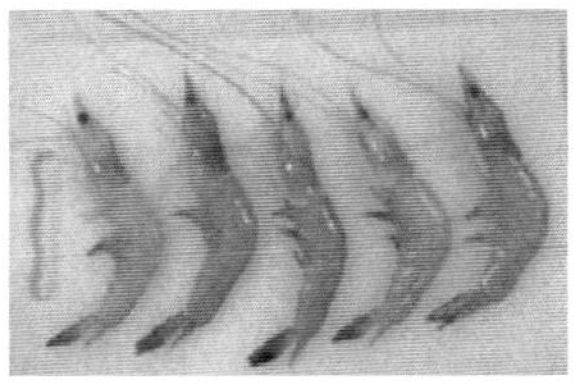

Whole fresh

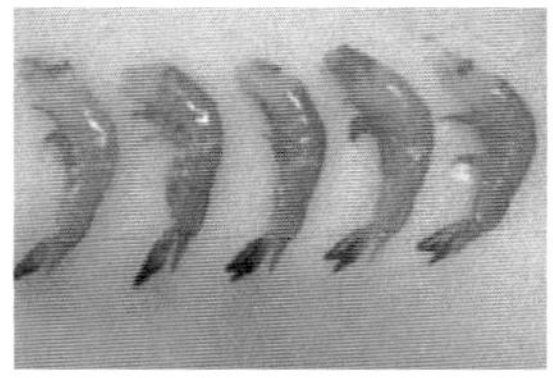

Headless

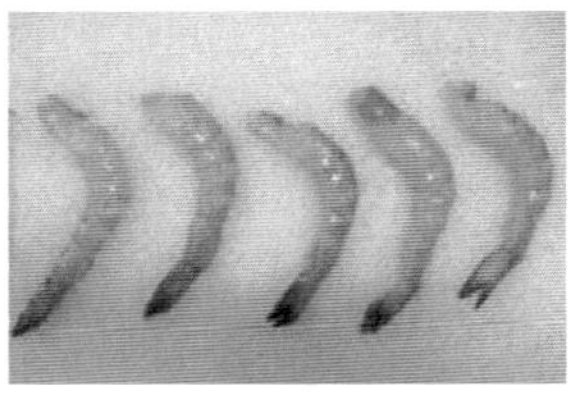

Peeled tail on

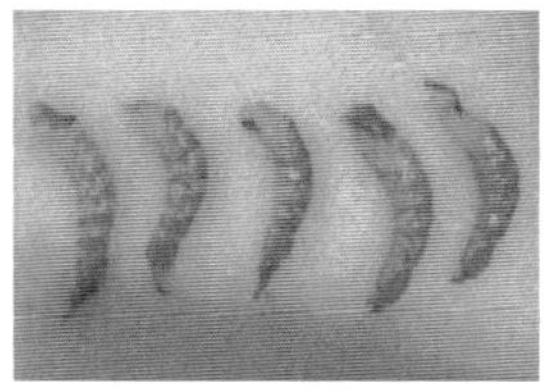

Peeled undevined

Figure 1. Forms of processed fresh shrimp.

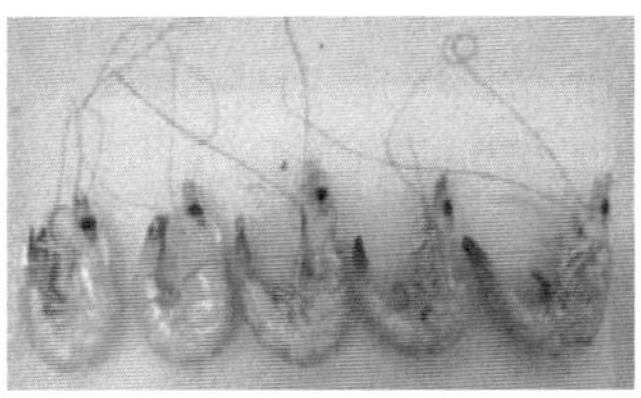

Whole steamed shrimp

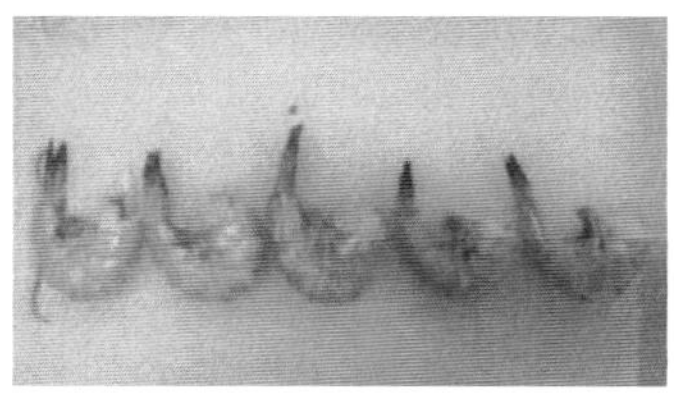

Headless steamed shrimp

Peeled tail on steamed shrimp

Peeled undevined steamed shrimp

Figure 2. Forms of steamed shrimp.

The value of yield was decreased both in the process of headless, carapace, and tail removal, as well as on the process of steaming shrimp vannamei (figure 3). The steamed shrimp has decreased in weight. This occurs because during the cooking process, the shrimp loses water levels as a result of the heating process that causes the occurrence of protein denaturation. Thus the ability to bind water has decreased, consequently there is water coming out of shrimp meat.

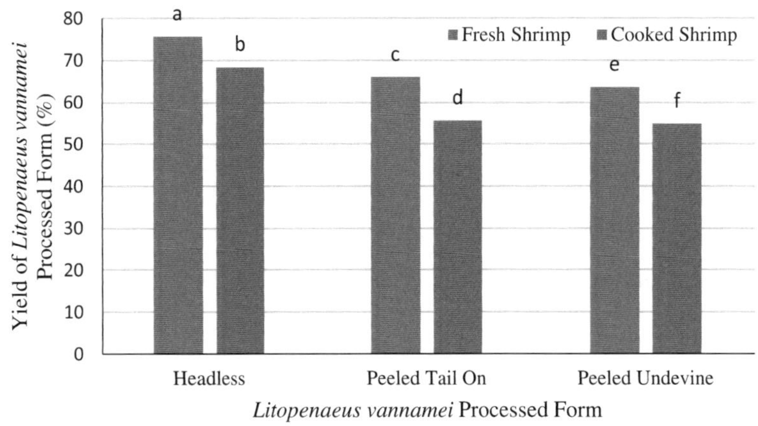

Figure 3. Yield of *Litopenaeus vannamei* processed form in fresh shrimp (blue bars) and cooked shrimp (red bars). Means followed by the same letter are not significantly different ($P < 0.05$) (*t* test).

There are several changes that occur in cooking shrimp. In general, shrimps show a loss of water content and shrink no more than 30% while thermal conductivity decreases no more than 5% and density increases not exceeding 4% of raw shrimp [8]. Examines the development of textures and histologic changes in collagen during thermal processing in shrimp. Thermal processing can increase the hardness and shrinkage deformation of the cooked shrimp [9].

Discoloration of the shrimp happen after cooking (figure 4). The colour was white initially, shrimp after cooking was transformed into the orange a bit red. Thermal causes denaturation of muscle protein, including carotenoprotein. Consequently, the carotenoid shows the original colour range from orange to red. Shrimp colour varies, depending on the species, seasons, food, and the environment [10].

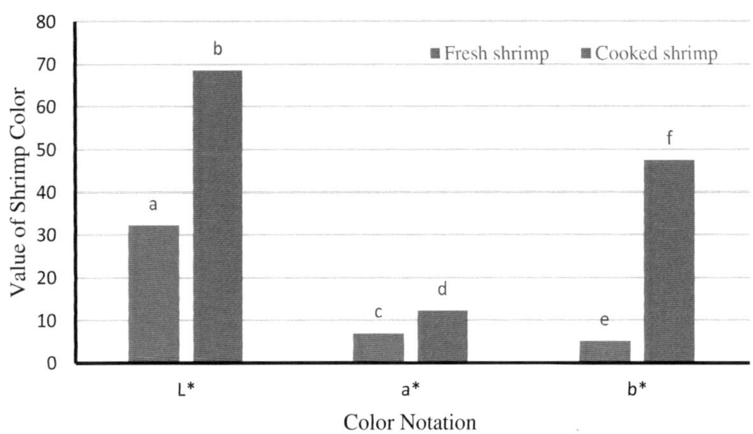

Figure 4. Value of fresh shrimp (blue bars) and cooked shrimp (red bars) colour. L* shows the brightness achromatic color, a* shows red green gradation chromatic color, and b* shows yellow blue gradation chromatic color. Means followed by the same letter are not significantly different ($P < 0.05$) (*t* test).

According to [4], with increasing temperature, water content of cooked shrimp meat was declining, while the fat content and the protein increases. The results of the Vannamei shrimp proximate analysis from Indramayu Waters are presented in table 1.

Table 1. Proximate composition of shrimp (*Litopenaeus vannamei*) from Indramayu waters.

Parameter	Raw shrimp[a] (before steaming)	Cooked shrimp[a]
Moisture (%)	77.59 ± 0.12[a]	69.74 ± 0.12[b]
Crude protein (%)	19.25 ± 0.32[c]	26.63 ± 0.07[d]
Crude fat (%)	0.30 ± 0.01[e]	0.82 ± 0.01[f]
Ash (%)	1.38 ± 0.11[g]	1.52 ± 0.01[g]

[a]Mean ± Standard Deviation, n = 4. Means followed by the same letter are not significantly different ($P < 0.05$).

The shrimp nutrients content was various in value depending on the habitat. Water is a major component of food. The content of water in the food ingredients also determine the acceptability, freshness and shelf life of foods [11]. Water content of fresh vannamei shrimp from Indramayu waters was 77.59%

The chemical composition of the most much reduced during cooking process is the water content. Water content in the flesh of steamed shrimp has decreased. A decrease in water content affected by the factors that cause the liquid from inside the meat seeped out. This decrease in water content resulted in the relative increase in protein, fat, and ash content. Decrease in water content caused by the occurrence of the protein denaturation process during steaming shrimp. The water is probably lost as a result of heat-induced denaturation of proteins during cooking of the meat. This causes less water to be entrapped within the protein structures held by capillary forces [12].

Thermal process causes the denaturation of muscle proteins [4]. This led to earlier denaturation of various kinds of shrimp proteins, that is, myosin, actin, sarcoplasmic proteins, and connective tissue (collagen is the main content of connective tissue). When protein denatured and coagulated myofibrillar proteins and collagen shrank and lost their water-holding capacity [2]. Protein aggregation induced by heating and also internal forces caused by pressure from connective tissue shrinkage [13].

The effect of steaming on changes in the chemical composition of fresh shrimp and cooked shrimp was compared using a *t*-test. The steaming process showed significantly different results on water, protein, and fat content of the shrimp.

4. Conclusion

Steaming process affect the physical and chemical characteristics of *Litopenaeus vannamei* shrimp. Yield in the form of processed peeled undevine cooked shrimp has the lowest value compare to the other of 54.78%. The Percentage of cooking loss during steaming process is 7.24%. After going through the process of ripening colour shrimp turns into a somewhat orange red. Chemical composition of white shrimp after steaming process decreased water content, while the protein, fat, and ash content increased.

References

[1] Mohan C O, Ravishankar C N, Bindu J, Geethalakshmi V and Gopal T K S 2006 Effect of thermal process time on quality of "shrimp kuruma" in retortable puoches and alumunium cans *Journal of Food Science* **71**(6) 496-500

[2] Niamnuy C, Devahastin S, Soponronnarit S and Raghavan G S V 2008 Kinetics of astaxanthin degradation and color changes of dried shrimp during storage *J. Food Eng.* **87** 591–600

[3] Rostini I 2011 *Pengembangan Edible Coating pada Udang Rebus Berbahan Dasar Surimi Limbah Filet Ikan Kakap Merah* (*Lutjanus* sp.) [The improvement of cooked shrimp's surimi based edible coating from red snapper fillet waste (*Lutjanus* sp.)] Bogor: Institut Pertanian Bogor

[4] Benjakul S, Visessanguan W, Kijroongrojana K and Sriket P 2008 Effect of heating on physical properties and microstructure of balck tiger shrimp (*Penaeus monodon*) and white shrimp (*Penaeus vannamei*) meats *International Journal of Food Science and Technology* **43** 1066-1072.
[5] [AOAC] Assosiation of Official Analytical Chemist 2005 *Official Methods of Analysis* Vol II 18th edn Gaithersburg: AOAC International
[6] Hadiwiyoto S 1993 *Teknologi Pengolahan Hasil Perikanan* [Fishery Product Processing Technology] Jilid I Yogyakarta: Liberty
[7] Maleki S J, Chung S Y, Champagne E T and Raufman J P 2000 The effects of roasting on the allergenic properties of peanut proteins *J. Allergy Clin Immunol* **106** 763–8
[8] Murakami E G 1994 Thermal processing affects properties of commercial shrimp and scallops *J. Food Sci.* **59** 237–41
[9] Mizuta S, Yamada Y, Miyagi T and Yoshinaka R 1999 Histological changes in collagen related to textural development of prawn meat during heat processing *J. Food Sci.* **64** 991-5
[10] Yanar Y, Celik M and Yanar M 2004 Seasonal changes in total carotenoid content of wild shrimps (*Penaeus semisulcatus* and *Metapenaeus monoceros*) inhabiting the Eastern Mediterranean *Food Chemistry* **88** 267-269
[11] Winarno F G 2008 *Kimia Pangan dan Gizi* [Food Chemistry and Nutrition] Jakarta: PT. Gramedia
[12] Aaslyng D M, Bejerholm C, Ertbjerg P, Bertram C H and Andersen J H 2003 Cooking loss and juiciness of pork in relation to raw meat quality and cooking procedure *Food Quality and Preference* **14** 277-288
[13] Erdogdu F, Balaban M O, Otwell W S and Garrido L 2004 Cooked-related yield loss for pacific white (Penaeus vannamei) shrimp previously treated with phosphates: effects of shrimp size and internal temperature distribution *Journal of Food Engineering* **63**(3) 279-300

IOP Conf. Series: Earth and Environmental Science **176** (2018) 012047 doi:10.1088/1755-1315/176/1/012047

Suitability of the various diversity index of phytoplankton in Garden House-Coastal Ornamental Lake, Pantai Indah Kapuk, North Jakarta

N T M Pratiwi*, Z Imran, S Hariyadi, R Meidwilestari

Department of Aquatic Resources Management, Faculty of Fisheries and Marine Science, Bogor Agricultural University
*e-mail: niken_tmpratiwi@yahoo.com

Abstract. Diversity is one of biotic community description within an ecosystem. A lot of diversity index has been developed. However, there is no consensus to suit the proper index to use. This study aimed to assess the suitability of the various diversity index of phytoplankton as community description, related to the water condition in coastal ornamental-lake of Garden House, Pantai Indah Kapuk, Jakarta. This research was conducted in July to December 2015 for sampling, laboratory analysis, and conformity assessment of diversity index (Margalef, Shannon, Simpson, McIntosh, and McIntosh evenness Index) that was based on the criterion of Discriminant Capacity approach. The results showed that there were 26 genera of phytoplankton from five classes, with density ranged from 8.9×10^3 to 7.6×10^9 cell/m^3, and was dominated by Cyanophyceae. Based on the criteria of Discriminant Capacity, McIntosh Index was the most appropriate diversity index to describe phytoplankton community in this lake. Furthermore, the low value of McIntosh index showed that the population of phytoplankton communities was not distributed homogeneously. Related to water quality, the phytoplankton community was closely correlated to pH, salinity, nitrate, nitrite, and orthophosphate.

1. Introduction

Garden House ornamental-lake is laid in Garden House Cluster of Bukit Golf Mediterania Residence, Pantai Indah Kapuk, northern coast of Jakarta. The waters are about 11615 m^2 in wide, and relatively shallow (70 cm). It is used as flood control and for aesthetics purpose, and also completed by sewage treatment plant (STP) for residence's waste water.

Meanwhile, a load of waste water, especially organic materials, sometimes exceeded the capacity of STP. This could lead the change of water quality of the lake, and also the coastal area when the outflow flooded into. The waste water could influence the condition of the waters, especially the high organic content that will transform into the nutrient. The nutrient augmentation and the change of nutrient proportion lead the change of community structure of phytoplankton.

Diversity is one of the biotic community, such as phytoplankton, that is used to describe the community stability within an ecosystem [1]. The determination of diversity index is also used in environmental monitoring or conservation. A lot of diversity index has been developed. However, there is no consensus to suit the proper index to use [2].

There are three approaches to determining the diversity, the richness index, species abundance models, and species proportional abundance index [3]. This study focused on the third approach that consists of many indexes. An evaluation is necessary since not all of the index could well-describe a community condition [4].

The species proportional abundance approach is relatively simple and applicable. Through this approach, the species richness and its evenness are well informed, that could vary the community description [5]. Several indexes of this approach are Shannon Index, Simpson Index, and McIntosh Index.

This study aimed to assess the suitability of the various diversity index of phytoplankton as community description, related to the water condition in coastal ornamental-lake of Garden House, Pantai Indah Kapuk, Jakarta.

2. Material and methods

There were five stations as sampling points in Garden House ornamental lake. The samples were taken monthly in six months, including plankton and water for water quality analysis [6]. The phytoplankton identification was conducted using identification book [7, 8, 9]. The cell counting was using one mL SRC [6].

Data analysis consisted of cluster analysis based on Canberra index [10] of water quality, several diversity index (based on species richness and proportional abundance approach), the evaluation of suitability of diversity index, PCA (Principal Component Analysis), and correlation between phytoplankton and water quality (Pearson's correlation) [11].

Those diversity indexes are Margalef Index, representatives the species richness approach; Shannon Index, Simpson Index, McIntosh Index, and McIntosh Evenness Index [3]. The effectiveness of those indexes was evaluated using discriminant capacity approach [4, 12].

3. Results

3.1. Phytoplankton abundance and composition

Phytoplankton species abundance and composition is necessary information to describe its community condition. The highest rate of abundance and number of species were found in Station 2, with about 661×10^6 cell/m^3 and 22 species (figure 1). The highest proportion of abundance was shown by Cyanophyceae and number of species was shown by Chlorophyceae (figure 2 and 3).

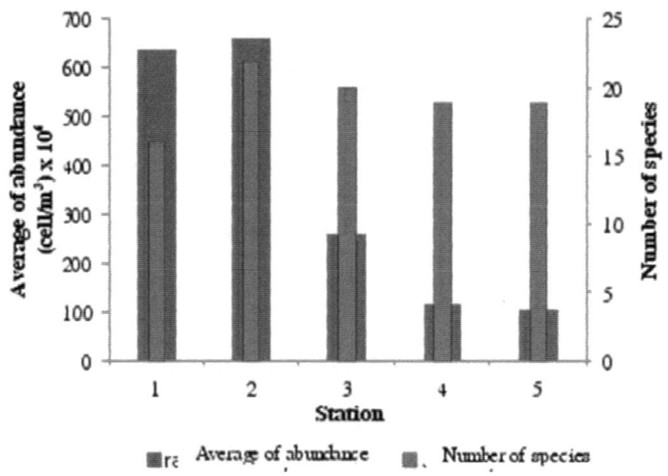

Figure 1. Average of abundance and number of phytoplankton in Garden House ornamental lake.

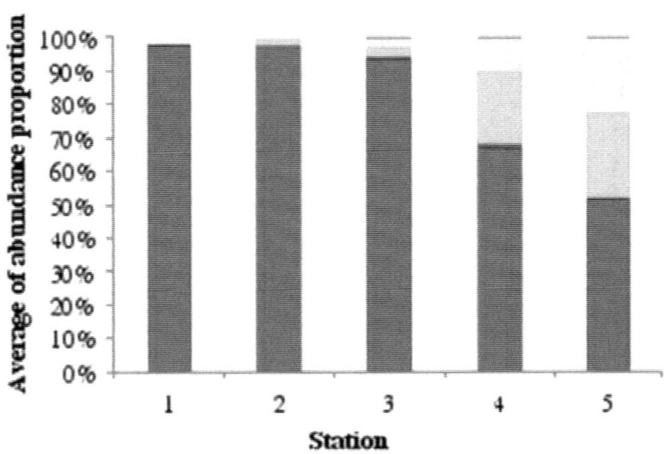

Figure 2. Proportion of abundance of phytoplankton in Garden House ornamental lake.

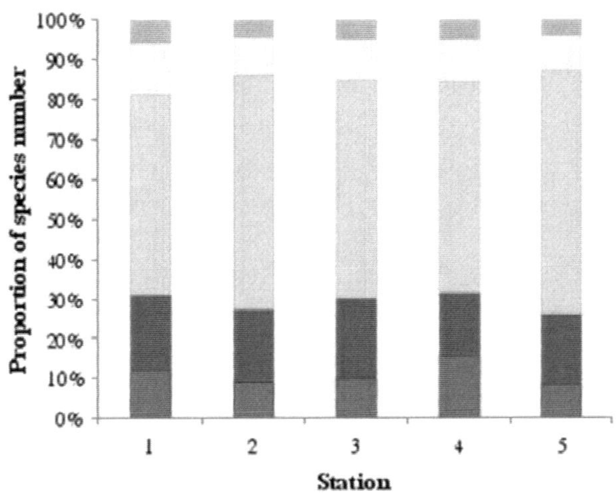

Figure 3. Proportion of number of phytoplankton in Garden House ornamental lake.

3.2. Water quality characteristic of Garden House ornamental lake

Bioavailable nutrients, such as ammonium, nitrite, nitrate, and orthophosphate, are essential to the growth and development of phytoplankton community. The concentration of those nutrients was fluctuated (figure 4). Furthermore, the physical characteristic indicated typically shallow water characteristic, with a very high TSS value at Station 5 (table 1).

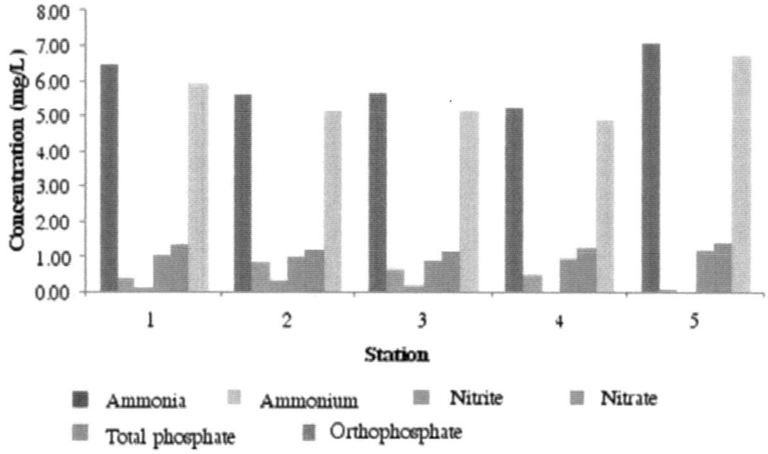

Figure 4. Bioavailable nutrients in Garden House ornamental lake.

Based on water quality characteristic, there was a determination of similarity index that was used as the base of cluster analysis. Figure 5 shows three clusters or groups of the station based on water quality of Garden House ornamental lake. Those were Group 1 (Station 1, 2, and 3), Group 2 (Station 4), and Group 3 (Station 5).

Table 1. Water quality characteristic of Garden House Ornamental Lake.

No	Parameters	Station				
		1	2	3	4	5
1	Depth (m)	0.75± 0.04	0.68± 0.30	0.45±0.18	0.79±0.07	0.64±0.11
2	Turbidity (NTU)	52.57±25.63	48.48±28.13	41.30±12.96	60.20±20.27	91.43±53.96
3	TSS (mg/L)	50.50±23.30	64.00±61.37	57.83±18.14	73.83±48.40	91.67±72.80
4	Temperature (°C)	29.82±0.78	29.22±0.90	29.32±0.90	29.52±0.90	29.78±0.97
5	pH	8.05±0.45	8.09±0.23	8.10±0.30	7.96±0.44	7.80±0.63
6	Salinity (‰)	1.62±0.34	1.60±0.36	1.62±0.38	1.53±0.36	1.57±0.28
7	DO (mg/L)	3.75±0.61	4.37±1.35	5.43±0.99	4.03±1.76	3.58±2.75
8	BOD_5 (mg/L)	15.75±9.07	15.25±8.37	16.07±8.82	14.70±9.12	18.27±11.28

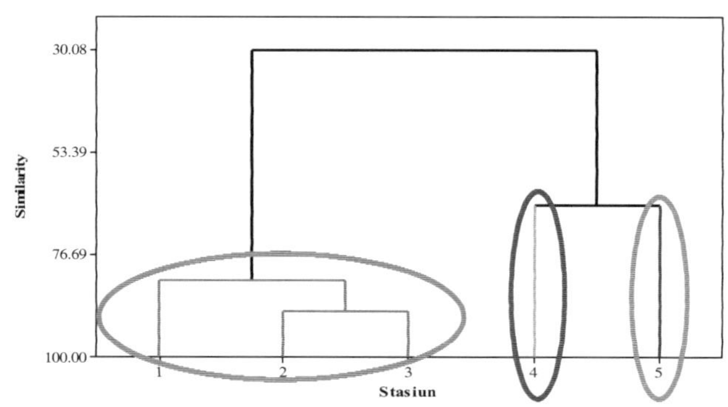

Figure 5. Groups of the station based on water quality.

3.3. The evaluation of diversity index

Discriminant capacity shows the suitability of the diversity index in describing the condition of phytoplankton community structure in Garden House. The value of the discriminant capacity of each phytoplankton diversity index is laid in Table 2. It is shown that the order from the highest to the lowest values, temporal and spatially. From both orders, it is concluded that the most suitable diversity index is McIntosh (Mc) with relatively high values at both approaches.

Table 2. Discriminant capacity of each diversity index.

Diversity Index	Value	
	Temporally	Spatially
Simpson index	61.471	25.784
McIntosh index (McE)	46.711	24.724
McIntosh index (Mc)	42.128	23.305
Shannon index (H')	34.600	18.312
Shannon index (E)	31.853	15.941
Margalef index	26.134	1.2320

3.4. Phytoplankton community structure in Garden House ornamental lake

Based on the best diversity index, there are values of McIntosh Index of phytoplankton in Garden House ornamental lake of each clustered station (Table 3). It seems that the diversity values are fluctuated along the observation.

Table 3. McIntosh Index of Beach group of station.

	Month					
	1	2	3	4	5	6
Group 1	0.0092	0.3207	0.2574	0.3435	0.2639	0.4090
Group 2	0.3925	0.4722	0.4226	0.4559	0.3919	0.4432
Group 3	0.2530	0.4458	0.2995	0.4188	0.4086	0.2949

3.5. The relationship between phytoplankton community and water quality

The principal component analysis was used to describe the relationship between phytoplankton community and water quality of each group of the station. Biplot of the relationship between phytoplankton and water quality is shown in figure 6.

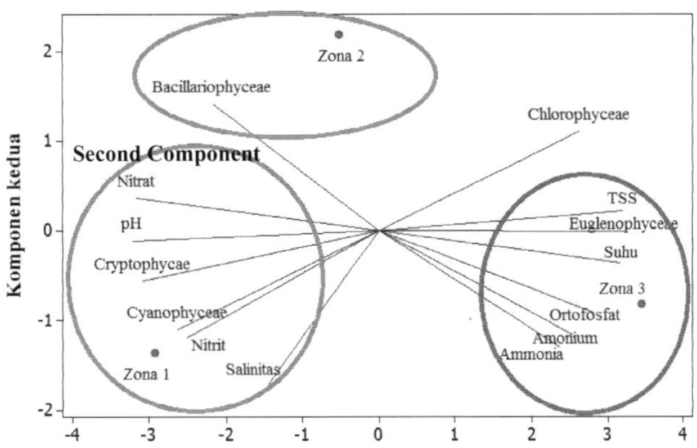

Figure 6. Biplot of the relationship between phytoplankton and water quality.

The description of the biplot was furthermore tested to determine their correlations. The result showed that most classes of phytoplankton were highly correlated to pH, salinity, nitrate, nitrite, orthophosphate, ammonium and temperature. Those results were shown in table 4.

Table 4. Correlation between phytoplankton and water quality.

Class	pH	Salinity	Nitrate	Nitrite	Orthophosphate
Cyanophyceae	0.849	0.881	-	0.998	-
Bacillariophyceae	-	-	0.803	-	-
Euglenophyceae	0.983	-	-	-	0.880
Cryptophycae	-	-	0.883	0.929	-

4. Discussion

Nutrient concentration in Garden House ornamental lake was relatively high with high TSS condition, especially Station 5 that located near from STP unit. The organic material that was decomposed into nutrients will lead abundance and composition of phytoplankton, proper to the concentration and composition. In this study, the most abundance of phytoplankton was Cyanophyceae, especially at Station 5. The least density was direct by the operational of the pedal wheel that station 5.

The high density of Cyanophyceae indicated that the waters were in eutrophic states. Cyanophyceae often found in organically polluted waters [13. 14]. Meanwhile, the number of species was relatively low. *Oscillatoria* sp., *Spirulina* sp., and *Merismopedia* sp. were found in relatively high density. Those species were often dominated by organically polluted waters [15], and indicated nutrient enrichment [16].

One of important test to determine the most effectiveness of diversity index is the calculation of discriminant capacity [3]. The higher the discriminant capacity value. The higher the effectiveness of diversity index. The McIntosh Index showed high discriminant capacity, both temporal and spatially. The value indicated that the McIntosh Index was suitable to describe phytoplankton condition at Garden House ornamental lake.

Diversity index with high discriminant capacity value could describe the difference of diversity index among stations with a relatively similar environmental condition or water quality vice versa [4,12].

In this study, the value of McIntosh Index of plankton community was close to 0, which indicated that organisms in the community were not homogenized distributed. When the value of McIntosh Index close to 1, it is indicated that organisms in the community are homogenized distributed both temporally and spatially [17]. The values of the diversity index fluctuated at both approaches.

Biplot of PCA shows that in Group 1. Cyanophyceae and Cryptophyceae were in high density and laid in the same radian with a high value of nitrate, pH, salinity and nitrite. Group 2 was characterized by high density of Bacillariophyceae. Furthermore. Group 3 was characterized by the high density of Euglenophyceae and high concentration of TSS, total ammonia, and phosphate.

As a whole, the density of phytoplankton was influenced by nutrients, pH, and salinity. Based on the correlation test, the phytoplankton density was highly correlated with nutrient, pH, salinity, and TSS. Nutrients have a role as limiting factors (nitrogen and phosphor) for the phytoplankton community [18]. The pH could influence the life of phytoplankton, which the low value of pH will decrease water quality that will also influence the living inside [19].

The correlation value between Cyanophyceae and pH, salinity and nitrite was relatively high ($r > 0.75$). A high pH and magnified nutrients (nitrate and nitrite) were suitable for Cyanophyceae. Besides the Cyanophyceae was also correlated to salinity [15]. Salinity is one of controlling factor for nitrogen fixation of Cyanobacteria [20]. Furthermore, Cryptophyceae was highly correlated to nitrate and nitrite. Nitrogen in the sort of nitrate, nitrite, and ammonium has a role in protein synthesis of phytoplankton [21].

The relationship between Euglenophyceae and nutrients (orthophosphate and ammonium) was relatively high ($r >0.75$). Orthophosphate was used to and influence the process and the development of phytoplankton. Phosphor transforms energy for phytoplankton [21]. Euglenophyceae was also highly correlated to TSS and temperature. TSS will influence light Transparency that used by phytoplankton in the photosynthesis process.

Garden House ornamental lake was assumed to receive a high organic material load that was indicated by the high density of Cyanophyceae. The McIntosh diversity index is the most suitable index for describing the condition of phytoplankton community structure, based on the high value of the discriminant capacity both temporal and spatially.

The less homogenous distribution of species density leads the low value of McIntosh index. The difference of species density distribution was assumed to be correlated to several characterized parameter that influences the density of phytoplankton. Furthermore, based on correlation test the density of phytoplankton was highly correlated to water quality such as pH, salinity, nitrate, nitrite, and orthophosphate. Those environmental parameters influence the biological process in phytoplankton community.

5. Conclusion

The most suitable diversity index to describe the phytoplankton Community structure in Garden House coastal ornamental lake. Pantai Indah Kapuk. Jakarta is McIntosh Index (McI). The phytoplankton community was highly correlated with salinity, pH, nitrite, nitrate and orthophosphate.

References

[1] Ismael A A and Dorgham M M 2003 Ecological indices as a tool for assessing pollution in El-Dekhaila Harbour (Alexandria. Egypt) *Oceanologia* **45**(1) 21–131
[2] Morris E K. Caruso T. Buscot F. Fischer M. Hancock C. Maier T S. Meiners T. Muller C. Obermaier E and Prati D 2014 Choosing and using diversity indices: insights for ecological applications from the German Biodiversity Exploratories *Ecology and Evolution* **4**(18) 3514-3524
[3] Maguran E A 1983 *Ecological Diversity and Its Measurement* (London: Cambridge University Press)
[4] Lexerød N L and Eid T 2006 An evaluation of different diameter diversity indices based on criteria related to forest management planning *Forest Ecology and Management* **222** 17-28
[5] Peet R K 1974 The measurement of species diversity *Annual Review of Ecology and Systematics* **5** 285-307
[6] [APHA] American Public Health Association 2012 *Standard Methods for Examination of Water and Wastewater.* 22nd ed (Washington: [APHA] American Public Health Association. [AWWA] American Water Works Association. [WEF] Water Environment Federation)
[7] Prescott G W 1970 *The Freshwater Algae* (Dubuque Lowa: Brown Company Publishers)
[8] Mizuno T 1979 *Illustrations of The Freshwater Plankton of Japan.* revised ed (Osaka: Hoikusha Publishing Co.. LTD)
[9] Yamaji I E 1979 *Illustration of The Marine Plankton of Japan* (Osaka: Hoikusha Publishing Co.. LTD)
[10] Krebs C J 1999 *Ecological Methodology* (California: Benjamin/Cummings Imprint)
[11] Walpole R E 1993 *Pengantar Statistika Edisi 3* Bambang Soemantri. penerjemah (Jakarta: PT Gramedia)
[12] Ravera O 2001 A comparison between diversity. similarity. and biotic indices applied to the macroinvertebrate community of a small stream: The Ravella River (Como Province. Northern Italy) *Aquatic Ecology* **35** 97-107
[13] Nandan S N and Aher N H 2005 The algal community used for assessment of water quality of Haranbaree Dam and Mosam River of Maharashtra *Journal of Environmental Biology* **26**(2) 223-227
[14] Wang X. Wang Y. Liu L. Shu J. Zhu Y and Zhou J 2013 Phytoplankton and eutrophication degree assessment of Baiyangdian Lake Wetland. China *The Scientific World Journal* 1-8
[15] Ganai A H and Saltanat P 2014 Effect of physicochemical conditions on the structure and composition of the phytoplankton community in Wular Lake at Lankrishipora. Kashmir *Biodiversity and Conservation* **6** 71-84

[16] Zargar S and Ghosh T K 2006 Influence of cooling water discharges from kaiga nuclear power plant on selected indices applied to plankton population of Kadra Reservoir *Environmental Biology* **27**(2) 191-198

[17] Harazika L P 2013 Diversity indices of macro invertebrates in the Satajan Wetland of Lakhimpur District. Assam *Annals of Biological Research* **4**(8) 68-72

[18] Dzialowskai A R. Wang S H. Lim N C. Spotts M W and Huggins D G 2005 Nutrient limitation of phytoplankton growth in central plains reservoirs. USA *Plankton Research* **27** 587-595

[19] Susana T 2009 Tingkat keasaman (pH) dan oksigen terlarut sebagai indikator kualitas perairan sekitar Muara Sungai Cisadane *Teknologi Lingkungan* **5**(2) 33-39

[20] Moisander P H. McClinton E and Paerl H W 2002 Salinity effect on growth. photosynthetic parameters. and nitrogenase activity in estuarine planktonic cyanobacteria *Microb Ecology* **43** 432-442

[21] Reynold C S 2006 The Ecology of Phytoplankton (Cambridge: Cambridge University Press)